Osswald, Turng, Gramann
Injection Molding Handbook

Tim Osswald
Lih-Sheng Turng
Paul Gramann

Injection Molding Handbook

2nd Edition

With constributions from

J. Beaumont, J. Bozelli, N. Castaño, S. Collins, B. Davis, M. DeGreiff, R. Farrell, P. Gramann, G. Holden, R. Lee, T. Osswald, C. Rauwendaal, A. Rios, M. Sepe, T. Springett, L. Turng. R. Vadlamudi, J. Wickmann, V. Yang

HANSER
Carl Hanser Publishers, Munich • Hanser Gardner Publications, Cincinnati

The Authors:
Prof. Dr. Tim A. Osswald, University of Wisconsin-Madison, Polymer Processing Research Group, Department of Mechanical Engineering , Madison, WI 53706, USA
Asst. Prof. Lih-Sheng Turng, University of Wisconsin-Madison, Polymer Processing Research Group, Department of Mechanical Engineering, Madison, WI 53706, USA
Dr. Paul J. Gramann, The Madison Group, Polymer Processing Research Corp., Madison, WI 53711, USA

Distributed in the Americas by:
Hanser Publications
414 Walnut Street, Cincinnati, OH 45202 USA
Phone: (800) 950-8977
www.hanserpublications.com

Distributed in all other countries by:
Carl Hanser Verlag
Postfach 86 04 20, 81631 Munich, Germany
Fax: +49 (89) 98 48 09
www.hanser-fachbuch.de

Library of Congress Cataloging-in-Publication Data
Injection molding handbook / [edited by] Tim Osswald, Lih-Sheng Turng, Paul Gramann ; Oswald et al. -- 2nd ed.
p. cm.
Includes index.
ISBN 978-1-56990-420-6 (hardcover)
1. Injection molding of plastics-Handbooks, manuals, etc. I. Osswald, Tim A. II. Turng, Lih-Sheng. III. Gramann, Paul J.
TP1150.I552 2007
668.4'12--dc22
2007035952

Bibliografische Information Der Deutschen Bibliothek
Die Deutsche Bibliothek verzeichnet diese Publikation in der Deutschen Nationalbibliografie; detaillierte bibliografische Daten sind im Internet über <http://dnb.d-nb.de> abrufbar.

ISBN 978-3-446-40781-7

unmodified reprint of the 2nd edition of 2008
Production Management: Jörg Steffen
Coverconcept: Marc Müller-Bremer, Rebranding, München, Germany
Coverdesign: MCP • Susanne Kraus GbR, Holzkirchen, Germany
Typesetting: Kösel GmbH & Co, Altusried-Krugzell, Germany

We dedicate this handbook to Professor Kuo-King (K. K.) Wang whose vision and pioneering contributions propelled the advancement of injection molding technology.

Tim A. Osswald, Paul J. Gramann, and Lih-Sheng (Tom) Turng

Experience is a good school, but the tuition is high.

Heinrich Heine

Preface

The injection molding manufacturing sector, with a total product value of almost $200 billion per year, is the fourth largest industry in the United States. Today, more than a third of all polymeric materials, approximately 15 billion pounds, are used by the injection molding industry annually.

The *Injection Molding Handbook* is primarily written for engineers, processors researchers, and other professionals with various levels of technical background. It not only serves as introductory reading for those becoming acquainted with injection molding, but also as an indispensable reference for experienced practitioners. The handbook presents a thorough, up-to-date view of injection molding processing equipment and techniques, with fundamental information on the chemistry, physics, material science, and process engineering. It also covers topics that directly affect the injection molding process, such as injection molding materials, process control, simulation, design, and troubleshooting.

The handbook presents a well-rounded overview of the underlying theory and physics that control the various injection molding processes, without losing the practical flavor that governs the manuscript between its covers. The carefully chosen contributing authors include experts in the field, as well as practitioners and researchers in both industry and academia.

The first three chapters of this handbook present the fundamental background, covering basic process principles and materials. Here, a unified approach is used by pulling in the influence of processing on the properties of a finished product. Chapters 4 through 6 present the injection molding machine, which includes the plasticating and clamping units, as well as the injection mold. Materials handling is introduced in Chapter 7 and statistical process control, as related to injection molding, is presented in Chapter 8. Chapter 9 gives an in-depth overview of special injection molding processes. Product design and injection molding simulation is presented in Chapters 10 and 11, respectively. The last two chapters present extensive process and material troubleshooting procedures that will be useful to anyone in the industry at any stage of process and product design.

It would be impossible to thank everyone who in one way or another helped us with this manuscript. Above all, we would like to thank the contributors to this handbook. They are John Beaumont, John Bozzelli, Scott Collins, Bruce Davis, Mauricio DeGreif, Robert Farrel, Lukas Guenthard, Geoffrey Holden, Chris Rauwendaal, Antoine Rios, Michael Sepe, Treasa Springett, Raghu Vadlamudi, Jerry Wickmann, and Venny Yang. They not only submitted quality contributions in a timely manner, but also served as sounding boards during all stages of the preparation. We are also grateful to Lynda Litzkow and Angela Maria Ospina for the superb drawing of some of the figures. We would also like to extend our appreciation to Wolfgang Glenz and Christine Strohm of Carl Hanser Verlag for their support throughout the book's development, and to Barbara Chernow and coworkers for copyediting and typesetting the final book. Our wives Diane Osswald, Stephanie Gramann, and Michelle Turng are thanked for their constant love and support.

Tim A. Osswald, Paul J. Gramann, and Lih-Sheng (Tom) Turng
Madison, WI
Summer 2007

Contents

Contributors

John Beaumont, Plastics Engineering Technology, Penn State Erie, Erie, PA 16563, USA

John Bozzelli, Injection Molding Solutions, Midland, MI 48640, USA

Nelson Castaño, IPIPC, Medellin, Colombia

Scott Collins, Collins Marketing Communications, Cleveland, Ott, USA

Bruce A. Davis, The Madison Group: PPRC, Madison, WI 53719

Mauricio DeGreiff, IPIPC, Medellin, Colombia

Robert Farrell, Plastics Engineering Technology, Penn State Erie, Erie, PA 16563, USA

Paul J. Gramann, The Madison Group: PPRC, Madison, WI 53711, USA

Geoff Holden, Holden Polymer Consulting, Inc., Prescott, AZ 86301, USA

Roger Lee, Colormax, AK-Tron Company, Telford, UK

Tim A. Osswald, Department of Mechanical Engineering, Polymer Engeneering Center, Madison, WI 53706, USA

Chris Rauwendaal, 12693 Roble Veneno Lane, Los Altos Hills, CA 94022, USA

Antoine C. Rios, The Madison Group: PPRC, Madison, WI 53719, USA

Michael Sepe, Sedona, AZ 86336

Lih-Sheng (Tom) Turng, Department of Mechanical Engineering, Polymer Engeneering Center, Madison, WI 53706, USA

Treasa Springett, Technical Center, Phillips Plastics Corporation, Prescott, WI 54021, USA

Raghu Vadlamudi, Technical Center, Phillips Plastics Corporation, Prescott, WI 54021, USA

Jerry Wickmann, Technical Center, Phillips Plastics Corporation, Prescott, WI 54021, USA

Venny Yang, Moldex 3D, Taipei, Taiwan

1 Introduction

P. J. Gramann, T. A. Osswald

Injection molding is the most important process used to manufacture plastic products. Today, more than one third of all thermoplastic materials are injection molded, and more than half of all polymer-processing equipment is for injection molding. The injection-molding process is ideally suited to manufacture mass-produced parts of complex shapes that require precise dimensions.

1.1 Historical Background

The development and refinement of plastics-processing equipment is considered by most people to be as important as the invention of plastics itself. Most of the important hurdles encountered throughout the history of the plastics industry were overcome by refining processing techniques and machinery. Although the invention of cellulose plastics, also known as Celluloid, Parkesine, Xylonite, or Ivoride, has been attributed to three people: the Swiss professor, Christian Schönbein; the English inventor, Alexander Parkes; and the American entrepreneur, John Wesley Hyatt. Hyatt also launched the enterprise that eventually became the plastics industry.

As the story goes, it all began in 1865 when billiard ball manufacturer Phelan & Collendar placed an advertisement that promised $10,000 to the person who would find a replacement for ivory in the manufacture of billiard balls. Elephants were being slaughtered at a rate of 70,000 per year, which would have led to the extinction of this great animal, exorbitant prices for the "white gold" from Africa, and reduced profits for the billiard ball industry. The $10,000 tag attracted the 28-year-old Hyatt's attention. After returning home from his job as a printer, he worked on this project every day until eventually, in 1869, he stumbled upon nitrocellulose, a material that Parkes and Schönbein had worked with before him. After mixing well the components, and allowing the solvents to evaporate from the mass completely, before solidification, he was soon manufacturing high-quality billiard balls. Instead of cashing in on the $10,000 prize, John Hyatt founded the Albany Billiard Ball Company with his brother Isaiah, becoming a direct competitor to Phelan & Collendar, marking the beginning of the plastics industry.

One of their first needs was to find a method to manufacture plate- and rodlike celluloid articles. Based on an 1870 patent to inject metal castings, by John Smith and Jesse Locke, the Hyatt brothers developed and patented the first plastics injection molding machine in 1872 [1] Three of these machines were in operation for many decades in an American celluloid factory, injection molding articles of relatively simple shape [2]. Their "stuffing machine," shown in Fig. 1.1, was the prototype of the plunger injection molding machine. The two drawings that show the detail of an inner mandrel with pegs are noteworthy. The purpose of this section was to increase contact

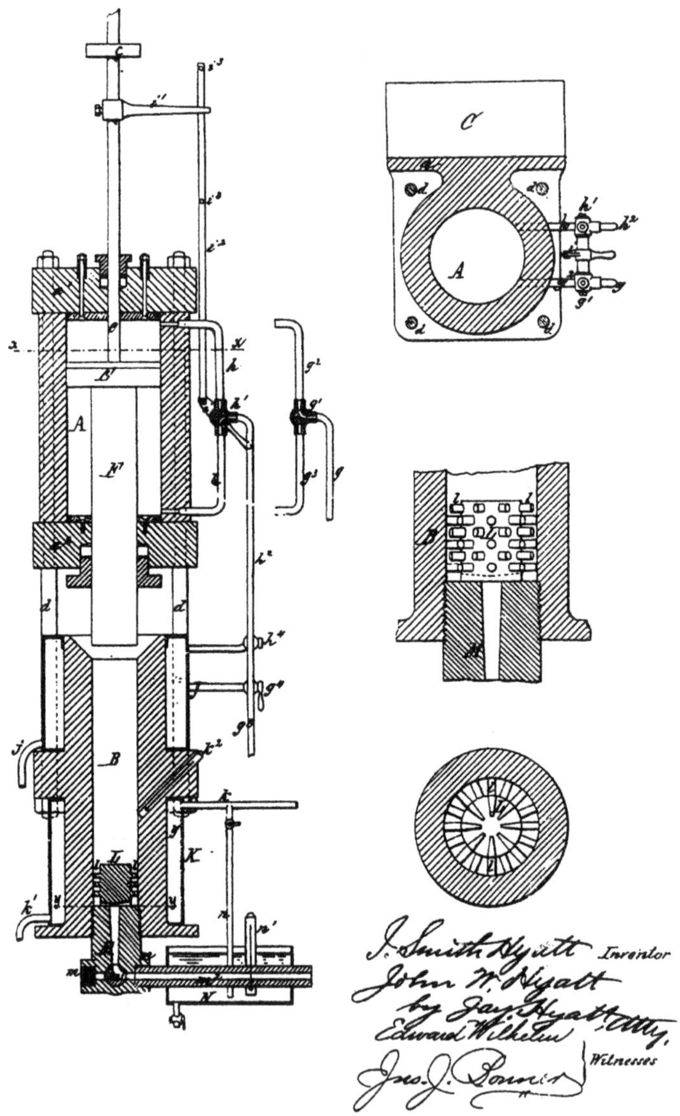

Figure 1.1 Original patent drawing of the Hyatt brothers' stuffing machine.

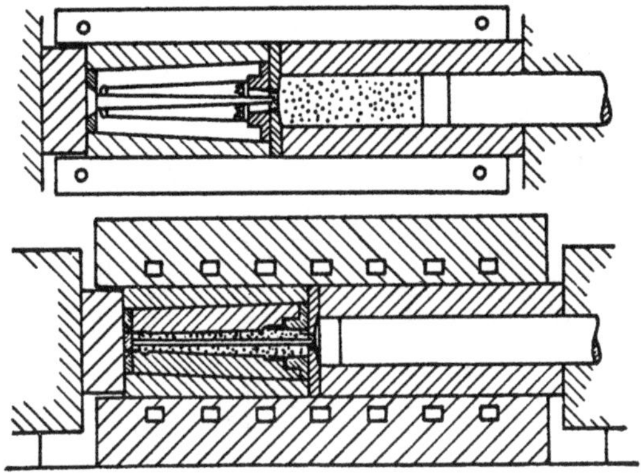

Figure 1.2 Gaylord's 1904 injection machine used to mold amber products.

area between the heating elements and the molding material, resulting in faster temperature rises during plastication. With no doubt, this is the predecessor of the torpedo in later injection molding machines. Because the only material available for injection was celluloid, not much changed in the industry for the next 50 years. In 1904, however, the Englishman E. L. Gaylord was able to actually patent the injection molding process itself. His injection molder is depicted in Fig. 1.2. He used his process to inject amber into high-quality finished products; however, the high cost of amber, which is very similar in properties to polystyrene, was the main reason why Gaylord's enterprise was not successful.

Finally, in 1919 the German A. Eichengrün was able to find the processing conditions that allowed him to injection mold celluloid parts of complex shape. Thus, for the fourth time, the injection molding process was invented. He refined the process and the equipment, and in 1923 filed for a patent in Germany, England, and the United States for the injection molding machine [3]. It took almost 8 years in the United States to issue his patent. His hand-operated machines, which were built by H. Buchholz, were an immediate success throughout the world. The Buchholz patent was assigned to the Grotelite Company in the United States. They automated the process and developed the first hydraulically operated injection molding machine in 1933. The validity of the Buchholz patent was challenged in 1934, however, which resulted in the sudden manufacture of numerous home made, hand-operated injection molding machines throughout the country. Finally, in 1942 the Buchholz patent was declared invalid using Hyatt's patent and the prior art as declared in the 70-year-old patent. Despite all of this, the Eichengrün-Buchholz machine, such as the one depicted in Fig. 1.3 [3], became the prototype for most injection molding machines that followed.

In 1926 L. E. Shaw invented the transfer molding process, which has since been a primary manufacturing process for thermosetting articles [3]. Here, as depicted in Fig. 1.4, the charge is placed inside an initial chamber, where it is heated and pres-

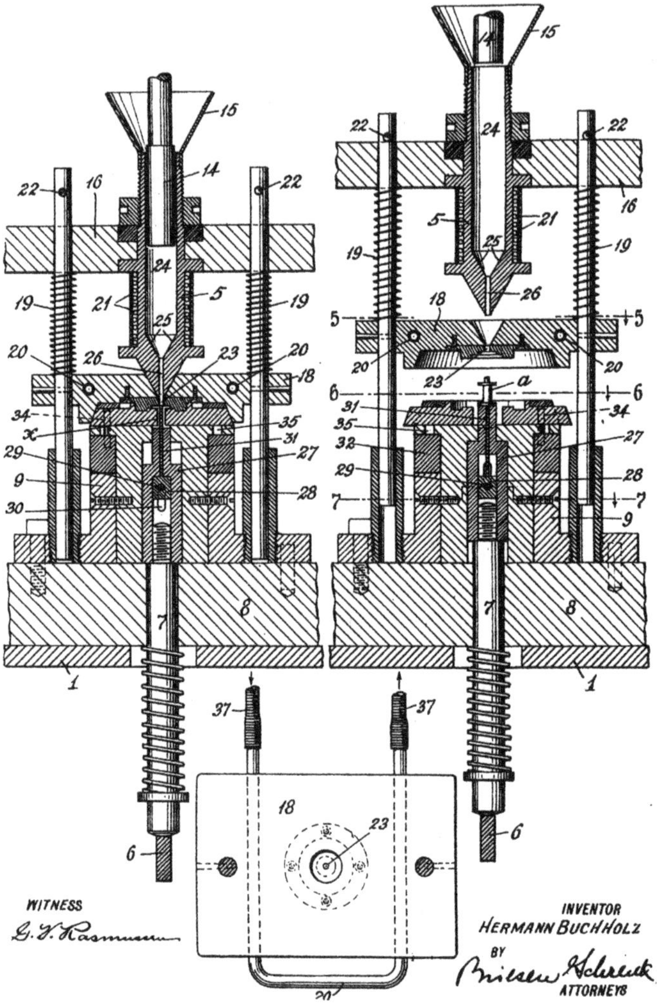

Figure 1.3 Eichengrün-Buchholz injection molder as depicted in a 1931 patent.

surized by a plunger, causing the softened material to flow through sprue, runners, and gates into the mold cavities. After Shaw's invention many compression molding machines were transformed into transfer molding type machines. In 1940, Shaw developed the jet-molding machine for thermosets, as depicted in Fig. 1.5 [3]. Here, a piston injected the thermosetting resin though a heated nozzle into the mold cavity. A heater was used to heat the nozzle rapidly as the resin was injected into the cavity, followed by water cooling that prevented the material to cure inside the nozzle while sitting idle during the curing process.

Through the 1930s and 1940s several companies manufactured injection molding machines in Europe and the United States. One major improvement came in 1932 when H. Gastrow introduced a torpedo in the melting region of the injection molding machine (Fig. 1.6). In a way, this invention was very similar to the mandrel present in the 1872 Hyatt patent. Gastrow's streamlined torpedo increased the plasticating capacity of injection molding machines by a measurable amount; however, the melting mechanisms of injection molding machines remained fairly inefficient even with a central core increasing the surface area and enhancing shear. For example, to increase the molding capacity of their equipment, the Hydraulic Press Manufacturing Company (HPM) increased the injection capacity in one of their machines to 36-oz by using four 9-oz pasticating cylinders, as shown in the photograph in Fig. 1.7.

It was clear early on that the screw extruder was a very effective plasticating unit. In fact, the use of screw extruders as injection molding machines goes back to the 1930s, when companies such as Eckart & Ziegler, Foster-Wucher, and Paul Troester used them to inject polymer melts into mold cavities directly. Unless fairly large extruders were used, however, the relatively low pressures and injection speeds generated by most systems were only sufficient to mold thick parts with large gates [4]. To increase the injection pressures and speeds to the levels required to inject thin parts, the Jackson and Church Company developed a two-stage injection molding

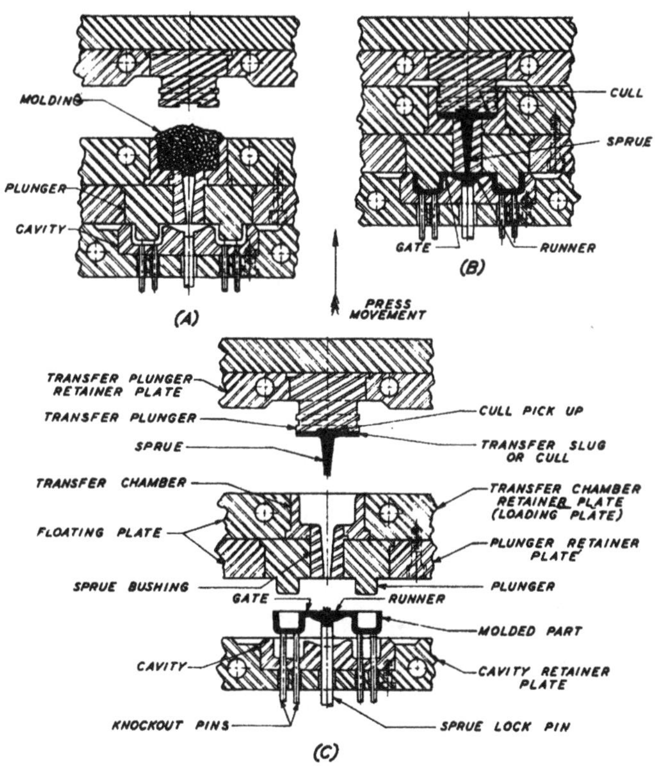

Figure 1.4 Shaw's 1926 transfer molding process.

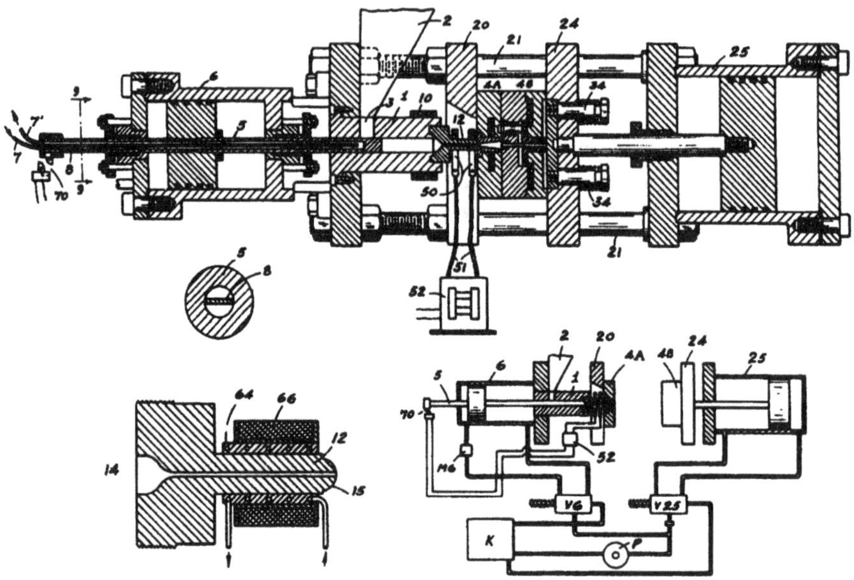

Figure 1.5 Shaw's 1940 jet molding machine.

machine in 1948 [3,4]. A two-stage system had a screw plasticating unit pumping the polymer melt into a piston injection unit. Their first products were 48-oz and 64-oz injection molding machines. In 1943, H. Beck of the I.G. Farbenindustrie in Germany filed for a patent that used the plasticating screw as the injection piston, as depicted in Fig. 1.8 [5]. The patent was granted in 1952. That same year the American W.H. Willert filed for a similar patent. A schematic of Willert's machine is shown in Fig. 1.9. In both inventions as the screw turns during the plasticating stage, it is displaced

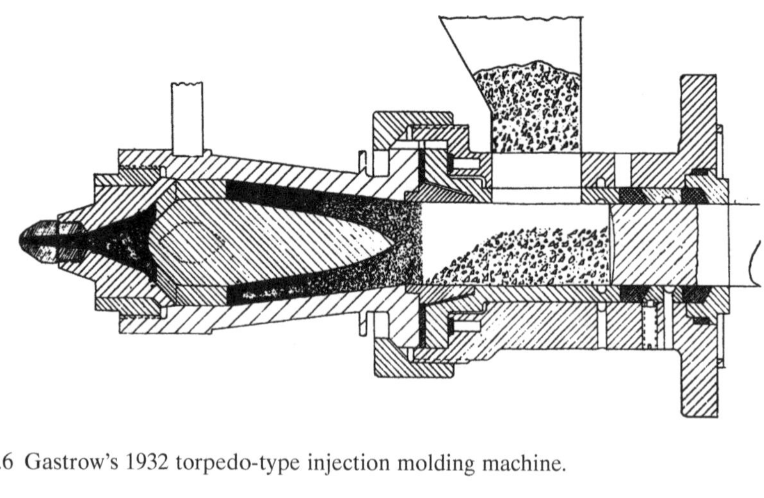

Figure 1.6 Gastrow's 1932 torpedo-type injection molding machine.

Figure 1.7 Hydraulic Press Manufacturing Company 36-oz torpedo-type injection molding machine with four 9-oz injection cylinders. (From Dubois, *Plastics History*, 1972, p. 224)

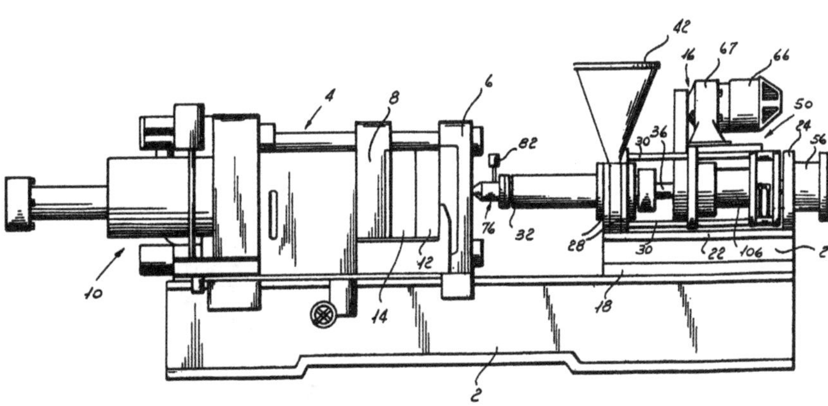

Figure 1.8 Beck's 1943 reciprocating screw injection molding machine as depicted in his 1952 patent.

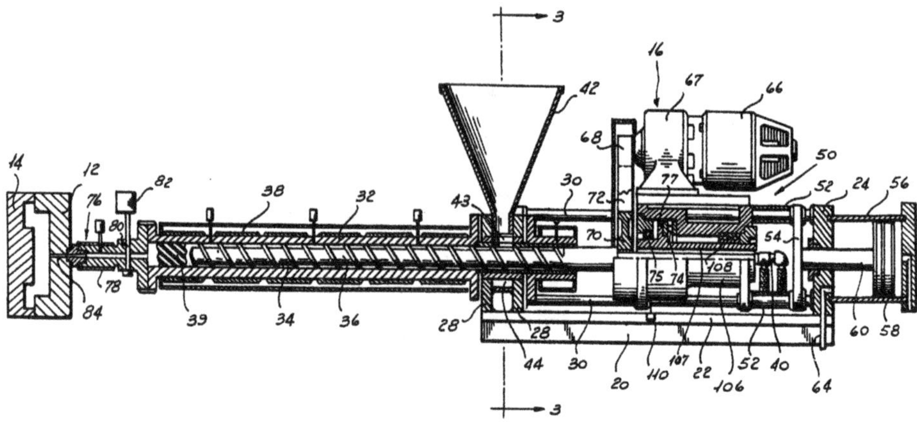

Figure 1.9 Willert's 1956 reciprocating screw injection molding machine.

back in the axial direction by the newly molten polymer melt. The screw plunges forward once enough material is plasticated, injecting the melt into the hollow cavity. The screw continues to force melt into the cavity until the gate freezes shut. Willert's patent was issued in 1956. Thus, for the second time the reciprocating screw injection molding machine was born. The first machine that used Beck's and Willert's systems was a 600-ton Reed Prentice injection molder in 1953. Despite debates and arguments that came early on from a fairly conservative plastics industry, this type of machine proved to be superior to the torpedo-plunger system, and has prevailed as the preferred technique to injection mold polymer articles.

Today's injection molding machines are direct grandchildren of Beck's and Willert's reciprocating screw injection molding machines. A modern injection molding machine with its most important elements is shown in Fig. 1.10. As shown in the figure, the main components of the injection molding machine are the plasticating unit, clamping unit, mold, and the hydraulic unit.

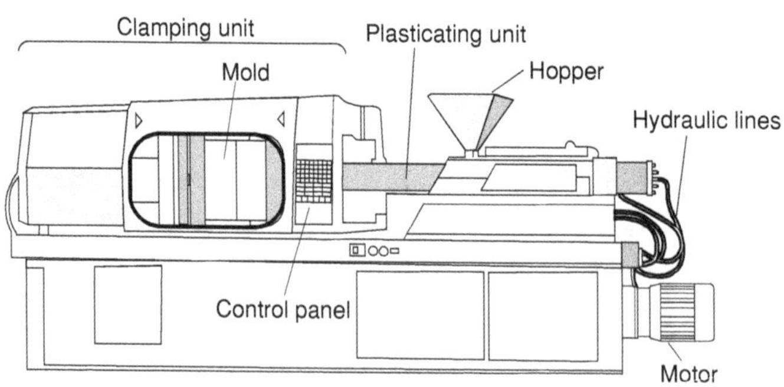

Figure 1.10 Schematic of an injection molding machine.

Injection molding machines today are classified by the following international convention*

$$\text{Manufacturer type } T/P$$

where T is the clamping force in metric tons and P is defined as

$$P = \frac{v_{max}\, p_{max}}{1000} \tag{1.1}$$

where v_{max} is the maximum shot size in cubic centimeters and p_{max} is the maximum injection pressure in bar. The clamping forced T can be as low as 1 metric ton for small machines, and as high as 11,000 tons.

1.2 The Reciprocating Screw Injection Molding Machine

1.2.1 The Plasticating and Injection Unit

A plasticating and injection unit is shown in Fig. 1.11. The major tasks of the plasticating unit are to melt the polymer, to accumulate the melt in the screw chamber, to inject the melt into the cavity, and to maintain the holding pressure during cooling.

The main elements of the plasticating unit follow:

- Hopper
- Screw
- Heating bands
- Check valve
- Nozzle

The hopper, heating bands, and the screw are similar to a plasticating single screw extruder, except that the screw in an injection molding machine can slide back and

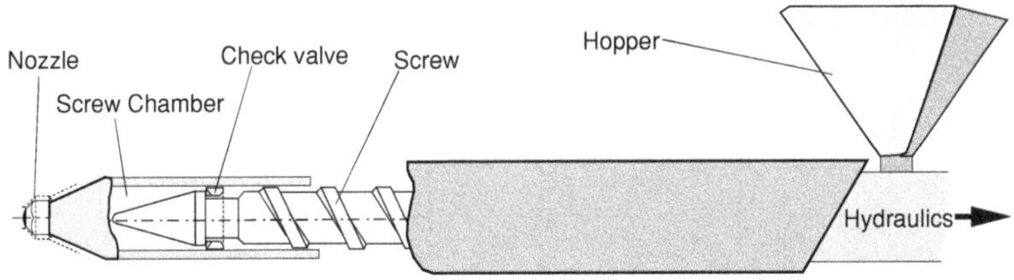

Figure 1.11 Schematic of the plasticating unit.

* The old U.S. convention uses MANUFACTURER T-v where T is the clamping force in British tons and v is the shot size in ounces of polystyrene.

forth to allow for melt accumulation and injection. This characteristic gives it the name *reciprocating screw*. The maximum stroke in a reciprocating screw is three times the screw diameter.

Although the most common screw used in injection molding machines is the three-zone plasticating screw, two-stage vented screws are often used to extract moisture and monomer gases just after the melting stage.

The check valve, or nonreturn valve, is at the end of the screw and enables it to work as a plunger during injection and packing without allowing polymer melt back flow into the screw channel. A check valve is depicted in Fig. 1.11, and its operation is described in the next section. A high-quality check valve allows less than 5% of the melt back into the screw channel during injection and packing.

The nozzle is at the end of the plasticating unit and fits tightly against the sprue bushing during injection. The nozzle type is either open or closed. The open nozzle is the simplest, rendering the lowest pressure consumption.

1.2.2 The Clamping Unit

The job of a clamping unit in an injection molding machine is to open and close the mold, and to close the mold tightly to avoid flash during filling and holding. Modern injection molding machines have three predominant clamping types: mechanical, hydraulic, and a combination of the two.

Figure 1.12 presents a toggle mechanism in the open and closed mold positions. Although the toggle is essentially a mechanical device, it is actuated by a hydraulic cylinder. The advantage of using a toggle mechanism is that as the mold approaches closure, the available closing force increases and the closing decelerates significantly; however, the toggle mechanism only transmits its maximum closing force when the system is fully extended.

Figure 1.13 presents a schematic of a hydraulic clamping unit in the open and closed positions. The advantages of the hydraulic system is that a maximum clamp-

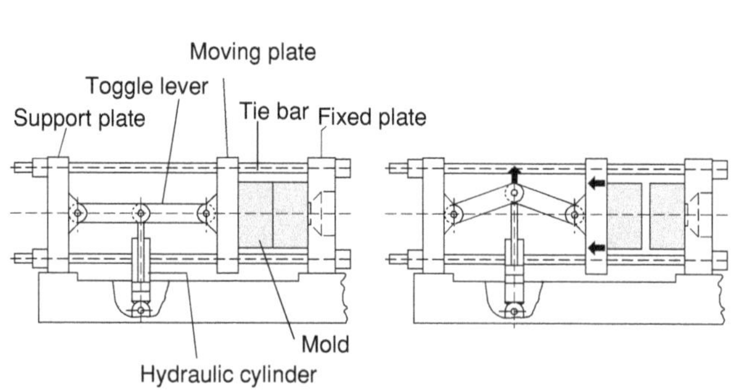

Figure 1.12 Clamping unit with a toggle mechanism.

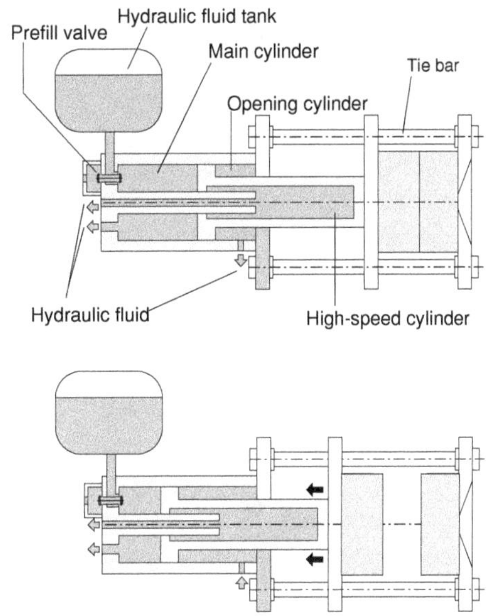

Figure 1.13 Hydraulic clamping unit.

ing force is attained at any mold closing position and that the system can take different mold sizes without major system adjustments.

1.2.3 The Mold Cavity

The central point in an injection molding machine is the mold. The mold distributes polymer melt into and throughout the cavities, shapes the part, cools the melt, and ejects the finished product. As depicted in Fig. 1.14, the mold is custom-made and consists of the following elements:

- Sprue and runner system
- Gate
- Mold cavity
- Cooling system (thermoplastics)
- Ejector system

During mold filling, the melt flows through the sprue and is distributed into the cavities by the runners, as in Fig. 1.15.

The runner system in Fig. 1.15(a) is symmetric whereby all cavities fill at the same time and cause the polymer to fill all cavities in the same way. The disadvantage of this balanced runner system is that the flow paths are long, leading to high material and pressure consumption. On the other hand, the asymmetric runner system shown in Fig. 1.15(b) leads to parts of different quality. Equal filling of the mold cavities can also be achieved by varying runner diameters. There are two types of runner

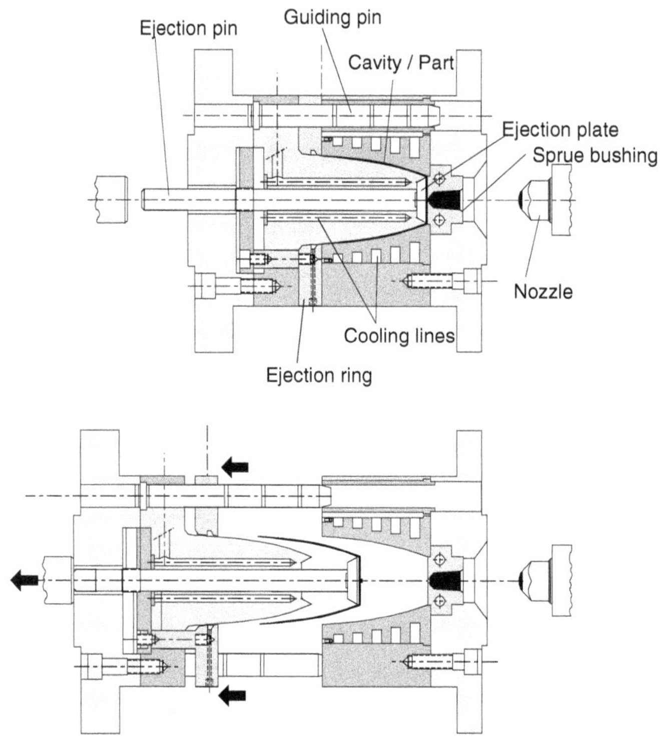

Figure 1.14 An injection mold.

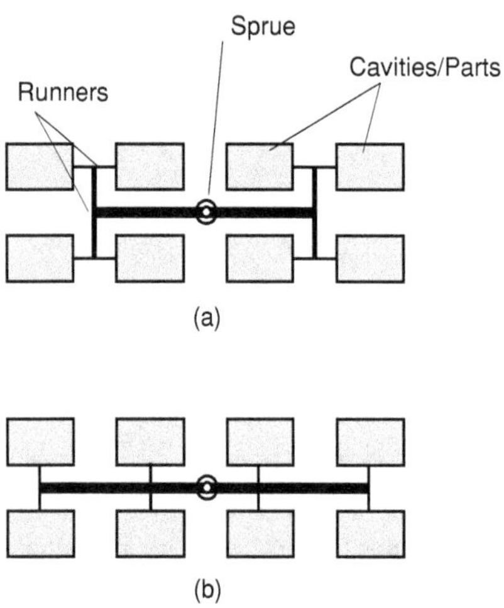

Figure 1.15 Schematic of different runner system arrangements.

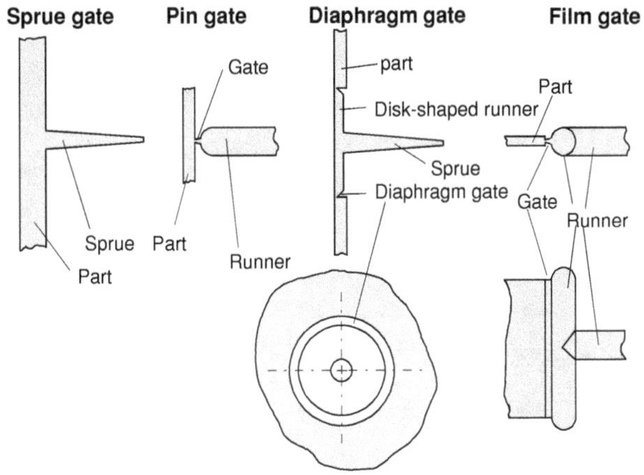

Figure 1.16 Schematic of different gating systems.

systems—cold and hot. Cold runners are ejected with the part, and are trimmed after mold removal. The advantage of the cold runner is lower mold cost. The hot runner keeps the polymer at or above its melt temperature. The material stays in the runners system after ejection, and is injected into the cavity in the following cycle. There are two types of hot runner system: externally and internally heated. Externally heated runners have a heating element surrounding the runner that keeps the polymer isothermal. Internally heated runners have a heating element running along the center of the runner, maintaining a polymer melt that is warmer at its center and possibly solidified along the outer runner surface. Although a hot runner system considerably increases mold cost, its advantages include elimination of trim and lower pressures for injection.

When large items are injection molded, the sprue sometimes serves as the gate, as shown in Fig. 1.16. The sprue must subsequently be trimmed, often requiring further surface finishing. On the other hand, a pin-type gate (Fig. 1.16) is a small orifice that connects the sprue or the runners to the mold cavity. The part is easily broken off from such a gate, leaving only a small mark that usually does not require finishing. Other types of gates, also shown in Fig. 1.16, are film gates, which are used to eliminate orientation, and disk or diaphragm gates, which are used for symmetric parts such as compact discs.

1.3 The Injection Molding Cycle

The sequence of events during the injection molding of a plastic part, as shown in Fig. 1.17, is called the *injection molding cycle*. The cycle begins when the mold closes, followed by the injection of the polymer into the mold cavity. Once the cavity is filled,

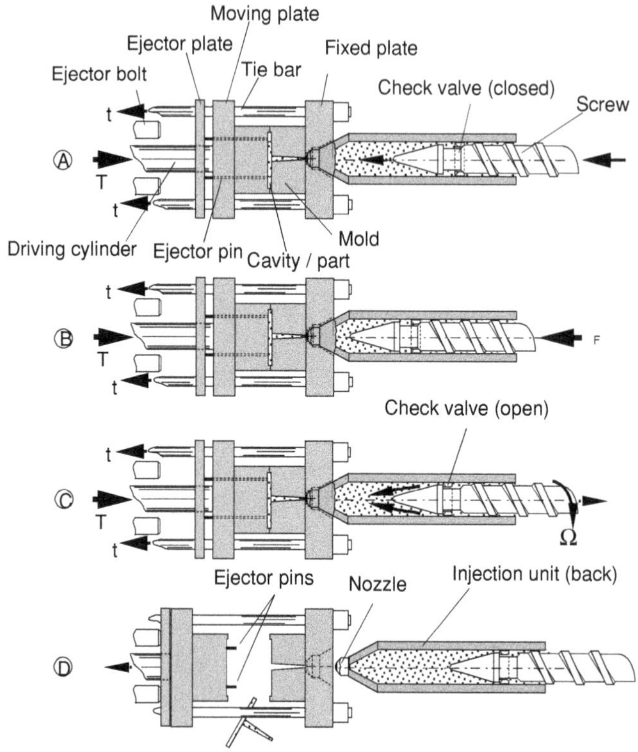

Figure 1.17 Sequence of events during an injection molding cycle.

a holding pressure is maintained to compensate for material shrinkage. In the next step, the screw turns, feeding the next shot to the front of the screw. This causes the screw to retract as the next shot is prepared. Once the part is sufficiently cool, the mold opens and the part is ejected. Figure 1.18 presents the sequence of events during the injection molding cycle. The figure shows that the cycle time is dominated by the cooling of the part inside the mold cavity. The total cycle time can be calculated using

$$t_{cycle} = t_{closing} + t_{cooling} + t_{ejection} \tag{1.2}$$

where the closing and ejection times, $t_{closing}$ and $t_{ejection}$, can last from a fraction of second to a few seconds, depending on the size of the mold and machine.

Using the average part temperature history and the cavity pressure history, the process can be followed and assessed using the *P-v-T* diagram as depicted in Fig. 1.19. [6,7]. To follow the process on the *P-v-T* diagram, we must transfer both the temperature and the pressure at matching times. The diagram reveals four basic processes: an isothermal injection (0–1) with pressure rising to the holding pressure (1–2), an isobaric cooling process during the holding cycle (2–3), an isochoric cooling

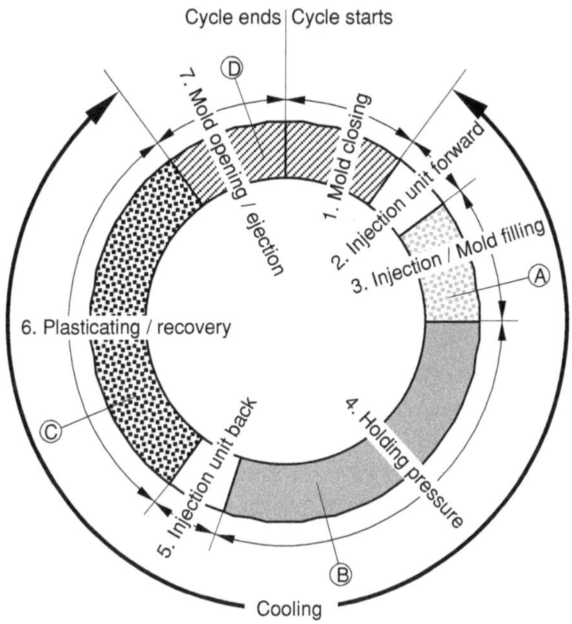

Figure 1.18 Injection molding cycle.

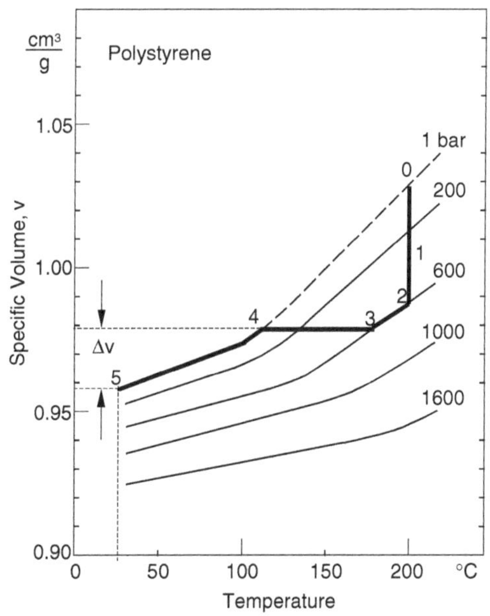

Figure 1.19 Trace of an injection molding cycle in a PvT diagram.

after the gate freezes with a pressure drop to atmospheric (3–4), and then isobaric cooling to room temperature (4–5).

The point on the PvT diagram where the final isobaric cooling begins (4) controls the total part shrinkage, Δ_v. This point is influenced by the two main processing conditions—the melt temperature, T_M, and the holding pressure, P_H, as depicted in Fig. 1.20. Here, the process in Fig. 1.19 is compared with one with a higher holding pressure and smaller part shrinkage. Of course, there is an infinite combination of conditions that render acceptable parts, bound by minimum and maximum temperatures and pressures. Figure 1.21 presents *the molding diagram* with all limiting conditions. The melt temperature is bound by a low temperature that results in a *short shot* or unfilled cavity and a high temperature that leads to material degradation. The hold pressure is bound by a low pressure that leads to excessive shrinkage or low part weight, and a high pressure that results in *flash*. Flash results when the cavity pressure force exceeds the machine clamping force, leading to melt flow across the mold parting line. The holding pressure determines the corresponding clamping force required to size the injection molding machine. An experienced polymer processing engineer or computer simulation program can usually determine which injection molding machine is appropriate for a specific application. For the untrained polymer processing engineer, finding this appropriate holding pressure and its corresponding mold clamping force can be difficult.

With difficulty, one can control and predict the component's shape and residual stresses at room temperature. For example, *sink marks* in the final product are caused

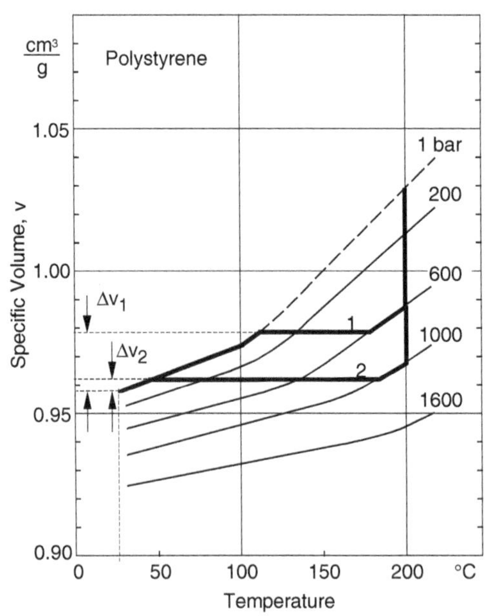

Figure 1.20 Trace of two different injection molding cycles in a PvT diagram.

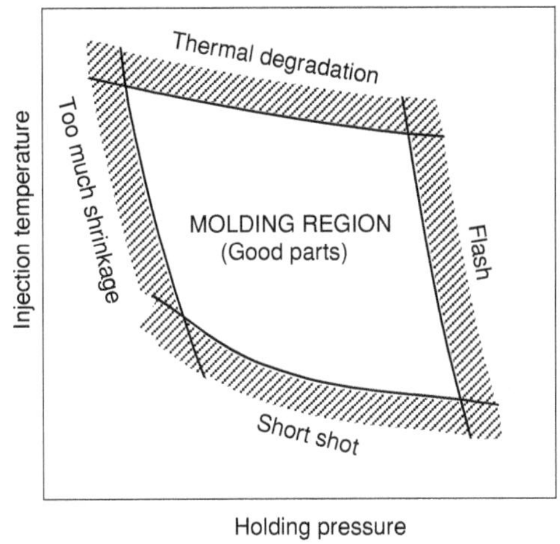

Figure 1.21 The molding diagram.

by material shrinkage during cooling, and residual stresses can lead to environmental stress cracking under certain conditions [8]. Warpage in the final product is often caused by processing conditions that lead to asymmetric residual stress distributions through the part thickness. The formation of residual stresses in injection molded parts is attributed to two major coupled factors—cooling and flow stresses. The first and most important is the residual stress formed as a result of rapid cooling.

1.4 Related Injection Molding Processes

Although most injection molding processes are covered by the conventional process description discussed earlier, there are several important molding variations including:

- Co-injection (sandwich) molding
- Fusible (lost, soluble) core injection molding
- Gas-assisted injection molding
- In-mold decoration and in-mold lamination
- Injection-compression molding
- Insert and outsert molding
- Lamellar (microlayer) injection molding
- Low-pressure injection molding
- Microinjection molding
- Microcellular molding
- Multicomponent injection molding (overmolding)

- Multiple live-feed injection molding
- Powder injection molding
- Push–pull injection molding
- Reaction injection molding
- Resin transfer molding
- Rheomolding
- Structural foam injection molding
- Structural reaction injection molding
- Thin-wall molding
- Vibration gas injection molding
- Water assisted injection molding
- Rubber injection
- Injection molding of liquid silicone rubber

These processes are all discussed in detail in Chap. 9 of this handbook.

References

1. Hyatt, I. S., Hyatt, J. W., Improvement in Processes and Apparatus for Manufacturing Pyroxyline, U.S. Patent No. 133,229, (1872).
2. Jacobi, H. R., *Kunststoffe* (1965), 55, 173.
3. Buchholz, H., Process of Making Molded Articles, U.S. Patent No. 1,810,126, (1931).
4. Dubois, J. H., *Plastics History U.S.A.* Chaners Books (1972), Boston.
5. Beck, H., *Kuststoffe* (1961), 51, 62.
6. Sonntag, R., *Kunststoffe* (1985), 75, 5.
7. Greener, J., *Polym. Eng. Sci.* (1986), 26, 886.
8. Michaeli, W., Lauterbach, M., *Kunststoffe* (1989), 79, 852.
9. Osswald, T. A., Menges, G., *Materials Science of Polymers for Engineers*, Hanser (1996), Munich.

2 Injection Molding Materials

T. A. Osswald*

Of all the materials that are injection molded, polymers** play the most significant role. As the word suggests, polymers are materials composed of molecules of high molecular weight. These large molecules are generally called *macromolecules*. The unique material properties of polymers and the versatility of processing methods are attributed to their molecular structure. The ease with which polymers and *plastics*† are processed makes them the most sought after materials today for many applications. Because of their low density and their ability to be shaped and molded at relatively low temperatures, compared with traditional materials such as metals, plastics, and polymers are the material of choice when integrating several parts into a single component—a design step usually called *part consolidation*. In fact, parts and components, which are traditionally made of wood, metal, ceramics, or glass, are frequently redesigned with plastics.

2.1 Historical Background

Natural polymeric materials such as rubber have been in use for thousands of years. Natural rubber, also known as *caoutchouc* (crying trees), has been used by South American Indians in the manufacture of waterproof containers, shoes, torches, and squeeze bulb pumps. The first Spanish explorers of Haiti and Mexico reported that natives played games on clay courts with rubber balls [1]. Rubber trees were first mentioned in *De Orbe Novo*, originally published in Latin by Pietro Martire d'Anghiera in 1516. French explorer and mathematician Charles Maria de la Condamine, who was sent to Peru by the French *Academie des Sciences*, brought caoutchouc from South America to Europe in the 1740s. In his report [2] he mentions several rubber items made by native South Americans, including a pistonless pump composed of a rubber pear with a hole in the bottom. He points out that the most remarkable property of natural rubber is its great elasticity.

The first chemical investigations on rubber or *gummi elasticum* were published by the Frenchman Macquer in 1761. It was not until the twentieth century, however,

* With Contributions from M. DeGreiff and G. Holden.

** From the Greek, *poli*, which means many, and *meros*, which means parts.

† The term *plastics* describes the compound of a polymer with one or more additives.

that the molecular architecture of polymers was well understood. Soon after its introduction to Europe, various uses were found for natural rubber. Gossart manufactured the first polymer tubes in 1768 by wrapping rubber sheets around glass pipes. Small rubber blocks were introduced during the same time period to erase lead pencil marks from paper. In fact, the word *rubber* originates from this specific application—*rubbing*.

These new materials slowly evolved from novelty status as a result of new applications and processing equipment. Although the screw press, which is the predecessor of today's compression molding press, was patented in 1818 by McPherson Smith [3]. The first documented *polymer processing* machinery dates to 1820, when Thomas Hancock invented a rubber masticator. This masticator, which consists of a toothed rotor in a toothed cylindrical cavity [4], was used to reclaim rubber scraps which resulted from the manual manufacturing process of elastic straps, perhaps the first recycling effort. In 1833 the development of the vulcanization process by Charles Goodyear [5] greatly enhanced the properties of natural rubber. In 1836 Edwin M. Chaffee invented the two-roll steam-heated mill, the predecessor of the calender, for continuously mixing additives into rubber for the manufacture of rubber-coated textiles and leathers. As early as 1845, presses and dies were used to mold buttons, jewelry, dominoes, and other novelties out of shellac and gutta-percha. *Gutta-percha* (rubber clump), which is a gum found in trees similar to rubber, became the first wire insulation and was used for ocean cable insulation for many years.

The ram-type extruder was invented by Henry Bewley and Richard Brooman in 1845. The first *polymer processing* screw extruder, the most influential equipment in polymer processing, was patented by an Englishman named Mathew Gray in 1879 for the purpose of wire coating. The screw pump is attributed to Archimedes, however, and the actual invention of the screw extruder by A. G. DeWolfe of the U.S. dates back to the early 1860s.

Cellulose nitrate plasticized by camphor, which was possibly the first thermoplastic, was patented by Isaiah and John Hyatt in 1870. Based on experience from metal die casting, the Hyatt brothers built and patented the first injection molding machine in 1872 to mold cellulose materials [6].

With the mass production of rubber, gutta-percha, cellulose, and shellac articles during the height of the industrial revolution, the polymer processing industry after 1870 saw the invention and development of internal kneading and mixing machines for the processing and preparation of raw materials [7]. A notable invention was the Banbury mixer, developed by Fernley Banbury in 1916. With some modifications this mixer is still used for rubber compounding.

Bakelite, developed by Leo Baekeland in 1907, was the first synthetically developed polymer. Bakelite, which is also known as phenolic, is a thermoset resin that reacts by condensation polymerization that occurs when phenol and formaldehyde are mixed and heated.

In 1924, Hermann Staudinger proposed a model that described polymers as linear molecular chains. Once this model was accepted by other scientists, the concept used to synthesize new materials was realized. Cellulose acetate and polyvinyl chloride (PVC) [8] were developed in 1927. Because of its higher wear resistance, polyvinyl chloride replaced shellac for phonograph records in the early 1930s. Wallace Carothers pioneered condensation polymers such as polyesters and polyamides. It was not until this point that the scientific world was finally convinced of the validity of Staudinger's work. Polyamides, first called Nylon, were set into production in 1938. Polyvinyl acetate, acrylic polymers, polystyrene (PS), polyurethanes, and melamine were also developed in the 1930s [9].

The first single-screw extruder designed for the processing of thermoplastic polymers was built circa 1935 at the Paul Troester Maschinenfabrik [10]. Roberto Colombo developed a twin-screw extruder for thermoplastics at around that same time.

World War II and the postwar years saw accelerated development of new polymeric materials. Polyethylene (PE), polytetrafluoroethylene, epoxies, and acrylonitrile-butadiene-styrene (ABS) were developed in the 1940s, and linear polyethylene, polypropylene (PP), polyacetal, polyethylene terephthalate (PET), polycarbonate (PC), and many more materials came in the 1950s. The 1970s saw the development of new polymers such as polyphenylene sulfide, and the 1980s beget liquid crystalline polymers.

Developing and synthesizing new polymeric materials has become increasingly expensive and difficult. Developing new engineering materials by blending or mixing two or more polymers or by modifying existing ones with plasticizers is now widely accepted.

The world's yearly production of polymer resins has experienced steady growth since the turn of the last century, with growth predicted well into the twenty-first century. Figure 2.1 [11] presents the world's yearly polymer production in millions of tons. In developed countries, the growth in annual polymer production has decreased; however, developing countries in South America and Asia are now starting to experience tremendous growth.

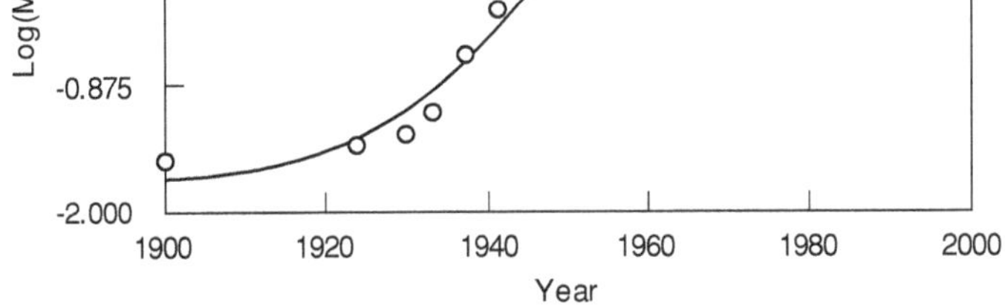

Figure 2.1 World yearly plastics production since 1900.

Of the more than 31 million tons of polymers produced in the United States in 1993, 90% were thermoplastics. As mentioned earlier, there are thousands of grades of polymers available to the design engineer. These cover a wide range of properties, from soft to hard, ductile to brittle, and weak to tough. Figure 2.2 shows this range by plotting important average properties for selected polymers. The abbreviations used in Fig. 2.2 are defined in Table 2.1. The values for each material in Fig. 2.2 are representative averages. Examples of various polymeric materials and their applications are given in Table 2.2.

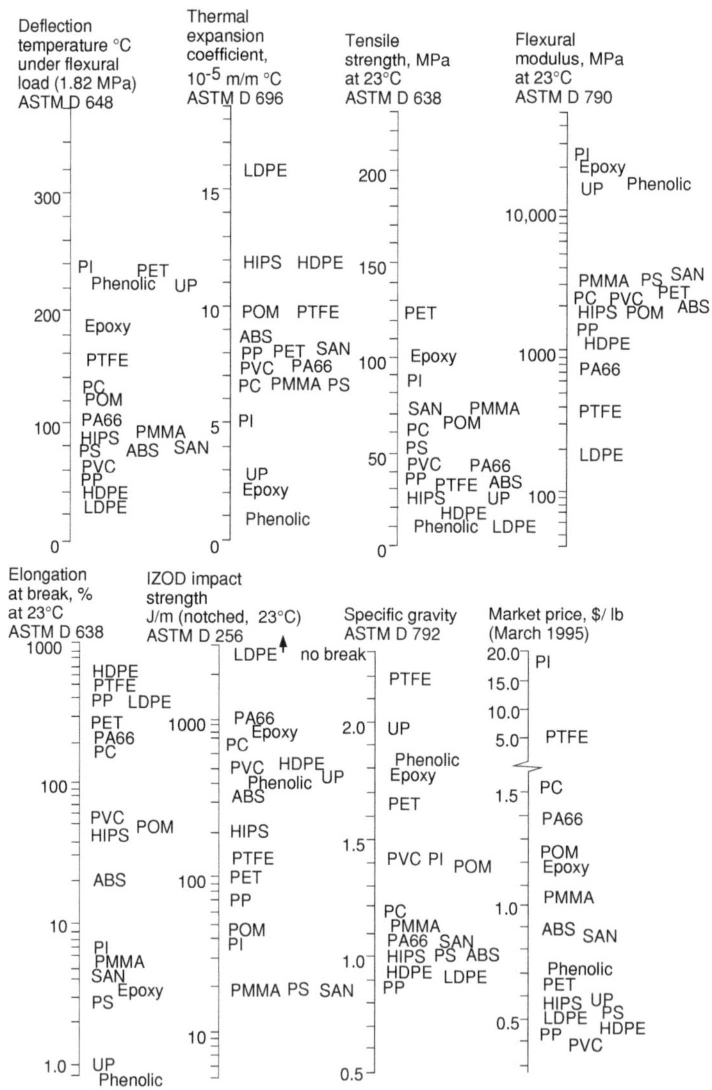

Figure 2.2 Average properties for common polymers.

Table 2.1 Abbreviation of Common Polymers

Polymer	ASTM D1600-93 Abbreviation
Acrylonitrile-butadiene-styrene	ABS
Epoxy	EP
High density polyethylene	HDPE
Impact resistant polystyrene	IPS (HIPS)
Low density polyethylene	LDPE
Phenol-formaldehyde (Phenolic)	PF
Polyacetal (polyoxymethylene)	POM
Polyamide 66	PA 66
Polycarbonate	PC
Polyethylene terephthalate	PET
Polyimide	PI
Polymethyl methacrylate	PMMA
Polypropylene	PP
Polystyrene	PS
Polytetrafluoroethylene	PTFE
Polyvinyl chloride	PVC
Styrene-acrylonitrile copolymer	SAN
Unsaturated polyester	UP

The relatively low stiffness of polymeric materials is attributed to their molecular structure, which allows relative movement with ease while under stress. The strength and stiffness of individual polymer chains, however, are much higher than the measured properties of the bulk. For example, polyethylene, whose molecules have a theoretical stiffness of 300,000 MPa, has a bulk stiffness of only 1000 MPa [12,13]. By introducing high molecular orientation, the stiffness and strength of a polymer can be substantially increased. In the case of *ultradrawn, ultrahigh molecular weight high-density polyethylene* (UHMHDPE) fibers can exceed a stiffness 200,000 MPa [13].

2.2 Macromolecular Structure of Polymers

Polymers are macromolecular structures that are generated synthetically or through natural processes. Cotton, silk, natural rubber, ivory, amber, and wood are a few materials that occur naturally with an organic macromolecular structure, whereas natural inorganic materials include quartz and glass. The other class of organic materials with a macromolecular structure are synthetic polymers, which are generated through addition polymerization or condensation polymerization.

In addition to polymerization, monomers are added to each other by breaking the double-bonds that exist between carbon atoms, allowing them to link to neighboring

Table 2.2 Common Polymers and Some of Their Applications

Polymer	Some Applications
Thermoplastics	
Amorphous	
Polystyrene	Mass-produced transparent articles, packaging, insulation (foamed)
Polymethyl methacrylate	Skylights, airplane windows, lenses, stop lights
Polycarbonate	Helmets, hockey masks, blinker lights, head lights
Unplasticized polyvinyl chloride	Tubes, window frames, siding, bottles, packaging
Plasticized polyvinyl chloride	Shoes, hoses, calendered films and sheets (floors and upholstery)
Semi-crystalline	
High density polyethylene	Milk and soap bottles, mass produced household goods
Low density polyethylene	Mass produced household goods, grocery bags
Polypropylene	Housings for electric appliances, auto battery cases
Polytetrafluoroethylene	Coating of cooking pans, lubricant-free bearings
Polyamide	Gears, bolts, skate wheels, pipes, fishing line, textiles, ropes
Thermoplastic Elastomers	
PVC-NBR-Plasticizer Blend	Overmolding handles or covers on PVC, SAN, ABS, and SMA parts
PA-Elastomer copolymers	Overmolding handles, covers or layers on PA, and PC parts
Thermosets	
Epoxy	Adhesive, matrix in fiber-reinforced composite parts
Melamine	Decorative heat-resistant surfaces for kitchens and furniture, dishes
Phenolics	Heat-resistant handles for pans, irons and toasters, electric outlets
Unsaturated polyester	Sinks and tubs, automotive body panels (with glass fiber)
Elastomers	
Polybutadiene	Automotive tires, golf ball skin
Ethylene propylene rubber	Automotive radiator hoses and window seals, roof covering
Natural rubber (polyisoprene)	Automotive tires, engine mounts
Polyurethane elastomer	Roller skate wheels, automotive seats (foamed), shoe soles (foamed)
Silicone rubber	Seals, flexible hoses for medical applications
Styrene butadiene rubber	Automotive tire treads

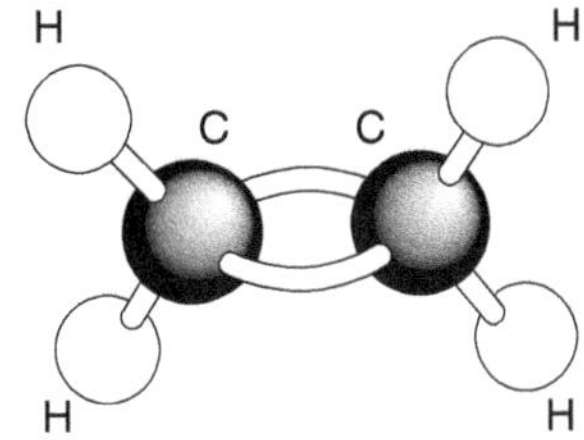

Figure 2.3 Schematic representation of an ethylene monomer.

carbon atoms to form long chains. The simplest example is the addition of ethylene monomers, schematically shown in Fig. 2.3, to form polyethylene molecules as shown in Fig. 2.4. The schematic shown in Fig. 2.4 can also be written symbolically as shown in Fig. 2.5. Here, the subscript n represents the number of repeat units which determines the molecular weight of a polymer. The number of repeat units is more commonly referred to as the *degree of polymerization*. Other examples of addition polymerization include polypropylene, polyvinyl chloride, and polystyrene, as shown in Fig. 2.6. The side groups, such as CH_3 for polypropylene and Cl for polyvinyl chloride, are sometimes referred to as the *X groups*.

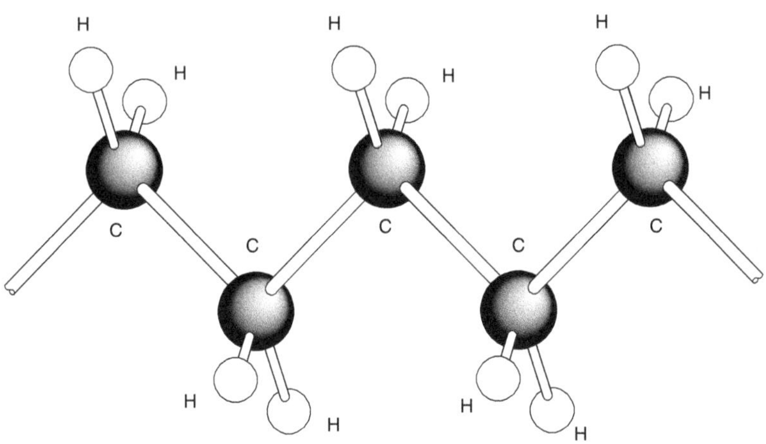

Figure 2.4 Schematic representation of a polyethylene molecule.

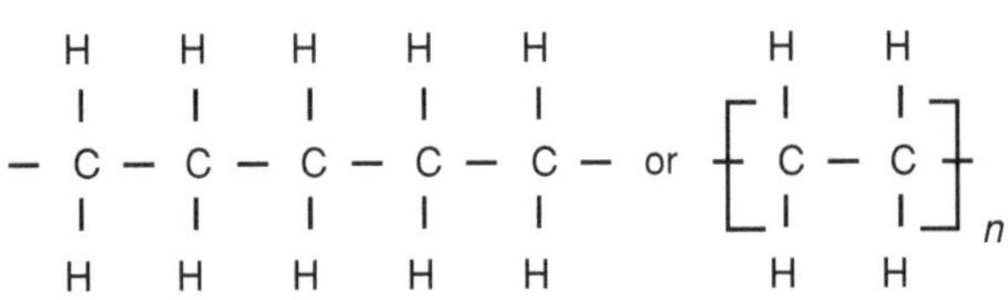

Figure 2.5 Symbolic representation of a polyethylene molecule.

Another technique to produce macromolecular materials is condensation polymerization. Condensation polymerization occurs when two components with end-groups that react with each other are mixed. When they are stoichiometric, these end-groups react, linking them to chains that leave a by-product such as water. A common polymer made by condensation polymerization is polyamide, whereas diamine and diacid groups react to form polyamide and water as shown in Fig. 2.7.

In the molecular level, there are several forces that hold a polymeric material together. The most basic force present are covalent bonds that hold together the backbone of a polymer molecule, such as the —C—C bond. The energy holding together two carbon atoms is about 350 KJ/mol, which would result in a polymer component strength between 1.4×10^4 and 1.9×10^4 MPa when translated. As will be seen in Chapter 8, however, the strength of polymers only lie between 10 and 100 MPa.

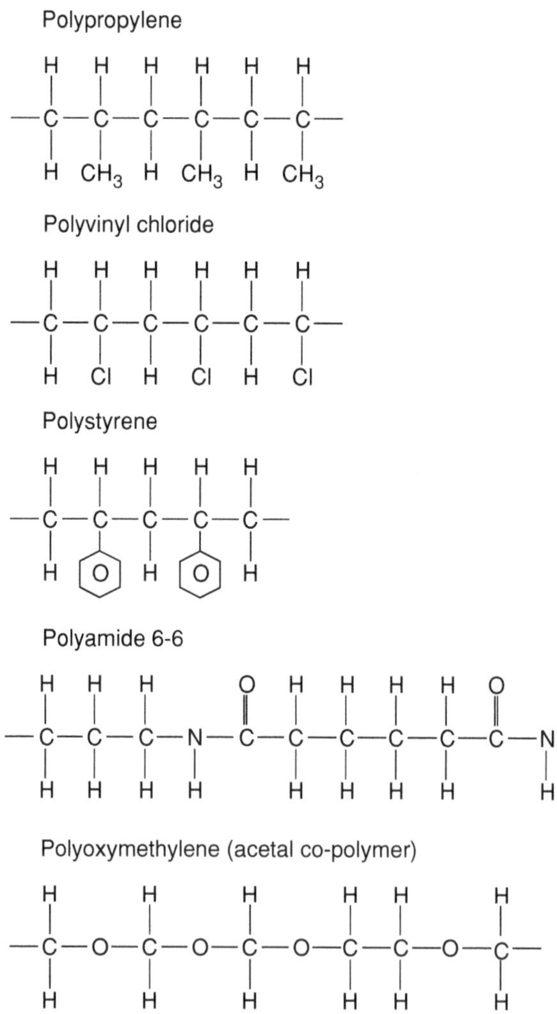

Figure 2.6 Symbolic representation of several thermoplastic molecules.

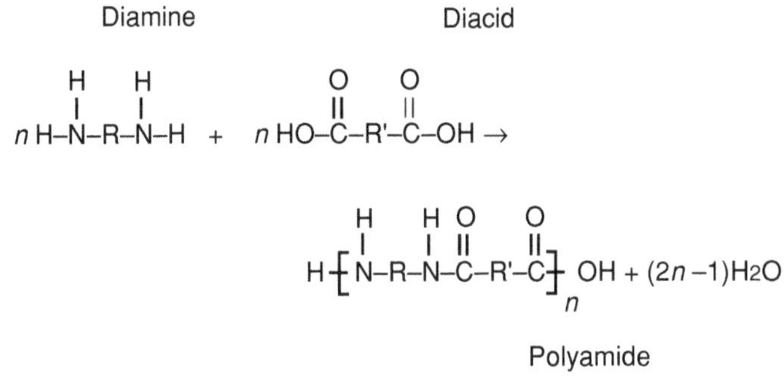

Figure 2.7 Symbolic representation of the condensation polymerization of polyamide.

Because of the comparatively low strength found in polymer components, it can be deduced that the forces holding a polymer component together do not come from the —C—C bonds, but instead from intermolecular forces, or the so-called Van-der-Waals forces. Thus, it becomes clear that as a polymer sample is heated, the distance between the molecules increases as the vibration amplitude of the molecules increases. The vibration amplitude increase allows the molecules to move more freely, enabling the material to flow in the macroscopic level.

Another important point is that as solvents are introduced between the molecules, their intermolecular separation increases, which leads to a reduction in stiffness. This concept can be implemented by introducing *plasticizers* into the material, thus lowering the glass transition temperature below room temperature and bringing out rubber elastic material properties.

2.3 Molecular Weight

A polymeric material may consist of polymer chains of various lengths or repeat units. Hence, the molecular weight is determined by the average or mean molecular weight which is defined by

$$\overline{M} = \frac{W}{N} \tag{2.1}$$

where W is the weight of the sample and N the number of moles in the sample.

The properties of polymeric material are strongly linked to the molecular weight of the polymer as shown schematically in Fig. 2.8. A polymer such as polystyrene is stiff and brittle at room temperature with a degree of polymerization, n, of 1000. Polystyrene with a degree of polymerization of 10 is sticky and soft at room temperature. Figure 2.9 shows the relation between molecular weight, temperature, and properties of a typical polymeric material. The stiffness properties reach an asymptotic

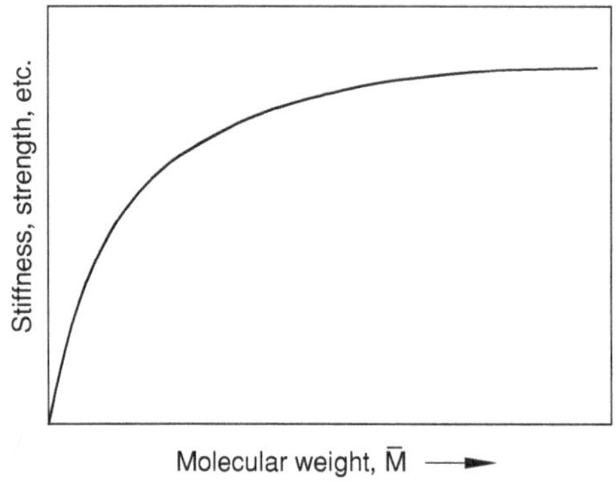

Figure 2.8 Influence of molecular weight on mechanical properties.

maximum, whereas the flow temperature increases with molecular weight. On the other hand, the degradation temperature steadily decreases with increasing molecular weight. Hence, it is necessary to find the molecular weight that renders ideal material properties for the finished polymer product, while having flow properties that make it easy to shape the material during the manufacturing process. It is important to mention that the temperature scale in Fig. 2.9 corresponds to a specific time scale, for example time required for a polymer molecule to flow through an injection molding runner system. If the time scale is reduced (e.g., by increasing the injection speed), then the molecules have more difficulty sliding past each other. This would require a somewhat higher temperature to assure flow. In fact, at a specific temperature, a polymer melt may behave as a solid if the time scale is reduced sufficiently. Hence, for this new time scale the stiffness properties and flow temperature curves must be shifted upward on the temperature scale. A limiting factor is that the thermal

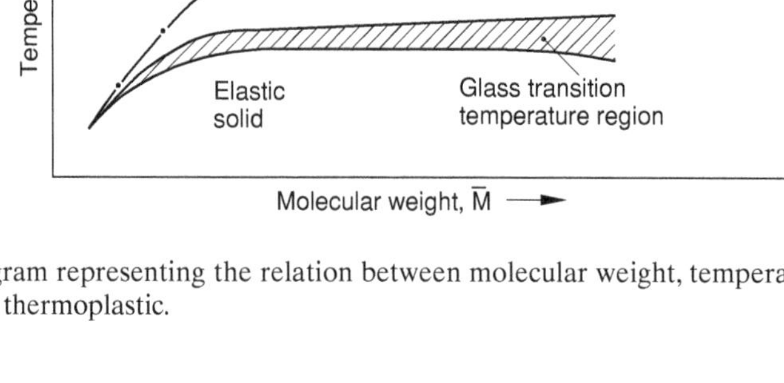

Figure 2.9 Diagram representing the relation between molecular weight, temperature, and properties of a typical thermoplastic.

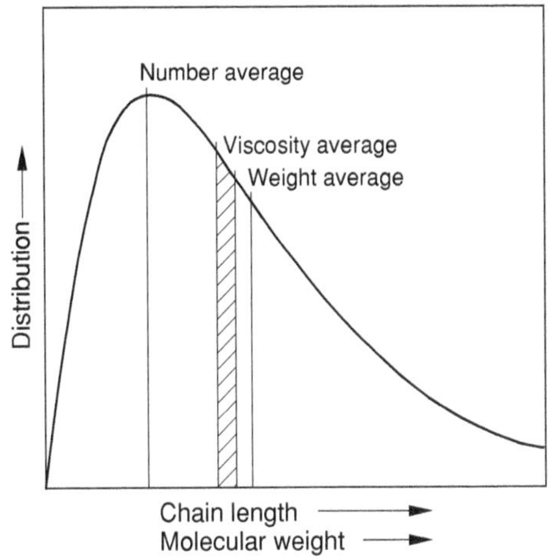

Figure 2.10 Molecular weight distribution of a typical thermoplastic.

degradation curve remains fixed, limiting processing conditions to remain above certain time scales. This relation between time, or time scale, and temperature is often referred to as *time—temperature superposition principle* and is discussed in detail in the literature [14].

With the possible exception of some naturally occurring polymers, most polymers have a molecular weight distribution such as shown in Fig. 2.10. We can define a number average, weight average, and viscosity average* for such a molecular weight distribution function. The number average is the first moment and the weight average the second moment of the distribution function. In terms of mechanics this is equivalent to the center of gravity and the radius of gyration as first and second moments, respectively. The *viscosity average* relates the molecular weight of a polymer to the measured viscosity such as is shown in Fig. 2.11.

Figure 2.11 [15] presents the viscosity of various undiluted polymers as a function of molecular weight. The figure shows how for all these polymers the viscosity goes from a linear (slope = 1) to a power dependence (slope = 3.4) at some critical molecular weight. The linear relation is sometimes referred to as *Staudinger's rule* [16] and applies for a perfectly *monodispersed polymer*† where the friction between the molecules increases proportionally to the molecule's length. The increased slope of 3.4 is due to molecular entanglement due to the long molecular chains.

A measure of the broadness of a polymer's molecular weight distribution is the *polydispersity index* defined by

$$\mathrm{PI} = \frac{\overline{M}_w}{\overline{M}_n} \tag{2.2}$$

* There are other definitions of molecular weight which depend on the type of measuring technique.

† A monodispersed polymer is composed of a single molecular weight species.

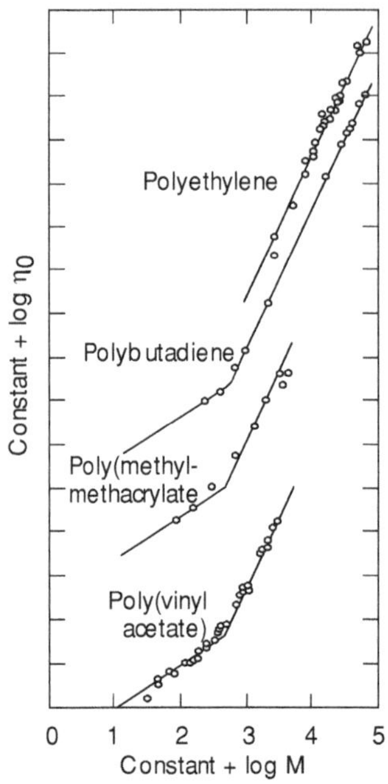

Figure 2.11 Zero shear rate viscosity for various polymers as a function of weight average molecular weight.

Figure 2.12 [17] presents a plot of flexural strength versus melt flow index* for polystyrene samples with three different polydispersity indexes. The figure shows that low polydispersity index—grade materials render high-strength properties and flowability, or processing ease, than high polydispersity index grades.

2.4 Conformation and Configuration of Polymer Molecules

The conformation and configuration of the polymer molecules have a great influence on the properties of the polymer component.

The conformation describes the preferential spatial positions of the atoms in a molecule. It is described by the polarity flexibility and regularity of the macromolecule. Carbon atoms are typically tetravalent, which means that they are surrounded by four substituents in a symmetric tetrahedral geometry.

The most common example is methane, CH_4, which is schematically depicted in Fig. 2.13. As the figure demonstrates, the tetrahedral geometry sets the bond angle at 109.5 degrees. This angle is maintained between carbon atoms on the backbone of

* The melt flow index is the mass (grams) extruded through a capillary in a 10-minute period while applying a constant pressure. Increasing melt flow index signifies decreasing molecular weight. The melt flow index is discussed in more detail in Chapter 3.

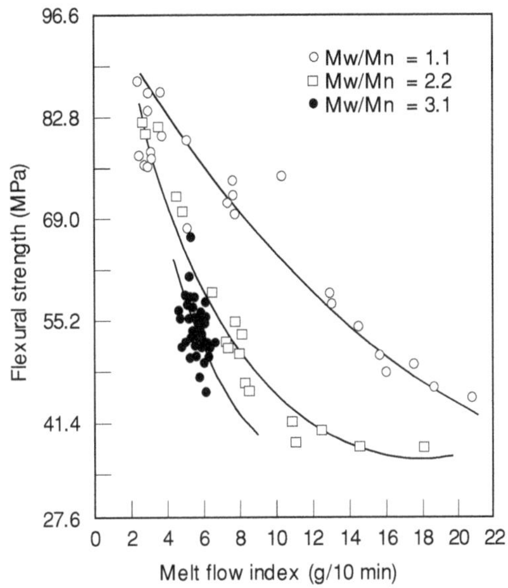

Figure 2.12 Effect of molecular weight on the strength-melt flow index interrelationship of polystyrene for three polydispersity indices.

a polymer molecule, as shown in Fig. 2.14. As shown in the figure, each individual axis in the carbon backbone is free to rotate.

The configuration gives the information about the distribution and spatial organization of the molecule. During polymerization it is possible to place the *X* groups on the carbon–carbon backbone in different directions. The order in which they are arranged is called the *atacticity*. The polymers with side groups that are placed in a random matter are called *atactic*. The polymers whose side groups are all on the same

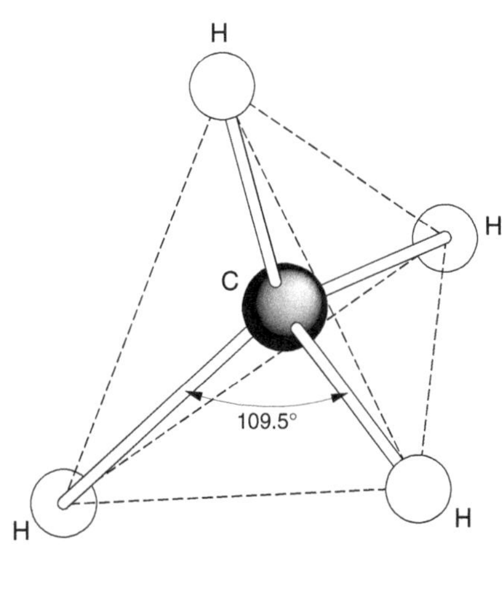

Figure 2.13 Schematic of tetrahedron formed by methane (CH_4).

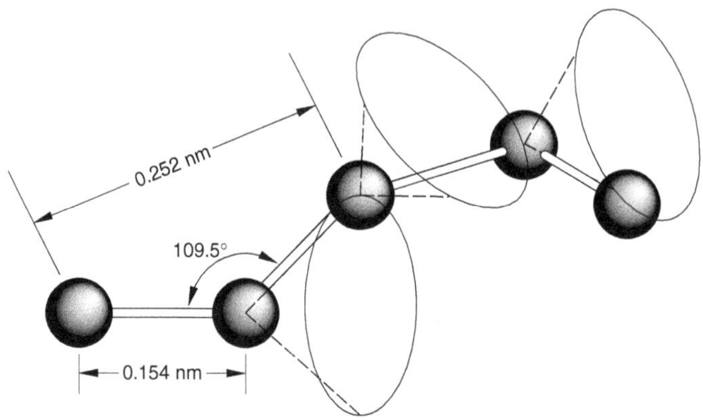

Figure 2.14 Random conformation of a polymer chain's carbon-carbon backbone.

side are called *isotactic*, and those molecules with regularly alternating side groups are called *syndiotactic*. Figure 2.15 shows the three different tacticity cases for

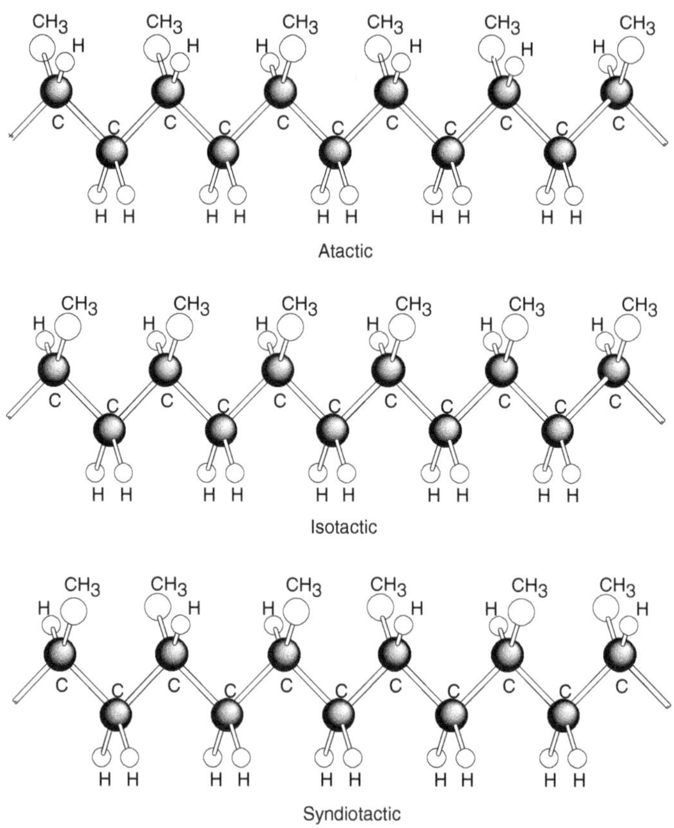

Figure 2.15 Different polypropylene structures.

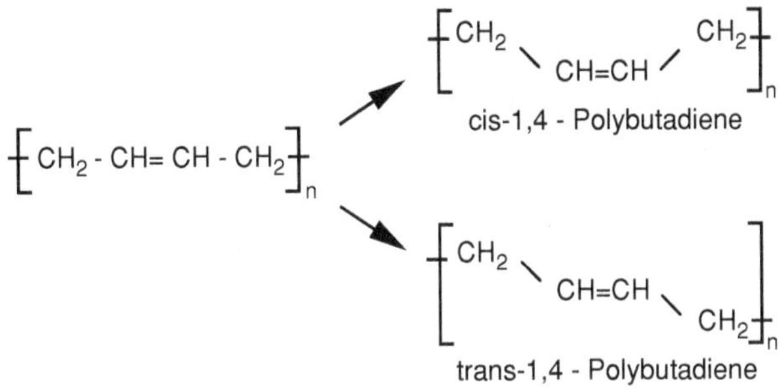

Figure 2.16 Symbolic representation of cis-1,4- and trans-1,4- polybutadiene molecules.

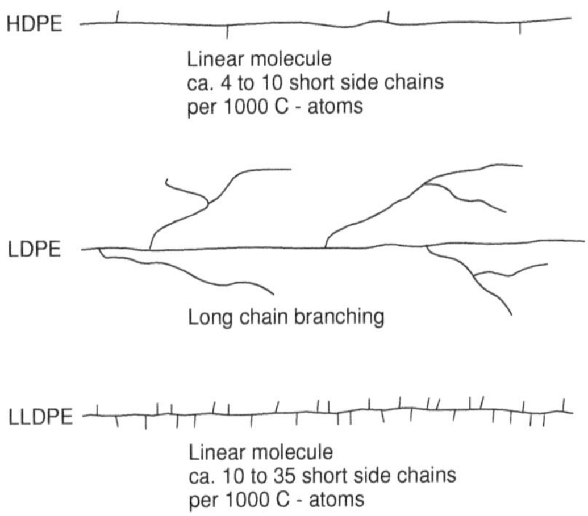

Figure 2.17 Schematic of the molecular structure of different polyethylenes.

polypropylene. The tacticity in a polymer determines the degree of crystallinity that a polymer can reach. For example, a polypropylene with a high isotactic content will reach a high degree of crystallinity and as a result be stiff, strong, and hard.

Another type of geometric arrangement arises with polymers that have a double bond between carbon atoms. Double bonds restrict the rotation of the carbon atoms about the backbone axis. These polymers are sometimes referred to as *geometric isomers*. The X groups may be on the same side (*cis-*) or on opposite sides (*trans-*) of the chain, as shown schematically for polybutadiene in Fig. 2.16. The arrangement in a cis-1,4- polybutadiene results in a very elastic rubbery material, whereas the structure of the trans-1,4- polybutadiene results in a leathery and tough material.

Branching of the polymer chains also influences the final structure, crystallinity, and properties of the polymeric material. Figure 2.17 shows the molecular architecture of high-density, low-density, and linear low-density polyethylenes. The high-density polyethylene has between 5 and 10 short branches every 1000 carbon atoms.

The low-density material has the same number of branches as HDPE; however, they are much longer and are themselves usually branched. The LLDPE has between 10 and 35 short chains every 1000 carbon atoms. Polymer chains with fewer and shorter branches can crystallize with more ease, resulting in higher density.

2.5 Thermoplastic Polymers

Thermoplastics are those polymers that solidify as they cool, restricting the motion of the long molecules. When heated, these materials regain the ability to "flow," as the molecules can slide past each other easily. Thermoplastic polymers are divided into amorphous and semi-crystalline polymers, and *thermoplastic elastomers* (TPEs). These TPE materials are quite soft and compliant, such as elastomers, but can be melted and processed like regular thermoplastics. Thermoplastic elastomers are covered separately later in this chapter.

The formation of macromolecules from monomers occurs if there are unsaturated carbon atoms (i.e., carbon atoms connected with double bonds), or if there are monomers with reactive end-groups. The double bond (e.g., in an ethylene monomer), is split, which frees two valences per monomer and leads to the formation of a macromolecule such as polyethylene. This process is often referred to as *polymerization*. Monomers (R) that possess two reactive end-groups (bifunctional) can similarly react with other monomers (R′) that also have two other reactive end-groups that can react with each other, which also leads to the formation of a polymer chain. Typical reactive end-groups are listed in Table 2.3.

2.5.1 Amorphous Thermoplastics

Amorphous thermoplastics, with their randomly arranged molecular structure, are analogous to a bowl of spaghetti. Due to their random structure, the characteristic size of the largest ordered region is of the order of a carbon–carbon bond. This dimension is much smaller than the wavelength of visible light and so generally makes amorphous thermoplastics transparent.

Figure 2.18 [18] shows the shear modulus*, G', versus temperature for polystyrene, one of the most common amorphous thermoplastics. The figure shows two

* The dynamic shear modulus, G', is obtained using the dynamic mechanical properties test described in Chapter 8.

Table 2.3 List of Selected Reactive End-Groups

Hydrogen in aromatic monomers	—H
Hydroxyl group in alcohols	—OH
Aldehyde group as in formaldehyde	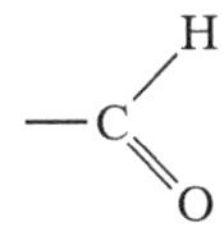
Carboxyl group in organic acids	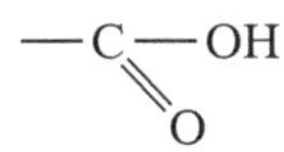
Isocyanate group in isocyanates	—N=C=O
Epoxy group in polyepoxys	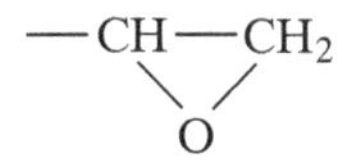
Amido groups in amides and polyamides	—CO—NH_2
Amino groups in amines	—NH_2

general regions: one where the modulus appears fairly constant*, and one where the modulus drops significantly with increasing temperature. With decreasing temperatures, the material enters the glassy region where the slope of the modulus approaches zero. At high temperatures the modulus is negligible and the material is soft enough to flow. Although there is not a clear transition between "solid" and "liquid," the temperature that divides the two states in an amorphous thermoplastic is referred to as the *glass transition temperature* T_g. For the polystyrene in Fig. 2.18 the glass transition temperature is approximately 110°C. Although data is usually presented in the form shown in Fig. 2.18, note that the curve shown in the figure was measured at a constant frequency. If the frequency of the test is increased—reducing the time scale—the curve is shifted to the right because higher temperatures are required to achieve movement of the molecules at the new frequency.

Figure 2.19 [19] demonstrates this concept by displaying the elastic modulus as a function of temperature for polyvinyl chloride at various test frequencies. A similar effect is observed if the molecular weight of the material is increased. The longer molecules have more difficulty sliding past each other, thus requiring higher temperatures to achieve "flow."

* When plotting G' versus temperature on a linear scale, a steady decrease of the modulus is observed.

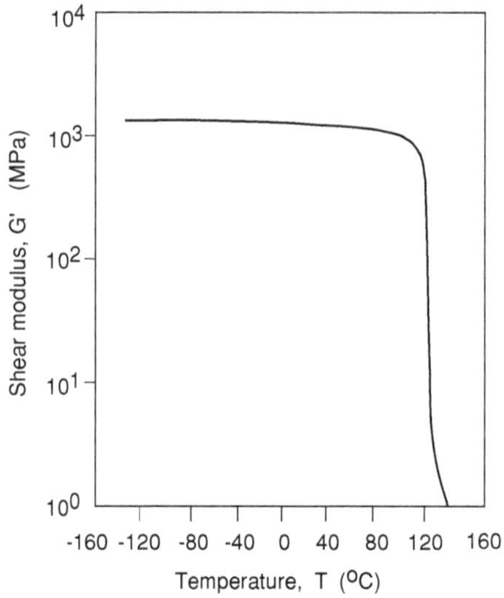

Figure 2.18 Shear modulus of polystyrene as a function of temperature.

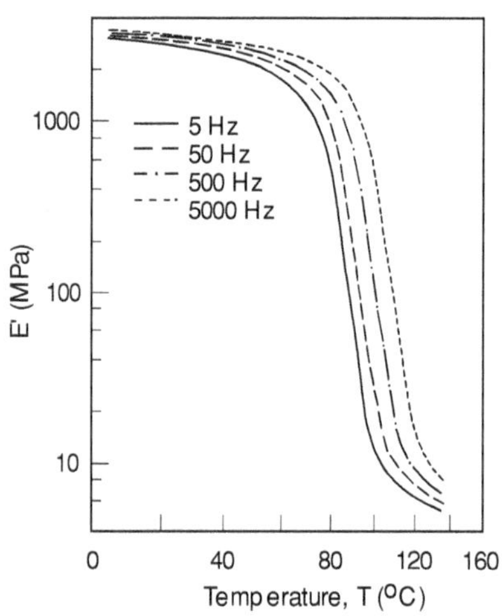

Figure 2.19 Modulus of polyvinyl chloride as a function of temperature at various test frequencies.

2.5.2 Semi-Crystalline Thermoplastics

The molecules in semi-crystalline thermoplastic polymers exist in a more ordered fashion when compared with amorphous thermoplastics. The molecules align in an ordered crystalline form as shown for polyethylene in Fig. 2.20. The crystalline structure is part of a *lamellar crystal*, which in turn forms the *spherulites*. The formation of spherulites during solidification of semi-crystalline thermoplastics will be covered in Chapter 7. The schematic in Fig. 2.21 shows the general structure and hierarchical arrangement in semi-crystalline materials. The spherulitic structure is the largest

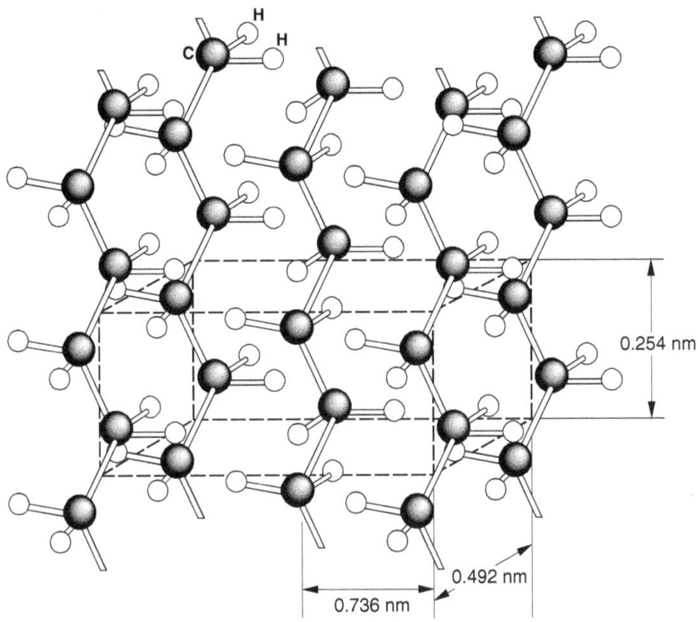

Figure 2.20 Schematic representation of the crystalline structure of polyethylene.

domain with a specific order and has a characteristic size of 50 to 500 μm. This size is much larger than the wavelength of visible light, making semi-crystalline materials translucent and not transparent.

The crystalline regions, however, are very small with molecular chains comprised of both crystalline and amorphous regions. The degree of crystallinity in a typical thermoplastic will vary from grade to grade. For example, in polyethylene the degree of crystallinity depends on the branching and the cooling rate. A low-density polyethylene (LDPE) with its long branches (Fig. 2.17) can only crystallize to about 40 to 50%, whereas a high-density polyethylene (HDPE) crystallizes to up to 80%. The density and strength of semi-crystalline thermoplastics increase with the degree of crystallinity as demonstrated in Table 2.4 [20], which compares LDPEs and HDPEs.

Table 2.4 Influence of Crystallinity on Properties for Low- and High-Density Polyethylene

Property	Low density	High density
Density (g/cm^3)	0.91–0.925	0.941–0.965
% crystallinity	42–53	64–80
Melting temperature (°C)	110–120	130–136
Tensile modulus (MPa)	17–26	41–124
Tensile strength (MPa)	4.1–16	21–38

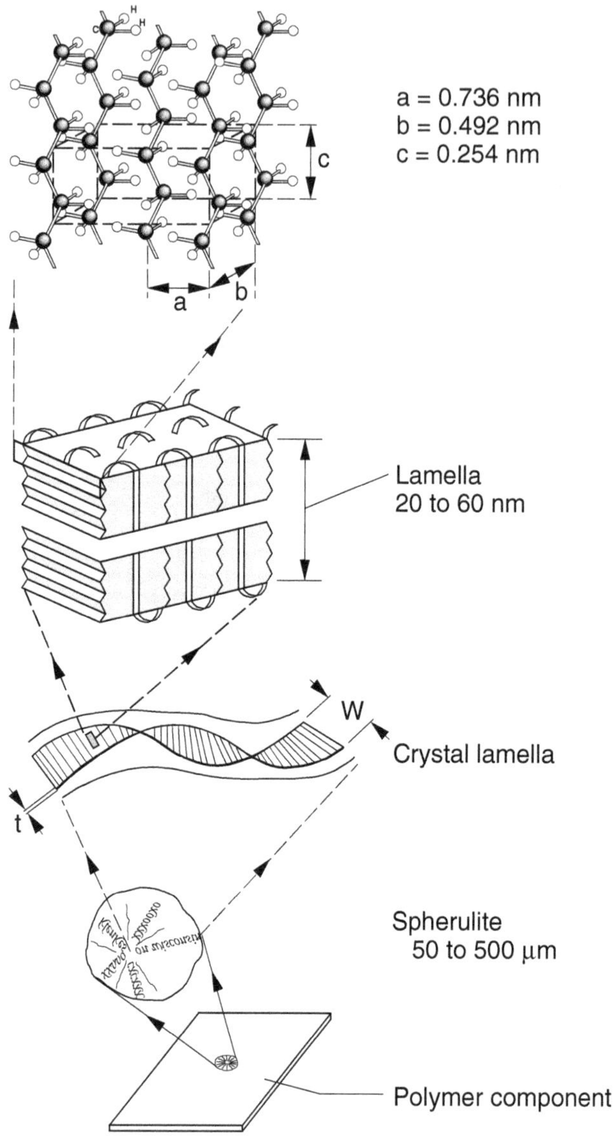

Figure 2.21 Schematic representation of the general molecular structure and arrangement of typical semi-crystalline materials.

Figure 2.22 shows the different properties and molecular structure that may arise in polyethylene plotted as a function of degree of crystallinity and molecular weight.

Figure 2.23 [18] shows the dynamic shear modulus versus temperature for an HDPE, which is the most common semi-crystalline thermoplastic. Again, this curve presents data measured at one test frequency. The figure clearly shows two distinct transitions: one at about –110°C, the *glass transition temperature*, and another near

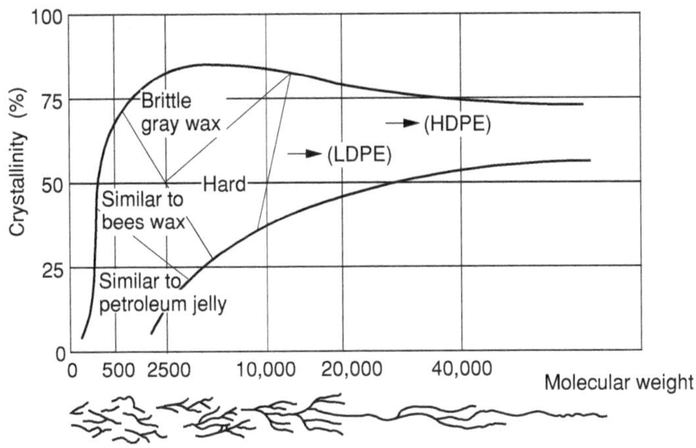

Figure 2.22 Influence of degree of crystallinity and molecular weight on different properties of polyethylene.

140°C, the *melting temperature*. Above the melting temperature, the shear modulus is negligible and the material will flow. Crystalline arrangement begins to develop as the temperature decreases below the melting point. The material behaves as a leathery solid between the melting and glass transition temperatures. As the temperature decreases below the glass transition temperature, the amorphous regions within the semi-crystalline structure solidify, forming a glassy, stiff, and in some cases brittle, polymer.

To summarize, Table 2.5 presents the basic structure of several amorphous and semi-crystalline thermoplastics with their melting and/or glass transition temperatures.

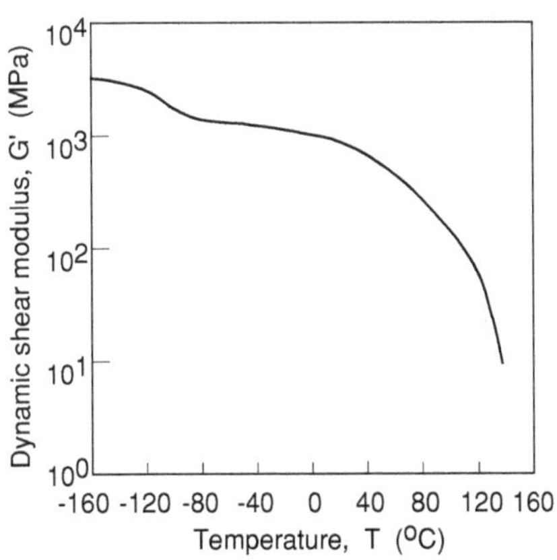

Figure 2.23 Shear modulus of a high-density polyethylene as a function of temperature.

Table 2.5 Structural Units for Selected Polymers with Glass Transition and Melting Temperatures

Structural Unit	Polymers	T_g(°C)	T_m(°C)
$—CH_2—CH_2—$	Linear polyethylene	–125	135
$—CH_2—CH(CH_3)—$	Isotactic poly-propylene	– 20	170
$—CH_2—CH(C_2H_5)—$	isotactic Polybutene	– 25	135
$—CH_2—CH(CH(CH_3)—CH_3)—$	Isotactic poly-3-methylbutene-1	50	310
$—CH_2—CH(CH_2—CH(CH_3)—CH_3)—$	Isotactic poly-4-methylpen-39 tene -1	240	
$—CH_2—CH(C_6H_5)—$	Isotactic polystyrene	100	240
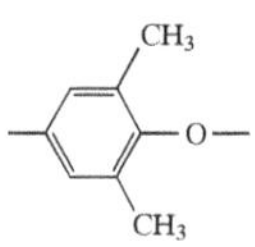	Polyphenylenether (PPE)	—	261
$—O—CH(CH_3)—$	Polyacetaldehyde	– 30	165
$—O—CH_2—$	Polyformaldehyde (polyacetal, polyoxymethylene)	– 85	178, 198
$—O—CH_2(CH_3)—CH—$	Isotactic polypropyleneoxide	– 75	75
$—O—CH_2—C(CH_2Cl)(CH_2Cl)—CH_2—$	Poly-[2.2-bis-(chlormethyl)-trimethylene-oxide]	5	181
$—CH_2—C(CH_3)(CO_2CH_3)—$	Isotactic polymethyl-methacrylate	50	160
$—C(Cl)(F)—C(F)(F)—$	Polychlortrifluorethylene	45	220

Table 2.5 ***(Continued)***

Structural Unit	Polymers	T_g(°C)	T_m(°C)
$—CF_2—CF_2—$	Polytetrafluorethylene	–113, +127	330
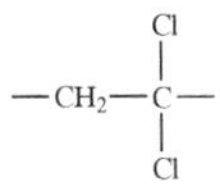	Polyvinylidenechloride	- 19	190
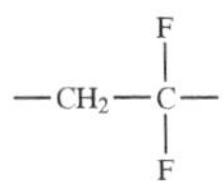	Polyvinylidefluoride	- 45	171
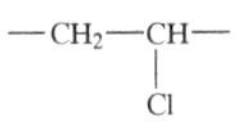	Polyvinylchloride (PVC) amorphous crystalline	80 80	— 212
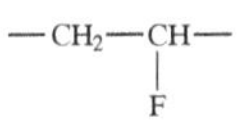	Polyvinylfluoride (PVF)	- 20	200
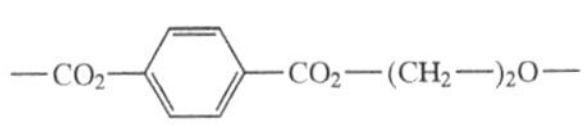	Polyethylene-terephthalate (PET) (linear polyester)	69	245
$—CO—(CH_2—)_4CO—NH—(CH_2)_6NH—$	Polyamide 66	57	265
$—CO—(CH_2—)_8CO—NH—(CH_2—)_6NH—$	Polyamide 610	50	228
$—CO—(CH_2—)_5NH—$	Polycaprolactam, Polyamide 6	75	233
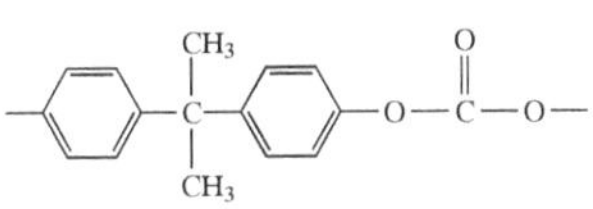	Polycarbonate (PC)	149	267
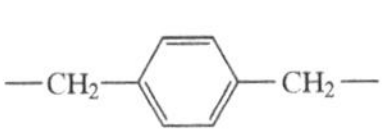	Poly-(*p*-xylene) (Parylene ‖ R)	—	400
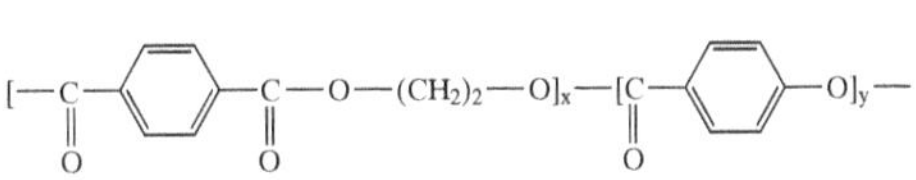	Polyethyleneterephthalate/ *p*.-Hydroxybenzoate-copolymers LC-PET,	75	280
|← 35% PET →|← 65% PHB →	Polymers with flexible chains		
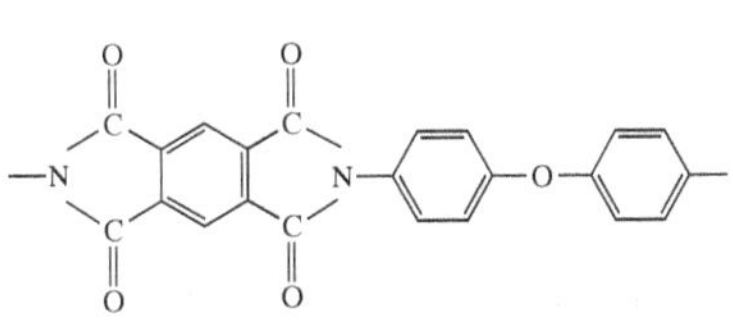	Polyimide (PI)	up to 400	

Table 2.5 ***(Continued)***

Structural Unit	Polymers	T_g(°C)	T_m(°C)
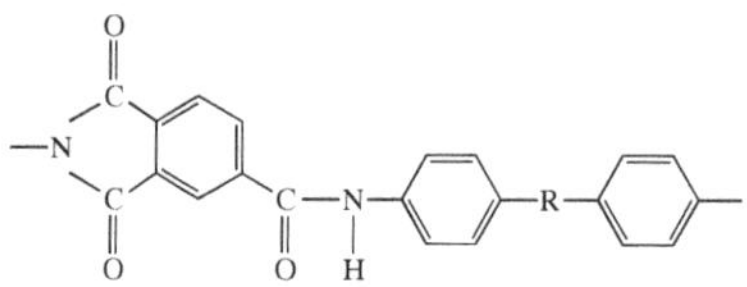	Polyamidimide (PAI)	≈260	
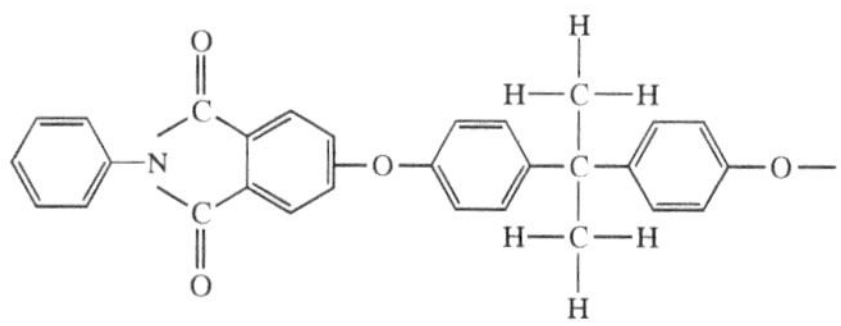	Polyetherimide (PEI) recommended:	 >300 <200	
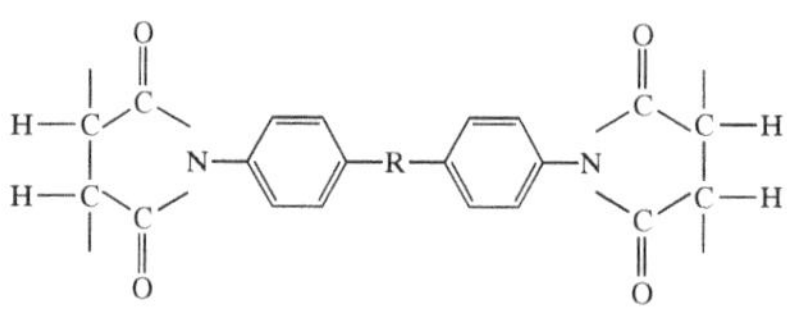	Polybismaleinmide (PBI)	≈260	
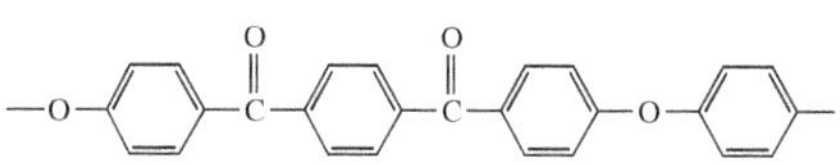	Polyoxybenzoate (POB)	≈290	
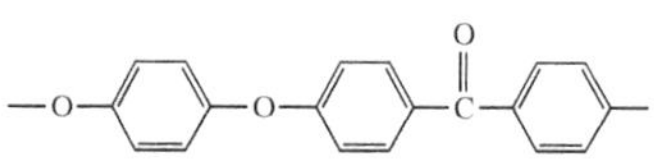	Polyetheretherketone (PEEK)	143	335
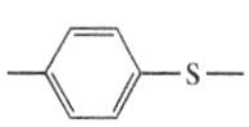	Polyphenylene-sulfide (PPS)	85	280
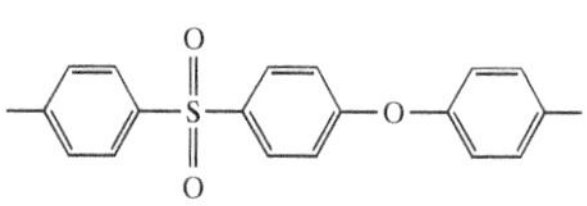	Polyethersulfone (PES)	~230	
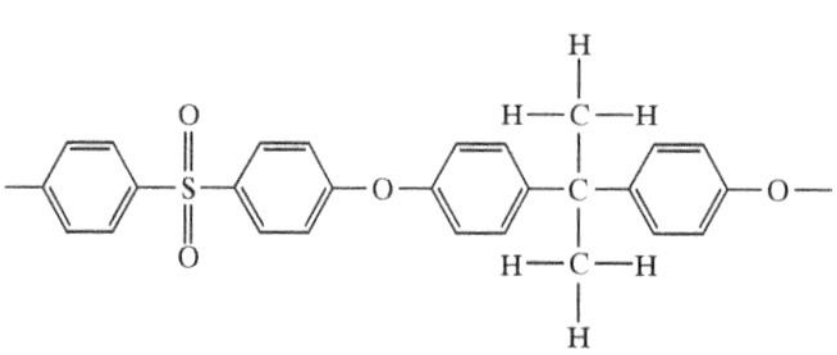	Polysulfone (PSU)	~180	

Furthermore, Fig. 2.24 [21] summarizes the property behavior of amorphous, crystalline, and semi-crystalline materials using schematic diagrams of material properties plotted as functions of temperature.

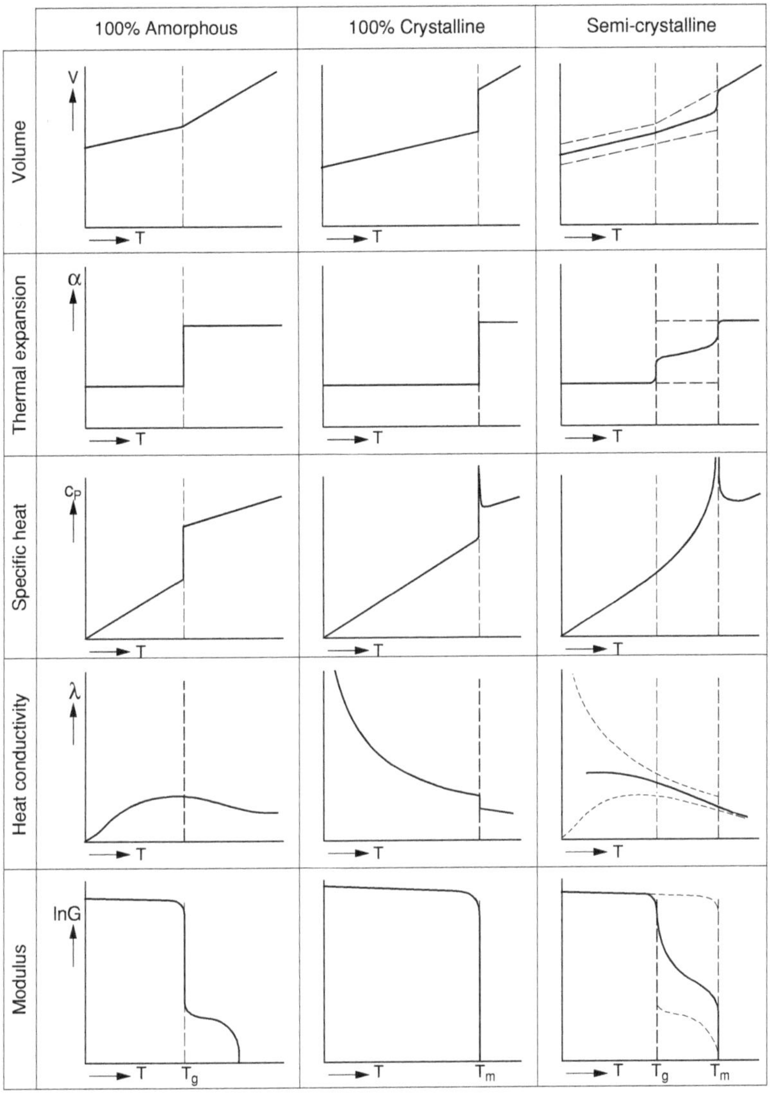

Figure 2.24 Schematic of the behavior of some polymer properties as a function of temperature for different thermoplastics.

2.5.3 Examples of Common Thermoplastics

Examples of various thermoplastics are discussed in detail in the literature [18] and can be found in commercial materials data banks [22,23]. The chemical structure and transition temperatures for many thermoplastic polymers are presented in Table 2.5. Examples of the most common thermoplastic polymers, with a short summary, are given in the following sections. Ranges of typical processing conditions are also presented, grade dependent.

Polyacetal (POM)

Polyacetal is a semi-crystalline polymer known for its high toughness, high stiffness, and hardness. It is also very sought after for its dimensional stability and its excellent electrical properties. It resists many solvents and is quite resistant to environmental stress cracking. Polyacetal has a low coefficient of friction. When injection molding POM the melt temperature should be between 200 and 210°C and the mold temperature should be greater than 90°C. Due to the flexibility and toughness of POM it can be used in sport equipment, clips for toys, and switch buttons.

Polyamide 66 (PA66)

Polyamide 66 is a semi-crystalline polymer known for its hardness, stiffness, abrasion resistance, and high heat deflection temperature. When injection molding PA66, the melt temperature should be between 260 and 320°C, and the mold temperature 80 to 90°C or more. The pellets must be dried before molding. Of the various polyamide polymers, this is the preferred material for molded parts that will be mechanically and thermally loaded. It is ideally suited for automotive and chemical applications such as gears, spools, and housings. Its mechanical properties are significantly enhanced when reinforced with glass fiber.

Polyamide 6 (PA6)

Polyamide 6 is a semi-crystalline polymer known for its hardness and toughness; however, it has a toughness somewhat lower than PA6. When injection molding PA6, the melt temperature should be between 230 and 280°C, and the mold temperature 80 to 90°C or more. The pellets must be dried before molding. Low-viscosity PA6 grades can be used to injection mold many thin-walled components. High-viscosity grades can be used to injection mold various engineering components such as gears, bearings, seals, pump parts, cameras, telephones, and the like. Its mechanical properties are significantly enhanced when reinforced with glass fiber.

Polycarbonate (PC)

Polycarbonate is an amorphous thermoplastic known for its stiffness, toughness, and hardness over a range from –150 to 135°C. It is also known for its excellent optical properties and high surface gloss. When injection molding PC, the pellets must be dried before molding for 10 hours at about 130°C. The melt temperature should be between 280 and 320°C, and the mold temperature 85 to 120°C. Typical applications for injection molded polycarbonate parts are telephone housings, filter cups, lenses for glasses and optical equipment, camera housings, marine light covers, safety goggles, hockey masks, and so on. Compact discs (CDs) are injection compression molded. PC's mechanical properties can also be significantly enhanced when reinforced with glass fiber.

Polyethylene (PE)

As mentioned in previous sections, the basic properties of PE depend on the molecular structure, such as degree of crystallinity, branching, degree of polymerization, and molecular weight distribution. Due to all these factors, PE can be a low-density polyethylene (PE-LD), linear low-density polyethylene (PE-LLD), high-density polyethylene (PE-HD), ultra-high-molecular-weight high-density polyethylene (PE-HD-HMW), and so on. When injection molding PE-LD, the melt temperature should be between 160 and 260°C, and the mold temperature 30 to 70°C, grade dependent. Injection temperatures for PE-HD are between 200 and 300°C, and mold temperatures between 10 and 90°C. Typical applications for injection molding PE-LD parts are very flexible and tough components such as caps, lids, and toys. Injection molding of PE-HD components include food containers, lids, toys, and buckets.

Polymethylmethacrylate (PMMA)

Polymethylmethacrylate is an amorphous polymer known for its high stiffness, strength, and hardness. PMMA is brittle, but its toughness can be significantly increased as a copolymer. PMMA is also scratch resistant and it can have high surface gloss. When injection molding PMMA, the melt temperature should be between 210 and 240°C, and the mold temperature 50 to 70°C. Typical applications for injection molded PMMA parts are automotive rear lights, drawing instruments, watch windows, lenses, jewlery, pipe fittings, and the like.

Polypropylene (PP)

Polypropylene is a semi-crystalline polymer known for its low density and its somewhat higher stiffness and strength than HDPE. PP, however, has a lower toughness than HDPE. Polypropylene homopolymer has a glass transition temperature of as high as –10°C, below which temperature it becomes brittle. When copolymerized with ethylene, however, it becomes tough. Because of its flexibility and the large range of properties, including the ability to reinforce it with glass fiber, polypropylene is often used as a substitute for an engineering thermoplastic. When injection molding PP, the melt temperature should be between 250 and 270°C, and the mold temperature 40 to 100°C. Typical applications for injection molded PP parts are housings for domestic appliances, kitchen utensils, storage boxes with integrated hinges (living hinges), toys, disposable syringes, food containers, and the like.

Polystyrene (PS)

Polystyrene is an amorphous polymer known for its high stiffness and hardness. PS is brittle, but its toughness can be significantly increased when copolymerized with butadiene. PS is also known for its high dimensional stability and its clarity, and it can have high surface gloss. When injection molding PS, the melt temperature should be between 180 and 280°C, and the mold temperature 10 to 40°C. Typical applications for injection molded PS parts are pharmaceutical and cosmetic cases, radio and television housings, drawing instruments, clothes hangers, toys, and so on.

Polyvinylchloride (PVC)

Polyvinylchloride comes either unplasticized (PVC-U) or plasticized (PVC-P). Unplasticized PVC is known for its high strength rigidity and hardness; however, PVC-U is also known for its low impact strength at low temperatures. In the plasticized form, the flexibility of PVC will vary over a wide range. Its toughness will be higher at low temperatures. When injection molding PVC-U pellets, the melt temperature should be between 180 and 210°C, and the mold temperature should be at least 30°C. For PVC-U powder the injection temperatures should 10°C lower, and the mold temperatures at least 50°C. When injection molding PVC-P pellets, the melt temperature should be between 170 and 200°C, and the mold temperature should be at least 15°C. For PVC-P powder the injection temperatures should 5°C lower, and the mold temperatures at least 50°C. Typical applications for injection molded plasticized PVC parts are shoe soles, sandals, and some toys. Typical applications for injection molded unplasticized polyvinylchloride parts are pipefittings.

2.6 Thermosetting Polymers

Thermosetting polymers solidify by a chemical cure. Here, the long macromolecules cross-link during cure, resulting in a network. The original molecules can no longer slide past each other. These networks prevent "flow" even after reheating. The high density of cross-linking between the molecules makes thermosetting materials stiff and brittle. The cross-linking causes the material to become resistant to heat after it has solidified; however, thermosets also exhibit glass transition temperatures that sometimes exceed thermal degradation temperatures. A more in-depth explanation of the cross-linking chemical reaction that occurs during solidification is in Chap. 3.

2.6.1 Cross-Linking Reaction

The cross-linking is usually a result of the presence of double bonds that break, allowing the molecules to link with their neighbors. One of the oldest thermosetting polymers is phenol-formaldehyde, or phenolic. Figure 2.25 shows the chemical symbol representation of the reaction, and Fig. 2.26 shows a schematic of the reaction. The phenol molecules react with formaldehyde molecules to create a three-dimensional cross-linked network that is stiff and strong. The by-product of this chemical reaction is water.

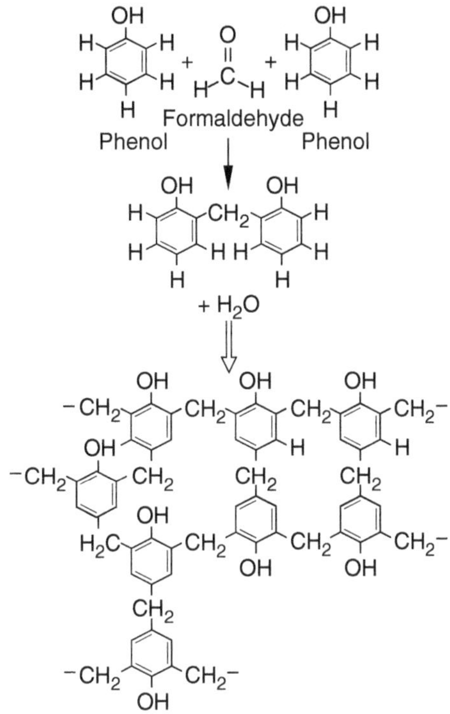

Figure 2.25 Symbolic representation of the condensation polymerization of phenol-formaldehyde resins.

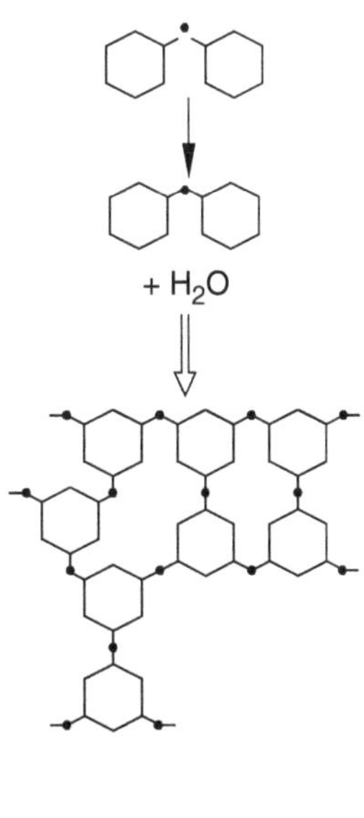

Figure 2.26 Schematic representation of the condensation polymerization of phenol-formaldehyde resins.

2.6.2 Examples of Common Thermosets

Examples of the most common thermosetting polymers, with a short summary, are given in the following.

Phenol Formaldehyde (PF)

Phenol formaldehyde is known for its high strength, stiffness, hardness, and its low tendency to creep. It is also known for its high toughness, and, depending on its reinforcement, it will also exhibit high toughness at low temperatures. PF also has a low coefficient of thermal expansion. PF is compression molded, transfer molded, and injection-compression molded. Typical applications for phenol formaldehyde include distributor caps, pulleys, pump components, handles for irons, and so on. It should not be used in direct contact with food.

Unsaturated Polyester (UPE)

Unsaturated polyester is known for its high strength, stiffness and hardness. It is also known for its dimensional stability, even when hot, making it ideal for under-the-hood applications. In most cases UPE is found reinforced with glass fiber. Unsaturated polyester is processed by compression molded, injection molding, and injection-compression molding. Sheet molding compound (SMC) is used for compression molding; bulk molding compound is used for injection and injection-compression molding. Typical applications for fiber-reinforced UPE are automotive body panels, automotive valve covers and oil pans, breaker switch housings, electric motor parts, distributor caps, ventilators, etc.

Epoxy (EP)

Epoxy resins are known for their high adhesion properties, high strength, and excellent electrical and dielectrical properties. They are also known for their low shrinkage, their high chemical resistance, and their low susceptibility to stress crack formation. They are heat resistant until their glass transition temperature (around 150 to 190°C), where they exhibit a significant reduction in stiffness. Typical applications for epoxy resins are switch parts, circuit breakers, housings, encapsulated circuits, and so on.

Cross-Linked Polyurethanes (PU)

Cross-linked polyurethane is known for its high adhesion properties, high impact strength, rapid curing, low shrinkage, and low cost. PU is also known for the wide variety of forms and applications. PU can be an elastomer, a flexible foam, a rigid foam, an integral foam, a lacquer, an adhesive, and so on. Typical applications for PU are television and radio housings, copy and computer housings, ski and tennis racket composites, and the like.

2.7 Copolymers and Polymer Blends

Copolymers are polymeric materials with two or more monomer types in the same chain. A copolymer that is composed of two monomer types is referred to as a *bipolymer*; one that is formed by three different monomer groups is called a *terpolymer*. One distinguishes between *random*, *alternating*, *block*, or *graft* copolymers depending on how the different monomers are arranged in the polymer chain. The four types of copolymers are schematically represented in Fig. 2.27.

A common example of a copolymer is an ethylene-propylene copolymer. Although both monomers would results in semi-crystalline polymers when polymerized individually, the melting temperature disappears in the randomly distributed copolymer with ratios between 35/65 and 65/35, resulting in an elastomeric material, as shown in Fig. 2.28. In fact EPDM* rubbers are continuously gaining acceptance in industry because of their resistance to weathering. On the other hand, the ethylene-propylene block copolymer maintains a melting temperature for all ethylene/propylene ratios, as shown in Fig. 2.29.

Another widely used copolymer is high impact polystyrene (PS-HI), which is formed by grafting polystyrene to polybutadiene. Again, if styrene and butadiene are randomly copolymerized, the resulting material is an elastomer called *styrene-butadiene-rubber* (SBR). Another classic example of copolymerization is the terpolymer acrylonitrile-butadiene-styrene (ABS).

Polymer blends belong to another family of polymeric materials which are made by mixing or blending two or more polymers to enhance the physical properties of each individual component. Common polymer blends include PP-PC, PVC-ABS, PE-PTFE, and PC-ABS.

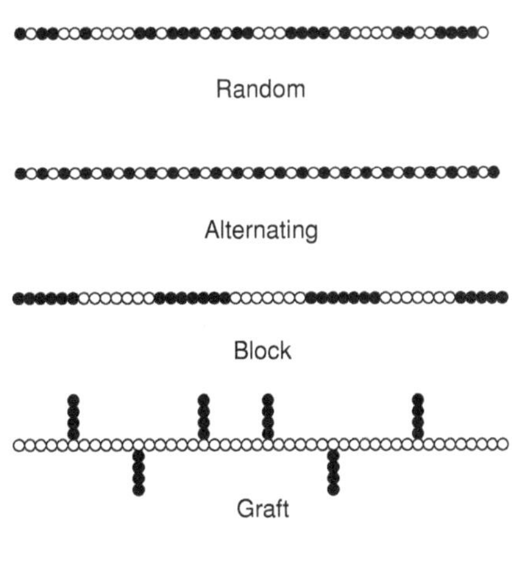

Figure 2.27 Schematic representation of different copolymers.

* The D in EP(D)M stands for the added unsaturated diene component that results in a cross-linked elastomer.

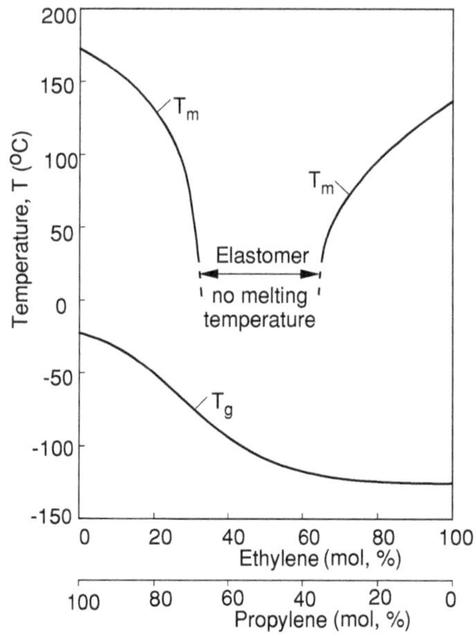

Figure 2.28 Melting and glass transition temperature for random ethylene-propylene copolymers.

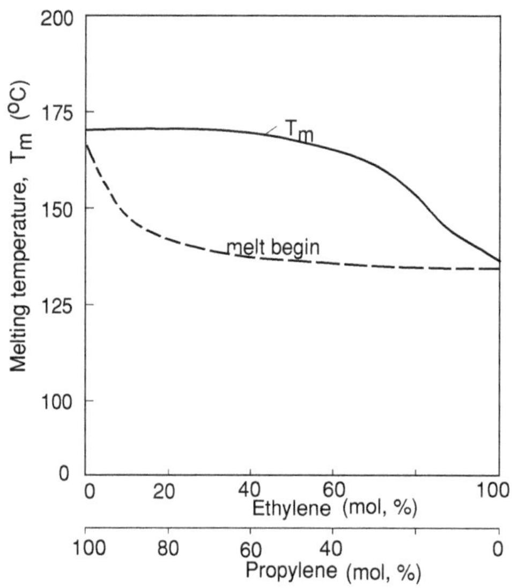

Figure 2.29 Melting temperature for ethylene-propylene block copolymers.

2.8 Elastomers*

A manufacturer transferring from a compression to an injection molding process may carry out the first trials fairly safely without modifying the compound, relying on adjustments of barrel temperature to obtain reasonable operating conditions.

Sorne typical formulations for NR and NBR polymers are listed in following table together wíth curing systems selected to offer a range of processing and cure requirernents. These are based either on MBTS, sulphenamides, or Sulfasan R because of the need for a certain minimum of scorch safety in the compounds.

Black NR Formulations

Natural Rubber	70
Whole tyre reclaim	60
Carbon black	75
Zinc oxide	40
Stearic acid	5
Paraffin wax	2
Antiozonant	1

Curing Systems	A	B	C	D	E
Sulphur	2.5	2.5	2.5	—	—
Dithiodimorpholine—DTM	—	—	—	1.4	1.2
Sulphenamide	1.2	1.2	—	1.4	1.2
MOR	—	—	1.2	—	—
Thiuram	0.3	—	—	0.2	0.5

A, B, and C are suitable for thin-section products and are in ascending order of scorch time. D and E are efficient vulcanizing systems suitable for thick sections. They give much reduced reversion and improved ageing resistance.

Nitrile Rubber Formulations (NBR)

Nitrile	100
Carbon black	80
Dioctyl phthalate	5
Zinc oxide	5
Stearic acid	1

* Contributed by M. DeGreiff.

Curing Systems	A	B	C	D
Sulphur	1.5	1.5	0.5	—
TMTD	0.5	—	3.0	—
MBTS	1.0	1.5	—	3.0

A and B are conventional curing systems which may be adequate where aging resistance is not a particular problem. C is a low sulphur system giving much improved aging but its scorch time is usually sufficient only for ram-type injection. D combines excellent ageing with a scorch time long enough for most applications.

2.9 Efficient Vulcanizing Systems

Efficient vulcanizing (EV) systems are defined as those where a high proportion of the sulphur is used for cross-linking purpose. These systems have two main advantages over conventional systems, giving vulcanizates with reduced reversion and better aging characteristics. In addition to these advantages, EV systems based on dithiodimorpholine (DTM) are very versatile, enabling a wide range of scorch times, cure rates, and states of cure to be chosen at will.

It is particularly important to avoid reversion for injection molding of thick sections, and EV systems give the complete answer to this problem. The conventional system (sulphur/MBTS/DPG) shows reversion immediately after the maximum modulus is reached, whereas the EV system (DTM/MBTS/TMTD) shows no reversion even after three times the optimum cure time. EV systems can be developed to give equivalent cure propoerties with much improved aging as compared with a conventional cure, even when antioxidants are omitted.

Accelerator systems for injection molding should be chosen to give adequate scorch time, fast cure without reversion, and appropriate product properties.

When molding thick scctions from polymers which revert (e.g., NR) EV systems should be used to minimize reversion. Combinations of a sulphenamide, dithiodimorpholine and TMTD are ideal, and the ratios can be varied to meet precise machine operating conditions and product requirements.

Where reversion is not a problem conventional sulphur/accclerator systems can be used and the following accelerators will give the best cure rates for each scorch time requirement:

	↓	
MOR		Decreasing
TBBS		scorch
CBS/TMTD		time

Accelerator loadings may be increased to give improved product properties or to counter the effect of oil addition.

2.10 Thermoplastic Elastomers*

Thermoplastic elastomers are a series of synthetic polymers that combine the properties of vulcanized rubber with the processing advantages of conventional thermoplastics. In other words, they allow the production of rubberlike articles using the fast processing equipment developed by the thermoplastics industry.

There are many different of thermoplastic elastomers, and details of their composition, properties, and applications have been extensively covered in the literature [22–29]. The commercially available materials used in injection molding can be classified into 10 types (Table 2.6). The commonly used abbreviations are listed in Table 2.7. The various themoplastic elastomers are discussed in more detail later in this chapter.

Before dealing with each type individually, we can consider some features that thermoplastic elastomers have in common. Most thermoplastic elastomers listed in Table 2.6 have one feature in common: They are phase-separated systems (i.e., the chlorinated olefin interpolymer alloys are the exception). One phase is hard and solid at room temperature in these phase-separated systems. The polymer forming the hard phase is the one listed first in this table. Another phase is an elastomer and fluid. The hard phase gives these thermoplastic elastomers their strength. Without it, the elastomer phase would be free to flow under stress and the polymers would be unusable. When the hard phase is heated, it becomes fluid. Flow can then take place, so the thermoplastic elastomer can be molded. Thus, the temperature at which the hard phase becomes fluid determines the processing temperature required for molding.

Table 2.6 Thermoplastic Elastomers Used in Injection Molding

1. Polystyrene/(S-B-S + Oil) Blends
2. Polypropylene/(S-EB-S + Oil) Blends
3. Polypropylene/(EPR + Oil) Blends
4. Polypropylene/(Rubber + Oil) Dynamic Vulcanizates
5. Polyethylene/(Polylefin Rubber) Block Copolymers
6. PVC/(NBR + Plasticizer) Blends
7. Chlorinated Olefin Interpolymer Alloys
8. Polyurethane/Elastomer Block Copolymers
9. Polyester/Elastomer Block Copolymers
10. Polyamide/Elastomer Block Copolymers

* Contributed by G. Holden.

Table 2.7 Abbreviations

ABS	Poly(acrylonitrile-co-butadiene-co-styrene)
ABS/PC	Blend of ABS and PC
BCP	Block Copolymer
DV	Dynamic Vulcanizate
EPR	Ethylene propylene rubber; i.e., poly(ethylene-co-propylene)
EPDM	Ethylene propylene diene rubber, an EPR that also contains a small number of out-of-chain double bonds
DOP	Dioctyl phthalate, a plasticizer for PVC
HIPS	High Impact Polystyrene
NBR	Nitrile butadiene rubber; i.e., poly(acrylonitrile-co-butadiene)
SAN	Poly(styrene-co-acrylonitrile)
PA	polyamide (e.g., nylon)
PC	Polycarbonate
PE	Polyethylene
PEst	Polyester
PET	Poly(ethylene terephthalate)
PP	Polypropylene
PS	Polystyrene
PU	Polyurethane
PVC	Polyvinylchloride
SBR	Styrene butadiene rubber; i.e., poly(styrene-co-butadiene)
S-B-S	Poly(styrene-b-butadiene-b-styrene) elastomeric block copolymer
S-EB-S	Poly(styrene-b-ethylene-co-butylene-b-styrene) elastomeric block copolymer
SMA	Poly(styrene-co-maleic anhydride)
TPE	Thermoplastic Elastomer

2.10.1 Service Temperatures

In these phase-separated systems, the individual polymers that constitute the phases retain many of their characteristics. For example, each phase has its own glass transition temperature (T_g), or crystal melting point (T_m) if it is crystalline. These determine the temperatures at which a particular thermoplastic elastomer goes through transitions in its physical properties. Thus, when the properties (e.g., modulus) of a thermoplastic elastomer are measured over a range of temperatures, there are three distinct regions (see Fig. 2.30). Both phases are hard at very low temperatures, so the material is stiff and brittle. At a somewhat higher temperature the elastomer phase becomes soft and the thermoplastic elastomer now resembles a conventional vulcanizate. As the temperature is further increased, the modulus stays relatively constant (a region often described as the "rubbery plateau") until the hard phase finally softens. At this point, the thermoplastic elastomer as a whole becomes fluid. Thus, thermoplastic elastomers have two service temperatures. The lower service temperature depends on the T_g of the elastomer phase, whereas the upper service temperature depends on the T_g or T_m of the hard phase. The difference between the upper and lower service temperatures is the service temperature range. Values of T_g and T_m

for the various phases in some commercially important thermoplastic elastomers are given in the literature [22–28]. The thermoplastic elastomers with polypropylene or polyethylene as the hard phase have excellent processing stability and can be left in the hot barrel of the injection molder for an hour or two without thermal degradation. There is also usually no need to purge the machine at shutdown. The other types are generally less forgiving. Some may degrade if left hot for more than about 30 minutes, and purging is often recommended.

2.10.2 Examples of Common Thermoplastic Elastomers

Features of each individual types are described in various publications [22–29], particularly [28], and in manufacturers literature. A short summary of their structure, properties and molding conditions is given later and in Chap. 3.

Polystyrene/(S-B-S + Oil) Blends

S-B-S and S-EB-S block copolymers form a physically cross-linked network (see Fig. 2.31). This network loses its strength at high temperatures, allowing the block copolymer to flow; however, at room temperature it has a combination of strength and elasticity similar to good quality vulcanizates [30,31]. Both S-B-S and S-EB-S block copolymers are difficult to injection mold as pure polymers, so they are always

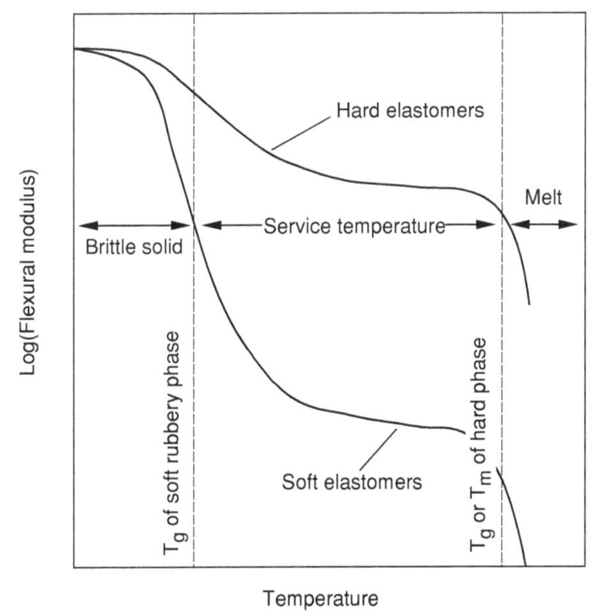

Figure 2.30 Stiffness of typical thermoplastic elastomers at various temperatures.

blended with other materials in this application. For S-B-S, the most important ones are polystyrene and mineral oils (quite large amounts of fillers can also be added, but they do not have much effect on physical properties). The structure of the polystyrene/(S-B-S + oil) blends is an interdispersed co-continuous network of the polystyrene hard and (S-B-S + oil) elastomer phases (Fig. 2.32). Both phases are strong materials at room temperature, so they can be blended together in almost any proportions, giving a wide range of product properties. Some general formulations are given in the literature [24].

Many of these blends are intended for footwear products, typically shoe soles. Because of the range of sizes and variety of styles, the use of tool steel molds is uneconomic in this application. Instead, aluminum molds are commonly used. These mold can only withstand low clamping pressure, which in turn restricts the injection pressure. Thus, blends intended for footwear products have relatively low melt viscosities and are molded on special low-pressure injection machines. There are three types: conventional reciprocating screw machines, fixed screw machines (essentially extruders that operate intermittently), and fixed screw machines that pump the melt into an accumulator, from which it is forced into the mold by the action of a piston. Despite the relatively low shear conditions, blends of the components can often be used with these machines and final compounding carried out during molding The polybutadiene segment in the S-B-S elastomer is unsaturated (i.e., it contains double bonds). If S-B-S is overheated, these begin to cross-link the polymer and so reduce its viscosity. Thus, although flow into the mold can be improved by raising a low melt temperature, at some point (about 225°C) this approach becomes counterproductive.

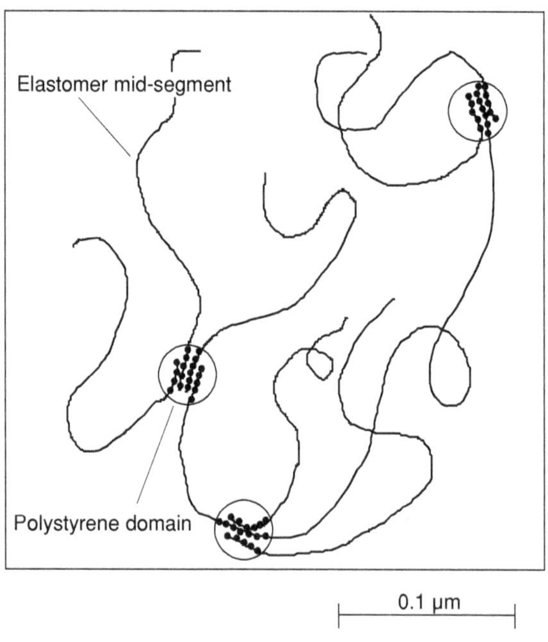

Figure 2.31 Morphology of styrenic block copolymers.

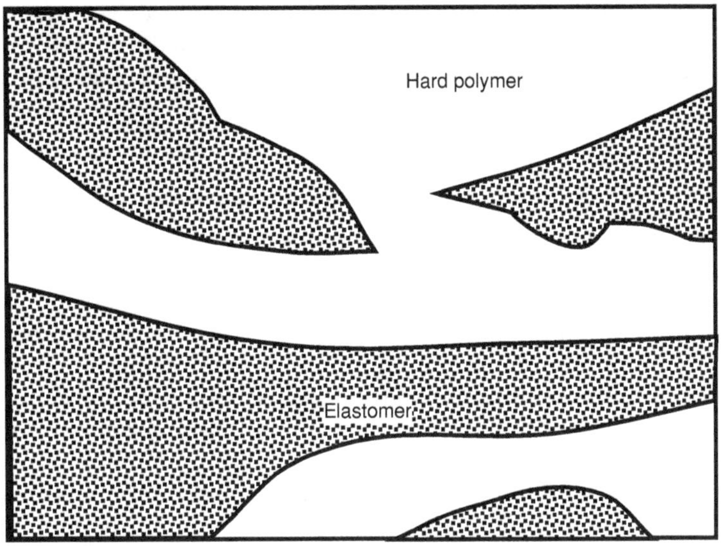

Figure 2.32 Morphology of hard polymer/elastomer blends.

Polypropylene/(S-EB-S + Oil) Blends

The structure of the polypropylene/(S-EB-S + oil) blends is similar to that of the polystyrene/(S-B-S + oil) blends [i.e., an interdispersed co-continuous network of the hard polypropylene phase and the strong, elastic (S-EB-S + oil) elastomer phase]. Again, the two phases can be blended together in almost any proportions and typical formulations are given in the literature [24]. These blends differ from the polystyrene/(S-B-S + oil) blends in several ways. Polypropylene is a relatively high melting material, so the blend requires higher processing temperature. The poly(ethylene-butylene) segment in the S-EB-S elastomer is completely saturated (i.e., it contains no double bonds). Thus, S-EB-S is more thermally stable (by about 75°C) than S-B-S and can tolerate the increase in processing temperatures. S-EB-S is extremely compatible with paraffinic mineral oils, so very soft blends (less than 10 Shore A) can be produced. Finally, the polypropylene/(S-EB-S + oil) blends have highly non-Newtonian viscosity. They are usually molded at high injection rates using small gates. This high injection rate requires good mold venting (vacuum assist is often helpful) and careful placement of the vents.

Polypropylene/(EPR + Oil) Blends

The structure of these blends is again an interdispersed co-continuous network—in this case, of the hard polypropylene phase and the EPR elastomer phase (which often contains oil). The difference is that in this case the EPR elastomer phase has little strength and can flow at room temperature. This limits the amount of the EPR phase that can be added to the polypropylene before there is a significant loss in strength. For this reason, these blends are relatively hard (greater than 60 Shore A). Similar blends can be made using polyethylene/(polylefin rubber) block copolymers in place of EPR. Polypropylene/EPR blends are injection molded under conditions generally similar to those used for pure polypropylene but with higher injection pressures.

Polypropylene/(Rubber + Oil) Dynamic Vulcanizates

The structure of the dynamic vulcanizates, such as PP/(Rubber + Oil) dynamic vulcanizates, is quite different. The important point is that although the hard polypropylene phase is continuous, the elastomer phase is a fine dispersion of strong vulcanized rubber particles (Fig. 2.33). Thus, quite large proportions of elastomer can be used without significant loss of strength, and products as soft as 35 Shore A are manufactured. EPDM is the most commonly used elastomer, but dynamic vulcanizates based on nitrile rubber (these have improved oil resistance), butyl rubber, and natural rubber are also available. All these dynamic vulcanizates can be overmolded against polypropylene. Grades suitable for overmolding against morepolar thermoplastics such as ABS, PC PA and PS have recently been announced [32]. All these products (especially those based on EPDM and butyl rubber) are thermally stable and degradation should not present a problem under the recommended processing conditions (see Chap. 3). Virgin product should be dried before molding, although this may not be necessary with the recycled scrap as long as it is used shortly after it is reground.

Polyethylene/(Polyolefin Rubber) Block Copolymers

These are multiblock copolymers consisting of alternating hard and soft segments of polyethylene and a poly(ethylene/α-olefin) copolymer, respectively. They form a two-

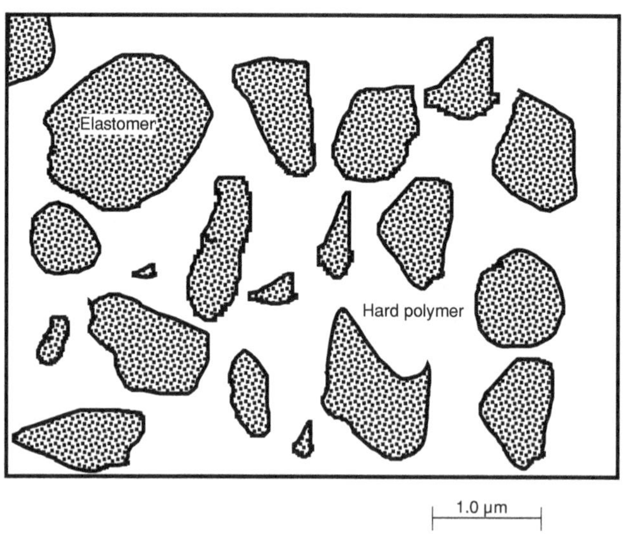

Figure 2.33 Morphology of dynamic vulcanizates.

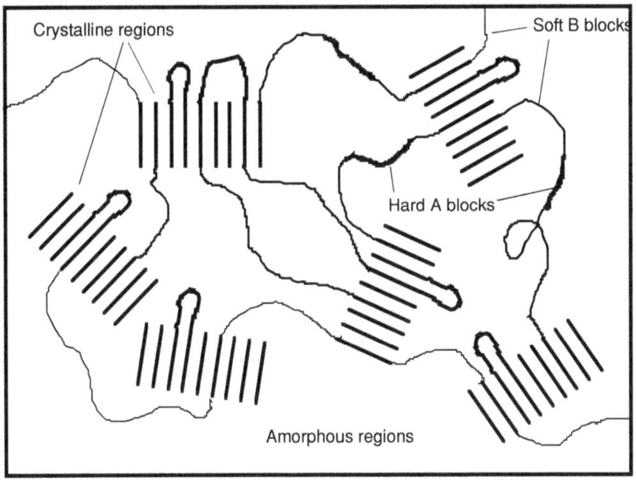

Figure 2.34 Morphology of multiblock polymers with crystalline hard segments.

phase network (see Fig. 2.34) similar in some ways to the physically cross-linked network formed by the S-B-S and S-EB-S block copolymers shown in Fig. 2.32. At high temperatures the network similarly loses its strength, allowing the material to flow. One difference is that the hard polyethylene phase is crystalline. Another is that in these and other multiblock copolymers, each polymer molecule can be part of several hard and soft regions. Polyethylene/(polyolefin rubber) block copolymers are transparent, low-density materials with excellent thermal stability. They have very non-Newtonian flow characteristics and are molded at high injection rates. The melting point of the polyethylene segment is low (from 50 to 100°C) [33,34]. Molded products must be cooled to well below this temperature range before they can be ejected from the mold, so these polymers require a very cold mold (10°C is recommended, which will require refrigerated coolant). Air-assisted ejection is also recommended. The polyethylene/(polylefin rubber) block copolymers can be added to polypropylene to improve impact resistance. [If enough is added, the products become relatively hard, heat resistant thermoplastic elastomers analogous to the harder versions of the polypropylene/(EPR + oil) blends.]

PVC/(NBR + Plasticizer) Blends

These blends present a somewhat different picture to the previous five types. They are much more polar, and therefore have improved oil and solvent resistance. PVC has limited thermal stability. If it degrades, then it can produce hydrogen chloride, which is both toxic and corrosive. Thus, in molding these materials, the same precautions must be used as with molding plasticized PVC. Melt temperatures must be below 200°C, and preferably below 190°C. Shear heating (by the use of high compression screws and/or high back pressures) must be avoided. The molten material must not be allowed to sit in the hot barrel of the injection molder and the machine should be purged at shutdown. Corrosion-resistant barrel liners and screws should be used. The user can buy precompounded products or make his own custom compounded material. In this case the PVC and the plasticizer (typically DOP) are mixed together in a tumble blender or high shear mixer until the plasticizer is completely absorbed, after which the powered NBR is added. Compounding can then be completed on the injection molding machine. Special types of NBR, with properties optimized for this end use, are commercially available.

Because they are polar materials, these blends cannot be overmolded against non-polar hard thermoplastics such as polypropylene. Instead, they are recommended for use with polar thermoplastics such as PVC, SAN, ABS, and SMA. Gradual migration of the plasticizer may present a problem, and overmolded products should be checked for this after a period of storage. The molded parts should not be allowed to come into prolonged contact with polystyrene or other materials whose properties are affected by DOP.

Chlorinated Olefin Interpolymer Alloys

The exact composition of these materials has not been disclosed. They are claimed to be single-phase systems, and in this they differ from the other thermoplastic elastomers. Like the PVC/(NBR + plasticizer) blends described earlier, they have limited thermal stability; thus, molding temperatures must be carefully controlled to avoid degradation, and the material must not be allowed to sit in the hot machine for extended times. Corrosion resistant barrel liners and screws similarly should be used and shear heating must be avoided. Melt temperatures should be less than 190°C. Resistance to flow is not much affected by temperature within the recommended range, so molding is not usually improved by increasing the processing temperatures. These interpolymer alloys have highly non-Newtonian flow characteristics, so they are also molded at high injection rates through small gates. They can be overmolded against polar thermoplastics such as PVC and polycarbonate/ABS blends, as well as against thermoplastic elastomers based on polyurethanes or polyesters, discussed later. Grades with adhesion to other hard thermoplastics are under development [35].

Polyurethane/Elastomer Block Copolymers

These polymers (as well as those discussed later) are structurally similar to the polyethylene/(polylefin rubber) block copolymers described in Section 2.3.5. Their morphology is shown in Fig. 2.34. In this case the crystalline hard segments are polyurethanes, and the soft segments usually are polyesters or polyethers. Melting temperatures for the polyurethane segments are high (typically greater than 180°C), so melt temperatures for injection molding must be more than this value. They are generally a little higher for the harder products. These polymers are quite susceptible to thermal degradation, which starts above 230°C, and gives foaming or bubbles in the final part. They should not be allowed to sit in the hot injection molding machine, and the machine should be purged with polystyrene or ABS before shutdown. Polyurethane/elastomer block copolymers readily absorb water, which causes severe degradation of the molten material. Both virgin polymer pellets and reprocessed scrap must be thoroughly dried (down less than 0.1% moisture) before being molded. Injection rates should be moderate, and fairly large gates with rounded edges are recommended. Further crystallization and crystal rearrangement will take place after molding is completed, and the properties of the product will improve with time for a period of about 6 weeks. The process can be speeded up by overnight annealing at about 115°C.

Polyester/Elastomer Block Copolymers

In these polymers the crystalline hard segments are polyesters and the soft segments are polyethers. They are more thermally stable than the polyurethane analogs described carlier, so they can be processed at somewhat higher temperatures, up to about 260°C. Again, processing temperatures should be higher for the harder products The melt viscosity changes significantly with temperature, so processing can be improved by increasing the melt temperature, although this will also increase the cycle time. The molten polymer can be allowed to sit in the hot injection molding machine for an hour or two. Purging before shutdown is generally not necessary, but if desired, it is best to use a thermally stable material such as polyethylene. As with the polyurethane/elastomer block copolymers, both virgin polymer pellets and reprocessed scrap must be thoroughly dried (less than 0.1% moisture) before use.

Polyamide/Elastomer Block Copolymers

In these polymers the crystalline hard segments are polyamides and the soft segments are usually polyethers. Processing temperatures depend on the melting temperature of the polyamide (e.g., nylon-6,6, nylon-6, nylon-11) and should be higher for the harder grades. Other processing conditions are essentially similar to the polyester analogs described earlier. Processing detail is given in Chapter 3.

References

1. Stern, H. J., *Rubber: Natural and Synthetic*, Maclaren and Sons LDT (1967), London.
2. de la Condamine, C. M., *Relation Abregee D'un Voyage Fait Dans l'interieur de l'Amerique Meridionale*, Academie des Sciences (1745), Paris.
3. DuBois, J. H., *Plastics History U.S.A.*, Cahners Publishing Co., Inc. (1972), Boston.
4. Tadmor, Z., Gogos, C. G., *Principles of Polymer Processing*, John Wiley & Sons (1979), New York.
5. McPherson, A. T., Klemin, A., *Engineering Uses of Rubber*, Reinhold Publishing Corporation (1956), New York.
6. Sonntag, R., *Kunststoffe* (1985), 75, 4.
7. Herrmann, H., *Kunststoffe* (1985), 75, 2.
8. Regnault, H. V. Liebigs Ann. (1835), 14, 22.
9. Ulrich, H., *Introduction to Industrial Polymers*, 2nd ed., Hanser Publishers (1993), Munich.
10. Rauwendaal, C., *Polymer Extrusion*, 2nd ed., Hanser Publishers (1990), Munich.

11. Utracki, L. A., *Polym. Eng. Sci.* (1995), 35, 1, 2.
12. Termonia, Y., Smith, P., *High Modulus Polymers*, A. E. Zachariades, Porter, R. S., eds., Marcel Dekker Inc. (1988), New York.
13. Ehrenstein, G. W., *Faserverbundkunststoffe*, Carl Hanser Verlag (1992), Munich.
14. Osswald, T. A., Menges, G., *Materials Science of Polymers for Engineers*, Hanser Publishers (1996), Munich.
15. Berry, G. C., Fox, T. G., *Adv. Polymer Sci.* (1968), 5, 261.
16. Staudinger, H., Huer, W., *Ber. der Deutsch. Chem Gesel.* (1930), 63, 222.
17. Crowder, M. L., Ogale, A. M., Moore, E. R., Dalke, B. D., *Polym. Eng. Sci.* (1994), 34, 19, 1497.
18. Domininghaus, H., *Plastics for Engineers*, Hanser Publishers (1993), Munich.
19. Aklonis, J. J., MacKnight, W. J., *Introduction to Polymer Viscoelasticity*, John Wiley & Sons (1983), New York.
20. Rosen, S. L., *Fundamental Principles of Polymeric Materials*, John Wiley & Sons (1993), New York.
21. van Krevelen, D. W., Hoftyzer, P. J., *Properties of Polymers*, 2nd. ed., Elsevier (1976), Amsterdam.
22. Holden, G., Legge, N. R., Quirk, R. P., Schroeder, H. E., eds., *Thermoplastic Elastomers*, 2nd ed., Hanser Publishers (1996), Munich.
23. Walker, B. M., Rader, C. P., eds., *Handbook of Thermoplastic Elastomers*, 2nd ed. Van Nostrand Reinhold (1998), New York.
24. Holden, G., "Thermoplastic Elastomers," In *Rubber Technology*, 3rd ed., Morton, M., ed., Van Nostrand Reinhold (1987), New York.
25. Holden, G., "Elastomers, Thermoplastic," In *Kirk-Othmer Encyclopedia of Polymer Science and Engineering*, 2nd ed., Kroschwitz, J. I., ed., John Wiley & Sons (1986), New York.
26. Holden, G., "Elastomers, Thermoplastic," In *Kirk-Othmer Encyclopedia of Chemical Technology*, 4th ed., Kroschwitz, J. I., ed., John Wiley & Sons (1988), New York.
27. Holden, G., "Thermoplastic Elastomers (Overview)," In *Polymeric Materials Encyclopedia*, Salamone, J. C., ed. CRC Press (1996), Boca Raton.
28. Holden, G., *Understanding Thermoplastic Elastomers*, Hanser Publishers (1999), Munich.
29. Holden, G., "'Thermoplastic Elastomers and Their Applications" in *Applied Polymer Science—21st Century*, Carraher, C., Craver, C. eds., American Chemical Society (In Press) Washington.
30. Holden, G., in Chap. 3 of Ref. 24.
31. Chap. 3 of Ref. 24.
32. Tan, O. H. C., Mehta, S., paper presented at the 158th meeting of the A.C.S. Rubber Division Cincinnati, OH October 17–19, 2000. See Rubber Chem. Technol. 74, p167.
33. Kresge, E. N., in Chap. 5 of Refe. 24.
34. Publication #H-69216, DuPont Elastomers L. L. C., Wilmington, DE 19809.
35. Personal Communication, Abel, W. R., III, Advanced Polymer Alloys, Wilmington, DE 19810.

3 Processing Fundamentals

T. A. Osswald*

The mechanical properties and the performance of a finished product are always the result of a sequence of events. Manufacturing of a plastic part begins with material choice in the early stages of part design. Processing follows this, at which time properties of the final part are made and frozen into place. During design and manufacturing of any plastic product one must always be aware that material, processing, and design properties all go hand-in-hand and cannot be decoupled. This approach is often referred to as the four Ps: polymer, processing, product, and performance. This chapter will cover the fundamentals of processing during injection molding. To cover the various phenomena that occur during the manufacturing process of an injection molded part, this chapter has been broken up into four distinct sections: processing data, rheology, anisotropy, and solidification. Throughout the chapter some simple methods are given to estimate, "on the back of an envelope," various parameters concerning process and part design. For more in-depth computations and predictions, the reader should refer to Chapter 11, which presents injection molding simulation techniques.

3.1 Processing Data

To arrive at high-quality products as quickly as possible, the machine settings at the beginning of the injection molding process optimization procedure should be as close as possible to optimal processing conditions. Although an educated guess of the processing conditions for a given material is always at hand, the final conditions are dependent on specific material grades, injection molding machine size, screw wear, part and mold design, and other material independent variables. Each processing condition setting is directly influenced by the following parameters:

- Desired properties of the finished product
- Injection molding machine
- Injection mold
- Injection molding material
- Part geometry

* With contributions from M. DeGreiff and G. Holden.

For example, using only a small amount of the total shot size, in every cycle, results in large residence times. This can be detrimental for certain materials. In addition, fast material throughput requires higher melting rates, which may be achieved by raising heater temperatures.

For most materials, however, a good starting point is always known and can be found in material data banks such as CAMPUS™ [1].

3.1.1 Temperature Settings

The material temperature is one of the main parameters that needs to be set on the injection molding machine. Table 3.1 presents processing temperatures for several thermoplastic polymers. The table presents the temperature settings for the heating zones in the plasticating unit of injection molding machines, as well as settings for the nozzle heater temperature and mold cooling temperatures.

As mentioned in Chapter 1 if the temperature setting is too high, the material can thermally degrade; if melt temperature is too low, the process can result in a short shot. The temperature settings must be high enough to complete melting, have low enough viscosity required for mold filling, and achieve sufficient homogeneity. It is important to note that the viscosity of amorphous thermoplastics is very sensitive to temperature changes when compared with semi-crystalline polymers.

During the plasticating process, the effects of viscous dissipation play a significant role. Here, the defining variable is the maximum screw speed that can be computed using

$$V_{screw} = R\Omega \tag{3.1}$$

where R is the radius and Ω is the screw speed in radians/second. The relation between speed and viscous dissipation is given by the Brinkman number and is discussed later in this chapter. Table 3.2 presents processing windows for various thermoplastic polymers with various heater temperatures and allowable residence times, along with recommended screw speeds.

3.1.2 Injection and Pack-Hold Pressure Settings

Injection speeds and pressures as well as pack-hold pressures are also very important controllable settings when manufacturing injection molded products. The injection pressure and speeds, and the pack-hold pressure are dependent on the type of material being molded. Table 3.3 presents some recommended values for the preceding parameters. For example, various materials require a high injection pressure and speed. High speeds are desired if a homogeneous filling is required and to be able to achieve an even degree of crystallinity when molding semi-crystalline polymer products. Low injection speeds and pressures are often desired

Table 3.1 Temperature Setting During Injection Molding of Thermoplastic Polymers [2,3]

Polymer	Abb.	Temperatures (°C)						
		Temp. range	Feed zone	Zone 1	Zone 2	Zone 3	Nozzle	Mold
Polyethylene	PE-LD	160–270	20–30	140–300	200–350	220–350	220–350	20–70
	PE-LLD	160–200						20–60
	PE-HD	200–300						10–90
	PE-UHMW	200–250						40–70
Polypropylene	PP	200–300	20–30	150–210	210–250	240–290	240–300	20–100
Styrenic polymers	PS	170–280	20–30	150–180	180–230	210–280	220–280	10–60
	SB	190–280						
	SAN	200–260	35–45					30–80
	ABS	200–260						
	ASA	220–260						
Polyvinylchloride	PVC-U	170–210	30–40	135–160	165–180	180–205	180–210	20–60
	PVC-P	160–190	40	125–150	150–175	160–200	150–200	
Polychlorotrifluoroethylene	PCTFE	270–280		250–280	290–320	340–370	340–370	80–130
Polymethylmethacrylate	PMMA	190–290	50–60	135–180	185–225	200–250	200–250	40–80
Polyoxymethylene	POM	180–230	30–40	150–180	180–205	195–215	190–215	40–120
Polycarbonate	PC	270–380	70–90	235–270	285–310	305–350	310–350	80–120
Polyamide	PA 6	240–290	60–80	210	230	240	230	60–100
	PA 66	260–300		265	260	280	280	40–90
Polyurethane	TPU	190–220		175–200	180–210	205–240	205–240	40–80
Cellulose-acetobutyrate	CAB	180–220	40	130–140	150–175	160–190	160–190	40–80
Cellulose-acetate	CA	180–220	40	135–165	140–185	165–200	185–200	40–80
Cellulose-propionate	CP	190–230	40	160–190	190–230	190–240	190–220	40–80
Polyphenylene-oxyde	PPO	230–300	40	260–280	300–310	320–340	320–340	80–105
Polyphenylene-ether	PPE							
Polysulfone	PSU	310–390		250–270	290–330	320–330	300–340	95–160
Ionomeres (Surlyn)		290–330		90–170	130–215	140–215	140–220	10–30
Polymethyl-pentene	PMP	260–320		240–270	250–280	250–290	250–300	60
Poly-4-methyl-pentene-1								
Thermoplastic polyesters	PET	260–280	60–80	240–250	245–255	250–260	250–260	90–140
	PBT							

Table 3.2 Processing Window for Various Thermoplastics Polymers

Polymer	Melt temperature (°C)	Residence time (minutes)	Screw flight speed (m/s)
ABS	220	4	0.8
	220	10	0.3
	260	4	0.3
PC	280	4	0.8
	280	10	0.3
	320	4	0.3
PA 6	260	4	0.8
	260	10	0.3
	280	4	0.3
PA 66	270	4	0.8
	270	10	0.3
	300	4	0.3

to achieve high-quality surfaces, and to avoid viscous dissipation and jetting during mold filling.

3.1.3 Drying

Another important aspect is the water content of polymers that are hygroscopic. Residual moisture content in those materials that were not properly dried can lead to bubbles inside the part, flash, and poor surface quailty. The moisture content can often be removed within the injection molding machine if a degassing stage is present in the plasticating zone of the machine; however, because most injection molding processes do not require such a feature, a degassing stage is often not at hand. For those situations, the polymer pellets must be dried before they are placed inside the injection molding machine's hopper. Table 3.4 presents recommended drying temperatures and times for some thermoplastic polymers. For each individual case, however, the resin supplier's recommendations should be followed.

3.1.4 Processing Data for Thermoplastic Elastomers*

The thermoplastic elastomers described as "Blends" in Table 2.6 can all be purchased as pre-compounded products. Standard compounds, optimized for particular end uses, can be purchased from suppliers in pelletized, ready-to-use form. In such cases the user has no extra costs or processing steps. Alternatively the user can often carry

* This section was contributed by G. Holden.

Table 3.3 Pressure Settings and Processing Data for Various Thermoplastic Polymers [3,7,10]

Polymer	Abb	Pressure			Injection volume (%)	Shrink (%)	Flow length* (mm)	Buffer (mm)
		Injection pressure bar	Pack hold bar					
Polyethylene	PE-LD	400–800	100–250		5–95	1.5–3	550–600	2–6
	PE-HD	600–1200				1.5–5	200–600	
Polypropylene	PP	1200–1800	100–200		10–90	1.2–2.5	250–700	2–6
Polystyrene and Copolymers	PS	600–1800	100–200		5–95	0.45–0.6	200–500	2–6
	SB							
	SAN							
	ASA							
	ABS	1000–1500	100–250 (20–50)		10–90 (20–80)	0.4–0.7	320	2–6
Polyvinylchloride	PVC-U	800–1600	Up to 500		20–85	0.5–0.7	160–250	Maintain low
	PVC-P	600–1000	50–100		20–85	1–2	150–500	Maintain low
Polymethylmethacrylate	PMMA	500–1200	100–400 (20–50)		20–75	0.4–0.8	200–500	2–6
Polyoxymethylene	POM	1200–1500	100–200	60–80	15–75	1.9–2.3	500	2–6
Polycarbonate	PC	1300–1800	50–150 (20–50)	40–60	15–85 (20–80)	0.7–0.8	150–220	2–6
Polyamide	PA 6	Not too high or long	50–150 (20–50)	Not too high or long	15–80 (20–80)	0.8–2.5	400–600	2–6
	PA 66						810	
Cellulose-acetobutyrate	CAB	800–1200	50–100	40–70	15–85	0.4–0.7	500	3–8
Cellulose-acetate	CA						350	
Cellulose-propionate	CP						500	
Polyphenylene-oxyde	PPO	1000–1200	30–50	50–70	15–85	0.5–0.7		3–5
Thermoplastic Polyester	PET	1200–1500	30–50	50–70	15–85	1.3–1.5	200–500	3–5
	PBT						250–600	

* Flow length with a 2-mm thick spiral mold with average processing conditions.

Table 3.4 Recommended Drying Parameters for Common Thermoplastics [3]

Polymer	Abb.	Temperature (°C)	Time (hours)
Polystyrene and Copolymers	PS	60–80	1–3
	SAN	70–90	1–4
	ABS	70–80	4
	ASA	80–85	2–4
Polymethylmethacrylate	PMMA	70–100	8
Polyoxymethylene	POM	80–120	3–6
Polycarbonate	PC	100–120	8
Polyamide	PA	80–100	16
Ethylen-Butyl-Acrylate	EBA	70–80	3
Polyether-Block-Amide	PEBA	70–80	2–4
Celluloseacetobutyrate	CAB	60–80	2–4
Celluloseacetate	CA		
Cellulosepropionate	CP		
Polyphenylenoxyde	PPO	100	2
Polyphenylene-ether	PPE		
Polysulfone	PSU	120–150	3–4
Thermoplastic polyester	PET	120–140	2–12
	PBT		
Ethylen-Vinylacetate	EVA	70–80	3
Polyarylate	PAR	120–150	4–8
Polyarylsulfone	PASU	135–180	3–6
Polyethersulfone	PES		
Polyphenylensulfide	PPS	140–250	3–6
Polyether-Etherketone	PEEK	150	8
Thermoplastic elastomers	TPE	120	3–4
Polyamidimide	PAI	150–180	8–16
Polyetherimide	PEI	120–150	2–7
Liquid crystal polymers	LCP	110–150	4–8
Thermoplastic Polyurethane	TPU	100–110	1–2

out his own compounding to produce custom products. Compounding can be carried out in a continuous process, e.g., an internal mixer followed by an extruder and pelletizer. This can be done well in advance of molding and the products stored. This option requires the purchase of mixing and pelletizing equipment. Another option is to dry blend the ingredients together and then carry out compounding in the barrel of the injection molding machine. This introduces several new requirements:

1. Both the hard polymer and the elastomer should have small particle size, typically less than 1 mm (0.04 in.).

2. The oil or plasticizer must be completely absorbed by the polymers before being fed into the injection molder. Heat speeds up this process, which can be carried out in a high shear mixer or a heated tumble blender.
3. The injection molding machine should have a fairly high compression screw, or one fitted with a mixing section.
4. The screw should be kept rotating for as long as possible. In a reciprocating screw machine, this can be done by adjusting the back pressure so that the screw returns to its starting position only a second or two before the mold opens.

In all these blends, an excellent dispersion of ingredients is vital. Any undispersed material will give flaws in the final product. Dispersion can be checked by compression molding the product into a thin film, which can then be examined for irregularities.

Many molded parts are colored. The color is often introduced into thermoplastic elastomers by blending them with pelletized mix (sometimes called a masterbatch) of the colorant with a base polymer. This base polymer must be suitable for mixing with the thermoplastic elastomer, otherwise it can give such problems as decomposition or delamination. Details of suitable base polymers for each type of thermoplastic elastomer are given in Table 2.2.

All the thermoplastic elastomer types listed in Table 2.6 have grades optimized for injection molding. These can be processed on conventional reciprocating screw machines using tool steel molds and high injection pressures (the polystyrene/(S-B-S + oil) blends are also molded on low pressure equipment to make footwear). Of course each type has different optimum molding conditions. As a generalization, molding conditions suitable for the thermoplastic used as the hard phase of the thermoplastic elastomer are a good place to start (the chlorinated olefin interploymer alloys can be processed under conditions similar to those used for plasticized PVC). Further details of appropriate processing conditions for each type are summarized in Table 2.2.

Thermoplastic elastomers are often insert molded or co-molded with hard plastics to give final parts with regions of different hardness/stiffness. Production in this way avoids subsequent part assembly using adhesives, clips or clamps. Of course, it is necessary that the thermoplastic elastomer forms a strong bond to the hard plastic. A familiar example is a kitchen spatula. These often have a stiff handle molded from polypropylene and a flexible blade molded from a polypropylene/(S-EB-S + oil) blend. This example illustrates a general rule—adhesion between the two pieces will be good if the hard plastic is similar in composition to the hard phase of the thermoplastic elastomer. This is especially true for the less polar thermoplastic elastomer—the more polar ones can often adhere to several other polar thermoplastics. Special grades of thermoplastic elastomers suitable for molding against dissimilar hard plastics are under development. This is an area in which there is a lot of developmental activity. The user should consult manufacturers of thermoplstic elastomers to obtain current information. Adhesion may often be improved by blending a small amount (say 5%) of the hard plastic into the thermoplastic elastomer, assuming that the two are technically compatible. Hard plastics that have been found satisfactory for overmolding against each type of thermoplastic elastomer are listed in Table 3.5.

Table 3.5 Injection Molding Conditions and Properties of Thermoplastic Elastomers*

	PS/ (S-B-S + Oil)	PP/ (S-EB-S + Oil)	PP/ (EPR + Oil)	PP/ (Rubber + Oil)	PE/ Polyolefin
Rubber	Blends	Blends	Blends	DVs	BCPs
Melt temperature (°C)	190–210	210–250	250	190–230[a]	180–290
Mold temperature (°C)	25	50	50	10–80[b]	10
Injection rate	medium	fast	fast	fast	fast
Thermal stability during molding	good	excellent	excellent	good-excellent[c]	excellent
Drying[d]	A	A	A	D	NN
Polymers for overmolding	PS, HIPS	PP	PP	PP	PE
Base for color concentate	PS, HIPS	PP	PP	PP	PE
Price range[e] (cents / lb)	90–150	185–280	80–120	165–300	80–100
Hardness range (Shore A or D)	40A–45D	5A–60D	60A–65D	35A–50D	65A–85A

	PVC (NBR + Plasticizer) Blends	Chlorinated Olefin Alloys	Polyurethane/ Elastomer BCPs	Polyester/ Elastomer BCPs	Polyamide/ Elastomer BCPs
Melt temperature (°C)	140–200	170–190	180–230[a]	195–250[a]	180–290[a]
Mold temperature (°C)	50	50	10–50[a]	45	10
Injection rate	medium	fast	slow/medium	fast	fast
Thermal stability during molding	adequate	adequate	adequate	good	good
Drying[d]	A	A	V	V	V
Polymers for overmolding	PVC, SAN, ABS, SMA	PC/ABS, PVC, PEst, PU-TPEs,	ABS, PC, PC/ABS, PVC, PEst	PEst, PC, ABS	PA, PC
Base for color concentate	Plasticized PVC	Plasticized PVC	Soft PU-TPEs ABS	Soft PEst-TPEs	Soft PA-TPEs
Price range[e] (cents / lb)	130–150	225–275	225–375	275–375	450–550
Hardness Range (Shore A or D)	50A–90A	45A–80A	70A[f]–75D	90A–80D	75A–70D

[a] Depends on hardness. Harder grades require the higher rmelt temperatures.
[b] Those based on NBR require higher (60 to 100°C) mold temperatures.
[c] Depends on the type of rubber. Those based on EPDM and butyl are the most stable.
[d] A = May improve appearance, especially in filled products, D = Desirable, NN = Not necessary, V = Vitally Important.
[e] For general purpose grades.
[f] As low as 60A when plasticized.
* The abbreviations used in this table are collected in Chapter 2.

Thermoplastic elastomers are also insert molded against metal, e.g., to produce handles for knives and kitchen utensils. Adhesion to metal is often poor, and so the metal insert is usually serrated and often heated before being placed in the mold. Another technique to improve adhesion is to solution coat the metal with an adhesive.

After molding is complete and the part has solidified, it must be removed from the mold without damaging or deforming it. This must be done in a reliable and consistent way, otherwise "hangups" will occur and stop production. Because thermoplastic elastomers are relatively soft materials, they have quite high surface friction and so they often stick to the mold surface. This problem becomes progressively worse as the products get softer. Molds with a high draft angle (at least 3°) on walls and cores give best results, especially with softer products. Vapor honing the mold surface or using a matte finish (such as #2 SPI) reduces the sticking problem. Another alternative is treating the mold surface with a permanent release coating, such as tungsten disulfide.

The softer thermoplastic elastomers may deform rather than eject if ejection methods appropriate for hard thermoplastics (e.g., small diameter ejection pins) are used. Stripper plates often give better results. If ejector pins are used, they should be as large as possible, and should push against sections of the part providing greatest support. Air or air assisted ejection is also useful, especially for deep draw parts. Even when mechanical means of ejection are used, incorporation of an air valve will serve to break a vacuum and assist in part ejection.

Drying thermoplastic elastomer pellets before molding is useful for some thermoplastic elastomers and vitally important for others. It will often improve part appearance by reducing splay marks. Of the various types of thermoplastic elastomers, drying is strongly recommended for the dynamic vulcanizates and absolutely essential for polyurethane, polyester and polyamide / elastomer block copolymers. For these three materials, the combination of water and heat in the barrel of an injection molder can cause polymer breakdown, foaming and sudden "spitting" of the molten polymer. The remaining types may not need drying, especially if used directly from sealed bags, but it is often desirable for those containing large amounts of filler. Sometimes, all that is necessary is the use of a vented two stage screw on the injection machine.

Manufacturers of thermoplastic elastomers often provide very detailed literature describing the appropriate molding parameters, and this should be consulted before use. Some of this information is summarized in Table 3.5.

3.1.5 Processing Data for Thermosets

The influence of temperature and temperature changes during processing of thermosetting polymers is significant. The material temperature, as well as the temperature of the cylinder, nozzle, and mold, influence competing mechanisms for mold filling and curing reaction. A rise in temperature will lower the viscosity, which will

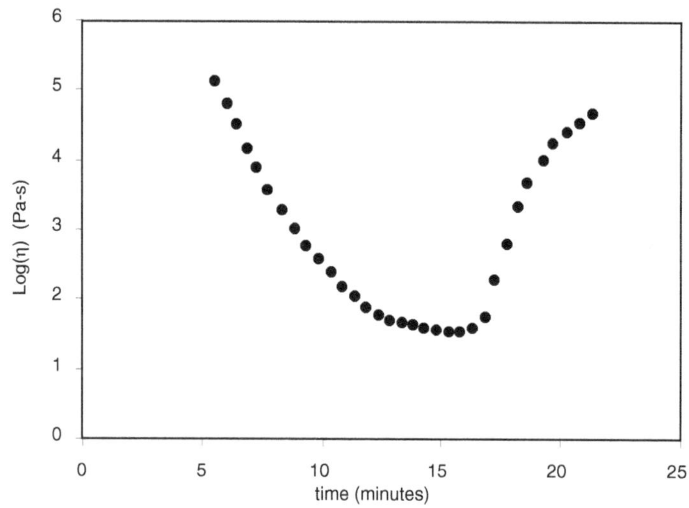

Figure 3.1 Viscosity development during cure for an epoxy resin as the temperature is ramped-up from 40°C (t = 0) to 180°C (t = 20 min).

ease the processability of the material; however, it will also speed the cross-linking reaction, which will in turn raise the viscosity and lead to processing problems. Figure 3.1 presents a predicted curve for the viscosity of epoxy as the temperature is ramped in time. Here, as expected, the viscosity drops as the temperature increases. As time progresses, the viscosity starts to rise due to the progression of the curing reaction. Hence, there is a processing window that would render optimal processing conditions. Table 3.6 presents suggested processing conditions for various thermosetting resins.

3.1.6 Processing Data for Elastomers

When injection molding elastomeric mixtures most of the heat required to raise the temperature to vulcanization temperatures results from the viscous dissipation during flow. When injection molding very thin products, cycle times of 1 minute or less can easily be achieved. Table 3.7 presents recommended processing temperatures for various elastomers.

3.2 Rheology of Polymer Melts

Rheology is the science of fluid behavior during flow-induced deformation. Among the variety of materials that rheologists study, polymers have been found to be the most interesting and complex. Polymer melts are shear thinning and viscoelastic, and

Table 3.6 Processing Conditions for Various Thermosets [2,3]

Polymer	Abb.	Temperatures (°C)				Pressures		Curing times
		Material	Cylinder	Nozzle	Mold	Injection bar	Pack bar	
Polyester, filled	UPE-GF		70–110	70–110				
Polyester, unfilled	UPE	100–140	30–60	80–110	155–175	500–2500		10–100
Phenolic	PF	120–140	65–85	85–110	155–190	1000–2500		15–80
Urea formaldehyde	UF		70–90	90–115	130–160	1000–2000	up 150	15–80
Epoxy	EP		50–60	70–100	180–190	800–1800	5–10	15–80
Polyester, filled	UPE-GF		70–110	70–110				
Polyester, unfilled	UPE	100–140	30–60	80–110	155–175	500–2500		10–100
Melamine	MF	120–140	65–85	80–110	155–190	1000–2500		15–80

their flow properties are temperature dependent. This section discusses the phenomena that are typical of polymer melts and covers the basic properties used to represent the flow behavior of polymers during injection molding. The section also briefly introduces rheometry. For further reading on rheology of polymer melts, the reader should consult References 4 to 9.

3.2.1 Shear Thinning Behavior of Polymers

Viscosity is the most widely used material parameter when determining the behavior of polymers during processing. Because the majority of polymer processes are shear rate dominated, the viscosity of the melt is commonly measured using shear deformation measurement devices. For example, the simple shear flow generated in the sliding plate rheometer [10], shown in Fig. 3.2, exhibits a deviatoric stress defined by

Table 3.7 Processing Conditions for Various Elastomers [3]

Polymer	Abb.	Temperatures °C			
		Material	Cylinder	Nozzle	Mold
Synthetic rubber	BR. NBR		70–110	70–110	170–200
Natural rubber	NR				
Silicon rubber	SiR				
Polybutadiene rubber	PB	240–280	60–100	60–100	40–80

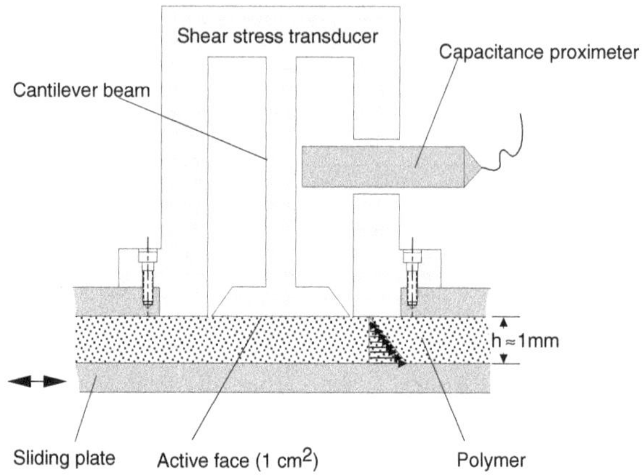

Figure 3.2 Schematic disgram of a sliding plate rheometer.

$$\tau_{xy} = \eta(T, \dot{\gamma})\dot{\gamma}_{xy} \tag{3.2}$$

where $\eta(T, \dot{\gamma})$ is the viscosity and $\dot{\gamma}_{xy}$ the shear rate defined by v/h. For the flow in Fig. 3.2, the magnitude of the rate of deformation tensor, $\dot{\gamma}$, is $\dot{\gamma}_{xy}$. There are also polymer processes, such as blow molding, thermoforming, and fiber spinning, which are dominated either by elongation or by a combination of shear and elongational deformation. In addition, some polymer melts exhibit significant elastic effects during deformation.

Most polymer melts are *shear thinning fluids*. The shear thinning effect is the reduction in viscosity at high rates of deformation, such as that shown in Fig. 3.3 for an LDPE resin. This phenomenon occurs because the polymer molecules are entangled at low rates of deformation, but stretched out and disentangled at high rates of deformation. The disentangled molecules can slide past each other more easily, thus lowering the bulk melt viscosity.

The power-law model proposed by Ostwald [11] and de Waale [12] is the simplest model that accurately represents the shear thinning region in the viscosity versus strain rate curve, but overshoots the Newtonian plateau at small strain rates. The power-law model is:

$$\eta = m(T)\dot{\gamma}^{n-1} \tag{3.3}$$

where m is called the *consistency index* and n the *power law index*. The consistency index may include the temperature dependence of the viscosity and can be represented as

$$m(T) = m_0 e^{-a(T-T_0)} \tag{3.4}$$

Power-law constants, as used in Eqs. (3.3) and (3.4), are presented in Table 3.8 for common thermoplastics.

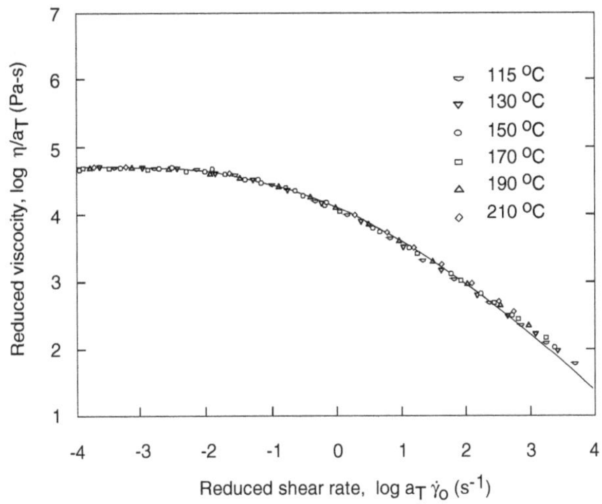

Figure 3.3 Reduced viscosity curve for an LDPE at a reference temperature of 150°C. Viscosity data measured at other temperatures have been shifted horizontally by a factor *aT* to construct the master curve.

3.2.2 Simplified Flows Common in Injection Molding

Many polymer processing operations can be modeled using simplified geometries and material models. This section presents several isothermal flows in simple geometries using a Newtonian viscosity and a power-law viscosity as described in Eq. (3.3).

Simple Shear Flow

Simple shear flows, such as that shown in Fig. 3.4, are common in polymer processing, such as inside extruders as well as the plasticating unit of an injection molding machine. The flow field in simple shear is the same for all fluids and is described by

Table 3.8 Power-Law Indexes, Consistency Indexes, and Temperature Dependence Constants for Common Thermoplastics

Polymer	m (Pa-s^n)	n	a (1/°C)	T_0(°C)
Polystyrene	2.80×10^4	0.28	–0.025	170
High density polyethylene	2.00×10^4	0.41	–0.002	180
Low density polyethylene	6.00×10^3	0.39	–0.013	160
Polypropylene	7.50×10^3	0.38	–0.004	200
Polyvinylchloride	1.70×10^4	0.26	–0.019	180

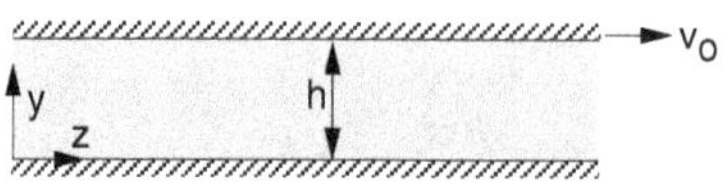

Figure 3.4 Schematic diagram of a simple shear flow.

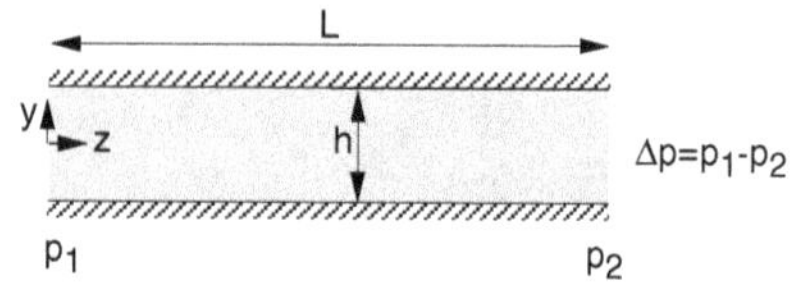

Figure 3.5 Schematic diagram of a pressure flow through a slit.

$$v_z(y) = v_0 \frac{y}{h} \tag{3.5}$$

$$Q = \frac{v_0 h W}{2} \tag{3.6}$$

Pressure Flow Through a Slit

The pressure flow through a slit, such as that shown in Fig. 3.5, is commonly encountered in flows through extrusion dies or inside injection molds. The Newtonian flow field is described using

$$v_z(y) = \left(\frac{h^2 \Delta p}{8\mu L}\right)\left[1 - \left(\frac{2y}{h}\right)^2\right] \tag{3.7}$$

$$Q = \frac{W h^3 \Delta p}{12\mu L} \tag{3.8}$$

When using the power law model [Eq. (3.3)] the flow field is described by

$$v_z(y) = \left(\frac{h}{2(s+1)}\right)\left(\frac{h\Delta p}{2mL}\right)^s \left[1 - \left(\frac{2y}{h}\right)^{s+1}\right] \tag{3.9}$$

$$Q = \frac{W h^2}{2(s+2)}\left(\frac{h\Delta p}{2mL}\right)^s \tag{3.10}$$

where $s = 1/n$.

Pressure Flow Through a Tube—Hagen-Poiseuille Flow

Pressure flow through a tube, such as that shown in Fig. 3.6, is encountered in the runner system in injection molds, in certain dies, and in the capillary rheometer. The Newtonian flow field is given by

$$v_z(r) = \frac{R^2 \Delta p}{4\mu L}\left[1 - \left(\frac{r}{R}\right)^2\right] \tag{3.11}$$

$$Q = \frac{\pi R^4 \Delta p}{8\mu L} \tag{3.12}$$

Using the power law fluid the flow field in a tube is described by

$$v_z(r) = \frac{R}{1+s}\left(\frac{R\Delta p}{2mL}\right)^s \left[1-\left(\frac{r}{R}\right)^{s+1}\right] \tag{3.13}$$

$$Q = \left(\frac{\pi R^3}{s+3}\right)\left(\frac{R\Delta p}{2mL}\right)^s \tag{3.14}$$

The preceding equations can be used when balancing the runner system inside a multicavity injection mold.

As an example, let us consider the multicavity injection molding process shown in Fig. 3.7. To achieve equal part quality, the filling time for all cavities must be balanced. For the case in question, we need to balance the cavities by solving for the runner radius, R_2. For a balanced runner system the flow rates into all cavities must match. For a given flow rate Q, length L, and radius R_1, we can also solve for the pressures at the runner system junctures. Assuming an isothermal flow of a non-Newtonian shear thinning polymer with viscosity η, we can compute the radius R_2 for a part molded of polystyrene with a consistency index of 2.8×10^4 Pa-s^n and a power law index (n) of 0.28.

The flow through each runner section is governed by Eq. (3.14), and the various sections can be represented using

$$\text{Section 1: } 4Q = \left(\frac{\pi(2R_1)^3}{s+3}\right)\left(\frac{2R_1(p_1-p_2)}{2mL}\right)^s$$

$$\text{Section 2: } 2Q = \left(\frac{\pi(2R_1)^3}{s+3}\right)\left(\frac{2R_1(p_2-p_3)}{2m(2L)}\right)^s$$

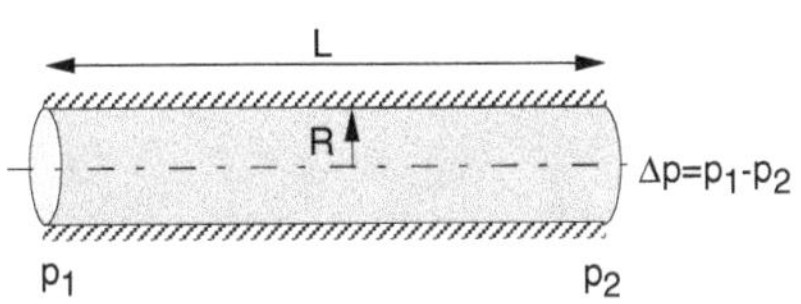

Figure 3.6 Schematic diagram of a tube flow.

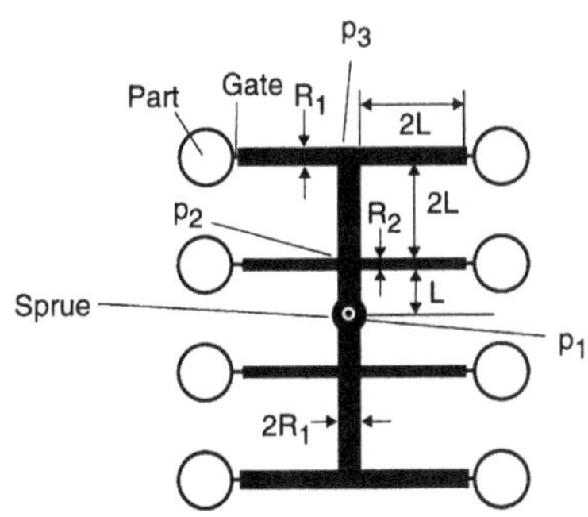

Figure 3.7 Runner system lay out.

$$\text{Section 3:}\ Q=\left(\frac{\pi R_2^{\ 3}}{s+3}\right)\left(\frac{R_2\,(p_2-0)}{2m(2L)}\right)^s$$

$$\text{Section 4:}\ Q=\left(\frac{\pi R_1^{\ 3}}{s+3}\right)\left(\frac{R_1\,(p_3-0)}{2m(2L)}\right)^s$$

Using values of L = 10 cm, R_1 = 4 mm, and Q = 20 cm^3/s, the unknown parameters, P_1, P_2, P_3, and R_2, can be obtained using the preceding equations. The equations are nonlinear and must be solved in an iterative manner. For the given values, a radius, R_2, of 3.4 mm would result in a balanced runner system, with pressures P_1 = 265.7 bar, P_2 = 230.3 bar, and P_3 = 171.9 bar. For comparison, if one had assumed a Newtonian model with the same consistency index and a power law index of 1.0 a radius, R_2, of 3.9 would have resulted, with much higher required pressures of P_1 = 13,926 bar, P_2 = 12,533 bar, and P_3 = 11,140 bar. The difference is due to shear thinning.

3.2.3 Estimating Injection Pressure and Clamping Force (Stevenson Model)

To aid the polymer processing engineer in finding required injection pressures and corresponding mold clamping forces, Stevenson [13] generated a set of dimensionless groups and corresponding graphs for nonisothermal mold filling of non-Newtonian polymer melts. Using the notation in Fig. 3.8 and a viscosity defined by $\eta = m_0 e^{-a(T-T_{ref})}|\dot{\gamma}|^{n-1}$, Stevenson defined four dimensionless groups that can be used in pressure and clamping force predictions:

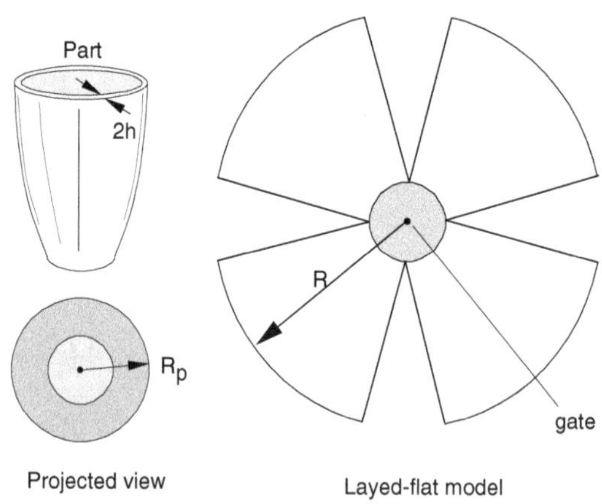

Figure 3.8 Layed-flat representation of an injection molded part.

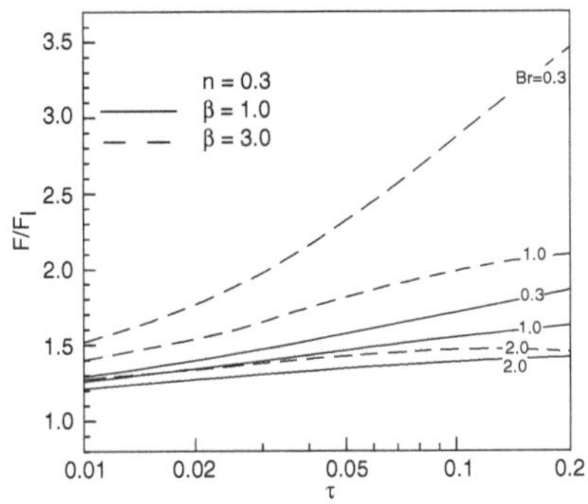

Figure 3.9 Dimensionless clamping force versus dimensionless groups.

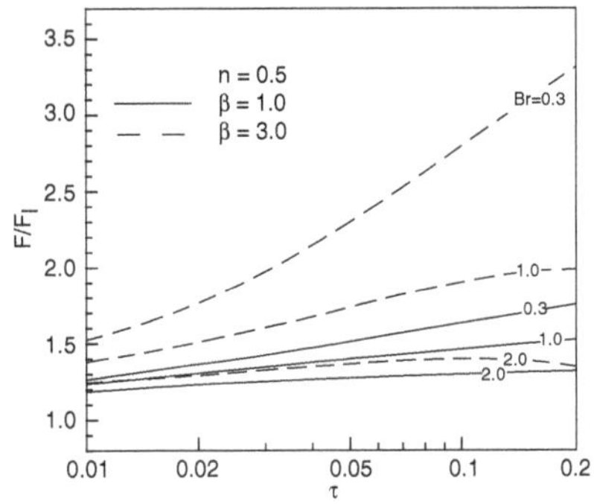

Figure 3.10 Dimensionless clamping force versus dimensionless groups.

- The dimensionless temperature β determines the intensity of the coupling between the energy equation and the momentum balance. It is defined by

$$\beta = a(T_i - T_m) \tag{3.15}$$

where T_i and T_m are the injection and mold temperatures, respectively.

- The dimensionless time τ is the ratio of the filling time, t_{fiil}, and the time for thermal equilibrium via conduction, defined by

$$\tau = \frac{t_{fill}k}{h^2 \rho C_p} \tag{3.16}$$

- The Brinkman number Br is the ratio of the energy generated by viscous dissipation and the energy transported by conduction. For a nonisothermal, non-Newtonian model it is

$$Br = \frac{m_0 e^{-aT_i} h^2}{k(T_i - T_m)} \left(\frac{R}{t_{fill} h} \right)^{n+1} \tag{3.17}$$

- The power-law index n of the Ostwald and de Waale model reflects the shear thinning behavior of the polymer melt.

Once the dimensionless parameters are calculated, the dimensionless injection pressures $\left(\frac{\Delta p}{\Delta p_I}\right)$ and dimensionless clamping forces $\left(\frac{F}{F_I}\right)$ are read from Figs. 3.9 to 3.12. The isothermal pressure and force are computed using

$$\Delta p_I = \frac{m_0 e^{-aT_i}}{1-n} \left[\frac{1+2n}{2n} \frac{R}{t_{fill} h} \right]^n \left(\frac{R}{h} \right) \tag{3.18}$$

and

$$F_I = \pi R^2 \left(\frac{1-n}{3-n} \right) \Delta p_I \tag{3.19}$$

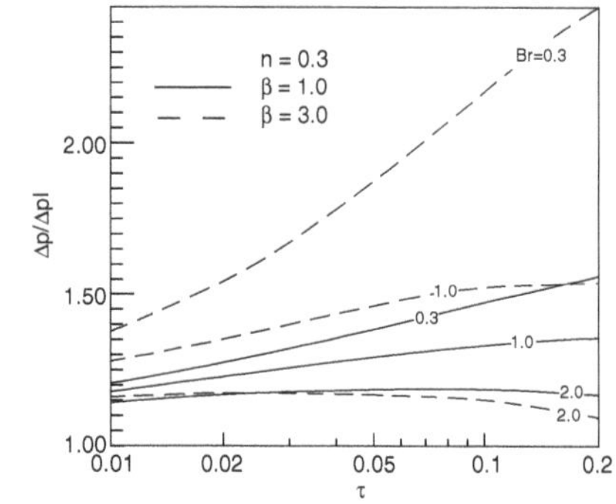

Figure 3.11 Dimensionless injection pressure versus dimensionless groups.

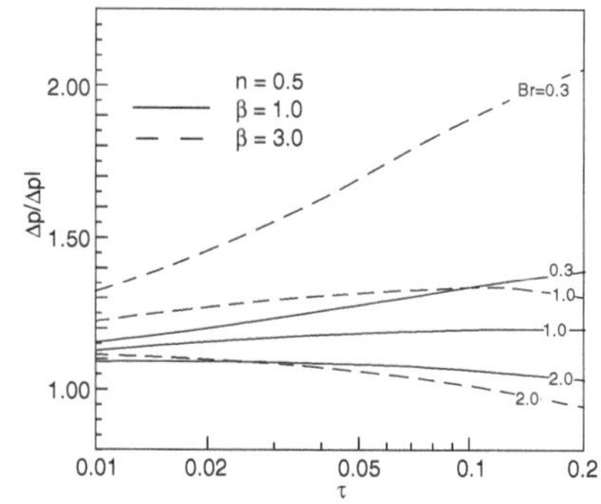

Figure 3.12 Dimensionless injection pressure versus dimensionless groups.

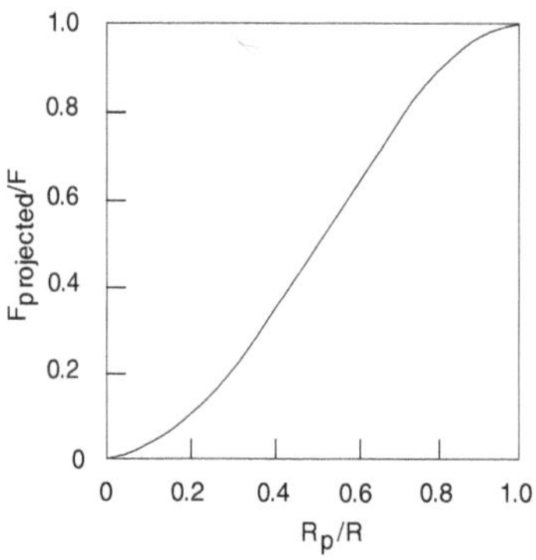

Figure 3.13 Clamping force correction for the projected area.

Because the part area often exceeds the projected area, Fig. 3.13 can be used to correct the computed clamping force.

As an example let us determine the maximum clamping force and injection pressure required to mold an ABS suitcase with a filling time, t_f, of 2.5 s. Use the dimensions shown in Fig. 3.14, an injection temperature, T_i, of 227°C (500 K), and a mold temperature, T_m, of 27°C (300 K). The properties necessary for the calculations are also given in the table.

Properties for ABS

$n = 0.29$	$\rho = 1020\,kg/m^3$
$m_0 = 29 \times 10^6\,Pa\text{-}s^n$	$C_p = 2343\,J/kg/K$
$a = 0.01369/K$	$k = 0.184\,W/m/K$

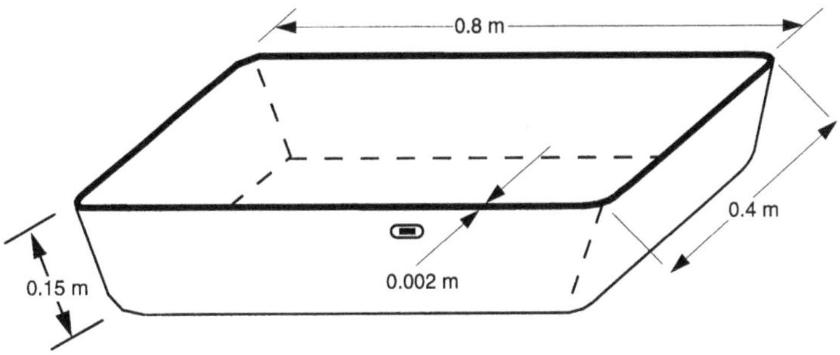

Figure 3.14 Suitcase geometry.

We start solving this problem by first laying the suitcase flat and determining the required geometric factors (Fig. 3.15). From the suitcase geometry, the longest flow path, R, is 0.6 m and the radius of the projected area, R_p, is 0.32 m. We can now compute the dimensionless parameters given in Eqs. (3.15 to 3.17),

$$\beta = 0.01369\,(500-300) = 2.74$$

$$\tau = \frac{2.5(0.184)}{(0.001)^2\,(1020)(2343)} = 0.192$$

$$Br = \frac{(29\times10^6)e^{-0.01369\,(500)}\,(0.001)^2}{0.184(500-300)}\left(\frac{0.6}{2.5(0.001)}\right)^{0.29+1} = 0.987$$

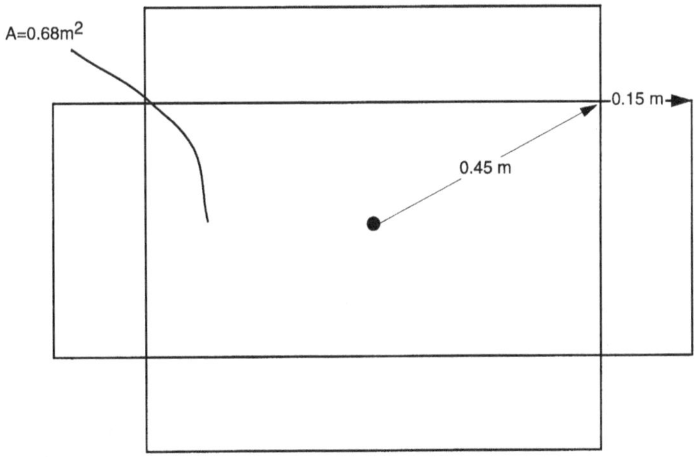

Figure 3.15 Layed-flat suitcase.

The isothermal injection pressure and clamping force are computed using Eqs. (3.18 to 3.20)

$$\Delta p_I = \frac{29 \times 10^6 e^{-0.01369(500)}}{1-0.29}\left[\frac{1+2(0.29)}{2(0.29)}\frac{0.6}{2.5(0.001)}\right]^{0.29}\left(\frac{0.6}{0.001}\right) = 171\,MPa$$

$$F_I = \pi(0.6)^2\left(\frac{1-0.29}{3-0.29}\right)(17.1 \times 10^7) = 50.7 \times 10^6\,N$$

We now look up $\Delta p/\Delta P_I$ and F/F_I in Figs. (3.9 to 3.12). Because little change occurs between $n = 0.3$ and $n = 0.5$, we choose $n = 0.3$. For other values of n, however, we can interpolate or extrapolate. For $\beta = 2.74$, we interpolate between 1 and 3 as

$$\beta = 1 \rightarrow \Delta p/\Delta p_I = 1.36 \text{ and } F/F_I = 1.65$$

$$\beta = 3 \rightarrow \Delta p/\Delta p_I = 1.55 \text{ and } F/F_I = 2.1$$

$$\beta = 2.74 \rightarrow \Delta p/\Delta p_I = 1.53 \text{ and } F/F_I = 2.04$$

$$\Delta p = (\Delta p/\Delta p_I)\Delta p_I = 262\,MPa = 2620\,bar$$

$$F = (F/F_I)F_I = 10.3 \times 10^7\,N = 10{,}300\ metric\ tons$$

The clamping force can be corrected for an $R_p = 0.32m$ using Fig. 3.13 and $R_p/R = 0.53$.

$$F_{projected} = (0.52)10{,}300 = 5356\ metric\ tons$$

For our suitcase cover, where the total volume is 1360 ml and total part area is 0.68 m^2, the preceding numbers are too high. A usefull rule-of-thumb is a maximum allowable clamping force of 2 tons/in^2. Here, we have greatly exceeded that number. Approximately 3000 metric tons/m^2 are normally allowed in commercial injection molding machines. For example, a typical injection molding machine with a shot size of 2000 ml has a maximum clamping force of 630 metric tons with a maximum injection pressure of 1400 bar. A machine with much larger clamping forces and injection pressures is suitable for much larger parts. For example, a machine with a shot size of 19,000 ml allows a maximum clamping force of 6000 metric tons with a maximum injection pressure of 1700 bar. For this example we must reduce the pressure and clamping force requirements. This can be accomplished by increasing the injection and mold temperatures or by reducing the filling time. Recommended injection temperatures for ABS are between 210 and 240°C, and recommended mold temperatures are between 40 and 90°C. As can be seen, there is room for adjustment in the processing conditions, so one must repeat the preceding procedure using new conditions.

3.2.4 Nonisothermal Flows in Polymer Processing

Although we simplify analyses of polymer processes by assuming isothermal conditions, most operations are nonisothermal because they include melting and viscous dissipation. Hence, the temperature of the polymer melt lies between the glass transition, for amorphous materials, T_g (or the melting temperature for semi-crystalline polymers T_m) and the heater temperature, T_h, and often exceeds T_h due to viscous dissipation. An estimate of the maximum temperature rise due to viscous heating in a simple shear flow is

$$\Delta T_{max} = \frac{\eta V_0^2}{8k} \tag{3.20}$$

As mentioned in the previous section, to estimate if viscous dissipation is important in a polymer process, the Brinkman number, Br, is often used. A simplified form of Eq. 3.17 is often used

$$Br = \frac{\eta V_0^2}{k(T_h - T_g)} \tag{3.21}$$

The choice of temperature difference, e.g., $(T_h - T_g)$, depends on the type of problem being analyzed.

3.2.5 Normal Stresses in Shear Flow

The tendency of polymer molecules to "curl-up" while they are being stretched in shear flow results in normal stresses in the fluid. For example, the shear flow presented in Eq. (3.2) has measurable normal stress differences, N_1 and N_2, which are called the *first* and *second normal stress differences*, respectively. The first and second normal stress differences are material dependent and are defined by

$$N_1 = \tau_{xx} - \tau_{yy} = -\Psi_1(\dot{\gamma}, T)\dot{\gamma}_{xy}^2 \tag{3.22}$$

$$N_2 = \tau_{yy} - \tau_{zz} = -\Psi_2(\dot{\gamma}, T)\dot{\gamma}_{xy}^2 \tag{3.23}$$

The material functions, Ψ_1 and Ψ_2, are called the *primary* and *secondary normal stress coefficients*, and are also functions of the magnitude of the strain rate tensor and temperature. The first and second normal stress differences do not change sign when the direction of the strain rate changes. This is reflected in Eqs. (3.22) and (3.23). Figure 3.16 [14] presents the first normal stress difference coefficient for the LDPE melt of Fig. 3.2 at a reference temperature of 150°C. The second normal stress difference is difficult to measure and is often approximated by

$$\Psi_2(\dot{\gamma}) \approx -0.1\Psi_1(\dot{\gamma}) \tag{3.24}$$

The normal stress differences play significant roles during processing. For example, the first normal stress difference is partly responsible for *extrudate swell* (Fig. 3.17)

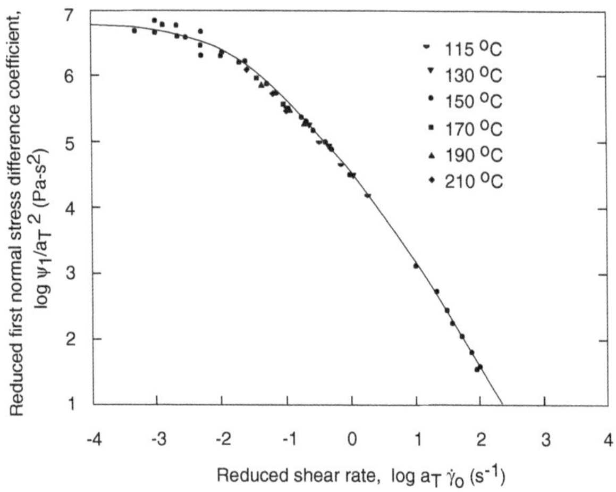

Figure 3.16 Reduced first normal stress difference coefficient for an LDPE melt at a reference temperature of 150°C.

at the exit of the die. The second normal stress differences help diminish the eccentricity of a wire in the die during the wire coating process [15].

3.2.6 Deborah Number

A useful parameter for estimating the elastic effects during flow is the Deborah number:

$$De = \frac{\lambda}{t_p} \tag{3.25}$$

where λ is the relaxation time of the polymer and t_p is a characteristic process time. For the extrusion case in Fig. 3.17, the characteristic process time can be defined by the ratio of characteristic die dimension and average speed through the die. A Deborah number of zero represents a viscous fluid, whereas a Deborah number of

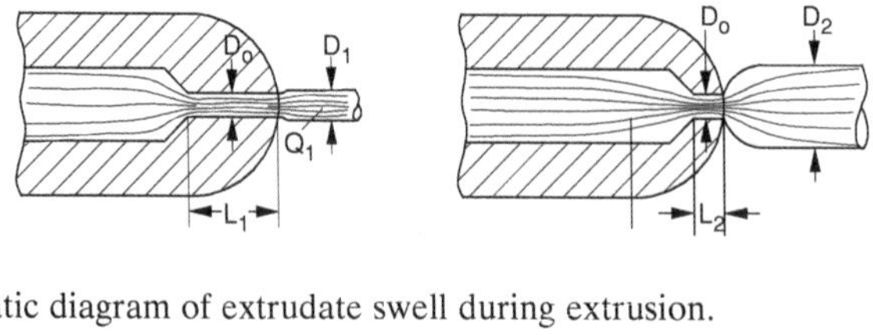

Figure 3.17 Schematic diagram of extrudate swell during extrusion.

∞ reflects an elastic solid. As the Deborah number exceeds 1, the polymer does not have enough time to relax during the process, resulting in *extrudate swell**, *shark skin* or even *melt fracture*.

Although many factors affect the amount of extrudate swell, fluid "memory" and normal stress effects are most significant; however, abrupt changes in boundary conditions, such as the separation point of the extrudate from the die, also play a role in the swelling or cross-section reduction of the extrudate. In practice, the fluid memory contribution to die swell can be mitigated by lengthening the land length of the die. This is schematically depicted in Fig. 3.17. A long die land separates the polymer from the manifold long enough to allow it to "forget" its past shapes.

Waves in the extrudate may also appear as a result of high speeds during extrusion, where the polymer cannot relax. This phenomenon is generally called *shark skin*. Polymers can be extruded at such high speeds that an intermittent separation of melt and inner die walls occurs. This phenomenon is often called *stick-slip* or *spurt flow* and is attributed to high shear stresses between the polymer and the die wall. This phenomenon occurs when the shear stress is near the critical value of 0.1 MPa [15–17]. If the speed is increased further, a chaotic pattern will eventually occur. This well-known phenomenon is called *melt fracture*. Shark skin is frequently absent, and spurt flow seems to occur only with linear polymers.

The critical shear stress has been reported to be independent of the melt temperature but inversely proportional to the weight average molecular weight [17,18]; however, Vinogradov et al. [15] presented results where the critical stress was independent of molecular weight except at low molecular weights. Dealy and co-workers [17] and Denn [18] give an extensive overview of various melt fracture phenomena. Both references are recommended.

3.2.7 Rheology of Curing Thermosets

A curing thermoset polymer has a conversion or cure dependent viscosity that increases as the molecular weight of the reacting polymer increases. For the vinyl ester whose curing history is shown in Fig. 3.18 [16], the viscosity behaves as shown in Fig. 3.19 [18]. Hence, a complete model for viscosity of a reacting polymer must contain the effects of strain rate, $\dot{\gamma}$, temperature, T, and degree of cure, c, such as

$$\eta = \eta(\dot{\gamma}, T, C) \tag{3.26}$$

* Newtonian fluids, which do not experience elastic or normal stress effects, also show some extrudate swell or reduction. A Newtonian fluid extruded at high shear rates reduces its cross-section to 87% of the diameter of the die. This swell is the result of inertia effects caused by the rearrangement from the parabolic velocity distribution inside the die to the flat velocity distribution of the extrudate. If the polymer is extruded at very low shear rates it swells to 113% of the diameter of the die.

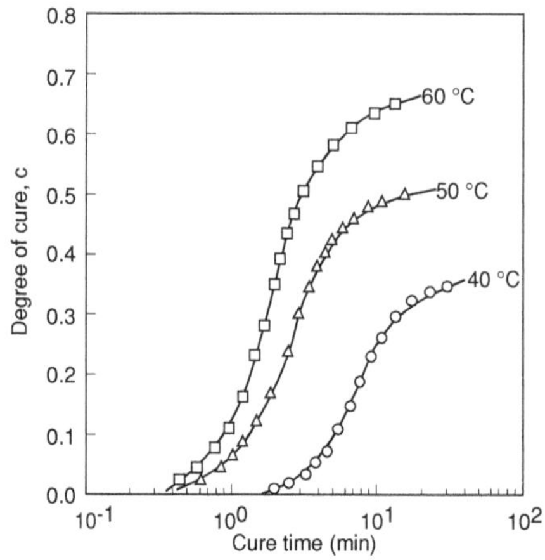

Figure 3.18 Degree of cure as a function of time for a vinyl ester at various isothermal cure temperatures.

There are no generalized models that include all these variables for thermosetting polymers; however, extensive work has been done on the viscosity of polyurethanes [19,20] used in the reaction injection molding process. An empirical relation that models the viscosity of these mixing activated polymers, given as a function of temperature and degree of cure, is written as

$$\eta = \eta_0 e^{E/RT} \left(\frac{c_g}{c_g - c} \right)^{C_1 + C_2 c} \tag{3.27}$$

where E is the activation energy of the polymer, R is the ideal gas constant, T is the temperature, c_g is the gel point, c the degree of cure, and C_1 and C_2 are constants that

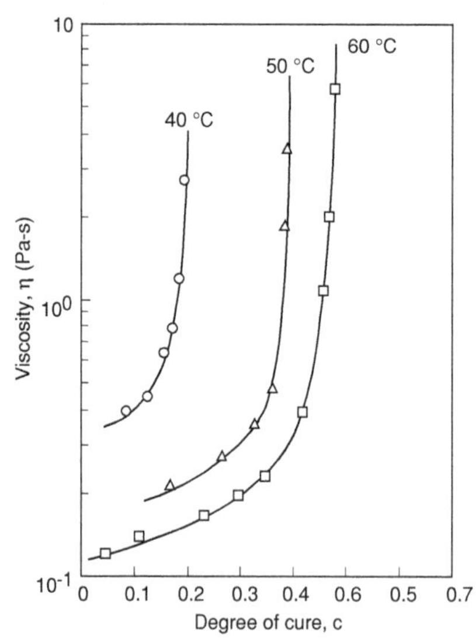

Figure 3.19 Viscosity as a function of degree of cure for a vinyl ester at various isothermal cure temperatures.

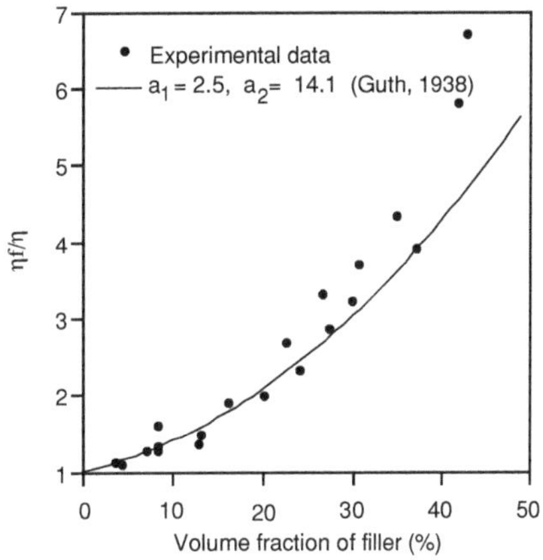

Figure 3.20 Viscosity increase as a function of volume fraction of filler for polystyrene and LDPE.

fit the experimental data. The gel point is the degree of cure when the molecular weight goes to infinity or when the molecules become interconnected.

3.2.8 Suspension Rheology

Particles suspended in a material, such as in filled or reinforced polymers, have a direct effect on the properties of the final article and on the viscosity during processing. The model that best fits experimental data is the one given by Guth [21]:

$$\frac{\eta f}{\eta_0} = 1 + 2.5\phi + 14.1\phi^2 \tag{3.28}$$

Figure 3.20 compares experimental data to Guth's equation. The experiments were performed on polyethylene and polystyrene containing different fill factors of spherical glass particles ranging between 36 μm and 99.8 μm in diameter. The model agrees well with the experimental data up to volume fractions of 30%.

3.3 Rheometry

In industry, there are various ways to qualify and quantify the properties of the polymer melt. The techniques range from simple analyses for checking the consistency of the material at certain conditions, to more difficult fundamental measurements to evaluate viscosity and normal stress differences. This section includes three such techniques to give the reader a general idea of current measuring methods.

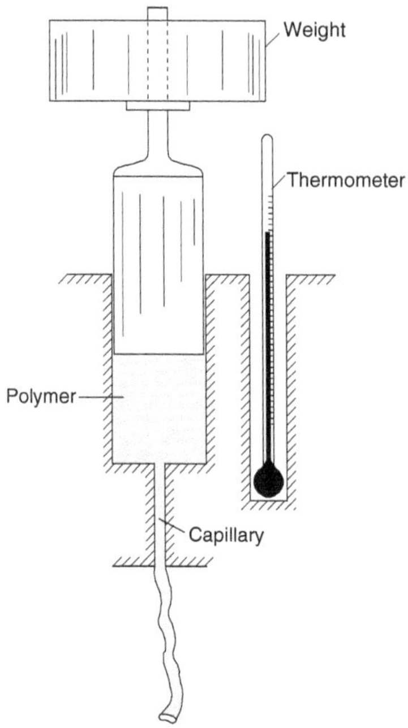

Figure 3.21 Schematic diagram of an extrusion plastometer used to measure melt flow index.

3.3.1 The Melt Flow Indexer

The melt flow indexer is often used in industry to characterize a polymer melt and as a simple and quick means of quality control. It takes a single point measurement using standard testing conditions specific to each polymer class on a ram-type extruder or extrusion plastometer, as shown in Fig. 3.21. The standard procedure for testing the flow rate of thermoplastics using an extrusion plastometer is described in the ASTM D1238 test [22]. During the test, a sample is heated in the barrel and extruded from a short cylindrical die using a piston actuated by a weight. The weight of the polymer in grams extruded during the 10-minute test is the melt flow index (MFI) of the polymer.

3.3.2 The Capillary Viscometer

The most common and simplest device for measuring viscosity is the capillary viscometer. Its main component is a straight tube or capillary, and it was first used to measure the viscosity of water by Hagen [23] and Poiseuille [24]. A capillary rheometer has a pressure driven flow for which the shear rate is maximum at the wall and zero at the center of the flow, making it a nonhomogeneous flow.

Because pressure-driven viscometers employ heterogeneous flows, they can only measure steady shear functions such as viscosity, $\eta(\dot{\gamma})$. They are widely used, however, because they are relatively inexpensive and simple to operate. Despite their simplicity, long-capillary viscometers give the most accurate viscosity data available. Another major advantage is that the capillary rheometer has no free surfaces in the test region, which is unlike other types of rheometers such as the cone and plate rheometer to be discussed next. When the strain rate–dependent viscosity of polymer melts is measured, capillary rheometers are capable of obtaining such data at shear rates greater than 10 second. This is important for processes with higher rates of deformation like mixing, extrusion, and injection molding. Because its design is basic and it only needs a pressure head at its entrance, the capillary rheometer can easily attach to the end of a screw- or ram-type extruder for online measurements. This makes the capillary viscometer an efficient tool for industry. The basic features of the capillary rheometer are shown in Fig. 3.22.

3.3.3 Viscosity from the Capillary Viscometer

A capillary tube of radius R and length L is connected to the bottom of a reservoir. Pressure drop and flow rate through this tube are used to determine the viscosity. At the wall, the shear stress is:

$$\tau_w = \frac{R}{2}\frac{(p_o - p_L)}{L} = \frac{R}{2}\frac{\Delta p}{L} \tag{3.29}$$

Equation (3.29) requires that the capillary be long enough to assure fully developed flow, where end effects are insignificant. Because of entrance effects, however, the actual pressure profile along the length of the capillary exhibits curvature. The effect

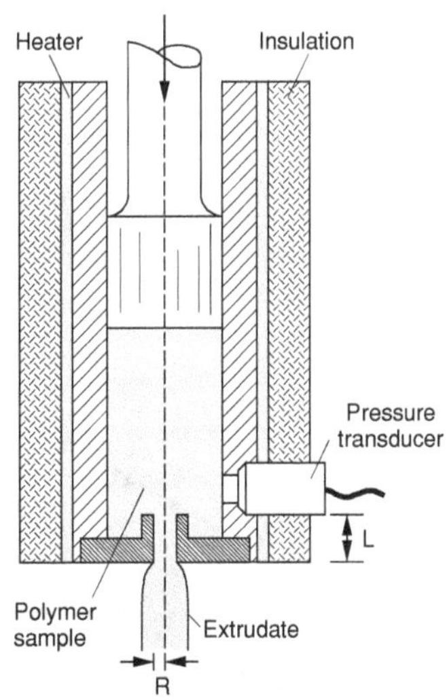

Figure 3.22 Schematic diagram of a capillary viscometer.

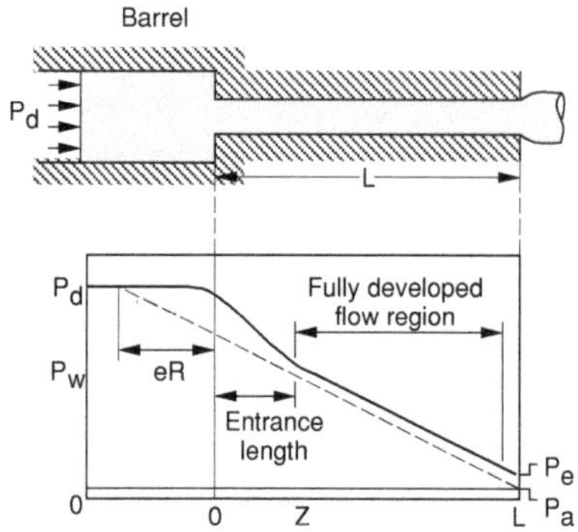

Figure 3.23 Entrance effects in a typical capillary viscometer.

is shown schematically in Fig. 3.23 [25] and was corrected by Bagley [26] using the end correction e:

$$\tau_w = \frac{1}{2}\frac{(p_o - p_L)}{(L/R + e)} \tag{3.30}$$

The correction e at a specific shear rate can be found by plotting pressure drop for various capillary L/D ratios as shown in Fig. 3.24 [25].

The equation for shear stress is then

$$\tau_{rz} = \frac{r}{R}\tau_w. \tag{3.31}$$

To obtain the shear rate at the wall the Weissenberg–Rabinowitsch [27] equation can be used

$$\dot{\gamma}_w = \frac{1}{4}\dot{\gamma}_{aw}\left(3 + \frac{d(\ln Q)}{d(\ln \tau)}\right) \tag{3.32}$$

where, $\dot{\gamma}_{aw}$ is the apparent or Newtonian shear rate at the wall and is written as

$$\dot{\gamma}_{aw} = \frac{4Q}{\pi R^3} \tag{3.33}$$

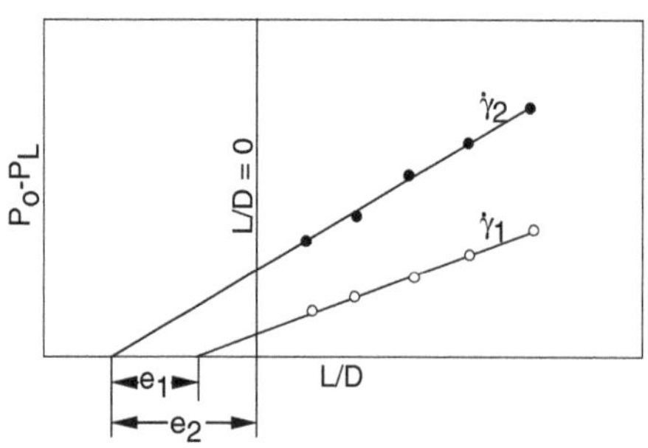

Figure 3.24 Bagley plots for two shear rates.

The shear rate and shear stress at the wall are now known; therefore, using the measured values of the flow rate, Q, and the pressure drop, $p_O - p_L$, the viscosity is calculated using

$$\eta = \frac{\tau_w}{\dot{\gamma}_w}. \tag{3.34}$$

3.4 Anisotropy Development During Processing

The mechanical properties and dimensional stability of a molded polymer part are strongly dependent upon the anisotropy of the finished part. The structure of the final part, in turn, is influenced by the design of the mold cavity, such as the type and position of the gate, and by the various processing conditions, such as injection speed, melt or compound temperatures, mold cooling or heating rates, and others. The amount and type of filler or reinforcing material also has a great influence on the quality of the final part.

This chapter discusses the development of anisotropy during processing of thermoset and thermoplastic polymer parts and presents basic analyses that can be used to estimate anisotropy in the final product.

3.4.1 Orientation in the Final Part

During processing the molecules, fillers, and fibers are oriented in the flow and greatly affect the properties of the final part. Because there are large differences in the processing of thermoplastic and thermoset polymers, the two will be discussed individually in the next two sections.

Processing Thermoplastic Polymers

When thermoplastic components are manufactured, the polymer molecules become oriented. The molecular orientation is induced by the deformation of the polymer melt during processing. The flexible molecular chains get stretched, and because of their entanglement they cannot relax fast enough before the part cools and solidifies. At lower processing temperatures this phenomena is multiplied, leading to even higher degrees of molecular orientation. This orientation is felt in the stiffness and strength properties of the polymer component. Orientation also gives rise to *birefringence* or *double refraction*. The various degrees of molecular orientation and the different main directions of orientation in the material introduce a variable refractive index field, $n(x,y,z)$, throughout the part. The value of the refractive index, n, depends on the relative orientation of the molecules, or the molecular axes, to the direction of the light shining through the part.

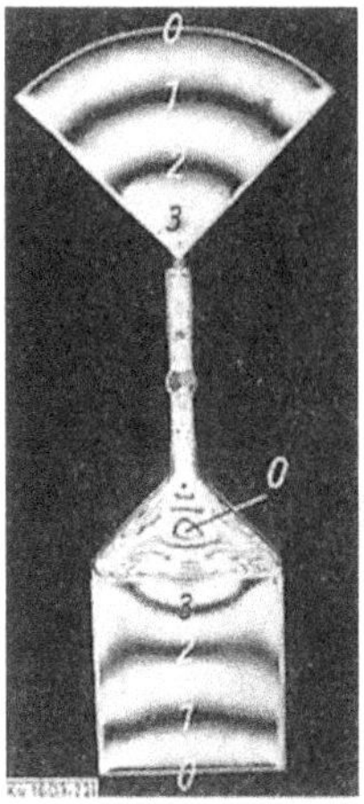

Figure 3.25 Isochromatics in a polycarbonate specimen of 1.7-mm-wall thickness.

As polarized light travels through a part, a series of colored lines called *isochromatics* become visible or appear, as shown in Fig. 3.25 [25]. The isochromatics are lines of equal molecular orientation and are numbered from zero, at the region of no orientation, up with increasing degrees of orientation. A zero degree of orientation is usually the place in the mold that fills last; the degree of orientation increases toward the gate. Figure 3.26 shows schematically how molecular orientation is related to birefringence. The layers of highest orientation are near the outer surfaces of the part with orientation increasing toward the gate.

The degree of orientation increases and decreases, depending on the various processing conditions and materials. For example, Fig. 3.27 [28] shows quarter disks of various wall thicknesses, molded out of four different materials: polycarbonate, cellulose acetate, polystyrene, and polymethyl methacrylate. We see that for all

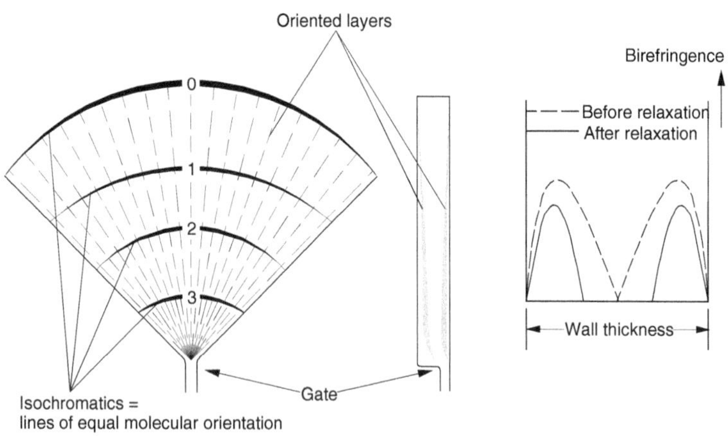

Figure 3.26 Orientation birefringence in a quarter disk.

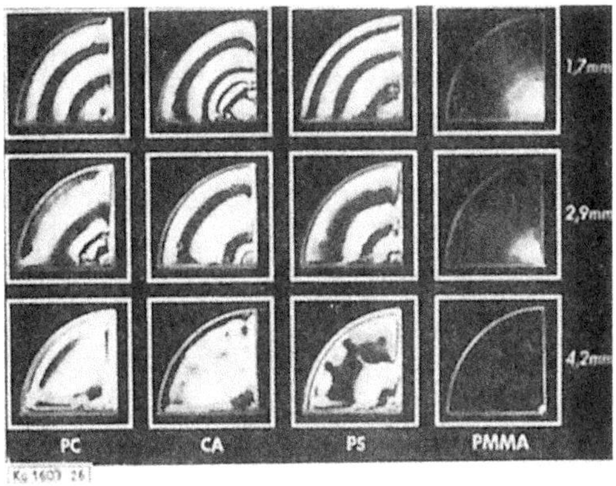

Figure 3.27 Isochromatics in polycarbonate, cellulose acetate, polystyrene, and polymethyl methacrylate quarter disks of various thicknesses.

materials the degree of orientation increases with decreasing wall thickness. An explanation for this is that the velocity gradients increase when wall thickness decreases. We will discuss how orientation is directly related to velocity gradients in subsequent sections of this chapter.

Orientation is also related to the process used to manufacture the part. For example, Fig. 3.28 [28] shows two injection molded polycarbonate parts molded with different injection molders: a piston-type and a screw-type machine. It is obvious that the cover made with the piston-type injection molder has much higher degrees of molecular orientation than the one manufactured using the screw-type injection molder. Destructive tests revealed that it was impossible to produce a cover that is sufficiently crack-proof when molded with a piston-type molding machine.

The articles in Figs. 3.25, 3.27, and 3.28 were injection molded—a common processing method for thermoplastic polymers. Early studies have already shown that a molecular orientation distribution exists across the thickness of thin injection molded

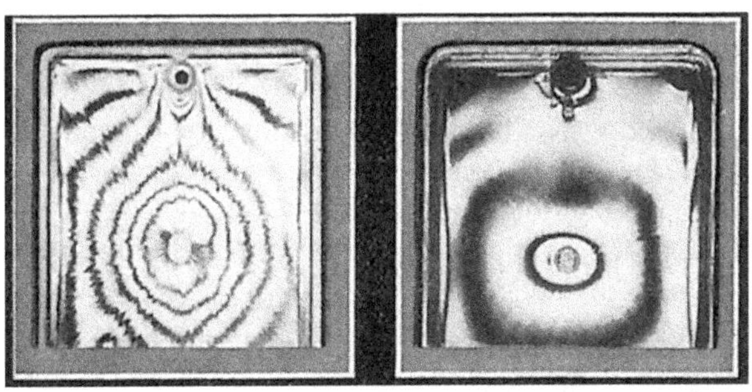

Figure 3.28 Isochromatics in polycarbonate parts molded with (left) piston-type and (right) screw-type injection molding machines.

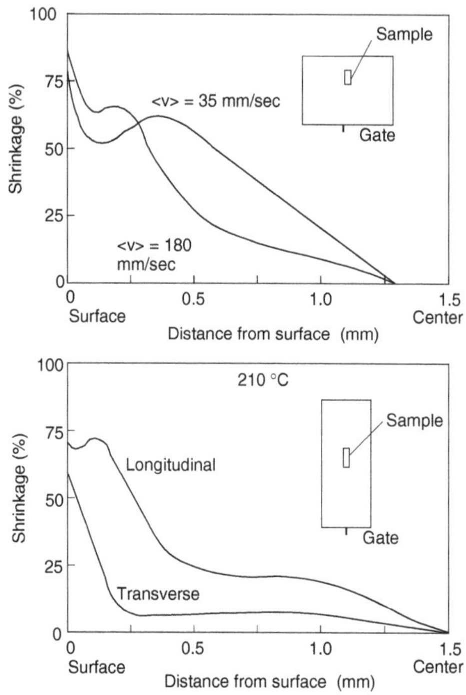

Figure 3.29 Shrinkage distribution of injection molded polystyrene plates.

parts [26]. Figure 3.29 [29] shows the shrinkage distribution in longitudinal and transverse flow directions of two different plates. The curves demonstrate the degree of anisotropy that develops during injection molding, and the influence of the geometry of the part on this anisotropy.

An example where the birefringence pattern of polymer parts can be used to detect severe problems is in the manufacture of polycarbonate compact disks [30,31]. Figure 3.30 shows the birefringence distribution in the rz-plane of a 1.2-mm-thick disk molded with polycarbonate. The figure shows how the birefringence is highest at the surface of the disk and lowest just below the surface. Toward the inside of the disk the birefringence rises again and drops somewhat toward the central core of the disk. A similar phenomena was observed in glass fiber reinforced [32–34] and liquid crystalline polymer [32] injection molded parts that show large variations in fiber and molecular orientation through the thickness.

These findings support earlier claims that molecular or filler orientation in injection molded parts can be divided into the seven layers schematically represented in Fig. 3.31 [35]. The seven layers may be described as follows:

- Two thin outer layers with a biaxial orientation, random in the plane of the disk
- Two thick layers next to the outer layers with a main orientation in the flow direction

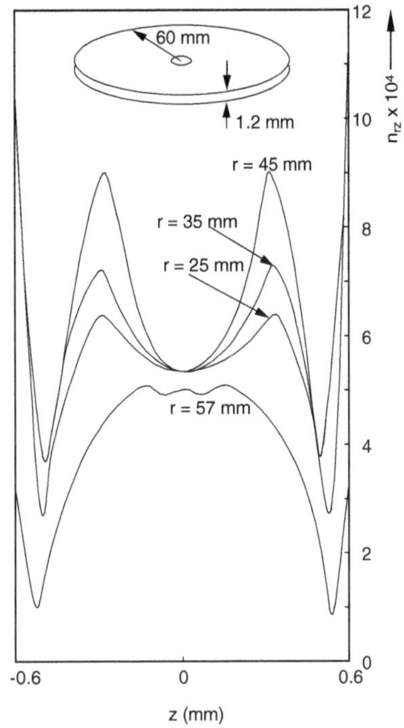

Figure 3.30 Birefringence distribution in the rz-plane at various radius positions. Numbers indicate radial position.

- Two thin randomly oriented transition layers next to the center core
- One thick center layer with a main orientation in the circumferential direction

There are three mechanisms that lead to high degrees of orientation in injection molded parts: fountain flow effect, radial flow, and holding pressure induced flow.

The *fountain flow effect* [36] is caused by the no-slip condition on the mold walls, which forces material from the center of the part to flow outward to the mold surfaces as shown in Fig. 3.32 [37]. As the figure schematically represents, the melt that flows inside the cavity freezes upon contact with the cooler mold walls. The melt that subsequently enters the cavity flows between the frozen layers, forcing the melt skin

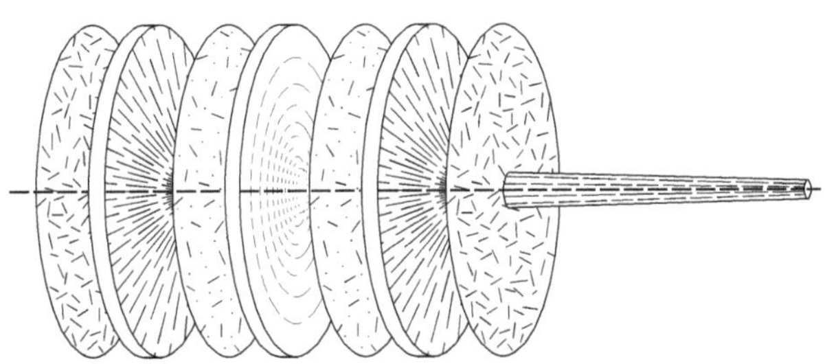

Figure 3.31 Filler orientation in seven layers of a centrally injected disk.

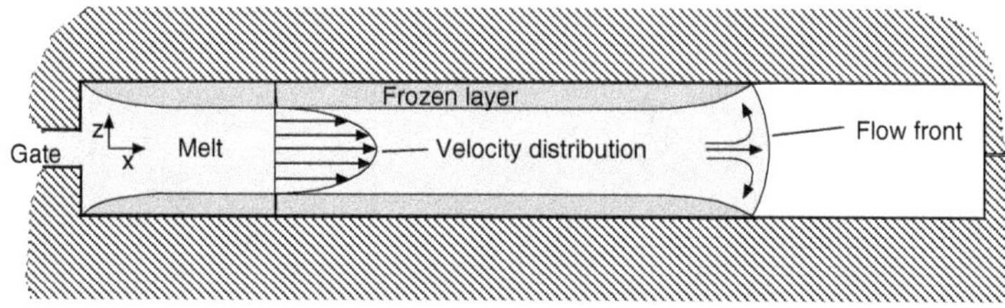

Figure 3.32 Flow and solidification mechanisms through the thickness during injection molding.

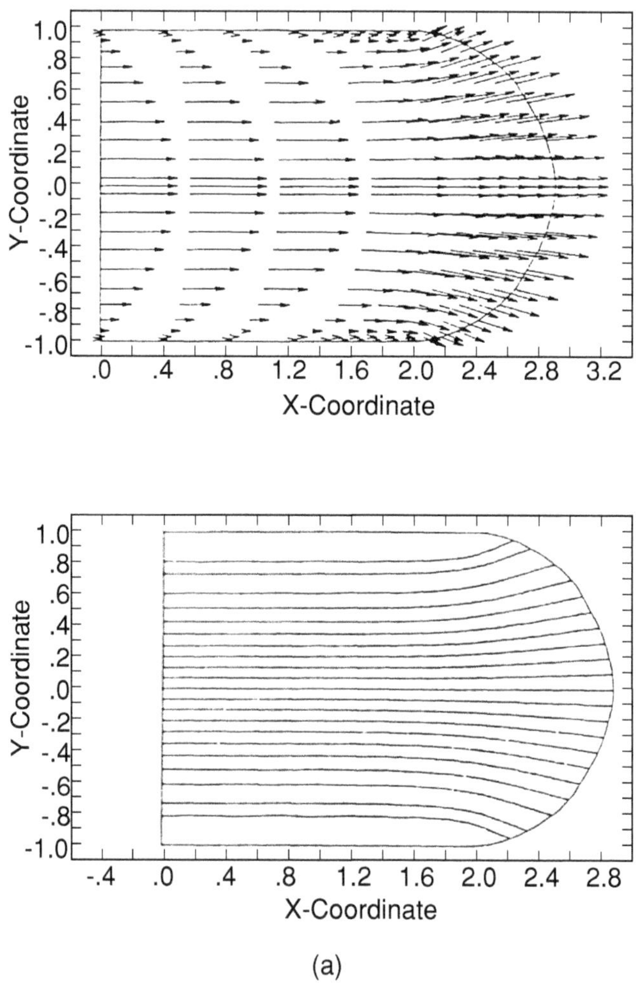

Figure 3.33 Fountain flow effect: (a) Actual velocity vectors and streamlines. Fountain flow effect: (b) relative to the moving front velocity vectors and streamlines.

at the front to stretch and unroll onto the cool wall where it freezes instantly. The molecules that move past the free flow front are oriented in the flow direction and laid on the cooled mold surface, which freezes them into place, yet allows some relaxation of the molecules after solidification. The fountain flow effect has been extensively studied in past few years using computer simulation [35]. Figure 3.33(a) [38] shows simulated instantaneous velocity vectors and streamlines during the isothermal mold filling of a Newtonian fluid,* and Fig. 3.33(b) shows the velocity vectors relative to the moving flow front. Figure 3.34 [39] presents the predicted shape and

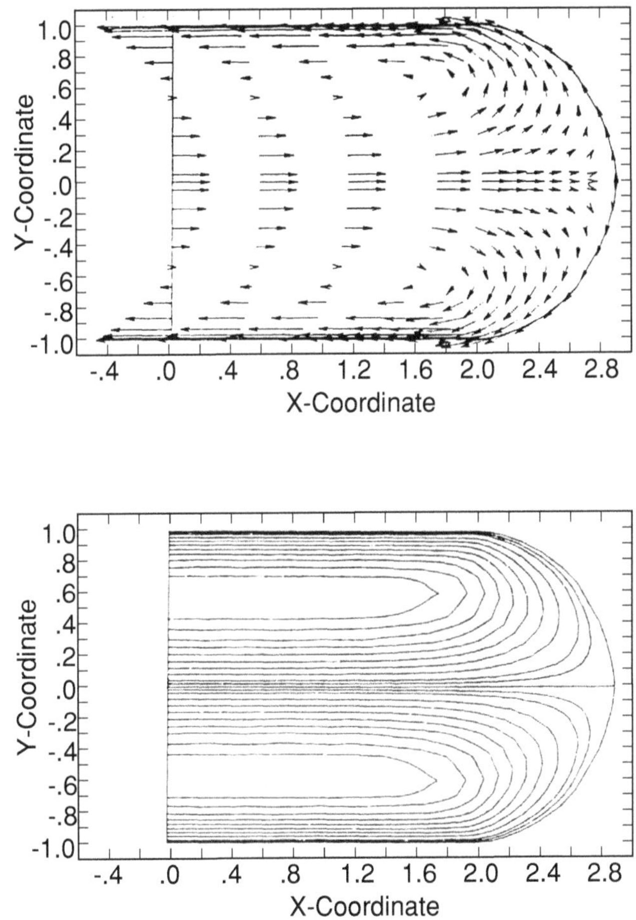

(b)

Figure 3.33 *(Continued)*

* The isothermal and Newtonian analysis should only serve to explain the mechanisms of fountain flow. The nonisothermal nature of the injection molding process plays a significant role in the orientation of the final part and should not be left out in the analysis of the real process.

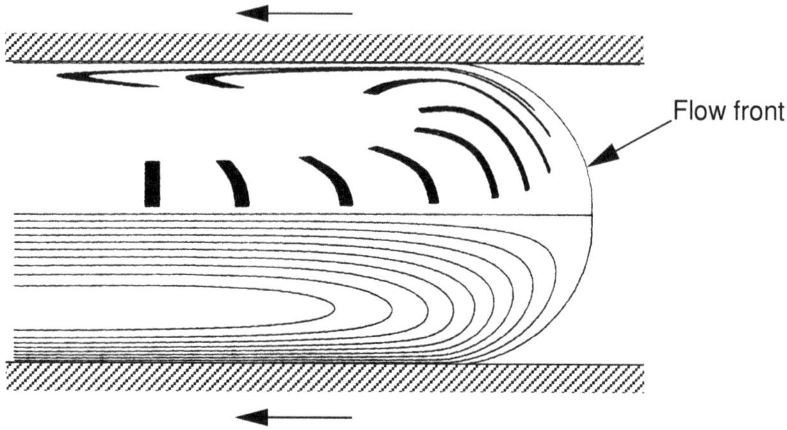

Figure 3.34 Deformation history of a fluid element and streamlines for frame of reference that moves with the flow front.

position of the tracer relative to the flow front along with the streamlines for a non-Newtonian nonisothermal fluid model. The square tracer mark is stretched as it flows past the free flow front, and is deposited against the mold wall, pulled upward again, and eventually deformed into a *V*-shaped geometry. The movement of the outer layer is eventually stopped as it cools and solidifies.

Radial flow is the second mechanism that often leads to orientation perpendicular to the flow direction in the central layer of an injection molded part. This mechanism is schematically represented in Fig. 3.35. As the figure suggests, the material that enters through the gate is transversely stretched while it radially expands as it flows away from the gate. This flow is well represented in today's commonly used commercial injection mold filling software.

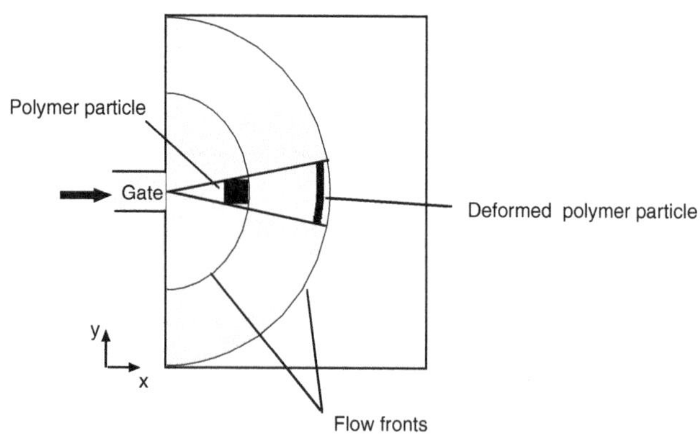

Figure 3.35 Deformation of the polymer melt during injection molding.

Finally, the flow induced by the holding pressure as the part cools leads to additional orientation in the final part. This flow is responsible for the spikes in the curves shown in Figs. 3.29 and 3.30.

Processing Thermoset Polymers

During the manufacture of thermoset parts, there is no molecular orientation because of the cross-linking that occurs during the solidification or curing reaction. A thermoset polymer solidifies as it undergoes an exothermic reaction and forms a tight network of interconnected molecules.

Many thermoset polymers, however, are reinforced with filler materials, such as glass fiber or wood flour. These composites are molded via injection molding, transfer molding, compression molding, or injection-compression molding. The properties of the final part are dependent on the filler orientation. In addition, the thermal expansion coefficients and the shrinkage of these polymers are highly dependent on the type and volume fraction of filler being used. Different forms of orientation may lead to varying strain fields, which may cause warpage in the final part. This topic will be discussed in Chapter 4.

During the processing of filled thermoset polymers the material deforms uniformly through the thickness with slip occurring at the mold surface, as shown schematically in Fig. 3.36 [40]. Several researchers have studied the development of fiber orientation during transfer molding, injection-compression molding of bulk molding compound (BMC), and compression molding of sheet molding compound (SMC) parts [41]. The fibers are usually between 10- and 25-mm long and the final part thickness is 1–5 mm. Hence, the fiber orientation can be described with a planar orientation distribution function.

Under certain circumstances, filler orientation may lead to crack formation as shown in Fig. 3.37 [42]. Here, the part was transfer molded through two gates that lead to a knitline and filler orientation shown in the figure. Knitlines are cracklike

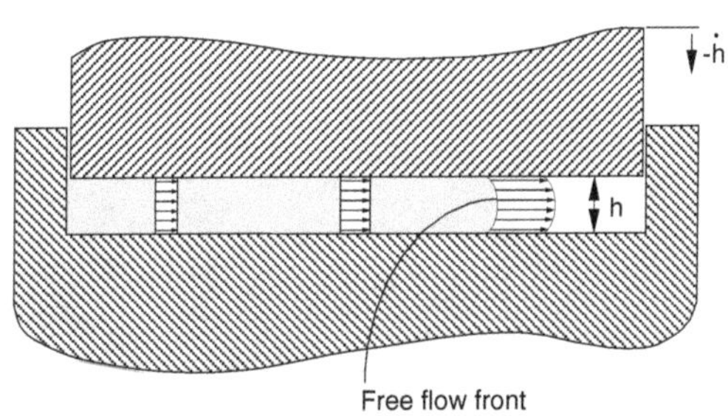

Figure 3.36 Velocity distribution during compression molding with slip between material and mold surface.

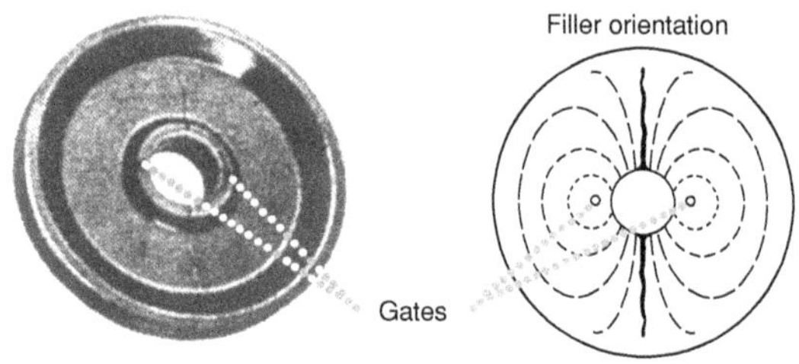

Figure 3.37 Formation of knitlines in a fiber filled thermoset pulley.

regions where few or no fibers bridge across, lowering the strength across that region to that of the matrix material. A better way to mold the part of Fig. 3.37 would be to inject the material through a ring-type gate, which would result in an orientation distribution mainly in the circumferential direction.

3.4.2 Fiber Damage

One important aspect when processing fiber reinforced polymers is fiber damage or *fiber attrition*. This is especially true during injection molding where normally high shear stresses are present. As the polymer is melted and pumped inside the screw section of the injection molding machine and as it is forced through the narrow gate, most fibers shorten in length, reducing the properties of the final part (e.g., stiffness and strength).

Figure 3.38 helps explain the mechanism responsible for fiber breakage. The figure shows two fibers rotating in a simple shear flow: fiber "a," which is moving out of its 0-degree position, has a compressive loading, and fiber "b," which is moving into its 0-degree position, has a tensile loading. It is clear that the tensile loading is not large enough to cause any fiber damage, but the compressive loading is potentially large

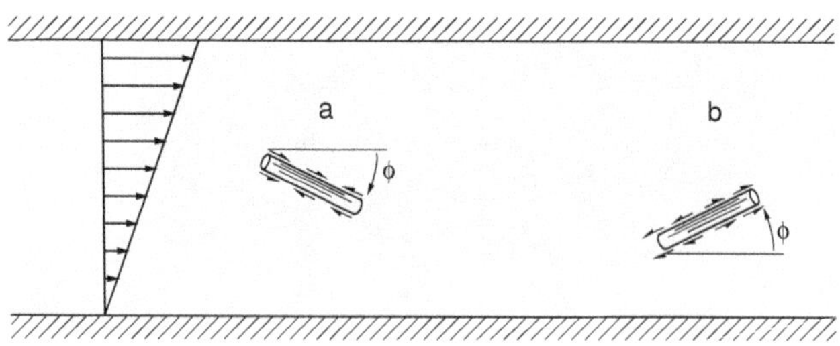

Figure 3.38 Fiber in compression and tension as it rotates during simple shear flow.

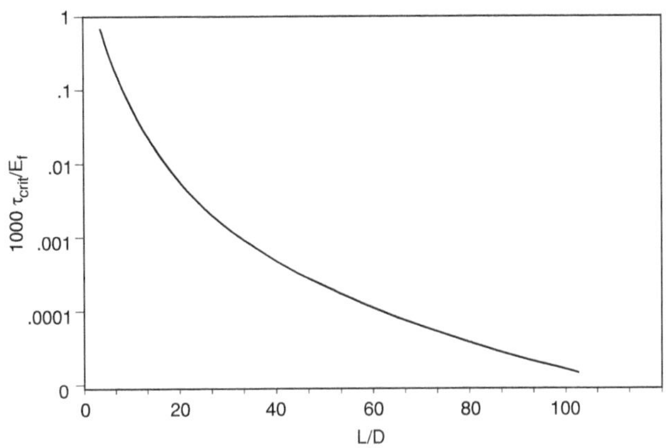

Figure 3.39 Critical stress, τ_{crit}, versus fiber L/D ratio.

enough to buckle and break the fiber. A common equation exists that relates a critical shear stress, τ_{crit}, to elastic modulus, E_f, and to the L/D ratio of the fibers:

$$\tau_{crit} = \frac{\ln(2L/D) - 1.75}{2(L/D)^4} E_f \tag{3.35}$$

where τ_{crit} is the stress required to buckle the fiber. When the stresses are greater than τ_{crit} the fiber L/D ratio is reduced. Figure 3.39 shows a dimensionless plot of critical stress versus L/D ratio of a fiber as computed using Eq. (3.35). It is worth pointing that although Eq. (3.35) predicts L/D ratios for certain stress levels, it does not include the uncertainty that leads to fiber L/D ratio distributions—very common in fiber-filled systems.

Figure 3.40 presents recent findings by Thieltges [43] where he demonstrates that during injection molding most of the fiber damage occurs in the transition section of the plasticating screw. Lesser effects of fiber damage were measured in the metering section of the screw and in the throttle valve of the plasticating machine. The damage observed inside the mold cavity was marginal. The small damage observed inside the mold cavity, however, is of great importance because the fibers flowing inside the cavity underwent the highest stresses, further reducing their L/D ratios. Bailey and Kraft [44] also found that fiber-length distribution in the injection molded part is not uniform. For example, the skin region of the molding contained much shorter fibers than did the core region.

Another mechanism responsible for fiber damage is explained in Fig. 3.41 [45], where the fibers that stick out of partially molten pellets are bent, buckled, and sheared-off during plastication.

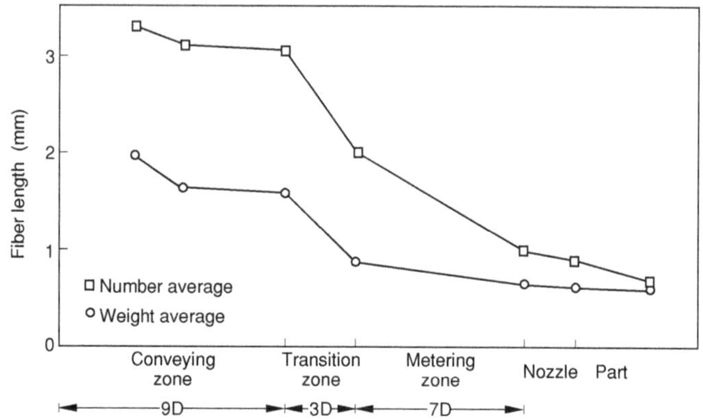

Figure 3.40 Fiber damage measured in the plasticating screw, throttle valve, and mold during injection molding of a polypropylene plate with 40% fiber content by weight.

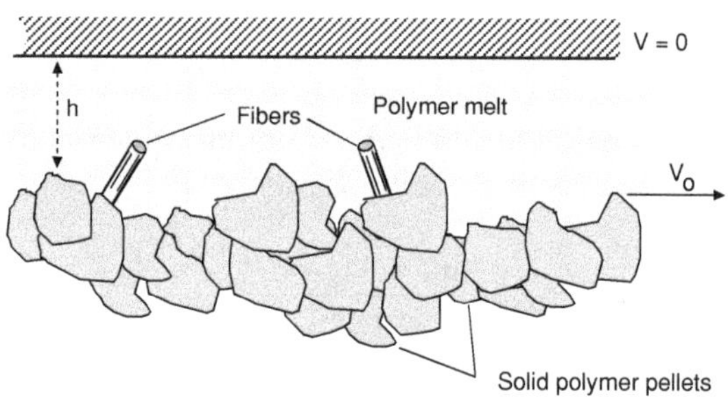

Figure 3.41 Fiber damage mechanism is the interface between solid and melt.

3.5 Solidification and Curing Processes

Solidification is the process in which a material undergoes a phase change and hardens. The phase change occurs as a result of either a reduction in material temperature or a chemical curing reaction. As discussed in previous chapters, a thermoplastic polymer hardens as the temperature of the material is lowered below either the melting temperature for a semi-crystalline polymer or the glass transition temperature for an amorphous thermoplastic. A thermoplastic has the ability to soften again as the temperature of the material is raised above the solidification temperature. On the other hand, the solidification of a thermosetting polymer results from a chemical reaction that results in the cross-linking of molecules. The effects of cross-linkage are irreversible and lead to a network that hinders the free movement of the polymer chains independent of the material temperature.

3.5.1 Solidification of Thermoplastics

The term *solidification* is often misused to describe the hardening of amorphous thermoplastics. The solidification of most materials is defined at a discrete temperature, whereas amorphous polymers do not exhibit a sharp transition between the liquid and the solid states. Instead, an amorphous thermoplastic polymer vitrifies as the material temperature drops below the glass transition temperature, T_g. A semi-crystalline polymer does have a distinct transition temperature between the melt and the solid state, the melting temperature, T_m.

Thermodynamics During Cooling

As heat is removed from a polymer melt, the molecules loose their ability to move freely, thus making the melt highly viscous. As amorphous polymers cool, the molecules slowly become closer packed, thus changing the viscous material into a leathery or rubberlike substance. Once the material has cooled below the glass transition temperature, T_g, the polymer becomes stiff and brittle. At the glass transition temperature, the specific volume and enthalpy curves experience a significant change in slope. This can be seen for polystyrene in the enthalpy–temperature curve shown in Fig. 3.42. With semi-crystalline thermoplastics, at a crystallization temperature near the melting temperature, the molecules start arranging themselves in small crystalline and amorphous regions, creating a very complicated morphology. During the process of crystalline structure formation, a quantum of energy, often called *heat of crystallization* or *heat of fusion* is released and must be conducted out of the material before the cooling process can continue. The heat of fusion is reflected in the shape of the enthalpy-temperature curve as shown for polyamide 6.6, polyethylene and polypropylene in Fig. 3.42. At the onset of crystalline growth, the material becomes rubbery, yet not brittle, because the amorphous regions are still greater than the glass transition temperature. As seen earlier, the glass transition temperature for some semi-crystalline polymers is far below room temperature, making them tougher than amorphous polymers. For common semi-crystalline polymers, the degree of crystallization can be between 30 and 70%. This means that 30 to 70% of the molecules form crystals, and the rest remain in an amorphous state. The degree of

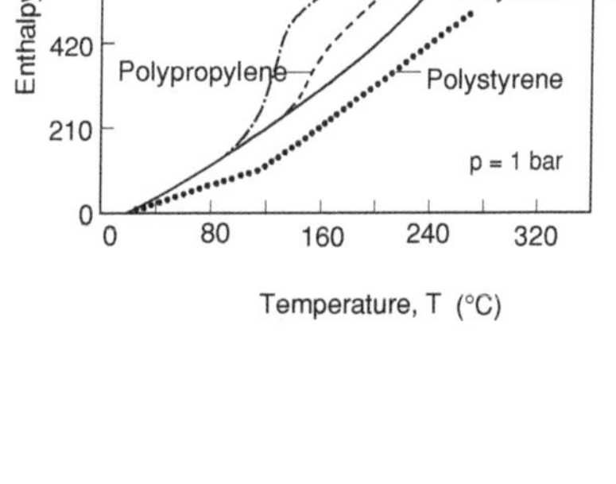

Figure 3.42 Enthalpy as a function of temperature for various thermoplastics.

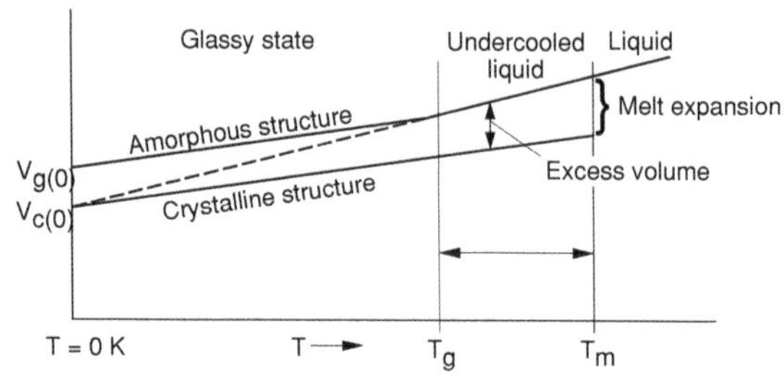

Figure 3.43 Thermal expansion model for thermoplastic polymers.

crystallization is highest for those materials with short molecules because they can crystallize faster and easier.

Figure 3.43 [46] shows the volumetric temperature dependence of a polymer. In the melt state, the chains have "empty spaces" in which molecules can move freely. Hence, undercooled polymer molecules can still move as long as space is available. The point at which this free movement ends for a molecule or segment of chains is called the *glass transition temperature* or *solidification point*. As pointed out in Fig. 3.43, the free volume is frozen-in as well. In the case of crystallization, ideally, the volume should jump to a lower specific volume. Even here, however, small amorphous regions remain that permit a slow flow or material creep. This free volume reduces to nothing at absolute zero temperature at which heat transport can no longer occur.

The specific volume of a polymer changes with pressure even at the glass transition temperature. This is demonstrated for an amorphous thermoplastic in Fig. 3.44 and for a semi-crystalline thermoplastic in Fig. 3.45.

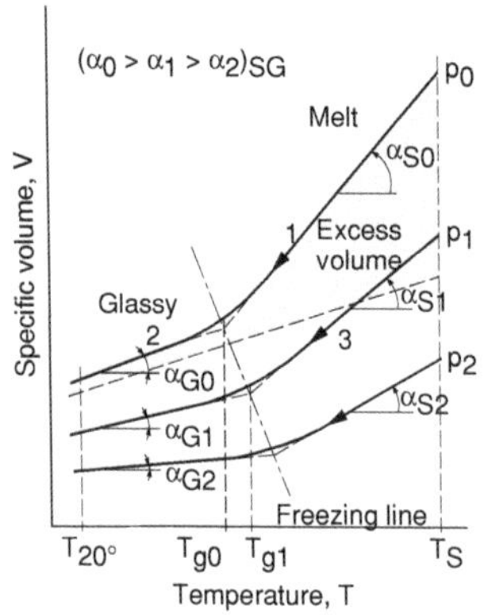

Figure 3.44 Schematic of a *p*-*v*-*T* diagram for amorphous thermoplastics.

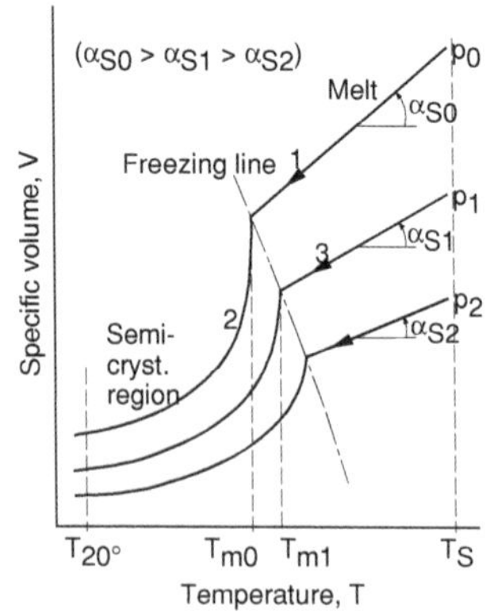

Figure 3.45 Schematic of a *p*-*v*-*T* diagram for semi-crystalline thermoplastics.

It should be noted here that the size of the frozen-in free volume depends on the rate at which a material is cooled; high cooling rates result in a large free volume. In practice this is very important. When the frozen-in free volume is large, the part is less brittle. On the other hand, high cooling rates lead to parts that are highly permeable, which may allow the diffusion of gases or liquids through container walls. The cooling rate is also directly related to the dimensional stability of the final part. The effect of high cooling rates can often be mitigated by heating the part to a temperature that enables the molecules to move freely; this will allow further crystallization by additional chain folding. This process has a great effect on the structure and properties of the crystals and is referred to as *annealing*. In general, this only signifies a qualitative improvement of polymer parts. It also affects shrinkage and warpage during service life of a polymer component, especially when thermally loaded.

All these aspects have a great impact to processing. For example, when extruding amorphous thermoplastic profiles, the material can be sufficiently cooled inside the die so that the extrudate has enough rigidity to carry its own weight as it is pulled away from the die. When heated above the melting temperature, semi-crystalline polymers with low molecular weights have a viscosity that is too low to be able to withstand their own weight as the extrudate exits the die. Temperatures below the melting temperature, T_m, however, cannot be used due to solidification or crystallization inside the die. Similar problems are encountered in the thermoforming process in which the material must be heated to a point so that it can be formed into its final shape, yet be able to withstand its own weight.

Semi-crystalline polymers are also at a disadvantage in the injection molding process. Because of the heat needed for crystallization, more heat must be removed to solidify the part; because there is more shrinkage, longer packing times and larger pressures must be employed. This implies longer cycle times and more shrinkage. High cooling rates during injection molding of semi-crystalline polymers will reduce the degree of crystallization; however, the amorphous state of the polymer molecules may lead to some crystallization after the process, which will result in further shrinkage and warpage of the final part. It is quite common to follow the whole injection molding process in the p-v-T diagrams presented in Figs. 3.44 and 3.45, and so predict how much the molded component has shrunk.

Morphological Structure

Morphology is the order or arrangement of the polymer structure. The possible "order" between a molecule or molecule segment and its neighbors can vary from a very ordered, highly crystalline polymeric structure to an amorphous structure (i.e., a structure in greatest disorder or random). The possible range of order and disorder is clearly depicted on the left side of Fig. 3.46. For example, a purely amorphous polymer is formed only by the noncrystalline or amorphous chain structure, whereas the semi-crystalline polymer is formed by a combination of all the structures represented in Fig. 3.46.

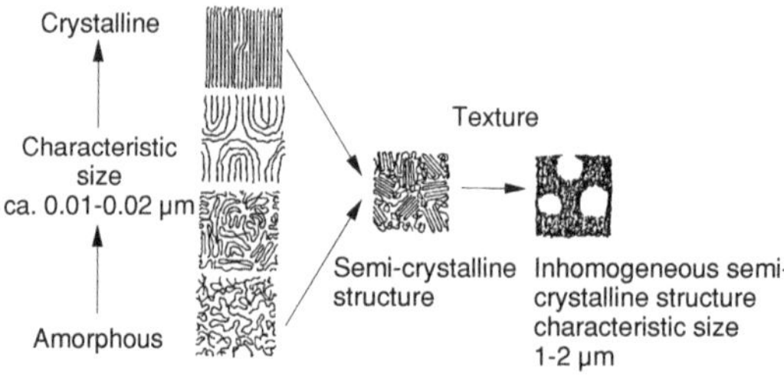

Figure 3.46 Schematic diagram of possible molecular structure that occurs in thermoplastic polymers.

The image of a semi-crystalline structure as shown in the middle of Fig. 3.46, can be captured with an electron microscope. A macroscopic structure, shown in the right-hand side of the figure, can be captured with an optical microscope. An optical microscope can capture the coarser macromorphological structure, such as the spherulites in semi-crystalline polymers.

An amorphous polymer is defined as having a purely random structure. It is not quite clear, however, if a "purely amorphous" polymer as such exists. Electron microscopic observations have shown amorphous polymers that are composed of relatively stiff chains show a certain degree of macromolecular structure and order (e.g., globular regions or fibrilitic structures). Nevertheless, these types of amorphous polymers are still found to be optically isotropic. Even polymers with soft and flexible macromolecules, such as polyisoprene, which was first considered to be random, sometimes show bandlike and globular regions. These bundlelike structures are relatively weak and short-lived when the material experiences stresses. The shear thinning viscosity effect of polymers is sometimes attributed to the breaking of such macromolecular structures.

Crystallization

Early on, before the existence of macromolecules had been recognized, the presence of highly crystalline structures had been suspected. Such structures were discovered when undercooling or when stretching cellulose and natural rubber. It was later found that a crystalline order also existed in synthetic macromolecular materials such as polyamides, polyethylene, and polyvinyls. Because of the polymolecularity of macromolecular materials, a 100% degree of crystallization cannot be achieved. Hence, these polymers are referred to as semi-crystalline. It is common to assume that the semi-crystalline structures are formed by small regions of alignment or crystallites connected by random or amorphous polymer molecules.

With the use of electron microscopes and sophisticated optical microscopes the various existing crystalline structures are now well recognized. They can be listed as:

- *Single crystals*. These can form in solutions and help in the study of crystal formation. Here, platelike crystals and sometimes whiskers are generated.
- *Spherulites*. As a polymer melt solidifies, several folded chain lamellae spherulites form that are up to 0.1 mm in diameter. A typical example of a spherulitic structure is shown in Fig. 3.47 [47]. The spherulitic growth in a polypropylene melt is shown in Fig. 3.48 [48].
- *Deformed crystals*. If a semi-crystalline polymer is deformed while undergoing crystallization, oriented lamellae form instead of spherulites.
- *Shish-kebab*. In addition to spherulitic crystals, which are formed by plate- and ribbonlike structures, there are also shish-kebab, crystals which are formed by circular plates and whiskers. Shish-kebab structures are generated when the melt undergoes a shear deformation during solidification. A typical example of a shish-kebab crystal is shown in Fig. 3.49 [49].

The crystallization fraction can be described by the *Avrami equation* [50] written as follows:

$$x(t) = 1 - e^{-Zt^n} \tag{3.36}$$

where Z is a molecular weight and temperature-dependent crystallization rate, and n is the Avrami exponent. Because a polymer cannot reach 100% crystallization, however, the above equation should be multiplied by the maximum possible degree of crystallization, x_∞.

Figure 3.47 Polarized microscopic image of the spherulitic structure in polypropylene.

Figure 3.48 Development of the spherulitic structure in polypropylene. Images were taken at 30-second intervals.

$$x(t) = x_\infty\left(1 - e^{-Zt^n}\right) \tag{3.37}$$

The Avrami exponent, n, ranges between 1 and 4 depending on the type of nucleation and growth. For example, the Avrami exponent for spherulitic growth from sporadic nuclei is around 4, disclike growth is 3, and rodlike growth is 2. If the growth

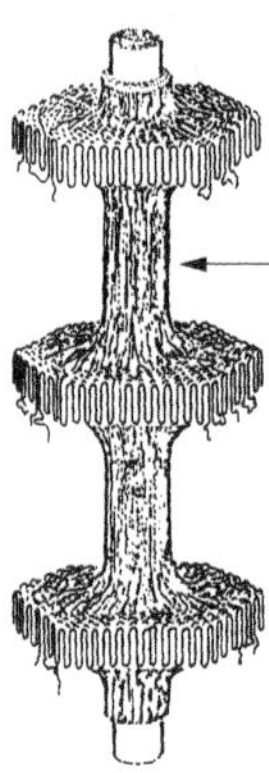

Figure 3.49 Model of the shish-kebab morphology.

Table 3.9 Maximum Crystalline Growth Rate and Maximum Degree of Crystallinity for Various Thermoplastics

Polymer	Growth rate (μ/min)	Maximum crystallinity (%)
Polyethylene	>1000	80
Polyamide 66	1000	70
Polyamide 6	200	35
Isotactic polypropylene	20	63
Polyethylene terephthalate	7	50
Isotactic polystyrene	0.30	32
Polycarbonate	0.01	25

is activated from instantaneous nuclei, the Avrami exponent is lowered by 1.0 for all cases. The crystalline growth rate of various polymers differ significantly from one to another. This is demonstrated in Table 3.9, which shows the maximum growth rate for various thermoplastics. The crystalline mass fraction can be measured experimentally with a differential scanning calorimeter (DSC).

A more in-depth coverage of crystallization and structure development during processing is given by Eder and Janeschitz-Kriegl [51].

Heat Transfer During Solidification

Because injection molded polymer parts are generally thin, the cooling process can be regarded as a one-dimensional problem, as shown in Fig. 3.50. During cooling, when the material's temperature drops below the glass transition temperature, T_g, it can be considered solidified. This is shown schematically in Fig. 3.51. Of importance here is the position of the solidification front, $X(t)$. Once the solidification front equals the plate's dimension L, the solidification process is complete. It can be shown that the rate of solidification decreases as the solidified front moves further away from the cooled surface. For amorphous thermoplastics, the well-known *Neumann solution* can be used to estimate the growth of the glassy or solidified layer. The Neumann solution is written as

$$X(t) \infty \sqrt{\alpha t} \tag{3.38}$$

where α is the thermal diffusivity of the polymer. It must be pointed out here that for the Neumann solution, the growth rate of the solidified layer is infinite as time goes to zero.

The solidification process in a semi-crystalline materials is a bit more complicated due to the heat of fusion or heat of crystallization, nucleation rate, and so on. When measuring the specific heat as the material crystallizes, a peak that represents the heat of fusion is detected.

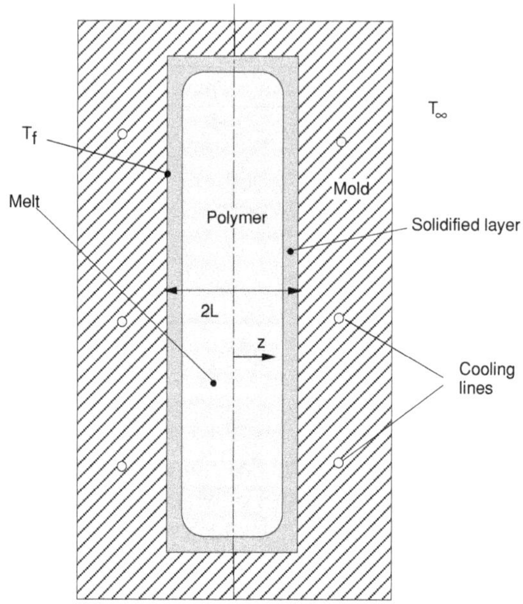

Figure 3.50 Schematic diagram of polymer melt inside an injection mold.

With semi-crystalline polymers the material that is below the melting temperature, T_m, is considered solid.* Experimental evidence [52] has demonstrated that the growth rate of the crystallized layer in semi-crystalline polymers is finite. This is mainly due to the fact that at the beginning the nucleation occurs at a finite rate. Hence, the Neumann solution presented in Eq. (3.38) as well as the widely used *Stefan condition* [53], do not hold for semi-crystalline polymers. This is clearly demonstrated in Fig. 3.52 [53], which presents measured thickness of crystallized layers as a function of time for polypropylene plates quenched at three different temperatures. For further reading on this important topic the reader is encouraged to consult the literature [54,55].

For most parts, the analysis can be simplified to a 1D problem, making it quite simple to predict cooling times. The cooling time for a platelike part of thickness *2L* can be estimated using

$$t_{cooling} = \frac{4L^2}{\pi\alpha} \ln\left(\frac{8}{\pi^2} \frac{T_M - T_W}{T_D - T_W} \right) \tag{3.39}$$

and for cylindrical geometries such as runners one can use

* It is well known that the growth of the crystalline layer in semi-crystalline polymers is maximal somewhat below the melting temperature, at a temperature T_c. The growth speed of nuclei is zero at the melting temperature and at the glass transition temperature.

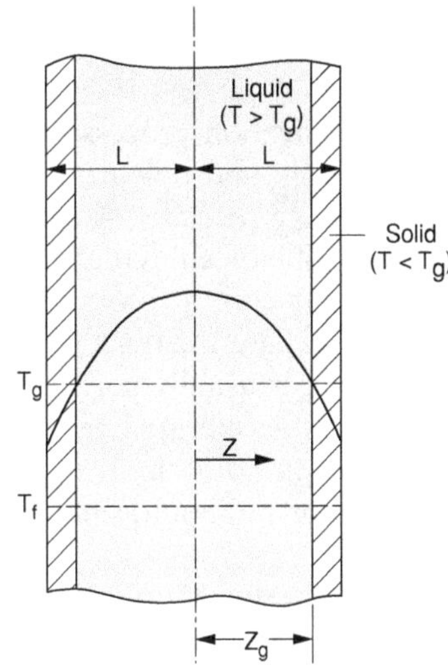

Figure 3.51 Schematic diagram of the cooling process of a polymer plate.

$$t_{cooling} = \frac{D^2}{23.14\alpha} \ln\left(0.692 \frac{T_M - T_W}{T_D - T_W}\right) \tag{3.40}$$

where D denotes the diameter of the runner, α represents thermal diffusivity, T_M represents the melt temperature, T_W the mold temperature, and T_D the average part temperature at ejection.

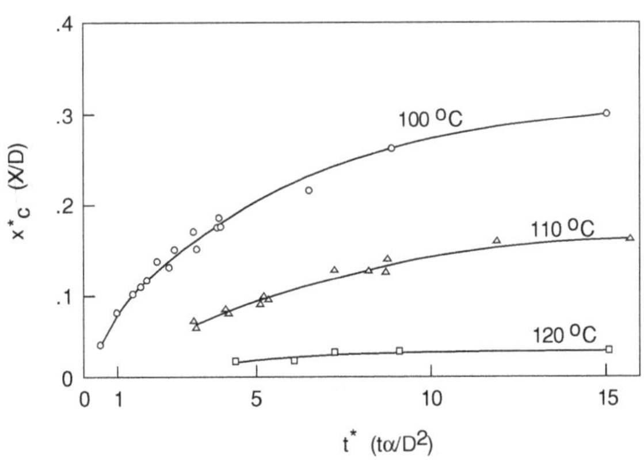

Figure 3.52 Dimensionless thickness of the crystallized layers as a function of dimensionless time for various temperatures of the quenching surface.

3.5.2 Solidification of Thermosets

The solidification process of thermosets, such as phenolics, unsaturated polyesters, epoxy resins, and polyurethanes, is dominated by an exothermic chemical reaction called *curing reaction*. A curing reaction is an irreversible process that results in a structure of molecules that are more or less cross-linked. Some thermosets cure under heat, others cure at room temperature. Thermosets that cure at room temperature are those where the reaction starts immediately after mixing two components, where the mixing is usually part of the process; however, even with these thermosets, the reaction is accelerated by the heat released during the chemical reaction, or the *exotherm*. In addition, it is also possible to activate cross-linking by absorption of moisture or radiation, such as ultraviolet, electron beam, and laser energy sources [56].

In processing, thermosets are often grouped into three distinct categories, namely those that undergo a *heat activated cure*, those that are dominated by a *mixing activated cure*, and those that are activated by the absorption of humidity or radiation. Examples of heat-activated thermosets are phenolics; examples of mixing activated cure are epoxy resins and polyurethane. Raising the processing temperature, however, accelerates even the mixing activated cure material's curing reaction.

Curing Reaction

In a cured thermoset, the molecules are rigid, formed by short groups that are connected by randomly distributed links. The fully reacted or solidified thermosetting polymer does not react to heat as observed with thermoplastic polymers. A thermoset may soften somewhat upon heating and degrades at high temperatures. Due to the high cross-link density, a thermoset component behaves as an elastic material over a large range of temperatures. It is brittle, however, with breaking strains of usually 1 to 3%. The most common example is phenolic, one of the most rigid thermosets, which consists of carbon atoms with large aromatic rings that impede motion, making it stiff and brittle. Its general structure after cross-linking is given in Figs. 2.25 and 2.26.

Similar to thermoplastics, thermosets can be broken down into two categories: thermosets that cure via *condensation polymerization* and those that undergo *addition polymerization*.

Condensation polymerization is defined as the growth process that results from combining two or more monomers with reactive end-groups, and leads to by-products such as an alcohol, water, and an acid. A common thermoset that polymerizes or solidifies via condensation polymerization is phenol formaldehyde, which was discussed in Chapter 2. The by-product of the reaction when making phenolics is water. Another well-known example of a thermoset that cross-links via condensation polymerization is the copolymerization of unsaturated polyester with styrene molecules, also referred to as free radical reaction as shown in Fig. 3.53. The molecules contain several carbon–carbon double bonds that act as cross-linking sites during

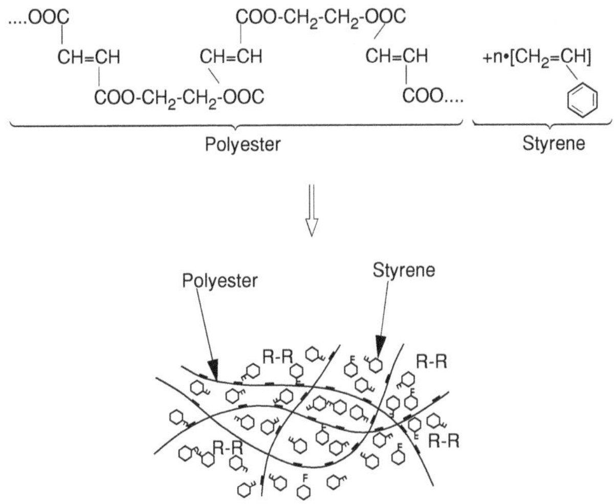

Figure 3.53 Symbolic and schematic representations of uncured unsaturated polyester.

curing. An example of the resulting network after the chemical reaction is shown in Fig. 3.54.

A characteristic of addition polymerization is that the molecules contain unsaturated double bonds and are combined into long chains or cross-linked systems. Examples of addition polymerization are polyurethanes and epoxies.

Cure Kinetics

As discussed earlier, in processing thermosets can be grouped into two general categories: *heat activated cure* and *mixing activated cure*. No matter which category a

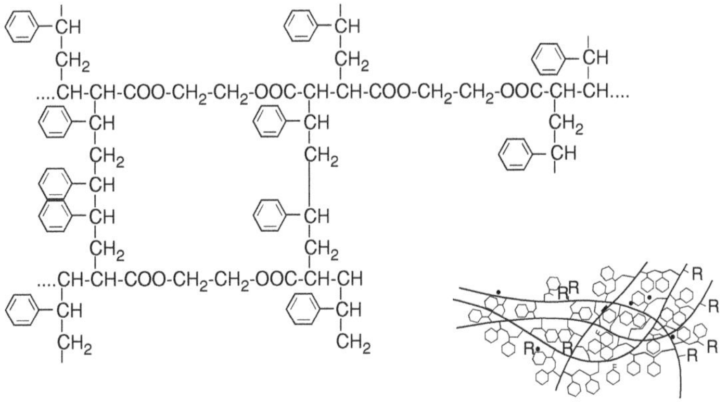

Figure 3.54 Symbolic and schematic representations of cured unsaturated polyester.

thermoset belongs to, however, its curing reaction can be described by the reaction between two chemical groups denoted by A and B that link two segments of a polymer chain. The reaction can be followed by tracing the concentration of unreacted As or Bs, C_A or C_B. If the initial concentration of As and Bs is defined as C_{Ao} and C_{Bo}, the degree of cure can be described with

$$C^* = \frac{C_{Ao} - C_A}{C_{Ao}} \tag{3.41}$$

The degree of cure or conversion, C^*, equals zero when there has been no reaction and equals 1 when all As have reacted and the reaction is complete; however, it is impossible to monitor reacted and unreacted As and Bs during the curing reaction of a thermoset polymer. It is known, however, that the exothermic heat released during curing can be used to monitor the conversion, C^*. When small samples of an unreacted thermoset polymer are placed in a DSC, each at a different temperature, every sample will release the same amount of heat, Q_T. This occurs because every cross-linking that occurs during a reaction releases a small amount of energy in the form of heat. For example, Fig. 3.55 [57] shows the heat rate released during isothermal cure of a vinyl ester at various temperatures.

The degree of cure can be defined by the following relation:

$$C^* = \frac{Q}{Q_T} \tag{3.42}$$

where Q is the heat released up to an arbitrary time τ, and is defined by

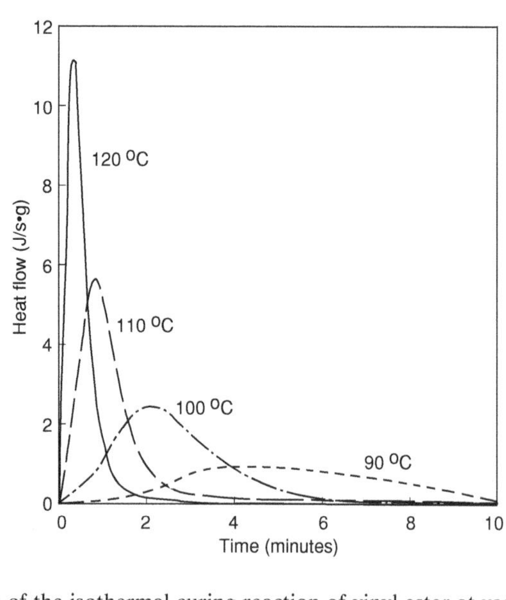

Figure 3.55 DSC scan of the isothermal curing reaction of vinyl ester at various temperatures.

$$Q = \int_0^{\tau} \dot{Q} dt \tag{3.43}$$

DSC data is commonly fitted to empirical models that accurately describe the curing reaction. Hence, the rate of cure can be described by the exotherm, $\dot{Q}$, and the total heat released during the curing reaction, Q_T, as

$$\frac{dC^*}{dt} = \frac{\dot{Q}}{Q_T} \tag{3.44}$$

With the use of Eq. (3.44), it is now easy to take the DSC data and find the models that describe the curing reaction. Two models that describe the curing reaction include that of unsaturated polyesters, which undergo a heat-activated cure, and of polyurethane that cure after its components are mixed at room temperature.

It is clear from Fig. 3.55 that the curing reaction rate is slow at first, then increases and slows down again toward the end of the reaction. It is also clear that higher temperatures accelerate the reaction. Hence, the curing kinetics for many heat-activated cure materials, such as vinyl esters and unsaturated polyesters, can be described fairly well by

$$\frac{dC^*}{dt} = k_0 e^{-E/RT} C^{*m} (1 - C^*)^n \tag{3.45}$$

where E is the activation energy, R the gas constant and k_0, m and n are constants that can be determined by curve fitting DSC data.

On the other hand, mixing activated cure materials such as polyurethanes will instantly start releasing exothermic heat after the mixture of its two components has occurred. The proposed *Castro-Macosko curing model* accurately fits this behavior and is written as [58]

$$\frac{dC^*}{dt} = k_0 e^{-E/RT} (1 - C^*)^2 \tag{3.46}$$

Heat Transfer During Cure

A well-known problem in thicksection components is that the thermal and curing gradients become more complicated and difficult to analyze because the temperature and curing behavior of the part is highly dependent on both the mold temperature and part geometry [59,60]. A thicker part will result in higher-temperatures and a more complex cure distribution during processing. This phenomenon becomes a major concern during the manufacture of thick components because high temperatures may lead to thermal degradation. A relatively easy way to check temperatures that arise during molding and curing or demolding times is desired. For example, a one-dimensional form of the energy equation that includes the exothermic energy generated during curing can be solved:

$$\rho C_p \frac{\partial T}{\partial t} = k \frac{\partial^2 T}{\partial z^2} + \rho \dot{Q} \tag{3.47}$$

Assuming the material is confined between two mold halves at equal temperatures, the use of a symmetric boundary condition at the center of the part is valid:

$$\frac{\partial T}{\partial z} = 0 \text{ at } z = 0 \tag{3.48}$$

and

$$T = T_m \tag{3.49}$$

at the mold wall.

With the use of the finite difference technique and a six-constant model that represents $\frac{dC^*}{dt}$, Barone and Caulk [61] solved Eqs. (3.47 to 3.49) for the curing of sheet molding compound (SMC). The SMC was composed of an unsaturated polyester resin with 40.7% calcium carbonate and 30% glass fiber by weight. Figures 3.56 and 3.57 show typical temperature and degree of cure distributions, respectively, during the solidification of a 10-mm thick part as computed by Barone and Caulk. In Fig. 3.56, the temperature rise that results from exothermic reaction is obvious. This temperature rise increases in thicker parts and with increasing mold tempera-

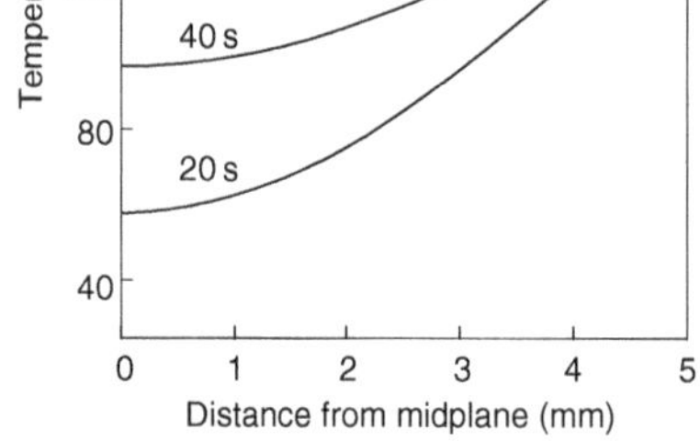

Figure 3.56 Temperature profile history of a 10-mm-thick SMC plate.

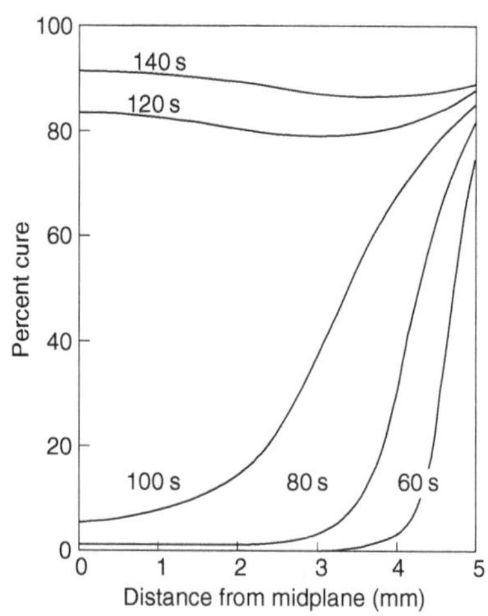

Figure 3.57 Curing profile history of a 10-mm-thick SMC plate.

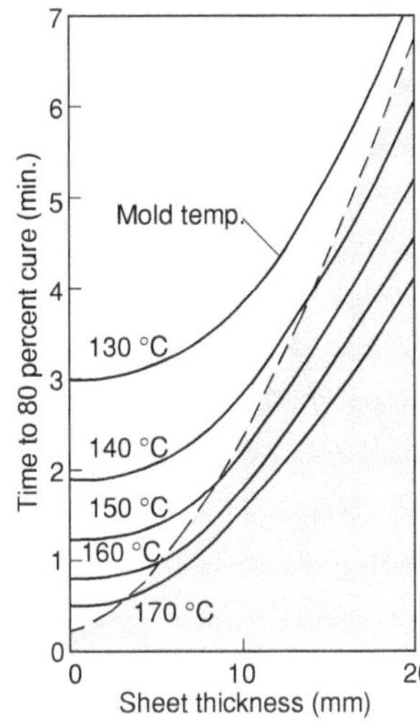

Figure 3.58 Cure times versus plate thickness for various mold temperatures. Shaded region represents the conditions at which thermal degradation may occur.

tures. Figure 3.58 is a plot of the time to reach 80% cure versus thickness of the part for various mold temperatures. The shaded area represents the conditions at which the internal temperature within the part exceeds 200°C because of the exothermic reaction. Temperatures greater than 200°C can lead to material degradation and high residual stresses in the final part.

Improper processing conditions can result in a nonuniform curing distribution that may lead to voids, cracks, or imperfections inside the part. It is of great importance to know the appropriate processing conditions that will both avoid the overheating problem and speed up the manufacturing process.

3.5.3 Residual Stresses, Shrinkage, and Warpage

Some major problems that occur when molding polymeric parts are the control and prediction of the component's shape at room temperature. For example, the resulting *sink marks* in the final product are caused by the shrinkage of the material during cooling* or curing. A common geometry that usually leads to a sink mark is a ribbed structure. The size of the sink mark, which is often only a cosmetic problem, is related to both the material and processing conditions as well as to the geometry of the part. A rib that is thick in relation to the flange thickness will result in significant sinking on the flat side of the part.

* In injection molding one can mitigate this problem by continuously pumping polymer melt into the mold cavity as the part cools until the gate freezes shut.

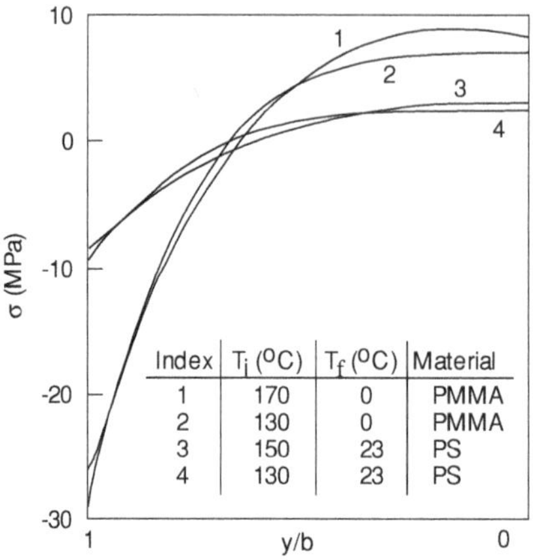

Figure 3.59 Residual stress distribution for 3-mm-thick PMMA plates cooled from 170°C and 130°C to 0°C, and for 2.6-mm-thick PS plates cooled from 150°C and 130°C to 23°C.

Processing conditions that cause unsymmetrical residual stress distributions through the thickness of the part often cause warpage in the final product. The Thermoplastic parts that will be most affected by residual stresses are those that are manufactured with the injection molding process. The formation of residual stresses in injection molded parts is attributed to two major coupled factors: cooling and flow stresses. The first and most important one is the residual stress that is formed because of the rapid cooling or quenching of the part inside the mold cavity. As will be discussed and explained later in this chapter, this dominant factor is the reason why most thermoplastic parts have residual stresses that are tensile in the central core of the part and compressive on the surface. Typical residual stress distributions are shown in Fig. 3.59 [62], which presents experimental* results for PMMA and PS plates cooled at different conditions.

Residual stresses in injection molded parts are also formed by the shear and normal stresses that exist during flow of the polymer melt inside the mold cavity during the filling and packing stage. These tensile flow-induced stresses are often very small compared with the stresses that build up during cooling. At low injection temperatures, however, these stresses can be significant in size, possibly leading to parts with tensile residual stresses on the surface. Figure 3.60 [60] demonstrates this concept with PS plates molded at different injection temperatures. The figure presents residual stress distributions through the thickness of the plate perpendicular and parallel to the flow direction. Isayev [62,63] has also demonstrated that flow stresses are maximum near the gate. The resulting tensile residual stresses are of particular concern since they may lead to *stress cracking* of the polymer component.

* The experimental residual stress distributions where directly computed from curvature measurements obtained from the layer removal method.

The development of models and simulations to predict shrinkage and warpage in the manufacturing of plastic parts is necessary to understand and control the complex thermomechanical behavior the material undergoes during processing. Shrinkage and warpage result from material inhomogeneities and anisotropy caused by mold filling, molecular or fiber orientation, curing or solidification behavior, poor thermal mold layout, and improper processing conditions. Shrinkage and warpage are directly related to residual stresses. Transient thermal or solidification behavior as well as material anisotropies can lead to the build up of residual stresses during manufacturing. Such process-induced residual stresses can significantly affect the mechanical performance of a component by inducing warpage or initiating cracks and delamination in composite parts. It is hoped that an accurate prediction of the molding process and the generation of residual stresses will allow for the design of better molds with appropriate processing conditions.

Estimating Part Shrinkage

To illustrate shrinkage and residual stress build-up during injection molding, the plate-shaped injection molding cavity shown in Fig. 3.50 is considered. As a first-order approximation, it can be assumed that a hard polymer shell forms around the melt pool as the material is quenched by the cool mold surfaces. Neglecting the packing stage during the injection molding cycle, this rigid frame contains the polymer as it cools and shrinks during solidification. The shrinkage of the polymer is, in part, compensated by the deflection of the rigid surfaces, which is a defor-

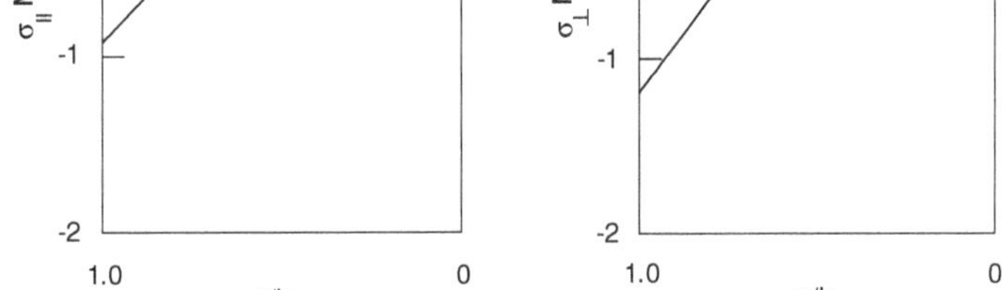

Figure 3.60 Residual stress distribution parallel and perpendicular to the flow direction for 2.54-mm-thick PS plate cooled from (1) 244°C and (2) 210°C to 60°C.

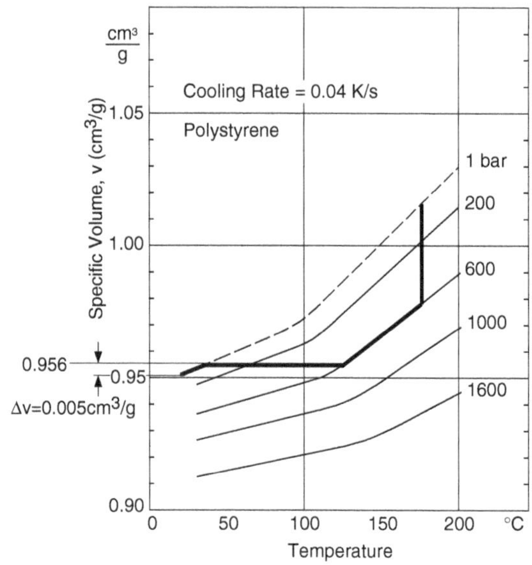

Figure 3.61 Trace of the given injection molding cycle in a polystyrene *P-v-T* diagram.

mation that occurs with little effort. In fact, if the packing stage is left out, it is a common experimental observation that between 85 and 90% of the polymer's volumetric changes are compensated by shrinkage through the thickness of the part [64].

The thickness change in a final part can be estimated with the use of a *P-v-T* diagram, as illustrated in Chap. 1. As an example let us consider a polystyrene component molded in a 1-mm-thick mold cavity. The melt is injected at 175°C to a maximum pack-hold pressure of 600 bar. This pressure is maintained until the gate freezes when the average part temperature has reached 125°C. At that point the pressure reduces to 1 bar and the temperature drops to 25°C. We can follow this process on a *P-v-T* diagram for polystyrene as illustrated in Fig. 3.61.

From the *P-v-T* diagram we can deduce that the part will have an average specific volume of 0.956 cm^3/g when the pressure reaches atmospheric pressure. The specific volume change from that point until the part reaches room temperature is 0.005 cm^3/g or 0.523%. This will lead to a final part thickness of 0.995 mm. It is important to point out that the cooling rate used to measure the *P-v-T* diagram is much slower than the cooling rates experienced during injection molding. This is especially critical with semi-crystalline polymers, where the degree of crystallinity, and therefore the shrinkage, will be overpredicted using the slow cooling rates experienced by the material during the *P-v-T* measurements.

Estimating Residual Stress Distribution

Solidification of an injection molded part starts during mold filling; however, flow continues during the postfilling or packing stage. This results in frozen-in flow stresses that are of the same order as the thermal stresses. Baaijens [65] calculated the residual stresses in injection molded parts, including the viscoelastic behavior of the polymer and the flow and thermal stresses. With his calculations, he demonstrated that the flow-induced stresses are significant and that a major portion of them stems back to the post-filling stage during injection molding. This is in agreement with experimental evidence from Isayev [60] and Wimberger-Friedl [66].

For a quick estimate of residual stress in an injection molded part, however, we can take the parabolic temperature distribution that is present once the part has solidified as a starting point. This will lead to a parabolic residual stress distribution that is compressive in the outer surfaces of the component and tensile in the inner core. Assuming no residual stress build up during phase change, a simple function based on the parabolic temperature distribution, can be used to approximate the residual stress distribution in thin sections [66]:

$$\sigma = \frac{2}{3}\alpha E(T_s - T_f)\left(\frac{6z^2}{4L^2} - \frac{1}{2}\right) \tag{3.50}$$

Here, T_s denotes the solidification temperature, T_f is the final temperature after cooling, α the thermal expansion coefficient, and E the modulus. The solidification temperature is the glass transition temperature for amorphous thermoplastics or the melting temperature for semi-crystalline polymers. Equation (3.50) was derived by assuming static equilibrium (e.g., the integral of the stresses through the thickness must be zero). Figure 3.62 [67] compares the compressive stresses measured on the surface of PMMA samples to Eq. (3.50).

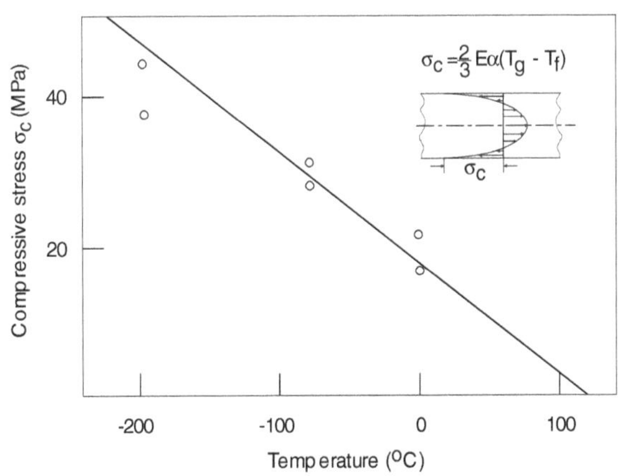

Figure 3.62 Comparison between computed, Eq. (50), and measured compressive stresses on the surface of injection molded PMMA plates.

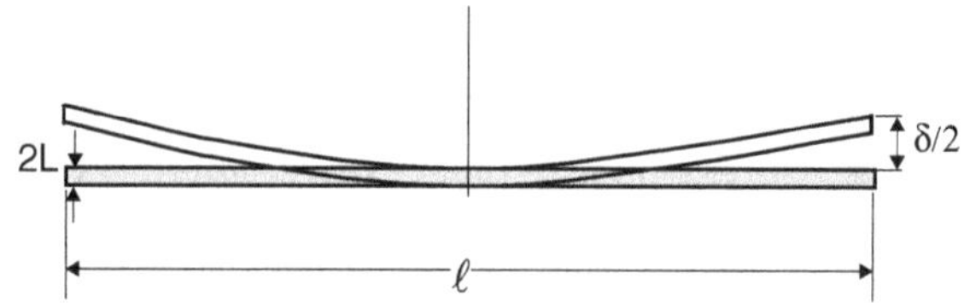

Figure 3.63 Nomenclature when computing part warpage.

Estimating Warpage due to Thermal Mold Imbalance

Thermal mold imbalances is a major cause of part warpage. These imbalance may be due to the location of the cooling lines and mold geometry. A rather simple but effective calculation is to predict warpage due to a temperature differential, ΔT, between the two mold halves. Using the nomenclature presented in Fig. 3.63 we can predict part warpage using the following relation

$$\delta = \frac{\alpha \ell^2 \Delta T}{4L} \tag{3.51}$$

To illustrate the preceding equation we can estimate how much a 30-cm, 1-mm-thick polystyrene ruler will warp due to a thermal mold imbalance of 5°C. Assuming a thermal expansion coefficient for polystyrene of 7×10^{-5} mm/mm/°C Eq. (3.51) gives a 15.7 mm warpage.

References

1. www.campusplastics.com
2. Saechtling, H., International Plastics Handbook, 3rd ed, Hanser Publishers, 1995, Munich.
3. Domininghaus, H., Plastics for Engineers, Hanser Publishers, 1995, Munich.
4. Macosko, C. W., *Rheology: Principles, Measurements and Applications*, VCH (1994), New York.
5. Dealy, J. M., Wissbrun, K. F., *Melt Rheology and Its Role in Plastics Processing*, Van Nostrand (1990), New York.
6. Tanner, R. I., *Engineering Rheology*, Clarendon Press (1985), Oxford.
7. Bird, R. B., Armstrong, R. C., Hassager, O., *Dynamics of Polymeric Liquids*, 2nd Ed., Vol. 1, John Wiley & Sons (1987), New York.
8. Gordon, G. V., Shaw, M. T., *Computer Programs for Rheologists*, Hanser Publishers (1994), Munich.
9. Bird, R. B., Wiest, J. M., *Annu. Rev. Fluid Mech.* (1995), 27, 169.
10. Giacomin, A. J., Samurkas, T., Dealy, J. M., *Polym. Eng. Sci.* (1989), 29, 499.
11. Ostwald, W., *Kolloid-Z.* (1925), 36, 99.
12. de Waale, A., *Oil and Color Chem. Assoc. Journal* (1923), 6, 33.
13. Stevenson, J. F., *Polym. Eng. Sci.* (1978), 18, 577.
14. Tadmor, Z., Bird, R. B., *Polym. Eng. Sci.* (1974), 14, 124.

15. Vinogradov, G. V., Malkin, A. Y., Yanovskii, Y. G., Borisenkova, E. K., Yarlykov, B. V., Berezhnaya, G. V., *J. Polym. Sci. Part A-2* (1972), 10, 1061.
16. Vlachopoulos, J., Alam, M., *Polym. Eng. Sci.* (1972), 12, 184.
17. Hatzikiriakos, S. G., Dealy, J. M., *ANTEC Tech. Papers* (1991), 37, 2311.
18. Denn, M. M., *Annu. Rev. Fluid Mech.* (1990), 22, 13.
19. Castro, J. M., Macosko, C. W., *AIChe J.* (1982), 28, 250.
20. Castro, J. M., Perry, S. J., Macosko, C. W., *Polymer Comm.* (1984), 25, 82.
21. Guth, E., Simha, R., *Kolloid-Zeitschrift* (1936), 74, 266.
22. ASTM, 8.01, Plastics (I), ASTM (1994), Philadelphia.
23. Hagen, G. H. L., *Annalen der Physik* (1839), 46, 423.
24. Poiseuille, L. J., *Comptes Rendus* (1840), 11, 961.
25. Dealy, J. M., *Rheometers for Molten Plastics*, Van Nostrand Reinhold Company (1982), New York.
26. Bagley, E. B., *J. Appl. Phys.* (1957), 28, 624.
27. Rabinowitsch, B., *Z. Phys. Chem.* (1929), 145.
28. Woebken, W., *Kunststoffe* (1961), 51, 547.
29. Menges, G., Wübken, W., *SPE, 31st ANTEC*, (1973).
30. Wimberger-Friedl, R., *Polym. Eng. Sci.* (1990), 30, 813.
31. Wimberger-Friedl, R., Ph.D. Thesis, Eindhoven University of Technology, The Netherlands (1991).
32. Menges, G., Geisbüsch, P., *Colloid & Polymer Science* (1982), 260, 73.
33. Bay, R. S., Tucker, C. L., III, *Polym. Comp.* (1992a), 13, 317.
34. Bay, R. S., Tucker, C. L., III, *Polym. Comp.* (1992b), 13, 322.
35. Menges, G., Schacht, T., Becker, H., Ott, S., *Intern. Polymer Processing* (1987), 2, 77.
36. Leibfried, D., Ph.D. Thesis, IKV, RWTH-Aachen, Germany (1970).
37. Wübken, G., Ph. D. Thesis, IKV, RWTH-Aachen (1974).
38. Mavrides, H., Hrymak, A. N., Vlachopoulos, J., *Polym. Eng. Sci.* (1986), 26, 449.
39. Mavrides, H., Hrymak, A. N., Vlachopoulos, J., *J. Rheol.* (1988), 32, 639.
40. Barone, M. R., Caulk, D. A., *J. Appl. Mech.* (1986), 361.
41. Lee, C.-C., Folgar, F., Tucker, C. L., III, *J. Eng. Ind.* (1984), 186.
42. Advani, S. G., Ph.D. Thesis, University of Illinois at Urbana-Champaign (1987).
43. Thieltges, H.-P., Ph.D. Thesis, RWTH-Aachen, Germany (1992).
44. Bailey, R., Kraft, H., *Intern. Polym. Proc.* (1987), 2, 94.
45. Mittal, R. K., Gupta, V. B., Sharma, P. K., *Composites Sciences and Tech.* (1988), 31, 295.
46. van Krevelen, D. W., Hoftyzer, P. J., *Properties of Polymers*, Elsevier Scientific Publishing Company, (1976), Amsterdam.
47. Wagner, H., Internal report, AEG (1974), Kassel, Germany.
48. Menges, G., Winkel, E., *Kunststoffe* (1982), 72, 2, 91.
49. Tadmor, Z., Gogos, C. G., *Principles of Polymer Processing*, John Wiley & Sons (1979), New York.
50. Avrami, M., *J. Chem. Phys.* (1939), 7, 1103.
51. Eder, G., Janeschitz-Kriegl, H., *Material Science and Technology, Vol. 18.*, Ed. H. E. H. Meijer, Verlag Chemie (1995), Weinheim.
52. Krobath, G., Liedauer, S., Janeschitz-Kriegl, H., *Polymer Bulletin* (1985), 14, 1.
53. Stefan, J., *Ann. Phys. and Chem., N.F.* (1891), 42, 269.
54. Eder, G., Janeschitz-Kriegl, H., *Polymer Bulletin* (1984), 11, 93.
55. Janeschitz-Kriegl, H., Krobath, G., *Intern. Polym. Process.* (1988), 3, 175.
56. Randell, D. R., *Radiation Curing Polymers*, Burlington House (1987), London.
57. Palmese, G. R., Andersen, O., Karbhari, V. M., *Advanced Composites X: Proceedings of the 10th Annual ASM/ESD Advance Composites Conference*, Dearborn, MI (1994), ASM International, Material Park.
58. Macosko, C. W., *RIM Fundamentals of Reaction Injection Molding*, Hanser Publishers (1989), Munich.
59. Bogetti, T. A., Gillespie, J. W., *45th SPI Conf. Proc.*, (1990).
60. Bogetti, T. A., Gillespie, J. W., *21st Int. SAMPE Tech. Conf.*, (1989).

61. Barone, M. R., Caulk, D. A., *Int. J. Heat Mass Transfer* (1979), 22, 1021.
62. Isayev, A. I., Crouthamel, D. L., *Polym. Plast. Technol.* (1984), 22, 177.
63. Isayev, A. I., *Polym. Eng. Sci.* (1983), 23, 271.
64. Wübken, G., Ph.D. Thesis, IKV, RWTH-Aachen, Germany (1974).
65. Baaijens, F. P. T., *Rheol. Acta* (1991), 30, 284.
66. Wimberger-Friedl, R., *Polym. Eng. Sci.* (1990), 30, 813.
67. Ehrenstein, G. W., *Polymeric Materials*, Hanser Publishers (2000), Munich.

4 Plasticating

C. Rauwendaal, P. J. Gramann

The plasticating unit of the reciprocating screw injection molding machine serves many important functions. The quality of the final molded part is strongly determined by the processes taking place in this part of the machine; however, the plasticating unit is often overlooked when molding problems occur. Its functionality is similar to the single screw extruder except for the fact that the screw moves axially in the barrel to inject the molten material into a mold. Its basic function is to accept and convey solid pellets and additives, perform melting, convey the melt, mix the plastic, possibly devolatilize the plastic melt, and generate pressure.

An important difference with extrusion is that the injection molding process is cyclic, whereas the extrusion process is continuous. To have consistency from cycle to cycle the conveying, heating, and mixing processes of the plasticating process must be repeatable and well understood. The goal of this chapter is to give the reader a good understanding of the plasticating unit.

This chapter will focus on what occurs in the plasticating unit of the injection molding machine by introducing its main parts, describing how they work, and indicating problems that may arise and some possible solutions, along with screw design issues. Models to predict solids conveying, melting, and melt conveying will also be given. Because *online* compounding is becoming more popular the chapter will also briefly discuss the theory of mixing. Several popular mixing sections will be shown with some new mixing devices that are successfully making their way into the injection molding industry.

4.1 The Plasticating Unit

The most common plasticating unit on injection molding machines is a single-screw extruder. This extruder is somewhat different from extruders used in continuous extrusion, where the screw rotates without interruption, but does not move forward or backward. The screw in an injection molding machine can rotate and can also move backward and forward. Such an extruder is called a *reciprocating screw extruder*; other terms used are ram-screw, recipro-screw, or RS extruder.

Another type of plasticating unit is the *ram* or *plunger extruder*. This is the earliest type of extruder used in injection molding. The advantage of the ram extruder is its simple design and excellent pumping characteristics. Important drawbacks of the ram extruder are very poor melting and mixing characteristics. With greater emphasis on quality and with increasingly large molded parts the ram extruder fell out of favor and was replaced by the reciprocating screw extruder. Today, nearly all injection molding machines use reciprocating screw extruders. Some plasticating units use a combination of a screw extruder and ram extruder; these units are called *ram-assisted screw extruders.*

4.1.1 The Ram Extruder

The overall objective of the plasticating unit is to take a solid pellet at its entrance and deliver a melt under high pressure to the mold at its exit. To do this, several different possibilities exist. The most basic and earliest design is the single-stage ram injection molding unit, which is rarely used today. In its most simple design, solid pellets are placed in a chamber and are melted by heating bands that are placed on the outside of the chamber.

The Two-Stage Ram Extruder

The limited melting capability of the single-stage ram extruder led to the development of the two-stage ram extruder. In this plasticating unit two ram extruders are used. The first ram extruder is used to melt the material and feed it to the second ram extruder. The second ram extruder is used to force the melted material into the mold.

The Ram-Assisted Screw Extruder

In this unit a conventional screw extruder is used to melt and mix the material. The melted material is then conveyed to a ram extruder. The ram extruder is used to push the melted material into the mold. Figure 4.1 shows an example of a ram-assisted screw extruder plasticating unit.

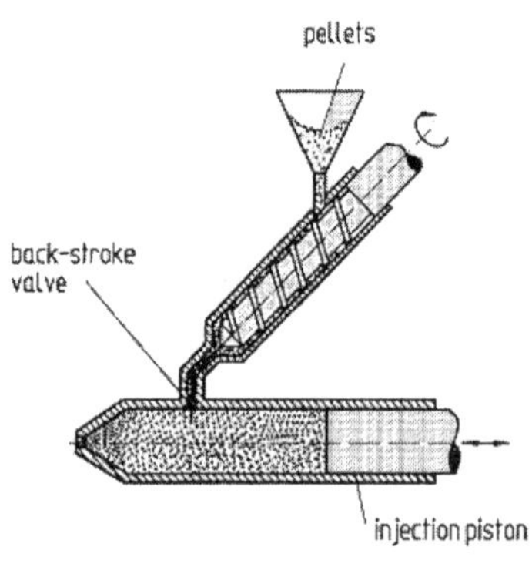

Figure 4.1 Ram injection molding unit.

The advantage of the ram-assisted screw extruder is that highly filled, viscous materials are more easily processed. This type of device is commonly used when processing bulk molding compound (BMC)—a blend of thermosetting polyester, fillers, and glass. Here, the extruder mixes the fillers and glass with the polyester and then feeds the chamber of the ram extruder.

4.1.2 The Reciprocating Screw

As mentioned earlier, the reciprocating screw is the most commonly used plasticating unit. Because of the importance of this type of unit it will be described in detail. The most important components are the screw, nonreturn valve, barrel, nozzle, and screw drive. The components will be discussed next.

The Plasticating Screw

The main components of the plasticating unit are shown in Fig. 4.2. The most important component is the screw [1]. It is a long cylinder with one or more helical flights wrapped around it and a nonreturn valve at the end (see Fig. 4.3). Such a device is often referred to as an *Archimedean screw*, after Archimedes, who developed the

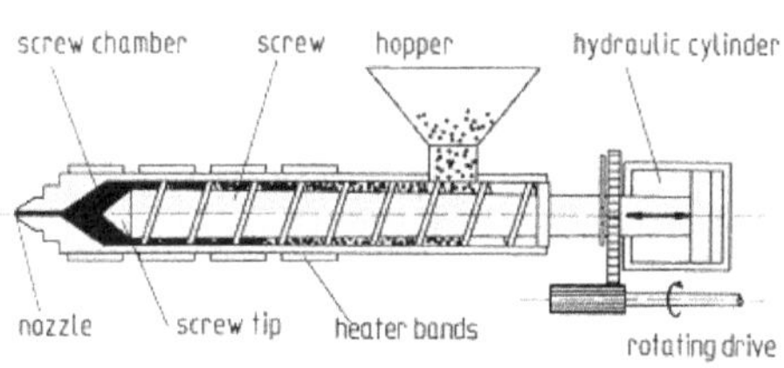

Figure 4.2 Reciprocating screw plasticating unit of the injection molding machine.

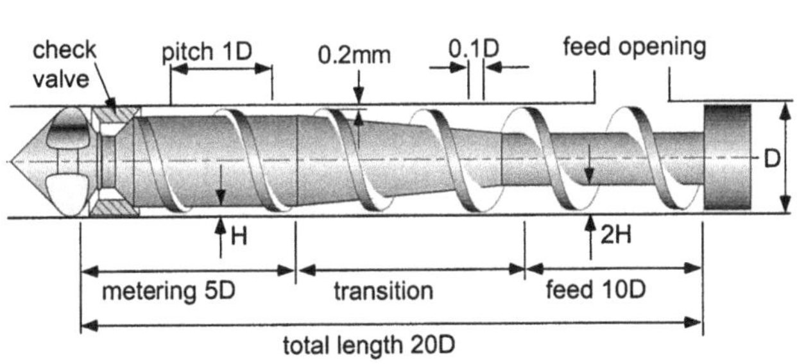

Figure 4.3 A typical extruder screw for injection molding.

basic screw conveyor thousands of years ago. The screw determines the conveying, the heating, the mixing, and in some cases the degassing of the plastic. Degassing or devolatilization is performed on a vented system; these are machines with a vent opening in the barrel through which volatiles can escape. Venting requires a special screw geometry, the so-called two-stage screw.

The main requirements of a reciprocating screw are to deliver a homogeneous and high-quality melt to the end of the screw within a cycle and from cycle to cycle. Homogeneous refers to uniformity of melt temperature as well as consistency. Good melt homogeneity is best achieved by using a distributive mixing section at the end of the screw [2]. Mixing will be discussed in more detail in a later section.

Screw Design for Injection Molding

It is generally recognized that proper screw design is of critical importance in the extrusion of sheet, film, pipe, tubing, and profiles using conventional, nonreciprocating extruders. In injection molding, however, where the plasticating units are usually reciprocating extruders, it is often assumed that screw design is of little importance. This assumption is unfortunately often incorrect. Screw design technology for nonreciprocating extruders has advanced considerably over the years, as evidenced by steady improvements in output and quality and the large number of articles devoted to the subject. On the other hand, screw design for reciprocating extruders has progressed relatively little, despite the fact that injection molding machines have become more and more sophisticated in terms of process monitoring and control, mold design, and the like.

The basic functions performed in a nonreciprocating extruder are solids conveying, melting, melt conveying, mixing, and sometimes devolatilization. The very same functions are performed in a reciprocating system with the difference being the screw is moving backward while these functions are performed, and the nature of the process is cyclic instead of continuous. If consistency from cycle to cycle is to be achieved, however, the conveying, heating, and mixing process should be highly repeatable. Thus, improvements in screw design developed for nonreciprocating extruders are largely applicable to reciprocating extruders as well.

The Standard Injection Molding Screw Injection molding screws often have similar characteristics to the extruder screw. A typical screw is shown in Fig. 4.3. The screw is single flighted with the flight pitch equal to the screw diameter and constant over the length of the screw. This is called a *square pitch geometry*, which is also frequently used in conventional extruders. The compression ratio usually varies from 2:1 for small machines to about 2.5:1 for large machines. The length of the feed section is about 50% of the total length, while the compression and metering section usually are 25%. This is different in a conventional extruder screw where the feed section is shorter and the compression and metering section longer. A typical screw for a regular extruder is shown in Fig. 4.4.

The width of the flight is about ten percent of the screw diameter. The channel depth in the feed section is deeper than in the metering section; the ratio of feed to

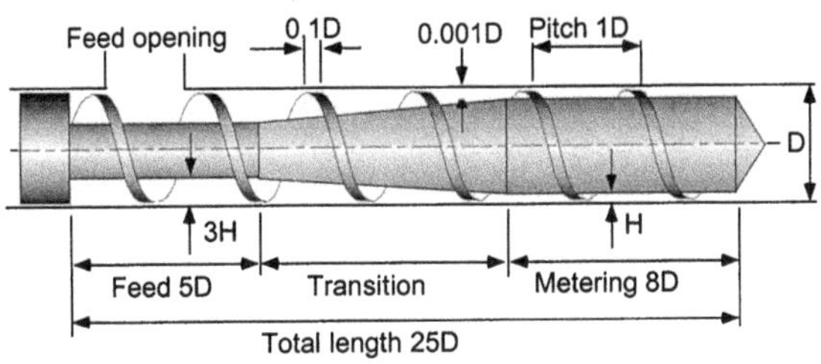

Figure 4.4 Typical screw for a regular extruder.

metering depth is called the *compression ratio*. The value of the compression ratio usually ranges from 2.5 to 3.0. In the modern extruder screw, mixing sections are used in most cases because the mixing capability of simple conveying screws is very limited.

Figure 4.5 shows some typical values of the channel depth in the metering section for reciprocating screws as described by Schwittay [3]. Within a diameter range of 30 to 120 mm, the channel depth increases approximately in a linear fashion with the diameter. Typical values of the compression ratio are shown in Fig. 4.6 [3].

For most thermoplastics the compression ratio increases from about 2:1 at small screw diameters to 2.5:1 for large screw diameters (100 mm or 4 inches). For rigid PVC, however, much lower values of the compression ratio are used, because PVC is quite susceptible to degradation, so using a lower compression ratio will deter this from happening. The radial flight clearance for reciprocating screws [3] is considerably larger than it is for regular screws [4], as shown in Fig. 4.7; the difference is a factor of about 2 to 3. This is mainly due to the fact that the reciprocating screw not only has an angular movement, but an axial movement. This action will wear the top of a flight quickly if it is not properly designed.

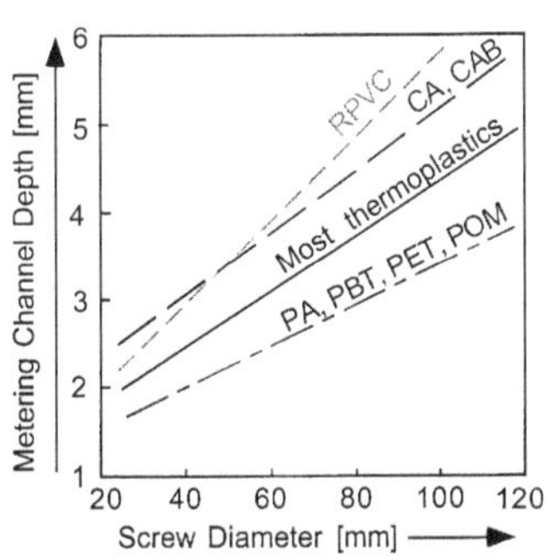

Figure 4.5 Typical values of channel depth in the metering section [3].

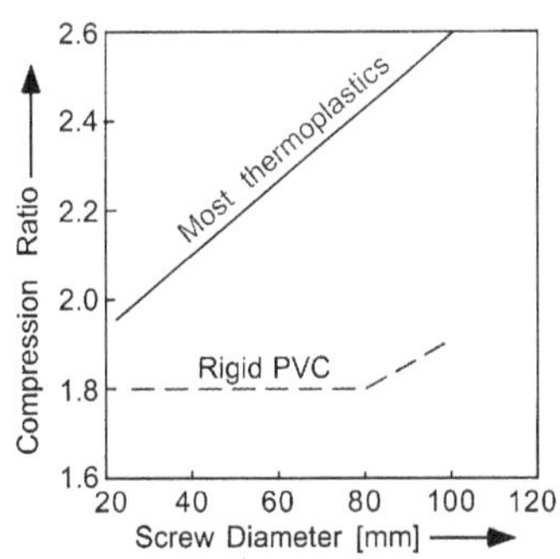

Figure 4.6 Typical values of the compression ratio.

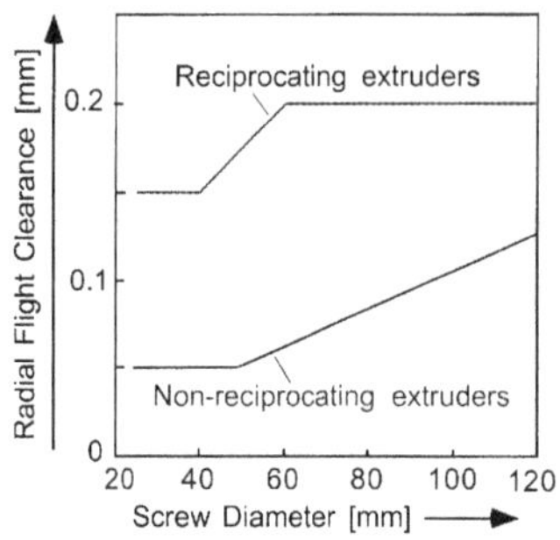

Figure 4.7 Flight clearance values for reciprocating and non-reciprocating extruders.

The relatively long feed section is largely due to the reciprocating action of the screw. The maximum stroke of the screw is usually about 3 to 4 diameters. When the screw is moving backward, the effective length of the feed section, as well as the overall length, reduces by the length of the stroke. The short metering section is used because the screw does not have to develop high pressure, as is common in continuous extrusion. The pressure that the screw needs to develop only needs to be high enough to push the screw backward during screw rotation. This pressure is usually only about 6 bar; however, the pressure can be intentionally increased by increasing the resistance to backward motion. This is done by raising the backpressure, usually by adjusting the flow control valve between the injection cylinder and the oil tank or by using a relief valve to control the pressure in that line.

Requirements for Reciprocating Screws The main requirements for a reciprocating screw are to deliver a homogeneous and high-quality melt to the end of the screw. Screws without mixing sections unfortunately have poor mixing ability. Good melt homogeneity is best achieved by using a distributive mixing section at the end of the screw [2]. It has been amply demonstrated in continuous extrusion that a good distributive mixing section at the end of the screw can improve extrudate quality and process stability. The same is clearly true in injection molding, even though relatively few publications have addressed this aspect of injection molding [5,6].

Desirable characteristics for mixing sections are [1,7]:

- streamlined flow
- minimum pressure drop
- barrel should be completely wiped
- operator friendly
- easy to machine

Important characteristics for distributive mixing sections are frequent splitting and reorientation of the fluid and substantial shear strain exposure. Various distributive mixing sections used in extrusion and injection molding are discussed in Ref. [2]. To achieve good dispersive mixing all material has to be exposed to a high and also uniform stress. This is generally accomplished by forcing the material through a small gap where it is exposed to high shear stresses. Different mixing sections will be discussed in a later section.

Solids Conveying In order to obtain a high degree of repeatability from cycle to cycle the conveying and heating process has to be quite consistent. The conveying process is driven by the frictional and viscous forces at the barrel surface. Frictional forces unfortunately can vary considerably in the extrusion process. This tends to destabilize the solids conveying process and can lead to surging. Several measures can be taken to improve the solids conveying process. The most potent measure is to machine grooves in the internal barrel surface in the region at the feed port. This will generally increase output and stability. Another measure is to apply a low-friction screw coating to the screw surface. This will reduce the retarding frictional force on the screw, thus improving solids conveying.

For steady conveying the solid material should move in plug flow (i.e., all particles should move at the same velocity). Hang-up of solid particles on the screw surface will reduce solids conveying rate and stability. The tendency for hang-up can be reduced by changing the flight geometry from a rectangular flight with a small radius to a flight with smoothly curved flight flanks [8]. This provides a gentle transition from the flight flank to the root of the screw. A double-flighted screw geometry in the feed section increases the screw surface area and reduces the channel volume; both these factors lead to a lower solids conveying rate. Thus, a double-flighted screw geometry is not recommended in the feed section, unless a high enough frictional force is available at the barrel surface.

The feed section length in a regular screw is typically about 5 diameters. This length generally results in a relatively steady solids conveying. The feed section length in a reciprocating screw has to be longer because the axial movement of the screw reduces the effective feed section length. As discussed earlier, in a standard reciprocating screw the feed section length is usually 50% of the total screw length. With the total length normally 20 diameters, the feed section is 10 diameters long. If one considers that the maximum stroke is usually 3 to 4 diameters, however, it would appear that there is no need to make the feed section longer than 7 to 8 diameters. The reduced feed section length is important in reciprocating screws because of the limited overall length. In other words, a shorter feed section allows for a longer transition section.

Screw design features that improve solids conveying:

- Flight helix angle of about 20 degrees
- Large flight flank radius (radius equals flight height)
- Single flighted geometry
- Low-friction coating on the screw surface

Plasticating In addition to solids conveying, the plasticating or melting process is critical in obtaining a high-quality and repeatable product. Because of the relative short length of most reciprocating extruders, melting deficiencies are more likely to occur in injection molding. In addition, the axial movement of the screw reduces the melting efficiency as compared with regular extruders [9]. Melting deficiencies fortunately do not directly affect product quality. There is a short time available for the melt in front of the screw to reach improved thermal uniformity, before the melt is injected into the mold; however, considering that the recovery time is often quite short, a few seconds, only limited improvement in melt temperature can occur. This is due to the very low thermal diffusivity of plastics.

The compression section length should preferably be equal to or larger than 7 diameters. A short compression section, as is commonly used, increases the chance of plugging and screw deflection; both are rather undesirable phenomena. Plugging occurs when the melting rate in the compression section cannot keep up with the reducing channel cross-sectional area. As a result, the solid bed will plug the screw channel. It will result in severe process instabilities. A rapid reduction in channel depth also increases the lateral forces acting on the screw because of the required deformation of the solid bed. The resulting deflection of the screw can cause rapid screw and barrel wear. This problem unfortunately occurs both in extrusion and in injection molding.

Reciprocating extruders equipped with rapid compression screws have been known to experience rapid and severe wear. In some cases the screw and barrel can be destroyed in a matter of hours. The use of hardfacing does not help much in these situations. It may slow down the rate of screw wear, but it will accelerate the barrel wear.

Barrier screws are often claimed to have improved melting performance relative to nonbarrier screws. This issue, however, is still being debated [10]; conclusive evidence has yet to be presented. From an engineering analysis [11] it is not clear that barrier screws offer better melting performance. Barrier screws can offer specific advantages, but there are some disadvantages as well that one needs to be aware of. The advantages are:

- No solids beyond the barrier section
- Barrier screws generally perform better than simple conveying screws
- The polymer melt is exposed to dispersive mixing as it passes over the barrier flight

Disadvantages of barrier screws are:

- Barrier screws perform no better than standard screws that incorporate efficient mixing sections
- Barrier screws are more expensive than standard screws with mixers
- Barrier screws are more susceptible to plugging

Barrier screws are based on melt separation, the melt collects in a separate melt channel. This means that the solid bed is restricted to the solids channel, which is only a fraction of the total channel. Plugging is therefore more likely to occur in a barrier screw than it is in a nonbarrier screw. One advantage of the barrier screw is that it does not allow unmolten particles to go beyond the barrier section. As a result, there is little chance that solid particles can make their way to the end of the screw. The same advantage, however, is obtained with a fluted mixing section, while it does not impose the space restriction on the solid bed.

Barrier sections can be advantageous in deep-flighted, large-diameter screws used for high-viscosity plastics. In deep-flighted, large-diameter screws there is more chance of insufficient melting capacity because a thick solid bed takes more time to melt. The advantage of a barrier screw in this case is that the depth of the solids channel can be reduced to essentially zero, whereas the melt channel can maintain substantial depth. This issue is more critical in large diameter ($D > 100\,\text{mm}$) extruders because heat transfer becomes more critical with an increasing screw diameter. The reason for this is that in scale-up the heat transfer area increases with the diameter squared, whereas the volume goes up with the diameter cubed.

Screw design features that favor high melting efficiency are:

- Multiple flights in the melting section
- Small flight clearance
- Long transition section (at least 5 diameters)
- Large flight helix angle

Melt Conveying The melt conveying requirements for reciprocating extruders are less severe than those for nonreciprocating extruders. The required pressure-generating capability in melt conveying is quite small. As a result, the metering section is usually quite short, typically about 5 diameters. In fact, it seems reasonable to choose an even shorter metering section. In some instances, screw designers have gone to the extreme of a zero-meter screw [12]. This, however, may reduce the melt conveying capability to the point where the operating window of the screw is reduced by the lack of pressure generating capability. Instead of a 5-diameter metering section, a better use of screw length would be a 2-diameter metering section followed by a 2- or 3-diameter mixing section.

Screw design features that favor a high melt conveying capacity are:

- Large channel depth
- Optimum helix angle for the plastic being processed
- Small flight clearance
- Large flight flank radius

Conclusions on Screw Design Screw design is important in extrusion as well as in injection molding. Even though some functional requirements may be less severe in injection molding than in extrusion, others may be more severe. Considering that the same functions are performed in both machines, there is no reason that the lessons learned in extrusion cannot be applied to injection molding. In doing this, one has to keep in mind the unique features and requirements for reciprocating extruders. Screw designs appropriate for conventional extruders are clearly not appropriate for reciprocating extruders; however, many features used advantageously on extruder screws can be used equally well on injection molding screws and result in improved performance.

Nonreturn or Check Valves

The screw of the injection molding machine produces a pool of polymer melt at the front of the screw as it rotates. This action forces the screw to move back away from the mold as the pool becomes bigger. When the chamber is filled to the specified amount the screw moves axially toward the mold and forces the polymer melt into the mold. Because of the relatively small channels and cavities that make up the runner system and the part being molded, a great deal of resistance must be overcome by the material to flow into the mold. This resistance could cause the material to flow back into the screw instead of into the mold during the injection and packing phases. To prevent this from occurring various shutoff systems have been used. The processor typically looks to have less than 5% of the melt flow back into the screw channel.

There are many different types of shutoff systems, generally termed *nonreturn valves* or *check valves*, that are available to the injection molder. The most common designs are the ball check valve and the sliding ring check valve. Both systems have their advantages over the other and are typically used for a general class of materials.

Ball Check Valve The ball check nonreturn valve works in a similar manner to the ball check valve commonly found in hydraulic systems, Fig. 4.8. The operation of a ball check valve is shown in Figs. 4.8 and 4.9. When the screw moves backward the ball is in the most forward position and the plastic melt can flow to the end of the screw (see Fig. 4.8). When the screw moves forward, the ball moves to the most rearward position against the ball seat forming a seal (see Fig. 4.9). In this position the valve is closed. The movement of the ball is due to the pressure difference across the ball. When the screw rotates and moves backward, the pressure behind the ball is higher and the ball is pushed forward. When the screw moves forward the pressure in front of the ball is higher and the ball is pushed backward.

The ball check nonreturn valve is usually less prone to wear than the sliding ring valve. The ball is free to turn, however, and material can leak through worn areas in

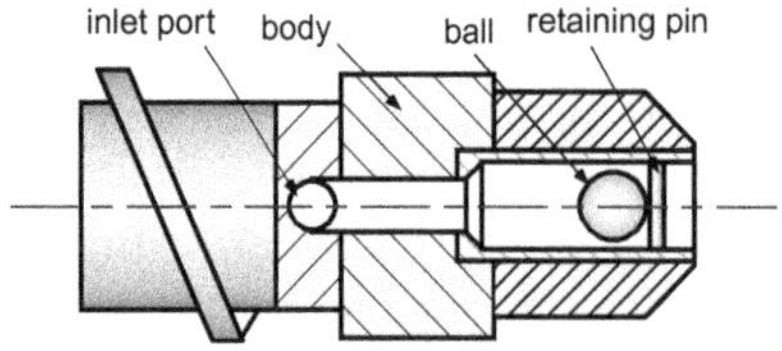

Figure 4.8 Ball check valve in open position.

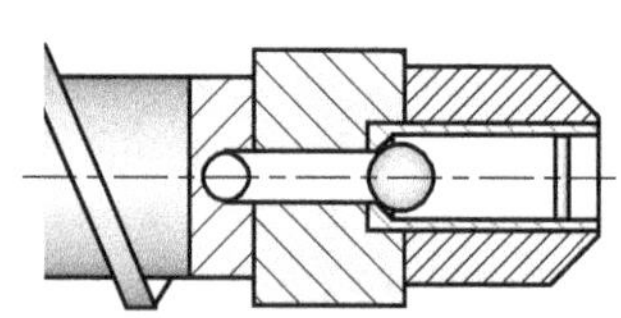
Figure 4.9 Ball check valve in closed position.

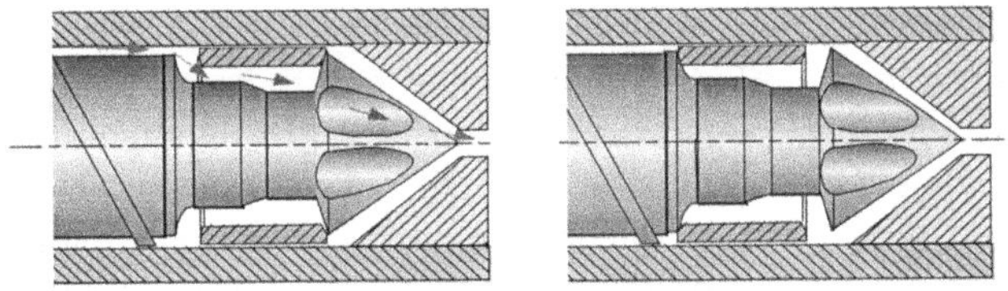

Figure 4.10 The ring valve in open (a) and closed (b) position.

the seal. Moreover, this device must leak in order for the ball to seat during injection—material must flow past the ball to suck it to the seat. This type of nonreturn valve is primarily used for low viscosity materials that are less prone to shear degradation.

Sliding Ring Check Valve The action of the sliding ring valve is illustrated in Fig. 4.10. Again, as the screw rotates plastic melt is created and accumulates at the discharge end of the screw. As the screw moves backward to make room for the plastic melt, the check ring is dragged to the most forward position against a stop at the end of the screw [see Fig. 4.10(a)]. The stop is usually a star-shape shoulder with four to six openings. When the check ring rests against this stop the plastic melt can flow through these openings.

As the screw moves forward to inject the plastic melt into the mold, the check ring is dragged to the most rearward position against the check ring seat, forming a seal [see Fig. 10(b)]. In this position the valve is closed and the plastic melt is thus prevented from leaking back into the screw channel during injection.

The sliding ring nonreturn valve characteristically has less leakage than the ball check nonreturn valve due to the large seating area, which reduces part-to-part variation. Because of the relative movement between the stop and the check ring, the stop will wear over time and will eventually need to be replaced. This device is primarily used for high-viscosity, shear-sensitive materials.

Design and performance criteria for both of these devices are: the valve must seat quickly, not leak, and seal effectively. The valve should not degrade the material, which could be caused by high shear forces or long residence times due to a non-streamlined flow path. Moreover, consideration should be given to the cost to manufacture the device. With both devices there is a loss of pressure because the polymer must flow around some constricting part that makes up the seal. As the material flows around or through the ball or sliding ring, it can be exposed to high degrees of shear, due to the high flow rates and small clearances. This of course manifests itself in viscous heating of the melt and ultimately higher melt temperatures.

As mentioned earlier, a popular method to reduce manufacturing costs is to perform on-line mixing. One example of this is in-house coloring, which is the blending of natural resin with a color concentrate, in the injection molding machine—essentially using mixing sections on the screw to promote distributive and dispersive mixing. This can be challenging, however, because most plasticating units for injection molding are relatively short, with the typical length-to-diameter ratio at 20:1. This does not leave much space to incorporate a mixing section. The nonreturn valve, specifically the sliding ring, lends itself to produce mixing. If designed properly, the common sliding ring nonreturn valve can be made to serve the dual purpose of being a valve and mixer, without affecting the melting and conveying capability of the plasticating unit. A mixing nonreturn valve can reduce the melt temperature nonuniformities significantly when compared with the standard nonreturn valve. Another advantage of such a valve is that mixing action can be achieved with a simple change to the screw geometry. An example of such a device is the CRD nonreturn valve [13], shown in Fig. 4.11. A significant advantage of creating mixing at the nonreturn valve is that the mixing takes place at the latest stage of plastication, just before the material enters the mold. An efficient mixing screw can achieve a good mixture at the end of the screw, but if coalescence takes place after the material leaves the screw, then the mixing quality of the final product may still not be acceptable.

In some cases a plain screw tip is used without an actual nonreturn valve. The most common application of such a screw is for high viscosity, thermally sensitive

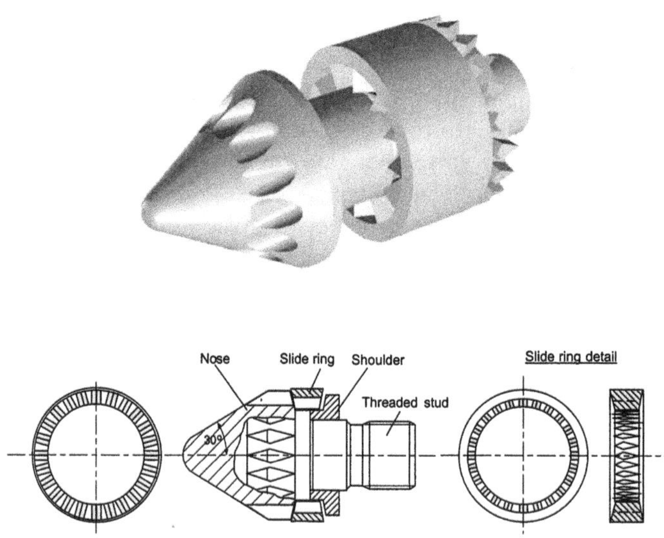

Figure 4.11 CRD nonreturn valve.

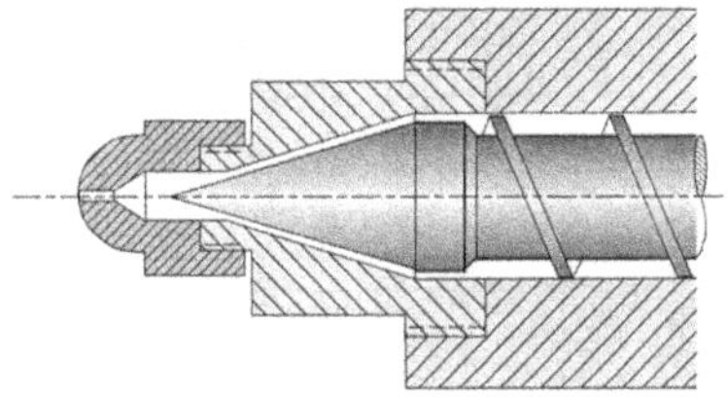

Figure 4.12 A plain screw tip as used for RPVC.

materials, such as rigid PVC (polyvinylchloride). Figure 4.12 shows a plain screw tip; it has a small gap between the tip and the barrel. The gap has to be small enough so that only a very small amount of plastic will leak back when the screw moves forward to inject the melt into the mold. The advantages of the plain tip are well-streamlined flow, no hang-up of plastic, and no need for maintenance.

The Barrel of the Plasticating Unit

The plasticating unit's barrel is a straight cylinder that closely fits around the screw. The radial clearance between the screw and barrel is typically 0.20mm or 0.008 inches; extruders 40mm in size and smaller have a clearance of about 0.15mm or 0.006 inches [3]. The barrel is often manufactured with a bimetallic liner that is centrifugally cast into the barrel. When the plasticating unit is vented, the barrel has a vent opening. At the feed end of the plasticating unit a feed opening is machined in the barrel. This opening is connected to the feed hopper, through which solid plastic particles are introduced to the plasticating unit.

The Nozzle of the Plasticating Unit

The nozzle of the plasticating unit is pushed against the sprue bushing. The nozzle tip usually has a radius that is slightly smaller than the mating radius of the sprue bushing to achieve a good seal; however, it also has to be large enough to avoid excessive wear. In the United States standard nozzle radii are 0.5 inches and 0.75 inches; in Europe the Euromap standards are 10, 15, 20, and 30mm. Nozzles can be either open or may contain a shut-off device. Open nozzles are recommended in the processing of thermally sensitive polymers and high-viscosity polymers, such as RPVC, thermosets, and elastomers. An example of an open nozzle is shown in Fig. 4.12.

Shut-off nozzles can be used to avoid drooling of molten plastic or stringing, they are also used to allow the plasticating unit to run with the nozzle retracted. Some nozzles are positively actuated during the cycle, others are controlled separately. An example of a sliding-bolt valve in the open position and in the closed position is shown in Figs. 4.13 and 4.14, respectively.

A commonly used nozzle valve is the needle valve shown in Fig. 4.15. The valve is open when the needle is retracted, as shown in Fig. 4.15. The valve is closed when the needle is pushed against the nozzle, as shown in Fig. 4.16.

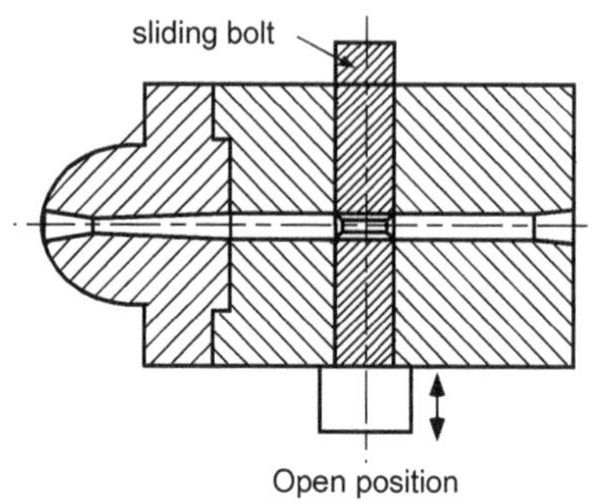

Figure 4.13 Example of a sliding-bolt shut-off nozzle in the open position.

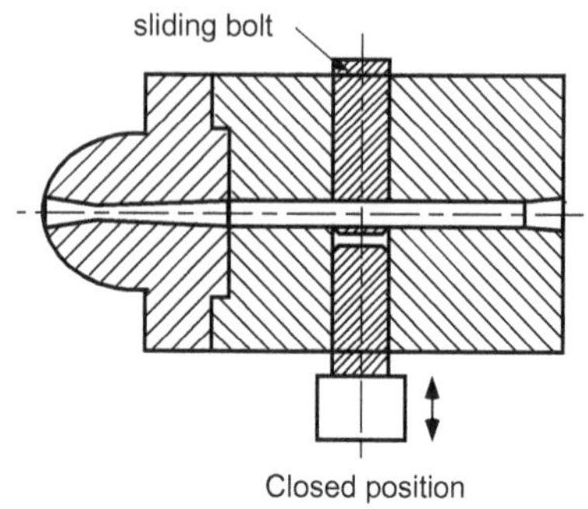

Figure 4.14 Example of a sliding-bolt shut-off nozzle in the closed position.

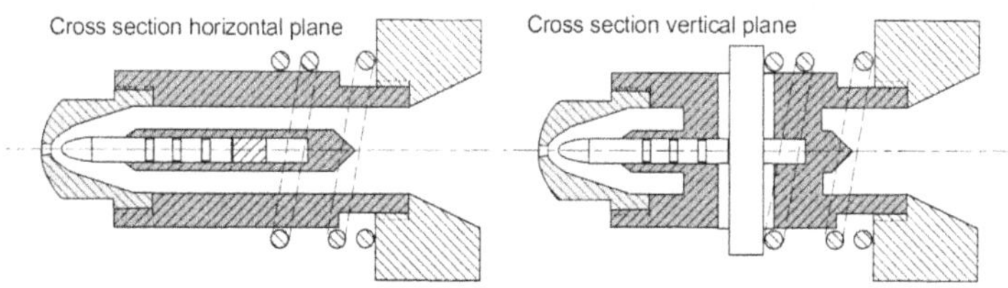

Figure 4.15 Schematic of a needle valve in the open position.

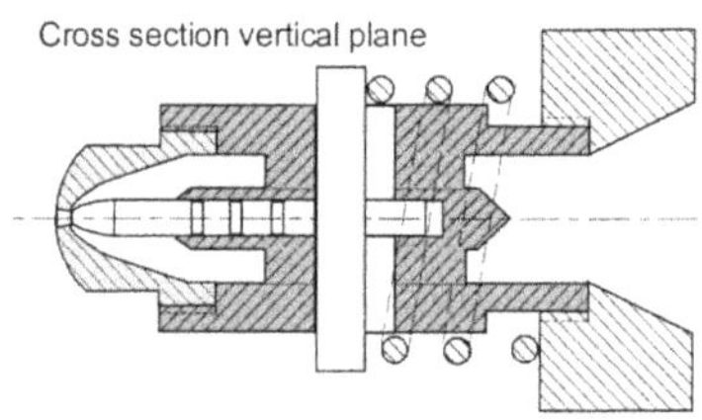

Figure 4.16 Schematic of needle valve in closed position.

The Feed Hopper and Feed Opening

The hopper should be designed to achieve a steady flow to the screw. The hopper cross section is preferably circular to minimize the chance of hang-up of material. The converging region should have a gradual compression to reduce the chance of bridging. Figure 4.17(a) demonstrates a poorly designed hopper, whereas Fig. 4.17(b) illustrates the desirable design features in a feed hopper.

Hopper design is critical when dealing with a feed with difficult bulk flow characteristics. Difficult materials are those with a wide range of particle size and/or

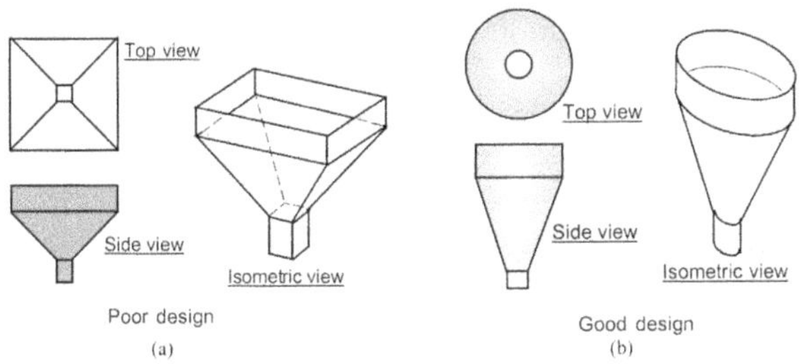

Figure 4.17 Common hopper designs. (a) poorly designed hopper (b) well designed hopper.

shape. Bulk materials that are highly compressible tend to cause problems in the hopper region. With such materials special features can be used to ensure steady flow, including vibrating pads, low friction coatings, internal cones, rotating wiper arms in the hopper, and so on.

It is important that the feed region of the plasticating unit be maintained at a low enough temperature to avoid sticking of the plastic to the metal walls. When plastic particles stick to the metal walls the size of the flow channel reduce; thus, the flow rate decreases. In an extreme case, flow can stop completely. Cooling capability is usually provided in the feed region so that sticking of the plastic particles can be avoided.

The Drive of the Plasticating Unit

The drive of the plasticating unit supplies the power to turn the screw and to push the screw forward during injection. Older injection molding machines use mostly hydraulic drives. The hydraulic system can be used to turn the screw, to push the screw forward, and to open and close the mold. Electric molding machines have more recently been introduced. These machines use electric motors to turn the screw, to push the screw forward, and to open and close the mold.

These all-electric molding machines were only available in small-to-midrange tonnage categories; however, as of the year 2000 all-electric machines are available up to 1100 tons and there are plans for machines up to 2500 tons. There are also a number of electric/hydraulic hybrid molding machines that combine electric with hydraulic drives.

In hydraulic molding machines the hydraulic drive consists of an electric motor and a hydraulic pump. The hydraulic pump converts the energy of the motor into hydraulic energy in the form of pressure. Both constant displacement pumps and variable displacement pumps are used. The pump efficiency should be about 90% at maximum operating pressure; the efficiency of the hydraulic system as a whole will be lower than the pump efficiency.

Hydraulic pumps should be located below the oil level, either at the bottom of the oil reservoir or below it. The drive unit has to be mounted so that noise propagation is minimized; this can be done by rubber support members and hose connections for all the pipelines.

4.2 Functions of the Plasticating Unit

Consistency from part to part must be maintained to create a high-quality product. To do this, the amount of mixing, heat, and shear that all material experiences must be kept the same throughout the process for days, weeks, and sometimes months. The cyclic process of injection molding must be made to be as repeatable as possible. Using a single screw with the ability to reciprocate has become the method of choice. The main functions of the plasticating unit are solids conveying, melting or plasticating, melt conveying, degassing, and mixing. These functions will be discussed in the following sections.

4.2.1 Solids Conveying

Solids conveying is the transport of particulate solid material in the injection molding machine. The particulate material usually consists of plastic pellets; however, it can also be granules, regrind, powder, or a combination of any of these. Solids conveying occurs both in the feed hopper and in the plasticating unit itself. The conveying in the feed hopper is driven by gravity; the flow resulting from this is called *gravity-induced conveying*. The weight of the plastic pellets causes them to move down the feed hopper and into the screw channel.

Starve Feeding

Most injection molding machines are flood fed. Thus, the screw takes in as much material as it can. In this case, the screw is essentially completely filled with plastic and the output is determined by the screw speed. Another approach is to use starve feeding.

In starve feeding the feed material is metered into the feed throat at a rate below the flood feed rate, Fig. 4.18. The output of the screw is no longer determined by its speed, but instead by the feed rate set for the feeder. In starve feeding the material does not build up in the feed hopper. As a result, this is one way to avoid flow problems in the feed hopper. Further, when using a vented barrel, starve feeding is way to keep from flooding the vent.*

In starve feeding the degree of fill is dependent on the feed rate and the screw speed. Increasing the screw speed reduces the degree of fill, whereas raising the output will increase the degree of fill. By changing feed rate or screw speed the effec-

* Personal communication with John Bozzelli of Injection Molding Solutions.

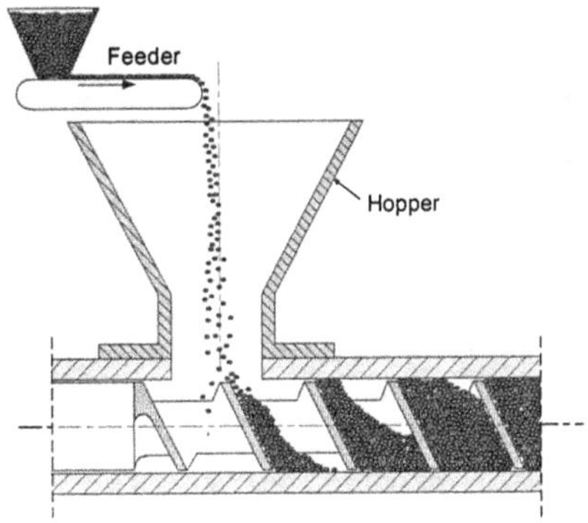

Figure 4.18 Metering of solids into the hopper to create a starve feed system.

tive *L/D* (length to diameter ratio) of the unit can be varied. A number of advantages can be obtained from starve feeding:

- The motor load and screw torque can be reduced
- The specific energy consumption, SEC, can be varied
- Eliminate leakage problems with venting
- Instabilities may be eliminated, e.g., solid bed break up
- Reduced agglomeration of feed stock ingredients

The SEC determines how much mixing and viscous heat generation occurs in the plastic. Starve feeding, therefore, allows an additional degree of process control. It can be very useful in cases where excessive motor load and/or excessive melt temperatures occur.

Drag-Induced Solids Conveying

Once the plastic particles drop into the screw channel the conveying mechanism changes to a drag-induced conveying. Gravity may still play a minor role; however, the drag forces generally dominate the gravitational forces in this part of the machine. There are two competing drag forces acting on the plastic. One is the frictional force between the plastic and the barrel. The other is the frictional force between the plastic and the screw. It is important to understand which of these two forces is the driving force for forward conveying and which one is the retarding force.

The forward conveying is due to drag forces between the plastic and the barrel. The drag forces between the plastic and the screw actually work against forward conveying. Without a drag force against the barrel, there is no forward conveying only rotation with the screw. For good, steady conveying, therefore, it is important to have a large drag force at the barrel and a small drag force against the screw.

The first comprehensive analysis of solids transport in conventional single screw extruders was made by Darnell and Mol [14]. The analysis developed here is an extended version of the Darnell and Mol analysis to include axial movement of the screw [15]. The following assumptions are made:

1. The particulate solids behave like a continuum
2. The solid bed is in contact with the entire channel wall
3. The channel depth of the screw is constant
4. The flight clearance can be neglected
5. The solids material moves in plug flow
6. The pressure is a function of down channel distance only
7. The coefficient of friction is independent of pressure
8. Gravitational forces are negligible
9. Density changes in the solid material are negligible
10. The screw channel curvature is negligible

Figure 4.19 shows the various forces acting on an element of the solid bed as well as a velocity diagram. To simplify the analysis the usual assumption is made that the screw is stationary and the barrel is rotating and moving axially. The barrel velocity v_b is the combination of the tangential barrel velocity v_{bt} and the axial barrel velocity v_{ba}. The angle α between v_b and v_{bt} is determined from:

$$\tan \alpha = \frac{v_{ba}}{v_{bt}} \tag{4.1}$$

The solid bed velocity v_{sz} in the down channel direction is determined by the flight helix angle ϕ. The relative velocity between the solid bed and the barrel, Δv, is determined by the vectorial difference between the barrel velocity v_b and the solids bed velocity v_{sz}. The solids conveying angle θ is the angle between Δv and v_b.

A force balance in down-channel and cross-channel direction yields an expression for the pressure profile along the screw channel:

$$P(z) = P_0 \exp\left(\left[\frac{f_b}{H}\cos(\theta + \phi - \alpha) - \frac{f_b f_s}{H}\sin(\theta + \phi - \alpha) - \frac{2f_s}{W} - \frac{f_s}{H}\right]z\right) \tag{4.2}$$

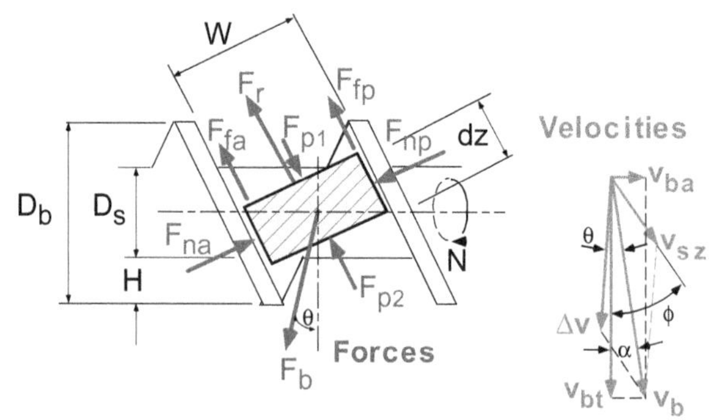

Figure 4.19 Forces and velocities in solids conveying.

where H is the channel depth, W the channel width, P_0 the pressure at $z = 0$, f_b the coefficient of friction at the barrel, and f_s the coefficient of friction at the screw. Equation (4.2) indicates that the pressure increases exponentially with down channel distance z. As a result, high pressures can be generated efficiently under certain circumstances. Equation (4.2) can be worked out to obtain a closed form expression for the solids conveying angle θ:

$$\theta = \arcsin\left[\frac{\left(1 + f_s^2 - k^2\right)^{1/2}}{1 + f_s^2}\right] - \phi \tag{4.3}$$

where:

$$k = \frac{H}{zf_s}\ln\frac{P}{P_0} + \frac{f_s}{f_b}\left(1 + \frac{2H}{W}\right)$$

The solids conveying rate relative to the screw is given by:

$$\dot{M}_{s1} = \rho_s HpWv_{sz} = \rho_s HpWv_b \frac{\sin\theta}{\sin(\theta + \phi - \alpha)} \tag{4.4}$$

where ρ_s is the density of the solids and p the number of parallel screw flights. The solids conveying rate relative to the barrel is given by:

$$\dot{M}_s = \rho_s HpWv_b\left[\frac{\sin\theta}{\sin(\theta + \phi - \alpha)} - \frac{\sin\alpha}{\sin\phi}\right] \tag{4.5}$$

From Eq. (4.5) it is clear that the net solids conveying rate, relative to the barrel, is reduced by the axial velocity component, v_a. The change in solids conveying rate with angle α is shown in Fig. 4.20.

The maximum axial barrel velocity is the one at which the net solids conveying rate becomes zero. This occurs when the axial component of the solid bed velocity equals the axial barrel velocity. The maximum axial velocity, therefore, becomes:

$$v_{ba-\max} = v_{sz}\sin\phi = \frac{v_b \sin\theta\ \sin\phi}{\sin(\theta + \phi - \alpha)} = v_b \sin\theta \tag{4.6}$$

The corresponding maximum angle α equals the solids conveying angle.

Improving Solids Conveying

In practical terms, the solids conveying rate can be improved by:

- reducing the friction at the screw surface
- increasing the friction at the barrel surface
- increasing the bulk density of the feed stock

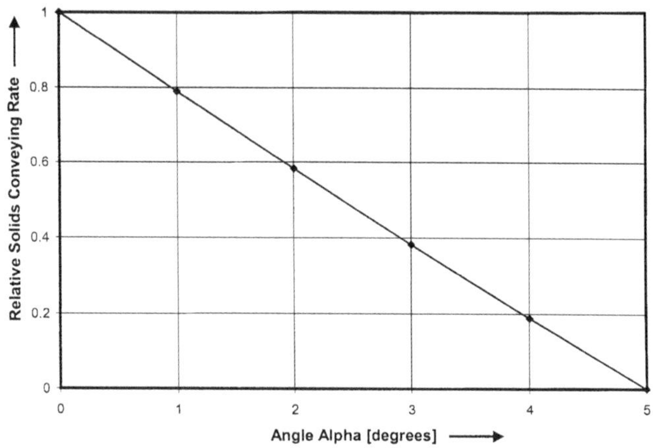

Figure 4.20 Solids conveying rate versus angle α.

Reducing the friction at the screw surface can be done by:

- applying a low-friction coating to the screw
- using a special surface treatment such as catalytic surface conversion (J-Tex) or Dyna-blue
- avoiding a multiflighted screw geometry in the feed section
- using a large flight flank radius along the screw
- avoiding a short flight lead in the feed section

Increasing the friction at the barrel surface can be done by:

- adjusting the barrel temperature
- increasing the surface roughness of the internal barrel surface
- machining grooves into the internal barrel surface

4.2.2 Melting or Plasticating

There are two sources of heat during plastication. One is the heat from heating elements placed along the barrel; the other is the frictional and viscous heat generation within the plastic. Frictional heating occurs when solid plastic particles slide past a metal surface or past one another. Viscous heating occurs when the plastic melt is sheared by the rotation of the screw. The viscous heating is primarily determined by the shear rate in the melt and the viscosity of the melt. The shear rate is a measure of the shearing action in a fluid, In the plasticating unit, the approximate shear rate in the screw channel is the circumferential velocity of the screw divided by the channel depth.

As an example, let us consider a 150-mm (6-inch) screw running at 90 rpm with a channel depth of 7 mm (0.28 inches).

The circumference of the screw is 3.14 * 150 = 471.24 mm.

The circumferential speed is (471.24 mm * 1.5 rev/s = 706.86 mm/s).

The approximate shear rate becomes 706.86/7 = 100.98 (1/s).

Typical shear rates in a screw channel range from 50 to 100 (1/s or s^{-1}). The shear rates in a plasticating unit increase with screw speed.

As a result of the high plastic melt viscosity, the viscous heat generation is often substantial. In fact, in many plasticating units, as well as extruders, most of the heat comes from frictional and viscous heat generation. This is particularly true at high screw speeds. As the screw speed increases the contribution from viscous heating increases and the contribution from the barrel heaters decreases. This is illustrated in Fig. 4.21. The barrel heaters usually contribute only a small amount of heat to the extrusion process. The frictional and viscous heat generation is essentially a transformation of mechanical energy from the screw drive into thermal energy to increase the plastic temperature.

When the plastic temperature reaches its melting point, melting will start. Melting usually starts at the barrel surface about 5 diameters from the feed opening. A melt film forms initially on the barrel surface. As the melt film grows thicker, a melt pool forms at the leading flank of the flight, pushing the solid bed against the trailing flank of the flight. This model, first described by Maddock [16], is shown in Fig. 4.22.

The two main sources of heat for melting are the heat from the barrel heaters and the viscous heat generation in the melt film between the solid bed and the barrel. For a high melting efficiency, it is important to keep the melt film as thin as possible. The melt film thickness is strongly determined by the screw flight clearance. The flight clearance, therefore, should be kept small. If the flight clearance increases (e.g., as a result of screw wear) the melting rate can reduce substantially.

The melting model shown in Fig. 4.22 is also called the *contiguous solids melting* (CSM) model. It has been observed in many experimental studies of melting in single

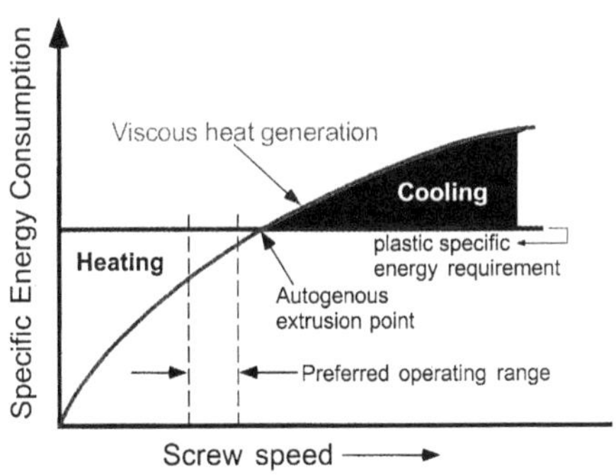

Figure 4.21 Specific energy consumption versus screw speed.

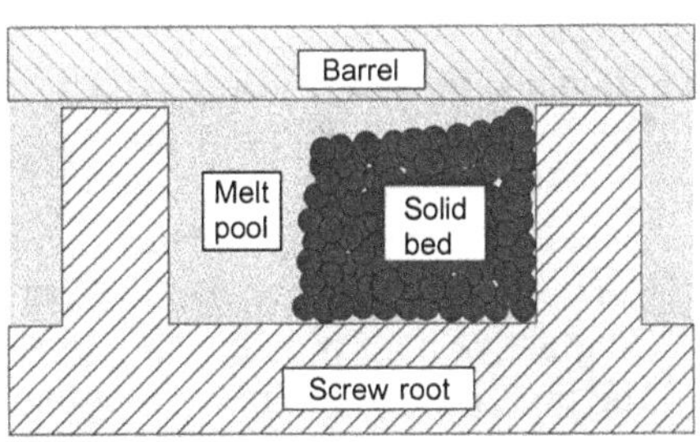

Figure 4.22 Maddock melting model.

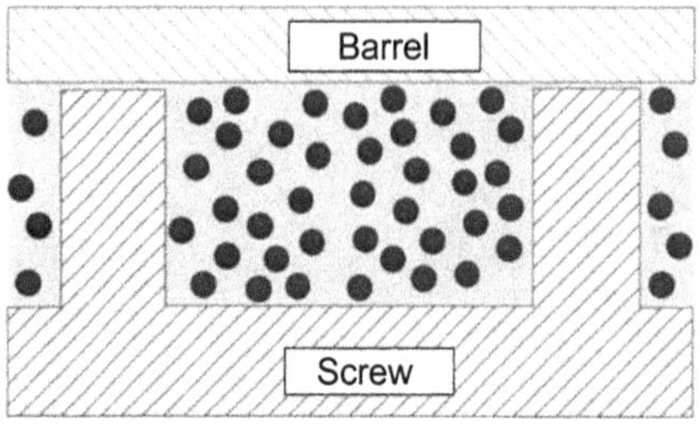

Figure 4.23 Dispersed solids melting model.

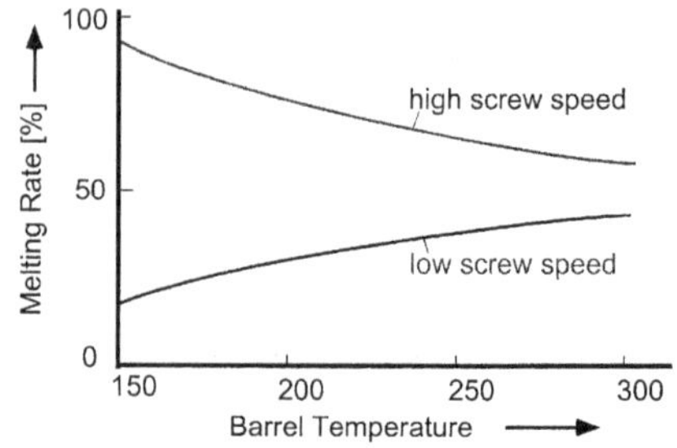

Figure 4.24 Effect of barrel temperature on melting rate.

screw extruders. In some cases, however, another melting is observed where the solid particles are discrete and floating in a melt matrix. This type of melting is called *dispersed solids melting* (DSM) [17] or dissipative melt mixing [18], see Fig. 4.23. DSM has been observed in twin screw extruders and reciprocating single-screw compounders. Melting by DSM occurs more efficiently than by CSM.

When the barrel temperature is increased, the amount of heat for melting from the barrel heaters increases. At the same time, however, the viscous heat generation in the melt film will reduce because a higher barrel temperature lowers the viscosity in the melt film. At low speeds, when the barrel heaters supply a major portion of the heat, higher barrel temperatures will increase melting (see Fig. 4.24). At high screw speeds, however, the major portion of the heat for melting comes from viscous heat generation. At high speed, therefore, increased barrel temperature can actually reduce melting (see Fig. 4.24).

The melting process in reciprocating screws can be analyzed by extending the theoretical analysis developed by Tadmor [19] for conventional extruders. The model used for analysis of melting is shown in Fig. 4.25.

The solid bed is pushed against the trailing flank of the flight, the melt pool against the leading flank of the flight, and the melt film is located between the barrel and the solid bed. The melt film thickness increases from the flight flank to the melt pool to accommodate the melt entering the melt film from the solid bed. Melting is assumed to take place primarily at the interface between the melt film and solid bed.

The following assumptions are made:

1. The melt density and thermal conductivity are constant
2. Convective heat transfer is neglected
3. Conductive heat transfer only in the radial direction
4. The melt flow is laminar (creeping flow)
5. Inertia and body forces are negligible
6. There is no slip at the walls
7. There is no pressure gradient in the melt film
8. The temperature dependence of the viscosity can be neglected

The temperature profile in the melt film is considered fully developed (i.e., no change in the down channel direction). The fully developed melt temperature profile for a Newtonian fluid in pure drag flow is a quadratic function of radial distance y:

$$T(y) = \left(\frac{-\mu \Delta v}{2k_m H_m^2}\right) y^2 + \left(\frac{2k_m \Delta T_b + \mu \Delta v^2}{2k_m H_m}\right) y + T_{mp} \tag{4.7}$$

From the temperature profile in the melt film the heat flux from the melt film to the solid bed interface can be determined. This heat flux can be expressed as:

$$-\dot{q}_m = \frac{2k_m \Delta T_b + \mu \Delta v^2}{2H_m} \tag{4.8}$$

The first term in the numerator, $2k_m\Delta T_b$, represents the conductive heating term. The thermal conductivity is represented by k_m and ΔT_b is the difference between the barrel temperature and the temperature at the interface. For semi-crystalline polymers the latter temperature is generally taken to be the crystalline melting point, T_{mp}.

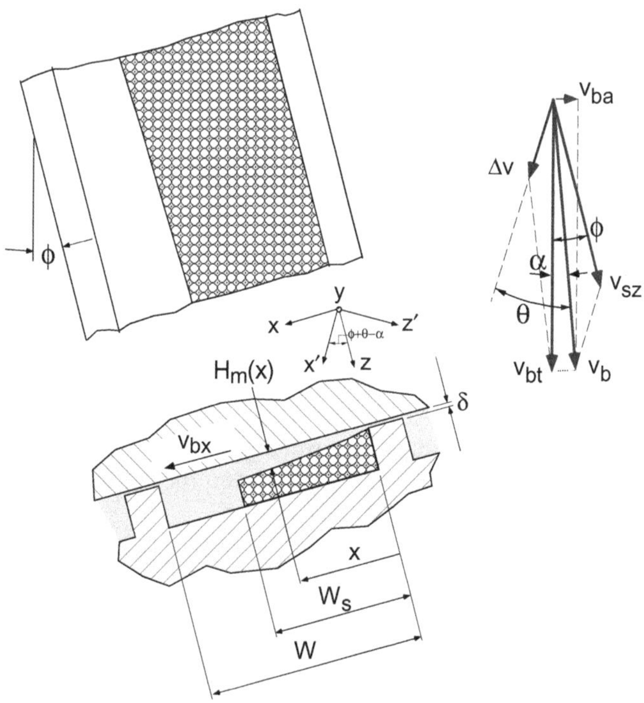

Figure 4.25 Melting model for reciprocating extruder.

The second term, $\mu\Delta v^2$, represents the viscous heat generation with μ the viscosity and Δv the relative velocity between the solid bed and the barrel.

If the temperature in the solid bed is assumed to be a function only of radial distance it can be written as:

$$T(y) = \Delta T_r \exp\left(\frac{y v_{sy}}{\alpha_s}\right) + T_r \tag{4.9}$$

where T_r is the reference temperature, v_{sy} the normal melting velocity, α_s the thermal diffusivity, and $\Delta T_r = T_{mp} - T_r$. From the temperature profile in the solid bed the heat flux from the interface into the solid bed can be determined. It can be expressed as:

$$-\dot{q}_{out} = \rho_s C_s v_{sy} \Delta T_r \tag{4.10}$$

where C_s is the specific heat of the solids. The melting velocity can be determined from a heat balance of the interface. The heat available for melting is simply the difference between the heat flux into and the heat flux out of the interface:

$$v_{sy} \rho_s \Delta H_f = \frac{2k_m \Delta T_b + \mu\Delta v^2}{2H_m} - \rho_s C_s v_{sy} \Delta T_r \tag{4.11}$$

From this expression the melting velocity can be determined:

$$v_{sy} = \frac{2k_m \Delta T_b + \mu\Delta v^2}{2\rho_s H_m \Delta H} \tag{4.12}$$

where ΔH is the sum of the latent heat of fusion ΔH_f and the specific heat from T_r to T_{mp}. The thickness of the melt film varies along the width of the solid bed. The thickness profile can be determined from a mass balance on an element of the solid bed. The melt film thickness can thus be written as:

$$H_m(x') = \left[\frac{4k_m \Delta T_b + 2\mu\Delta v^2}{\rho_m \Delta v \Delta H} x' + \delta^2\right]^{1/2} \tag{4.13}$$

where δ is the radial flight clearance. Coordinates x' and z' are determined by the direction of Δv, which is the relative velocity between the solid bed and the barrel as shown in Fig. 4.25. This relative velocity is:

$$\Delta v = \frac{v_b \sin(\phi - \alpha)}{\sin(\theta + \phi - \alpha)} \tag{4.14}$$

The melting rate can be determined by integrating the melting velocity over the width of the solid bed. Assuming that the flight clearance can be neglected, the melting rate becomes:

$$\frac{d\dot{M}_p}{dz} = \left[\frac{(2k_m \Delta T_b + \mu \Delta v^2)\rho_m v_b \sin(\phi - \alpha)}{2\Delta H}\right]^{1/2} \sqrt{W_s} = \Omega \sqrt{W_s} \tag{4.15a}$$

The melting rate expressed in terms of the tangential and axial components of the barrel velocity can be written as:

$$\frac{d\dot{M}_p}{dz} = \left[\frac{(2k_m \Delta T_b + \mu \Delta v^2)\rho_m (v_{bt} \sin\phi - v_{ba} \cos\phi)}{2\Delta H}\right]^{1/2} \sqrt{W_s} = \Omega \sqrt{W_s} \tag{4.15b}$$

It is clear from Eqs. (4.15a) and (4.15b) that the melting rate is reduced by the axial velocity component. This is graphically illustrated in Fig. 4.25.

The change in width of the solid bed along the length of the screw channel can be determined from a mass balance in down channel direction:

$$d\dot{M}_p = -\rho_s v_{sz} d(H_s W_s) \tag{4.16}$$

If it is assumed that dH/dz equals the taper of the screw channel, A_z, the solid bed width at time $t = 0$ becomes:

$$W_s(z) = W\left[\frac{2H_f}{z_0 A_z} - \left(\frac{2H_f}{z_0 A_z} - 1\right)\left(\frac{H_f}{H_f - A_z z}\right)^{1/2}\right]^2 \tag{4.17}$$

where:

$$z_0 = \frac{2\rho_s v_{sz} H_f \sqrt{W}}{\Omega}$$

In this equation W is the width of the solid bed at $z = 0$, H_f the channel depth at $z = 0$, and z_0 the total melting length required to complete melting in a constant depth screw. Equation (4.17) can only be used as long as the melting occurs in the compression section of the screw. The total length required to complete melting can be determined from Eq. (4.17) by setting W_s equal to zero. This results in the following expression for the helical melting length:

$$z_t = z_0\left(1 - \frac{z_0 A_z}{4H_f}\right) \tag{4.18}$$

This expression clearly shows that the melting length can be reduced significantly by using a positive taper in the melting zone of the screw. The maximum taper that can be used before plugging occurs is $2H_f/z_0$. At this value $z_t = 0.5z_0$ and the solid bed width becomes independent of the down channel distance z. This situation obviously cannot be achieved in practice unless there is another channel into which the melt can flow.

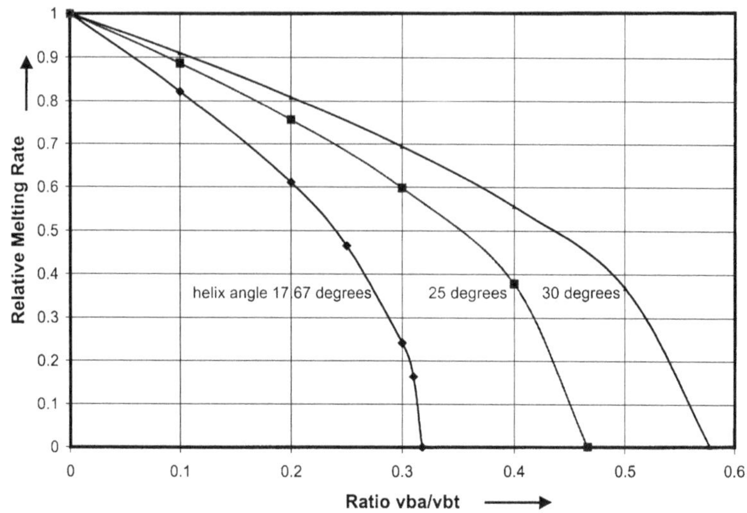

Figure 4.26 The relative melting rate versus the ratio of axial to tangential barrel velocity.

The helical length z_t will increase when the axial velocity increases. For a zero taper (constant depth) screw the increase in melting length with axial velocity is shown in Fig. 4.26.

From this figure it is clear that melting is strongly affected by the axial velocity component. At the common helix angle of 17.67 degrees the maximum possible axial velocity is about 32% of the tangential velocity. This value, however, cannot be reached in practice because the melting length becomes excessive. It is also clear from this figure that melting performance can be improved by using a larger helix angle.

In a reciprocating screw the point where $W_s = W$ will move down the screw because of the axial motion between the screw and barrel. If it is assumed that melting starts at a fixed distance from the feed opening the location where melting starts will move down the screw at velocity v_{ba}. This means that the solid bed width is both a function of down channel distance as well as a function of time. The solid bed width as a function of down channel distance and time can be written as:

$$W_s(z, t) = W_s(z - t v_{ba} \csc \phi, t = 0) \tag{4.19}$$

There are a number of simplifying assumptions behind Eq. (4.19). One is that the change in solid bed width in down channel direction is not affected by the shift in the onset of melting. This is a reasonable assumption as long as the onset of melting occurs in the compression section of the screw. When the melting occurs both in the feed section and in the compression section, however, this assumption may not be correct. In this case the melting in the feed section should be separated from the transition section, which means that simple closed form expressions are more difficult to obtain.

Another assumption is that the solid bed profile at the beginning of screw recovery is determined by Eq. (4.17). This is reasonable when $v_{bt} >> v_{ba}$ and when screw recovery is a major part of the total molding cycle. If this is not the case one should analyze melting during the ram forward portion of the cycle when $v_{bt} = 0$ and $v_{ba} < 0$ and the stationary melting when $v_{bt} = v_{ba} = 0$.

How to Improve Melting

The melting capacity of the plasticating unit is determined by the processing conditions and the design of the screw. Screw design issues have been covered in the previous section. The melting capacity can be improved by:

- Preheating the plastic before it enters the plasticating unit
- Increasing the barrel temperature when operating at low screw speed
- Lowering the barrel temperature when operating at high screw speed
- Making sure that the screw flight clearance is small

4.2.3 Melt Conveying

The melt conveying in reciprocating screws can be analyzed using the standard flat plate approximation [4]. In this analysis the stationary screw channel is unrolled onto a flat plane and the barrel is represented by a flat plate moving relative to the screw (see Fig. 4.27).

The barrel velocity has components both in the tangential (v_{bt}) and axial direction (v_{ba}). The axial barrel velocity equals the backward recovery speed of the screw. The barrel velocity is determined by vectorial addition of the tangential and axial components:

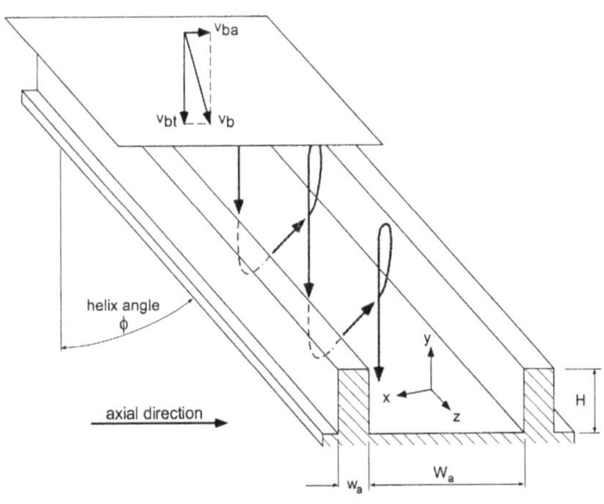

Figure 4.27 Flat plate model and coordinate system.

$$v_b = \left(v_{bt}^2 + v_{ba}^2\right)^{1/2} \tag{4.20}$$

The barrel velocity can be written as:

$$v_b = v_{bt}\left(1 + \tan^2\alpha\right)^{1/2} \tag{4.21}$$

The barrel velocity component in the direction of the screw channel is given by:

$$v_{bz} = v_b \cos(\phi - \alpha) \tag{4.22}$$

If the polymer melt is assumed Newtonian and isothermal the down channel velocity as a function of normal distance, y, can be written as [4]:

$$v_z(y) = \left(\frac{v_{bz}}{H} - \frac{g_z H}{2\mu}\right) y + \frac{g_z}{2\mu} y^2 \tag{4.23}$$

where μ is the melt viscosity, g_z the pressure gradient in down channel direction, and H the channel depth. The flow rate down the screw channel is obtained by integrating the down channel velocity over the cross channel area of the screw channel:

$$\dot{V}_s = \int_0^H pWv_z(y)dy = 0.5\,pWHv_{bz} - \frac{pWH^3 g_z}{12\mu} \tag{4.24}$$

where p is the number of flights and W the perpendicular channel width.

In Eq. (4.24) it is assumed that the effect of the flight flanks can be neglected. In other words, the down channel velocity is considered to depend only on normal distance, y, and not on cross channel distance, x. It is also assumed that leakage flow through the flight clearance is negligible. The first term on the right-hand side of Eq. (4.24) is the drag flow rate and the second term is the pressure flow rate.

It should be realized that the flow rate in this equation is the flow rate down the screw channel (i.e., the flow rate relative to the screw surface). It is more interesting, however, to determine the flow rate relative to the barrel surface because this is the amount that actually flows out of the system. This net melt flow rate can be written as:

$$\dot{V} = 0.5\,pWHv_{bz} - \frac{pWH^3 g_z}{12\mu} - \frac{pWH^3 v_{ba}}{\sin\phi} \tag{4.25}$$

If the drag flow rate is defined as the flow resulting from the relative motion between the screw and the barrel, the drag flow rate becomes:

$$\dot{V}_d = 0.5pWHv_{bt}\cos\phi\left[1 + \tan\phi\,\tan\alpha\left(1 - 2\csc^2\phi\right)\right] \tag{4.26}$$

The expression before the square brackets is the drag flow rate of a nonreciprocating screw. The bracketed term indicates how much the drag flow rate is affected by the axial movement of the screw. When the axial velocity is zero ($\tan\alpha = 0$), the bracketed term equals unity. When the axial velocity is greater than zero the drag flow rate will reduce with the axial velocity.

At a certain axial velocity the drag flow rate will become zero. This situation can occur in reciprocating screw systems. During recovery, when the screw is rotating and moving backward, there is typically no net flow out of the plasticating unit. If it is assumed for the time being that the pressure flow can be neglected, then the axial screw speed at which the net output becomes zero can be determined from Eq. (4.26). This maximum axial speed can be expressed as:

$$v_{ba-\max} = v_{bt}\,\frac{\sin\phi\,\cos\phi}{2-\sin^2\phi} \tag{4.27}$$

Thus, the maximum axial speed as a fraction of the tangential speed v_{bt} is only a function of the helix angle ϕ. This relationship is shown in Fig. 4.28.

At the commonly used square pitch helix angle, $\phi = 17.67$ degrees, the maximum axial speed is rather small, about 15% of the tangential speed. The maximum axial speed increases with the helix angle up to about 55 degrees where the highest value occurs. The highest value of the maximum axial speed that can be obtained is about 35% of the tangential speed.

From Eq. (4.27) the minimum recovery time can be determined. The recovery time is the stroke L_s divided by the axial speed (recovery speed). The minimum recovery time can be written as:

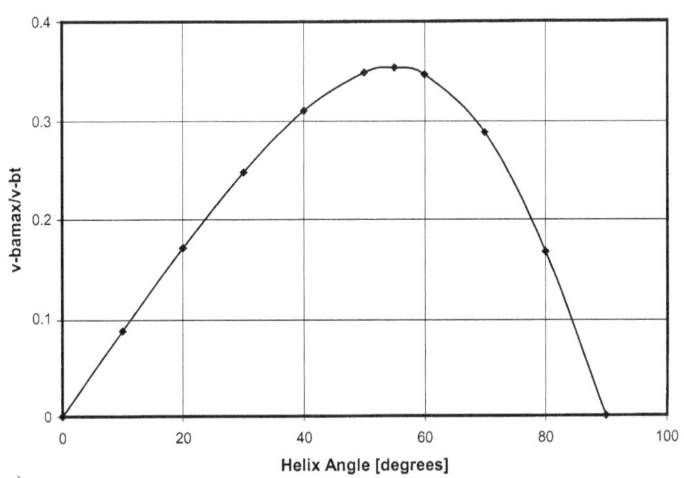

Figure 4.28 Maximum axial speed versus screw flight helix angle.

$$L_{t\min} = \frac{L_s(2 - \sin^2\phi)}{v_{bt} \sin\phi \cos\phi} \tag{4.28}$$

If the stroke is given as a percentage of maximum stroke ($L_s = sL_{max}/100$), then the minimum recovery time becomes:

$$t_{r\min} = \frac{sL_{\max}(2 - \sin^2\phi)}{100\, DN \sin\phi \cos\phi} \tag{4.29}$$

Using typical values of $L_{max} = 3.5D$ and the square pitch flight helix angle the minimum recovery time becomes:

$$t_{r\min} = 0.07349 \frac{s}{N} \tag{4.30}$$

The minimum recovery time, therefore, is primarily determined by the percentage of maximum stroke and the screw rotational speed. This relationship is shown in Fig. 4.29.

The minimum recovery time is directly proportional to the stroke and inversely proportional to the screw speed. In Eqs. (4.27 to 4.30) it is assumed that the pressure flow is negligible. This is a reasonable assumption as long as the back pressure during recovery is low. This is usually the case in injection molding with typical recovery pressures of about 10 to 20 bar (150 to 300 psi); however, if the pressure flow does play a role of importance the expressions for maximum axial speed and minimum recovery time become more complex. The maximum axial velocity can be written as:

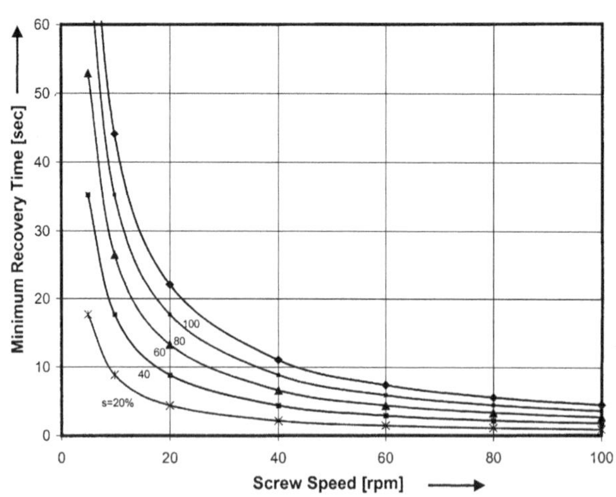

Figure 4.29 Minimum recovery time versus screw speed.

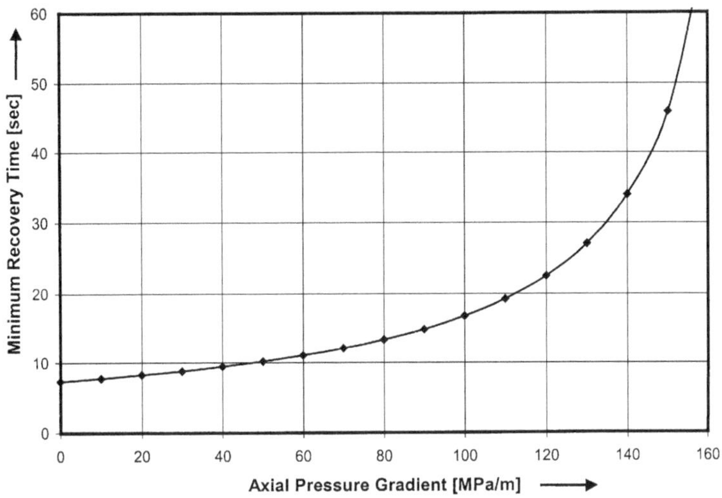

Figure 4.30 Minimum recovery time versus pressure gradient.

$$v_{ba\max} = v_{bt}\frac{\sin\phi\cos\phi}{2-\sin^2\phi}\left(1-\frac{H^2 g_a}{6\mu v_{bt}}\tan\phi\right) \tag{4.31}$$

Thus, when the axial pressure gradient is positive, $g_a > 0$, the maximum axial velocity will be reduced. This effect will be more pronounced with large values of channel depth and helix angle and with a low polymer melt viscosity. A positive pressure gradient in the melt conveying zone of the screw indicates that the melt conveying section has to develop pressure, resulting in reduced throughput. With Eq. (4.31) the minimum recovery time can be expressed as:

$$t_{r\ \min} = \frac{0.12 s L_{\max}(2-\sin^2\phi)}{\sin 2\phi(6v_{bt} - \tan\phi H^2 g_a/\mu)} \tag{4.32}$$

The effect of the pressure gradient on minimum recovery time is shown in Fig. 4.30.

The example illustrated in Fig. 4.30 is for a 63.5-mm-screw running at a screw speed of 60 rev/minute with a channel depth of 4.6 mm, a helix angle of 17.67 degrees, a melt viscosity of 1000 Pa-s, and a maximum stroke of 222 mm. It is clear in this example that the effect of the pressure flow is minor when the axial pressure gradient is below 20 MPa/m (72.5 psi/inch). At higher values of the pressure gradient, however, the minimum recovery time increases rapidly.

The actual velocity in axial direction (recovery speed) is directly related to the flow rate down the screw channel because this flow rate equals the axial velocity multiplied with the cross-sectional area of the screw ($\pi D^2/4$). From this relationship the axial velocity can be expressed as:

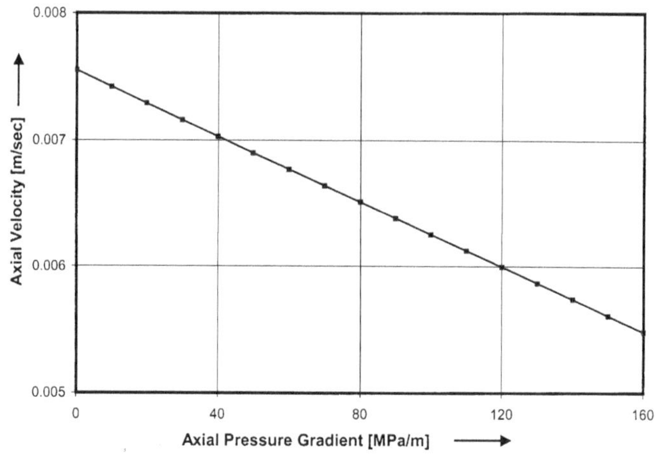

Figure 4.31 Axial velocity versus axial pressure gradient for 63.5-mm screw.

$$v_{ba} = \frac{2pWHv_{bt}\cos\phi - pWH^3 g_a \sin\phi/(3\mu)}{\pi D^2 - 2pWH\sin\phi} \tag{4.33}$$

Eq. (4.33) clearly shows the main factors that determine the recovery speed: the number of flights (p), the channel width (W), the channel depth (H), the screw diameter (D), the screw speed (v_{bt}), the flight helix angle (ϕ), the melt viscosity (μ), and the axial pressure gradient (g_a). Figure 4.31 shows the relationship between axial velocity and axial pressure gradient as determined from Eq. (4.33) and using the data for the 63.5-mm-screw running at 60 rpm from the previous example.

There is a linear reduction in the axial velocity with the axial pressure gradient. The actual values of the axial velocity are only a small percentage (2 to 5%) of the tangential barrel velocity.

The expression can be simplified if the channel width is taken as $W = \pi pD\sin\phi$ and the second term of the denominator is neglected. This leads to:

$$v_{ba} \approx H\left(N\sin\phi\cos\phi - \frac{H^2 g_a \sin^2\phi}{3\pi D\mu}\right) \tag{4.34}$$

where N is the rotational speed of the screw and g_a the pressure gradient in axial direction. The highest axial speed will obviously result in the shortest possible recovery time. When the pressure flow is small relative to the drag flow the recovery time can be reduced most effectively by:

- Increasing the channel depth in the metering section
- Increasing the screw speed
- Increasing the flight helix angle (up to about 50 degrees)

There are obviously limits to how far the channel depth can be increased. If the channel depth is made too large the melting capacity of the plasticating unit will become insufficient to handle the large conveying rate of the screw. This problem can be counteracted by increasing the length of the plasticating unit.

Melt conveying is strongly influenced by screw design. The optimum geometry of the melt conveying section of the screw depends on the polymer melt viscosity and the operating conditions. The details of screw optimization are discussed elsewhere [4]. There are several operational factors that can improve melt conveying. They are:

- Reduce barrel temperature
- Increase screw temperature
- Reduce flight clearance
- Reduce back pressure
- Increase screw speed

4.2.4 Degassing or Devolatilization

In some molding operations a vent opening in the barrel is used to remove volatiles from the plastic. Venting requires special screws designed to create a zero pressure region under the vent port; this is necessary to keep the plastic melt from flowing out of the vent port—a condition called *vent flow*. The screw that is used in a vented system is called a *two-stage screw* (see Fig. 4.32).

Vented systems are often used with hygroscopic plastics to lower the moisture level to a level where it does not cause problems in the extruded product. Other volatiles can be removed, too, including monomers, solvents, and air. In many cases, removing volatiles through venting is more cost effective than a drying–separation operation before extrusion.

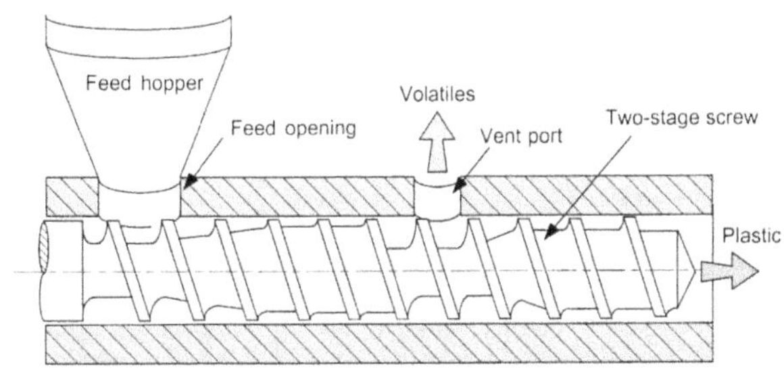

Figure 4.32 A two-stage screw.

A number of conditions have to be fulfilled to achieve efficient devolatilization:

- The plastic has to be melted completely by the time it reaches the extraction section under the vent port. Thus, the vent port cannot be too close to the feed port.
- The pressure under the vent port has to be zero; therefore, the extraction section of the screw has to have large conveying capability and be only partially filled.
- The screw section just before the extraction section has to be fully filled with plastic melt. This is achieved by incorporating a screw section with low conveying capability.
- The screw section downstream of the extraction section (the second stage) has to have larger conveying capability than the screw section upstream of the extraction section (the first stage). This can be achieved by increasing the channel depth and/or pitch or the second stage relative to the first stage.

4.2.5 Mixing

The demands on the plasticating unit of the injection molding machine have constantly increased over the years. This primarily relates to three factors: the quality of the part produced, the economy of production, and versatility of the process. These factors are closely related to one another. When investigating the reciprocating screw of the injection molding machine and how it influences these three factors, one main theme arises: the lack of good mixing of the reciprocating screw frequently limits the quality of the molded parts.

Due to its flexibility, the three-zone screw has become the de facto general purpose screw. With the demands of on-line compounding in the injection molding unit, however, the limits of this screw configuration have quickly been reached. It is nearly every processor's goal to processes various materials without the changing the screw. To obtain this type of screw the following demands need to be met:

- Fine break up of secondary components
- Homogenous mixture of primary and secondary components
- Thermally homogenous melt
- High flow rate
- Minimum shear stress induced into material
- Self-cleaning capabilities
- Minimum wear

Types of Flow

To obtain such a screw that meets many of these objectives, the principles of mixing that take place in the screw must be well understood. Before this can occur, the type of flows that are generated by the screw need to be specified. Flows are generally

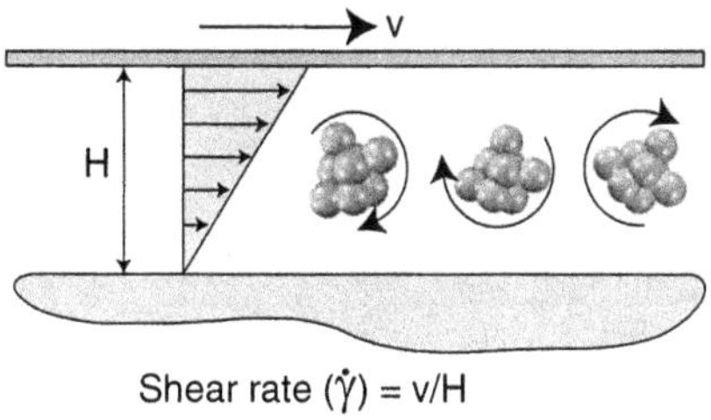

Figure 4.33 Shear flow.

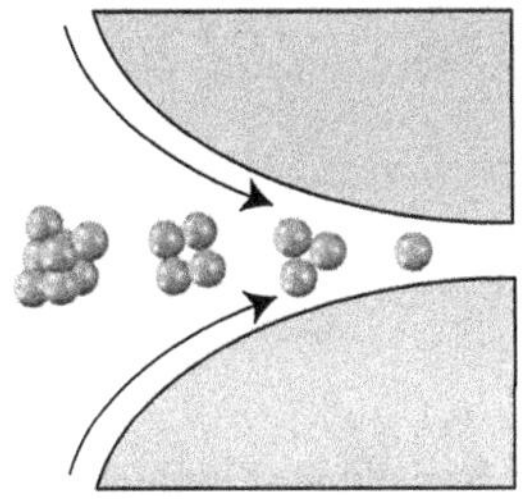

Figure 4.34 Irrotational or elongational flow.

classified as rotational, shear, or elongational. As the name implies, rotational flows create an environment that rotates the fluid elements. Because these flows do not produce any significant mixing they should be avoided. Shear flows are the most common type of flow created in polymer processing, represented in Fig. 4.33. This diagram is a simplistic model of flow in the screw channel where the top sliding wall represents the barrel and the stationary wall the screw. Elongational flows, sometimes referred to as *irrotational flows*, create a flow field that stretches or pulls apart the fluid elements, as shown in Fig. 4.34. These flows are generally difficult to produce and to sustain at a high elongational rate.

Distributive Mixing

When performing a mixing task to create a homogeneous compound or blend, the polymers and/or additives being processed can have widely different properties (e.g., viscosity), which may cause significant mixing problems. The mixing demands of the system are increased when this occurs. In polymer processing, the mixing action that takes place can be broken down into two major categories: distributive and dispersive. *Distributive mixing* involves the increase in spatial distance between solid agglomerates or droplets without their reduction in size. This distribution is achieved by imposing large strains on the system such that the interfacial area between the two or more phases increases and the local dimensions, or striation thickness, of the secondary phases decrease. Figure 4.35 shows a simplified schematic of distributive

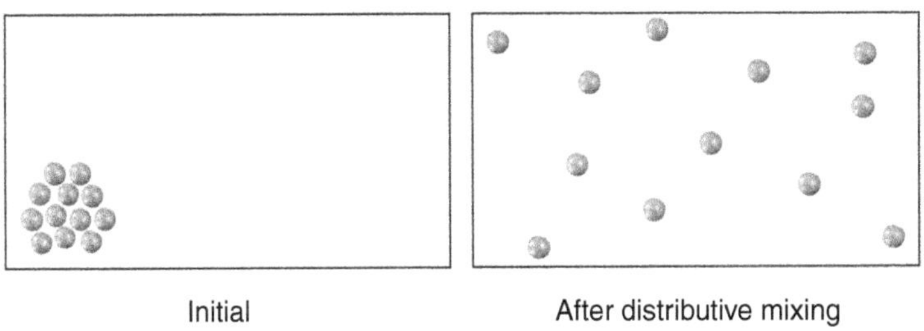

Figure 4.35 Schematic of distributive mixing.

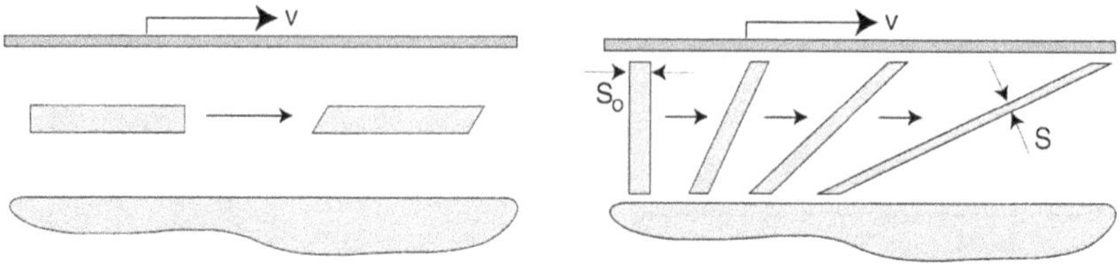

Figure 4.36 Reduction of striation thickness (S) through shear. (a) Horizontal secondary phase. (b) Vertical secondary phase.

mixing. Here, a group of cohesionless particles are grouped together in the lower left corner. The flow of the primary phases displaces the particles throughout the region. Because there are no cohesive bonds between the particles they are free to move virtually independent of one another. The final state of the particles shows a nearly random distribution of particles throughout. Segregation of particles is another form of distributive mixing that may take place. Here, particles with different characteristics will preferentially move to certain regions [20].

Imposing large strains on a mixture does not always guarantee a homogeneous blend. This is especially true in polymer processing where the mixing of liquids typically involves viscosities ranging between 100 to 100,000 Pa/s. With viscosities this high, the Reynolds number (ratio of inertia forces to viscous forces) is usually under 1.0, indicating laminar flow. With this type of flow the natural mixing of diffusion and the exponential rate of mixing found in turbulent flow is not present. This means that the mixing that takes place in polymer processing takes a great deal of effort. In addition, the initial orientation of the secondary phase relative to the flow direction becomes significant in the amount of total strain generated. Figure 4.36 shows a simple example of mixing in laminar flow with two fluid elements in the same shear flow. The element shown in Fig. 4.36(a), which is parallel to the direction of flow, is between two streamlines that are near each other, causing little deformation. On the other hand, in Fig. 4.36(b), where the fluid element is perpendicular to the direction flow and cuts across many streamlines, a more deformed fluid element is produced.

The deformation that occurs in this example can be analyzed with equations developed by Erwin [21]. To quantify the mixing that takes place the interfacial area between the secondary component (fluid element) and the primary component can be calculated. The expression that relates the growth of the interfacial area between the two components is given as:

$$\frac{A}{A_o} = \gamma \cos \alpha \tag{4.35}$$

where, A_o is the initial interfacial area, A is the current interfacial area, γ is the total strain, and α the angle that defines the orientation of the surface with respect to the direction of flow, i.e., in Fig. 4.36(a) α equals 90 degrees and in Fig. 4.36(b), zero degrees. The higher surface increase implies better mixing. Another approach to look at mixing is to monitor the reduction in the striation thickness [21]. Here, a smaller striation thickness indicates a higher degree of mixing. It turns out that the striation thickness is inversely proportional to the interfacial area increase described earlier. For simple shear flow the striation thickness expression becomes

$$\frac{S}{S_o} = \frac{1}{\sqrt{1+\gamma^2}} \tag{4.36}$$

Figure 4.37 shows a graph that displays the striation thickness as a function of shear. This clearly shows that the striation thickness reduces rapidly during the initial stages of deformation, which indicates that efficient mixing is taking place. However, the rate of the reduction in striation thickness slows down markedly after the first few units of strain. Here, the striation thickness reduces more in the first five units of strain than in the next 1000 units. This decrease in mixing efficiency occurs as the fluid element orients away from the optimal orientation needed for effective area increase. As a result, the mixing efficiency reduces with shear strain. A more efficient mixing environment would be realized if the optimal orientation for mixing were to be maintained. This is possible through the use of mixing sections. In these devices, the fluid element is occasionally repositioned to a more optimal orientation for shear. It is possible to achieve an increase in the interfacial area that is described by Ref. [21,22] if n-1 mixing devices are used:

$$\frac{A}{A_o} = \left(\frac{\gamma}{n}\right)^n \tag{4.37}$$

Here, it is assumed that the mixing section produces a randomly oriented fluid element (secondary component), and the shear strain imposed by the mixing section is insignificant. Figure 4.38 shows the results of this expression. There is a dramatic

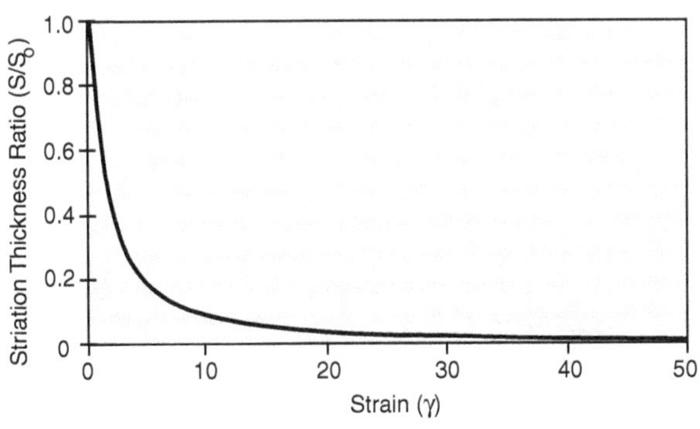

Figure 4.37 Striation thickness versus shear strain.

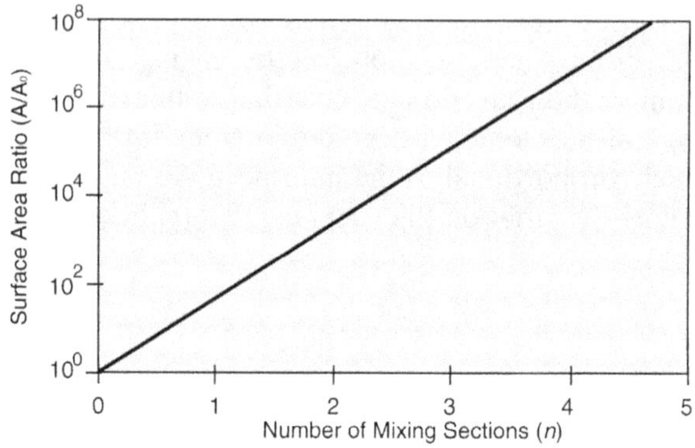

Figure 4.38 Surface area increase of secondary phase with increase in number of mixing sections.

increase in the mixing that takes place as the number of mixing events are increased. This figure clearly demonstrates the significance of using mixing sections during polymer processing.

A maximum mixing capability can be achieved if the surfaces are maintained optimally oriented at tall times. The maximum possible area increase, usually referred to as *Erwin's Ideal Mixer*, is given by

$$A_n = A_o\left(\frac{\gamma_1}{2}\right)^n \tag{4.38}$$

Up to this point only shear flows have been taken into account. It should be noted, however, that if elongational flows are considered, *Erwin's Ideal Mixer* is automatically achieved. In this mixer, the interfacial area increases exponentially. Figure 4.39

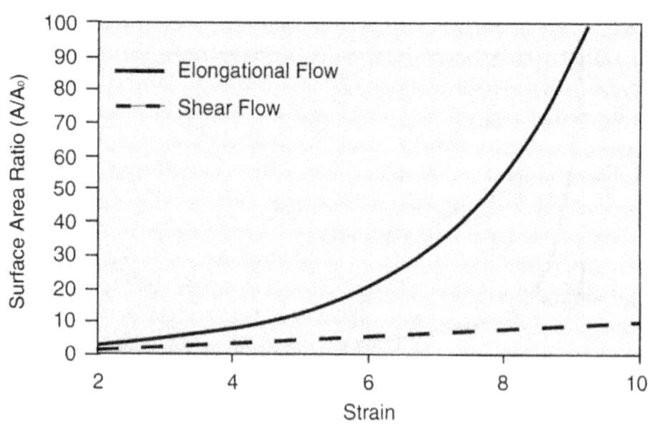

Figure 4.39 Surface area increase of secondary phase in an elongation and shear type flows.

shows the area increase for shear and elongational flows as a function of strain. It is obvious from the figure that elongational flows are more effective than shear flows. As explained earlier, however, it is difficult to generate elongational flows and even more so for a long period of time.

Demixing Creating motion of fluid or solids does not guarantee distributive mixing. In fact, distributive demixing may occur. One such occurrence appears in the channel of the screw. Figure 4.40 illustrates a fluid element that is being deformed by shear due to cross-channel flow. If we follow this fluid element, starting in the lower left root of the screw, it is deformed by a positive shear. As the fluid element moves through the upper region of the channel the fluid element is exposed to a negative shear, causing a deformation in the opposite direction. It should be noted that this demixing effect only appears in the outermost region of the channel. The shear deformation in the inner region of the channel is the same, creating a continuous mixing condition.

Another occurrence of demixing is called *segregation*. With solid particles this can occur when particles with different characteristics preferentially move to certain regions of a mixture. A common place for this to occur is in the hopper or resin silo. Here, as the mixture of resin pellets, additives, and fillers of various sizes sit in the hopper or silo, vibrations and flow of material allow small particles to slip down in-between larger particles. This lets the small particles move downward while large particles rise upward. The mechanism is called percolation and is common in many industries [1]. This commonly occurs when processing highly filled PVC. Here, the filler can vary by several percentages due to segregation in the silo. A common method to reduce segregation is to use a wetting agent (e.g., water), that creates a cohesive bond between particles and thus reduces their movement.

From the preceding discussion, the following should be followed to achieve efficient and effective distributive mixing:

- Reorientation of the flow field
- Elongational flows are preferred over shear flows
- Avoid conditions where segregation may appear

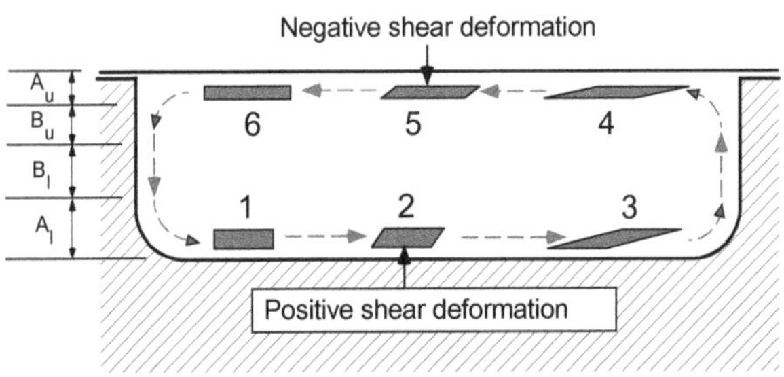

Figure 4.40 Shear deformation by cross-channel flow in different regions of the screw channel.

The most important condition for efficient mixing is frequent reorientation. Adding a mixing section that can reorient the secondary component numerous times will dramatically increase the mixing ability of the system. The optimal flow field is one that creates a sustained or multiple events of elongational flow. This type of flow is also preferred for dispersive mixing, which is discussed next.

Dispersive Mixing

Dispersive mixing entails the break up of liquid or solid agglomerates to produce a finer level of dispersion. Figure 4.41 demonstrates the break up of an agglomerate of solid particles that are held together by a cohesive force. Notice that there is a small amount of distributive mixing that takes place, which can be seen by the small increase in distance between particles. More distributive mixing obviously needs to take place after this initial break up to make a homogeneous mixture. The break up of a secondary liquid (polystyrene) in a continuous phase (polyethylene) is shown in Fig. 4.42 [23]. Here, breakup of the polystyrene threads into droplets takes place after an interfacial force is overcome. Distributive mixing must now take place to completely randomize the droplets.

A size reduction in the secondary phase will decrease the variance in the mixture and the chance of segregation will be lowered. During dispersive mixing a hydrodynamic force is applied to the agglomerate or droplet by the flow of material. Similar to distributive mixing, the type of flow that is created is a major factor in determining how effective the device will be in breaking up the secondary component. Again, elongational flows will be preferred over shear flows.

The break up of an agglomerate of solid particles involves overcoming cohesive forces that hold together the individual particles. The most common example of this is the dispersion of carbon black into a rubber compound. The magnitude of force that can be applied by shear to separate two particles is given by [24]:

$$F_{\text{shear}} = 3\pi\eta_s\dot{\gamma}r^2 \tag{4.39}$$

where η_s is the shear viscosity of the carrier fluid, $\dot{\gamma}$ the magnitude of the strain rate tensor, and r the radii of the particles. If pure elongational flow is generated, the magnitude of the force between two particles is given as:

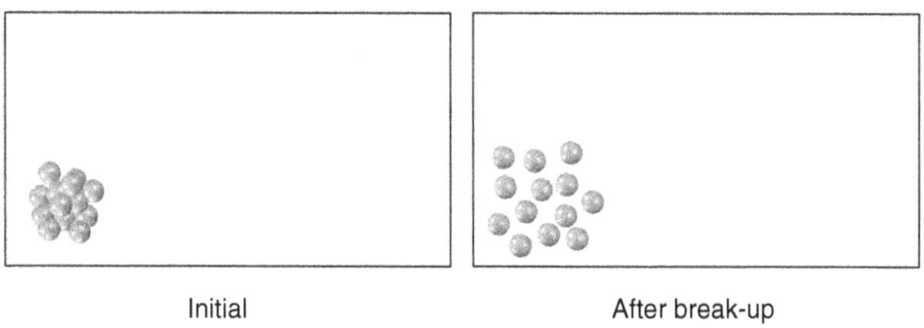

Figure 4.41 Schematic of dispersive mixing.

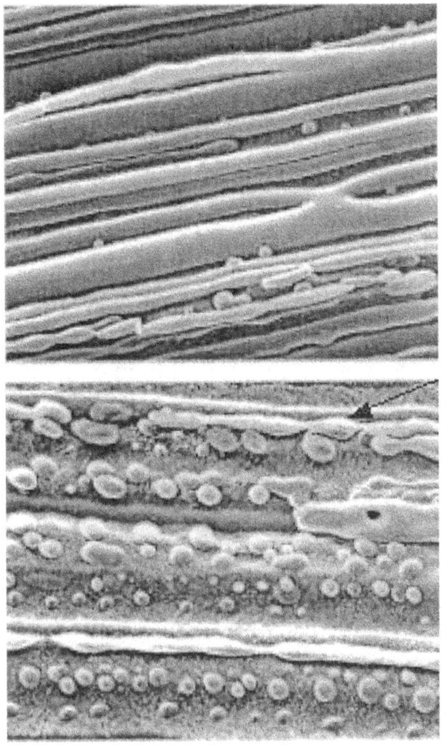

Figure 4.42 Breaking polystyrene threads in a high density polyethylene matrix. (*From*: Mixing and Compounding of Polymers by Ica Manas-Zloczower and Zehev Tadmor, Hanser, p. 136).

$$F_{\text{elong}} = 6\pi\eta_e\dot{\varepsilon}r^2 \tag{4.40}$$

which is twice as large as the maximum force generated by shear. In addition, the elongational viscosity, η_e, is at least three times greater than the shear viscosity and in many cases much more, which means the force by elongation is at least six times higher than by shear. It should also be pointed out that the agglomerates in elongational flows are always oriented in the direction of maximum force generation, whereas in simple shear flow the agglomerate tumbles quickly through the maximum position of force.

Solid particles are not the only ingredients that need to be dispersed; liquid secondary components also need to be broken up. The droplets of the liquid secondary component tend to stay spherical due to surface tension, which tries to maintain the lowest possible surface to volume ratio. As the droplets flow through the mixer a force is applied that causes them to deform. If the force is large enough and applied for long enough the droplet will deform into a thread. The striation thickness, or diameter, of the thread will reduce until it becomes unstable (Rayleigh disturbances [25]) and break up into small droplets will occur. This break up is governed by the viscosity ratio of the droplet and the continuous phase, surface tension, the size of the droplet, and the type and strength of flow [25]. It is well known [26] that when the viscosity ratio becomes higher than 4.0 shear flows are not able to break up secondary liquid components. This, of course, is an important fact when mixing polymers with different viscosities. Elongational flows fortunately do not have this same limit and are able to create dispersive mixing at both high- and low-viscosity ratios. Furthermore, elongational flows are able to break up droplets to a smaller size than shear flows.

Some of the favorable characteristics of a dispersive mixer are:

- Generation of high stress
- Elongational flows preferred over shear flows
- Multiple passes through high stress region
- Similar mixing for all fluid elements—uniform mixing

Economics of Mixing

An important aspect that needs to be considered during a mixing analysis is the economics of mixing. For example, is it more economical to replace an existing screw with a high-performance mixing screw than it is to buy a preblended polymer and use a screw with poor mixing performance. If compounding is done in-house there is more control of the process and materials. This typically makes it easier to control quality and to perform statistical analysis on the process.

Another aspect of the economics of mixing is energy consumption. High energy consumption increases the costs to produce the blend or compound. Furthermore, this energy is typically dissipated into the system as heat, which can lead to polymer degradation and a longer cooling cycle. Again, we can see large differences when we analyze the two dominant flows that produce mixing. The amount of power required for shear and elongational flows can be expressed as [25]:

$$\text{Power}_{\text{shear}} = \frac{4\eta_s}{t_0}\left(\frac{A}{A_0}\right)^2 \tag{4.41}$$

$$\text{Power}_{\text{elong}} = \frac{12\eta_e}{t_0}\left[\ln\left(\frac{5A}{4A_0}\right)\right]^2 \tag{4.42}$$

where, η_s is the shear viscosity, η_e is the elongational viscosity, A is the current interfacial area, A_o the initial interfacial area, and t_o the time for mixing. Figure 4.43 shows the ratio of power consumption during shear mixing to power consumption during elongational mixing. From this figure it is clear that the power needed for shear is always greater than that needed for elongation. When the interfacial area is greater than 1.0, the power required for shear becomes dramatically more than that for elongation. Again, this directly relates to increase in cost to manufacture a product due to higher energy costs, more waste due to material degradation, and a longer cycle time as a result of higher temperatures.

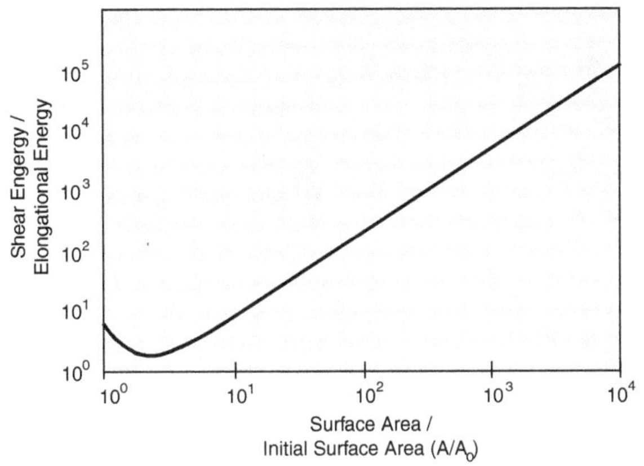

Figure 4.43 Ratio of shear energy to elongation energy versus area increase of secondary phase.

Coalescence

The final morphology after mixing is a result of the competing processes of break up and coarsening of dispersed particles and droplets. Similar to the situation found with distributive mixing, demixing is possible with dispersive mixing where solid particles join to form a large agglomerate and liquid droplets coalesce to form a large droplet. When particles join together it is called *flocculation*, and when droplets come together to form a larger droplet it is called *coalescence.*

Coalescence occurs in nearly every mixing process, even with only a few percent of the dispersed phase. The process of coalescence entails the following:

1. Approach of two or more droplets toward one another
2. Drainage of fluid between them
3. Rupture of the film between the droplets
4. Combining of both droplets to from one larger droplet

This coalescence process is shown in Fig. 4.44 [27]. For this complete process to occur several important aspects of flow must be present:

- High probability that two or more droplets will collide
- Enough force to drain and rupture the film between the droplets
- Enough time for the coalescence process to transpire.

Stagnant regions in a system can satisfy these conditions. By definition, this is a slow moving flow, which gives a relatively long period of interaction time. A simple example of a stagnant region is shown in Fig. 4.45(a). A system or mixer that is not

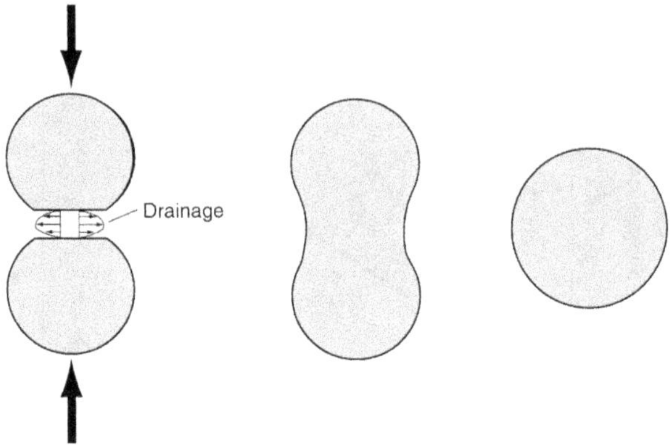

Figure 4.44 Schematic of coalescence of two droplets [26].

streamlined will incur stagnant regions. Even with a great amount of mixing, coalescence can reverse the mixing generated and create an unacceptable product.

Two straightforward methods are possible to reduce coalescence. The first is to insure a streamlined system, including the screw, mixing sections, shutoff valves, nozzles, and gating, among others. Figure 4.45(b) shows the streamline alternative to Fig. 4.45(a). The second method is to use a compatiblizer to alter the final morphology of the mixture by controlling the sizes of the domains of a multiphase blend. There are two mechanisms present that control this: the interfacial tension between phases is reduced and agglomeration of the domain is reduced [28]. Figure 4.46 shows how the use of compatibilizers can alter the domain size of the blend. The graph shows the domain size of a TPU/PE blend (PE is the dispersed phase) with and without a compatibilizer for an increasing dispersed phase content. Two interesting phenomena are shown here. The first is that the domain size of the PE is increasing as its content increases—higher probability for collision. Second, the domain size is significantly reduced with the use of a compatibilizer. Another important benefit of compatibilizers is the increase in adhesion between phases and the consequential

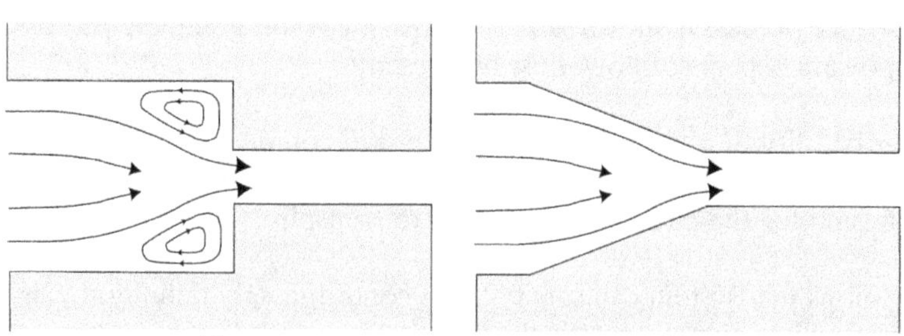

Figure 4.45 (a) Nonstreamlined system with stagnant regions. (b) Streamlined system.

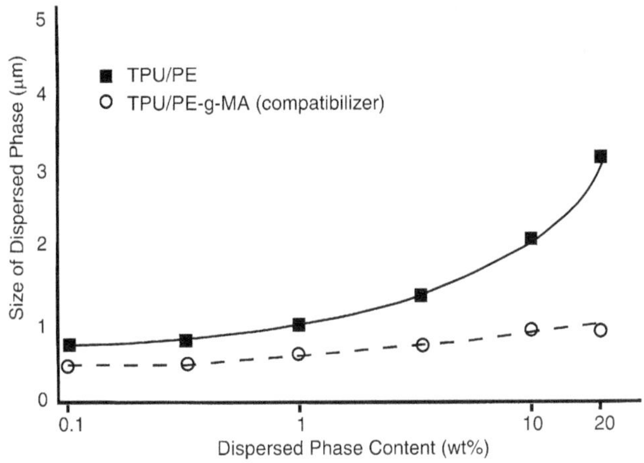

Figure 4.46 Increase in the size of dispersed phase as amount of dispersed phase is increased.

increase in strength of the blend [28]. Figure 4.47(a,b) shows the fracture surface of two TPU/PE samples with and without a compatibilizer, respectively. The fracture surface of the uncompatibilized blend occurs at the interface of the PE and TPU, indicating a poor adhesion between the two. The fracture surface of the TPU/PE blend with a compatibilizer is not at the surface of the PE, but through the PE signifying good adhesion between the TPU and PE.

Mixing Sections

The flow in the screw of the injection molding machine creates some mixing by the cross-flow components that were discussed earlier. The best method to improve

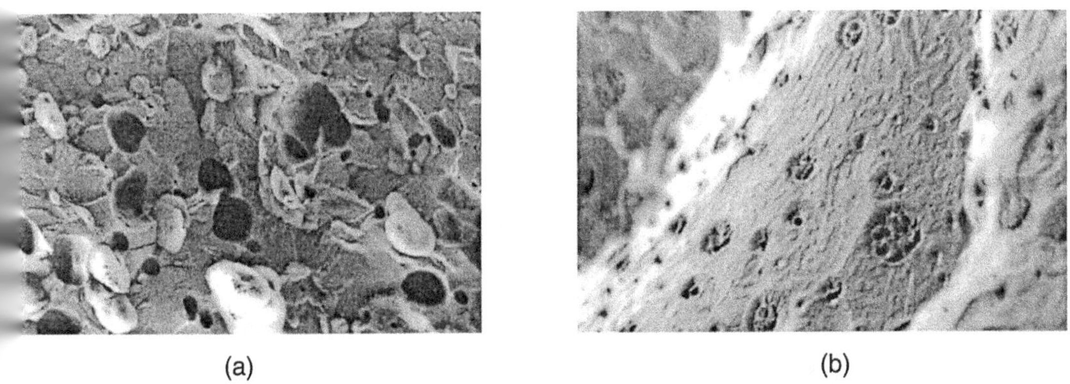

Figure 4.47 SEM micrographs of fracture surfaces. (a) 80% TPU, 20% PE with fracture at surface of PE (b) 80% TPU, 20% PE-g-MA (compatibilizer) with fracture through PE [28].

mixing is to incorporate mixing sections to the screw. These mixing sections are useful both in reducing the domain size of the secondary (dispersed) phase and distribute it randomly throughout, as well as in reducing large variations in melt temperature before the polymer enters the mold. Mixing sections should have the following characteristics to make them attractive for injection molding [29]:

- Minimum pressure consumption; preferably, a forward pumping capability
- Streamlined flow
- Barrel surface should be wiped by mixer to produce a self-cleaning effect
- Easy to assemble, install, run, clean and disassemble
- Easy to manufacture and reasonably priced

As stated earlier, critical for distributive mixing is reorientation of the fluid elements and significant strain. Some of the common distributive mixers used are the rhomboid mixer, slotted flight mixer, variable depth mixer, and Twente mixing ring.

The rhomboid mixer is manufactured in various shapes and sizes by controlling the two helical angles of the grooves that make up the teeth. The most popular rhomboid mixer is the pineapple mixer, as shown in Fig. 4.48. The angle of the two helical cuts are identical, but opposite in direction, giving a symmetric diamond tooth shape. The teeth of the rhomboid mixer create splitting and reorientation and, as a result, create very good distributive mixing. These devices are self-wiping systems where the barrel is wiped by the mixer teeth. One drawback to these mixers is that they are typically pressure consuming devices. There is some flexibility in controlling pressure consumption by varying the angle of the helical cut. Another drawback is that a stagnant region can appear on the trailing side and root of the rhomboid tooth [30].

As the name implies, the slotted flight mixer has open slots in the screw flight. In fact, the most basic method to create this type of mixer is to simply machine slots into the flight of a standard screw. A commonly used mixer, the Saxton mixer [31], is shown in Fig. 4.49. This mixer provides frequent splitting and reorientation, resulting in effective distributive mixing, while also providing a self-wiping effect. It also has a benefit of forward pumping due the orientation of the flights.

Figure 4.48 Pineapple mixing section.

Figure 4.49 Saxton mixing section.

The variable depth mixer is typically a multiflight section where the channel depths vary periodically out of phase with one another (Fig. 4.50). This causes the material to move from one channel to the other. In this mixer the splitting and reorientation mechanisms are not strong, making the distributive mixing of this device moderate.

The Twente Mixing Ring has proven to be an excellent distributive mixer for the extrusion industry and is now being used successfully in many injection molding applications, especially when online compounding is being performed. This device consists of a screw section that has hemispherical cavities and a free-floating ring with holes (Fig. 4.51). As the polymer melt travels through the mixer it must travel from the ring to the cavity and back to the ring, continuously. This action provides for excellent splitting and reorientation. The drawback to this device is that there is no forward pumping capability; thus, it is a pressure-consuming mixing device. It is also not well streamlined, which can cause difficulties with product change over and with thermally sensitive materials.

To a certain degree, some dispersive mixing can be generated in a distributive mixer, and vice versa. When break-up is critical for an acceptable blend, however, a mixer designed specifically for dispersive mixing should be used.

The most basic dispersive mixer is the blister ring, as shown in Fig. 4.52. This device is an extra shoulder added to the screw that causes the material to experience

Figure 4.50 Channels of a wave-type screw.

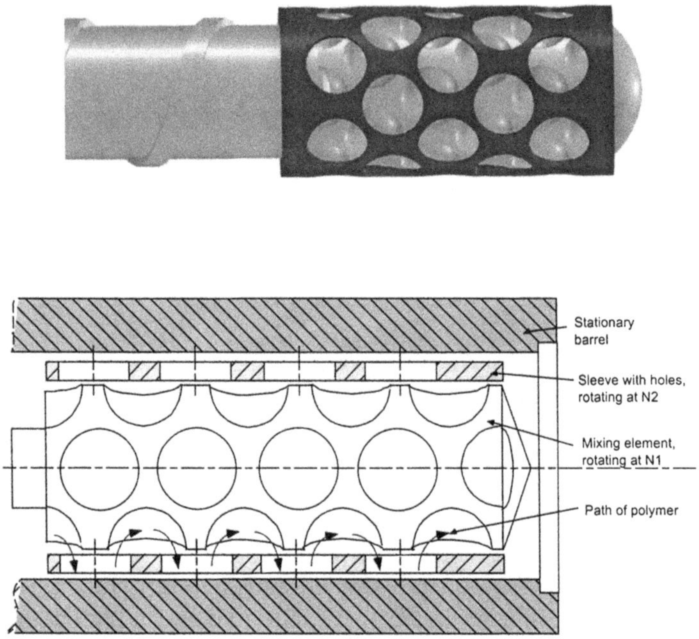

Figure 4.51 Twente mixing section.

a high shear flow. The benefits of this device are that it is relatively easy to incorporate, the number of high stress events are easily dictated by the number of rings used, and the residence time of the polymer under high stress is easily controlled by the length of the mixer. This is a high pressure consuming device and care must be taken to avoid degradation of the polymer.

The most common dispersive mixer is probably the fluted mixing section. This device has inlet channels and outlet channels that are connected by a small land region, commonly called the *barrier flight*. As the polymer flows from the inlet to outlet channel, over the barrier flight, a high shear event is encountered. This is where the dispersive mixing takes place. In addition, cross-flow takes place as the polymer flows in the channels, generating some distributive mixing. One of the benefits of this

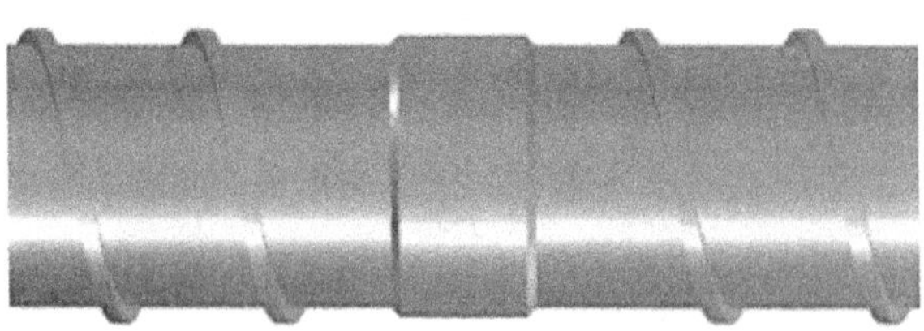

Figure 4.52 Blister ring.

Figure 4.53 Maddock/LeRoy mixing section.

mixer is that it is difficult for large particles to make their way past the mixer due to the small gap between the barrier flight and barrel. In addition, all material will experience one high dispersive mixing event. The most popular fluted mixing section is the LeRoy or Maddock mixing section [32], as shown in Fig. 4.53. Because the flights or channel walls, as well as the barrier flight, of the LeRoy are parallel to the axis of the screw, 90-degree helix angle, the device is highly pressure consuming. This high pressure consumption can increase the melt temperature and residence times, along with the possibility of material degradation. The helix angle can be modified to create a neutral or pressure-generating device, as shown in Fig. 4.54. For most polymers, the optimum helix angle is around 45 degrees [29]. A disadvantages to most fluted mixers is that only one dispersive mixing event takes place and that areas of stagnation may exist.

From the preceding discussion, the optimal mixing device is one that generates distributive mixing with many chances for splitting and reorientation, as well as many dispersive mixing events, preferably using elongational flows. The device should also be self-cleaning, easy to incorporate into an existing system, and economical to make. A couple of mixing devices that attempt to incorporate all these features are the elongational pin mixer (EPM) and the CRD mixing section [33,34] (Figs. 4.55 and 4.56).

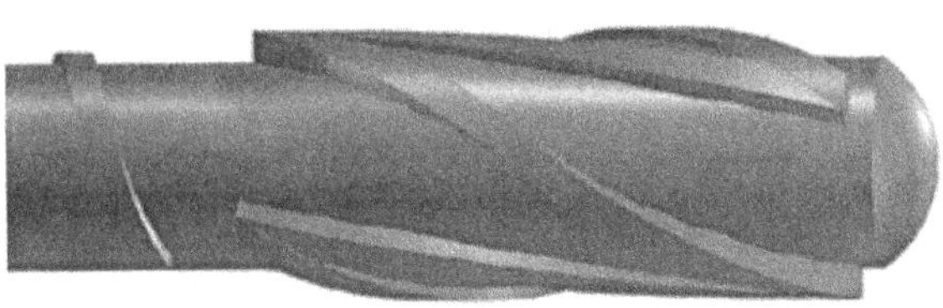

Figure 4.54 Z-shaped fluted mixer (Zorro mixer).

Figure 4.55 Elongational pin mixing section.

Figure 4.56 CRD mixing section.

An interesting aspect to these mixer is that they were designed using numerical methods. Here, the boundary element method [35] was used to optimize the shape and configuration of the wedges and slots to insure excellent dispersive and distributive mixing. These devices incorporate wedge-shaped geometries that create an elongational flow. The EPM does this by placing diamond-shaped pins circumferentially around the screw, ensuring that all material experiences the dispersive mixing event. This is a relatively simple device to incorporate with the option of using multiple rings in series to increase the number of dispersive mixing events. This device is also a good distributive mixer that acts similarly to the pineapple mixer, while also creating good backmixing [36]. The EPM is a pressure-consuming mixer; however, the wedge-shaped pins can be placed in a helical fashion around the screw to create a less pressure consuming device.

The CRD is a mixing section [37] that looks like a multiflighted screw; however, there are distinct characteristics that make it an efficient dispersive and distributive mixer. The mixing flights of the CRD are made to have a large clearance so that the polymer can back flow over them. This increases distributive mixing by increasing the backmixing [36]. In addition, the pushing flight flank is slanted or curved so that a strong elongational flow is produced as the polymer flows over the flight, as shown Fig. 4.57. To insure that self-wiping is maintained, some flights with little clearance are positioned along the mixer. To further increase the dispersive and distributive mixing ability of the CRD tapered slots are machined into the flight. This allows for more distributive mixing while increasing the dispersive effect by incorporating wedge-shaped slots that create elongational flow. Because this system is basically a multiflighted screw with special additional features, it has forward pumping capabil-

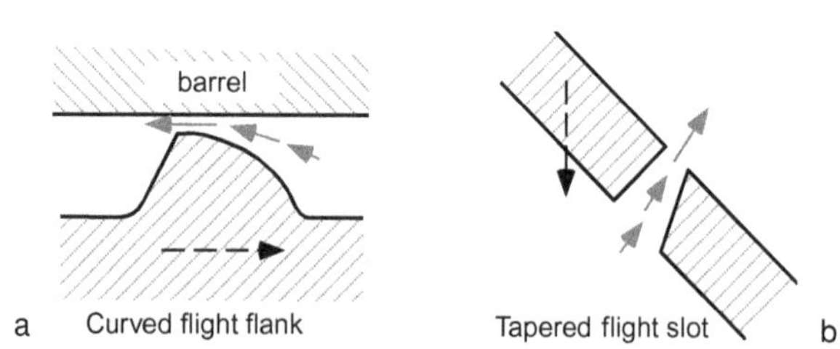

Figure 4.57 Elongational flow generation in the CRD mixing section. (a) Flow of the modified flight. (b) Flow through the tapered slot.

Figure 4.58 Combination of the CRD and elongational pin mixing sections.

ity, is easy to incorporate into an existing system, and relatively inexpensive. If more backmixing [36] is needed the helix of the CRD flights can be reversed and an EPM can be added (Fig. 4.58).

Another option available to the injection molder is to incorporate a static mixing device between the plasticating unit and the mold. Putting a mixing element here ensures mixing at nearly the latest point in the process. As the name implies, the static mixer does not move; the mixing action is created as the polymer flows through the device. The static mixer has the advantages of being easy to incorporate, relatively inexpensive, and their modularity provides flexibility in specifying the amount of mixing to take place. Because the devices are not dynamic they are pressure consuming and care must be taken that stagnation zones do not exist. Static mixers of varying geometry are available from a number of different suppliers [4].

Compounding machines that use barrel pins intermeshing with the extruder screw provide very efficient distributive mixing, for example, the Buss kneader. The high cost of these machines has limited commercial applications. REE, Inc., has developed a floating sleeve mixing device that achieves highly effective distributive mixing (38, 39). This patented (40) mixer uses pins attached to a floating sleeve. The pins intermesh with a slotted screw to create a vortex-like mixing action, hence, the name vortex intermeshing pin (VIP) mixer.

Figure 4.59 shows one configuration of the VIP mixer. It consists of a floating sleeve similar to the Twente mixing ring (41, 42). The sleeve has internal pins that fully intermesh with the screw. The screw flights have circumferential slots that allow the pins to travel through the slots. The pins reach almost to the root of the screw. As a result, there is an effective intermeshing action.

The vortex intermeshing pin mixer can be used in existing extruders and injection molding machines without barrel modification. The low cost of the VIP mixer allows it to be used in many applications. In injection molding the VIP mixer can be incorporated into the non-return valve (NRV). Injection molders can significantly improve the mixing capability of their injection molding machines simply by replacing the standard NRV with a VIP NRV. The VIP mixing technology has been licensed to Sulzer Chemtech of Winterthur, Switzerland. The VIP mixer has been commercially available since 2004.

One advantage of the VIP mixer is that it achieves a full intermeshing action over the full circumference of the screw. In a twin screw extruder, the intermeshing action of pin mixing elements only occurs in a small fraction of the extruder, as shown in

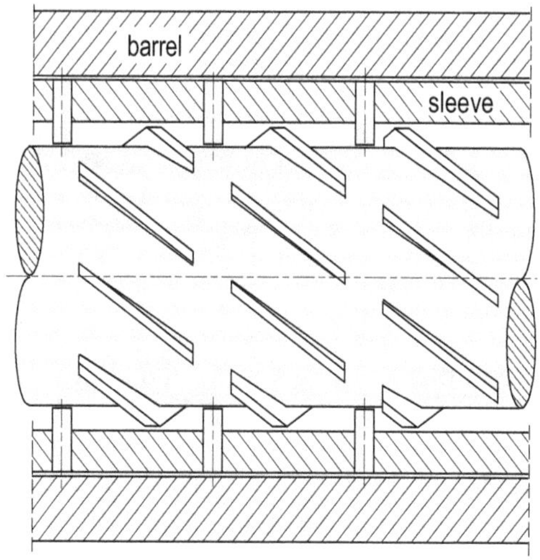

Figure 4.59 VIP mixer with slotted screw flights

Fig. 4.60. As a result, the VIP mixer in a single screw extruder achieves more effective distributive mixing than a similar device in a twin screw extruder. Thus, the distributive mixing action of the VIP mixer is significantly better than that of a twin screw extruder using pin mixing elements.

The intermeshing action of the pins achieves effective reorientation of the interfaces resulting in efficient distributive mixing. The VIP mixer can achieve good mixing over short axial distances. This makes this mixer particularly suitable for applications where limited space is available to incorporate mixing devices. A good example is the plasticating unit of most injection molding machines. The plasticating unit is generally a reciprocating single screw extruder (RSSE) as discussed in Section 4.1.2. The screw in a RSSE is usually short (20D to 25D) with little space available to incorporate mixing elements.

As a result, one of the most common applications of the VIP mixer is the VIP NRV. This non-return valve allows for efficient mixing in the NRV without adding

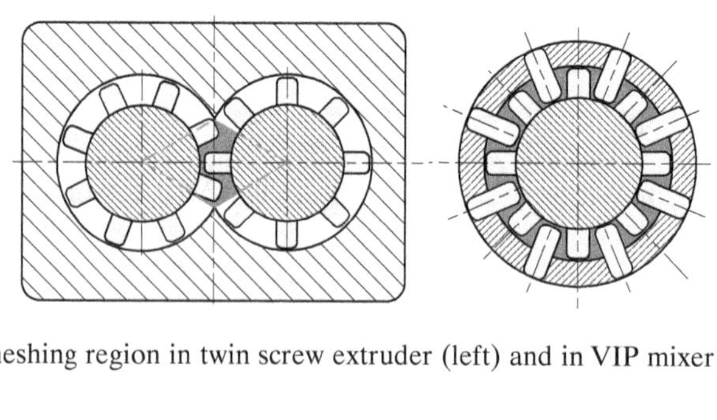

Figure 4.60 Intermeshing region in twin screw extruder (left) and in VIP mixer (right)

Figure 4.61 VIP non-return valve

length to the screw. In order to achieve efficient mixing, the pins of the sleeve intermesh with pins on the rotor. This allows a maximum number of reorientation events over a short distance. Figure 4.61 shows an example of a VIP NRV. The male part of the valve has a threaded connection to the injection molding screw.

The VIP mixer has proven itself in very demanding applications. In one application, a low viscosity liquid color was injected in a high viscosity fluid; the viscosity difference was about five orders of magnitude (100,000:1). Such extreme viscosity differences make mixing very difficult and conventional mixing devices (e.g., twin screw extruders) were unable to achieve acceptable mixing. A VIP mixer with a length of only 1D was able to achieve good mixing with the colorant injected just before the mixer. The mixer has also been successful in reactive extrusion applications where the viscosity differences ranged from five to six orders of magnitude.

4.3 Conclusion

The quality of the final molded part is strongly influenced by the processes taking place in the plasticating unit of the injection molding machine. The basic functions of the plasticating unit are to accept and convey solid pellets and additives, perform melting, convey the melt, devolatilize gases, achieve a level of mixing, and generate pressure. Maintaining consistency with each of these functions from cycle to cycle is of extreme importance.

This chapter focused on what takes place in the plasticating unit of the injection molding machine by introducing its main parts, describing how they work, showing the importance of proper screw design, and introducing problems that may arise and some possible solutions. Models to predict the solids conveying, melting, and melt conveying were also given. The importance of mixing was discussed with several examples of common mixing sections used to help create a more homogenous product.

With the information from this chapter it is hoped that reader gained more insight to what transpires in the plasticating unit, what problems may arise, options that may be taken to solve these problems, and how to assure that they will not appear in the future.

References

1. Rauwendaal, C., Ed., *Mixing in Polymer Processing*, Marcel Dekker (1991), New York.
2. Rauwendaal, C., *Plastics World*, (1990) November, 45–49.
3. Schwittay, D., *Spritzgießschnecken für die Thermoplastverarbeitung*, Company Literature Bayer AG (1978).
4. Rauwendaal, C., *Polymer Extrusion*, 4th ed. Hanser Publishers (2001), Munich.
5. Peischl, G. C., Bruker, I., *Polym. Eng. Sci.* (1989), 29, 202.
6. Fraser, K., Coyle, D. J., Bruker, I., "Evaluation of an Injection Molding Mixing Screw," SPE ANTEC (1989), Dallas, TX, 214–219.
7. Rauwendaal, C., *Plastics World* (1991), 43–47.
8. Spalding, M., Dooley, J., Hyun, K. S., "The Effect of Flight Radii Size on the Performance of Single Screw Extruders," SPE-ANTEC (1999).
9. Rauwendaal, C., *International Polymer Processing VII* (1992), 1, 26–31.
10. Leonard, L., *Plastics Machinery and Equipment* (1990), October, 43–44.
11. Rauwendaal, C., *Polym. Eng. Sci.* (1986), 6, 1245.
12. Miller, B., *Plastics World* (1982), March, 34–38.
13. Rauwendaal, C., Gramann, P. J., "Non-Return Valve with Distributive and Dispersive Mixing Capability," SPE-ANTEC, Orlando, FL. (2000).
14. Darnell, Mol, *SPE J.* (1956), 12, 20.
15. Rauwendaal, C., *Intern. Polym. Processing*, VII (1992), 1, 26.
16. Maddock, B. H., *SPE J.* (1967), July, 23–29.
17. Rauwendaal, C., *Adv. Polym. Tech.* (1996), 15, (2), 135.
18. Tadmor, Z., Cogos, C., *Principles of Polymer Processing*, John Wiley & Sons (1979), New York, p. 279.
19. Tadmor, Z., *Polym. Eng. Sci.* (1966), 6, 185.
20. Mohr, W. D., Saxton, R. L., Jepson, C. H., *Ind. Eng. Chem* (1957), 49, 1855.
21. Erwin, L., Polym. *Eng. & Sci.* (1978), 18, 572.
22. Erwin, L., In: *Mixing in Polymer Processing*. Rauwendaal, C. (Ed.) Marcel Dekker (1991), New York.
23. Meijer, H. E., Janssen, J. M. H., "Mixing of Immiscible Liquids," In: *Mixing and Compounding of Polymers* Manas-Zloczower, I., Tadmor, Z. (Eds.) Hanser (1994), Munich, p. 136.
24. Tadmor, Z., Klein, I., *Engineering Principles of Plasticating Extrusion*, Van Nostrand Reinhold Company (1970), New York.
25. Osswald, T. A., Menges, G., *Materials Science of Polymers for Engineers*, Hanser (1995), New York.
26. Grace, H. P., *Chem. Eng. Commun.* (1982), 14, 225.
27. Jenssen, J. M. H., Ph.D. Thesis, Eindhoven University of Technology, (1993), The Netherlands.
28. Datta, S., Lohse, D. J., *Polymeric Compatibilizers*, Hanser (1996), Munich.
29. Rauwendaal, C., *Understanding Extrusion*, Hanser (1998), Munich.

30. Gramann, P. J., Ph.D. Thesis, University of Wisconsin-Madison (1995), Madison, WI.
31. Saxton, R. L., U.S. Patent 3,006,029.
32. LeRoy, G., U.S. Patent 3,486,192.
33. Rauwendaal, C. J., Gramann, P. J., Davis, B. A., Osswald, T. A., U.S. Patent 6,136,246.
34. Rauwendaal, C. J., U.S. Patent 5,932,159.
35. BEMflow, Fluid Dynamics Simulation Program, The Madison Group: Polymer Processing Research Corporation.
36. Rauwendaal, C., Gramann, P. J., Backmixing in Screw Extruders, SPE ANTEC, Orlando, FL (2000).
37. Rauwendaal, C. J., Osswald, T. A., Gramann, P. J., Davis, B. A., "New Dispersive Mixers for Single Screw Extruders," SPE ANTEC, 277–283 (1998).
38. M. Scheuber, R. Maurer, and C. Rauwendaal, "New Intermeshing Pin Mixer for Injection Molding," Technical Papers of Annual Technical Conference of the Society of Plastics Engineers (2003).
39. R. Maurer, M. Scheuber, and C. Rauwendaal, "New Intermeshing Pin Mixer for Extrusion," Technical Papers of Annual Technical Conference of the Society of Plastics Engineers (2003).
40. C. Rauwendaal, "Intermeshing Element Mixer," US Patent 6,709,147 B1 (2004).
41. G. Semmekrot, US Patent 5,013,233 (1991).
42. A. J. Ingen Housz and S.A. Norden, Int. Polym. Process., 10, 120 (1995).

5 Clamping Unit

R. Farrell

Many of the processes that shape plastics require molds that open and close and are held closed with considerable force. Examples of these processes are injection molding, compression/transfer molding, and blow molding. Here, we will focus on injection molding.

The clamping process usually occurs in two phases. First is the traversing phase, in which the mold is opened or closed. This is usually done at a relatively fast speed, but at a low force. Second is the holding phase, in which very little if any motion occurs, but where high forces are usually involved. During this phase plastic melt is injected into the closed mold at considerable pressure. After the part has cooled (cured), the traversing phase is entered again and the mold is opened and the part ejected.

Because of the high cyclic stresses experienced by the various elements of the clamp, one of the major modes of failure in clamping systems is metal fatigue. For this reason, a review of metal fatigue is given in the following section.

5.1 Metal Fatigue and Its Importance in Clamp Design

5.1.1 Importance in Clamp Design

Why should a "plastics engineer" need to know about *metal* fatigue? The answer is that the tooling and machinery used to process plastics are subjected to repeated loading. In the case of injection molding, the molds and machines may cycle millions of times to complete a production run. Metal fatigue results from cyclic loading.

The plastics engineer may be involved in the design of machines. If the machine is not properly designed then it will fail. The insidious nature of metal fatigue is that the failure may not occur for months or years. By that time hundreds or thousands of the improperly designed machines may have been manufactured and sold. Correcting the design flaw now becomes *very* expensive.

The plastics engineer may be involved in plastics processing. It is important to know that operating the equipment above its intended loading may cause a fatigue failure. If the process engineer unwittingly modifies the equipment, a stress riser may also be introduced that will result in a fatigue failure.

5.1.2 A Brief History of Metal Fatigue

Metal fatigue has been under study for more than 150 years [1]. Cannons that had fired successfully for thousands of rounds suddenly exploded, causing great consternation amongst early cannoneers. Fragments of the cannons showed evidence of "crystallizing," which is why early investigators thought these failures were a result of a metallurgical change with time. The most intense and significant research in metal fatigue was conducted by the railroad industry. As the railroads began their rapid expansions worldwide in the mid-to-late nineteenth century, failures of the axles caused many fatal accidents and prompted a more intense study of metal fatigue. August Wohler, chief locomotive engineer of the Royal Lower Silesian Railways in Germany, was one of the more famous investigators of that time, and his work led to the establishment of the endurance limit concept for design. A classic paper entitled "The Fracture of Metals Under Repeated Alterations of Stress" was published by Ewing and Humphrey in 1903. Jenkins established the importance of cyclic deformation in 1923. During that period, his student, Griffith, wrote his classic work on fracture mechanics. In 1927, Moore and Kommers published their book, *The Fatigue of Metals*. Shortly thereafter, an ASTM Committee on Fatigue Research was formed. The SAE Committee on Fatigue Design and Evaluation was also formed during this period. Much of the research centered on the effects of various factors associated with endurance strength during the 1930s and 1940s. During the 1950s Coffin and Manson studied the effects of cumulative damage where the stress amplitude may very from cycle to cycle. Irwin, Paris and others developed fracture mechanics as a practical tool during the 1960s, and the two fields of study began to merge. By the 1970s both fields were established as engineering tools. Since that time, the effort has been to get the knowledge from the minds of the "experts" into the hands of the designers and engineers.

5.1.3 The Three Phases of Metal Fatigue

When a static load is applied to a part, the yield or tensile strength of the material determines if the part will fail. If the part is loaded repeatedly, then the stress at which the part will fail is much lower and is perhaps only one forth the stress required for static rupture. To understand why the allowable repeated stress is so much lower one must understand fracture mechanics. A flaw will form a crack and grow if the stress intensity at the notch of the flaw reaches a critical level for a critical size flaw. Flaws can be a result of an inclusion, a void, or a slip in the lattice structure. Once a critical flaw exists within a critical stress field, the crack will grow and transfer energy each time a load is applied. The rate of crack propagation will depend upon the stress magnitude and on the specific fatigue properties of the material. As the load cycling continues, the area remaining to carry the load will be reduced as the crack grows. As the area become smaller, the stress intensity increases, and so does the rate of crack growth. During the final load application the stress intensity exceeds the critical stress for controlled crack growth, and the unstable crack growth results in complete and dramatic brittle failure. An examination of a part that has failed in fatigue exhibits a "crystalline" appearance over the final rupture surface and a smoother surface over the area that experienced the repeated crack growth. "Beach-marks" can frequently be seen representing each cycle of crack growth and forming a radial pattern from the originating site of the crack. Thus, fatigue failures have three phases; crack initiation, crack growth, and final rupture.

5.1.4 Determination of Design Stress for Metal Fatigue

Definitions

At this point some definitions are in order. All stresses are engineering stresses based upon the original cross sectional area. Tension stresses are assigned a positive sign and compressive stresses are assigned a negative sign. Mean stress (S_m) is equal to half the sum of the maximum plus the minimum stress. Alternating stress (S_a) is the absolute value of half the difference between the maximum minus the minimum stress. The stress range (S_r) is the absolute value of the difference between the maximum and the minimum stress. Ultimate tensile strength (S_u) is the maximum load sustained by a specimen divided by the original cross-sectional area. Fracture strength (S_{fr}) is the maximum load divided by the actual area at fracture. The reduction in area (RA) along with load can be used to calculate the fracture strength value. Fatigue strength (S_e) is the maximum alternating stress a material will endure without failure. *S–N* curves are plots of fatigue strength verses the log of cycles (N), as shown in Fig. 5.1. For ferrous metals, the *S–N* curve usually levels at between 2 and 10 million cycles without failure. It is generally agreed that if a part survives 10 million cycles at a particular stress level, then it will never fail at that stress level. For nonferrous metals, such as aluminum, the *S–N* curve does not level at 10 million cycles, but continues downward, so both the fatigue strength and the cycles to failure must be reported.

Scope of this Discussion

This discussion will be limited to the fatigue of steel in the 60,000 to 180,000 psi tensile strength range because most machine parts are constructed from steels that have strengths in this range. In addition, low cycle fatigue (less than 100,000 cycles) will not be discussed. All of the references cited at the end of this chapter have good discussions on low cycle fatigue as well as nonferrous metals. Although screws and barrels on injection molding machines usually operate between 350 and 650°F, most components of the clamping system operate at or near room temperature. It will only be noted here that for elements that may be exposed to high temperatures, the tensile strength of steel usually peaks at about 10% greater than its room temperature strength when a temperature of about 500°F is reached.

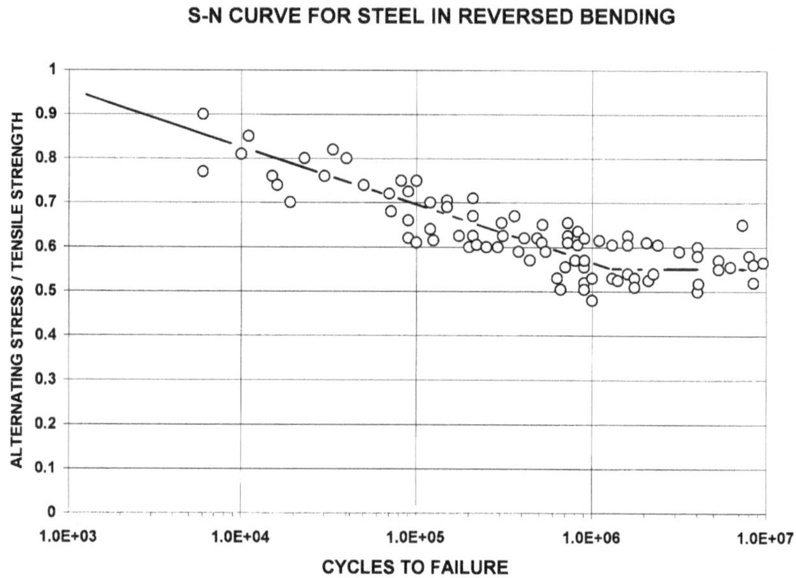

Figure 5.1 *S–N* curve for steel showing scatter of data.

Obtaining Fatigue Data

Fatigue data for many metals is available in the literature and should always be used whenever possible. Published fatigue data is gathered by at least two different methods and it is important to know how the data being used was gathered. One of the most frequently used methods is the rotating beam tester shown in Fig. 5.2. This test applies a bending load to a rotating beam specimen using a four-point loading to apply a constant moment (bending stress) to a specimen. The specimen is usually rotating at 1750 rpm. As the specimen rotates, each point on the surface goes through equal alternating tension and compression stresses as shown in Fig. 5.3. The stress at the surface is calculated using Eq. (5.1), which is known as the simple beam equation.

$$S = Mc/I \tag{5.1}$$

where S = tensile stress (psi)
M = bending moment (lb.—in)
c = distance from neutral axil to extreme fiber (in)
I = moment of inertia (in^4)

Because of the very low hysteresis in steel, the rate of cycling has little effect because very little heat is generated, even at speeds as high as 10,000 rpm [2]. This is not true for more pliable materials such as plastics, which must be tested at much slower rates to minimize heat generation.

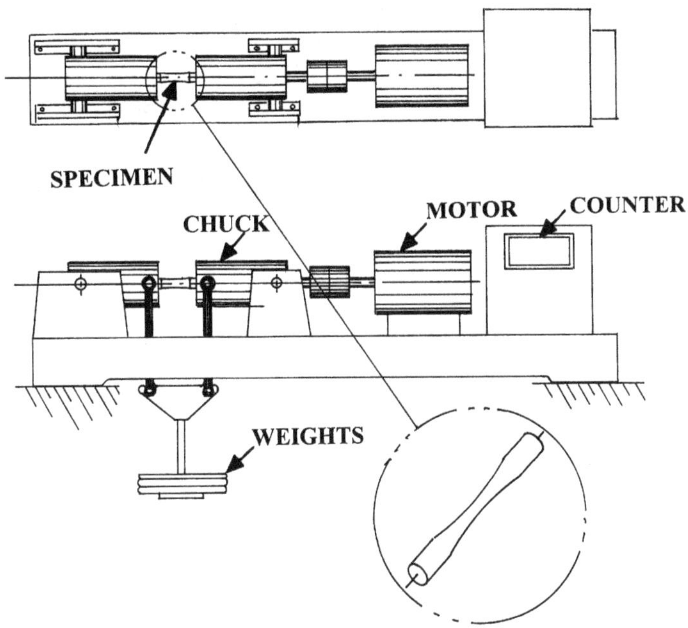

Figure 5.2 Rotating beam fatigue testing machine and specimen.

There is good correlation between ultimate tensile strength and fatigue strength for steel in the 60,000 to 180,000 psi range. The fatigue strength in bending is usually reported at between 40 and 60% of the tensile strength, with 50% often used for approximations when published data are not available. The same correlation exists

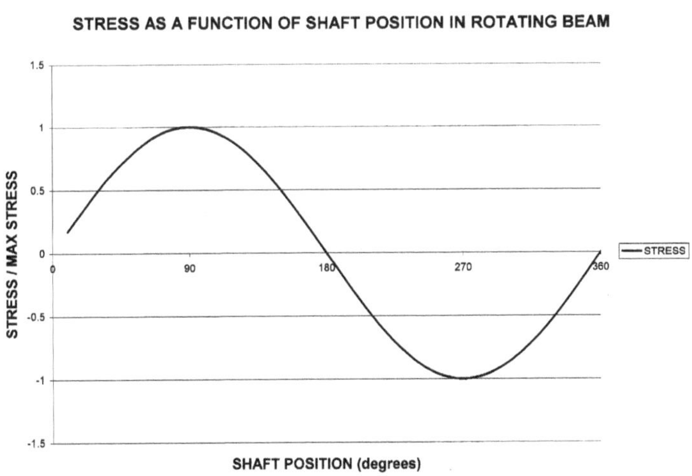

Figure 5.3 Stress history of a point on a rotating beam specimen as it rotates from tension when at the bottom to compression stress when at the top.

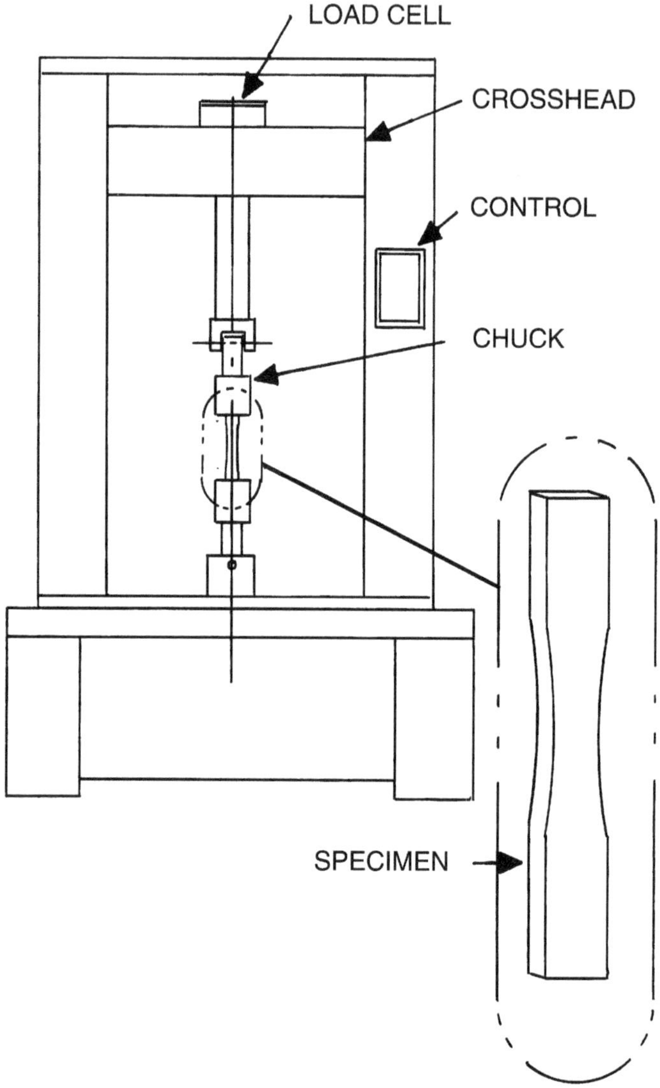

Figure 5.4 Uniaxial fatigue testing machine with the ability to cycle the load from tension–tension, to tension–compression, to compression–compression, thus allowing for testing the effect of mean stress.

between Brunelle hardness number (BHN) and fatigue strength, with the latter being 250 times the BHN. This rule of thumb is useful because the Brinell hardness of a part can be measured nondestructively.

A second method for gathering fatigue data is shown in Fig. 5.4. The cross-head can be cycled between two load or strain ranges and loads the specimen in uniaxial tension and/or compression. With this method the magnitude of tension and compression do not have to be equal, as shown in Fig. 5.5.

It is important to know that the specimens are usually between 0.25 and 0.30 in in diameter at the critical point, and that they have a large radial taper from the grip diameter down to the test diameter, so a stress riser is not introduced. The surface of the specimen is very carefully ground and then polished to a mirror finish. Great care is taken to avoid generating heat at the surface because this is likely to affect the surface properties.

Published Fatigue Data Must be Adjusted

When using published fatigue strength data there are five adjustments that must be made to the fatigue strength in order to determine a proper design stress. It is important to note that the upcoming factors are only to be applied to fatigue data gathered from rotating beam testing, which is the most common method. For information on data from other test methods, please refer to Refs. [1 to 5].

Loading Factor (LF) The adjustment for loading factor (LF) is to correct for the type of load the part will experience in service [2]. When a specimen is tested in a rotating beam tester, the surface at any point around the test diameter is exposed to equal alternating tension and compress stresses. The maximum stress (which results in failure) occurs at the small volume of material at the surface. In bending, the stress goes from a maximum value at the surface to zero at the center of the bar. A "flaw" of a critical size must exist for the bar to fail at a given stress level. The larger the volume of material exposed to a given stress level, the greater the probability that such a flaw will exist. For a given material, a volume exists such that the proba-

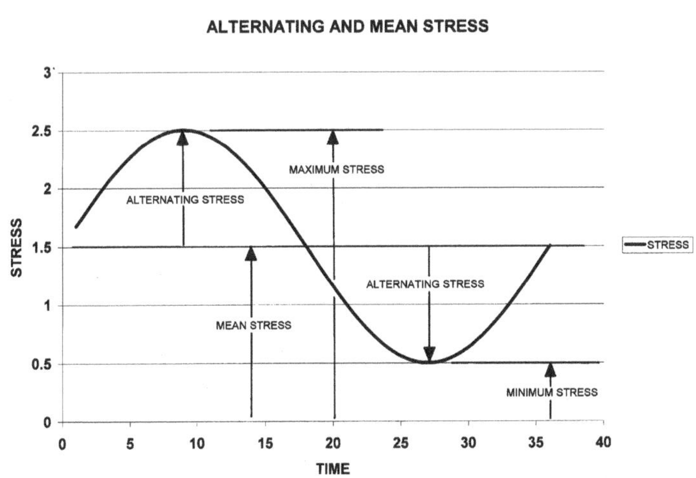

Figure 5.5 Stress history of a point on a uniaxial fatigue tester with the loading set for tension–tension.

bility of finding a critical flaw is nearly 1.0. If the applied load in service is bending and has a volume of material at the critical point similar to that near the surface of the 0.3-in test specimens, then no adjustment needs to be made to the test data and LF = 1.0. If the applied load is direct tension or compression, the entire cross-section at the critical point is exposed to the same stress and, unless the cross-section is very small (i.e., <<0.25 in), the probability of finding a critical flaw is nearly 1.0 and LF is equal to 0.85. Once a critical flaw is guaranteed to be found, then no further downward adjustment needs be made. Whether the critical volume is large enough to contain one critical flaw or many, it makes no difference. Crack initiation will occur.

If the part is loaded in torsion, then the LF is equal to 0.577. Based upon the maximum-distortion-energy theory [6], the maximum shear stress that exists in the test specimen cycled to failure on the rotating beam tester is 0.577 times the principle stress. It is worth noting at this time that machine elements that are cyclically stressed in compression (direct or bending), may fail in fatigue if the alternating principle shear stress at the critical point exceeds the shear endurance limit of the material. One example of this is the toggle links in a mechanical clamping system. This author has observed cracks along a 45-degree angle through failed toggle links.

Size Effect Factor (SEF) Another adjustment that must be made to rotating beam fatigue data is an adjustment, based on part size, to account for the probability of finding a critical flaw within the stressed volume of material at the critical point of the part. As stated earlier, once the probability of finding a flaw is equal to 1.0, then no further downward adjustment is required. Thus, for parts larger than 0.30 in and uniaxially loaded, the LF of 0.85 is the only reduction required and an SEF of 1.0 can be used. For parts loaded in bending, an adjustment is required. As the size increases to greater than 0.30 in, the probability of finding a flaw approaches 1.0 and SEF approaches 0.85 as a limit. The literature is filled with various empirical fits to size effect data. Reference [1] gives Eq. (5.2) for diameters between 0.30 and 10.0 in. For diameters less than 0.30 in, it is recommended to use an SEF of 1.0. I have had good results using Eq. (5.3) for diameters between 0.25 and 12.0 in. The logic behind this equation is that at 0.25 in the SEF should equal unity and at large diameters the SEF should approach 0.85. Equation (5.2) may be too conservative at large diameters and includes other factors, such as lower mechanical properties.

$$\mathrm{SEF} = 0.869\, d^{(-0.097)} \tag{5.2}$$

where d = section diameter (in) in bending

$$\mathrm{SEF} = 0.85 + 0.0375/d \tag{5.3}$$

The tensile strength of the metal at the critical point of the part is not always equal to the nominal strength of the part. In general, a uniform microstructure cannot be obtained throughout very large sections of heat-treated parts. For forgings, the more it is reduced in size from the original billet, the better the mechanical properties become because of grain refinement. In large forgings, the material near the surface may have higher mechanical properties than near the core because of more grain refinement near the surface. For large castings the mechanical properties near the surface may be higher than near the core because of the faster cooling near the surface as the casting is poured. If the part is used in the "as cast" state, this may be partially offset by the poor surface condition. Thus, it is important to use the tensile strength of the material at the critical point when performing an analysis. The critical point on most parts is fortunately near the surface. Equation (5.3) is in good agreement with the experience I have had in designing large tie rods in the 12-in diameter range. It is worth noting that threaded rods have the critical point near the surface and not at the core or center line of the rod.

Surface Finish Factor (SFF) Another adjustment that must be made to rotating beam fatigue data is to account for a surface finish that may differ from the carefully polished surface used on the rotating beam test specimens. The amount of reduction is a function of both degree of roughness and the tensile strength of the steel [1–3]. For surfaces that are machined or ground, I have found Eq. (5.4) to give realistic results for steels with tensile strengths between 60,000 and 180,000 psi.

$$\text{SEF} = 1.0 - S_u/(S_u + 5.5*10^6\sqrt{(1/f)}) \qquad (5.4)$$

where S_u = tensile strength in psi
f = rms machined finish in micro-inches within the range of 4 to 500

Surface Treatment Factor (STF) Surface treatment should not be confused with surface finish. Surface *treatment* involves effects from such treatments as carburizing, decarburizing, plating, or nitriding. References [1–5] should be consulted for adjustments that should be made for these surface treatments. In general, treatments that put the surface at the critical point in compression are beneficial. Such treatments as cold working (shot peening) or diffusion processes (nitriding or carburizing) are examples of beneficial surface treatments. Anything that decarburizes the surface or causes corrosion is detrimental. Surface treatments can have a significant effect on fatigue strength and should not be overlooked.

Failure Rate Factor (FRF) When fatigue tests are run, many samples are run at each loading and the cycles to fail are recorded. There is usually a fair amount of scatter in the data; however, the *S–N* curve usually has a best fit through the mean value. Thus, at any given life, 50% of the specimens failed at stresses greater than the mean, and 50% failed at less than the mean value. If the distribution is normal (Gaussian), then 68.3 % of all failures fall within 1 standard deviation (SD), 95% lie within 1.96 SD, 99.7% lie within 3.0 SD, and 99.97% lie within 3.72 SD. It is common practice in aircraft design to use the 99.7% value.

For most ductile steels 1 SD of fracture strength ranges from 1 to 3%, with 2% being the norm. The RA for most steels ranges between 45 and 65%, with the average at 55%. As a rule, as tensile strength increases, RA decreases, so that a lower RA occurs for high-strength steels, and a higher RA occurs for low strength steels. At 55% RA, the average fracture strength is about 2.22 times the ultimate tensile strength. Thus, each standard deviation equals approximately 4.44% of the ultimate strength. If the designer uses 3.72 SD, then there is a 99.97% probability of survival by using 3.72*4.44% or 16.5% lower than the published mean fatigue strength. For the lower strength steels, where RA = 65%, then 3.72*5.71% or 21.2% reduction would be in order. For high-strength steel 13.5% reduction might be appropriate. I have had good results with Eq. (5.5) for determining the failure rate factor FRF.

$$\mathrm{FRF} = 1.0 - 0.02 * SD * (1.0/(1.0 - \mathrm{RA}/100)) \tag{5.5}$$

where SD = the number of standard deviations below the published mean value required to give the desired probability of survival.

The Adjusted Fatigue Strength (S_e') The adjusted fatigue strength (S_e') is then equal to the published fatigue strength (S_e) based upon rotating beam tests times each of the reduction factors and is shown in Eq. (5.6). Whenever possible, S_e should be taken from published values. For approximating S_e when published values are not available, half of the tensile strength may be used. This is reasonably close for steel with tensile strengths in the range of 60,000 to 180,000 psi.

$$S_e' = \textit{Fatigue strength} * \mathrm{LF} * \mathrm{SEF} * \mathrm{SFF} * \mathrm{STF} * \mathrm{FRF} \tag{5.6}$$

Example 1 Determine the adjusted fatigue strength for 99.7% survival (3.0 SD) for a 5.50-in-diameter tie rod used in a 500-ton injection molding machine. The rod is made of AISI C11L37 with an ultimate tensile strength of 96,000 psi and RA of 48%. An eight-pitch thread is made on the ends of each rod using a carbide insert having a 0.018 in tip radius and producing a surface finish of 32 rms. The minor diameter of the thread is 5.365 in. No surface treatments are applied.

Use $S_e = S_u/2 = 96{,}000/2 = 48{,}000$ psi because no published value is given.

Loading factor (LF) = 1.0 since the primary stress is bending at the root of each thread

Size effect factor (SEF) (Eq. 5.2) = 0.85 + 0.0375/5.365 = 0.857

Surface finish factor (SFF) (Eq. 5.3) = $1.0 - 96{,}000/(96{,}000 + 5{,}500{,}000 * \sqrt{(1/32)}) = 0.910$

Surface treatment factor (STF) = 1.0 since no surface treatments are applied

Failure rate factor (FRF) (Eq. 5.4) = 1.0 − 0.02 * 3.0 (1.0/(1.0 − 48/100)) = 0.885

$$S_e' = 48{,}000 * 1.0 * 0.857 * 0.910 * 1.0 * 0.885 = 33{,}129\ \text{psi}$$

5.1.5 Determination of Survival Factor (SF)

The analysis of a design requires three steps: First, the design stress must be determined as shown earlier; second, the theoretical stress at the critical point must be calculated, including the geometric stress concentration factor; and finally, the design and theoretical stresses must be compared using the appropriate design equation to determine the probability that the design will survive.

Very early research into metal fatigue revealed that the mean stress can affect the survival rate of a part. It has been shown [1–5] that as the mean stress increases in tension, the allowable alternating stress must be reduced. As the mean stress increases in compression, the allowable alternating stress can also be increased. This can be used to explain why there are beneficial effects from surface treatments that produce a residual compressive stress. Figure 5.6 shows the approximate relationship between alternating and mean stress for ductile steels. Because the curves follow parabolas, W. Gerber [6] of Germany proposed in 1874 that a parabolic equation be used. Equation (5.7) is widely used in Europe, whereas the more conservative Goodman or Soderberg equations are more widely used in the United States. I have had good results with the Gerber equation. In Eq. (5.7), K_f is referred to as the fatigue stress concentration factor and is equal to the geometric (material independent) stress concentration factor K_t adjusted for "notch sensitivity" of the material. There is much discussion in the literature [1–5,7] about how to determine notch sensitivity; however, I favor the method proposed by Heywood [4] in Eq. (5.8). In earlier work, Heywood [5] also suggests that K_f given in Eq. (5.8) is for infinite life situations and that for finite life K_f, approaches unity as the cycles to failure approach 1. Equation (5.9) represents Eq. (10.13) from Heywood and reflects this effect.

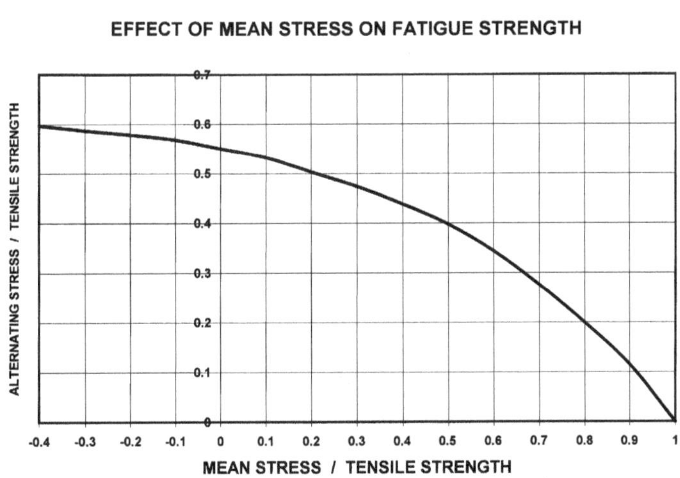

Figure 5.6 Fatigue strength relationship between alternating and mean stress for steel showing how a compressive (negative sign) mean stress increases the allowable alternating stress.

$$1/SF = (S_m / S_u)^2 + K_f * S_a / S'_e \tag{5.7}$$

where SF = the survival factor and must be greater than 1.0 for part survival beyond 2 million cycles.

$$S_m = (S_{max} + S_{min})/2$$

$$Sa = (S_{max} - S_{min})/2$$

$$K_f = K_t/(1 + 2a(K_t - 1)/(K_t * S_u * \sqrt{(r)})) \tag{5.8}$$

where r = notch radius (in) and a is a factor as follows:
a = 5000 for a transverse hole
a = 4000 for a shoulder
a = 3000 for a groove

$$K_{f,N} = 1 + (K_f - 1)/(0.915 + 200/n^4) \tag{5.9}$$

Where $K_{f,N}$ = fatigue s.c.f. for failure in N cycles.
$n = \text{Log}_{10}(N)$

Example 2

Consider the survival factor for 99.7% survival for the tie rod in Example 1. Analysis of the thread load distribution and stress in the thread fillets is complex and will be discussed later; however, the analysis results in a nominal combined stress of 28,700 psi and a K_t of 2.03 in the fillet of the thread closest to the pressure face of the tie rod nut. Using Eq. (5.8) with r = 0.018 in and a = 3000 for a groove results in a calculated value for K_f of 1.642. Using Eq. (5.7), results in 1/SF = 0.734. Thus, the survival factor is about 1.363. This may seem high, but it must be considered that the loading in the rods is typically balanced to no better than 5% and that there is a bending moment applied (typically about 20%) at the rod ends from bending in the stationary platen. When these additional considerations are factored in, the survival factor is reduced to only 1.08. This does not leave much room for errors in stress calculations, overloading from mold flashing, or poor surface finish (tool marks) in the thread fillets. This is why tie rods do fail from time to time.

Example 3

Consider an 8.00-in-thick stationary die plate made of low carbon hot rolled steel with an S_u = 55,000 psi and an RA = 50%. The surface adjacent to the injection unit is machined to 250 rms finish. A stress analysis indicates a hoop stress of 12,500 psi around the nozzle hole when the smallest recommended mold is being used. Located on this same surface are 3/8-in taped holes for mounting a purge guard. Determine the survival factor to have 99.7% (SD = 3.0) survive.

$$\text{Let } S_e = 55{,}000/2 = 27{,}500\text{psi}$$

LF = 1.0 because the platen is bending around the mold.

$$\text{SEF} = 0.85 + 0.0375/8 = 0.855$$

$$\text{SFF} = 1.0 - 55{,}000/(55{,}000 + 5.5 * 10^6 \sqrt{(1/250)}) = 0.86$$

STF = 1.0 because of no surface treatment.

$$\text{FRF} = 1.0 - 0.02 * 3.0 * (1.0/(1.0 - 50/100)) = 0.88$$

Thus,

$$S_e' = 27{,}500 * 1.0 * 0.855 * 0.86 * 1.0 * 0.88 = 17{,}794\,\text{psi}$$

From Ref. [7] K_t is found to be 2.90 for a hole, then

$$K_f = 2.90/(1.0 + 2 * 5{,}000(2.90 - 1)/(2.90 * 55{,}000 * \sqrt{(0.1875)})) = 2.274$$

$$1/\text{SF} = (6{,}250/55{,}000)^2 + 2.274 * 6{,}250/17{,}794 = 0.812$$

Thus, the survival factor is 1.23, which explains why platens almost never fail in fatigue.

5.1.6 Conclusion of Discussion of Metal Fatigue

It is hoped that this review has given the reader a good overview of metal fatigue. Injection molding machines are high production equipment that experience repeated loading millions of times and have metal fatigue as a major mode of failure. For the processing engineer, the information given here may be sufficient to make them aware of the damage that can be done by abusing the equipment through overloading. This review should only serve as an introduction for the machine designer. It is strongly suggested that the references cited at the end of this chapter be consulted to insure a complete analysis is done to prevent a fatigue failure.

5.2 Functions of the Clamping System

There are three major functions of clamping systems for injection molding machines.

- Traverse the mold to the closed and to the opened position

The first is to support the mold and traverse it between the open and closed positions. This function requires only enough force to accelerate the moving mass and to overcome friction. Guidance of the moving mold half is important to prevent excessive wear in the mold leader pins.

- Hold mold closed during part formation

The second is to hold the mold closed during the injection of melt into the cavities. This may require considerable clamping force as the melt pressure within the mold is often very high.

- Provide means of ejecting parts

Once the parts have cured (cooled) and the mold has opened, a means must be provided to eject the parts. The force required for this is relatively small compared with the clamping force.

5.3 The Three Types of Clamping Systems

Clamping mechanisms for injection molding usually fall into three categories: hydraulic (or pneumatic), hydromechanical, and mechanical.

5.3.1 Hydraulic

Hydraulic (or pneumatic) clamps are distinguished from the other types by the following:

- The load path passes through the fluid (oil or air) used to pressurize the clamping cylinder(s) (Fig. 5.7)
- The length of the column of fluid being pressurized is equal to or greater than the stroke of the clamp. This fact separates this class of clamps from hydromechanical clamps.
- The clamping force is in direct proportion to the pressure applied to the clamping cylinder.
- Available clamp stroke is a function of maximum open daylight and the shut height of the mold.

An example of a hydraulic clamp is shown in Fig. 5.8. These clamps are often favored for their simplicity and ease of setup.

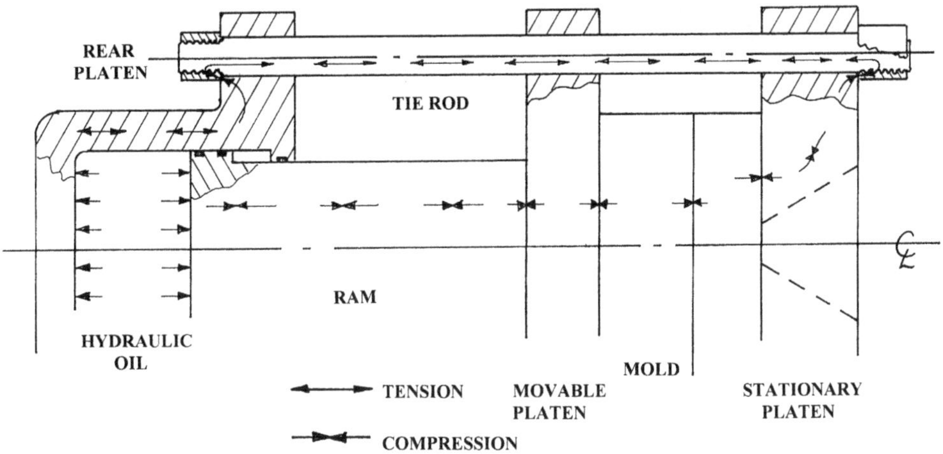

Figure 5.7 Load path for a hydraulic clamp showing only the top half. Note that the load path passes through a long column of relatively "flexible" oil. (Note the convention for tensile and compressive stresses, where opposing arrowheads imply compression and arrowheads pointing away from each other imply tension.)

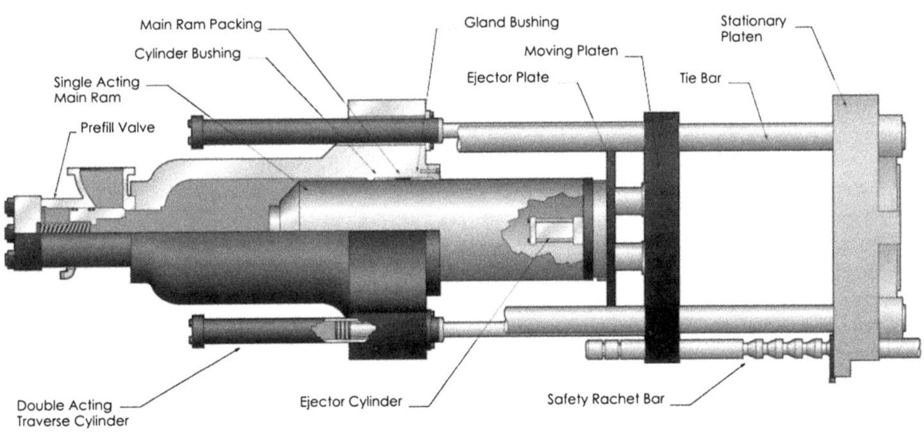

Figure 5.8 Typical hydraulic clamp. (Courtesy Van Dorn.)

5.3.2 Hydromechanical

Hydromechanical clamps are distinguished from the other types by the following:

- The load path passes through the fluid (oil or air) used to pressurize the clamping cylinder(s) (Fig. 5.9).

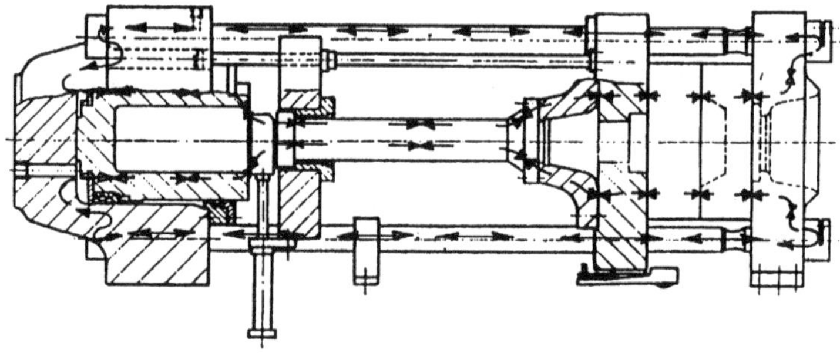

Figure 5.9 Load path for an hydro-mechanical clamp.

- The length of the column of fluid being pressurized is independent of and much less than the stroke of the clamp. Typically the height of the oil column is less than 2 cm (3/4 in). This fact separates this class of clamps from hydraulic clamps.
- The clamping force is in direct proportion to the pressure applied to the clamping cylinder.
- Available clamp stroke is a function of the traversing actuator stroke with provisions for varying mold shut heights typically being through tie rod nut adjustments on the rear platen.

An example of a hydromechanical clamp is shown in Fig. 5.10. A wide variety of novel approaches have been used in the design of this class of clamps.

5.3.3 Mechanical

Mechanical clamps are typically referred to as *toggle clamps* and are distinguished from the other types by the following:

- The load path does not pass through the hydraulic/pneumatic cylinder(s) or electrically driven actuator(s) (Fig. 5.11).
- The clamp force is not in direct proportion to the force of the actuator. Instead, it is a complex relationship between the available actuating force during toggle-up and the stiffness of the clamp. This will be discussed later.
- Available clamp stroke is a function of the linkage system with provisions for varying mold shut heights typically being through tie rod nut adjustments on the rear platen. The pin-to-pin center line length of the rear link is the dominant factor in determining clamp stroke. An example of a mechanical clamp is shown in Fig. 5.12.

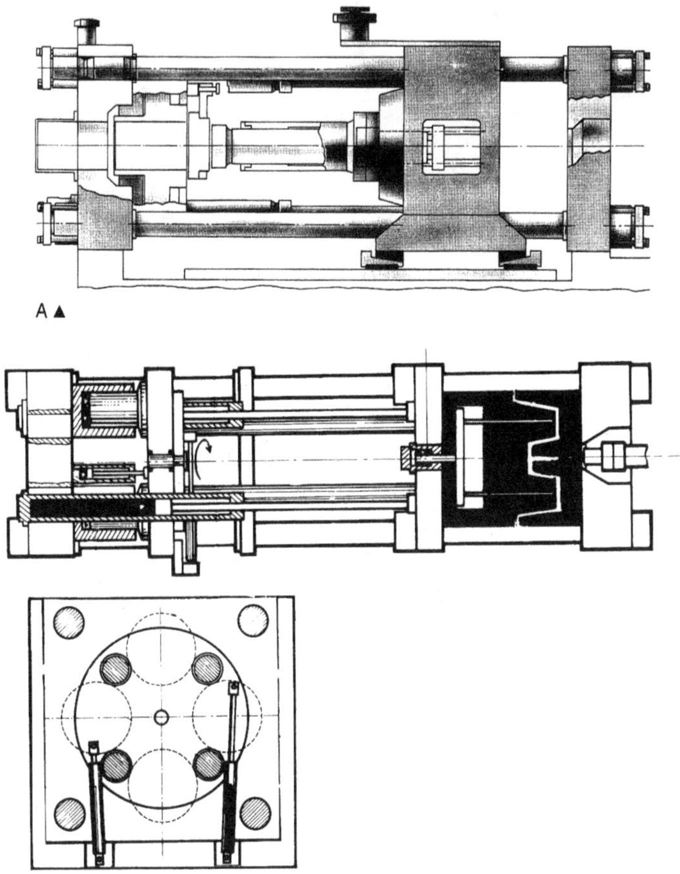

Figure 5.10 Typical hydromechanical clamp. (Courtesy Battenfeld (top) and Billion (center and bottom).)

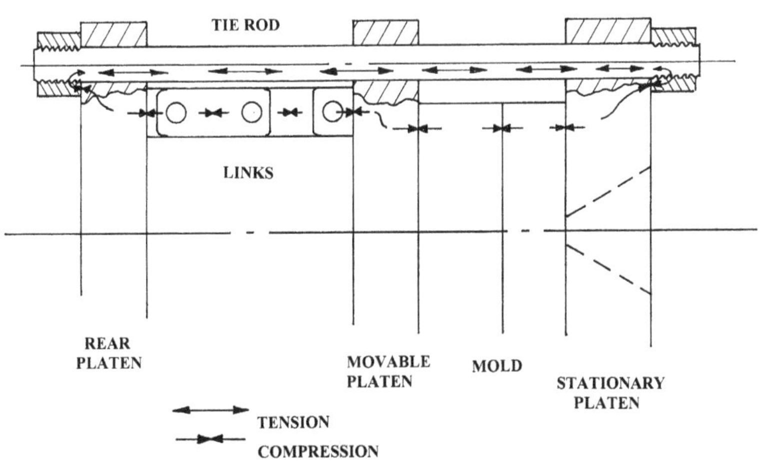

Figure 5.11 Load path for a mechanical (toggle) clamp.

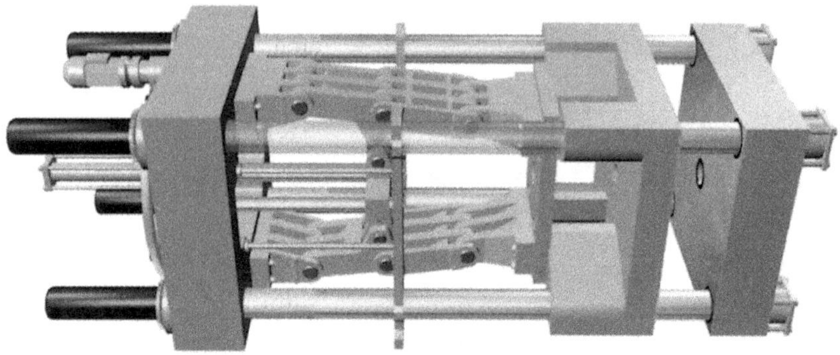

Figure 5.12 Typical mechanical (toggle) clamp (Courtesy Engel.)

5.3.4 Types of Toggle Systems

There are numerous variations on toggle clamp design. Most toggle clamps are hydraulically driven, but the trend today is to use electric drive systems to reduce the oil leakage problem and to increase the operational efficiency of the clamp. The toggle links are usually laid out symmetrically about the clamp center line either top-to-bottom or front-to-back with either four-point (four pin) or five-point (five pin) toggle links (Fig. 5.13). The difference between the two approaches is in the

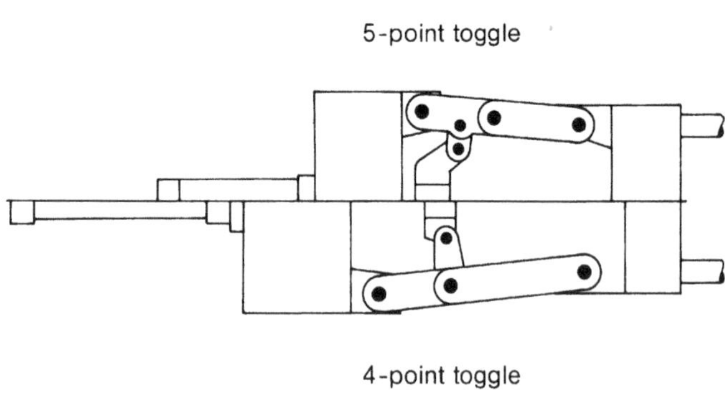

Figure 5.13 Five-point (pin) toggle compared to four-point (pin) toggle. (Note pin count is for only half of the clamp and actual total pin count is twice this.)

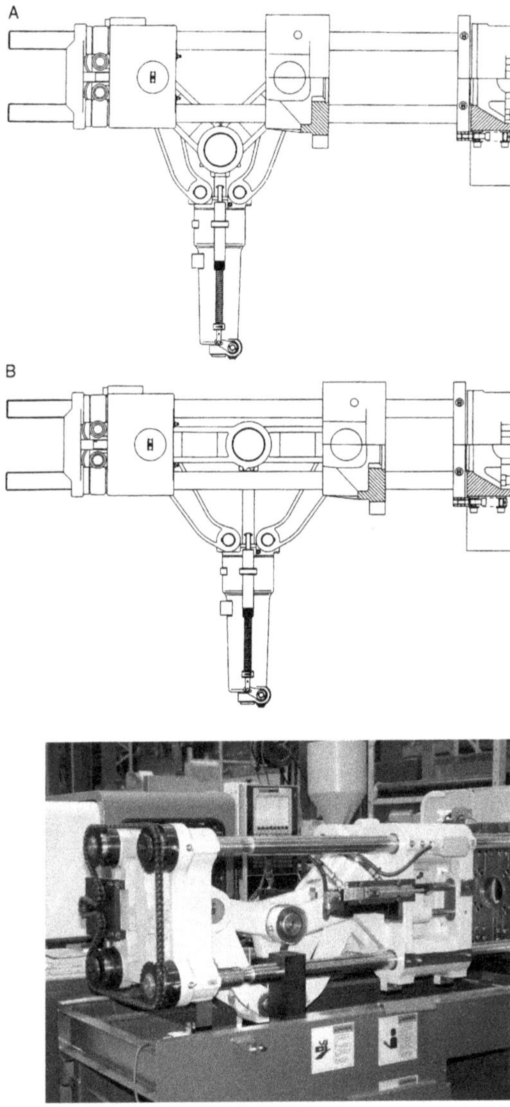

Figure 5.14 Monotoggle clamp shown open in top and bottom view and closed in center view. (Top and center view courtesy Arburg and bottom view courtesy Milacron.)

motion and force curves. It should be noted that even though reference is made to four or five points (pins), there are actually twice that many link pins in a complete toggle system. A few machine manufacturers use a "mono-toggle" design in which the driving actuator is positioned either above or below the clamp center line (Fig. 5.14).

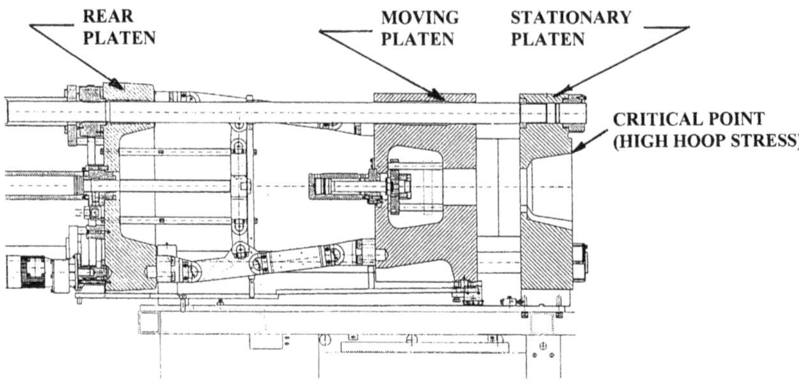

Figure 5.15 Typical arrangement of platens for a toggle clamp. (Courtesy Engel.)

5.4 Key Elements of a Clamp

5.4.1 Platens

Clamps for horizontal injection molding machines usually have three platens; a stationary, movable, and a rear platen (Fig. 5.15). These are usually made of mild steel plate or cast steel or ductile iron, and they are designed so that the maximum stress is less than 12,000 psi. This maximum design stress is selected because it is low enough to prevent a fatigue failure of mild steel even if a stress riser in the form of a small hole is present at the critical point.

Stationary Platen

The stationary platen is the one closest to the injection unit and has provisions for the injection unit nozzle to pass through it at the center line of the machine. The thickness of stationary platens vary with manufacturers, but they usually have a thickness equal to about 25% of the width or height of the platen. The mold mounting face of the stationary platen will have either tee slots or a multitude of tapped holes that should be located in a standard pattern. In the United States this is prescribed by *Recommended Guidelines Interchangeable Mold Mounting Dimensions*, developed by The Society of the Plastics Industry, Inc., (SPI). Some manufactures also provide a hardened centering ring for locating the mold on its sprue bushing. The stationary platen may be either stress or deflection limited depending on the size of the mold. The platen is usually deflection limited with a large mold and stress limited with a small mold. This means that the platen thickness is designed to limit the maximum relative deflection between the center line of the machine and the corner

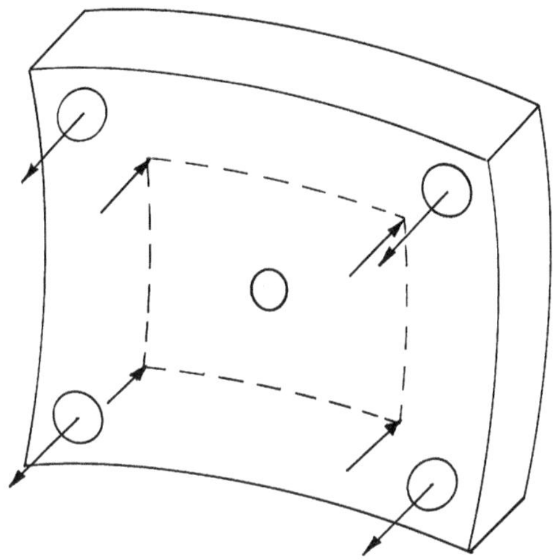

Figure 5.16 Deflection of stationary platen showing peak contact stress from mold occurring at the mold corners.

of the largest mold that will fit into the press. As clamp force builds, the stationary platen tends to deflect and wrap around the mold, forming a somewhat spherical concave surface (Fig. 5.16). To calculate the amount of deflection that occurs, the most accurate method is to use a finite element program. A good approximation can be made, however, by using beam equations with adjustments made for plane strain. The beam equations should be applied in two directions and the results superimposed (Fig. 5.17). Thus, the total deflection at the platen center-line relative to the corners

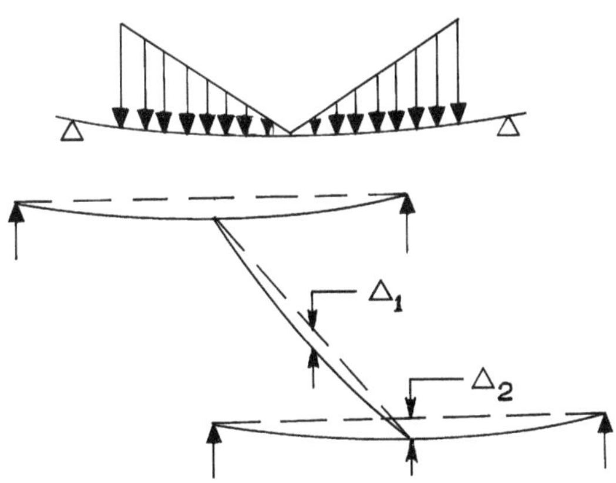

Figure 5.17 Superposition of simple beam equation solution yields approximate platen deflection by adding Δ_1 and Δ_2. Note that the distributed load on the platen is assumed to be maximum at the edges and zero at the center.

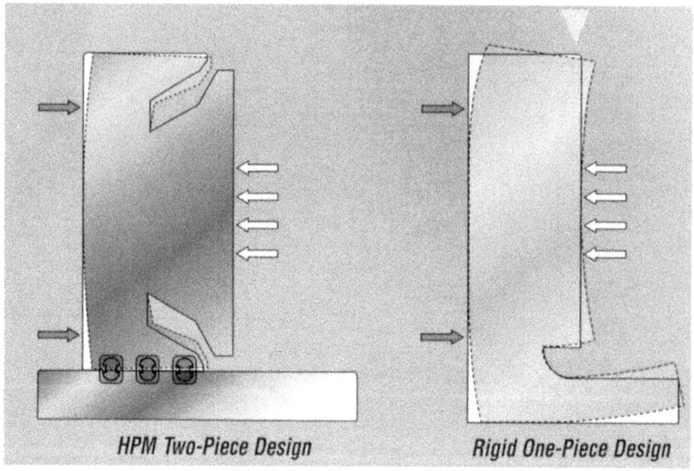

Figure 5.18 "Parabolic" platen that allows the mold mounting face to remain flat as the platen deflects under load. (Courtesy HPM.)

of the mold would be the sum of Δ_1 and Δ_2. The magnitude of this deflection can be normalized by dividing the deflection at the center-line by the diagonal of the mold mounting face. There is no published allowable value for this in the United States, but good practice seems to be between 0.0001 and 0.0002 mm/mm (in per in) of mold diagonal. Thus, if a mold which measures 24 in across the diagonal is installed, the platen deflection at the center of the mold relative to the corners of the mold would be about 0.003 in. Most molds are compliant enough that they can conform to the platen deflection. Whether they actually do depends on how the clamp force is transmitted through the movable platen. Because the platen tends to take on a concave surface when clamp force is applied, many machinery manufacturers machine the mold mounting face to a slight convex shape (crowned) so that the center of the mold is better supported. The crown is usually very small and is on the order of only a few thousandths of an inch. Other techniques have also been used to counteract platen deflection (Fig. 5.18 to Fig. 5.20). In these approaches, the mold mounting surface is kept flat by allowing other portions of the platen to deflect without distorting the mounting surface.

Once the platen thickness has been determined based on deflection limitations, the next consideration is to determine the minimum mold dimensions that can be accommodated without exceeding fatigue stress limitations. The critical point is usually the hoop stress around the cored or machined pocket provided for the injection nozzle (Fig. 5.21). Some manufacturers design their platens so that at maximum clamp force the maximum nominal tensile stress is less than 12,000 for a mold that has a length and width equal to half the tie rod spacing. This hoop stress is on the face nearest the injection unit. Because this portion of the platen is the highest stress area, the manufacturer should be consulted before drilling any holes there. It is also a good idea to consult the machine manufacturer to determine the smallest mold that

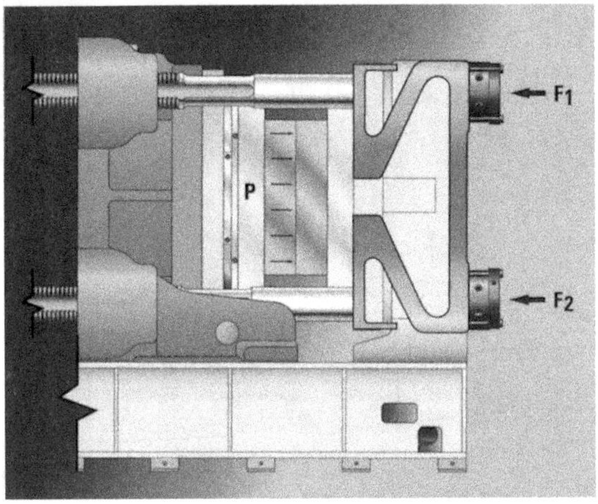

Figure 5.19 "Reflex" platen that allows the mold mounting face to remain flat as the platen deflects under load. (Courtesy Husky.)

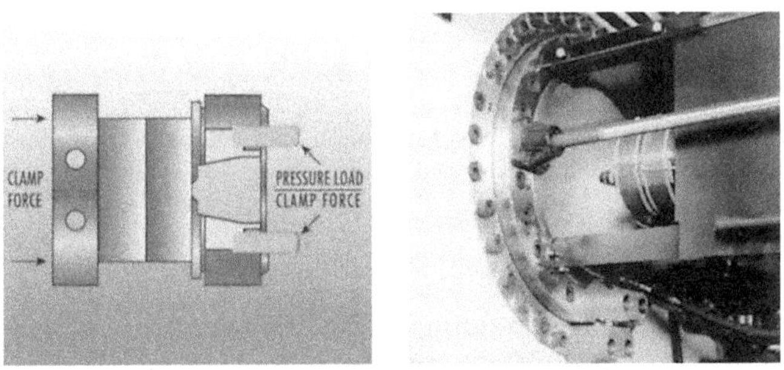

Figure 5.20 Stationary platen deflection control showing clamp force applied near the center of the mold mounting surface. (Courtesy Milacron.)

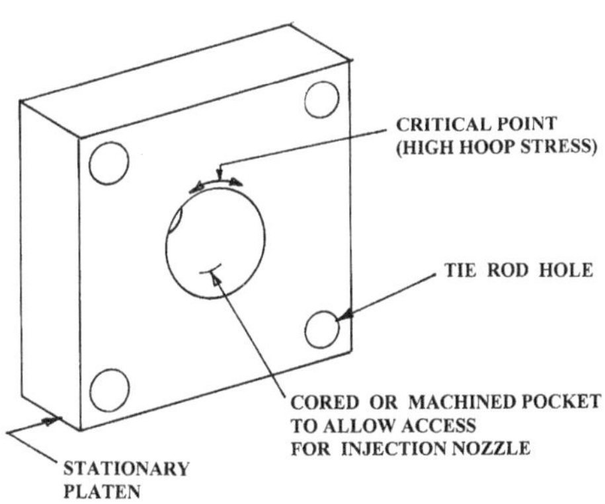

Figure 5.21 Critical point on stationary platen near nozzle access pocket where any mounting holes may be located in the hoop stress around the edge of the hole.

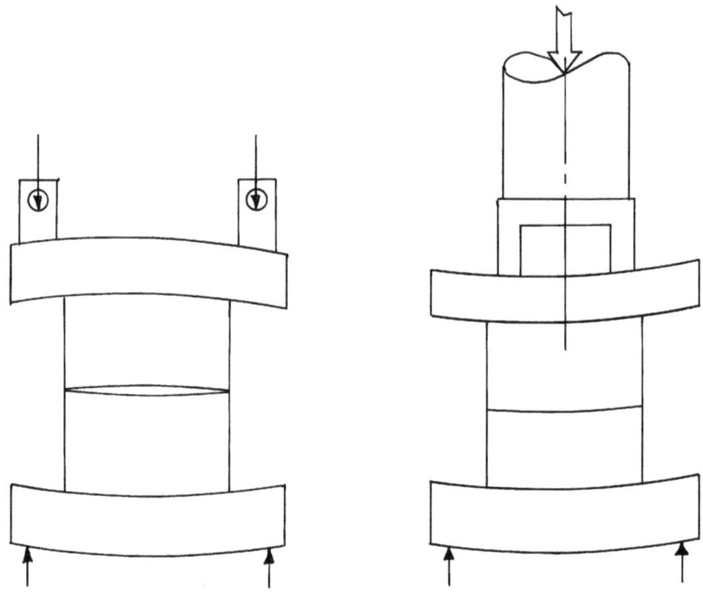

Figure 5.22 Deflection of moving platen as a function of load path with right view showing clamp load applied near the center of the moving platen and left view showing clamp load applied near the edges of the moving platen.

can be run at the rated clamp force. There is risk of a fatigue failure, as well as of hobbing the mold into the mold mounting face of the platen. Even though the proper size mold is run in a machine, it is not unusual to see witness marks on the platen(s) at the locations of the mold corners when it is removed after an extended run. This is simply a result of fretting corrosion as a result of the very small shear strain occurring between the mold surface and the platen as the platen "wraps" around the mold as clamp force is applied. It is prudent to clean both the platen and the mold surface to remove the corrosion before installing a new mold.

Moving Platen

The moving platen supports the movable mold half. It is usually not as thick as the stationary platen because the clamping force is often applied closer to the machine center-line instead of at the corners where the tie rods are located. The deflection of the moving platen can be either concave or convex, depending on how close to the center the clamp force is applied (Fig. 5.22). For instance, if the clamp force is applied at the center line then the deflection will be convex. This would then tend to cause the mold to go into compliance with the deflection of the stationary platen and results in the least probability of mold flashing at the center of the mold. If the clamp force is applied at the corners (near the tie rods), the deflection will be concave. This will result in the least favorable mold clamping condition and have the highest probability of mold flash at the center of the mold. The mold mounting face has the same provisions for attaching the mold as the stationary platen. An additional set of holes is provided, however, to allow ejector rods to pass through the platen. These ejector rod holes should also be sized and located according to some standard pattern. For machines used in the United States, this should be the SPI pattern. This will insure that molds can be moved from one machine to another with confidence that both the mold mounting holes and the ejector holes will be aligned.

Parallelism between the moving platen and the stationary platen is important to reduce wear on the mold leader pins and to insure proper tie bar loading. Parallelism is usually measured under two different conditions. The first measurement is made when the clamp is closed (toggled) without the mold halves touching (zero clamp force). The second condition is when the clamp is closed at the rated clamp force. For small to medium size machines, the platens should typically be parallel within 0.25 mm/m (0.003 in per foot) of mold diagonal at zero clamp force and half of this value at full clamp force. Machines greater than 500 tons should have only two thirds of these values.

Rear Platen

Not all horizontal injection molding machines have a third (rear) platen, but for those that do, the primary purpose is for supporting the clamp actuators and completing the load path around the mold. These platens are usually designed based upon stress limitations. There are at least two critical areas that are highly stressed for a hydraulic press (Fig. 5.23). These can fail in fatigue if not properly designed or if the foundry

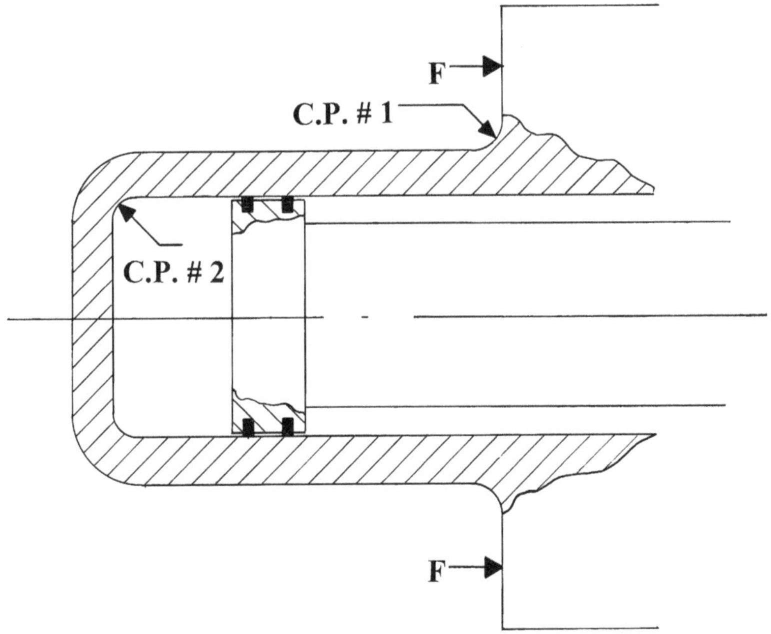

Figure 5.23 Two of the critical points on a rear clamp cylinder made from a single casting. *Note*: The size and finish of each radius is important in order to extend fatigue life.

does not exercise proper quality control, especially if it is a large ductile iron casting. This author has experienced failures where improper inoculation techniques resulted in a "fading" of the melt that left the thick portions of the clamp cylinder castings with properties closer to grey iron than to ductile iron. This error was not detected until 5 years after the machines were placed into operation. They began to fail in fatigue at Critical Point 2 (Fig. 5.23) after accumulating nearly 1 million cycles. Other machine manufacturers have experienced fatigue failures at Critical Point 1. Some machine manufacturers have eliminated these problem areas by making the rear platen an assembly (Fig. 5.24).

Two Platen Machines

The trend today for large molding machines is to eliminate the third (rear) platen and have only two platens. In addition to requiring less floor space, other advantages include lower manufactured cost and a more rigid clamp. Two platen machines are not new. Improved Machinery Company manufactured two platen machines as far back as 1968 with the introduction of their 1200-ton machine and in 1971 with their 2500 ton machine (Fig. 5.25). The moving platen of the 2500-ton machine was suspended on overhead rails. Coupled with the ability to retract the tie rod partially into the moving platen, this made it much easier to change very large molds. These molds often weigh more than 40 tons and this concept allowed for the use of a special mold

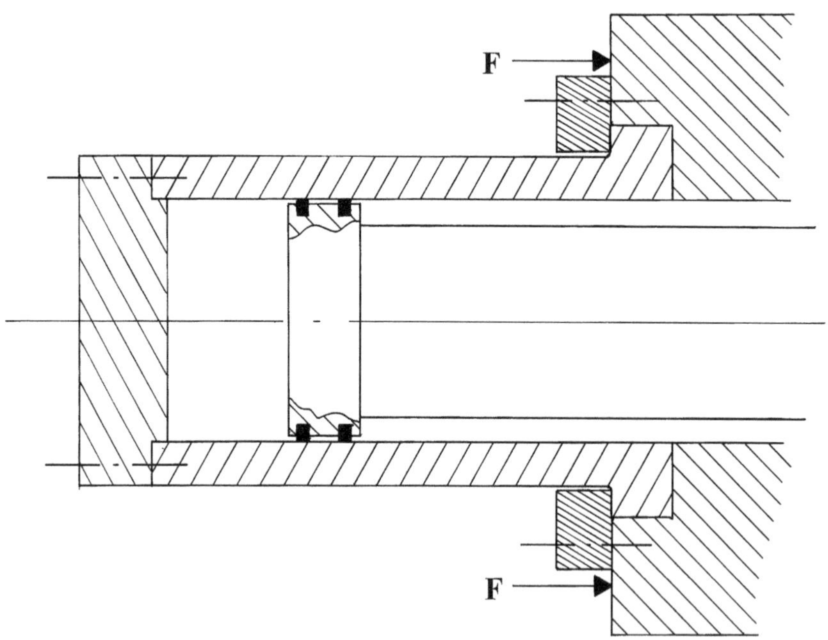

Figure 5.24 Elimination of critical points by using a composite design on the rear clamp cylinder.

Figure 5.25 *Top*: Two platen 2500 ton machine with "breech lock" system manufactured by Improved Machinery (IMPCO) in 1973. Note overhead suspension of moving platen, which allows for easy access of mold cart to facilitate mold changes. Automobile is shown for effect *Bottom:* Two platen IMPCO 1200 ton machine with "breech lock" system manufactured be Improved Machinery in 1968. (Courtesy Van Dorn.)

Figure 5.26 IMPCO 1200 ton machine viewed through rear guard. (Courtesy Van Dorn)

cart that could drive completely through the mold mounting area for ease of mold installation. Figure 5.26 shows the 1200 ton machine as seen from the back side and showing how the tie rods could be retracted for ease of mold change. Two platen machines usually have a means for disengaging the tie bar nuts from the tie rods so that the clamp can be opened. Two different concepts are often used on today's machines. One concept is to use split nuts that either translate open and closed as shown in the top two photo's of Fig. 5.27 or swing in a clamshell fashion as shown in the bottom photo. The second approach as shown in Fig. 5.28 is to use a nut that has half of the threads removed. Thus, after the clamp closes and the tie rod has entered the nut, the nut can rotate 90 degrees to engage the threads. Figures 5.29 through 5.32 show several of the newer two platen clamps.

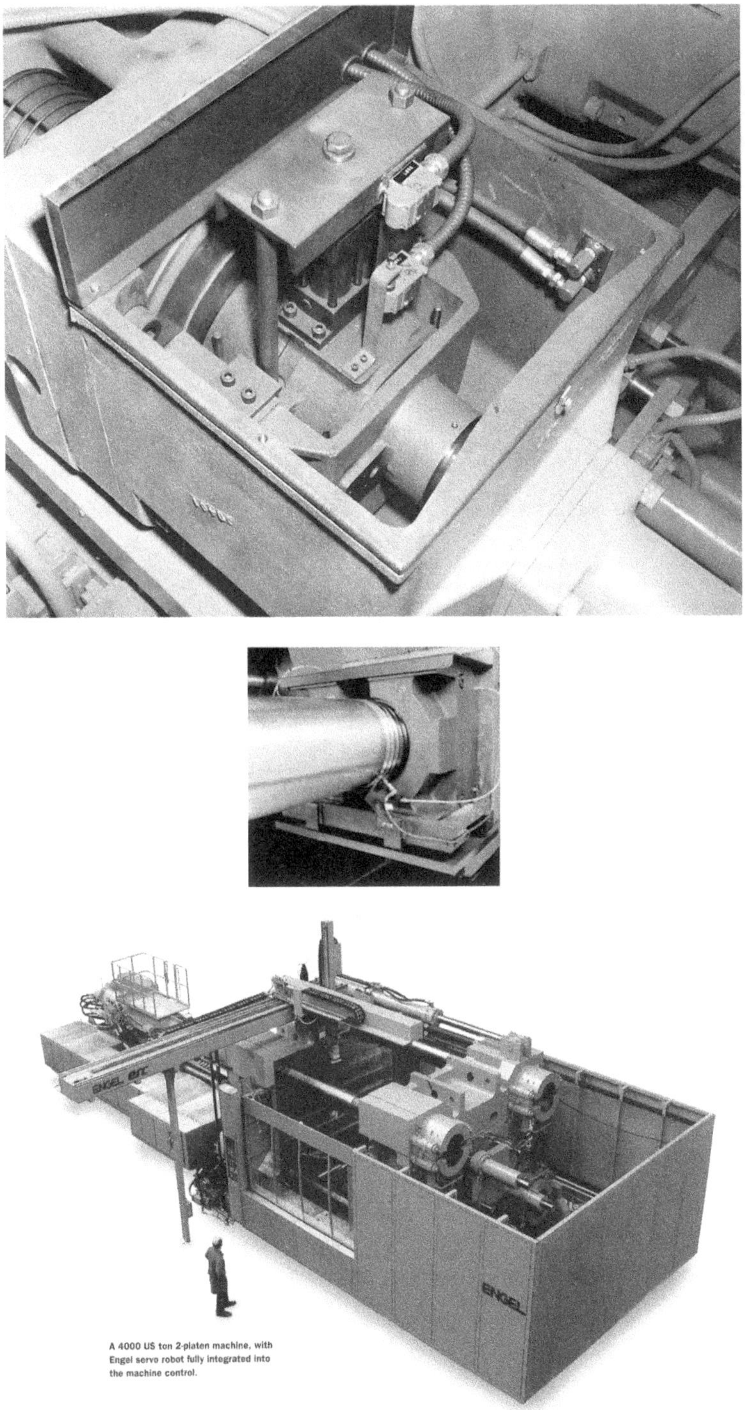

Figure 5.27 Split nut designs used on two platen clamps. Top view is the translating split nut used on IMPCO's 1200-ton "Breech Loc" system. Middle view is the translating split nut used on Milacron's MAXIMA. Bottom view is Engel's hinged split nut used on their large tonnage machines shown in Fig. 5.29.

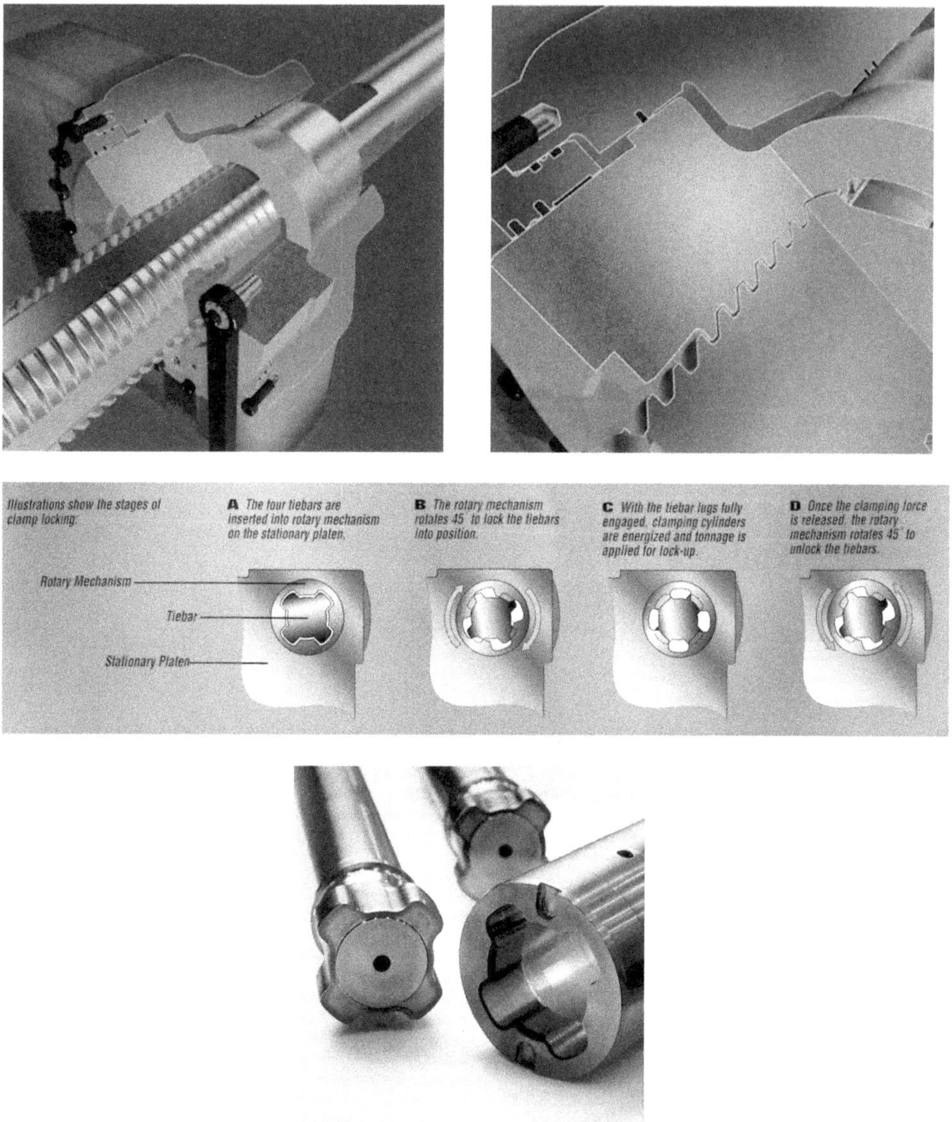

Figure 5.28 Top views are Husky's rotating nut design used on their two platen E-Series machines shown in Fig. 5.30. Bottom views HPM's rotating mechanism used in their two platen machine shown in Fig. 5.31. (Courtesy Husky and HPM.)

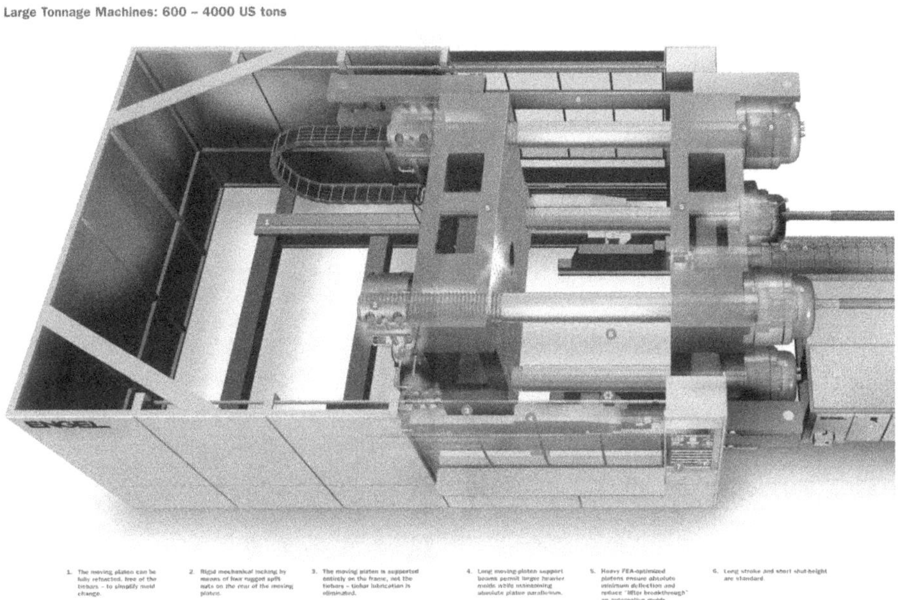

Figure 5.29 Engel's two platen Large Tonnage Machines. (Courtesy Engel.)

Figure 5.30 Husky's two platen E-Series Machines. (Courtesy Husky.)

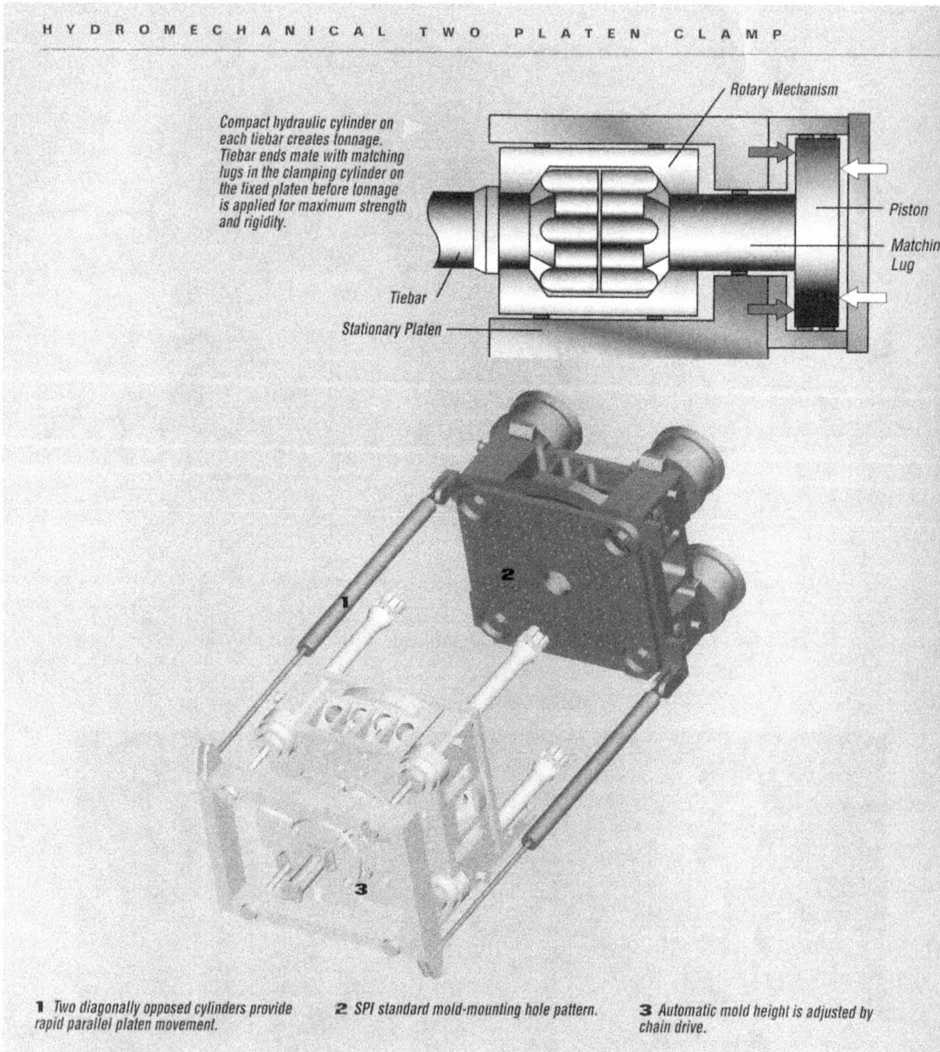

Figure 5.31 HPM's two platen machines. (Courtesy HPM.)

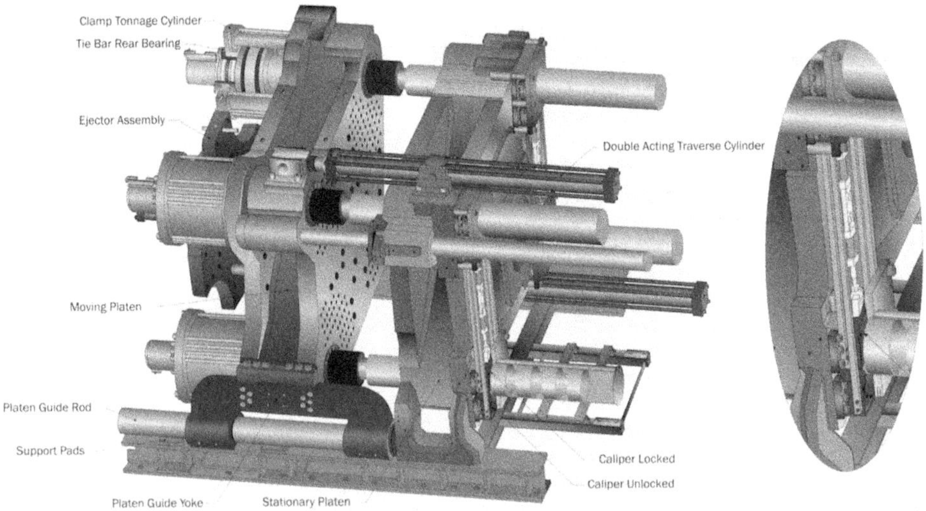

Figure 5.32 Van Dorn's two platen CALIBER SERIES.

5.4.2 Tie Rods and Nuts

Tie rods (sometimes called *tie bars* or *strain rods*) are the major tension element for completing the load path around the mold. A second function sometimes performed by tie rods is to provide guidance and support for the moving platen. This is often true for small machines, although not for large machines, because of the relative flexibility of larger (longer) tie rods.

In the United States, tie rods usually have a Unified, Acme, or Buttress thread form, depending upon the application. These threads have flank angles of 30, 14.5, and 7 degrees, respectively. Unified threads are the most common but Acme are sometimes used on the adjustable end of tie rods for machines that change shut height by adjusting the nuts. The lower flank angle of Acme threads reduces the total contact stress applied at the pitch line and thus reduces the probability of galling. Buttress threads are ideal for use on clamps that use split nuts such as two platen machines. Because buttress threads have a very low flank angle, the radial component of the tie rod load is much lower and the force required to hold the nut halves together is very low. The Unified thread form is most often used for clamps that use more conventional nuts.

Tie rod nuts are made of steel, ductile iron, cast iron, or bronze, depending on the application. Steel or ductile iron are the two most common materials used, especially on the fixed end of the rod. These are frequently used in the cast form to allow for cast-on projections that can be used to torque the nut for the proper preload. Bronze or cast iron nuts are sometimes used on the adjustable end. Either material has superior wear properties against the steel tie rod.

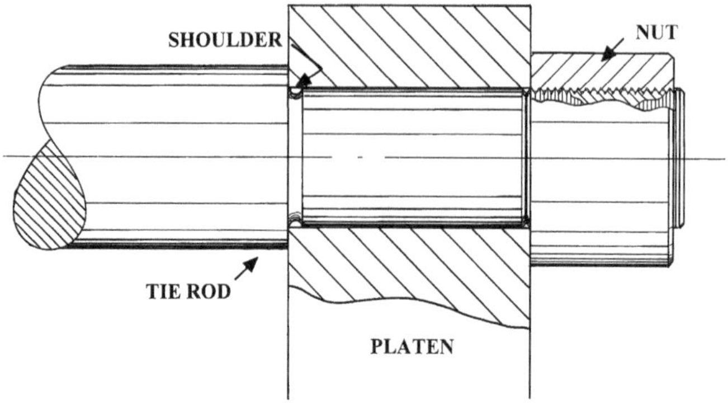

Figure 5.33 Tie-rod and nut system in which a shoulder is provided on the rod onto which the nut can be properly preloaded to resist clamp opening force. Nut must be properly tightened to prevent it from loosening during clamp cycling.

Methods to Retain the Tie Rod Nut

Preload of the tie rod nut is required to prevent the load bearing face of the nut from separating from the platen during mold opening. When the mold is being held closed by the clamp, the tie rod is in tension and the tie rod nut is being pulled against the platen. When the mold is being opened, there may be some tendency for parts to stick in the cavity, especially for deep parts with a shallow draft. The force required to open the mold puts the tie rod into compression and tends to push the nut face away from the platen. The magnitude of the force that is developed depends on the degree to which the parts stick and on the capacity of the clamp to develop an opening force. Typical hydraulic and hydromechanical clamps can develop, opening forces equal to about 5 to 10% of their nominal clamp force rating (this will vary from manufacturer to manufacturer). Typical toggle clamps can develop opening forces greater than 25% of their nominal clamping force rating (this also will vary with each manufacturer). Thus, a significant preload may be required to keep the tie-rod-to-platen joint from opening.

To keep the tie rod nuts in place as the machine cycles, some method must be used to retain the nut. There are three methods commonly used. One method is to have a reduced threaded diameter so that a shoulder exists onto which the rod can be tightened (Fig. 5.33). An alternative to this is to use a split ring to provide a shoulder for tightening (Fig. 5.34). The third method for retaining nuts is shown in Fig. 5.35. This

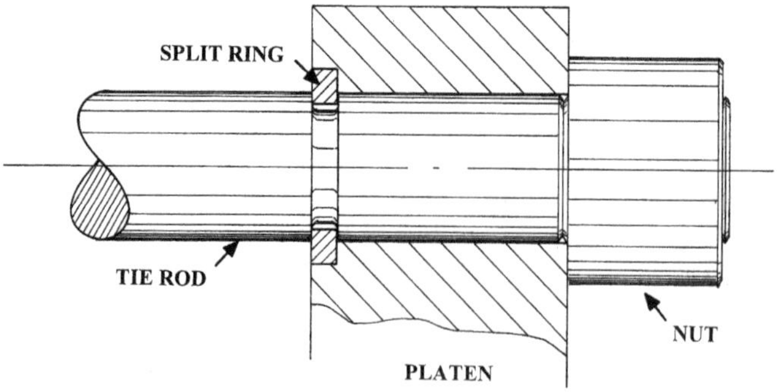

Figure 5.34 Tie-rod and nut system in which a split ring is used to provide a positive retention that will resist clamp opening force. Tie-rod nut must be tightened properly to prevent the nut from loosening.

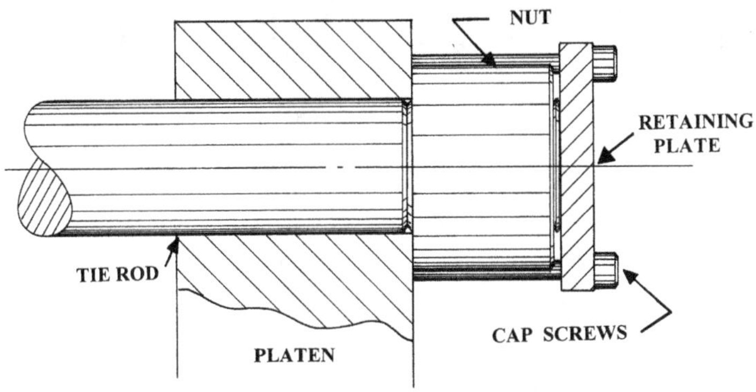

Figure 5.35 Tie-rod and nut system in which a retaining plate is used to capture the nut between the platen and the end of the rod. Note that much less bolt torque is required for proper preload.

uses a retainer plate placed over the end of the rod and three or more relatively small cap screws to clamp the nut against the face of the platen. The last method has the advantage that a more accurate preload can be obtained with a much smaller torque value on the fastener. For example, consider the torque required to preload the nuts on a 200-ton toggle clamp that can develop 50 tons of opening force. The preload must be *at least* 25,000 pounds for each tie rod. Consider first the method shown in Figs. 5.33 and 5.34. The torque, T, required can be approximated by Eq. (5.10).

$$T = K * D * F/12 \qquad (\text{Ft} - \text{Lb}) \tag{5.10}$$

Where F = 25,000 lbs. (minimum)
D = nominal fastener dia. = 3.5 in (tie-rod diameter)
K = 0.15 to 0.20 depending on lubrication.
Thus: $T = 25{,}000 \times 3.5 \times 0.2/12 = 1458$ ft. lb.!

A considerable torque that may be difficult to accomplish in the field! Consider now the method shown in Figure 5.35 where four 5/8 socket head cap screws might be used to retain each nut. Thus: $T = (25{,}000/4) \times (5/8) \times 0.2/12 = 65$ ft. lb. This is much easier to deal with in the field.

Tie-Bar–Less Machines

A discussion of clamp design would not be complete without a discussion of machines without tie-bars (Figs. 5.36 and 5.37). These machines offer the distinct advantage that mold changes can be made without interference by the tie rods. Since there are no tie-rods, the load path around the mold is completed by using a C-frame that is usually an integral part of the clamp base. The deflection that occurs in the C-frame under load can be analyzed today by using finite element methods (Fig. 5.38). The end of the C-frame that supports the stationary platen tends to deflect around the large radius in the notch of the frame. This results in the platen tipping toward

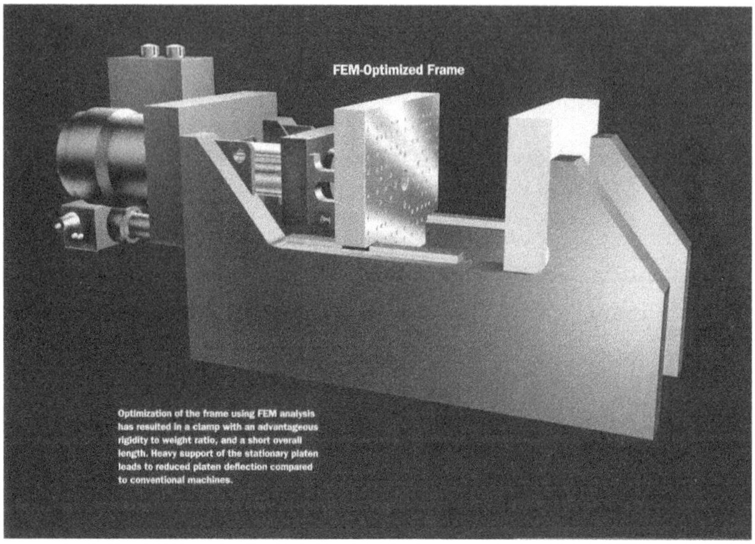

Figure 5.36 Tie bar-less clamp to provide easy access for mold changes. (Courtesy Engel.)

Figure 5.37 Top view is Milacron's PROWLER tie-bar-less clamp. Bottom view is HPM's ACCESS tie-bar-less clamp.

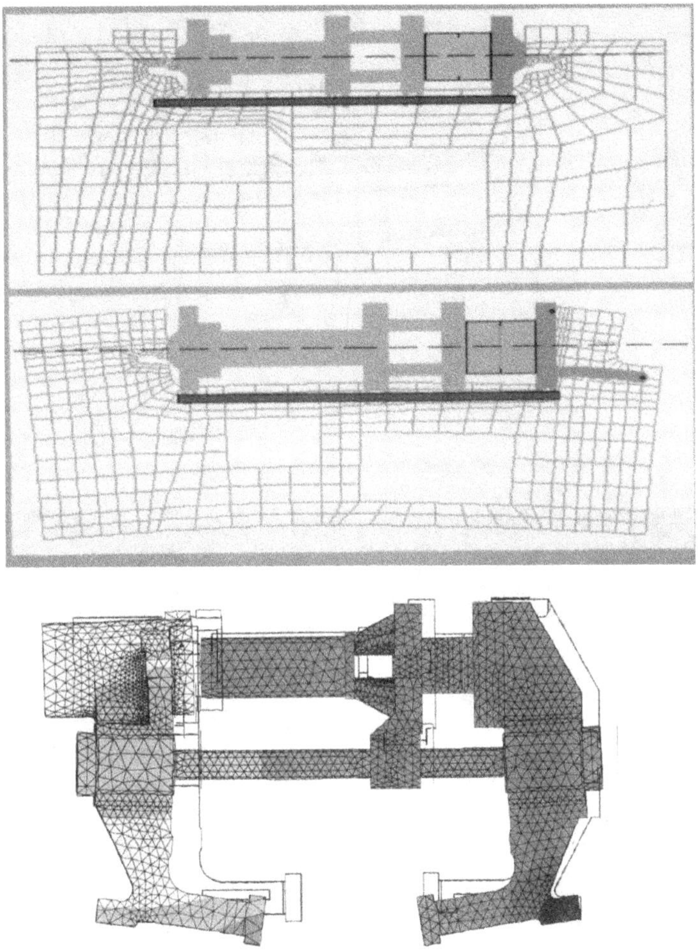

Figure 5.38 Finite element analysis of C-frame showing areas of deflection. (Top view courtesy of HPM and bottom view courtesy of Milacron.)

the injection unit. If the moving platen and mold are not allowed to "follow" the stationary platen then misalignment between the two mold halves occurs. Several methods have been used to provided the required "flexibility." Figure 5.39 shows the use of a pin joint (top) and flexible element (bottom).

5.4.3 Toggle Pins and Bushings

For toggle clamps, link pins and bushings must be designed to operate with minimum wear under high contact stresses. The typical nominal contact stress between the pins

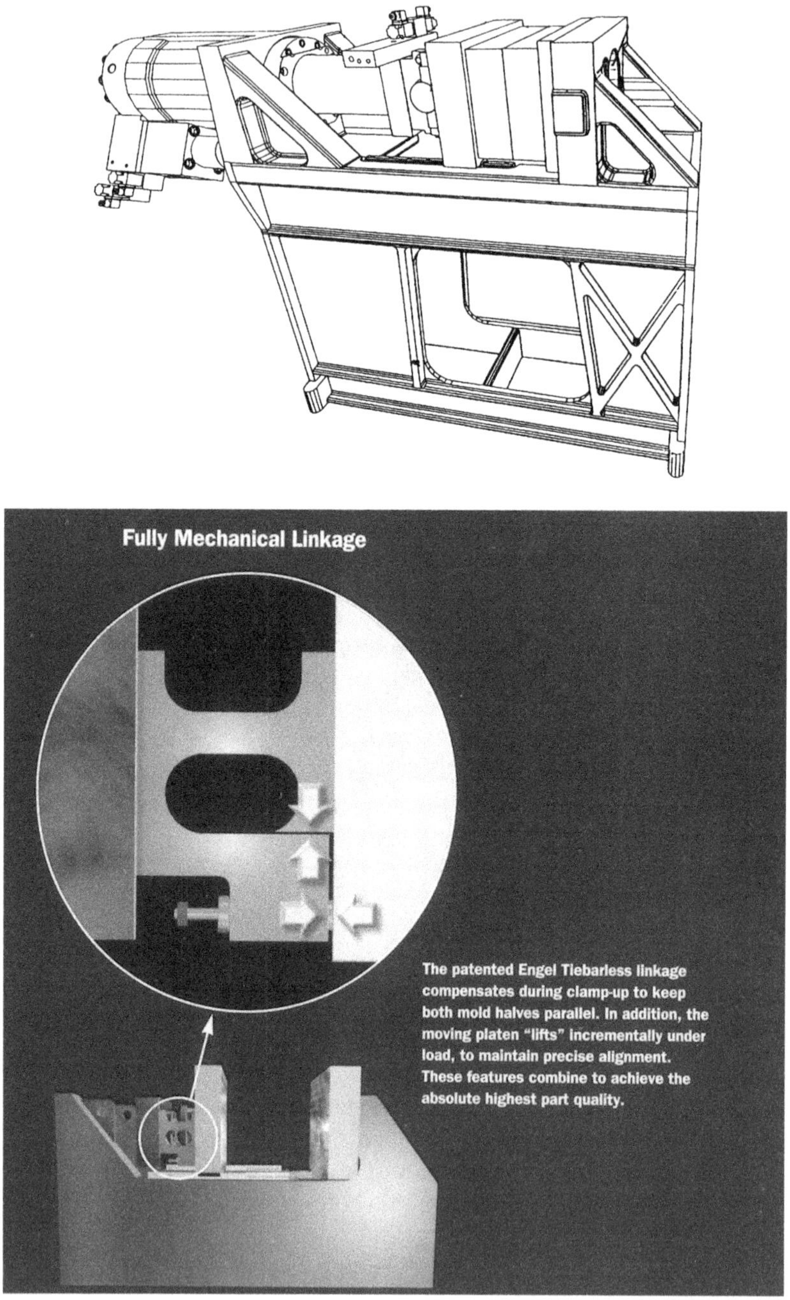

Figure 5.39 Two methods used to provide flexibility required so platens stay aligned as frame deflects. Top view shows pin joint behind moving platen to allow the moving platen to align with the stationary platen as the C-frame deflects. Bottom view shows the use of a flexing element to allow for alignment. (Courtesy Engle.)

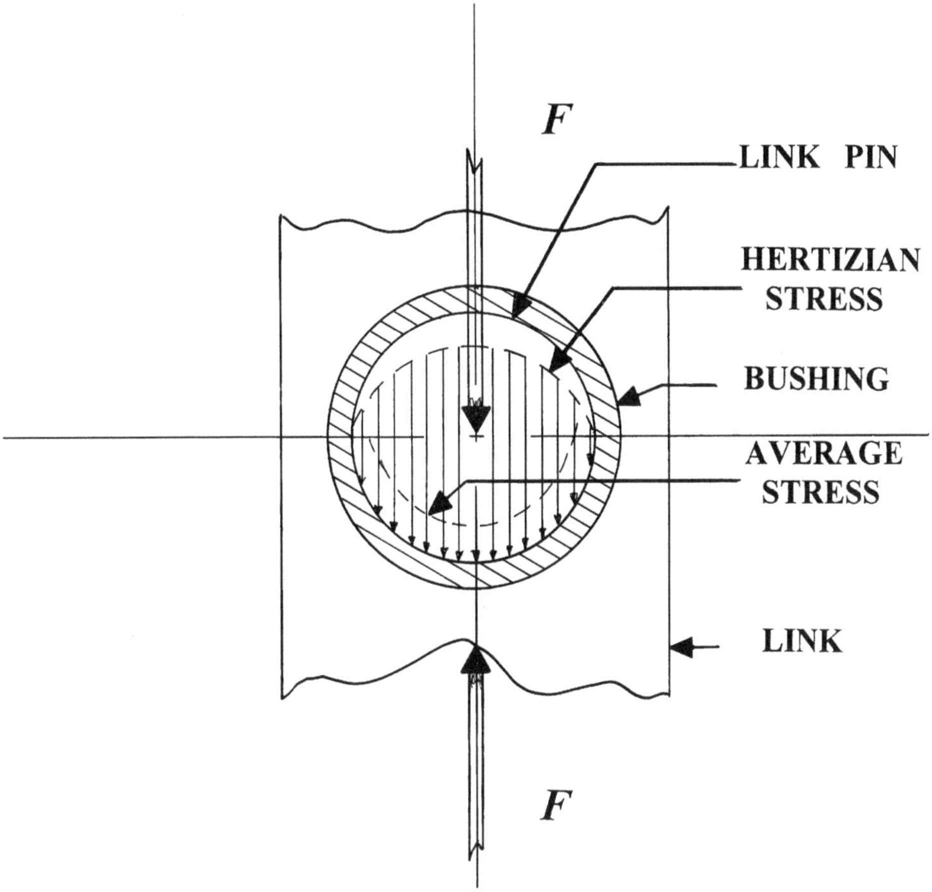

Figure 5.40 Toggle link pin and bushing combination showing the Hertzian contact stress as contrasted to the average contact stress between them.

and bushings is in the 10,000 to 12,000 psi range. Because of the clearance between the pin and bushing, the maximum contact stress is many times higher than the nominal stress (Fig. 5.40). It is for this reason that the link pins and bushings are usually made of carburized or nitrided steel and lubricated with either extreme pressure grease or heavy oil. Some machine are designed with bushings that require lubrication only once every several hundred thousand machine cycles. Even though this may not be the most conservative design, it saves the costs associated with an automatic lube system and eliminates the problem of the accumulation of grease or oil on the clamp end. This can be a great advantage in a medical or clean-room molding application.

Bushings are usually installed with a shrink fit. The resulting diametrical clearance between the pin and bushing is typically between 0.001 and 0.002 mm/mm (in per in) of pin diameter. For lubricated bushings, grooves are provided for the lubricant.

The most conservative design allows the link pins to rotate freely and independently of the links. By allowing the pin to rotate freely (or float), the entire pin circumference provides a wear surface and the pin wears uniformly. This results in a much longer wear life. If the pin is prevented from rotating, the same small area is stressed during the final 5 to10 degrees of link rotation as lock up occurs. This causes the pin to wear egg-shaped. Badly worn link pins and bushings can result in slop in the toggle mechanism and can result in alignment problems as the mold is closed.

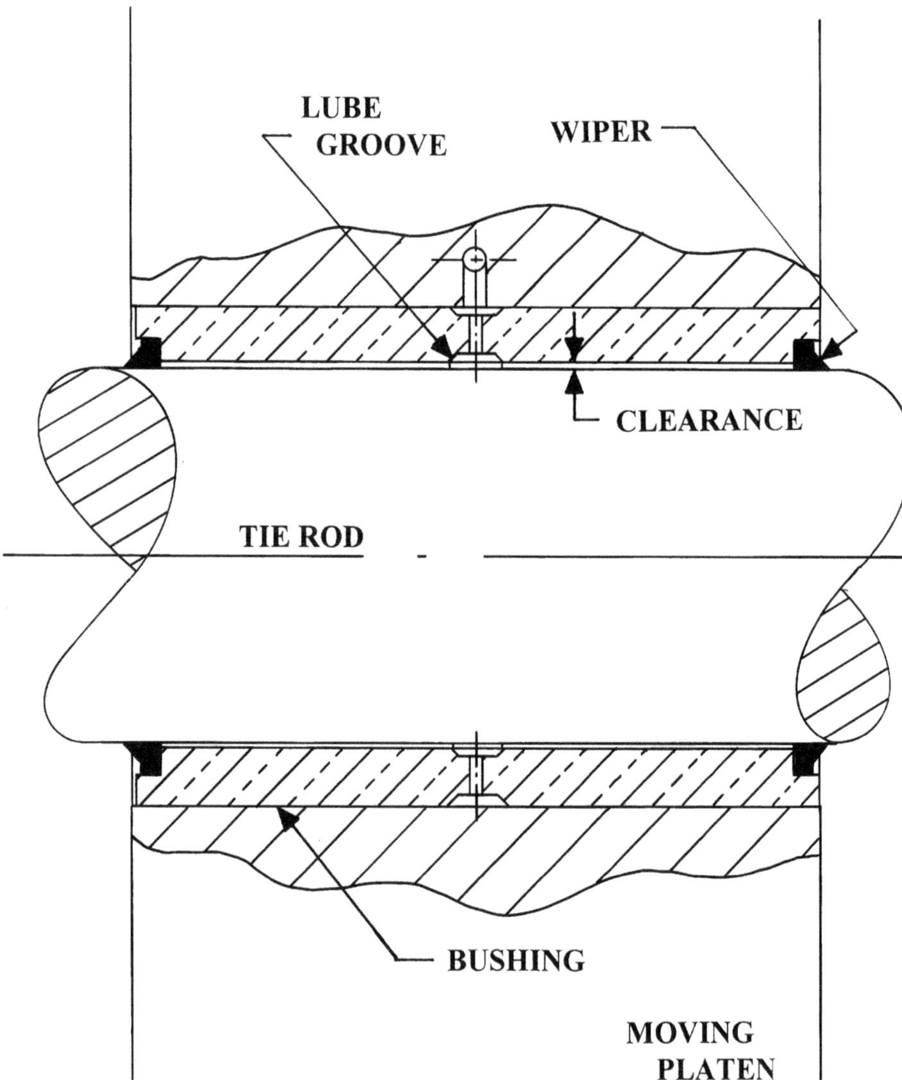

Figure 5.41 Tie-rod bushings installed in moving platen. Note relatively large running fit with the tie rods and the use of a wiper to exclude foreign material generated from the molding operation.

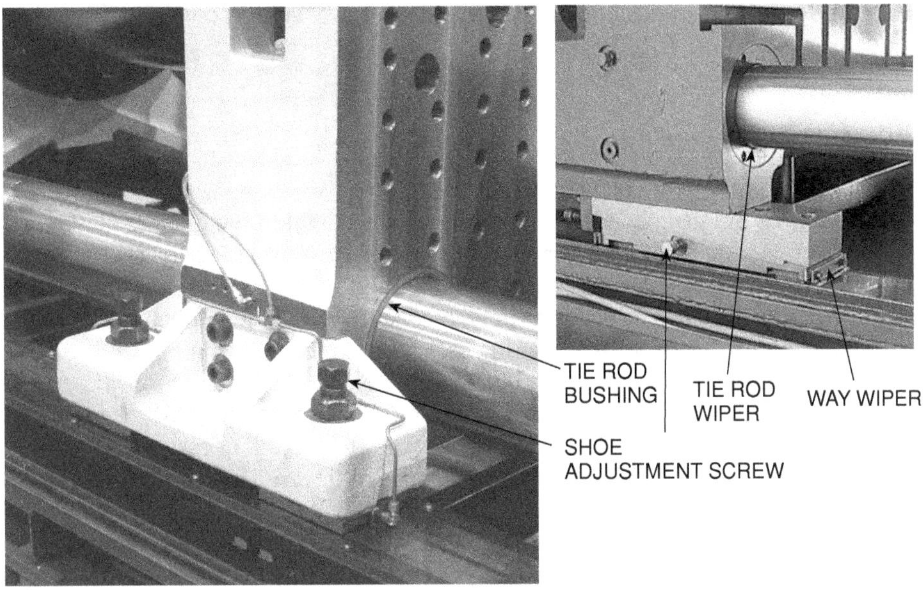

Figure 5.42 Moving platen support shoes with adjusting means for wear compensation. (Left is courtesy of Milacron. Right is courtesy of Sandretto.)

These alignment problems result early wear in mold leader pins and perhaps in tie rod failure due to unbalanced loading.

5.4.4 Tie-Rod Bushings

Bushings (Fig. 5.41) are usually provided in the moving platen. These bushings are used to help guide the moving platen laterally as it opens and closes the mold. A loose running fit of 0.002 to 0.003 mm/mm (in per in) of rod diameter is typically used between the rod and bushing. Leader pins in the mold provide the final (and most accurate) guidance once the mold is closed enough to engage them. A loose running fit is used in the tie rod bushing to allow for the tolerance stack-up that occurs. Allowance must be made for tolerance in the tie rod hole location, tie rod diameter, lack of tie rod straightness, and sag in the tie rod due to its own weight. This last point is often overlooked, but tie-rod sag on a 750-ton machine, for example, can exceed 0.005 in at midspan.

Bronze bushings with oil or grease lubrication are the most common design. Some machines have plastic, wood, or composite bushings that require little or no lubrication. This provides a cleaner mold area that can be of great importance in medical and clean-room molding applications. Unless the tie rods are chrome plated (which is rare), periodic application of a thin film of lubricant may be specified by the machine builder to prevent the tie rods from corroding.

Tie-rod wipers (Fig. 5.42) are often installed on the mold mounting side of the moving platen to exclude any foreign matter that may collect on the tie rods during operation. Provisions are usually made to replace these wipers conveniently as they wear.

5.4.5 Moving Platen Support

Tie-rod bushings are usually not used to support the weight of the moving platen and mold except on some small machines. Instead, shoes are provided for that purpose (Fig. 5.42). These shoes usually ride on a smooth and lubricated surface (way) made of high carbon spring steel attached to the machine base. Most manufacturers provide adjustable shoes with replaceable wear pads. The need for adjustment of these wear pads is evidenced by indications that the movable platen is beginning to ride on the tie rods. Follow the manufacturer's recommendations for periodic inspection and adjustment of these shoes. The longer the shoe, the better it resists the tipping moment imposed by the movable mold half that is cantilevered from the mold mounting face of the moving platen. In addition, as the moving platen is approaching the closed position, it tends to tip forward during the deceleration. Some machine designs have other means for resisting the tipping moment. Hydraulic clamps use the cylinder rod and some hydromechanical clamps use guide rods projecting from the rear surface of the moving platen into guide bushings in the rear platen. A large moment of inertia of the rod(s) and a rigid joint between the rod(s) and the back of the platen is important to resist tipping.

All parts are occasionally not ejected from the mold as it opens. If the mold is closed on these parts with a high force, mold damage can occur. To provide the maximum mold protection, it is important to reduce the drag force during the final closing as the mold halves come together. Some machinery manufacturers use shoes with rollers or ball bearings (Fig. 5.43) running on hardened surfaces (ways) in an effort to reduce closing frictional drag. This is very effective. Because of the high Hertzian contact stress between the roller and the way, however, spalling (pitting) problems can be encountered if the system is not properly designed. In addition, there can be a problem with resin fines and stringers that tend to fall onto the ways unless way wipers are used. Another method for reducing drag on the ways is to inject compressed air into the platen shoes so they act as air bearings. This reduces the contact stress on the ways, but it may introduce an oil mist into the mold area.

5.4.6 Shut Height Adjustment

Most hydraulic clamps use spacer blocks to adjust for significant reductions in mold shut height (Fig. 5.44). Toggle and some hydromechanical clamps use adjustable tie-rod nuts adjacent to the rear platen to make adjustments for different shut heights. The two most common ways of adjusting these nuts is by either a central ring gear (Fig. 5.45) or by a chain and sprocket drive (Fig. 5.46). For chain drive systems, care must be taken to provide *at least* 90 degrees of chain wrap-around on the nut

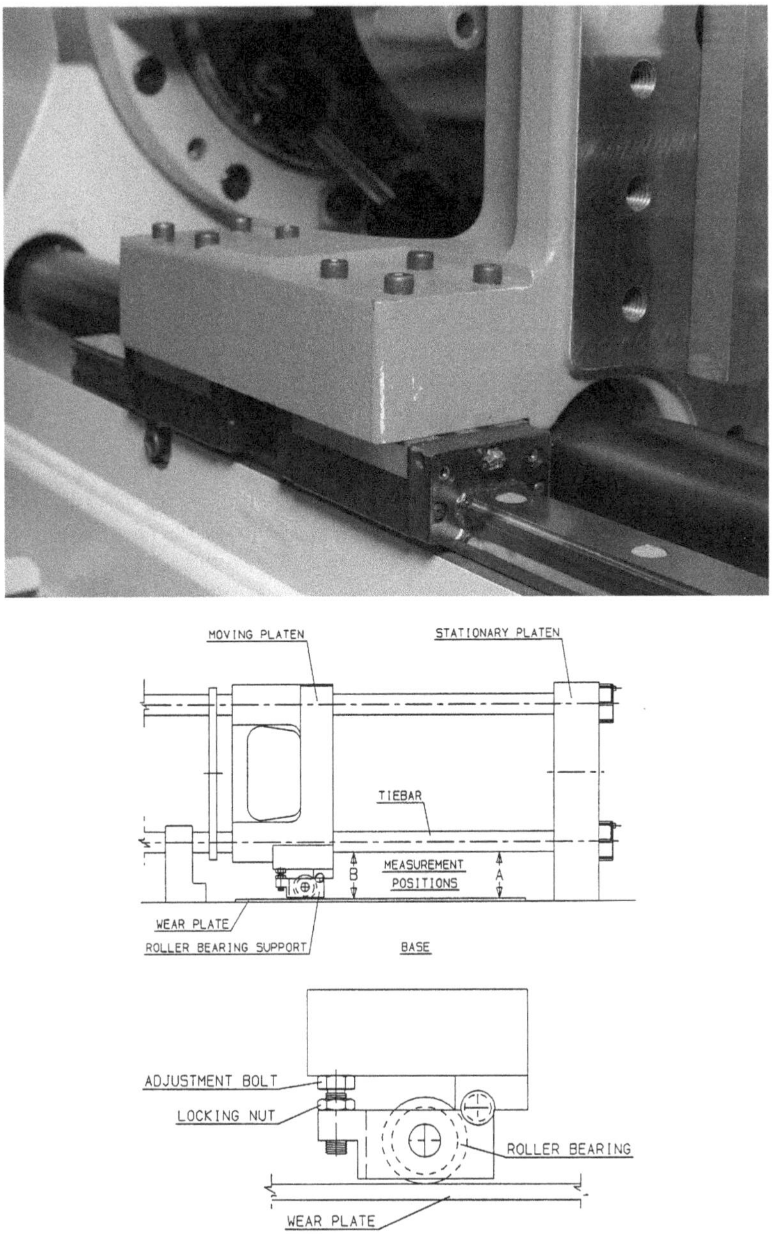

Figure 5.43 Moving platen support shoes with rollers to reduce drag force during clamp closing. (Top is courtesy of Husky. Bottom is courtesy of Engel.)

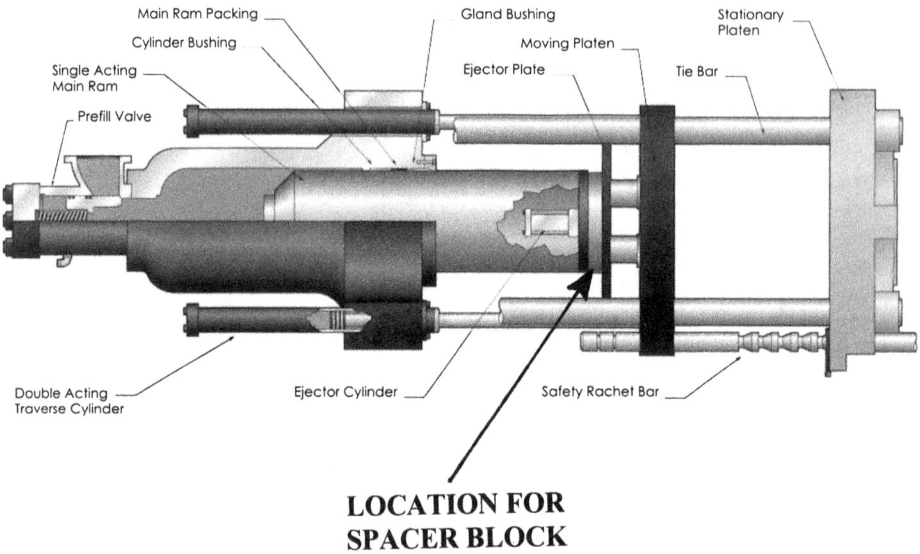

Figure 5.44 Hydraulic clamp with provisions to add a spacer block for molds with a small shut height. Spacer block is removed for large molds. (Courtesy Van Dorn.)

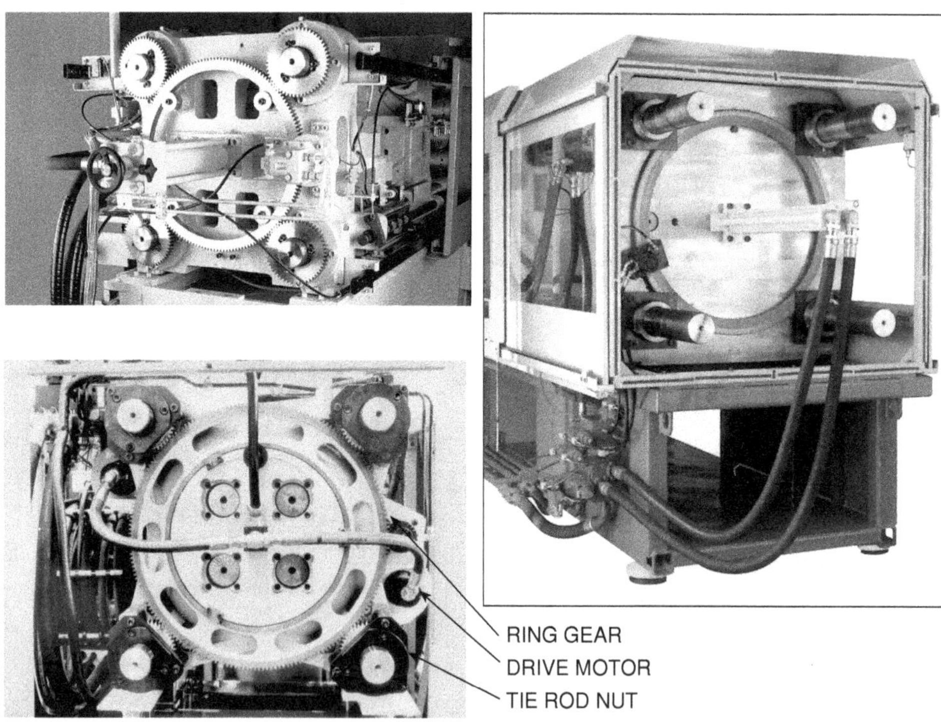

Figure 5.45 The use of a ring gear driven by a hydraulic motor to drive the adjustable (rear) nuts in a synchronized manner for clamp shut height adjustments. (Courtesy Sumitomo [top-left], Husky [bottom-left], and Van Dorn [right].)

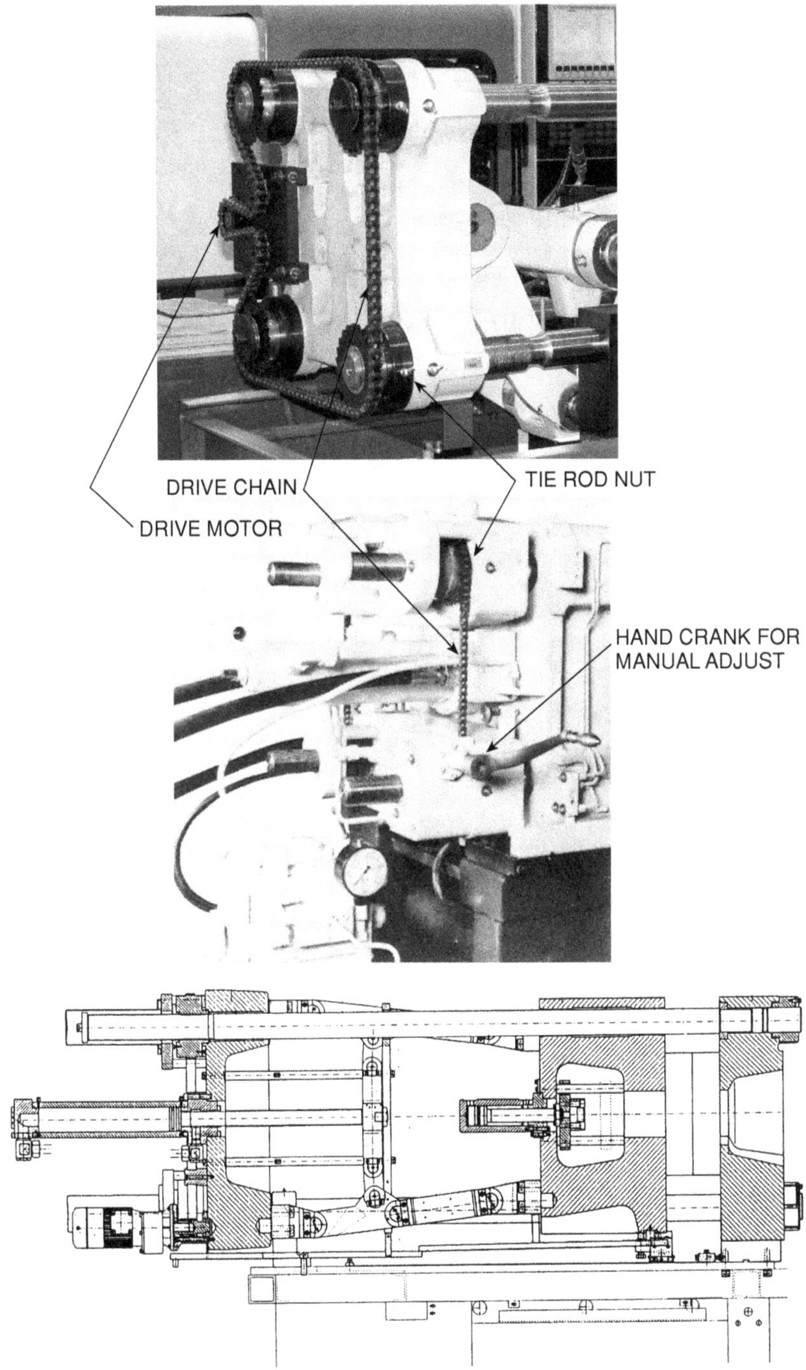

Figure 5.46 Chain and sprocket drive. Top view shows a system with hydraulic drive motor. (Courtesy Milacron.). Center view shows an optional manual (hand crank) drive. (Courtesy Engel.)

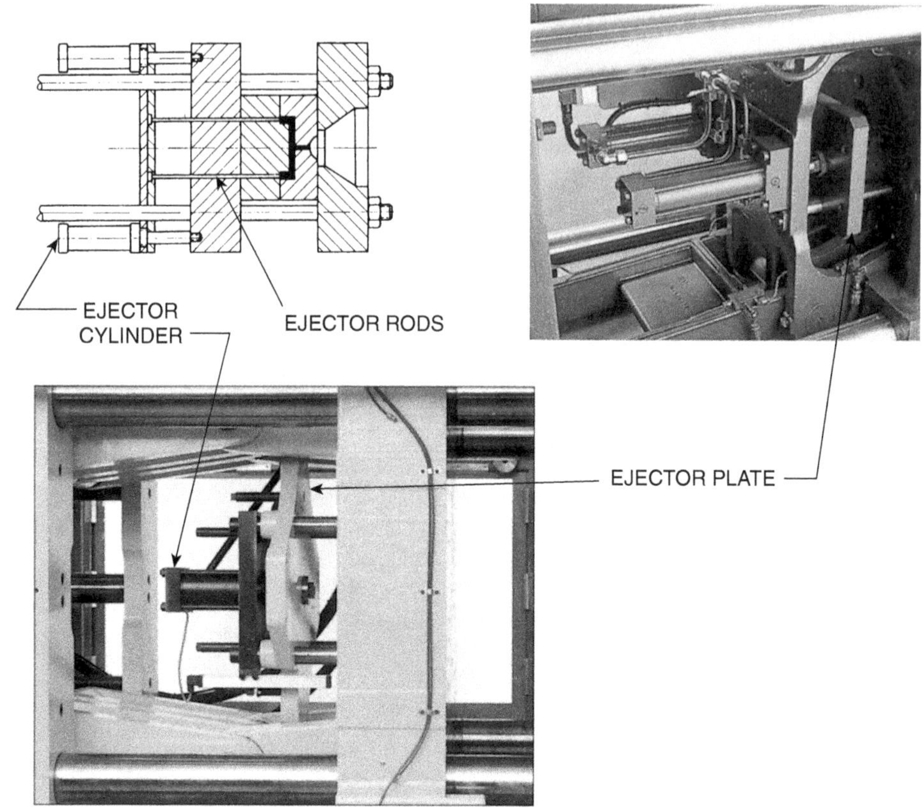

Figure 5.47 Top left view shows a typical ejector system. Top right shows an ejector system for a toggle clamp in which the moving platen is cast with a surface for mounting the ejector hydraulic cylinders. (Courtesy Sandretto.). Bottom left shows an ejector system for a toggle clamp in which the hydraulic cylinder mounting plate is attached to the rear of the moving platen on stand-offs. (Courtesy Van Dorn.)

sprockets so that the chain does not jump teeth under load. In either case it is important to note the location of "timing marks" between the rear platen and the nuts so that if the nuts are removed for any reason they can be reinstalled with the proper tooth alignment. This will insure that the clamp is square and that the tie-rod loading is balanced. These timing marks are usually made by the machinery manufacturer after a clamp squaring and balancing procedure has been completed. The drive motor for the nuts is usually either electric or hydraulic.

5.4.7 Ejector Systems

As mentioned earlier, one of the functions of the clamp is to eject the parts after the mold opens. On machines used in the United States (especially medium and large machines), an ejector plate is designed into the machine and is usually just behind the moving platen (Fig. 5.47). This plate drives a series of customer-supplied ejector rods (often referred to as “knock out” rods) that pass through holes in the moving platen and into the back of the mold to engage the mold’s ejector plate. The ejector plate is usually driven by a hydraulic cylinder(s) (or electric actuator) to come forward as the parts are to be ejected from the mold. The required ejector force is usually small (a few tons) compared with the clamping force. It is important that the ejector plate has good guidance and that the ejector rods be matched in length (to within a few thousandths of an inch) so that excessive bushing wear does not occur. Standards have been established to allow the molder to move molds from machine to machine that prescribe the size and location of ejector holes for a range of machine sizes (Fig. 5.48). The standard for machines manufactured in the United States was developed by SPI.

5.5 A Special Discussion of Tie-Rod Design

5.5.1 Why This Element Is So Important

For mechanical and hydromechanical clamps, the tie rod is the largest contributor to clamp deflection and thus has a significant influence on clamp stiffness. In addition, even though the nominal stress in most tie rods is only 10,000 to 13,000 psi, there are mitigating factors that result in the actual stresses reaching close to 60,000 psi. Medium carbon steel such as AISI 1045 or alloy steels such as AISI 4140 or even AISI 4340 are sometimes used because tie rods are so highly stressed. Of all the elements in the clamp, the tie rod is the most likely one to fail in fatigue.

5.5.2 Thread Load Distribution

Fatigue failure usually occurs in the fillet of first thread engaged by the tie-rod nut on the side of the nut adjacent to the stationary platen (Fig. 5.49). The reason for this is that there is a nonuniform thread load distribution. In other words, the threads do not each carry an equal share of the total load. When the threads in both the nut and the rod are cut, the thread pitch is identical in both so there is uniform contact between the threads. When a load is applied to the rod, the rod stretches under a tension load, whereas the nut compresses (Fig. 5.50). Thus, the first thread nearest to the loaded face of the nut may carry two or three times the average thread load. If, for example, a rod loaded to 100 pounds has 10 threads engaged with a nut, then each thread should carry 10 pounds of force if there is a uniform thread loading; however, because the first thread may carry two to three times the average, it will carry 20 to 30 pounds of force. Figure 5.51 shows the thread load distribution for a 5.00-in tie rod used in a 400-ton machine. This figure was produced using methods from Ref.

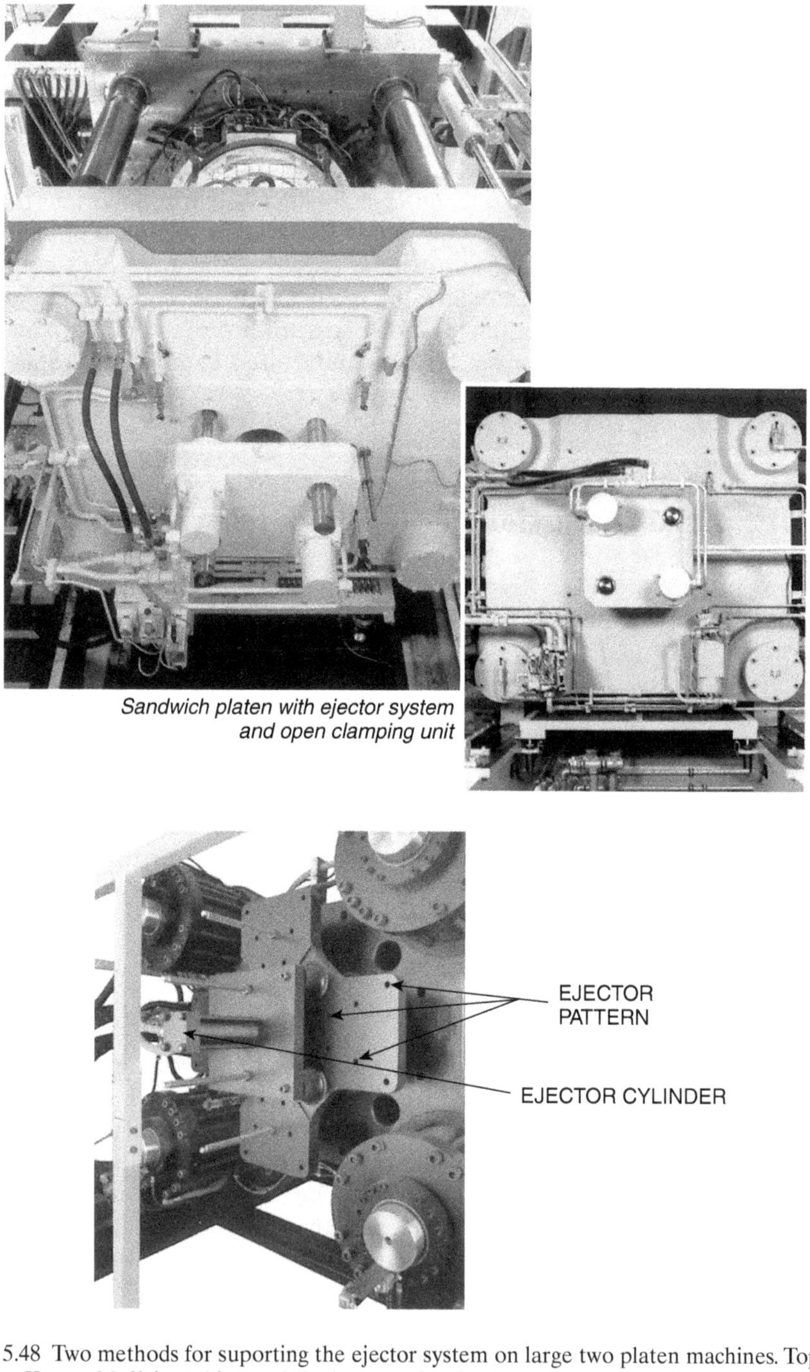

Figure 5.48 Two methods for suporting the ejector system on large two platen machines. Top views are for a Krauss Maffei machine and the bottom view is of a Van Dorn machine.

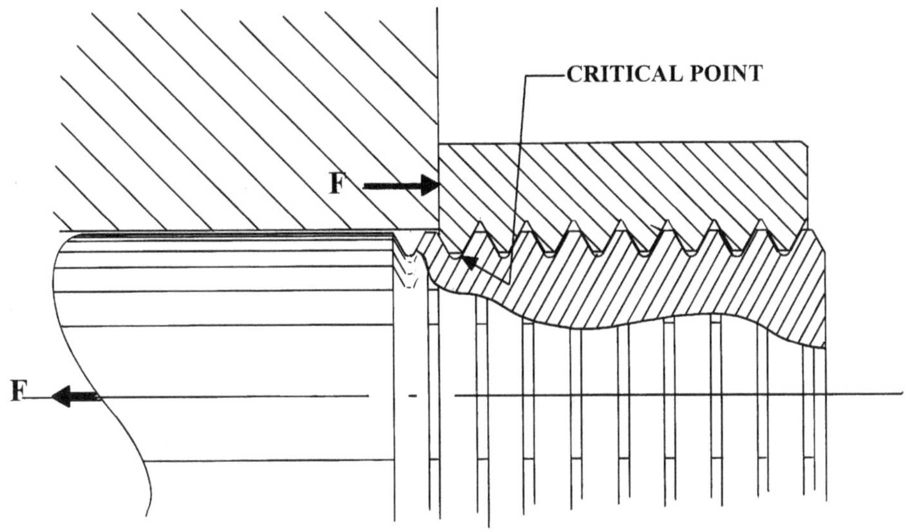

Figure 5.49 Location of the critical point for a tie rod.

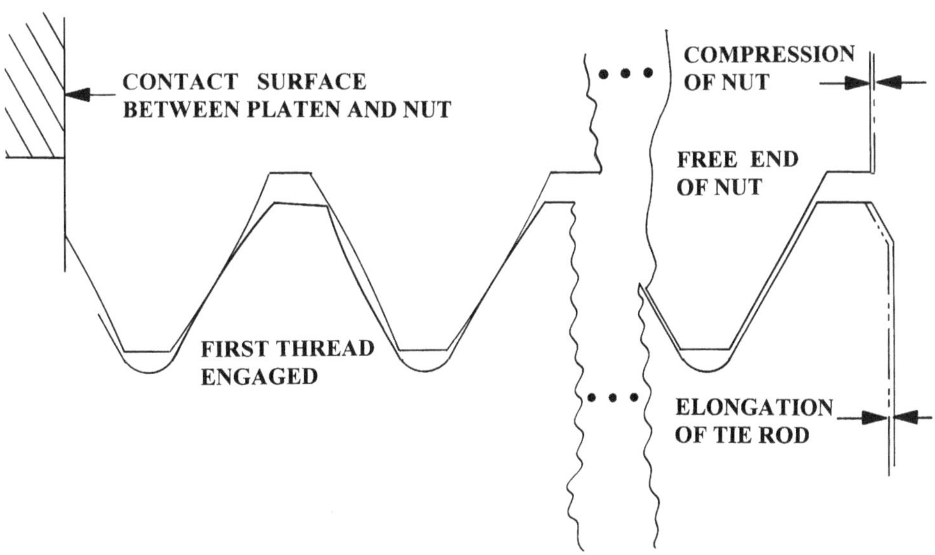

Figure 5.50 Nonuniform thread load distribution between a tie rod and nut resulting from the tie-rod elongating as the nut compresses.

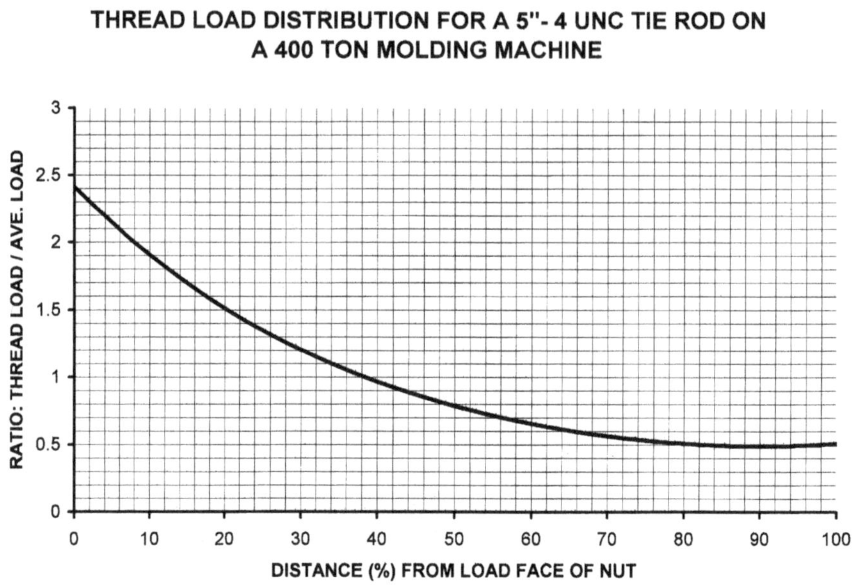

Figure 5.51 Thread load distribution as a ratio of average load for a tie rod on a 400-ton injection molding machine. Note that the first thread adjacent to the platen caries nearly 2.5 times the average load.

[9]. The resulting thread load distribution causes the first thread engaged to carry nearly 2.5 times the average load. The remaining threads then each carry a smaller proportion of the total load. The length of typical tie-rod nuts ranges from 75 to 100% of the nominal thread diameter. Increasing the length of the tie-rod nut does not solve the thread load distribution problem. The outside diameter of the nut is usually 1.5 time as large as the nominal rod diameter. Increasing this has only a small effect in improving the load distribution.

5.5.3 Thread Bending Stress

The load on each thread is applied at the pitch diameter with the maximum bending stress occurring in the fillet area of the thread. Various empirical relationships have been developed to predict this stress. Roark [10] offers an equation from Heywood (Ref. 49 in Roark) and modified by Kelly and Pedersen (Ref. 74 in Roark). Figure 5.52 shows a typical thread profile labeled to coincide with Eq. (5.11).

$$S_b = K_{\text{t-bend}}(W/t)[(1.5a/e^2) + \cos(\beta)/2e + 0.45/\sqrt{(be)}] \tag{5.11}$$

Where: S_b = the nominal bending stress in the thread fillet
t = the thread thickness

a = the normal distance to the load line
e = the normal distance to the thread axis
β = as shown in the figure
b = the distance between the critical point and the load point
$K_{t\text{-bend}}$ = the geometric stress riser = $[1 + 0.26(e/r)^{0.7}]$

Although this thread profile is not in actual use, it is included here simply to provide an example for explaining how the factors used in Eq. (5.11) are determined. For the unified thread form, for example, α and γ have values of 30 and 60 degrees, respectively. For the buttress thread used in the United States, α and γ have values of 7 and 45 degrees, respectively. The lines A and B are formed by the intersection of the beam axis with the line joining the point of maximum tensile bending stress with the corresponding point on the compression side of the thread profile. For symmetrical thread profiles such as Acme and Unified threads, A and B are equal and γ is bisected. Heywood recommends making θ equal 30 degrees and, as the ratio of b to e increases to more than 3, reducing θ to 12 degrees for very long beams. W/t is the load per inch of thread thickness.

$$W = \text{the load acting on the thread flank} = H * F/(N_t * \cos(\alpha - \Phi)) \tag{5.12}$$

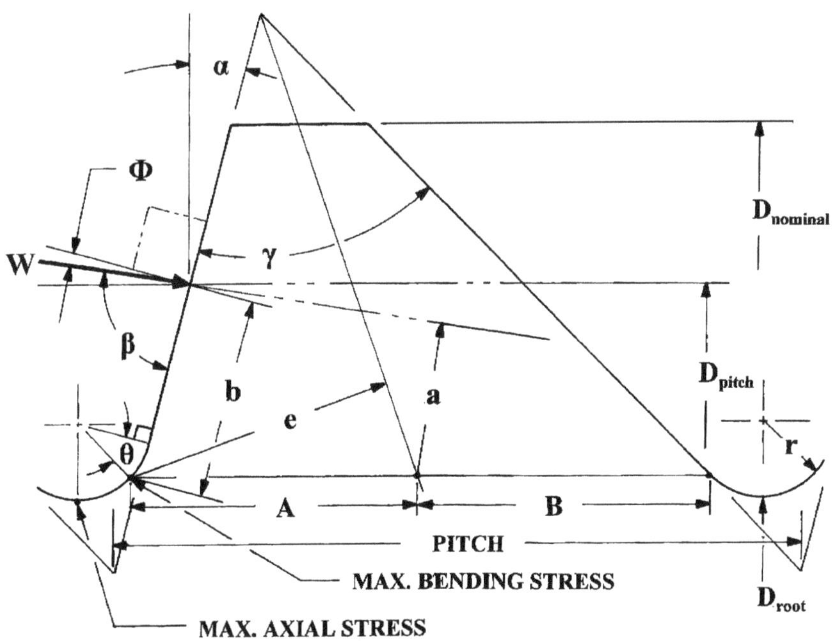

Figure 5.52 A typical thread profile showing dimensions used to calculate the maximum bending stress located in the fillet radius. Note that the maximum stress is the combination (not algebraic) of the maximum bending stress with the maximum axial stress at the base of the fillet.

H = is the thread load factor
F = is the total load on the tie rod
N_t = is the number of threads carrying the total load
α = is the flank angle
Φ = is the friction angle = arctan(coefficient of friction)

Equation (5. 12) is used to find W as a function of total load, F, carried by N_t threads with a thread load distributuion factor, H, and a friction angle, Φ. Typical values for friction coefficient range between 0.15 and 0.2 (Φ between 8.5 and 11.5 degrees) depending on the degree of lubrication. It can be seen from Eq. (5.11) that as friction is reduced, β approaches 90 degrees and a is reduced. Both of these reduce the bending stress. As the fillet radius is increased, the geometric stress concentration factor, $K_{t\text{-bend}}$, also decreases.

5.5.4 Thread Axial Stress

The axial stress is the local thread load plus the load from the threads "up stream" passing under the thread divided by the root area. For the first thread this load is the total load applied to the rod. A stress concentration factor [6, 7, 8] must be applied to this value. For a well-designed thread profile having a liberal fillet radius, a typical value for $K_{t\text{-axil}}$ is between 2.5 and 3.5. To improve fatigue resistance, it is important that the fillet radius be as large as possible with a smooth finish having no tool marks.

5.5.5 Combined Stresses

The combined stress is the result of the bending stress (including $K_{t\text{-bend}}$) superimposed with the axial stress (including $K_{t\text{-axil}}$) at that thread location. The two stresses occur at slightly different locations, as seen in Fig. 5.52, and cannot be added algebraically. Through photoelastic studies, Heywood [5] has developed am empirical relationship for combining these two stresses as shown in Eq. (5.13).

$$S_{\text{combined}} = S_b + S_a/[1 + c(S_b/S_a)] \tag{5.13}$$

Where $S_a = K_{t\text{-axial}}$ times, the nominal axial stress in the root diameter

$$c = [(60 - \alpha)44]^2 \tag{5.14}$$

The first thread engaged by the nut has the highest combined stress because the thread loading is highest on that thread and the axial stress is equal to the total load on the rod. This is why fatigue failure usually occurs here. Figure 5.53 shows a plot of the combined stress at various distances along the length of engagement. The lighter line marked "FATIGUE" represents the combined stress after adjustments for the notch sensitivity of the tie rod metal.

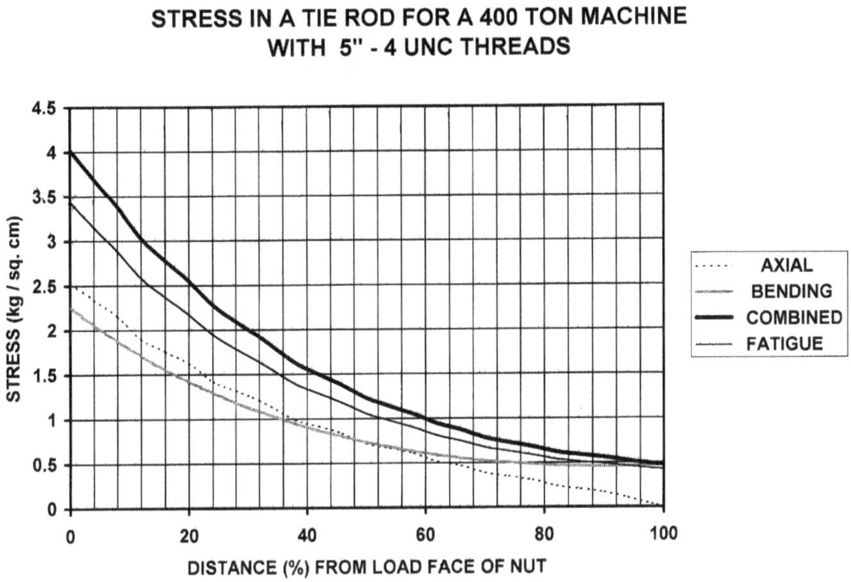

Figure 5.53 Graph of maximum thread-bending stress, axial stress, and combined stress in a tie rod. Note that these include the stress concentration factors.

5.5.6 Mitigating Factors

The combined stress at the first thread is usually increased in actual service due to several mitigating factors. One factor is the deflection of the stationary platen around the mold. The platen bearing surface under the nuts takes on a slope as a result of this platen deflection. This causes a bending moment to be superimposed onto the axial load applied to the rod. As a result, there can be a 10 to 20% increase in stress over that caused by the nominal axial load alone. This percentage will be even higher if the stationary platen is made too thin. I have observed cases where tie rod failures have occurred when a special wide platen machine was designed without increasing the platen thickness. The increase in platen slope at the tie rod intersection lead to fatigue failures of the tie rods.

Another factor is the imbalance in load among the four tie rods. For various reasons, it is very difficult to get the tie rods perfectly balanced. Thus, a 5% imbalance (which is not uncommon) will raise the peak stress by that amount. It can be seen that the combination of the 20% bending and the 5% imbalance can nearly eliminate the 1.3 safety factor that is typical of most machines. This is why tie-rod failure can occur in machines running molds that are not parallel, do not have a balanced layout of the cavities, or are overpacked and flash frequently.

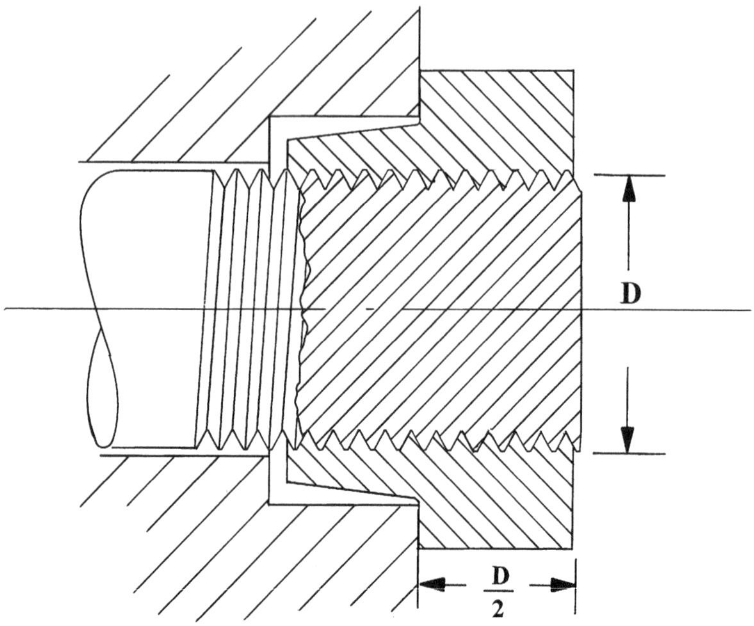

Figure 5.54 Tension nut that allows the first few threads engaged with the tie rod to elongate with the rod and thus reducing the load on the first few threads.

5.5.7 Ways to Improve the Design

There are several ways to improve tie-rod design. Improving thread load distribution so the first thread carries a lower load is very effective. Two methods used to accomplish this are tension nuts and nuts with tapered threads. Of these two methods, tension nuts are the most common. This design results in the first portion of the nut engaging the rod to be in tension rather than compression (Fig. 5.54). Because this portion of the nut is stretching (as is the rod), the threads in the nut are in better alignment with the threads on the rod, and the thread loading is more uniform. This results in a lower bending stress at the base of the first thread and thus a lower combined stress. Care must be taken in the design of the portion of the nut that is in tension so a fatigue failure does not occur there.

The second method for improving the thread load distribution is with nuts that have the pitch diameter tapered so that less penetration into the pitch diameter of the tie rod occurs at the first thread engaged near the load face of the nut. Because these threads are engaged near their tips, they bend more and carry a lower load (are more compliant). This passes more of the total load to the threads "downstream." By using the proper taper, the thread load distribution can be reversed so that the threads nearest the end of the rod have the highest loading, as seen in Fig. 5.55. Because the axial loading is lowest at the end of the rod, the combined axial and bending stress can be made more uniform as seen in Fig. 5.56. By comparing Fig. 5.56

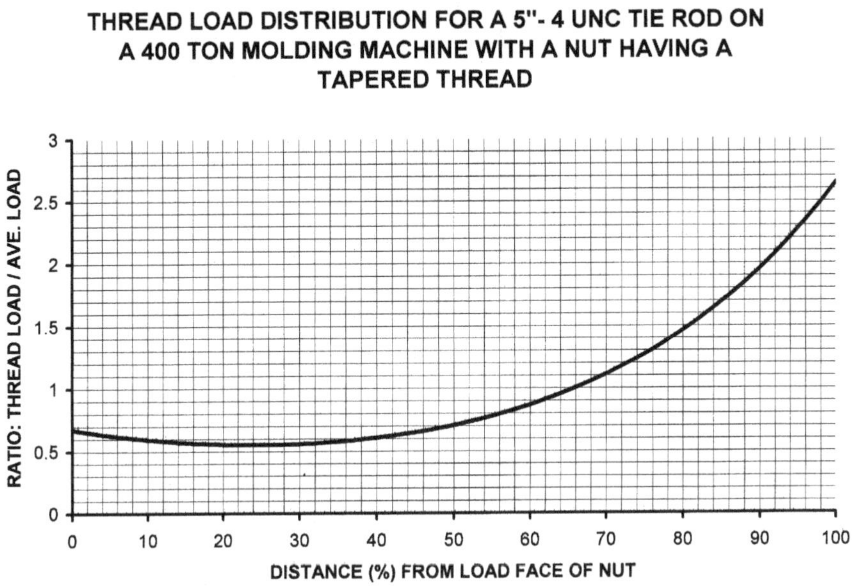

Figure 5.55 Thread-load distribution for a nut with a tapered pitch diameter such that the pitch diameter nearest the pressure face is the greatest.

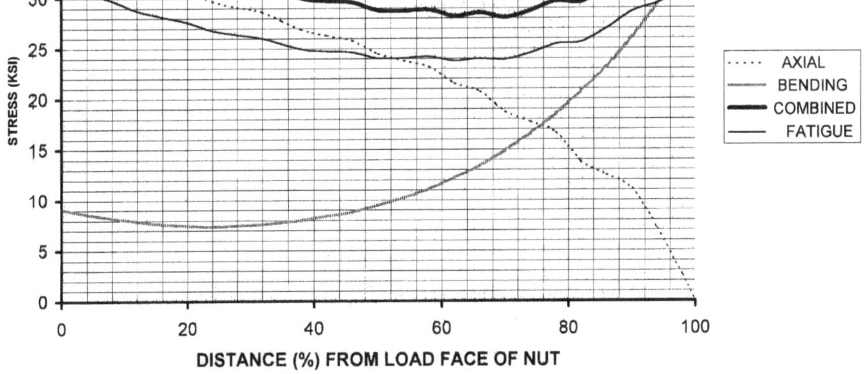

Figure 5.56 Graph of maximum thread-bending stress, axial stress, and combined stress for a nut with a tapered pitch diameter. Note dramatic decrease in the combined stress as compared with Fig. 5.53. Fatigue stress curve shows an adjustment downwards to account for "notch sensitivity".

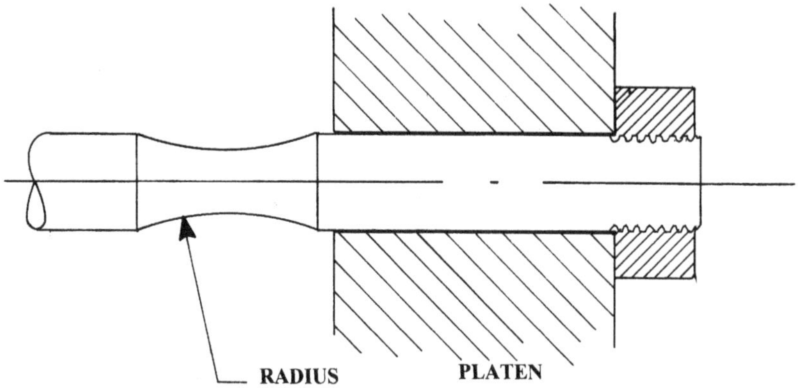

Figure 5.57 Reduction of tie-rod diameter to add some flexibility to the rod and thus reduces the superimposed bending stress caused by platen deflection. For those rods that incorporate a split collar for retention, this radius also reduces the stress in the retaining ring groove.

with Fig. 5.53, one can see the advantage of using a *properly* tapered thread. A 30% improvement in fatigue safety factor is not uncommon. This may not sound like much, but when the typical fatigue safety factor is only about 1.3, increasing the safety factor to 1.9 is dramatic! Care must be taken when using tie rod nuts with tapered threads to insure that they do not get installed backward. This *will* result in a premature fatigue failure.

On machines that use a split collar on the tie rods, a gradual reduction in tie rod diameter is usually made adjacent to the groove in the rod (Fig. 5.57). This both reduces the stress riser caused by the groove and makes the rod slightly more compliant with the platen as it bends around the mold. This reduces the bending moment transferred to the tie rod and thus the resulting superimposed stresses in the threads. Rolled threads also help because rolling the threads (cold flowing) causes the material in the fillet area to be left in compression.

Most fatigue failures occur on the end of the tie rod adjacent to the stationary platen rather than the adjustable end. One theory for this is that fatigue cracks may start on both ends of the rod, but when a different mold is installed, the adjustable nut is moved to a new location. Thus, that crack on the adjustable end stops growing. The stationary end of the rod is not so lucky because that nut is never moved. The crack that may start there will continue to grow until failure occurs.

5.6 Understanding Clamp Spring Rate (Stiffness)

Every element in the load path within a clamping system strains (deflects) as the load increases. Because the machine elements are usually designed to operate within the elastic range of the material, the deflection of each element is a linear relationship with the applied load. The energy stored in each element as it deflects is simply half of the product of the force times the deflection. The amount of stored energy is the same whether the element is compressed or elongated by the load. Thus, as an injection molding machine develops its clamp force, the total energy stored in clamp is the sum of the energy stored in each deflecting element of the clamp system. This includes the mold. Because the total stored energy is simply the sum of the stored energy in each element, it reasonable to replace the clamping system with an equivalent spring that has a spring rate equal to the total load divided by the sum of the absolute values of all the deflections of elements in series with the load.

5.6.1 How to Determine Clamp Spring Rate

The spring constant of any clamp system can be calculated in the same way. To start with, it is necessary to trace the load path through a clamping system. Figure 5.11 shows how one might analyze the load path in a toggle machine. The load and deflection in each element in that path can then be calculated. The governing equation for most of the required deflection calculations is: Deflection = Stress (S) * Free Length (L) / Young's Modulus (E).

Most machine elements are made of steel ($E = 29.5$ million psi) and are designed to a stress of between 10,000 and 12,000 psi. The mold deflection can be estimated assuming an average stress of 5,000 psi. The free length will vary for each element. Platen deflections may be estimated assuming a deflection of 0.0001 mm/mm to 0.0002 mm/mm (in per in) of diagonal distance between tie-rod center lines. In any case, the deflections of the platens are usually small compared with those of the tie-rods and links. The deflection within the tie-rod nuts can be estimated using the same method used for the tie rod, but using a free length of only 25 to 50% of the tie-rod diameter. This deflection is also small compared with that of the tie rod. Once each deflection is calculated, they are summed to get the total deflection that an equivalent spring would experience under the same load. Finally, the spring rate can be calculated by dividing the total clamp force by the total deflection calculated earlier. The units can be expressed as (tons per in) for use in a later section on understanding toggle clamps. Figure 5.7 shows how one might analyze the load path for a hydraulic clamp. For a hydraulic clamp the deflection of the large cylinder rod and the ejector box is substituted for the deflection of the links in a toggle clamp. The equivalent deflection of the large column of oil behind the ram at the time the mold is closed must also be added to the total deflection. This last "deflection" is usually several times as large as all the other deflections combined. This is because hydraulic oil is much more compressible than steel. The bulk modulus of oil is such that, at 2000 psi, the oil compresses about 1%. For example, if the column of oil behind the ram is 50 in long, then the equivalent deflection of that column is 0.50 in. The stiffness of a hydraulic clamp, such as the one shown in Fig. 5.8, is usually only about one fifth that of an equivalent sized toggle or hydromechanical clamp.

Because the mold is in the load path, its deflection must be considered in calculating the spring constant. If there is proper match between the mold and the chosen clamp size to run the mold, its contribution to the total deflection is usually only a few percentages of the total. If a mold is installed that was designed for a smaller machine, however, the mold deflection can have a considerable impact on overall stiffness. For example, if a mold that usually runs in a 100-ton machine is installed into a 300-ton machine, then its deflection may considerably reduce the system stiffness. The resulting stiffness may be so reduced that the 300-ton machine may only be able to develop 200 tons or less!

5.6.2 The Importance of Clamp Spring Rate

A stiff clamp has several advantages for the molder.

- Stiff clamps resist flashing better

This can be understood by realizing that for the mold to flash, it must uncompress to the point that the mold halves separate at the parting line. This requires the clamp "spring" to stretch even further than it did while developing the initial clamp force. Thus, the more "stiff" the clamp is, the more injection pressure it takes to flash the mold. Although it is true that if the mold flashes, a stiffer clamp will impose a higher additional force on the tie rods, it is still better to design a clamp that is less apt to flash.

- A lower initial clamp force can be used

Because a clamp with a high spring constant has a higher "holding" capacity, the initial clamping force can be reduced. This saves wear and tear on the clamp and the mold.

- A stiff clamp requires less energy

A clamp with a high spring constant uses less energy to clamp than does a clamp with a low spring constant running at the same clamp force. The energy to develop clamp force is represented by the area under the curves in Fig. 5.58. As can be seen, the area under the toggle clamp curve is much less (typically only one fifth) than is the area under hydraulic clamp curve for the same developed clamp force.

When the part has cooled and the clamp force is released so the clamp can be opened, the stored energy in the spring system must be dissipated. This energy usually shows up in the hydraulic oil. The additional cost to cool the oil adds to the expense of operation.

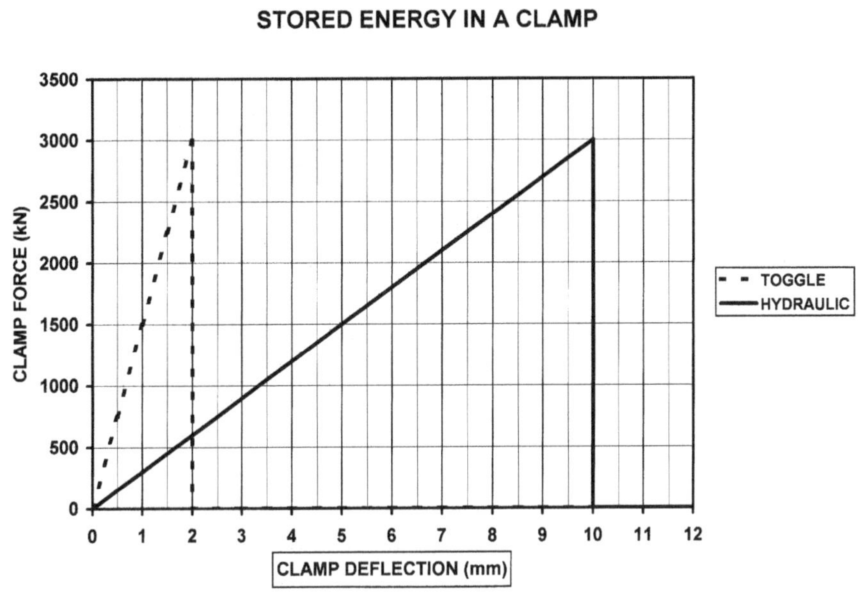

Figure 5.58 Energy required to deflect a clamping system. Note that the area (energy) under the toggle clamp is much less than the area under the equivalent-sized hydraulic clamp for the same clamp force.

5.7 Math Model for a Toggle Clamp

Mathematical relationships can be developed that describe the kinematic relationship between the toggle link positions and the available closing force. This closing force is a function of pin center-to-center distances for each link; the diameter of each pin; the coefficient of friction at each pin; the frictional resistance between the moving platen and the base ways; and the driving force of the actuator. Figure 5.59 shows a symmetrical toggle system and the defining geometry used in Eqs. (5.15 to 5.25).
α and γ as a function of β in Fig. 5.59:

$$Y_1 = Y_4 + L_{24}\cos(\beta) + L_{12}\sin(\alpha) \tag{5.15}$$

$$\text{Thus, } \alpha = \sin^{-1}\{(Y_1 - Y_4 - L_{24})/L_{12}\} \tag{5.16}$$

$$Y_3 = Y_4 + L_{24}\cos(\beta) + L_{24}\sin(\gamma) \tag{5.17}$$

$$\text{Thus, } \gamma = \sin^{-1}\{(Y_3 - Y_4 - L_{24})/L_{23}\} \tag{5.18}$$

$$X_3 = X_1 + L_{12}\cos(\alpha) + L_{24}\cos(\gamma) \tag{5.19}$$

Friction angles in pins in Fig. 5.59:

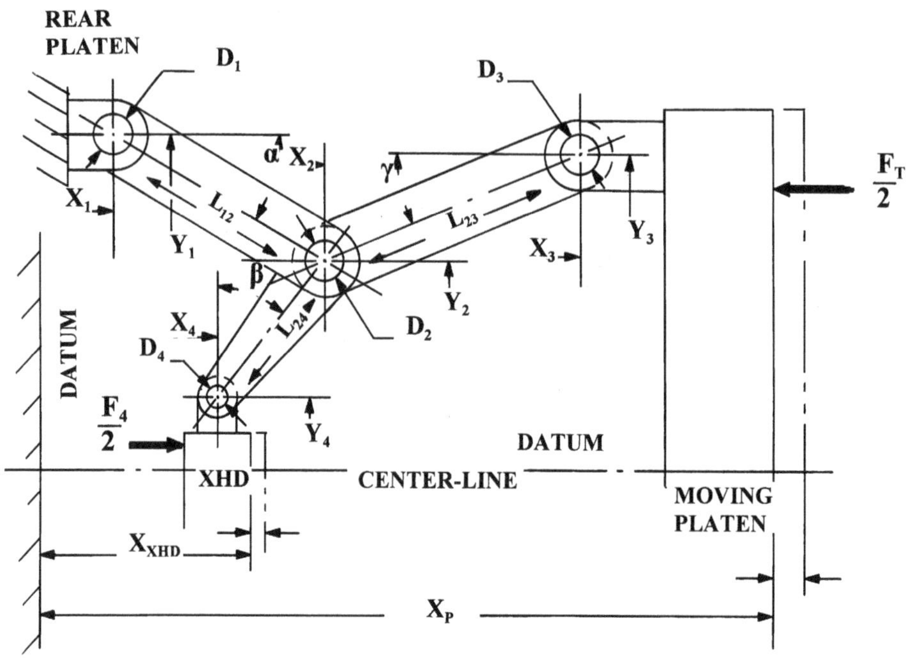

Figure 5.59 Typical four-point toggle system showing the notation for the lengths and angles used to derive the kinematic relationship for predicting available clamp force.

$$\alpha' = \alpha + K * \tan^{-1}\{\mu(D_1 + D_2)/(2L_{12})\qquad (5.20)$$

$$\gamma' = \gamma + K * \tan^{-1}\{\mu(D_2 + D_3)/(2L_{23})\}\qquad (5.21)$$

$$\beta' = \beta + K * \tan^{-1}\{\mu(D_2 + D_4)/(2L_{24})\}\qquad (5.22)$$

Clamping Force (F_T) in Fig. 5.59:

$$F_T = F_4\{\cos(\gamma')/\sin(\alpha')\} * \zeta_1/\zeta_2\qquad (5.23)$$

Where $$\zeta_1 = \cos(\beta') * L_{12}\cos(\alpha') + \sin(\beta') * L_{12}\sin(\alpha')\qquad (5.24)$$

$$\zeta_2 = \cos(\gamma') * L_{12}\cos(\alpha') + \sin(\gamma') * L_{12}\sin(\alpha')\qquad (5.25)$$

Equations (5.15 through 5.19) are simply based on geometry. The lengths L_{ii} are the distances between pin center lines for each link. Equations (5.20 through 5.22) are to account for the friction in the pins. The variable, K, takes on a value of (+1) when the cross-head is driving and a value of (–1) when the platen is driving (as during the untoggle motion). Equation (5.23) gives the available clamping force as a function of driving force. Figure 5.60 shows a plot of the available closing force for a 250-ton toggle machine. The force is plotted as a function of the change in position of the moving platen relative to the stationary platen. Notice that when the two platens are close, the available force is much higher than 250 tons. Also note the effect that the coefficient of friction in the pins has on available force at any position. At a higher coefficient of friction, the curve slips to a lower position.

Knowing the available clamping force is only half the answer in determining the actual clamp force that a clamp design can develop. The other half of the answer lies in the spring rate (stiffness) of the clamp/mold system. This can be calculated using the methods described earlier in Section 5.6.1. In Fig. 5.61 the clamp spring rate is superimposed onto a plot of the available clamp force. To lay in the spring rate curve (in this case 3800 tons/in), select a convenient point on the horizontal axis (in this case 0.1 in) and connect a line to the clamp force developed when the equivalent spring is deflected by that amount. In this example, 380 tons is developed at 0.10 in of deflection. The maximum force a particular clamp can develop is determined by making the spring rate curve tangent to the available clamp force curve. The point where this tangent line intersects the vertical axis represents the maximum clamp force. The point where this tangent line intercepts the horizontal axis is where the mold halves make initial contact when running at maximum clamp force. This is often referred to as the *kiss point* because it is where the mold halves actually make contact as the clamp closes. The distance from the kiss point to the vertical axis is the interference and represents the total of all the deflections of all the elements in a load path.

As long as the available force curve is above or to the left of the spring rate line (tangent line), the clamp will operate. If the coefficient of friction in the toggle pins

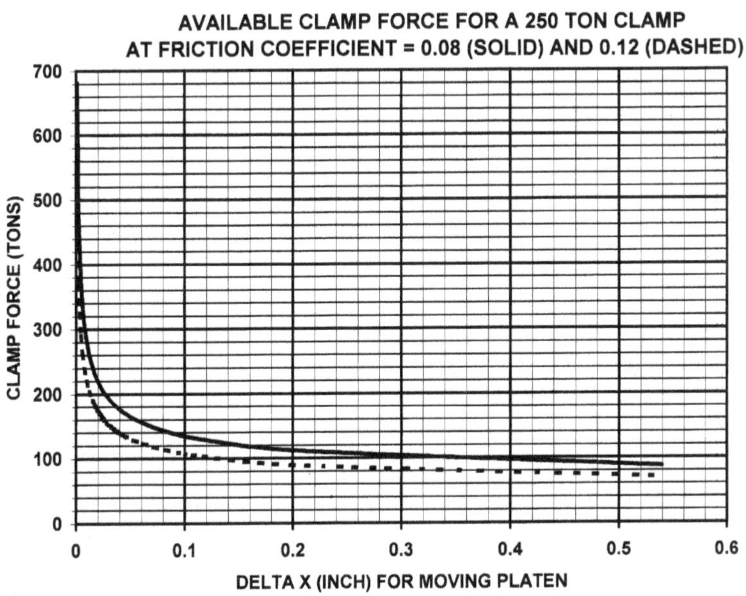

Figure 5.60 Graph of available clamp force as a function of moving platen position relative to the closed position. Note that as the clamp cycles thousands of times, lubrication is gradually squeezed out of the pin joints and friction increases, resulting in lower available clamp force.

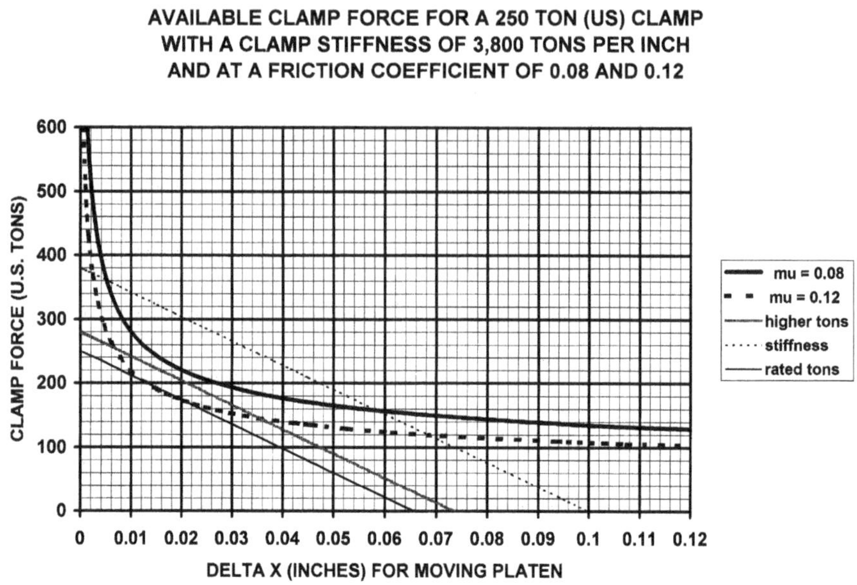

Figure 5.61 Graph of the spring rate line superimposed to the curve of the available clamping force. The maximum clamp force is where the spring rate line intersects the vertical axis and where it intersects the horizontal axis is the kiss point that representing where the mold halves make contact and the clamp begins to be deflected. The clamp "stalls" any time the available clamp force curve falls below the spring rate line.

and bushings goes up (due to a need for lubrication), then the available force curve will shift to the right and down. This will cause that curve to intercept the spring rate line. It is at this intercept point that the clamp will stall and not complete the toggling process. Because the available clamp force curve drops lower as friction goes up, it is best not to operate the clamp at this level because, as the clamp cycles, the lubricant gradually migrates out of the pin joints and friction increases.

The question of lubrication strategy bears a closer look. It is better to program the lubrication system to lubricate with a smaller amount of lubricant at a greater frequency than to lubricate infrequently with a large amount of lubricant. The clamp stroke used during molding can also influence the required frequency of lubrication. This is because of lubrication dynamics. If the pin joint rotates through a larger angle (longer stroke), then a thicker layer of lubrication is developed as the relative motion occurs between the pin and the bushing. Thus, for short clamp strokes, more frequent lubrication is required than for longer strokes.

There is one word of caution. If the clamp is left toggled for a very long time, the lubricant will be squeezed out of the pin joint and friction may become so high that the clamp will not be able to untoggle. Thus, it is very important not to shut down a toggle machine in the toggled position for a long period of time. The migration of lubricant from the joints can be observed by strain gaging the tie rods and observing the gradual drop in clamp force over time after the clamp is toggled. Depending upon the clearance between the pins and bushings, the squeezing out of lubricant can result in as much as a 1% drop in clamp force over the first second after the initial toggle. This normally does not pose a molding problem. In fact, most molders are not even aware that such an event occurs.

When the clamp is untoggled, the stored energy in the deflected elements will be released. It is very important that the actuator, whether it is a hydraulic cylinder or an electrically driven ball screw, or rack and pinion, controls the release of this energy to prevent the clamp from "snapping" open. If an hydraulic cylinder is used to drive the clamp, then the oil leaving the cylinder is usually metered out at a controlled rate. The clamp energy then shows up as an increase in oil temperature. It is important to note that in the event the clamp cannot be untoggled, for the reason discussed earlier, *great care must be taken* because the cross-head is jacked away from its toggled position. At a certain point in this process the cross-head will tend to "snap" open and could *cause serious injury* to anyone who has their hands in the toggle mechanism.

If one were wishing to design a toggle system, there are a few design guides and constraints. It can be seen from Eqs. (5.20 through 5.22) that it is desirable to keep the pin diameters as small as possible in order to keep the friction angle small. The limiting factor here is the maximum allowable shear stress on the pin. For hardened steel pins, 12,000 psi in double shear works well. The width of the bushing is based on the allowable average contact stress. For hardened steel pins and bushings with good lubrication, 12,000 psi works well.

The factor that has the greatest effect on maximum clamp stroke is the length, L_{12}, of the rear link. The constraint here is that the rear links usually swing through more than 90 degrees of travel and must not collide as they make their closest approach to the machine center line ($\alpha = 90$ degrees). The distance Y_3 (Fig. 5.59) is usually limited by the vertical platen dimension, so in order to have a competitive stroke, Y_1 is frequently greater than Y_3. As Y_1 extends beyond the tie-bar center line, a moment is imposed onto the rear platen. The constraint here is the stress limitation on the rear platen. Because Y_1 is greater than Y_3, there is a resulting splay in the links at the toggled position. It is best to design the linkage so that the rear and front links do not go completely into alignment; otherwise, it may be difficult to untoggle the clamp. On the other hand, if the links are too far out of alignment, the clamp will not be "self-locking" and will snap open as soon as the cross-head driving force is removed; approximately 1 degree works well. In addition, the cross-head link, L_{24}, should not go completely vertical at the toggled position for the same reason. A good minimum value for β is about one-half degree. It is easy to see that designing toggle systems is an iterative process.

5.8 The Farrell Square Root Rule

Nature operates in a well-behaved manner. Even though we have not always learned to recognize the patterns, it is this well-behaved and continuous nature that allows for scale models to be created that provide methods for scaling up or down from existing examples. Thus, for example, wind or water tunnel tests on model aircraft or water vessels can produce data that is useful in designing full-size craft. In addition, some humans are able to visualize events on a microscopic scale by relating to other events or experiences that occur on a larger scale. It is this ability to scale up or down that is important in the development of relationships presented in the following sections.

I have designed an entire line of injection molding machines based on a scale factor model. This model is based on the fact that certain constraints exist on the design of machines that are constant throughout the world. For example, the melt pressure tending to open the clamp is usually on the order of 2 tons/inch^2 of projected area. Every corresponding element in every molding machine is usually made of the same or similar material. Low-carbon steel is usually used for platens, and medium-carbon steel is usually used for tie rods, injection barrels, and screws. Each of these materials has an allowable design stress that is used by most designers around the world. Thus, for example, the tie rods for a machine made by almost every machine manufacturer are designed to a nominal stress of about 11,000 psi. If one were to design the clamping system for a range of machine sizes, then the cross-section of the tie rods would be in proportion to the clamp force. Thus, the diameter of the rods would follow the square root of the ratio of the clamp force. I referred to this as the *Farrell Square Root Rule*, and it can be expressed as Eq. (5.26).

Square Root Rule:

$$L_1 + L_0 \times \sqrt{(F_1/F_0)} \tag{5.26}$$

where L_1 = any dimension on the machine in question
L_0 = the corresponding dimension on the model machine
F_1 = the clamp force
F_0 = the clamp force of the model machine

Even though this is a very subtle rule, it is extremely powerful if the overall machine specifications are generated by this rule. As it turns out, the industry specifications for injection molding machines fortunately do loosely follow the specifications generated by this rule. Thus, a designer can design a complete line of machines scaled around one lead prototype machine. For example, if the prototype machine were a 250-ton machine, then a 1000-ton machine would be twice as large in every direction. The layout and all the detail drawings for all of the components on the 250-ton machine essentially could be used for the 1000-ton machine by multiplying all of the dimensions by two. This has tremendous implications on cost to engineer an entire line of machines. It also helps the engineer make size and weight estimates on machine sizes not designed yet.

5.8.1 Relationships that Follow from the Farrell Square Root Rule

- Dimensions of corresponding elements follow the square root

Between any two machine sizes, the size ratio of any corresponding component is in proportion to the square root of the ratio of clamp forces.

- Corresponding angles are equal

Angles of machine elements such as toggle links are always the same for corresponding events in the machine cycle.

- Stresses in corresponding elements are identical

As it turns out, the stress is identical at the critical point in every corresponding element at every corresponding time in the machine cycle. Thus, once the lead machine has completed fatigue testing, the other machine sizes do not require this validation.

- To achieve equal velocities of corresponding elements at corresponding portions of the clamping cycle requires oil flow rates in proportion to the ratio of the clamp forces.
- Volumes and weights follow the clamp force ratio raised to the 1.5 power

Because similar materials are used for corresponding machine elements, the densities of the material are identical. Because the weight of corresponding elements is proportional to the volume, and because the dimensions follow the square root ratio, the ratio of weights for corresponding elements is equal to the ratio of the clamp forces raised to the 1.5 power.

- Clamp dry cycle times are in proportion to the square root of the clamp force ratio

Because the clamp closing and opening speeds are nearly the same throughout the size range, the dry cycle time follows the ratio of clamp strokes, which is equal to the square root of the clamp force ratio.

- Corresponding weight supporting bearings are stressed in a ratio equal to the square root of the ratio of clamp forces

Thus, if moving platen support shoes are designed to the Square Root Rule, then the bearing stress between the bottom of the shoe and the ways is in proportion to the square root of the ratio of the clamp forces. For example, if the shoes for a 1000-ton machine are designed according to the Square Root Rule, then the dimensions of the 1000-ton shoe would be twice as large as those for the 250-ton shoe. The bearing area would be four times the bearing area for the 250-ton machine. Because the moving weight of the 1000-ton machine will be eight times that of the 250-ton machine, the contact stress under the 1000-ton shoe would be twice that of the 250-ton machine. This may not create a problem, however. The bearing material used on the shoes for all sizes will have an equal *PV* rating where *P* is the bearing stress and *V* is the velocity. Because *V* is the same for all sizes and the contact stress for the 1000-ton is double, the wear of the shoe pad will be twice a great as that for the 250-ton machine after traveling the same distance. Even though the distance traveled per cycle will be twice as great for the 1000-ton machine, it will typically cycle at half the rate. Thus, after 1 year of use, the 1000-ton machine will have the same travel, but twice the wear, on its pads. Because all dimensions on the 1000-ton machine (including bearing clearances) are twice those of the 250-ton machine, then twice the wear may be tolerated.

5.8.2 The Whole Machine Can Follow the Square Root Rule

Even the injection end of the machine can follow this rule. According to the Square Root Rule, shot size (volume) follows the ratio of clamp force raised to the 1.5 power. A quick check of published machine specifications [12] from a number of machine manufacturers shows that there is reasonable correlation between the average shot size for various clamp sizes and the shot size predicted by the Square Root Rule. For example, one manufacturer of toggle machines has an average shot size of 12.7 ounces for their 170-ton machine. The average shot size is 26.8 ounces on their 300-ton machine. The Square Root Rule predicts 29.8 ounces. The average is 64.95 ounces on their 500-ton machine (64.0 is predicted), and the average is 112.9 ounces on their 730-ton machine (113 is predicted).

If the Square Root Rule is followed throughout the entire machine, then the layout for an entire 250-ton machine, for example, only needs the scale doubled in the title box of the drawing to make it a layout for a 1000-ton machine. The square root rule is thus a very powerful design tool that can save *tremendous* amounts of design hours, it can also make estimates easier for hypothetical situations that marketing people always encounter.

References

1. Bannantine, J. A., Comer, J. J., Handrock, J. L., *Fundamentals of Metal Fatigue Analysis*, Prentice Hall (1990), Englewood Cliffs, NJ.
2. Juvinall R. C., *Engineering Considerations of Stress, Strain, and Strength*, McGraw-Hill, Inc. (1967), New York.
3. Osgood, C. C., *Fatigue Design*, Wiley-Interscience, (1970), New York.
4. Heywood, R. B., *Designing Against Fatigue of Metals*, Jarrold & Sons Ltd. (1962), Norwich, Great Britain; Barnes & Noble, New York, United States.
5. Heywood, R. B., *Designing by Photoelasticity*, Chapman & Hall Ltd. (1951), London.
6. Timoshenko, S., *Strength of Materials*, Third ed., Van Nostrand Reinhold Company (1958), New York, pages 454–475.
7. Peterson, R. E., *Stress Concentration Factors*, John Wiley & Sons, Inc. (1974), New York.

8. Roark, R. J., *Formulas For Stress And Strain*, fifth ed. McGraw-Hill Book Company, (1975), New York.
9. Sopwith, D. G., *Proc. Inst. Mech. Eng.* (1949), 160, 124.
10. Heywood, R. B., "Tensile Fillet Stresses In Threaded Projections", *Proc. Inst. Mech. Eng.* (1948), 159.
11. Stoeckly, E. E., Macke, H. J., "Effect of Taper on Screw-Thread Load Distribution", ASME (1951), paper No. 51–S45.
12. *Plastics Technology Processing Handbook & Buyer's Guide 2000/2001*, Gardner Publications Inc. (2000), Ciucinnati.

6 Mold Design

J. Beaumont

Every new injection molded part must have a mold custom designed and built. The mold is comprised of a mold body that includes a cavity that forms the features of the plastic part it will produce and has a means to deliver the melt to the cavity, a means of cooling the molten plastic delivered to the cavity, and a means of ejecting the solidified part from the mold cavity. Each time a part is formed, this is called the "molding cycle." To increase productivity, a mold may contain multiple cavities so that multiple parts are produced during each cycle. Injection molds commonly consist of 1, 2, 4, 8, 16, 32, 64, and 128 cavities. Even though each new product requires unique core and cavity sets to produce the parts, there are "modular molds" that permit the exchange of core and cavity inserts, allowing different products to be formed in a common mold frame. In addition, there are "family molds" that include different cavities capable of producing different parts in a single molding cycle. Injection molds are often built with tolerances as little as +/– 0.005 mm (0.0002 inches), and must be durable enough to withstand thousands, and even millions, of molding cycles.

The basic requirements of an injection mold are:

- Contain a core and cavity set(s) that defines the features of the part to be molded
- Provide means for molten plastic to be delivered from the injection molding machine barrel to the part forming cavities
- Act as a heat exchanger, which will cool the part rapidly and uniformly
- Provide the means to eject the molded part from the mold
- Have a structure that will resist internal melt pressures, which can potentially exceed 200 MPa (29,007 psi) and compressive forces from the molding machines clamp, which can commonly reach thousands of tons. In addition, the pressure and forces acting on an injection mold are cyclic, which will significantly reduce the life of any structural component
- In multicavity molds, provide uniformity to each cavity through steel dimensions, melt delivery, and cooling

Although nearly all molds are custom designed and built, some components and mold bases may be purchased, then machined and finished by a mold builder. Figure 6.1 is a standardized mold base that could be purchased from a catalogue. A mold builder would then add the custom designed cavity geometry, ejection, runners, gates, and coolant channels.

There are numerous types of injection molds and means of identifying them. These mold types are generally identified by a description that might include the number of cavities, the melt delivery system, and the ejection means used to remove

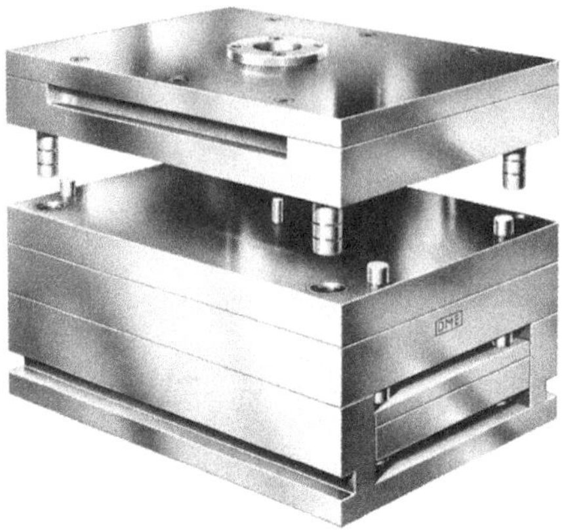

Figure 6.1 A standardized mold base. A mold builder adds cavity geometry, ejection, runners, gates and coolant channels to the mold (DME Company).

the molded part. As a mold becomes more specialized, the descriptions become more diverse. This chapter includes descriptions of some of the various types of molds and their components. It will also describe some of the advantages and disadvantages, particularly with regard to the melt delivery system, provided by each mold type. We will first review the construction of a common two-plate cold runner mold and then present some of the other common mold types and some of their key features.

6.1 Standard Mold Assembly

Although there is no one standard mold assembly, the assembly shown in Figure 6.2 provides a representation of a common construction. This Figure is a simple cross section capturing some of the more critical features within the mold. These illustrations show the same mold closed and then open with the ejection assembly forward. Figure 6.2B shows a runner and two halves of the mold in an open position ejecting molded part. The top half of the mold is commonly referred to as the 'top half', 'A half', 'cavity half', or 'stationary half' of the mold. The bottom half is commonly referred to as the 'bottom half', 'B half', 'core half', 'movable half', or 'ejector half'. The top half is attached to the stationary platen of the molding machine and the bottom half is attached to the movable platen. The *core* usually refers to the portion of the two mold halves where there are protrusions on which the forming plastic part will shrink and adhere to during mold opening. The part is then usually pushed off the core by some mechanical ejection means. In some cases, this may be assisted with air being blown between the core and the molded part. There are some molds where the cavity and core halves, are switched and provide ejection from the A half.

It is common that the A half of the mold body includes a clamp and cavity retainer plate, as can be seen in Fig. 6.20. The cavity is commonly inserted from the back of the retainer plate and clamped in place by bolting the two plates together. For high-production molds it is strongly recommended that cooling be placed directly in the cavity inserts. If this is not done, the heat from the plastic must be extracted through the interface between the insert and retainer plate. This interface can significantly restrict heat transfer, which is controlled by contact pressure and surface finish. As contact pressure and surface finish may vary from cavity to cavity, this can also result in variations in cooling between the molded part cavities. In single cavity molds, the cavity may sometimes be machined directly into a cavity plate, which would replace the retainer plate.

A *sprue bushing* provides a path for the melt to travel from the injection unit to the parting line where it may feed either directly to a cavity (in a single cavity mold) or to a runner in a multicavity mold. The sprue bushing is normally hardened and has a polished, tapered, conical flow path to facilitate ejection. Because of the high cost of machining this component, it is normally purchased from a supplier of standard mold components. A designer should specify a standard sprue size available from the supplier's catalogue when designing a mold. During mold opening, the plastic sprue is pulled from the sprue bushing by either an undercut in the core retainer plate, as used in Fig. 6.2, or an undercut in the *sprue puller*. The sprue puller

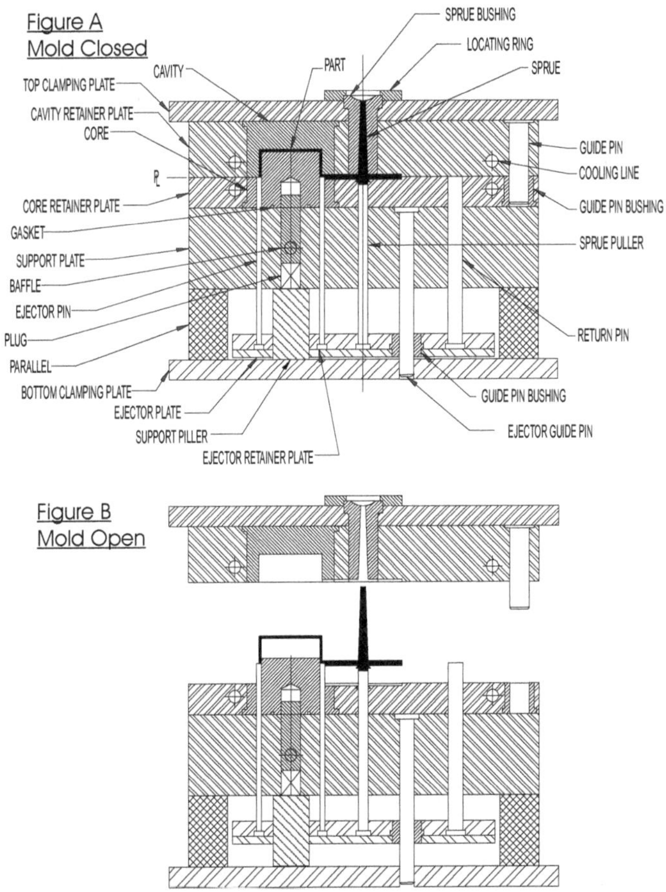

Figure 6.2 Cross-sectional view of a common mold assembly. (A) Mold closed. (B) Mold open.

is actually an ejector pin that, when moved forward [Fig. 6.2(b)], will either drive the sprue out of the undercut in the core retainer plate or, when pushed out of the retainer plate, will expose an undercut allowing the sprue to fall freely.

The "B" half of the mold body normally contains the core and the ejection assembly. Space is usually provided to allow movement of an ejector plate to which ejector pins, or other ejector means, are attached. The ejector plate is commonly attached to a hydraulic cylinder, which will move it forward and backward. Return pins are always used as a safety to assure that the ejector plate is returned during mold closing. If the ejector system is not returned during mold closing, the ejector return pins will hit a noncritical area on the retainer plate of the opposing mold half and drive the plate and ejector pins back. Without the ejector return pins, the mold could close with the ejector pins in their forward position. The forward positioned pins would then hit on the closing cavity components, causing damage.

During mold filling and packing, pressures within a cavity can easily reach 70 MPa (10,152 psi) with potential to reach to greater than 138 MPa (20,015 psi). These pressures will act to drive the core and cavity inserts backward. The core inserts and their back-up support plate are suspended over the ejector housing by a set of *parallels* or rails. To resist deflection, the support plate must be sufficiently thick, and possibly further supported with "support pillars". The support pillars are steel columns that are positioned between the clamp plate and the support plate passing through holes in the ejector plate.

Molds are normally described by a combination of their melt delivery means and ejection methods. This chapter will focus more on the classification and discussion of molds relative to their melt delivery systems because these are factors that are fundamental to the formation of a plastic part.

In a very broad sense, molds can be classified as either "cold runner" or "hot runner" molds. Cold runner molds can be broken down further as either two-plate or three-plate molds. The description of hot runner molds is more complex and generally includes a description of their two primary components—their manifolds and drops. These regions, or components, are either heated, heat conducting, or insulated. The following will describe the basic operation of both types of molds and their primary benefits.

6.2 Cold Runner Molds

With thermoplastic materials, a *cold runner mold* refers to a mold in which the runner is cooled, solidified, and ejected with the molded part(s) during each molding cycle.

6.2.1 Two-Plate Cold Runner Mold

A two-plate cold runner mold is by far the most basic and most common type of mold. It is the simplest, lowest cost to construct, and the easiest to operate and maintain compared with a hot runner mold. Color changes are easily accommodated because the runner system and part are cooled and ejected each cycle, leaving no material in the mold.

The mold in Fig. 6.2 has a single parting line, or plane, that opens during each molding cycle to allow removal of both the runner and molded part(s). The parting line exists between the core and cavity halves of the mold. A *sprue* is shown that provides a passage for the melt to travel from the injection unit, through the mold cavity half, to the parting line. At the parting line the sprue connects to the primary runner, which either feeds the melt directly to one part forming cavity or branches into additional runner sections, which will feed the melt to multiple cavities or multiple positions on a single cavity.

With a single-cavity mold, the cavity is generally placed in the center of the mold and the sprue delivers the melt directly to the center of the cavity. This would be typical of a part like a bucket. In a single-cavity mold, one might be able to envision how the cavity could be gated nearly anywhere by offsetting either the sprue or the cavity.

With a multicavity mold, the sprue delivers the melt to a runner, which in turn delivers the melt to the part cavities. As the cavities and the runner are along the same parting plane, gating is limited to the cavity perimeter. Some specialized gates, such as jump gates, provide some limited flexibility to gate off of the immediate perimeter of the part.

Cold runner systems have several advantages over hot runner systems. Because of their simplicity, they are cheaper to machine. The cost to maintain a cold runner system is less because there are no heaters, heater controllers, or thermocouples to maintain. Operation is also much simpler as there is no need to tend to the various heater controllers or deal with the many potential problems, such as gate drool, freeze or clogging, leaking, material degradation, or hang ups in the runner manifold.

Disadvantages of the cold runner system include the fact that the unwanted frozen runner must be dealt with. This requires the need to separate the runner from the molded parts and then sell throw out or grind the runner for reuse. This step of regrinding introduces additional potential for material contamination, and the need for granulators and their maintenance. The reground material can often be fed back in, at a controlled ratio, with the virgin material and remolded. If the molded part requirements will not allow reground materials, the runner must be sold or thrown away. As the reprocessed material will differ somewhat from the virgin material, it can be expected that this method of adding back regrind can alter the molding process and the properties of the molded part. This may or may not be acceptable depending on the sensitivity of the plastic material, requirements of the molded part, and the ease of molding it.

Reselling, or discarding, the reground runner is normally required when medical, optical, and high-tolerance parts are being molded. The reground material may alter flow and mechanical properties. It can also introduce unknown contaminates. Using a hot runner system helps eliminate these concerns. In addition, a hot runner [or hot sprue] can reduce the cycle time when the cooling time of the sprue and runner control the cycle time. The required cooling for the thick sprue or runner system may be longer than what is required for the thin-wall part. A hot runner, or hot sprue, eliminates the need for this delay, reducing the overall cycle time.

Another major limitation of the two-plate cold runner mold is that it offers the least flexibility in gating location. Gating in two-plate cold runner molds must be on the edge, or perimeter, of the molded part (except in the case of a single-cavity, sprue-gated mold). This limitation in gating location can sometimes lead to core deflection, gas traps, or undesirable weld lines.

6.2.2 Three-Plate Cold Runner Mold (Fig. 6.3)

The primary advantage of the three-plate cold runner mold over the two-plate cold runner mold is that gating is no longer limited to the perimeter of the part cavity. Compared with hot runner systems, three-plate molds are low cost, relatively easy to operate, have fast startups, require less skill to operate, and provide for easy color changes.

The three-plate mold has a second parting plane located behind the cavity plate. The second parting plane, between the cavity plate and top clamp plate, provides for a runner to travel under the mold cavity to any position relative to the part cavity. A secondary sprue then transfers the melt from the runner, through the mold cavity insert, to a desired location on the part cavity. The secondary sprue is attached to the part by a small diameter *pin gate*. Due to the increased flexibility in gating locations, the three-plate cold runner mold might be used in multicavity molds producing parts like a cup, where gating in the center of the cavity would be desirable.

Figure 6.4 illustrates one variation of a three-plate cold runner mold. Here the ejection of the part and runner begins with the mold first opening at a primary parting line, defined between the core and cavity plates (Fig. 6.4, Step 2). At a position where the part has been fully retracted from the cavity, a pull rod (A) will begin to pull a floating cavity plate to open the mold at a second parting line (Fig. 6.4, Step 3). As the secondary parting line opens, a stationary sprue puller, with an undercut, holds the base of the secondary sprue such that the small diameter gate tears away from the molded part as the cavity moves away. Once the mold has opened sufficiently for the secondary sprue to be fully relieved, the runner stripper plate moves forward to strip the secondary sprues from the stationary sprue puller, allowing the runner to be ejected (Fig. 6.4, Step 4). In the design shown in Figure 6.4, both the molded part and the runner are ejected using a stripper plate. The stripper plate ejecting the molded part is activated by a push rod (C). The stripper plate ejecting the runner is activated by a second puller pin (B), which pulls the plate forward as the mold is opening.

The disadvantages of the three-plate compared with the two-plate cold runner mold includes the added complexity of the mold, which create maintenance and operations problems. Molds must provide the proper opening, and the sequencing of the opening, of the primary and secondary parting lines, as well as the activation of ejection of both the runner and the molded parts. The cavity region must now be able to float on guide pins as the mold opens and closes. This increased number of moving

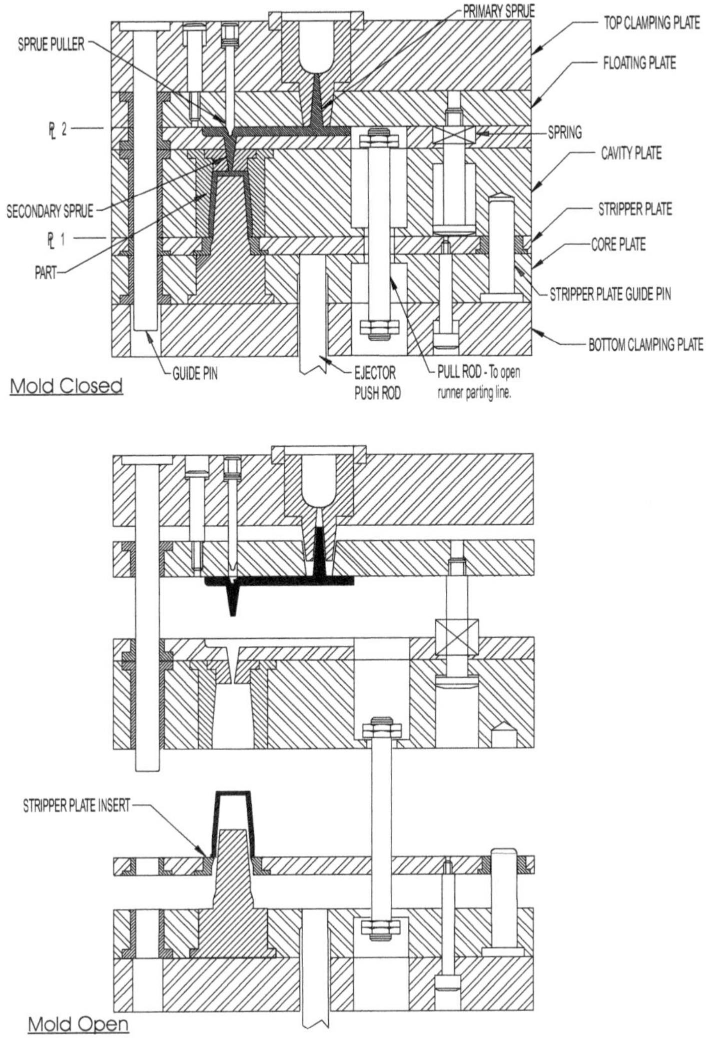

Figure 6.3 Typical three-plate cold runner mold.

parts increases wear related problems. The addition of the second parting line requires that the mold opens further, which will increase cycle time and mold opening requirements. There is also increased potential for runners hanging up during ejection. As the opening distance of the secondary parting line is limited, the runner may bridge across the mold halves as it attempts to fall free during ejection.

As stated earlier, the three-plate cold runner offers the same advantage of providing flexibility in gating locations as a hot runner; however, the three-plate has the same disadvantage as the two-plate cold runner in dealing with the frozen runner. In

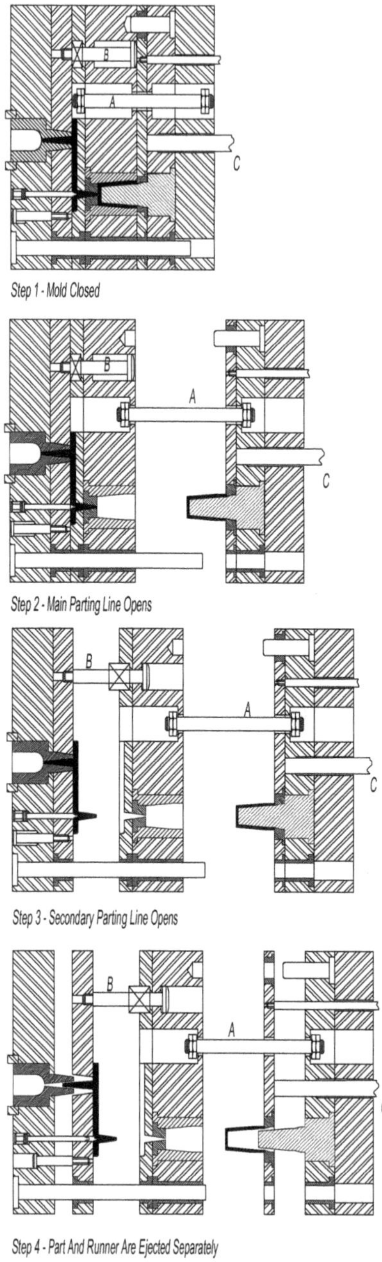

Figure 6.4 Opening and ejection action of a three plate mold.

addition, the three-plate will normally have a higher pressure drop during mold filling than any of the other molds, particularly hot runner molds. The fact that the runner must pass under the cavity plate and then up through the plate usually results in a longer runner than that is required in the two-plate cold runner mold. In contrast to a hot runner mold, this runner must have a relatively small cross-section to conserve on regrind and to assure that the runner does not extend the cooling cycle of the process. This is not a concern with a hot runner mold, which can therefore be a larger, more freely flowing channel. Finally, gate designs in a three-plate cold runner mold are limited to small, restricted gates, which must be designed to allow them to be torn away from the part during mold opening.

6.3 Hot Runner Molds

A hot runner mold (Fig. 6.5) refers to a mold in which the runner stays molten and is not ejected during the molding cycle. The most recognizable and accepted advantage of hot runner systems over cold runners is the elimination of a runner that must be filled and discarded each cycle and requires subsequent handling, storage, or disposal. Without a runner, shot size, plastification time (time required to create the next shot), runner cooling time, and required mold opening stroke decreases. In addition, the clamp tonnage may be reduced due to the elimination of the reactive force created by a cold runner. In some cases, these benefits allow use of smaller injection molding machines.

The hot runner design is similar to the cold runner in a three-plate mold in that the runner travels behind the cavity plate providing flexibility in gating locations. Flexible gating is achieved, but without the high pressure drops experienced in the three-plate cold runner systems due to the larger flow channels normally used. Hot runners also nicely accommodate "stack molds," which allow the molding of twice the number of parts in a mold without increasing the required clamp tonnage.

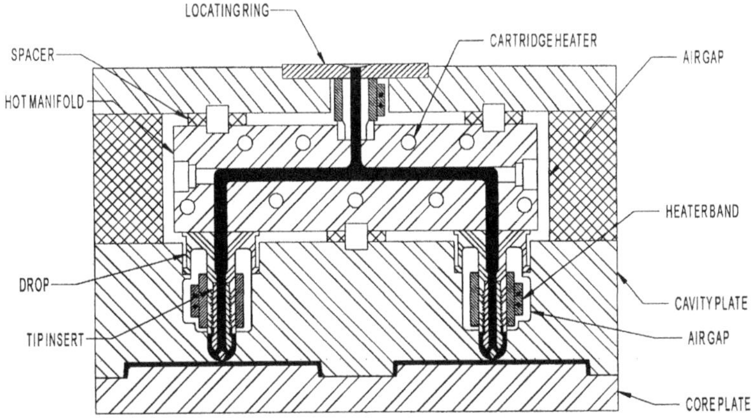

Figure 6.5 Externally heated manifold and drop.

The description and the designs of hot runner molds are more complex. The hot runner system is comprised of two primary components: the manifold and the drop. The manifold, contained within the mold body of the molds A half, delivers the melt from the machine nozzle to selected positions behind the cavity plate. Hot drops then provide passage for the melt from the manifold directly to the part-forming cavity or indirectly by connecting to a cold runner that might feed multiple part-forming cavities (indirect feed). The drop is generally positioned 90 degrees to the manifold and travels through the cavity plate.

The manifold in a hot runner system is either heated or insulated. The drops are either heated, heat conducting, or insulated. A heated manifold and each drop has its own individual heater and temperature controller. A heat-conducting drop is made of a high thermal conductivity material, like beryllium copper, and conducts its heat from the manifold. An insulated hot runner system has no direct heat source other than the molten plastic flowing through it. Each of these systems will be described in more detail in the following sections.

There are several combinations of manifolds and drops used to create the various hot runner systems. Each combination, rather than each separate type of manifold and drop, will be discussed in the following sections. Table 6.1 illustrates the more common combinations of manifolds and drops.

Due to the complexity of the hot runner systems, most of them are produced by companies specializing in their design and manufacture. A company building a hot runner mold normally purchases standard components, or an entire hot runner system, from one of these specialized companies. As there are numerous competing companies, variations between each system offered exist. The following examples present some of these variations. Many of the disadvantages of the various systems are known by their suppliers who continue to apply their best efforts to address them. The following can be used in discussions with these companies while allowing them to present their solutions to problems as well as their counterarguments.

Table 6.1 Common Hot Manifold and Hot Drop Combinations

Manifold		Drop
Externally heated	→	Externally heated
	↘	
Internally heated	→	Internally heated
	↗	
Insulated	→	Insulated

6.3.1 Externally Heated Manifold and Drops (Fig. 6.5)

Externally heated systems have the ability to provide the lowest pressure drop of all molds. The flow channels are cylindrical in cross-section and generally have a larger diameter than a cold runner mold. The cylindrical flow channel is the most efficient shape for flow. As cooling of the runner and regrind are not an issue with hot runners, a larger diameter is permissible. Both the larger diameter and the fact that there is no growing frozen layer in the runner system contribute to the relatively low pressure drop of these types of molds.

Of the various hot runner molds, externally heated manifolds and drops (also referred to as hot nozzles) are recommended for thermally sensitive and high-viscosity materials due to their free flowing open channel. In addition, because the heating of the manifold and drop is external, a frozen layer of plastic along the outer channel does not develop as it does with internally heated, or insulated, systems. Without this frozen layer, a larger flow channel will be available and a color change can be achieved more easily than it can with an internally heated hot runner system.

Disadvantages of an externally heated runner system include the potential for leaking of molten plastic, and the amount and location of the required heat from the heaters. Improper design or operation can result in plastic leaking between the drop and the manifold. This leaking plastic can engulf the manifold, entombing it and destroying heaters, wires, and thermocouples. In addition, the external heat source is in direct conflict with the cooling of the mold. The hot drops must often be surrounded by an air space, which helps insulate them from the cavity. The air space will require additional space in the mold. Proximity of drops approaching a cavity both introduce localized hot spots and limit positioning of cooling lines. Externally heated drops also have more limited control of the critical gate tip of the hot drop. Heat-conducting gate tips are often used, which can minimize these problems when properly designed. The tip conducts heat from the heated outer body of the drop. The gate tip temperature is being controlled by the same heaters controlling the outer body temperature of the drop. This restricts the control and may result in a compromise of temperatures for each of these two regions of the drop.

6.3.2 Externally Heated Manifold with Internally Heated Drops (Fig. 6.6)

The internally heated drop can potentially provide better gate tip control (less freezing, drooling, stringing, etc.) because heat can be applied more directly to the tip. In addition, because the heater is surrounded by plastic, the heat from the heated center probe is more naturally isolated from the mold cavity, where the part is trying to

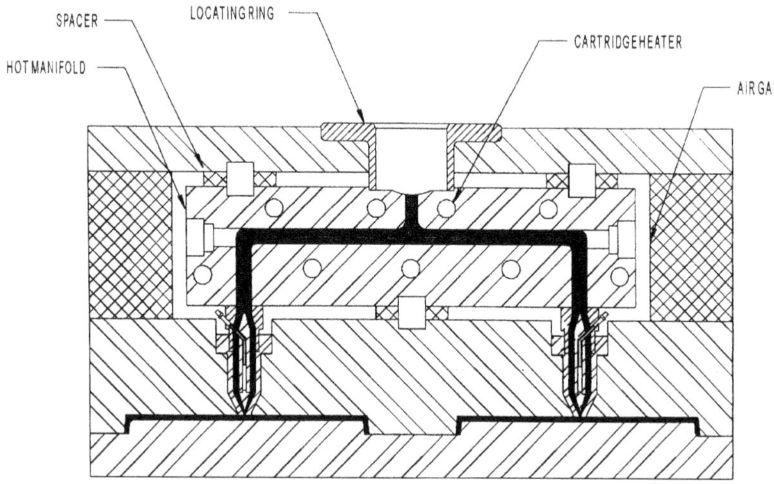

Figure 6.6 Externally heated manifold with internally heated drop.

freeze. There are designs that have the probes passing from the manifold through the drop. A frozen layer will develop around the perimeter of the flow channel. This frozen layer helps act as a seal around the flow channel and reduces the potential of leaking at the juncture of the manifold and the drop.

Potential problems created by the internally heated drop include the fact that because the heater is located in the middle of the flow channel it causes a restriction to flow. This location can also result in stagnant flow regions, depending on how the heater is located in the melt stream. These factors can combine to increase the degradation of material, problems with color changes, increased pressure during mold filling, and undesirable temperature variations in the melt (created between the heated probe and the cold mold wall). During normal color changes, the material in the frozen layer will never be flushed and can continue to bleed into the cavity. Suitable color changes can often be achieved, by first running hot, fast cycles to melt and reduce the thickness of the frozen layer of the first colored material. The second colored material is then added as the process is returned to normal. The result is that the frozen original color is covered by a frozen layer of the new color. If no color changes are required, this is not an issue.

Due to the conditions in the flow channel of an internally heated drop, they generally cannot be used with thermally sensitive materials. They should also be used with caution when molding clear and light colored materials. If stagnation points exist, with time many materials will degrade in these locations, break away, and result in undesirable streaking or black specs in lighter-colored parts.

The internal heater creates an inefficient annular-shaped flow channel. The result is that the pressure drop through an internally heated drop can be significantly higher than a full round drop with an equivalent amount of plastic material. This may not have much significance if there is sufficient pressure to fill the mold cavities. Although the diameters of the flow channels can be increased to decrease filling pressure, one must be cautious of the resulting increase in material residence time.

6.3.3 Internally Heated Manifold and Internally Heated Drops (Fig. 6.7)

This combination eliminates most leaking problems, naturally provides good isolation of the heater from the surrounding mold, and provides for gate tip control by the heated probe. Because heaters are internal, there is no need for a separate manifold block which must be heated and insulated from the surrounding mold with an air space. Plastic around the perimeter of the flow channel freezes against the colder mold, solidifies, and provides a thermally insulating boundary of plastic. This reduces the challenge with externally heated hot runner systems of insulating the heat from the part-forming cavity where you are trying to freeze the plastic. In addition, without the air gap surrounding the manifold in an externally heated system, leakage concerns are virtually eliminated.

Many of the concerns with the internally heated manifold and drop are similar to those discussed previously regarding the use of the internally heated hot drops. Some of the problems are further complicated from the increased use of internal heaters and the required crossing arrangement of these heaters (see Fig. 6.7). This crossing arrangement of internal heaters significantly increases the opportunity of stagnant flow regions where material can hang up and degrade. These systems, therefore, are not recommended for use with thermally sensitive materials. In addition, the pressure required during mold filling will be the highest of any of the hot runner molds per amount of material in the runner. The flow cross-section can be increased to

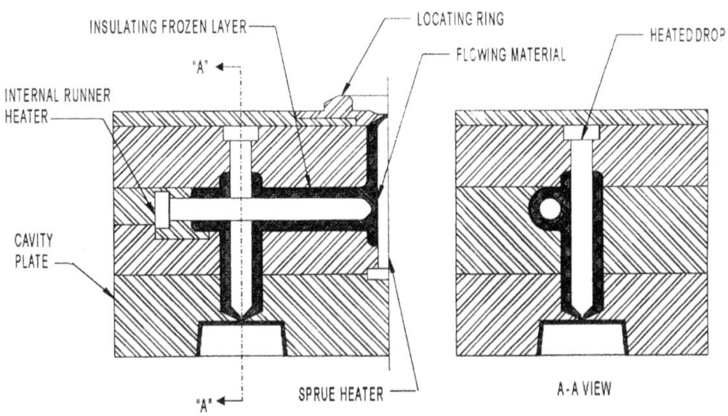

Figure 6.7 Internally heated manifold and drop.

reduce the pressure, but it will result in increased residence time for the molten material. The frozen layer development around the perimeter of the flow channel will create a challenge for color changes.

6.3.4 Insulated Manifold and Drops (Fig. 6.8)

This type of runner has no heat introduced other than that from the plastic melt flowing through it. For this reason, the insulated system must be run with fast and regular cycles to ensure that the melt in the manifold and drops does not solidify. A very large diameter flow channel is generally used (often greater than 30 mm). The first plastic injected into the mold fills the runner system. The plastic material along the perimeter of the cooler flow channel freezes acting as a thermal insulator. As the mold cycle continues, new material passes between these frozen outer insulating layers. Keeping this flow channel open depends on material moving through the runner fast enough so that the molten material in the center is continually replaced and does not have a chance to freeze completely.

The main advantages of the insulated systems over the heated systems are low cost, the ability to have thorough color changes, low pressure loss due to the large diameter runners, and that it introduces less heat to the part-forming cavities.

The molds are designed such that a seccond parting line is located along the melts flow channel. This parting line can easily be opened while the mold is still in the molding machine. Once the melt has frozen, this parting plane can be opened and the runner completely removed. This technique provides for very thorough color changes as well as for servicing the mold if the runner inadvertently freezes off.

This type of hot runner is rarely used due to the many drawbacks related to its lack of temperature control. Any interruption or variation in the cycle can cause significant variations in flow and melt conditions, as well as the potential for the runner to freeze off completely. In addition, there is an inherent variation in melt temperature across the flow channel, which can cause unacceptable results with most materials. If this system is used, it is normally limited to low tolerance parts and commodity

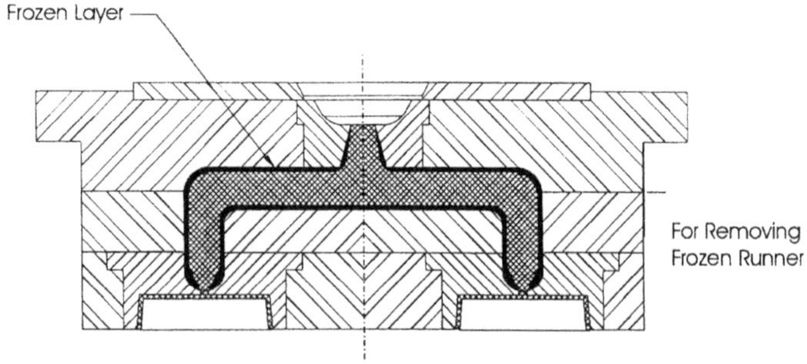

Figure 6.8 Insulated manifold and drop.

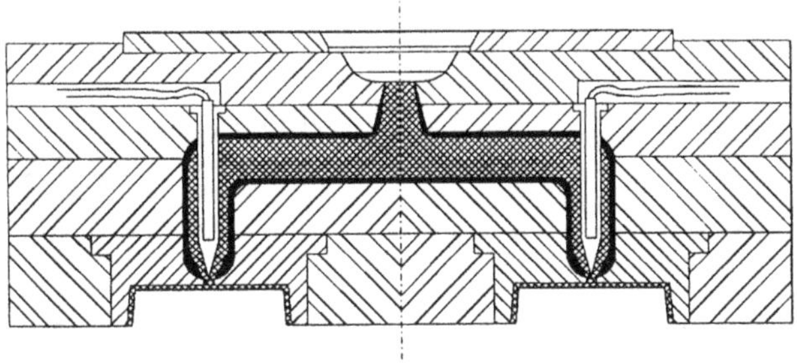

Figure 6.9 Insulated manifold with internally heated drop.

plastics like polyethylenes, polypropylenes, and polystyrenes. Internally heated drops are often used in conjunction with an insulated manifold to increase temperature control of the gate tips (see Fig. 6.9). Insulated runner systems are not sold by companies that specialize in designing and building hot runners. The designs, and use, of these systems are more of an art developed by a handful of companies.

6.3.5 Miscellaneous Hot Runner System Design

There are a few specialized systems that are unique from the more common designs discussed previously. One system of interest, produced by Gunther, achieves a full round flow channel while eliminating the need for an air barrier and the associated leakage space and leakage issues (see Fig. 6.10). The flow channel is a heated tube

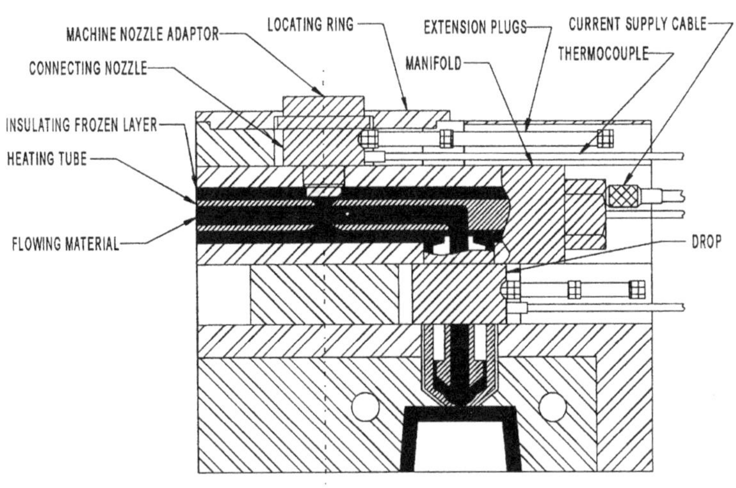

Figure 6.10 Gunther hot runner system.

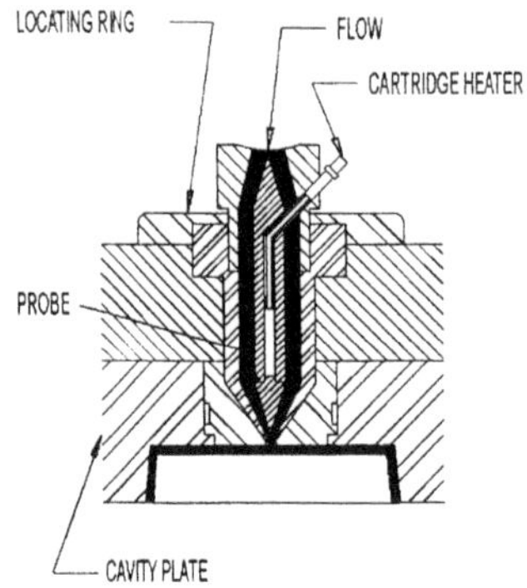

Figure 6.11 Internally heated hot sprue.

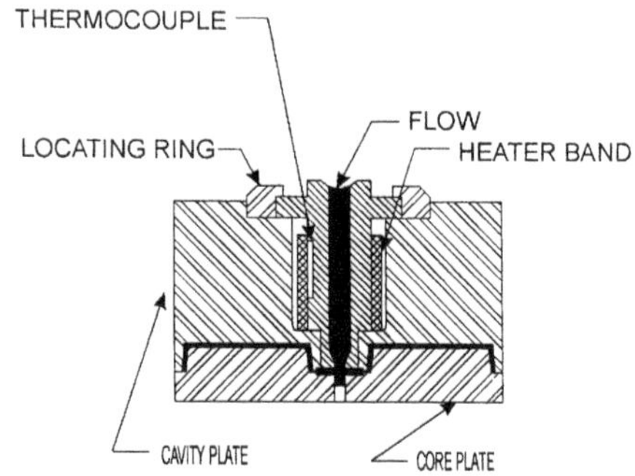

Figure 6.12 Externally heated hot sprue.

suspended in a channel machined in the mold. Material injected into the system initially fills both the flow channel and the gap between the suspended heated tube and the mold walls. The material in the gap remains stationary, acting as an insulator while material is injected through the externally heated full round tube. This special design conquers some of the leakage and thermal concerns of normal externally heated systems. The external plastic insulator resists leaking plastic and heat loss while still providing a full round flow channel.

6.3.6 Hot Sprues

In some applications the manifold is eliminated and a single drop, called a *hot sprue*, is used. The hot sprue directs the melt from the machine nozzle directly into a single cavity (Fig. 6.11), or into a cold runner that might either feed multiple cavities (Fig. 6.12), or multiple gates feeding a single cavity. In a single-cavity mold, the hot sprue might be able to eliminate the need to cut the more common cold sprue from the molded part manually, or the need to deal with the left over sprue. In thin-walled parts, a large-diameter cold sprue would most likely extend the required cooling time. Because a cold sprue must be tapered for ejection purposes, the intersection of the sprue and runner may be quite thick. Given an entry diameter of 3.175 mm (0.125 in) on a 76.2-mm (3-in) long sprue, the diameter of the sprue would increase to about 9.525 mm (0.375 in), where it intersects with the cold runner. This is also the location of the "sprue puller," which is required to pull the cold sprue from the sprue bushing during mold opening. If this thick region is not frozen sufficiently at the time of mold opening, it will break and the sprue will remain lodged in the sprue bushing. When this happens, the cooling cycle must be extended so that the sprue is sufficiently solidified to allow for it to be extracted from the sprue bushing without breaking. This additional time may far exceed the time required to cool the part. By using a hot sprue this problem can be eliminated and a faster cycle can be achieved.

6.3.7 Hot Drops (Nozzles)

Hot drops are the part of the hot runner system that delivers the melt from the hot manifold to the part-forming cavity. There are a number of basic requirements hot drops must meet:

- Conduct heat to the gate (prevent gate from freezing)
- Provide thermal separation between the hot drop and cold part-forming cavity
- Provide clean separation of the melt and the frozen part (satisfy gate vestige requirements)
- Minimize flow restrictions or areas of material hang up
- Provide good temperature control of the melt

In addition to satisfying the preceding requirements, there are some common problems associated with heated and heat-conducting drops:

- Thermal expansion of the manifold, causing misalignment and/or damage to the drop and gate tip
- Premature start-up of the hot runner system, before thermal expansion has provided a secure seal between the manifold and the drop, is a common cause of melt leakage and results in damage to externally heated systems
- The length of the drop is limited in heat conducting nozzles as heat must be conducted along its length from the heated manifold

Externally Heated Drops with Gate-Tip Inserts (Figs. 6.13 and 6.14)

A gate-tip insert allows variability in gate design and control with externally heated drops. Many suppliers of standardized hot drops provide optional gate-tip designs,

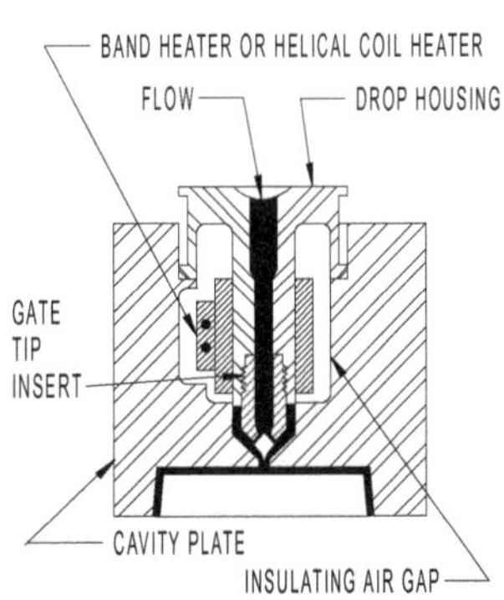

Figure 6.13 Externally heated drop with gate tip insert.

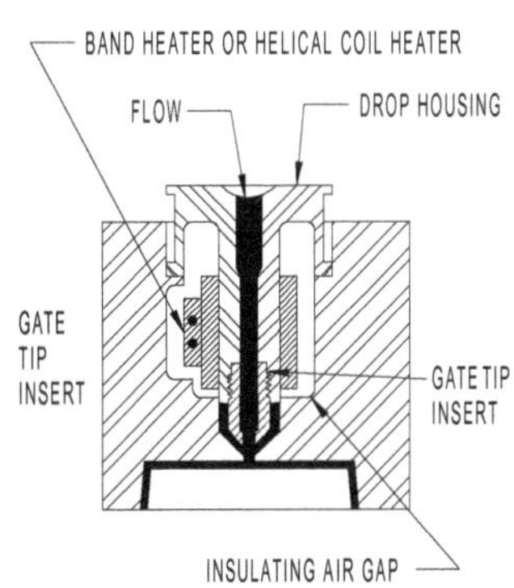

Figure 6.14 Externally heated drop with open flow tip insert.

which can be changed by simply screwing them into the drop. Hot drop gate tips must provide ease of flow and temperature control, and limit gate vestige.

Shutt-Off Nozzles

There are two primary types of shut-off nozzles; mechanical and thermal. The mechanical type can be spring (spring or torsion bar), hydraulically or pneumatically actuated (Fig. 6.15). The most common reason for using a mechanical shut-off nozzle is to create a flawless gate with no drooling and stringing. This is often a desirable characteristic for parts to be used for medical applications. A flawless surface will help assure that surgical gloves and other protective clothing will not tear. Another advantage to the mechanical shut-off nozzle is that large gate openings are possible. This makes processing at high flow rates, with shear-sensitive and fiber-filled materials, much easier. In addition, shut-off nozzles can be used to prevent backflow from the cavity to the runner when pressure is eliminated at the end of packing. This backflow can result in undesirable high shrinkages in the gate region of the part.

The prices of mechanical shut-off nozzles are high, expensive to maintain, and increase the sophistication of the mold set-up and operation (more functions to control). These systems are also prone to leakage and failures due to the interaction of moving parts.

The use of mechanical shut-off nozzles has seen increased use due to their ability to sequence the opening and closing of multiple gates in a given cavity. In large or thin-walled parts, which require multiple drops to fill, weld lines can be virtually eliminated by strategically sequencing the opening and closing of drops. Melt can be fed to the cavity by a first drop while all other drops are closed. As the melt from the first drop passes over a second drop, the second drop can be opened and begin filling. This will eliminate a weld line between drops.

Thermal shut-off nozzles have many of the same advantages of the mechanical shut-off nozzles, but without the mechanical problems. These drops have a small separate gate-tip heater. This heater is timed such that it turns on to allow mold filling

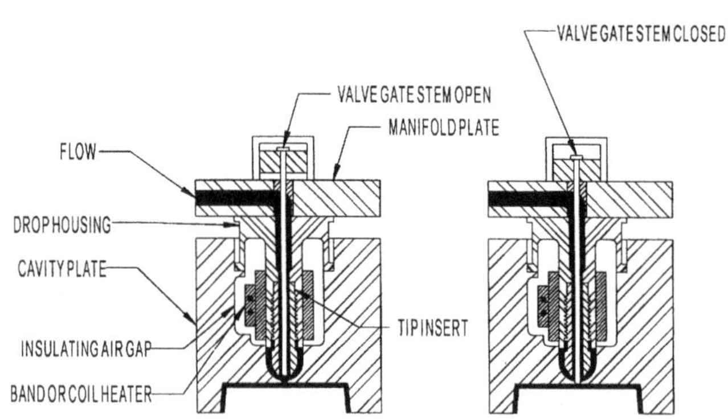

Figure 6.15 Shut off nozzle [valve gate shown] in open and closed position.

and packing, and shuts off to allow the gate to freeze prior to ejection. This requires excellent cooling near the gate tip so that the melt is assured of freezing in the tip. Setting up the process, to time the cyclic freezing and melting of material in the gate tip properly, creates a challenge requiring a skilled operator.

Thermal shut-off nozzles increase control at the gate tip. As a result, cooling lines can be placed very close to the gate without concern of premature freeze off. The gate-tip heater can also allow for very small gates, which might otherwise freeze off. The ability to cycle the heat off can reduce the potential for stringing. Though these drops improve control of the gate, they cannot be expected to eliminate gate vestige with the large gates that mechanical shut-off nozzles can provide.

Multi-tip Hot Drops

Hot drops can be made to feed multiple parts or gates from a single drop. The same concept allows them to be designed to edge gate a part when desirable. This approach could also be used to gate in multiple locations in the center of a hub of a large gear or other cylindrical part.

Heat Conducting Nozzle

Heat-conducting nozzles are not heated by their own dedicated heaters; rather, they draw their heat from the heated manifold. One type uses a bushing made of a high thermally conducting material, usually beryllium copper (BeCu), which conducts heat from the manifold to the gate. The bushing is designed so that plastic fills and surrounds the drop during the first shots. The surrounding material remains as an insulating layer, ensuring that little heat will be lost to the surrounding mold. This design provides a full round flow channel with heat being provided by the bushing. The advantages of using this special type of drop are that it is relatively inexpensive (no heaters or controllers needed) and it requires little space in the mold. The primary limitation of this type of drop is its lack of individual temperature control as all heat is conducted from the manifold. In addition, the length of the drops is limited because the heat must be conducted along their entire length to the gate. Additional disadvantages include possible degradation of the stagnant plastic surrounding the bushing. Further, the BeCu that is commonly used to conduct the heat must not come in contact with polypropylene or nylon because a chemical reaction will occur. Nickel plating can be used to protect the BeCu, although the plating is prone to wear, which will expose the BeCu. Finally, air borne particles developed during machining of BeCu presents a health risk to some individuals and is therefore avoided by many machine shops.

Another type of heat conducting nozzle uses conducting probes placed in the middle of the melt stream. This method provides better gate tip control. These probe-type heat-conducting nozzles (internally heating the melt) can be used in conjunction with a bushing or without a bushing. Without a bushing, the probe-type heat-conducting nozzle is very compact. It is suggested, that an externally heated nozzle with no flow restrictions be used for high flow rates with reinforced plastics.

6.3.8 Special Considerations in the Operation of Hot Runner Molds

Along with the many advantages of using a hot runner system there are a number of potential problems that should be recognized. As might be expected, there are differences in the processing characteristics of a cold and a hot runner mold.

The negative aspects of hot runners include the high cost of engineering, building, and maintaining them. These high costs can often be easily justified in high-production molds, but can be a serious draw back in lower production tooling. Operating problems can include failed heaters or controllers, difficulties in correcting filling imbalances, and leaking manifolds, which can seriously damage the heaters and their connecting wires. Higher-skilled operators are needed to deal with the increased number of process issues, such as temperature control and balances, startup procedure, heater failure, gate-tip drool, freeze off, or clogging.

Heat Load and Distribution

The effect of the additional heat load created in the mold by the heated system must be considered during design. Special attention must be directed toward the fact that a hot drop leads directly to the part that requires cooling. This can both increase the heat load in the mold and limit the placement of cooling lines in the location of the hot drop. The result can be variations in cooling across regions of a part, and/or from one side to the other side of a cavity wall at the location of the drop. This can affect cycle time and increase potential for residual stresses and warpage in plastic parts.

Startup Procedures

Start-up procedures for most hot runner molds must be strictly adhered to in order to avoid costly problems. Because rubber gaskets might be used for sealing water lines, the water must be turned on before the hot runner; otherwise, the gaskets can overheat and the water seal can be lost. With most externally heated systems, the hot runner must be heated and then allowed to soak for a period of time until all components are fully heated. The heat soak is critical because the seals between the hot manifold and hot drops are often dependent on thermal expansion of these components. Improper start-up procedure, or a heater failure, might result in an incomplete seal. The plastic material can then leak into the surrounding insulating air gap and wire channels, entombing the manifold and wires in plastic.

Pack Phase Characteristics

As the runner and gate do not freeze during the molding cycle, pack pressure can be higher and will be more consistent through the pack stage of the molding cycle than with cold runner systems. In a cold runner, the freezing runner and gate will increase the pressure loss with time as material is fed into the cavity during the pack, or compensation, stage. This will create a degrading pressure profile until the gate freezes. The degrading pressure can sometimes be beneficial because it acts to naturally profile the pressure and reduce over pack of regions near the gate relative to regions away from the gate. What is occuring during the packing phase must first be understood before the benefits of profile packing can be.

During packing the regions near the gate will normally freeze last. This results from the fact that, during packing, material is flowing from the gate region to replenish shrinking material throughout the part. This is called *compensating flow*, and is referred to as the *compensation or packing phase* of the injection phase of the molding cycle. The highest volume of flow is therefore nearest the gate, with nearly zero flow occurring in the extremities. The high flow in the gate region results in a greater amount of fresh material being fed through it, and a higher amount of frictional heating occurring there. This combination will cause the gate region of a part to be hotter longer than the extremities of the part.

During packing of a mold cavity, the flow of material will result in a pressure drop between the gate and regions further away. For example, if the pressure in the immediate vicinity of the gate is 60 MPa (8700 psi), pressure in regions away from the gate might be only 40 MPa (5800 psi). If pack pressure is held at a constant 60 MPa until the regions near the gate freeze, then the regions away from the gate would have been packed at 40 MPa, whereas those near the gate at 60 MPa. This will result in variations in shrinkage and residual stresses in molded parts, which is the fundamental cause of warpage. Therefore it may be desirable to profile the packing pressure so that all regions of the part freeze at a pressure closer to 40 MPa. If the pressure initially near the gate is 60 MPa, and the pressure drop across the part is 20 MPa, then one might want to reduce the pressure gradually as the part progressively freezes. If this is timed correctly, when the regions near the gate are freezing, the pressure near the gate should be only 400 Bar.

With a cold runner, a reduction of pack pressure will naturally occur as the runner and gate freeze.Therefore, it will naturally provide a degrading pressure profile. Though the hot runner does not naturally accomplish a pack profile, it does provide a better potential capability to control the pack profile if desired. With a cold runner, to control the pressure profile one must deal with the complication of the naturally degrading pressure present in the runner system. In a hot runner system, however, setting up a pressure profile in the part cavity should be easier because there is no freezing runner to complicate this process. This will allow a more direct and known control of the pressure profile.

Because the gate does not freeze in a hot runner system, there is also the need to consider backflow in the gate region of the part. When pack pressure is terminated, any remaining unfrozen material midwall in the gate region of the part will be at a high pressure and will depressurize by driving material back through the gate. This can increase sinks in the gate region, which will already be relatively hot due to its close proximity to the heated drop.

Hot Runner Diameter

Because there is no concern with regrind or lengthy cooling time, as with a cold runner, a hot runner can have a larger diameter flow channel. The larger diameter will reduce the pressure required to fill the mold. This can be important in parts that require significant fill pressure. There are a number of considerations that have the potential to limit the diameter. Increasing the runner diameter will increase residence time of the melt in the heated manifold and reduce flow rate through it. The increased thermal exposure in the manifold may be detrimental to some materials. The slower flow rate could complicate flushing material from any low flow regions and complicate color change.

In addition, as molten plastic is compressible at the high pressures used during injection molding, the increased volume of a large-diameter runner will act like a spring. As the melt fills a mold under high pressure it becomes compressed, with the highest compression occurring in the injection unit and the hot runner, where pressures are the highest. As the injection screw moves forward at a constant rate, the melt in the manifold will compress and begin feeding material to the cavity. As the pressure rapidly builds, the molten material in the manifold will act more rigidly as it continues to drive the material forward. If the screw stops moving prior to the cavity being filled, the melt front will continue to move forward for some distance as the melt decompresses and expands. The larger the volume of material in the barrel and the hot runner, the further the melt may advance on its own during this decompression and the more spring like melt behavior.

The following example contrasts the potential performance characteristics of a 16-mm (0.625-in) diameter versus a 22-mm (0.875-in) diameter hot runner.

$$\text{Pressure: Given } \Delta P = \frac{8Q\eta L}{\pi r^4} \tag{6.1}$$

If we assume injection flow rate (Q), viscosity (η), and flow length (L) are the same when comparing these runner diameters, pressure is found to be a function of the fourth power of the runner radius (r).

$$\frac{8Q\eta L}{\pi} = K \tag{6.2}$$

$$\text{Therefore } \Delta P = \frac{K}{r^4} \tag{6.3}$$

In this example, a 40% increase in runner diameter reduces fill pressure to only 26% of the original pressure. In reality, the effect would not be quite as dramatic as viscosity is affected by shear rate, which is effected by runner diameter assuming the constant flow rate. The larger diameter would decrease shear rate, which would increase viscosity.

Melt Compression

The 40% increase in runner diameter results in nearly doubling the volume of material in the manifold (100% increase). A given single hot runner manifold section, 254-mm (10.0-inch) long by 22.225-mm (0.975 in) diameter, contains a volume of 98.5 cm^3 (6.01 in^3) of molten plastic material. If the material were a Montell, PRO-

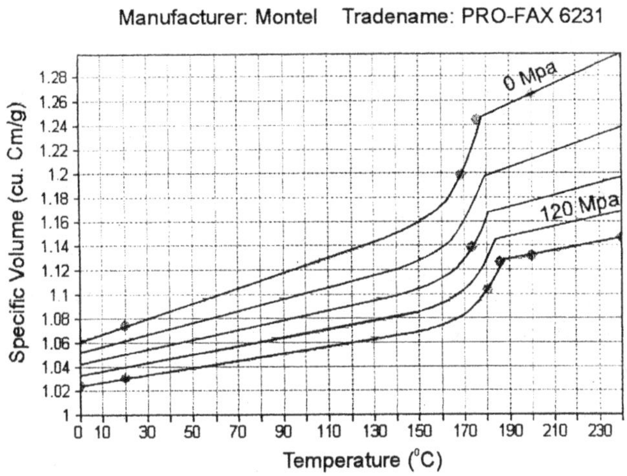

Figure 6.16 PVT Chart for Montel PRO-FAX 6231 Polypropylene to determine specific gravity of the material at specified temperatures and pressures.

FAX 6231 polypropylene, according to the PVT data from Fig. 6.16, the molten material being processed will compress by more than 10% at a pressure of 120 MPa (17,404 psi). As pressure is relieved, a volume of 6.6 cm^3 of material could expand forward into the cavity versus only 3.3 cm^3 with the smaller diameter runner channel (pressure loss from the expanding melt and freezing of the advancing melt would reduce this somewhat). In a part with a 1.5-mm wall thickness, this would be a 21.6 cm^2 section of the part. In a screw with a 3.175-cm diameter, this is equivalent to 0.41 cm of screw travel. Given a second section of the same hot runner manifold feeding a second cavity, the additional compression would result in screw travel of 0.82 cm.

The flow behavior of the melt is like a spring, where the first end is attached to a driving force and the second end to a movable object. When the driving force is quickly applied to the first end, the spring will initially compress before moving the object. As the spring compresses, it will develop sufficient energy to move the object (flow front) forward. When the driving force is removed, the first end of the spring will stop, whereas the second end will continue to move the object forward as it decompresses. Note that the second end may not fully decompress because the object is still providing a resistance.

6.3.9 Stack Molds

Stack molds are specially designed molds that allow doubling the number of cavities in a mold with very little increase in the required clamp tonnage. These molds accomplish this by including a second parting line, or plane, where cavities are positioned back to back to those along the first parting line. The two parting lines are indicated by A and B on Figure 6.17. By this means, a second set of cavities does not increase

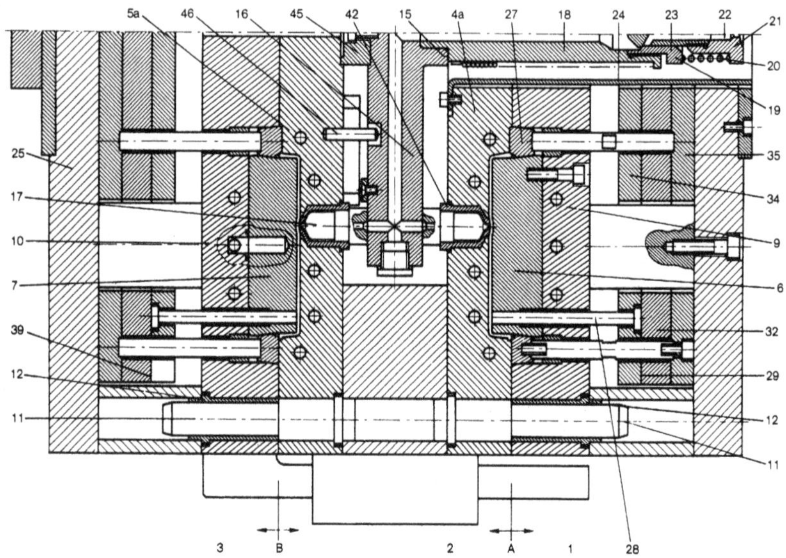

Figure 6.17 Stack mold with hot manifold. *1*, *2*, *3* plate sets, *4a*, *5a* cavity plates, *6*, *7* core inserts, *9*, *10* core carrying plates, *11* leader pins, *12* leader bushings, *16* hot manifold, *17* heated drops (nozzles), *18* hot sprue bar, *19* sliding shut-off nozzle, *20* compression spring, *21* sprue bushing, *22* spreader, *23* connector, *24* protecting tube, *25* clamping plate, *27* ejector strip, *28* ejector pins, *29*, 39 ejector retainer plate, *32* ejector plate, *34*, *35* ejector plates, *42* reversed-sprue bushing, *45* locating ring, *46* dowel pin.

the projected area for which the melt pressure will act on the clamp to open the mold. In addition, by stacking the cavities, a much smaller machine platen can be used, but this often requires specially designed machines with longer clamp strokes.

Most stack molds are fed through use of a hot runner system. The melt is fed from the machine nozzle through a “sprue bar” to a hot manifold positioned between the two parting lines. The sprue bar passes across the first parting line to the center section of the mold, where the hot manifold is positioned. The manifold distributes the melt to the cavities along both parting lines. During mold opening, the sprue bar separates from the injection nozzle as it remains attached to the manifold contained in the moving center section of the mold.

6.4 Runner Design

The runner is possibly one of the most important, yet least appreciated and understood, features in successfully molding plastic parts. This is particularly true in multicavity molds, where it is desirable that the delivery of melt to each cavity is identical. A significant percentage of rework in new molds is the result of improperly designed runners. Some of this rework is correctly diagnosed as originating in the mold cavity. However a significant amount of rework on molds, resulting from problems developed in the runner, is misdiagnosed as cavity, gate, cooling, material, or process problems.

Consider that the plastic forming of each molded part must first pass through the runner. The design of this runner can determine whether the part can be molded at all. It will also influence the size, shape, and mechanical properties of the molded part. The influence of the runner can go beyond this and even affect the materials fundamental characteristics, including molecular weight, conductivity, color, and flow. Despite its importance, the runner is often viewed simply as "that thing which gets the melt from the molding machine to the cavity". In a $100,000-cold runner mold, many mold builders leave machining of the runner to the apprentice rather than to their skilled toolmakers. The cross-sectional size is often dictated by standard cutter sizes rather than an evaluation of flow requirements. Tolerances are rarely specified for either the size of the runner or for the match up of the two halves of a full round runner.

The following contrasts the potential impact of slight variations in runner diameters commonly encountered from lack of tolerances provided when machining runners:

1. Given a 3.175-mm (0.125-in) runner diameter. What is the effect of a 0.0254 mm (0.001 in) variation in runners diameter?
   - given pressure drop along a round channel is $\Delta P = \dfrac{8Q\eta L}{\pi r^4}$ (Eq. 6.2).
   - if only the effects of changes in the runners radius are being contrasted and Q, η, and L are considered constant, then $\Delta P = \dfrac{K}{r^4}$.

   A variation of only 0.0254 mm (0.001 in) would result in more than a 3% variation in pressure flowing through a 3.175-mm (0.125-in) diameter runner. In a runner requiring 34.5 MPa (5000 psi) to fill, this would result in about a 1 MPa (150 psi) variation in pressure loss along the runner's length.
2. It is not unusual to see the two halves of a full round runner mismatched by 0.254 mm/side (0.010 in/side). This would reduce the effective channel diameter, in this example, by nearly 0.508 mm (0.020 in). Again, using the preceding simplified approach it can be found that the pressure drop through a given length of the runner would nearly double.

Evidence of the lack of understanding of runners include the fact that the impact of shear-induced flow imbalances in runners was not documented, or clearly explained, until 1997 [1]. For the first time it became widely publicized that the industry standard "naturally balanced" runners were creating significant imbalances. Imbalance ratios ranging from 65:35 to as high as 95:5 were found to develop in a standard *H* patterned geometrically balanced runner used in a four-cavity test mold. This phenomenon was being overlooked by nearly the entire molding industry in both cold and hot runner molds. In addition, the industry's leading state-of-the-art mold filling simulation programs had been developed without the realization of the shear-induced imbalance. As a result, these programs did not predict the imbalance and thereby left the analyst with false impressions of a good design. The shear-induced runner imbalance will be presented in Section 6.4.2.

Runners are designed as either a cold runner or a hot runner. Hot runners are most often made up of standardized components, developed by companies that specialize exclusively in hot runners. A company can purchase these components and include them in their own design or contract to have the hot runner company design and build a complete system for them. An entire assembled "hot half" can be purchased, which would include a complete assembly of hot runner components in a frame with all wiring in place. This can then be directly assembled to the mold with little additional work. In some cases, the supplier of the hot runner systems can be contracted to develop the entire mold. Due to the complexity of the hot runner system and their potential problems, most mold builders will purchase standard components and assemblies in order to minimize their start-up and maintenance problems. In addition, if problems develop, the faults can more easily be isolated.

6.4.1 Cold Runner Design

Unlike most hot runners, which are designed by companies with engineers who specialize in the runner design, engineering is rarely applied in the design of cold runners. Rather, they are simply laid out in some geometrically balanced pattern and assigned a diameter based on standard radiused cutter tools. In a small percentage of designs, a mold filling analysis may be applied and the runner diameters optimized.

Runners are used to deliver the melt from the injection nozzle of the molding machine, through the mold, and to the cavities. The runner should satisfy the following requirements:

- The cross-section should be large enough so that the pressure to fill the runner is not excessive. If the cross-section of a runner is designed too small, it may result in incomplete cavity filling or premature freezing during the packing stage
- The cross-sections should be as small as possible to reduce waste, minimize handling of regrind, minimize variations created through use of regrind, and avoid extending the cooling times waiting for the runner to solidify sufficiently for ejection
- Provide balanced flow to all cavities in a mold
- Maximize the efficiency of the flow channel by minimizing the ratio of surface area to cross-sectional area. This is best achieved with a full round runner channel

A variety of cross-sectional designs of runners are shown in Fig. 6.18. The ideal runner will have a full round cross-section. This will have the lowest ratio of surface area to cross-sectional area. The disadvantage to this geometry is that both halves of the mold must be machined with a half round channel that must match up when the mold is

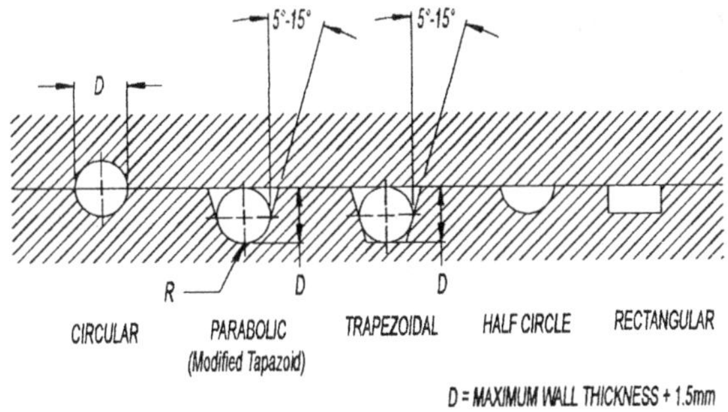

Figure 6.18 Runner-cross section design.

closed to form the desired full round channel. Alignment of the two half runners should be specified to within 0.05 mm (0.002 in) to avoid common misalignment related problems. It is common that a $100,000 mold with cavity dimensions specified to within 0.005 mm (0.0002 in) has no specified tolerance on runner alignment. The negative impact of misaligned runner halves can be significant and is often overlooked.

The trapezoidal runner cross-section is the most common alternative to the full round runner. It provides the advantage of being machined in only one mold half. This reduces machining costs and avoids problems related to misalignment found in full round runners. The parabolic cross-section, a variation of the more common trapezoid, provides the best alternative to the full round runner. This cross-section provides the same advantage as the trapezoidal shape while reducing the material wasted in the less-efficient trapazoid shape. The only disadvantage is that the required radiused cutter is not as common and is generally custom made. However, this additional cost may be less than $50.

The rectangular and half round shapes, seen in Fig. 6.18, should never be used in thermoplastic molds. These are extremely inefficient shapes for a flow channel, resulting in the highest pressure drop for the amount of material used. However, the rectangular shape can have an advantage, when it is used with some *thermosetting* materials, particularly with thick-walled parts. With thermosetting materials, it is beneficial to be able to heat the material as quickly as possible. The high pressure required to push the thermoset material through this shape results in energy being put into the material as heat. In addition, the rectangular shape increases the surface area of the runner channel, which exposes more of the melt to the heat from the hot mold. These two factors combined increase the temperature of the thermoset material as it flows through the runner. The hotter material entering the cavity will both improve the flow of the melt through the cavity, and reduce the cure time of the thermosetting material. The rectangular-shaped runner should be used with caution because it can increase damage to long glass-fiber reinforcements and significantly increase the pressure required to fill the mold; however, it is worth considering in some applications.

6.4.2 Runner Balancing in Geometrically Balanced Runners (Cold and Hot Runners)

One of the primary objectives of the runner in a multicavity mold is to deliver identical melt conditions to each of the cavities in the mold. The geometrically balanced runner (where all runner branches are the same length with symmetrical cross-sectional shapes) has always been thought to provide the optimum "natural balance" which is desired for consistent conditions between cavities during molding (Fig. 6.19[a–d]). This same concept of a naturally balanced runner system is also often applied to multiple runner branches, which may be feeding a single part (Fig. 6.19[e]). Despite the geometrical balance, there is nearly always some variation between cavities nearer to the center of the mold (cross-hatched cavities/regions in diagram) versus cavities further away from the center. In most cases, this imbalance does not occur until there are more than four cavities in the mold. The imbalance, actually depends on the number of branches in the runner and can occur in molds with as little as a single cavity, dependant on the layout of the runner system.

In the most common *H* pattern runner layouts, which might be found in an eight-cavity mold (Fig. 6.19[a]), the parts formed in the inner most cavities (those nearest the sprue) are generally larger and heavier. It can also be expected that there will be a variation in the mechanical properties of the parts formed in these different regions of the mold. This is particularly true when using a fiber-reinforced material, where variations in fiber length of 2:1 and significant variations in impact strength have been found as a direct result of the shear variation in the runners. In addition, it is common that the inner cavity may be flashing while trying to properly pack out cavities in the outer extremities of a mold. For years, this mold imbalance had been *incorrectly diagnosed* as being caused by variations in mold temperature, dimensional variations in cavities, gates or runners, and/or deflection of mold plates.

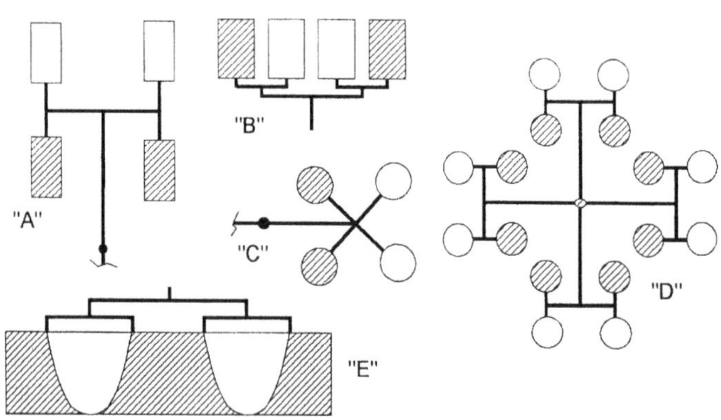

Figure 6.19 Sample runner layouts illustrating regions of high sheared material (shaded regions).

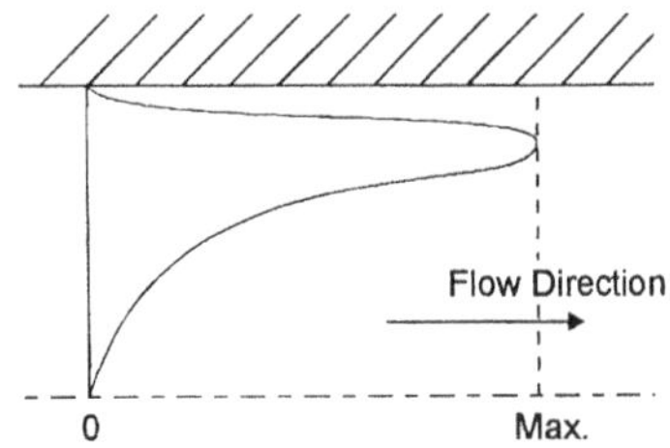

Figure 6.20 Shear rate distribution profile across the runner-centerline to outer wall.

The imbalance resulting in these conventional geometrically balanced runner systems has become more obvious as the demands on tolerances of plastic parts have become higher and attention to quality more prevalent. The missed diagnosis of the problem has been supported by the inability of the most commonly used commercial mold filling analysis programs to analyze the problem. This flaw in flow simulation technology was a result of the industry not recognizing the contribution of the runner to the imbalance.

Flow of Plastics Through a Runner and the Development of Asymmetric Melt Conditions

The flow of plastic in a runner is quite complex as the shear rate and temperature, and therefore the viscosity, vary both along and across the runner channel. At all flow rates the shear rate is highest just inside the outermost regions of the flow channel and is zero at the center (see Fig. 6.20). It can also be expected that the shear rate will be highest in the runner and gates of a mold where the velocity of the melt is normally at its highest.

The high shear rate region near the outer wall has a combined effect on viscosity. Viscosity in this region is reduced because of (1) its non-Newtonian characteristics and (2) the resultant frictional heating. The resultant frictional heating will cause the melt in these outer layers to be well above the temperature of the melt in the center of the runner channel. This higher temperature in the outer layers will nearly always exist despite some heat loss to the cold mold wall through conduction. In hot runners, and particularly with thermoset injection, or transfer molding, the frictional heating in the outer laminates is compounded by the heat gained from the heated mold wall.

Flow and Material Imbalances Resulting from the Asymmetrical Shear Distribution Across the Runner

Flow imbalances between cavities are most commonly recognized when there are more than two branches in the runner system. As the polymer melt flows down the primary runner it develops a high sheared region around its outer perimeter (cross-section "C-C" in Fig. 6.21; Note: For simplification, the frozen layer developed at the channel wall is not shown. The figure illustrates conditions of the flowing material only). When the melt stream is split at a runner branch, the high sheared (hotter)

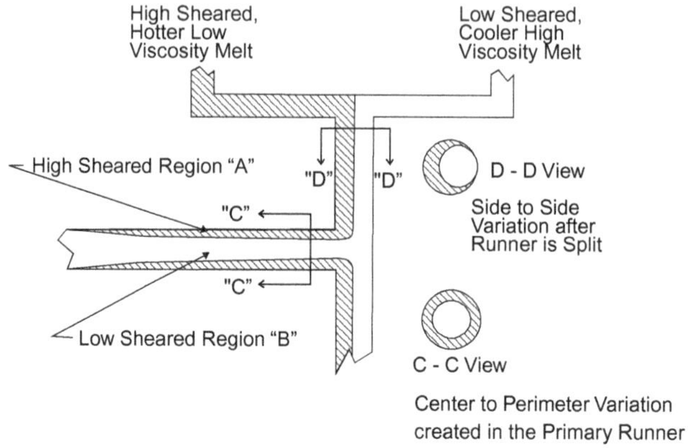

Figure 6.21 Cross-sectional view of high and low sheared regions showing: (1) side-to-side (D-D view), and (2) center to perimeter variation (C-C view) in the primary runner system.

outer laminates on one side of the primary runner (region "A" in Fig. 6.21) will flow along the left wall of the left-branching secondary runner. The low sheared (cooler) center laminate from the primary runner (region "B" in Fig. 6.21) will go to the opposite, right, side of the left-turning secondary runner. The high sheared (hotter) outer laminates of the primary runner will similarly follow along the right side of the right-turning secondary runner and the low sheared (cooler) center laminate will follow to the opposite, left side of the right-turning runner. The result is that one side of the secondary runners will consist of material with different properties than the other side (cross-section "D-D"). If the melt experiences a second flow branch by flowing into a tertiary runner, an imbalance will be created between the cavities being filled by these branches. The high sheared (hotter) material from the secondary runner will follow the tertiary branch on the inner side of the mold. The low sheared (cooler) material will follow the tertiary branch on the outer side. The result is that the material traveling down each of these branches will be of a different viscosity, temperature, and flow rate.

The imbalance described above can be seen by molding a series of short shots. Figure 6.22 illustrates the development of two different flows and the short-shot pattern expected in a conventional geometrically balanced eight-cavity mold. Flow #1 cavities (the inner four) will normally fill first, and be larger and heavier than the parts formed from Flow #2 (the outer four cavities). If there are additional branches in the runner system, the melt variations continue into the further branches. The imbalance can sometimes be reversed due to hesitation of the melt in Flow #1 when it hits a restrictive gate, cavity, or feature within the cavity. Comparative short shots are best taken at normal processing conditions and when the best filling cavities are approximately 80% full [1].

It can also be expected that the high shear region near the perimeter of the flow channel may adversely affect material properties. The effects of the high shear can also be traced visually when samples are molded of flexible vinyls. The effect of the

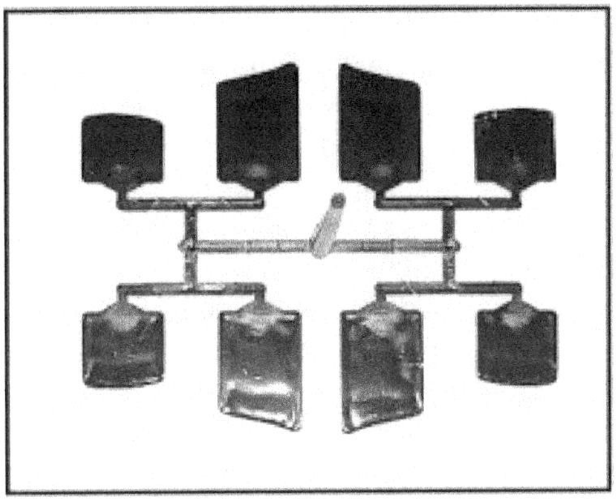

Figure 6.22 Typical "naturally balanced" runner system showing significant shear induced imbalances.

high shear in the perimeter of the runner accumulates to a point where the material can begin to break down in the Flow #1 runner branches.

Though the imbalance is most recognized in molds with two or more branches, variations can occur from one side of the cavity to the other in molds with only one or even no branches. In a mold with a single branch in the runner, the velocity, shear rate, temperature, and viscosity distribution across the branching runner will become nonsymmetrical from side to side in the secondary runner (see Fig. 6.23). At this point the secondary runner branches may be balanced to one another. Therefore, the flow rate will be the same in each branch. However, the side-to-side variation in the runner will continue into the part being molded. This can create dimensional variations between the two parts. This is a common condition in molds with parting line injec-

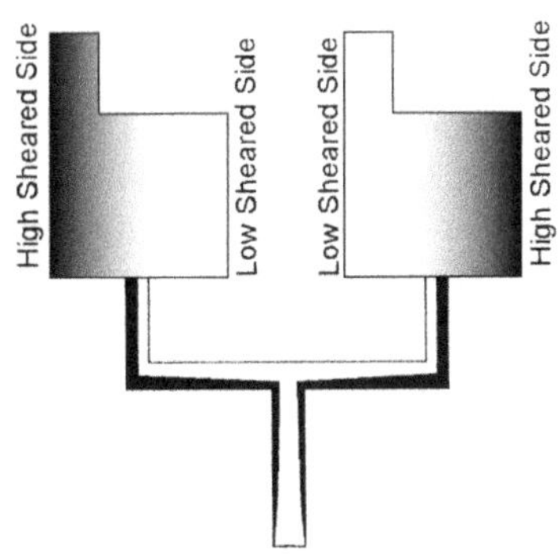

Figure 6.23 Illustration of the development and placement of high sheared material in a geometrically balanced runner system.

tion such as electrical connectors. This same condition can be seen in standard four-cavity molds, where every other part will be the same (two families of parts).

The imbalance created in the runner system is affected by runner geometry, material, and the process. In general, a larger-diameter runner decreases the variations in shear that is developed in the runner. The result is that the larger the runner diameter the less the imbalance. This creates a problem in both cold and hot runner molds. In a cold runner mold, the larger diameter increases the amount of runner scrap that must be handled, and potentially increases the required cooling time. Most plastic parts are relatively thin and take only a few seconds to cool. Though the runner does not need to be cooled to the same degree as the molded part, it still needs to be sufficiently frozen so that it does not stick in the mold or stick to the molded parts when it is ejected with them. Increased handling may include regrinding, storing, and reselling. In many cases, a certain amount of the material can be fed back into the process from which it originates. Though the ability to reuse the material may seem attractive, there are many hidden problems introduced with this, including controlling its reuse and its related variation in the molding process, the possibility of contamination during handling, increased need of granulators and maintenance, storage of unused materials, increased labor, and cleanliness.

Hot runner molds commonly have larger-diameter runners because they are not concerned with runner regrind or effect on cooling time. Despite the larger runner cross-section, these systems still regularly cause shear-induced runner imbalances. Unlike the cold runner, there is no cold runner channel that might help to offset the frictional heating developed in the melts perimeter; rather, the hot runner channel allow the heat that is developed to remain in the runner longer. In addition, there is a limit to the increase in the diameter that is acceptable. As the diameter increases the residence time of the plastic material at the elevated runner temperature decreases flushing capubilities and the spring reaction of the runner discussed earlier. Despite the ability to decrease the imbalance with increasing runner diameter, it should be realized that increasing the runner diameter does not get rid of the problem entirely.

Studies on different materials have found a relationship between filling imbalances and a material's viscosity. A material with a high viscosity will experience greater filling imbalances than a lower viscosity material. It has also been found that the greater the sensitivity of a material's viscosity to temperature, the greater the imbalance [2].

Fill speed is the most significant process variable affecting the imbalance. The imbalance is most dramatic at slow injection rates, which is in contrast to the expected behavior. At low fill rates, the shear difference is decreased, but the filling is affected by other factors such as thermal hesitation and a materials elastic characteristics that increases the imbalance. Fast fill rates will normally reduce the imbalance initially, and then increase it again as the fill rate continues to increase. Despite this sensitivity to fill rate, there is no fill rate that achieves a balance.

Addressing Runner Imbalances

The runner design should be balanced to allow each cavity in the mold to fill at the same time and with the same material conditions. Traditionally, it has been thought that the geometrical balanced, or artificially balanced, runner designs were providing these conditions. Artificial balancing has been applied to both nongeometrically balanced runners and geometrically balanced runners. This method modifies runner or gate diameters, or uses flow restrictors, in an attempt to balance mold filling variations. Although this may reduce some problems, it is fairly sensi-tive to process and material variations. Even if a filling balance is achieved, product variations will still exist due to the remaining differences in packing pressure, melt temperature and material shear history between parts. Artificial balancing of nongeometrically balanced runners is discussed further in Section 6.4.3.

The geometrically balanced runner is preferred. The geometrically balanced runner provides for a better uniformity between cavities, which will increase the molding process window. Radial-patterned runners provide a near ideal flow balance. The primary challenge with the radial patterns are providing uniform cooling of the cavities, increased spacing requirements in the mold, and increased runner volume. The *H*-patterned geometrically balanced runner provides for a more compact mold, which has less runner volume and more traditional cooling. One should be aware, however, of the shear-induced imbalances developed in this runner layout. New melt rotation technology addresses this imbalance. The alternative to this is artificial balancing of the geometrically balanced runners. Artificial balancing is commonly done in response to an observed filling imbalance and is based on a trial and error method of changing runner or gate diameters. Artificially balanced runners require that runner diameters feeding different cavities be different. The Flow #1 runner branches must be restricted. This artificial balance cannot be assisted through the use of the standard 1-dimensional beam elements used in most of the popular mold filling simulation programs. Additional drawbacks of artificial balancing include the fact that it is trying to correct a flow, shear, and thermal imbalance developed earlier in the primary runner. Despite achieving a limited flow balance, there will still be variations in melt temperatures, packing pressure, and other shear-affected properties in the material. In addition, this type of flow balance will be very sensitive to process and material variations (i.e. any shift in material or process will disrupt the balance).

A more recent development for addressing the shear induced runner imbalance is the MeltFlipper™ developed in 1997 [3]. The MeltFlipper, which is positioned at the intersection of the primary and secondary runners in a standard *H* pattern runner, rotates the melt approximately 90 degrees in the secondary runner, thereby strategically rearranging the shear-induced variation across the melt. The hotter shear-thinned laminates in the primary runner, which normally transfers to the inside surfaces (as shown in the cross section "B-B", Fig. 6.24[a]) of the secondary runners, are rearranged so that they end up on the bottom (or top relative to the "left" or "right" side of the runner; see cross-section "B-B," Fig. 6.24[b]) of the secondary runner. The relatively cooler nonshear-thinned regions in the center of the melt stream, which normally transfers to the outer surfaces (right side—see cross-section "B-B," Fig. 6.24[a]) of the secondary runner, are rearranged so that they end up on the opposite top (or bottom relative to the "left" or "right" side of the runner) of the secondary runner. Even though there is still a nonsymmetrical melt distribution in

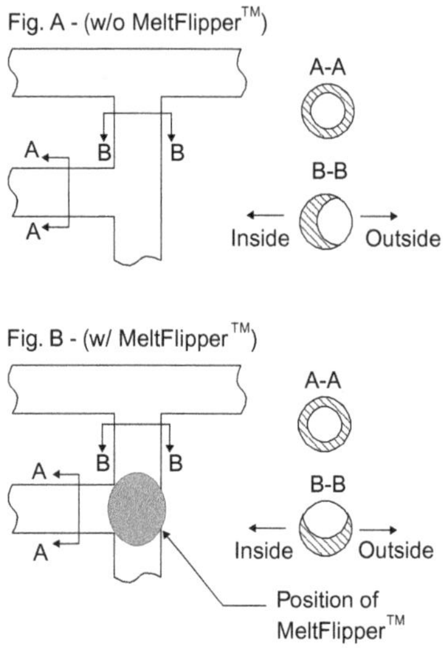

Figure 6.24 Effect of melt flipper on temperature distribution across the flow channel.

the secondary runner, the nonsymmetry is from top to bottom rather than left to right; therefore, when the melt splits left and right at the tertiary runner, the melt conditions entering each of these branches are now the same. Flow to all cavities will thereby be balanced. When there are more than three branches in a mold, as might occur in a standard 16- or 32-cavity mold, it is necessary first to rotate the melt approximately 90 degrees at the intersection of the primary and secondary runners. The melt will then need to be rotated again at later runner sections at angles less than 90 degrees. This method provides a true balance in filling and material properties. The amount of rotation is most commonly achieved through a controled elevation change at the intersection of branching runners. This simple method provides for use with both cold and hot runners without introducing any flow restrictions or ejection problems.

Figure 6.25 illustrates the effect of using the melt rotation technology in an eight cavity mold. The short-shot molded parts on the left side the photograph show the characteristic filling imbalance found with a conventional "H" patterned geometrically balanced runner. The partially molded parts on the right side were molded with the same mold, material and process using sets of inserts containing the patented melt rotation technology. The original imbalance is virtually eliminated.

Shear-Induced Imbalances with Hot Runner Design

The hot runner can provide more design flexibility than a cold runner. If residence time, color change, material degradation and precise shot control are not an issue, a relatively large diameter runner can be used. The larger runner diameter will decrease the required pressure and reduce the shear-induced runner imbalance. If a

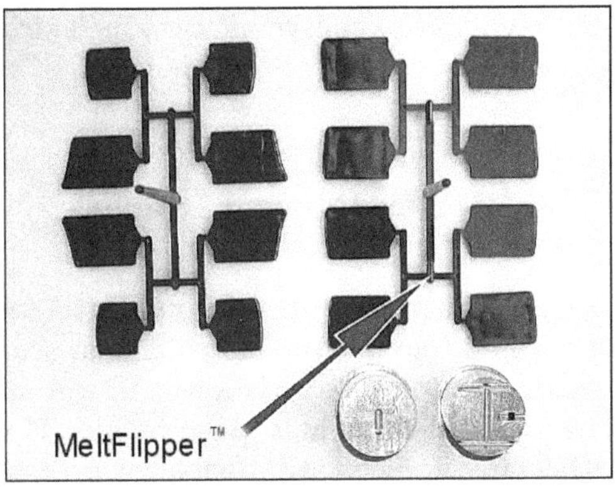

Figure 6.25 The effect of melt rotation technology is shown by contrasting the left (without melt rotation) and right (with melt rotation) short-shots. Both of these moldings were produced under identical conditions except for the addition of inserts containing the patented MeltFlipper™ technology.

small diameter or long channel is desired, problems associated with shear-induced imbalances will increase. Various methods of the melt rotation technology are also available for improving the balance in hot runners. Though these can sometimes be retrofited, it is prefered to be machined into the original manifold. The designs do not add any flow restrictions nor do they require any operator intervention.

The hot runner also provides more options for runner layout. With a large flow-channel diameter, "tree," or "fishbone," -style runner layouts are less of a problem than with smaller diameter cold runners. These designs, however, should only be used with the understanding that imbalance will still exist from the geometrical flow variations plus the effect of shear-induced imbalances.

Shear-Induced Imbalances in Stack Molds

Stack molds provide further complication in providing balanced filling to all cavities. In addition to the potential for imbalance created between cavities along a given parting line, or parting plane, most stack molds will have a shear-induced imbalance between each of the parting planes. Shear is developed as melt flows along the extended sprue bar used in stack molds. As the melt splits in the manifold, the high shear laminates will become positioned along the injection side of the mold. These laminates will end up feeding the cavities nearest the injection units (first parting plane), whereas the low-shear material will feed the cavities positioned along the further parting plane (second parting plane). The result is that shear-induced non-symmetry in the melt can be created in two planes. The previous melt rotation methods for solving shear-induced imbalances only addresses imbalances along a single plane. A new melt rotation means has been developed to eliminate this imbalance [4]. This method recombines nonsymmetrical melt streams that have been rotated in order to re-create full symmetry of the melt.

6.4.3 Nongeometrically Balanced Runner Layouts

Nongeometrically balanced runners are not recommended for most applications. These runners are laid out in a tree or fishbone pattern as seen in Fig. 6.26. These layouts are the most sensitive to process variations and will normally result in the largest variation in filling and melt conditions to the cavities. However, these layouts are still used today, particularly in high-cavitation, low-tolerance parts. Figures 6.27 and 6.28 contrast a geometrically and nongeometrically balanced runner layout used in a 64-cavity mold. The geometrically balanced runner would require six progressive branches. The optional fishbone runner layout would reduce the amount of required runner material and possibly reduce the required pressure to fill. Furthermore, the shear-induced runner imbalances in the geometrically balanced runner would decrease much of its expected benefit unless melt rotation technology is applied.

One strategy in designing the fishbone-type runner is to design the primary runner as large as possible. This causes the runner to act as a manifold, filling quickly and

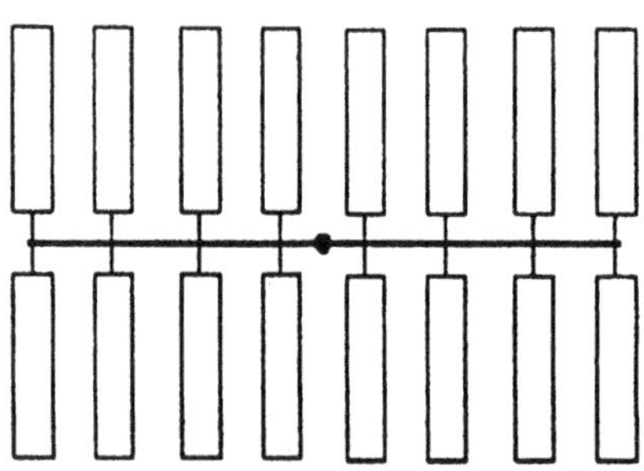

Figure 6.26 Nongeometrically balanced runner layout. Not recommended for most applications.

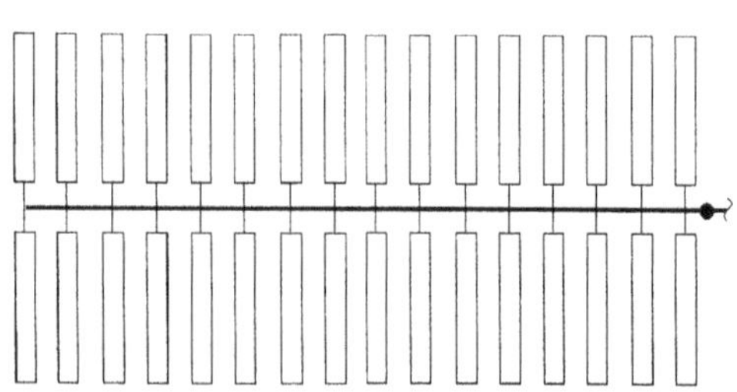

Figure 6.27 Unbalanced 64-cavity mold.

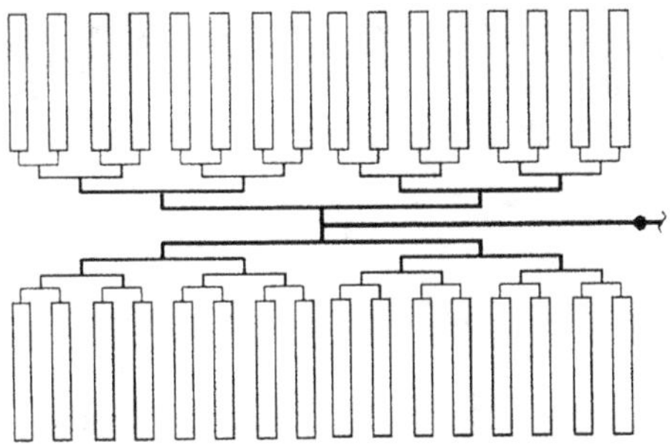

Figure 6.28 Balanced 64-cavity mold.

minimizing the variation in pressure drop between cavities. The problems that the melt reaching the first gates and cavities will experience a hesitation effect, possibly freezing off before all cavities can be filled. In addition, this may result in an excessively large primary runner and reduce cycle time. A smaller-diameter primary runner will develop more resistance and will help force the melt into the cavities closest the sprue as it fills. Even though there may be less chance of hesitation as these cavities nearest the sprue fill out, as soon as they fill, the melt pressure in them will become quite high and approach a hydrostatic condition. Meanwhile, as each progressive cavity fills, the flow rate in the remaining flow front will increase, thereby further increasing fill pressure. This will compound the high pressure condition in the early filling cavities. This high pressure can result in flash during mold filling. In either of these cases, the size of the cavity and gate will affect how these runners fill.

As with the geometrically balanced layout, these runners can sometimes be artificially balanced (see Fig. 6.29). Again, this would require that the runner diameters feeding the different cavities vary. Runner branches feeding the cavities closest to the sprue would be restricted and become larger as the material progresses down the primary runner. The resultant balance will be particularly sensitive to the process and the material, leaving a very small process window. If artificial balancing is required, it is suggested that the ratio of the shortest to the longest runner branch be kept as small as possible. In some cases this will require that the shorter runner branches, closest to the sprue, be purposely lengthened. In industry, artificial balancing is commonly attempted by modifying the gates because the gates can be more easily modified with a small hand grinder while still in the molding machine. This approach should be discouraged. Balance through adjusting the cross-sectional size of a short gate is even more sensitive to process and material variations than the longer runner branches. In addition, even if a filling balance is achieved with the gate, it can be expected that the different gate sizes will cause each of the parts to pack out differently as gate freeze characteristics will be altered. Gates should be left for their intended purpose.

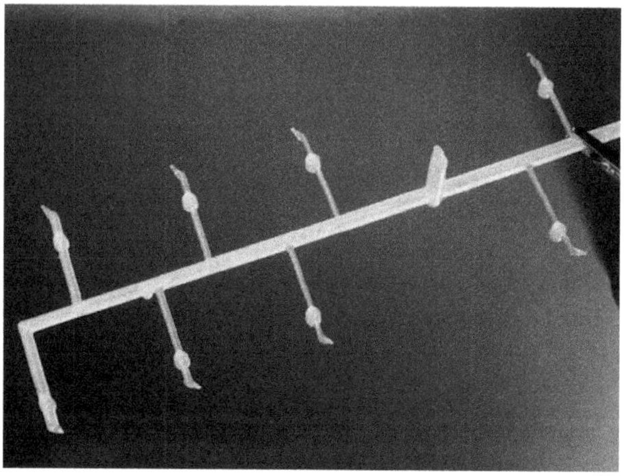

Figure 6.29 Artificially balanced runner system.

6.5 Gate Design

The gate is the link between the part and the runner system. It is normally a restricted area that facilitates separation of the runner from the part. The size, shape, and placement of the gate can significantly affect the ability to successfully mold a product. The key feature of the gate is to allow for easy, or automatic, separation of the part from the runner system while allowing for filling and packing of the part.

It is desirable that the gate design allows for easy removal from the part. This implies that the cross-section of the gate needs to be relatively small. A gate design that is too small might inhibit packing of the part, cause overshearing of the material, jetting, and other gate-related defects.

It is commonly recommended that the thickness of a gate, or the diameter of a gate, be 30 to 70% of the wall thickness of the part to which it is attached. The smaller cross-section allows for easier gate removal, but also increases the potential for the molding problems mentioned earlier.

6.5.1 Gate Types

Sprue Gate (Fig. 6.30)

Sprue gating is different from almost any other type of gating and refers to the cases where there is no traditional runner system or conventional gate. The part is gated directly from the sprue. It is used in single-cavity molds and provides for the melt

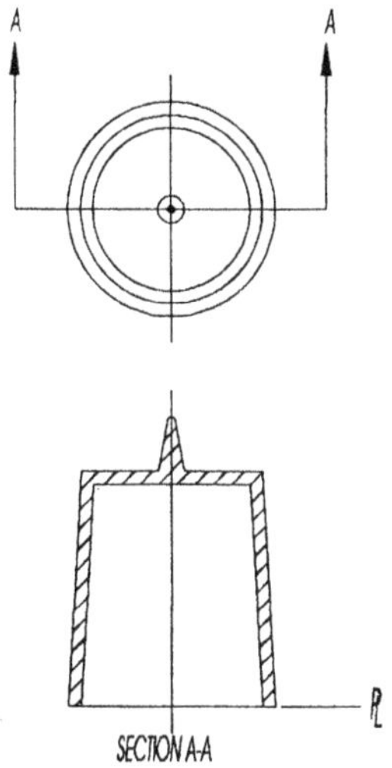

Figure 6.30 Sprue gate.

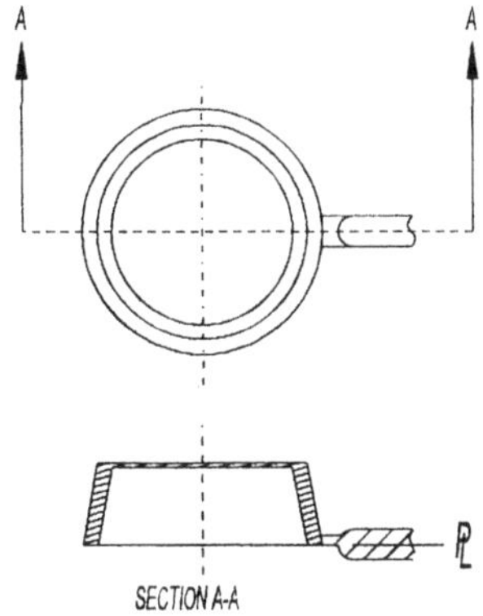

Figure 6.31 Tab gate.

to be delivered to the center of the cavity. This is ideal for many cylindrical or symmetrically shaped parts, such as buckets, tubs, helmets, boxes, cups, and disk-shaped parts. With a cold sprue there is no traditional gating. The sprue is tapered to facilitate ejection with the molded part and connects directly to the part at its widest point. The sprue must be manually removed from the molded part after molding. With a hot sprue, a small diameter, or valve gate, can be used and will result in automatic degating of the part from the sprue.

Tab Gate (Fig. 6.31)

Tab gates are the most basic type of gate. They are normally rectangular in cross-section and attach to the part, along its perimeter, at the parting line of the mold. They are used when automatic degating is impractical or undesirable. A tab gate would be desirable in a multicavity mold where parts are to be positioned for automated postmolding assembly. The tab gate will maintain the molded parts position and orientation on the runner, which will provide for easy postmold handling, such as assemblies, decoration, or inspection. The primary disadvantage of the tab gate is the need for manual degating when control of postmolding positioning is not required.

Variations of the tab gate include the lapped tab gate and the notched tab gate (see Fig. 6.32). The lapped tab gate reduces the possibility of jetting. The melt traveling through the gate will be directed downward opposite the gate into an opposing wall. The impingement on the wall will create a flow resistance, which will inhibit jetting. The major drawback is that the lapped gate cannot easily be cut off [cleanly] in a single operation and may actually require machining to be removed cleanly. To

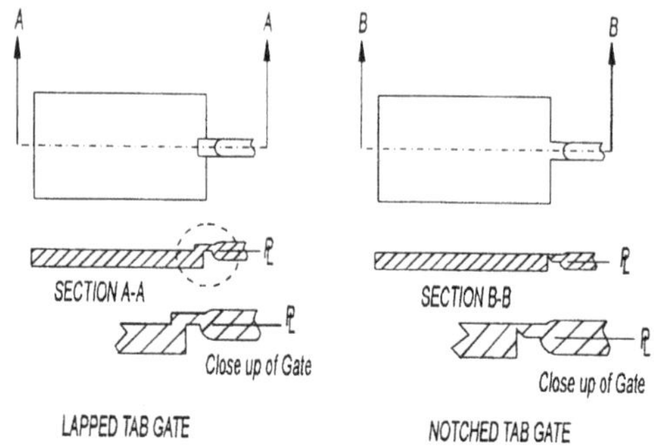

Figure 6.32 Variations of tab gate.

avoid degating problems, placement should be in a location where the lapped region will not need to be removed; however, the resultant flow through the cavity from this location must always be considered.

The notched tab gate is much like a standard tab gate, but can simplify postmold degating or even provide for automatic degating when brittle materials are being molded. The notch creates a focused fracture point, which will cause brittle materials to break when flexed. This flexing could be performed after demolding or while still in a mold that features two-stage ejection. With two-stage ejection an ejector pin firsts acts on the runner and gate, raising it at an angle, relative to the stationary part, that is sufficient to break it at the notch. Following this, a second stage in the ejection will push the part from the mold. If two-stage ejection is not used, the notch may assist in postmold degating because it can easily be broken off rather than cut off. Again, this requires a brittle material.

Fan Gate (Fig. 6.33)

Fan gates are similar to a tab gate in that they are attached to the part at the parting line and require manual degating. The difference is that the fan gate expands out from the runner in the shape of a fan with its widest end opening to the cavity. The fan region can be relatively thick and feeds a thin gate land, which is attached directly to the part. This design spreads and slows the melt as it enters the cavity. The benefits of the slower flow, and the broad uniform melt flow front includes improved melt orientation, reduced chance of jetting, reduced stresses in the gate region of the part, and reduced shear rates through the gate. In addition, the fan gate may replace several more restricted gates, eliminating weld lines formed between them. The major disadvantage is that the fan gate's width causes a problem with degating. Manual degating may require fixtures, shearing devices, or machining. Variations of fan gate designs are illustrated in Figs. 6.34 and 6.35. These designs are used to help improve the distribution of the melt across the thin gate land prior to entering the part-forming cavity.

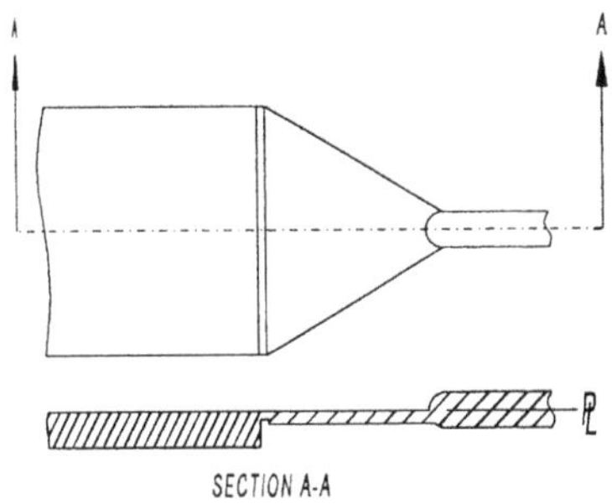

Figure 6.33 Fan gate.

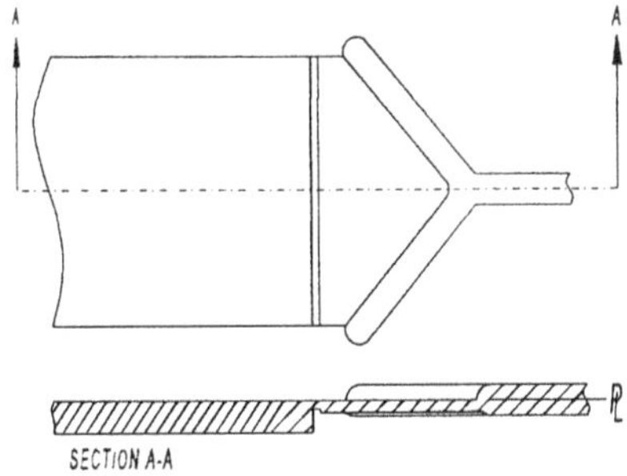

Figure 6.34 Variation #1 of fan gate.

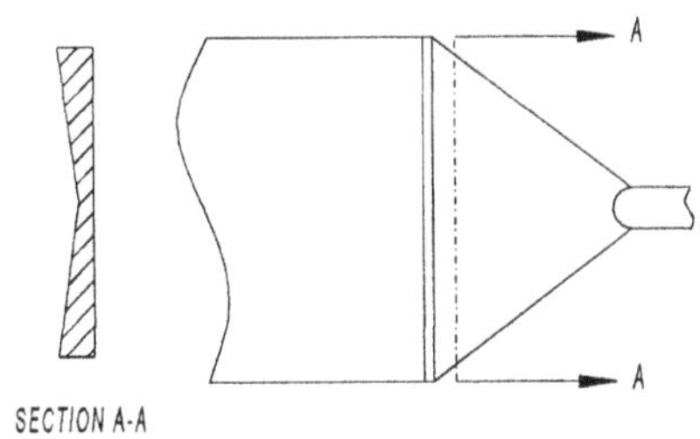

Figure 6.35 Variation #2 of fan gate.

Film, ring, and diaphragm gates are all gates that provide similar advantages to those of a fan gate, such as slowing the flow and providing a uniform flow front in a part. Most of these also have several variations, which serve particular needs.

Film Gate (Fig. 6.36)

The film gate attempts to capture the advantages of the fan gate, but it utilizes less space and material. Here, the runner attaches to a gate manifold that distributes the melt along a broad thin gate land that is attached directly to the part. The disadvantage compared with the fan gate is that the flow of melt through it is less predictable. The melt entering the manifold from the runner can hesitate at the gate land in that immediate location. It then can proceed down the manifold and enter the part at the ends of the manifold where there will be no hesitation. Increasing injection rate can potentially reverse this situation. The result is that filling from this gate can be somewhat sensitive to process variations. Film gates normally work best at fast fill rates where hesitation is minimized.

Ring Gate (Fig. 6.37)

Ring gates are essentially film gates that have been wrapped around a cavity. They are used for external gating on cylindrical parts. They are usually used with two-plate cold runner systems on cylindrical parts that are open on either end (like a tube).

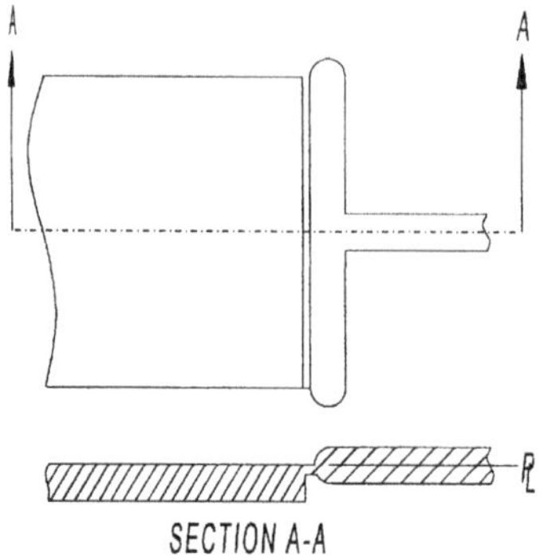

Figure 6.36 Film gate.

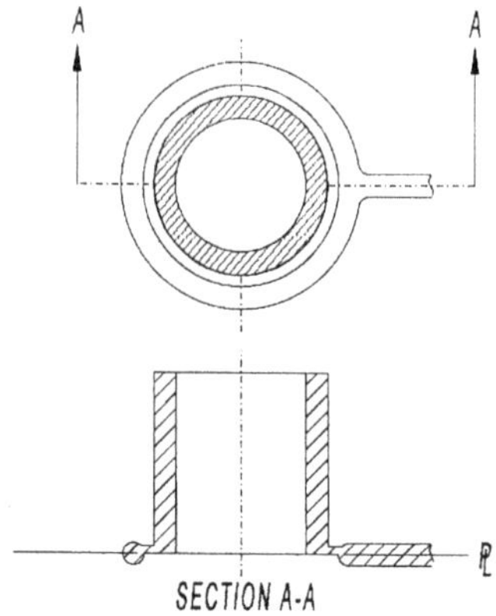

Figure 6.37 Ring gate.

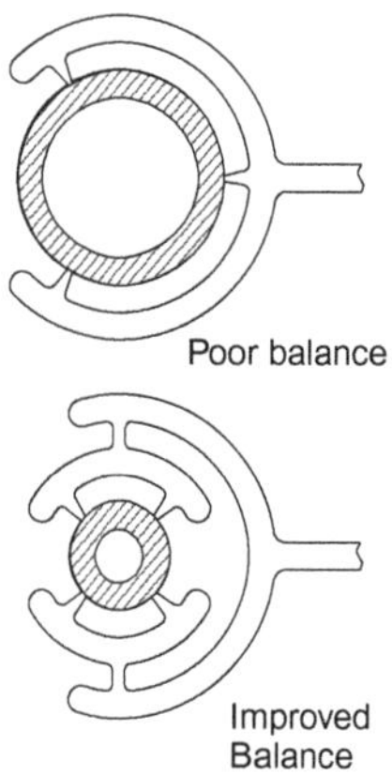

Figure 6.38 Variations of ring gate.

Their objective is to eliminate weld lines, provide uniform flow, and resist core deflection. The disadvantage is that the gate is difficult to remove and can cause similar unpredictable and unbalanced flow as with the film gate.

Two variations of ring gates can be seen in Fig. 6.38. These variations allow the use of tunnel gates for automatic degating. The second variation provides for an improved flow balance. This will minimize potential for core deflection and provide good concentricity to the part. The major disadvantage of the variations is that they will result in weld lines being formed on the part.

Diaphragm Gate (Fig. 6.39)

Diaphragm gates are used on cylindrical shaped parts that are normally open on either end (like a tube). They are used in three-plate cold runner, hot runner, or single-cavity sprue-gated molds. The advantages include no weld lines, minimal

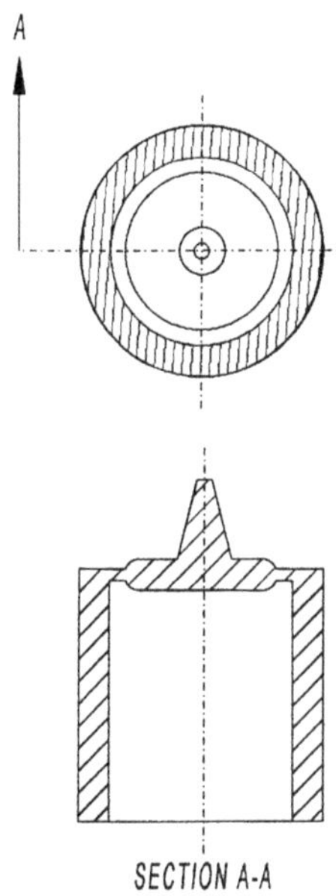

Figure 6.39 Diaphragm gate.

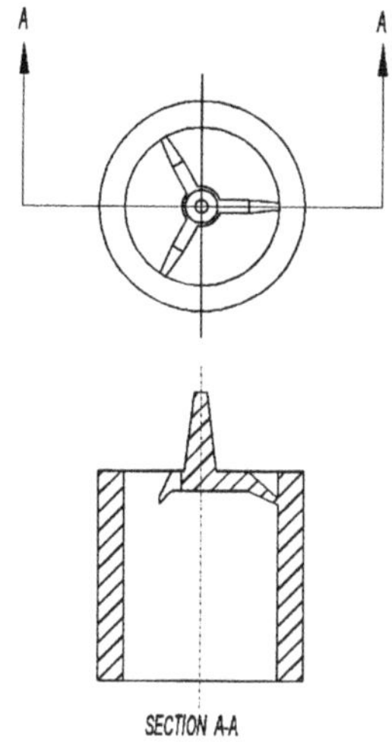

Figure 6.40 Variations of diaphragm gate.

potential for core deflection, and development of an ideal flow pattern for predictable shrinkage and minimal unwanted warpage. From a flow standpoint, this is the ideal gate for most cylindrical parts. The primary disadvantage is the difficulty of removing the gate.

The runner layout shown in Fig. 6.40 provides an option to the diaphragm gate. This design uses short runner lengths that radiate out from a central location, connecting to the cavity through either tab or tunnel gates. This design provides for balanced filling of cylindrical parts while simplifying the degating operation. It is recommended that three or four uniformly spaced gates be used to minimize the potential for core deflection and to improve concentricity of the molded part.

Tunnel Gate (Fig. 6.41)

A tunnel gate is typically conical in shape, with the smallest end of the cone attached to the part. The gate is cut in the mold such that it tunnels from the runner to the cavity below the parting line of the mold. During ejection, the gate is sheared, or torn away, from the part as it is pulled through the tunnel. It is most commonly used to provide automatic degating in standard two-plate cold runner molds; however, tunnel gates can be used in combination with a variety of mold and runner types and configurations providing automatic degating. Its primary advantage is automatic degating in cold runner systems.

The design of a tunnel gate is more critical than it is for most other gate types. If the gate diameter is too small, it will result in overshearing of the material and pre-

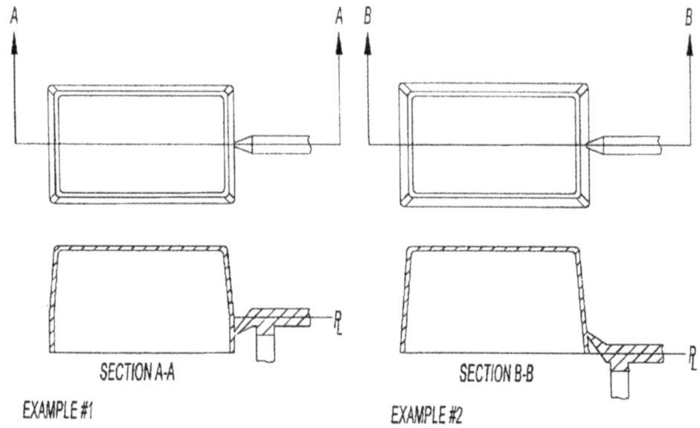

Figure 6.41 Tunnel gate (submarine).

mature freeze-off, preventing proper packing of the cavity. If it is too large, it may tear out sections of the part's wall during ejection. The gates are tapered so that they can be ejected from the mold. The gate-tip diameters typically vary between 30 and 70% of the wall thickness of the part, but they are material and application dependent. It is preferred that the plastic material being used is not too brittle because the gate and runner must distort enough to be retracted from the tunnel. The angle at which the gate tunnels toward the part and the taper on the gate will affect how much it is distorted during ejection. Increasing the thickness of the main body of the gate will cause the gate to be warmer and more flexible during ejection.

Variations in gate tip design can affect the gate vestige, shearing, packing, and pressure drop experienced by the melt. Design A in Fig. 6.42 is the most common tip design. Design B shown in Fig. 6.42 is less common and has a D-shaped opening to the cavity. This design can improve packing when used with semi-crystalline materials that have sharp melt temperature transitions. As melt passes through the gate, material will be trapped at the base of the gate, insulating the flowing material from the cold steel at the bottom side of the gate tip. This gate can also reduce gate vestige. The drawback of this design is that it is more difficult to maintain consistent gate sizes between gates. In addition, with the D-shaped tip, a small amount of wear will increase the opening size more quickly compared to the more traditional, full conical tunnel gate. In multicavity molds, this will result in gate-size variations that will cause fill and pack differences within the various cavities.

Cashew or Banana Gate (Fig. 6.43)

Cashew or banana gates are very similar to tunnel gates except that they can provide gating in regions that cannot be reached by the standard tunnel gate. The primary limitation is that the curved shape requires that the material in the gate go through considerable distortion during ejection. This generally requires use of low modulus materials that can go through significant strain without breaking. Some high modulus

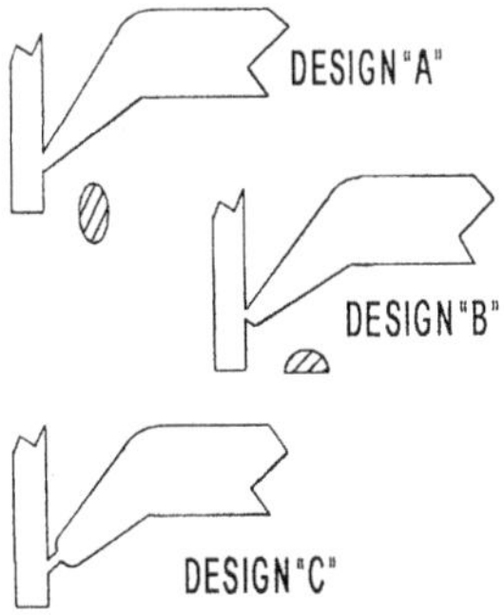

Figure 6.42 Tunnel gate tip design.

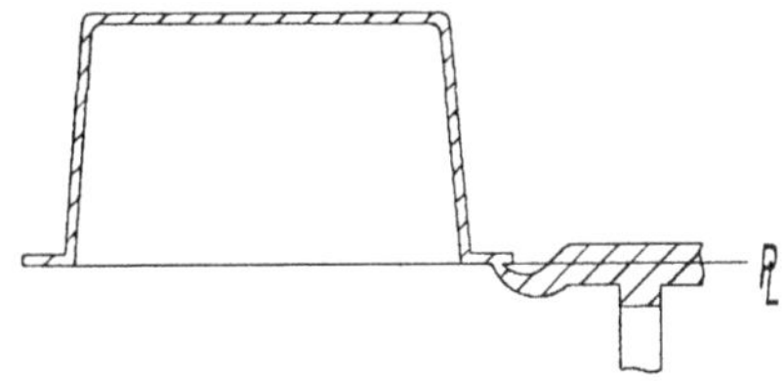

Figure 6.43 Cashew, or banana, gate.

amorphous materials can be used with this type of gate if properly designed with a thick main gate body. The thicker cross-section will keep the gate hot and more flexible while it is ejected.

Jump Gate (Fig. 6.44)

Jump gates can be used to eliminate gate vestige from the outside of a part. The jump gate is another variation of a tunnel gate, but here the tunnel gate jumps past the outer wall of the part into some internal feature on the core half of the part. This is used when automatic degating is required in a two-plate cold runner mold, but there is a cosmetic or physical concern with the gate being attached to the exterior of the part. The feature to which the tunnel gate jumps is often added to the part solely for the purpose of providing a connection to the gate. This feature can either be manually removed after ejection, or designed such that it will not interfere with the normal operation or the assembly of the molded part.

Pin Point Gate (Fig. 6.45)

A pin point gate is a restricted gate used in three-plate cold runner molds. This gate must be small enough to be torn away from the part wall without damaging the wall. These gates are generally less than 50% of the wall thickness of the part to which they are attached. The term *pin point gate* may also be applied to the gate tips in hot drops.

Chisel Gate (Fig. 6.46)

A chisel gate is a cross between a tunnel gate and a fan gate. Like a tunnel gate, the chisel gate tunnels into the part and is torn or sheared off during ejection. Instead of having a circular cross-section, however, the chisel gate has a flat profile. To eliminate an undercut, the chisel gate is widest where it attaches to the runner. It then tapers with a decreasing width and thickness as it tunnels toward the cavity wall.

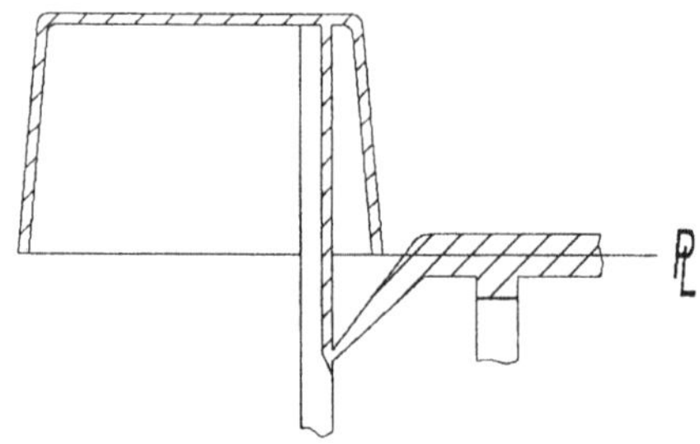

Figure 6.44 Jump gate.

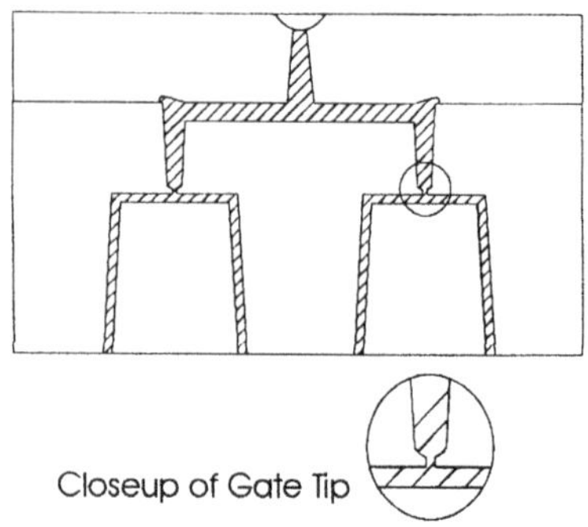

Figure 6.45 Pinpoint gate.

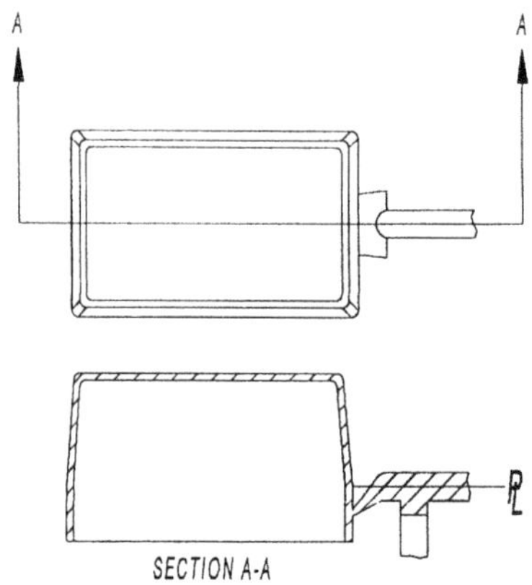

Figure 6.46 Chisel gate.

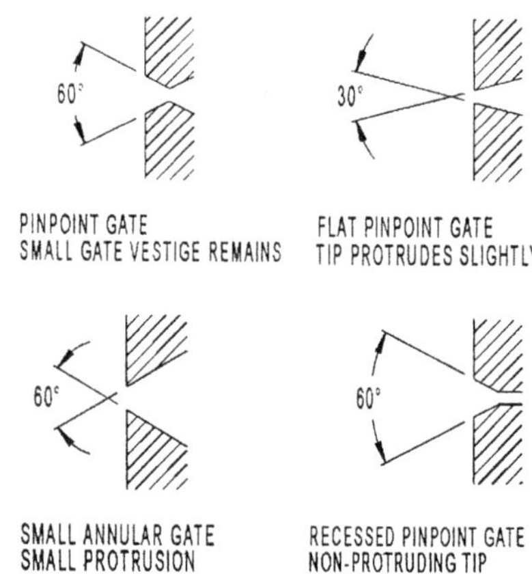

Figure 6.47 Gate tip designs for most hot runner systems.

Hot Runner Gates

Hot runner gates are all small diameter restrictive gates usually less than 2 mm in diameter. Larger gates will increase gate vestage and may require a mechanical shut off. Designs vary depending on the material and concerns with gate vestige. Figure 6.47 illustrates a few variations of gate tip designs.

6.5.2 Positioning Gates

The following are important considerations in the positioning of gates:

- Ability to reach gating location with the desired runner system/mold type: A gating location on the perimeter of the mold cavity can use a standard two-plate cold runner system. A gate location that is inside the perimeter is more restrictive and will require a three-plate cold runner system or a hot runner mold.
- Gate types available for the selected location, mold type, and material: Use of a diaphragm gate would require a three-plate cold runner or hot runner mold. A cashew gate would be limited by the type of plastic material being molded.
- Consideration of wall thickness variations in molded parts: If there are variations in wall thickness, the gate should be located in the thickest wall section. Gating into thinner wall sections will restrict the control of the packing of the thicker region. This can result in excessive shrinkage, warpage, sinks, and voids.
- Effect of gating location(s) in creating weld lines and their effect on the location of weld lines: This can be both a cosmetic and structural concern.
- Potential cosmetic and structural concern at the gate location on the part: The gate region of a part may have problems with gate blush and gate vestige, and may contain high-residual stresses.
- Effect on flow pattern and its influence on shrinkage: Nonuniform shrinkage will also create a potential for warpage and residual stresses.
- Length of flow through the required runner and part cavity: Flow length will affect the pressure to fill the mold and potentially influence the required runner diameter or part thickness requirement.
- Effect on clamp tonnage. Gate location on a part can increase clamp tonnage with increasing flow length, by gating into a region with a large projected area versus a region with a smaller projected area, and by creating unbalanced filling within the cavity.
- Effect on core deflections: Unbalanced filling around a core can cause it to deflect.
- Effect on venting: Will the resultant flow pattern allow for venting along the parting line or will it require special venting? Also consider the potential for gas traps.
- Jetting: Jetting is a common problem caused by restricted gates or poor placement of gates in a mold. The problem results from the inertial effect of the melt entering a cavity at a high velocity. If the melt does not impinge on anything as it enters the mold, it will jet across the cavity in a stream (Fig. 6.48). When a normal flow front is developed, this stream of material will eventually be pushed

Figure 6.48 Jetting shown in cavity short shot.

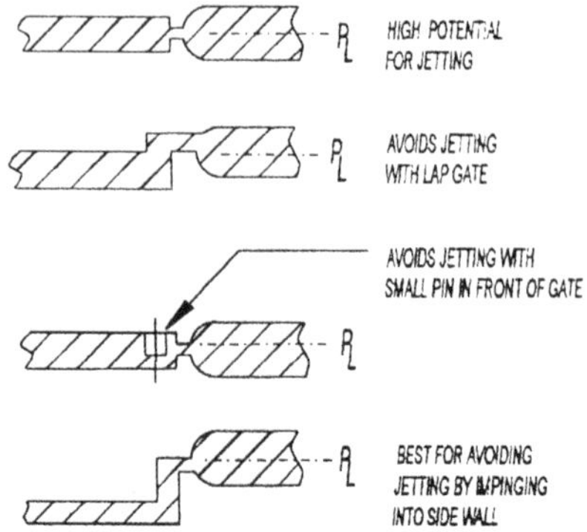

Figure 6.49 Gate designs to avoid jetting.

up against the cavity wall and will cause cosmetic as well as mechanical flaws. Figure 6.49 illustrates gate placements on a part and how they may address jetting.

6.6 Structural Design of a Mold for Long Life and Rigidity

6.6.1 Mold Material Selection

The materials used for molds must satisfy numerous requirements. They must withstand the significant forces and pressures developed during the molding process, have good wear resistance, provide good heat transfer, and good machineability. It is common that different components within the same mold are produced from different steels. The core and cavity inserts must be made from premium steel, which is void free. This steel must also have the ability to be hardened in order to resist the abrasion of the flowing plastic material over millions of cycles. The mold frame can be made out of less-specialized steels because it needs only to have good strength and machinability. If the mold includes components that will slide across one another, the steels should be of different types and hardness in order to resist galling. The least costly of the moving parts are made of softer steel; therefore, the more expensive component will not wear as readily and will less likely need replacement. It is sometimes desirable to use steel with high thermal conductivity, which can help reduce cycle time and provide more uniform cooling. Softer steels, which may reduce machine time and cost, are less wear resistant and will not polish as well.

Table 6.2 identifies some of the more common mold steels, their applications and properties.

Table 6.2 Identifies Some of the More Common Mold Steels, Their Applications and Properties

Steel Type	AISI Designation	Hardness (Rockwell C)	Thermal conductivity [W/(m °C)]	Common Applications
Prehardened	P-20	30–36	29	Mold plates
	4130/4140	30–36	46.7	Mold plates
Prehardened Stainless	420 SS	30–35		Mold plates
Stainless	420 SS	50–52	23	Cavities and cores
Air Hardened	H-13	50–52	24.6	Cavities and cores, gate inserts
	S7	54–56		Stripper rings
	A2	52–58		Slides, stripper rings
Maraging	350	52–54		Small cavity inserts with changing thickness
Beryllium-Copper	BeCu	28–32	130	Cavity and core inserts
Aluminum			220	Prototype and low production tooling
Copper			385	

Molders' desired characteristics of mold material

- Wear resistance for abrasive materials or necessary metal-to-metal contact
- Impact strength (toughness) for small, thin, or unsupported inserts
- Compressive strength to withstand mold clamping forces
- Hot hardness for high-temperature operation
- Corrosion resistance for use with corrosive materials, in high humidity, or for protecting cooling lines
- High thermal conductivity for high production molds

Mold builders' desired characteristics of mold material

- Machinability
- Polishability
- Heat-treating dimensional stability
- Weldability
- Nitriding ability

6.6.2 Fatigue

Mold failures are most commonly a result of metal fatigue. When designing injection molds, it is important to know what fatigue is and how it can affect the life of a mold.

Fatigue results from cyclical, or repeated, loads and will result in steel failure at loads far lower than its normal yield strength. One of the worst loading cases is the type experienced by an injection mold where each cycle consists of the steel being fully loaded during clamp up and injection, and then the load being fully relieved as the mold opens and the part ejected. Fatigue failure can be avoided by designing the mold within the steel's fatigue stress limits and by minimizing stress concentrators. Corrosion and galling (developed from rubbing of mating surfaces) will significantly decrease a steel's fatigue limit. Finishing processes alter the surface of a part as well as the fatigue life. Polishing a component can reduce surface defects that initiate small cracks. Product shape can contribute greatly to the length of a mold's life. Sharp corners or rough surfaces on a part should be areas of concern.

For the ferrous metals normally used for molds, it has been found that if a mold component can withstand more than 2 million cycles at a given load, it is expected to last virtually forever. As a result of fatigue failure, the design stress for mold steels is commonly only 10% of its yield strength.

6.6.3 Deflection of Side Walls

Sidewalls must resist excess deflection from the melt pressures developed during mold filling. Excessive deflection can result in the steel cavity walls springing back after filling and seizing the molded part such that the mold cannot be opened. Either a thicker cavity wall or a tight fit of the cavity insert in the retainer plate will reduce the potential for excessive deflection.

6.6.4 Core Deflection

Core deflection is caused by unbalanced pressures developed on a core from the melt. The location of gates, therefore, can have a significant effect on core deflection. The two gating locations indicated in Fig. 6.50(a) will result in a high pressure developing on the gate side of the core, causing it to bend away from the gate. Cores that are taller than 2.5 times their diameters are particularly susceptible to bending. The gating location shown in Fig. 6.50(b) is always preferred. This location will both minimize the potential for core bending and help to avoid problems, such as gas traps, weld lines, and nonconcentricity that otherwise might develop.

Even with the ideal gating shown in Fig. 6.50b, there is still the potential of core deflection with longer cores. This can result from shear-induced melt variations from the runner, nonhomogenous melt temperatures, dimensional variations in the mold cavity sizes, inconsistencies in mold cooling, etc. Two design variations are presented in Chapter 10 that can help reduce core deflection (see Fig. 10.36).

The deflection of a core results from both shear and bending. Core deflection can be affected by how the core is mounted to the retainer plate or how it interacts with

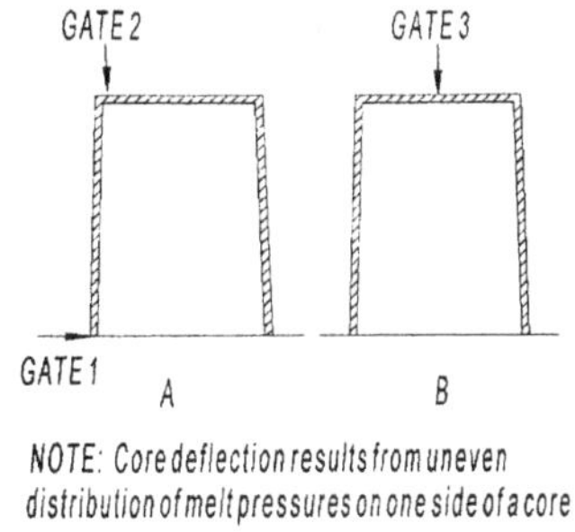

Figure 6.50 (A) Poor gate location resulting in deflection of core. (B) Preferred gate location minimizing potential for core deflection.

the corresponding cavity half of the mold. Means of reducing the potential of core deflection include:

- Increasing the length and diameter of the portion of the core that is mounted in the retainer plate (Fig. 6.51).
- Use of interlocks: Interlocks may be permanently mounted (Fig. 6.52) or retractable. Retractable interlocks are mechanically retracted after the melt has formed around the core. After the cores are retracted, the melt continues to flow and fill in the region previously occupied by the retracted interlock.

Ring and disk gates are commonly used with long cores in order to minimize core deflection and other filling-related problems.

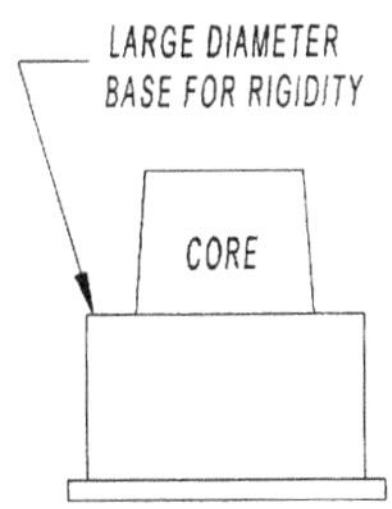

Figure 6.51 Increasing base diameter decreases overall deflection.

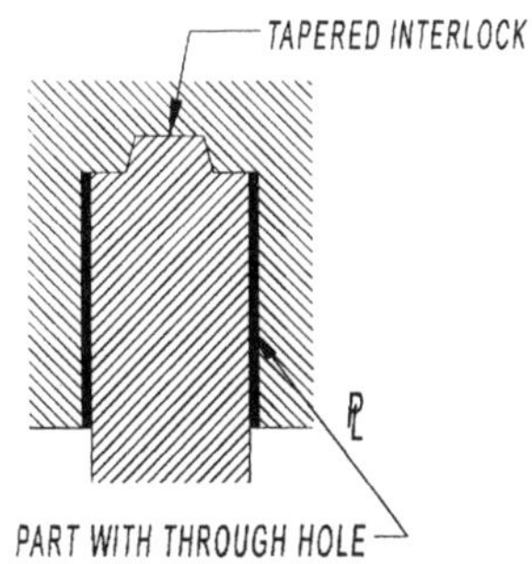

Figure 6.52 Use of interlocks to resist deflection.

6.6.5 Deflection of Support Plates

The support plate on the ejector half of the mold must resist the high forces developed from clamping and from the injection pressures. The support plate is particularly susceptible to deflection because it is suspended over the open space of the ejector housing. This deflection will result in flashing and irregularities in the molded part. It is important that the plate must be sufficiently thick and rigidly supported by support columns to resist deflection.

The practice in industry is to overdesign the structure of the mold using excessive plate thickness and support. This is an accepted practice because the cost for the thicker mold plates, support columns, and related machining is relatively small. A structural analysis would ideally be performed, although it is rarely done. The following is a simplified method that can be used to help determine the required thickness of the support plate. It uses standard beam equations. Judgment must be made as to whether the ends of the support plate should be treated as fully supported or freely supported. This example assumes a fully supported plate because the ends of the plates are reasonably fixed when the machines clamp is closed.

Example 6.1

The objective is to determine how thick the support plate needs to be to resist deflection in excess of 0.005 mm. The assumption is that there are no support pillars, that the two sides of the plate are fully constrained, and that there is a distributed load [cavities are well distributed across the plate (see Fig. 6.53)] of 200 metric tons (200,000 kg.). The 200 tons is based on the machines rated clamp. The clamp should be the maximum load acting to deflect the plates. Modulus of the support plate steel is 206,897 MPa (2,109,770 kg/cm^2)

Given the loading case shown, the plate deflection (y) can be determined as follows.

$$y = \frac{Fl^3}{384\,EI} \tag{6.2}$$

$$I = \frac{bd^3}{12} \tag{6.3}$$

$$y = \frac{Fl^3\,12}{384\,E(bd^3)} \tag{6.4}$$

where d = plate thickness
b = plate width (see Fig. 6.54)
l = plate length

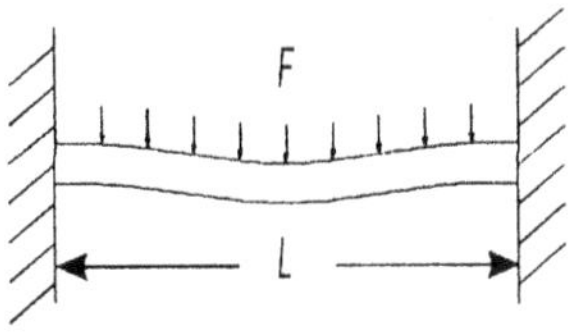

Figure 6.53 Distributed load with both ends fixed.

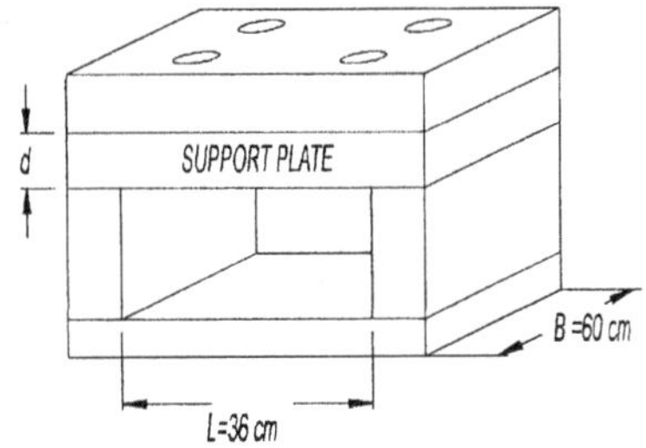

Figure 6.54 Plate deflection.

From Eq. 6.4, the required plate thickness can be found by manipulating the equation to solve for d (plate thickness).

$$d = \sqrt[3]{\frac{Fl^3 12}{384Eby}} \tag{6.5}$$

$$d = \sqrt[3]{\frac{(200{,}000 \text{ kg})(36 \text{ cm})^3(12)}{(384)(2{,}109{,}770 \text{ kg}/\text{cm}^2)(60 \text{ cm})(5.0\times10^{04}\text{ cm})}}$$

= 16.64 cm plate thickness required to maintain a deflection of 0.005 mm.

- To reduce this plate thickness, support pillars can be placed along the plate's centerline. This effectively reduces the unsupported area to less than one half. If 3.8-cm diameter pillars are spaced every 5.0 cm, the unsupported length of the support plate is reduced to approximately 16.1 cm. Solving for thickness (d), given a length (l) of 16.1 cm, gives a plate thickness of only 5.91 cm.
- The required plate thickness can also be affected by modulus of the support plates.

It is also important to check that the design stresses of the steel are not exceeded. The stress developed in the plates during loading for the preceding example are found as follows. Again, this is based on simple-beam equations using the previous assumptions.

$$\sigma = \frac{Fl}{12Z} \quad Z = \frac{I}{d/2} = \frac{bd^2}{6} \tag{6.6}$$

$$Z = \frac{(60 \text{ cm})(5.91 \text{ cm})^2}{6} = 349.3 \text{ cm}^3$$

$$\sigma = \frac{(100{,}00 \text{ kg})(16.1 \text{ cm})}{12(349.3 \text{ cm})^3}$$

$$\sigma = 384.1 \,{}^{\text{kg}}\!/_{\text{cm}^2} = 37.7 \text{ MPa}$$

The flexural stress of 37.7 MPa is acceptable, as the stress on the plate is below the material's fatigue yield stress limit of 69 MPa (this includes a safety factor).

6.7 Mold Cooling

In addition to all of its other tasks, the injection mold acts as a heat exchanger. The design of a molds cooling system requires consideration of:

1. Coolant efficiency: Coolant efficiency is significantly affected by the coolants flow rate through the cooling channels. The flow rate should be sufficient enough that there is a good turbulence in the water. Turbulent flow is much more efficient in extracting heat as compared with laminar flow. With laminar flow, heat must be primarily conducted away from the boundary of the cooling channel through the coolant. As water is a relatively good insulator, this results in the outer laminate of the flow channel being quite warm relative to the center of the coolant flow channel. In circular cooling channels turbulence begins at a Reynolds Number of approximately 2300. As turbulence increases, there will be a continued improvement in coolant efficiency. A Reynolds Number of 10,000 or better is preferred to assure efficient cooling.
2. Coolant temperature: Coolant temperature for a particular part and material must balance consideration of cooling time and part quality. These two factors are usually in contrast to one another. A warmer mold will usually result in the molded part having less residual stresses and a better surface finish; however, economics favors lower molding temperatures in order to reduce cycle time.
3. Pressure drop: Excessively long or small cross-sections in a cooling circuit will result in high pressures. If pressures are excessive, they will reduce the flow rate. Additional factors increasing pressure drop are sharp bends and pipe fittings used to connect to external water hoses.
4. Mold material: There can be significant variations in thermal conductivity for different mold materials (see Table 6.2).
5. Spacing of cooling lines: Spacing should provide for uniform cooling from one region of the part to another as well as from one side of a surface to the other.
6. Layout of cooling lines: Layout of cooling lines must consider physical restrictions within the mold. In addition, it is important to evaluate whether a series or a parallel circuit should be used. These decisions can affect the balance of flow through the circuit, coolant temperature variations, exceeding pressure limits, and maintenance of the cooling system.
7. Coolant: Coolants may include water, water glycol solutions of various ratios, and oil. The coolant with the best thermal conductivity should be matched to the required temperature. Water is the best media for heat extraction, but is limited to temperatures between freezing and boiling. For personal safety reasons and avoiding damage of equipment, water is not normally used for temperatures lower than 5°C and greater than 90°C. Glycol solutions are normally used at less than 5°C. Oil is most commonly used for temperatures greater than 90°C.

6.7.1 Practical Considerations

Pump Capacity

Pumps should be sized in order to provide sufficient flow rate and pressure for a given mold. On the other hand, it is more common that a mold must utilize the available pumps within a molding shop; therefore, the molds cooling channel design should accommodate the flow rate and pressure limits of these pumps.

Flow Control Valves

Flow control valves with sight glasses should be made available at each molding press. These provide individual control of the flow rate to various circuits within a mold as well as a means of monitoring the flow rate through these circuits.

Inlet Temperature

The cooling of a mold is often restricted by the available cooling systems in a plant. Most plants will have a central closed loop recirculating water cooling tower system with separate portable temperature controllers at each molding machine. The central tower system cools water through an evaporative heat exchange system using outside air. The major weakness with this method is that water temperature is dependent on outside temperatures and will vary day-to-day and season-to-season. The colder the day, the cooler the water. It can be expected that water temperature can vary between 15 and 30°C during the year.

The individual press side temperature controllers can be of three basic types. The most common is a mold temperature controller that can only heat the water. The mold temperature controller continuously circulates water through the mold at a set coolant temperature. It uses the central water as a heat exchange media and will raise the temperature to a desired level. Problems arise when a mold is set up to run within the temperature ranges between which the tower system fluctuates. Consider a mold set up to run with 20°C water during the winter and the central water system providing water at 15°C. The mold temperature controller can easily raise the water to an individual mold by the required 5°C; however, during the summer months, the water from the central system may already have reached 30°C. Most mold temperature controllers can do nothing to lower the water temperature; therefore, the cooling will drift out of control, varying the cycle and the molded parts.

A second press side temperature controller is a chiller that can only cool the water. These systems chill the water and use either the central water system or shop air as a heat exchange medium. Again, these systems continuously circulate a coolant (commonly a water glycol solution) through the mold at a set temperature. A chiller system uses a refrigerant and can provide coolants to temperatures below freezing. Glycol must be added to the water as temperatures approach freezing. The colder the coolant, the more glycol is required. Though less common, some plants will use a central chiller system. This provides maximum control when supplemented with individual press side mold temperature controllers.

A third method does not continually circulate water through the mold, nor does it attempt to control water temperature. Rather, it intermittently feeds, or pulses, water from the central water source to a mold in response to strategically placed thermocouples in the mold and an operator set mold temperature. This method allows the molten plastic material to provide the required heat. When the thermocouples indicate a need for coolant, the controller allows water to flood the mold. If there is not sufficient heat provided by the molten plastic, auxiliary electric heaters can be built into the mold. Though this system is less common, it has been documented in some application to have reduced cycle times, energy usage, molded in stresses, and warpage in molded parts. The standard commercial mold cooling analysis programs are currently based on the assumption that water is continuously circulated through the mold. This limits the use of these programs in analyzing this type of cooling method; however, the programs can still provide some helpful information regarding the flow balance of the cooling system, pressure losses, and uniformity of cooling.

Fouling of Water Line

Fouling of water lines can occur from either minerals in the water being deposited on the inside of the coolant channels or from corrosion and rusting of the steel. This fouling both builds up and restricts water flow and acts as an insulator. One millimeter of scale build up can have the equivalent thermal resistance of 50 mm of steel. This can create a potentially significant process variable to which many molders are not sensitive. A process may be established when a mold is first commissioned. With time, scale will build up that will affect the cooling and the related part formation. This can cause molded part characteristics to drift with time. Regular treatments of flushing the cooling lines with muriatic acid solutions will help to eliminate the amount of mineral build up. Stainless steels are also used for molds to reduce fouling problems. Even though the stainless steel has a lower thermal conductivity than many mold steels, it can provide improved and more consistent cooling performance in the long run. Molds built with most other steels should have their cooling lines plated to minimize corrosion.

Positioning of Water Lines

The ideal layout of cooling lines is often restricted due to part geometry and ejection requirements. The designer must be careful not to overcompensate for cooling on one side of a cavity wall if cooling cannot be placed on the other side. This could potentially cause the part to warp toward the hotter side.

Positioning of Coolant Inlet and Outlets on a Mold

- The bottom or the sides of the mold are preferred for positioning inlets and outlets. If there is any leaking at the fittings, the water will drip into the cavity and potentially rust the critical core and cavity inserts and otherwise negatively affect the molding.

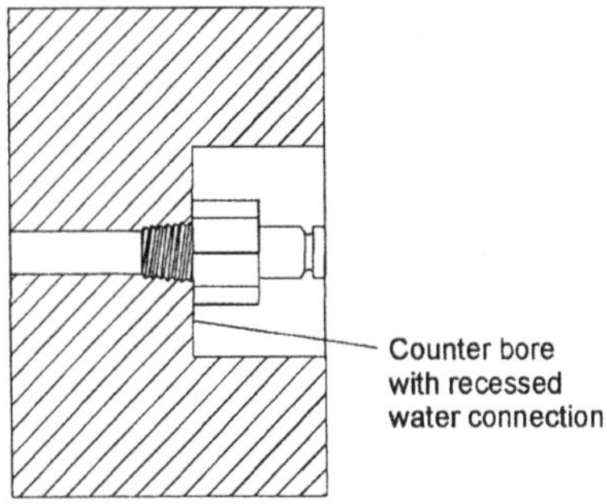

Figure 6.55 Illustration showing recessed coolant connection.

- Recess all coolant connections (Fig. 6.55). This will eliminate the possibility of damage to, and the need to remove, fittings when a mold is being removed from the molding machine.

6.7.2 Thermal Expansion

As a mold changes temperature, the steel will expand and contract. This generally does not create a problem as long as the whole mold is at approximately the same temperature; however, there are cases when a molder may adjust a mold half-temperature to address certain molding problems, such as ejection. This can create problems due to misalignment of the mold halves, which can affect part dimensions and damage interfacing parts of the mold such as leader pins or interlocks. A hot manifold may also create a similar variation if the thermal imbalance is not properly compensated for by the cooling system.

6.7.3 Parallel versus Series Cooling Circuits

Within a mold, cooling lines can be laid out in a series or parallel circuit (see Fig. 6.56). Each has its advantages and disadvantages, although it is common that these are often misunderstood and misapplied.

Parallel Circuit

In a parallel circuit, water is fed to multiple parallel branches from a single source or manifold. This manifold may be externally mounted or built into the mold. The ideal case is that the water is equally divided into each of the branches at the same temperature and flow rate. The primary advantage of a parallel circuit is that because its layout results in shorter flow lengths, it has less chance of exceeding the pump's pressure limits. As a result, the full flow rate from the pump is more assured. It is also perceived that the water temperature will be more uniform than it is in a series circuit.

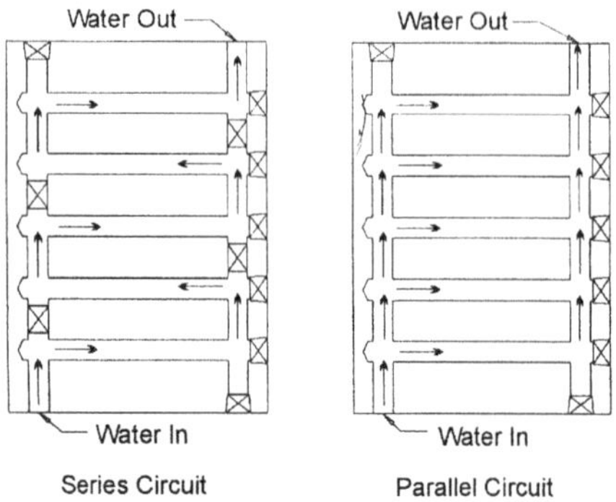

Figure 6.56 Series vs. parallel coolant circuit.

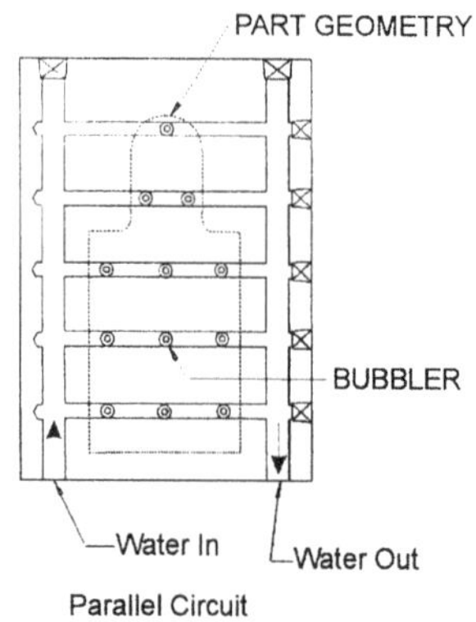

Figure 6.57 Parallel circuit with variation in bubblers causing unbalanced water distribution.

This is not always the case. The perception is that the water will fill the main feed branch and flood across all the secondary branches with the same water temperature and flow rate. There are several potential problems with this.

First, to get this to work reasonably well, the main feed branch must be large enough so that there is very little pressure drop along its length. Second, if there is any variation in the flow resistance in any of the secondary parallel branches, the flow rate will vary and the desired uniform cooling will be lost. This variation could result from the nature of the parallel circuit design, from variations in mold set ups, or developed with age. In some molds the cooling lines within each of the parallel branches could have various lengths, bends, bubblers, baffles, and diameter (see Fig. 6.57). With external manifolds, variations could be created by various water hose lengths, diameters, or connectors feeding each branch. This could vary every time you hook up the mold. With age, scale and rust buildup can create variations in the diameters and thereby the flow resistance.

Finally, by distributing the water between each of the parallel branches, the flow rate through each of these branches can only be a fraction of the total. This can potentially decrease the efficiency of the cooling by reducing the level of turbulence. If a Reynolds number of 10,000 cannot be obtained in each of the parallel branches, the designer should consider modifying the circuit design.

Mold cooling simulation programs provide the best means to design parallel circuits to help assure that all channels have the same flow rate and that flow is turbulent.

Series Circuit

In a series circuit there is a single inlet and a single outlet with no branching. This results in a relatively long flow path for the water. Advantages of the series circuit include:

1. The water flow rate through a mold will be higher than it is with a parallel circuit if the pumps pressure limit is not exceeded. The increased water turbulence will result in more efficient cooling.

2. Incontrast to the common externally manifolded parallel network designs, the fewer inlets and outlets in a series circuit will simplify mold set up. Set up is faster and has less chance of incorrect water hook up.
3. Flow rate is more assured to be constant along the entire length of the circuit because it is not divided into various branches that could have flow imbalances.
4. If a circuit were to clog, the clog would be obvious because there would be no water flow. A clogged branch in a parallel circuit is not easily detected because effect on flow in and out of the mold would difficult to detect. Undetected clogs will affect cycle and molded product.

The two major concerns of the series circuit are that the pressure drop through the circuit will be too high and that the water temperature rise through the long circuit will be excessive. If pressure exceeds the pressure limit of the water pump, the flow rate will be reduced and negatively affect the efficiency of the cooling. Excessive water temperature rise will result in variations in cooling. Variations in cooling within a single cavity will contribute to residual stresses and warpage in a molded part. Variations in cooling between cavities in a multicavity mold will result in variations between the molded parts; however, if pressure is not exceeded, the high flow rate through the mold minimizes the potential for excessive temperature build up through the length of the circuit. The advantages of the series circuit make it worth considering. Many smaller molds can easily be designed entirely with series circuits.

Combining Circuits in a Mold

It is important when designing a mold that the balance of all circuits to each other be considered. Each of the individual circuits in a mold are often fed from a single external water manifold. The mold may include combinations of highly restrictive circuits and nonrestrictive circuits. Circuits that are cooling critical areas of the mold might include restrictive features such as bubblers and baffles. Other circuits may be much less restrictive and serve only secondary roles such as cooling a retainer plate or stripper plate. In some cases a circuit might feed a single slide in a mold with side action. When these different circuits are fed from a single water manifold, a majority of the water will flow through the less-restrictive circuit. This will leave the restrictive circuits starved of water. These restrictive circuits, which include bubblers and baffles, are often supposed to be providing critical cooling to the mold; therefore, sections of a parallel circuit which have low flow restriction should be combined in series or restricted by some means such that they are better balanced to the highly restrictive sections of the circuit.

6.7.4 Baffles and Bubblers

Baffles (Fig. 6.58) and bubblers (Fig. 6.59) are generally used to cool inside restrictive cores. Multiple baffles or bubblers are often used to cool the inside of larger cores or to get closer to corners in box shaped parts. Running numerous baffles or bubblers up from the bottom of a core eliminates the need to drill cross-channels in a normal circuit arrangement. The baffles and bubblers can be angled in nearly any direction within the mold.

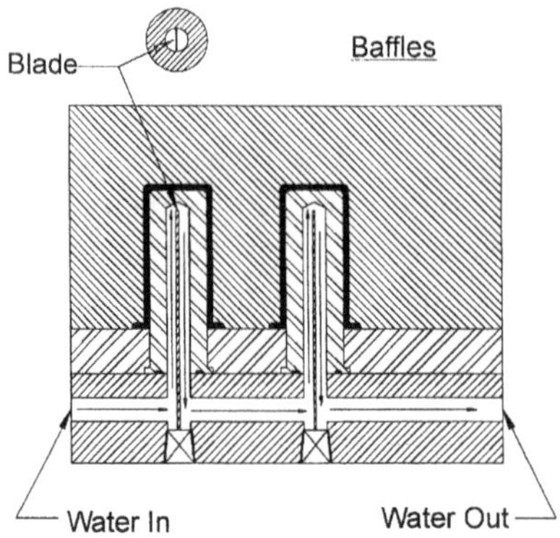

Figure 6.58 Baffle design.

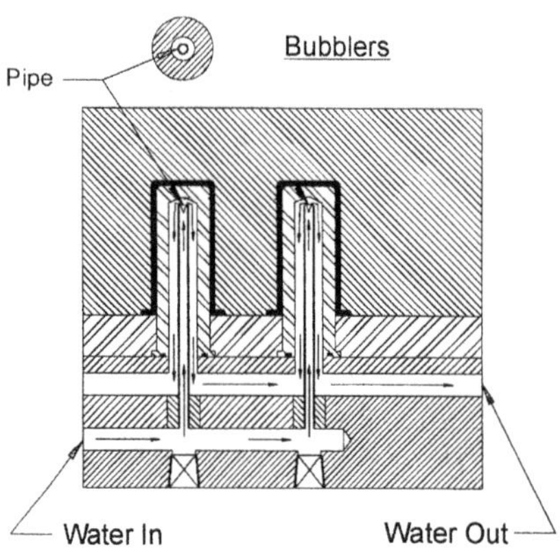

Figure 6.59 Bubbler design.

A baffle is comprised of a hole drilled across a water line. A thin metal blade is inserted in the hole such that it interrupts the water line. The blade diverts all the water up to the top of the baffle and it returns on the opposite side of the blade, returning to the same main water line. Numerous baffles can be on a single water line. This is essentially a series circuit.

A bubbler requires a feed line and a separate return line. In this case a hole is drilled through both of the feed and return lines. The lower feed line feeds water up through a tube to the top of the bubbler hole. The water emerges from the tube and returns down the outside of the tube to the return line. This arrangement is essentially a parallel circuit that has separate feed and return lines.

6.8 Mold Ejection Systems

6.8.1 Basic Ejection Problems

Ejection of plastic parts from a mold can present a significant challenge. It is critical when designing a plastic part, which is to be injection molded, that the part be designed considering how it is to be ejected from the mold. In other words, the part's design is strongly influenced by this requirement. Basic considerations include location of the parting line of the mold relative to the part during mold opening, tapering vertical surfaces that must be drawn from a core or cavity, and undercuts that will prevent the part from being ejected.

Consider even a simple five sided square box [see Fig. 6.60(a)]. The location of the parting line along the free edge of this box is somewhat obvious. This location allows the mold to open freely and the part to be pushed off the core; however, the straight walls of the part present a problem. After plastic is injected into the cavity, it begins to cool and shrink. This shrinking plastic develops a significant pressure on the mold's core. The part must be pushed off the core by some means. If the pressure is excessive, the force required to eject the part may result in damaging it. It is important, therefore, to minimize the negative effect of the pressure created by the shrinking plastic. This is done by tapering the inside walls of the part. This is referred to as *draft*. The larger the draft angle, the easier the ejection is, and the less potential there is for damaging the part.

In addition to the pressure created on the sidewalls of the core, the ejection of this part will result in the development of a vacuum created in the growing space between the base of the box and the top of the core [see Fig. 6.60(b)]. When on the core, no air exists between the part and the core. As the part is ejected from the core, more space is created between the part and core, but no air is allowed to enter the space. This creates a vacuum between the core and the part. The further the part is pushed along the core, the greater the vacuum and the greater the force needed to eject the part completely. Again, tapered sidewalls almost immediately solve the problem, allowing air to enter freely [see Fig. 6.60(c)] as the part is pushed off the core.

If the inside surface of the part is tapered, the outside surfaces of the same walls should be designed with the same taper. This will minimize problems created from variations in wall thickness.

There are many means of applying the ejection force required to eject this part from the core. The simplest, least expensive, and easiest means would be the use of ejector pins. The two potential regions to place these pins would be through the core to push on the inside base of the box or along the free edges of the box. As can be seen in Fig. 6.61, placing the pins up through the core results in the sidewalls being drawn in and the part being distorted at its base. The recommended position of the ejector pins is along the side edges of the part. These edges are rigid and more likely to be able to withstand the forces of ejection. Corners and ribs also provide rigid

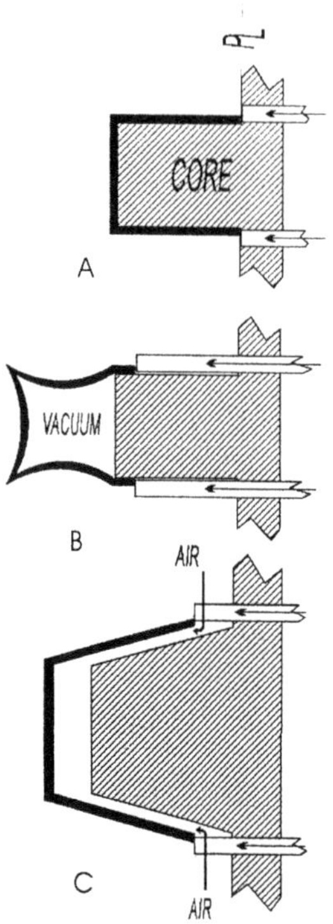

Figure 6.60 Five sided box. (A) Square design. (B) Ejection difficult due to vacuum formation. (C) Remedied by tapering inside and outside walls.

structure from which to eject a part. Because of the possibility of impression or witness lines/circles created by ejector pins, a designer must carefully place the ejector pins where the flaw would not be objectionable.

Polishing the core in the direction of draw (the direction in which the part is being ejected from the core) can decrease the required ejection force. Polishing the core transverse to the direction of draw can have the opposite effect. When polished against the direction of draw, the tiny grooves from polishing actually create undercuts and thereby hold the part on the core. It is normally expected that a finer polish

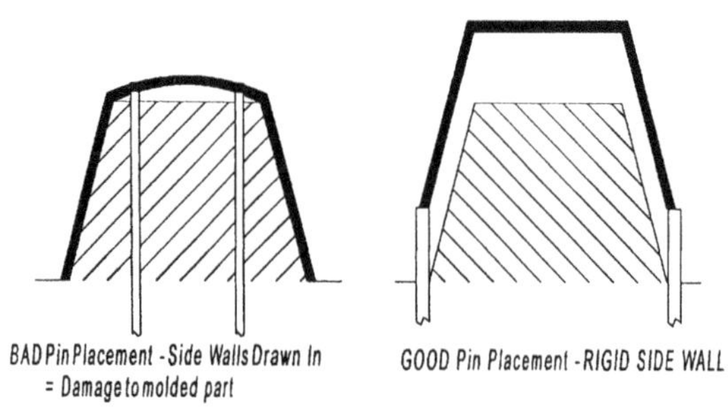

Figure 6.61 Ejector pin location.

Table 6.3 (Effect of Surface Finish on Ejection)*

Rating: 1 = Good; 2 = Average; 3 = Not Recommended (may) be acceptable)	Spifinish A-1 Diamond #3	A-2 Diamond #6	A-3 Diamond #15	B-1 Paper 600	B-2 Paper 400	B-3 Paper 320	C-1 Stone 600	C-2 Stone 400	C-3 Stone 320	D-1 Glass #11 Blast (Dry)	D-2 Oxide #240 Blast (Dry)	D-3 Oxide #24 Blast (Dry)
Plastic												
ABS	3	3	2	2	2	1	1	1	1	1	1	2
Acetal	3	3	2	1	1	1	1	1	1	1	1	N
Acrylic	1	1	1	2	2	2	2	2	2	3	3	3
Nylon	3	3	2	2	1	1	1	1	1	1	1	2
Polycabonate	3	2*	1*	2	2	3	3	N	N	1*	N	N
Polyester	3	3	2	1	2	2	1	2	N	1	1	N
Polyethylene	N	3	3	2	2	1	1	1	1	1	1	1
Polystyrene	3	3	2	1	1	1	1	1	1	1	1	2
Crystal Polystyrene	2	1	1*	2	2	3	3	N	N	1*	N	N
Polypropylene	N	3	3	2	2	1	1	1	1	1	1	1
Polyurethane	N	N	N	N	3	3	2	2	2	1	1	2
PVC	N	N	N	N	3	3	2	2	2	1	1	2

* DIAMOND and GLASS by CRYSTAL MARK (Tech Mold Development).
N indicates a nonacceptable application.

will reduce ejection force. However, this is material dependant. Some materials, like polyethylene and polypropylene, will be easier to eject when a courser finish is applied to the steel. Table 6.3, produced by the Society of Plastics Industry (SPI), describes the affect of surface finish on ejection of a plastic parts with various families of material.

For parts such as a five-sided square box, ejection problems can be resolved rather painlessly; however, all plastic part designs are not so simple. Consider the same box design but with the addition of a flange and an extending surface (see Fig. 6.62). The position of the parting line is the first obstacle encountered. If the parting line is placed at the free edge of the box, the flange creates an undercut that would trap the part in the cavity upon mold opening [see Fig. 6.62(a)]. When the parting

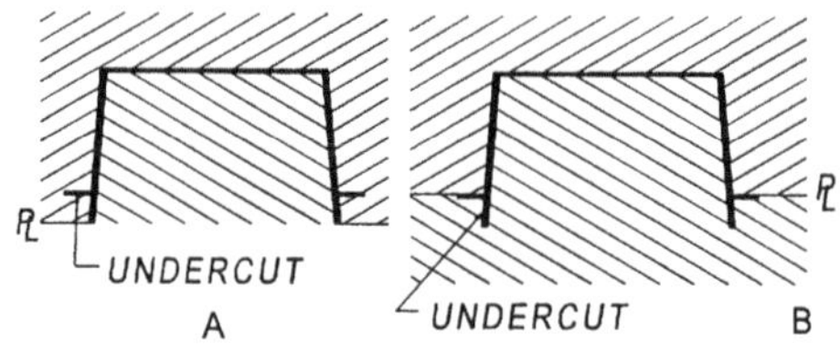

Figure 6.62 Part design's dependence on ejection.

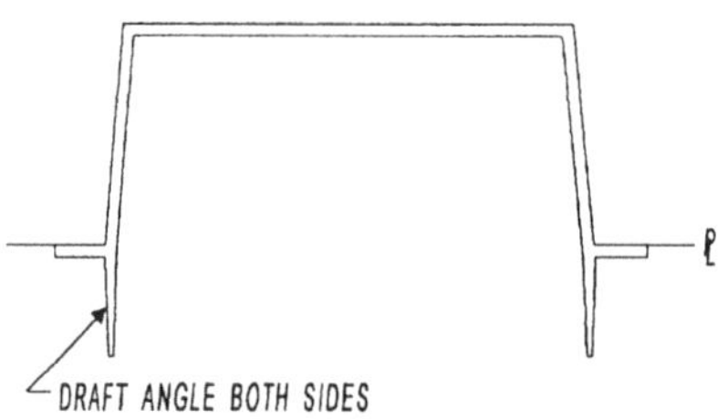

Figure 6.63 Part design change to avoid undercut and accommodate ejection.

line is positioned at the top free edge of the flange, the outward taper of the extending surface creates an undercut that would trap the part on the core [see Fig. 6.62(b)]. To accommodate ejection, the part design should be altered. The extending surface should be redesigned to include a draft angle on either side (see Fig. 6.63). This modification will allow the parting line to be placed along the top free edge of the flange and the extending surface to be extracted from the core side during ejection. In this case the part could be ejected using an ejector blade or two stage ejection.

6.8.2 Means of Ejection

Pins

Using pins for ejection is the most common and least expensive means of ejection. Figure 6.2 illustrated a typical pin ejection system. Here, the pins would be driven forward by the ejector plate. Return pins are included to drive the plate back by the cavity plate when the mold closes. Return pins are included as either a primary means of returning the ejector pins or as a back up if a hydraulically driven system fails to return the ejector plate and pins. Use of pins sometimes results in excessive stress because of their small contact area with the part. One method of increasing the contact area of an ejector pin is to use the edge of an oversized pin, as shown in Fig. 6.64.

Ejector Sleeves

Ejector sleeves are used for small parts with cylindrical cores (see Fig. 6.65) or for round, internal features on a part such as a hollow boss. Although this type of ejec-

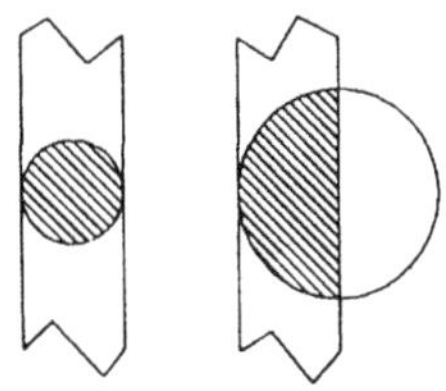

Figure 6.64 Increase in contact area by utilizing an oversized ejector pin.

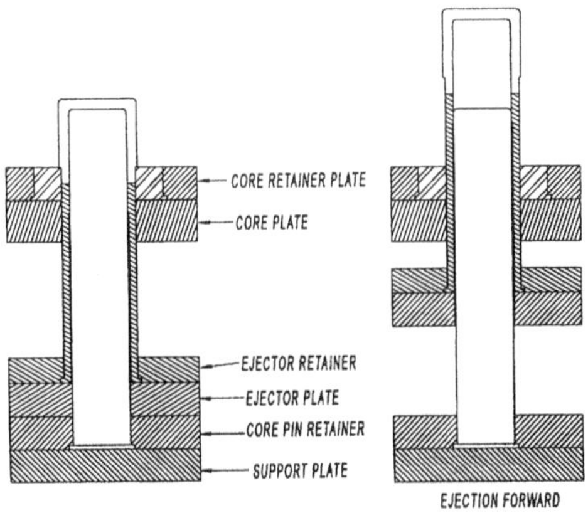

Figure 6.65 Core pin with stripper sleeve.

tion is more expensive, it provides an excellent positive force on the part. The core being stripped by an ejection sleeve must extend down through the ejector plate to the clamp plate. The sliding of the stripper sleeve on the core, during ejection, will create wear and with time will increase potential for flashing between the core and sleeve. Most common ejector sleeves are limited to small cylindrical parts with less than 50-mm diameter.

Stripper Plates

Stripper plates provide the same positive features of ejector sleeves but can easily accommodate larger parts. Figure 6.3 showed the application of a stripper plate in a three-plate cold runner mold. Unlike an ejector sleeve, stripper plates are not limited to cylindrical shapes. In addition, the core is not required to extend down through the ejector plate and can be used to eject multiple parts. The plate normally includes stripper inserts that contact the core and the plastic part. These stripper inserts are made of a harder mold steel to withstand the wear associated with the ejection process. During ejection the plate is driven forward by ejector rods that are either connected to the ejector plate or directly to the machine's hydraulic ejection system. When using a stripper plate, cooling the stripper plate should be taken into consideration to avoid variations in thermal expansion. Unlike the ejector sleeve, the stripper plate cannot be used to eject internal features like a boss.

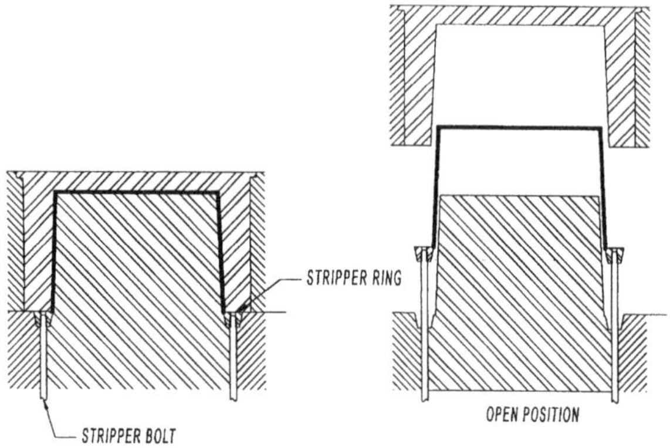

Figure 6.66 Stripper ring ejection.

Stripper Rings

Stripper ring ejection is similar to stripper plate ejection except for the shape of the stripper (Fig. 6.66). Instead of a bulky plate, the ring only borders the edges of the part. This feature is usually used with a large cylindrical part in a single cavity mold.

Blade Ejection

Blade ejectors are often applied to straight edges of parts where an increase in contact area is necessary and cannot be achieved by pins (Fig. 6.67). The blade can be a flat section machined from a round pin.

Air Ejection

Air ejection is yet another means of ejection that can be used alone or in combination with another type of ejection. This method injects air between the core and the molded part to push the part off of the core.

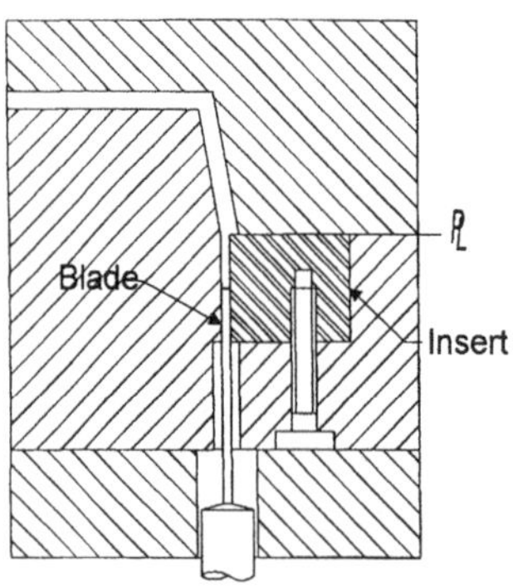

Figure 6.67 Blade ejection.

6.8.3 Ejection Considerations

The ejection system of a mold is typically in the moveable half. This requires that the parts adhere to the movable mold half. This is accomplished by several methods:

- Shrinkage onto projections/core features
- Use of undercuts
- Temperature
- Vacuum created between the part and core steel

Typical ejection problems generally arise with thin-walled parts, brittle materials, or highly elastic materials. A thin-walled part formed from a brittle material is more susceptible to fracturing under the force of ejection. Elastic materials or elastomers tend to distort too much during ejection, and the high coefficient of friction holds the parts to the core. High and long packing pressures will also increase the required ejection force. Finally, when running a mold with a thin-walled cavity, the spring back force of the deflected steel cavity wall can actually clamp the part in the cavity and prevent the mold from opening. This is particularly a problem with parts that have little to no draft.

Ejection must ultimately overcome the forces developed from three factors: (1) the shrinkage of the plastic onto the core, (2) the expansion or spring back force of the core and cavity from the injection pressure, and the resultant pressures acting on the frictional resistance of the steel/plastic interface, and (3) the vacuum developed between the ejecting part and the core. There are several ways to minimize the influence of the ejection force:

- Use release agents (silicone, PTFE, etc.)
- Polish in the direction of ejection
- Taper walls
- Decrease cooling time
- Possibly decrease injection rate
- Decrease hold pressure
- Decrease hold pressure time
- Decrease ejection rate
- Use a mold plating with low friction surface
- Use of air poppets to relieve vacuum

Release agents should be used with caution. They will be deposited on the part surface and interfere with secondary operations, including painting, printing, and use of adhesives for assemblies. It is important to have some ejection force pushing on the deepest features of the part geometry such as bosses and ribs. This is easily accomplished on bosses by using ejector sleeves. However, it can be difficult with ribs to find enough surface area to place sufficient ejection. Figure 6.68 shows a rib design that allows for more surface area, where an affective ejector pin or sleeve can be used. This change must always be approved by the part designer.

6.8.4 Ejection Design

When designing the ejector housing and ejector plate, the following must be considered: the placement of the support pillars for rigidity to withstand deflection of the plates on the moveable mold half, the use of ejector guide pins and bushings to ensure alignment, and the use of return pins to protect the cavity from damage. The actuation of the ejector plate can be achieved by three methods:

- Mechanical push—retraction of machine platen and mold plate against a stationary push rod
- Mechanical pull—ejection from stationary mold half (see Fig. 6.69)
- Hydraulic/pneumatic action—allows timed, and multiple pulse ejection. This is the most common method used today.

The return motion of the ejector plate can also be activated by hydraulic cylinders or springs. A three-plate mold generally uses the concept of stationary mold half ejection. The runner and secondary sprues are held to the stationary half of the mold by multiple sprue pullers while both the primary and secondary parting lines open. When the mold is fully open, the parts are ejected from between the primary parting line, whereas the runner is ejected from between the secondary parting line. Ejection of the runner is activated by a stripper plate, which is connected to the moving plates by a bolt, or external latch. This bolt, or latch, then pulls the runners stripper plate, stripping the runner from the sprue pullers (see Fig. 6.3).

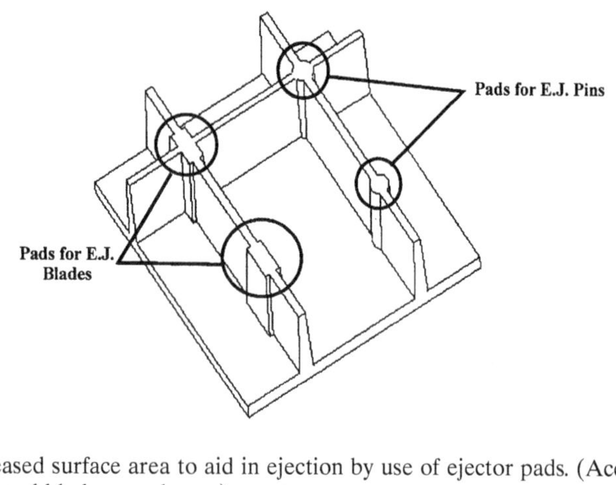

Figure 6.68 Increased surface area to aid in ejection by use of ejector pads. (Accommodations for both ejector pins and blades are shown.)

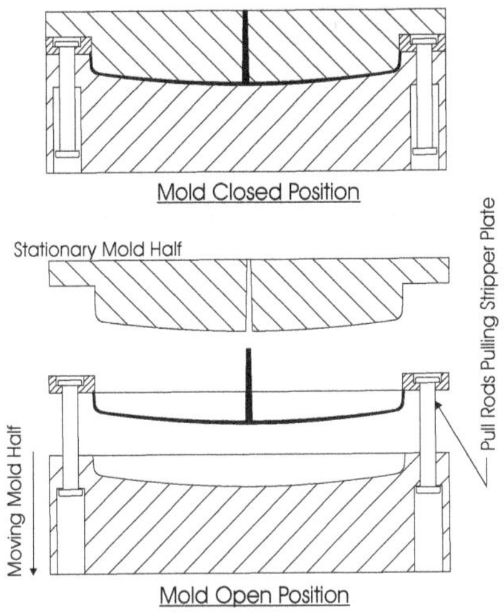

Figure 6.69 Ejection off of a stationary core.

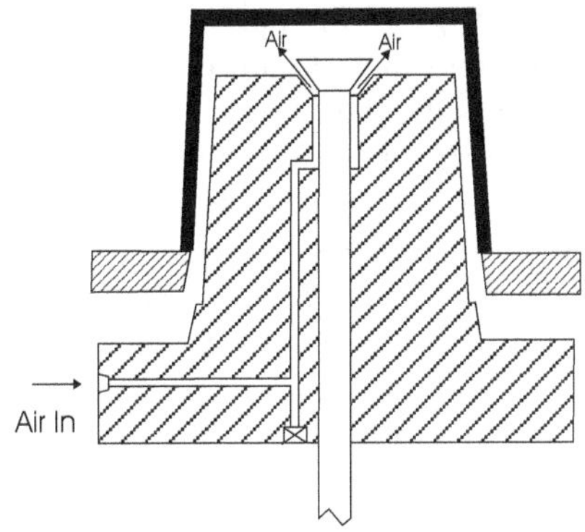

Figure 6.70 Air poppet valve used to break the vacuum and ease ejection.

Some special types of ejection include double-stage ejection, which can combine stripper pins, plates, cams, hydraulic/mechanical combinations, compressed air, and mechanical/air combinations. Figure 6.70 illustrates a simple case where an air poppet valve is used to allow air to enter the space created as the stripper plate drives the part off the core. Without the air poppet, a vacuum would result and complicate the ejection.

Ejection systems cannot be designed without first considering undercuts created from the part design. A designer should always ask several questions about undercuts:

1. Can the part be redesigned to eliminate the undercut?
2. Can the part be pushed, or stripped, off with the undercut without it being damaged?
   - The temperature of the part at time of ejection can have a big effect on potential ejection problems. A warmer part will be more ductile
   - The maximum allowed undercut can be calculated from permissible strain values if available
3. Is it possible to sequence the ejection to relieve the undercut?
4. Will the undercut require that the mold have side action?
   - Split cavity or split core
   - Side core or side cavity
5. Will the part require unthreading from the mold?

Internal Undercuts. Internal undercuts on a part typically require one of four methods to eject it from the mold. Choosing the appropriate means for removing the undercut is dependent on part design. The choices are as follows:

1. *Stripping.* Stripping is commonly applied to cylindrical parts, such as caps, which might include an undercut for snapping onto a can. The undercut should be designed with rounded or tapered surfaces so that it is not damaged during ejection (Figure 6.71). The material must also have good ductility so that it can be deflected during ejection. If threads are to be stripped, the undercut developed by the threads must be tapered to allow relief during ejection. Stripper plates or rings are generally required to assure positive ejection without damaging the part.

2. *Collapsible Cores.* Collapsible cores allow for molding nontapered or nonradiused undercuts and are generally used with relatively small diameter cylindrical parts. Standard collapsible cores can be purchased from suppliers of mold components. The operation of the collapsible core is shown in Figure 6.72:

- Step 1. Spring action of collapsing core segments are held open during molding by a stationary support core
- Step 2. Ejector plate moves forward, driving collapsing core segments forward and leaving the stationary support core behind. Without the support of the stationary core, the core segments collapse away from the inside of the molded part and relieve the undercuts.
- Step 3. Stripper plate, or other means, eject the part from the collapsed core segments.

3. *Lifters.* (Figure 6.73) Lifters are commonly used with larger parts having localized undercuts. As the ejector plate moves forward it drives the angled core, or core section, forward. The angle of the core section causes the core to move away from the inside of the part, thus relieving the undercut.

4. *Unscrewing Molds.* Unscrewing molds are used for both internal and external threads. An electrical motor, mechanical rack and pinion, hydraulic rack and pinion, or a lead screw can drive the unscrewing mechanism. These are usually very complex molds requiring significant experience for their successful design and manufacture.

External Undercuts. External undercuts are generally relieved with the use of side cores, split cavities, or lifters. Activation of side cores and cavities or split cavities can be achieved by several means, including mechanical (cam pins or lifters), hydraulic, or pneumatic. Choosing the most applicable means for ejection of external undercuts requires careful evaluation of the following methods:

1. Side Cores or Split Cavities. Figure 6.74 illustrates the movement of a side core being activated by cam pins during the normal mold opening motion. Hydraulic or pneumatic cylinders could also activate the same side motion. Hydraulic or pneumatic cylinders provide an advantage when a long stroke is required or if it is desireable to time the core movement independent of the mold opening.

2. Lifters. As with internal undercuts, the lifter utilizes the movement of the ejector plate to drive a cavity section at an angle, away from the external undercut.

3. Unscrewing Molds. Unscrewing molds are sometimes used with external threads, but are only used when the highest tolerances are required. Due to the rela-

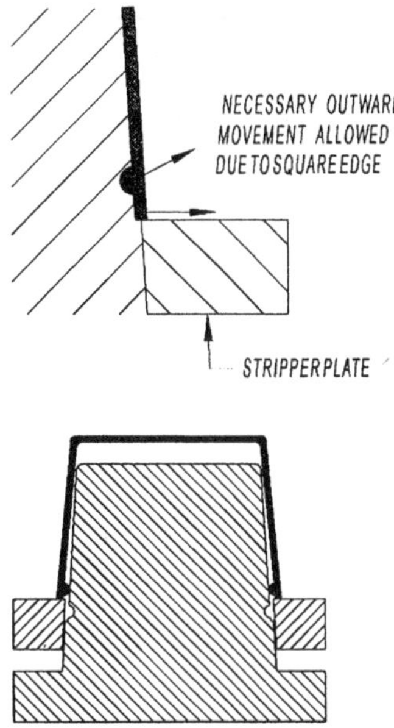

Figure 6.71 Undercut design.

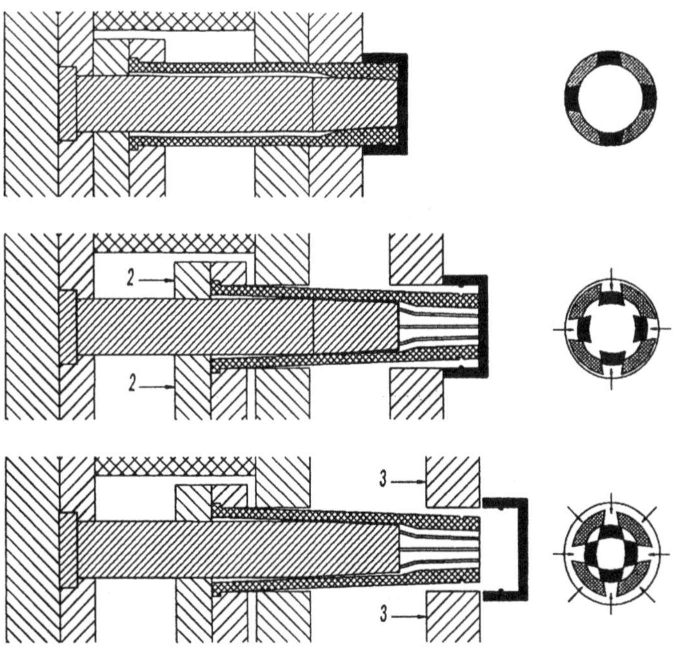

Figure 6.72 Collapsible core.

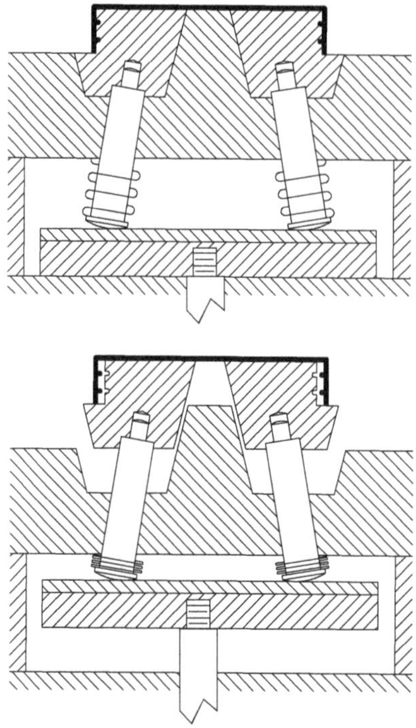

Figure 6.73 Lifter (split-core) used to relieve an internal undercut.

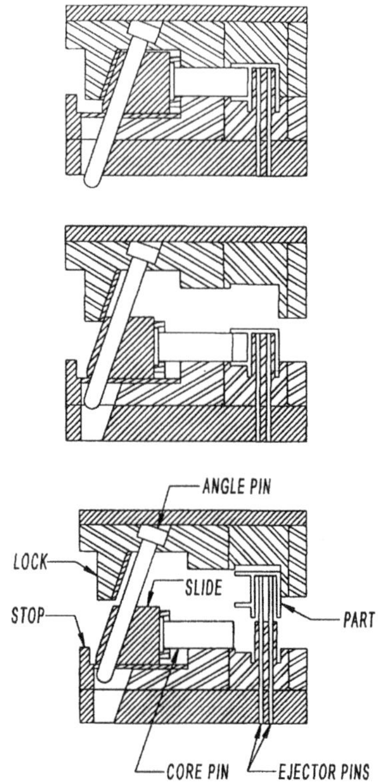

Figure 6.74 Side core using an angle pin and cam block.

tively low cost, the preferred method of ejection for external threads is the use of side action. It is also possible to produce external threads by positioning the threads along the parting line of the mold so that the two mold halves form them. Stripping generally is not possible with external threads.

6.9 Vent Design

Prior to melt being injected into a mold cavity, the cavity is occupied by the room air that was trapped there when the mold closed. As the melt enters the cavity through the gate, the air must have a means to escape. If the air cannot escape, it will be compressed by the flowing plastic. This compression of the trapped air will result in a number of negative effects. At the high melt-injection pressures experienced during mold filling, the trapped air can be compressed to a point where it can experience dieseling. The resultant hot compressed air can both burn the plastic material and even potentially damage the surface of the mold steel. In addition, as the air is compressed it will develop a pressure and create a resistance against the plastic melt trying to fill the mold. This pressure can become significant enough to affect the filling pattern and, in some cases, prevent the cavity from filling. This is a phenomenon not recognized by most mold-filling simulations. However, this effect is negligible if the mold is properly vented.

To prevent these problems, vents are placed in certain regions of the mold that will allow the air to escape. This requires consideration of the gate location to help assure that the air is not trapped by the advancing melt front in a region of the cavity that cannot be easily vented. The most desirable location for vents is the parting line of the mold. The parting line vent is easy to machine and, if clogged, can easily be wiped clean. Vents should be placed at the last place in the mold expected to fill. Additional vents along the perimeter of the part at the parting line of the mold are recommended in order to maximize the ability of the air to escape.

Parting line vents are normally ground into the parting line surface. They normally consist of two regions (see Fig. 6.75). The initial vent land at the perimeter of the cavity must be very shallow so that air can escape, but the molten plastic cannot. The depth is dependent on the material, but it is commonly only about 0.03 mm (0.0012 in) for high-viscosity amorphous materials and 0.015 mm (0.0006 inches) for low-viscosity semi-crystalline materials. In order to minimize the resistance to the escaping air, the length of this vent land is kept relatively short (approximately 2 mm). A vent relief is then provided. The depth and length of this relief is not nearly as critical, but it is commonly about 0.4-mm deep. The length must be sufficient to allow the escaping air to vent to the atmosphere.

Fast injection rates, which are common with thin-walled parts, require more liberal venting. This could include:

- continuous perimeter venting
- increased venting in regions other than end of fill to relieve gas pressure
- venting of runners
- vacuum venting

Vacuum venting can be used in cases where the air cannot be relieved by conventional means. This may be required in cases experiencing high-speed injection or cases

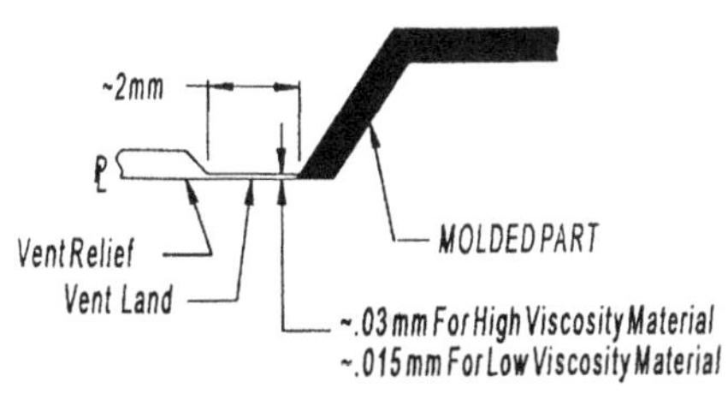

Figure 6.75 Vent design.

where the air will be trapped in some region of the cavity that cannot practically be vented by other means. With vacuum venting, the cavity region of the mold is commonly surrounded by a gasket. The normal vents in the cavity are channeled to some common location from which a vacuum hose can be attached. When the mold closes, a vacuum is applied that draws the air out of the cavity. This can be complicated by air being drawn up through the ejector pins or any other features internal to the cavity.

Though the parting line location is ideal for venting, it is not always accessible. It is common that parts have ribs or other regions that cannot be vented out the parting line. If a region of the cavity cannot be vented at the parting line of the mold, venting should be attempted at other moving parting lines, or seams, in the mold. This could include parting lines on slides, retracting cores, and ejector pins. It is important that the vent area is accessible so that it can be cleaned in the event that it is clogged. If it is not readily accessible, then an attempt should be made to provide venting through some moving feature within the mold. This movement provides the possibility of any residue clogging the vent to be wiped off during the movement. A common case of venting through a moving feature that is wiped clean during movement is an ejector pin. As with a parting line vent, provisions should be made to allow the gas to escape, but not the plastic melt (see Fig. 6.76). When a stationary vent is required that cannot easily be cleaned, the mold can be designed so that an air blast can be applied through the stationary vents during mold opening in order to blow out any residue build up.

Even though the use of an ejector pin may seem convenient, it usually does not provide sufficient area to properly vent. In addition, the position of the trapped air may move if the melt flow front changes such that gas trapped in the cavity does not end up at the small vent pin. An alternative is the use of a larger stationary vent. In the case of a free-standing rib that requires venting, a split cavity may be used (see Fig. 6.77). Large venting areas can be accommodated using a laminate structure (see

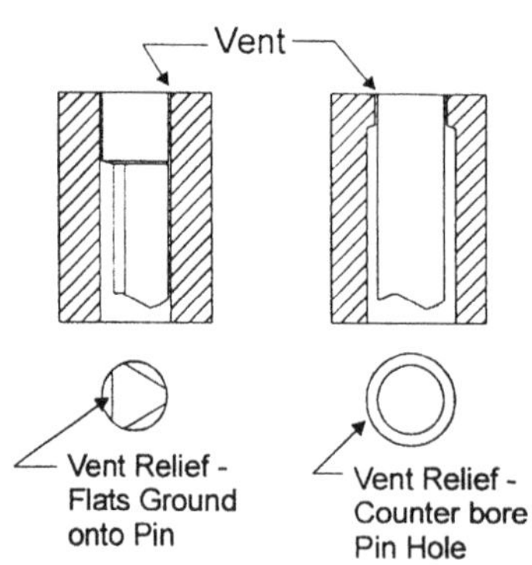

Figure 6.76 Common venting designs on pins.

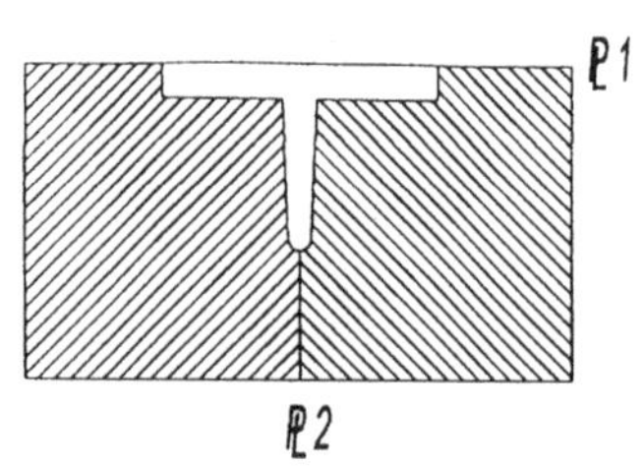

Figure 6.77 Use of split cavity insert to vent a freestanding rib.

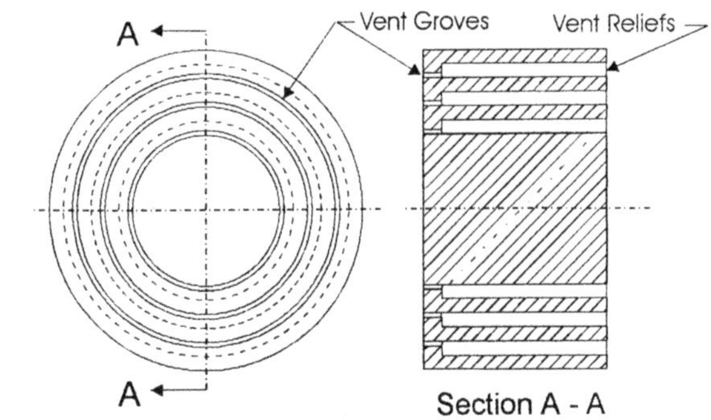

Figure 6.78 Laminated cylinders used for venting.

Fig. 6.78). Porous sintered metal inserts can also be used in areas that require venting. It should be realized that all of these stationary vents are troublesome and will clog with time. How quickly the vent will clog will depend on factors such as material characteristics, melt temperature, mold temperature, location in the cavity, and so on. It is common that these venting means can also negatively affect cooling because they may restrict heat transfer in their location.

References

1. Beaumont, J., Ralston, J., Shuttleworth, J., Carnovale, M. *Journal of Injection Molding Technology* (1999) 3(2): 1–11.
2. Beaumont, J. P. and Young, J. H. *Journal of Injection Molding Technology* (1997) 1(3): 1–11.
3. U.S. Patent 6,077,470, "Method and Apparatus for balancing Injection Molds," June 20, 2000.
4. Provisional Patent, "Runner Balancer for Hot Runner Stack Molds," June 2000.

7 Material Handling and Auxiliary Equipment

S. Collins

The plastics industry worldwide is conducted on a massive scale, with raw materials produced, shipped, and handled in extremely large volumes—often in railcar or truckload quantities. When these materials arrive at an injection-molding plant, processors must be in position to unload, segregate, and properly store each resin as needed, whether in silos, bins, or gaylords. The processor's material handling system may be required to handle everything from commodity plastics like polypropylene and polyethylene, to hygroscopic (water-absorbing) engineering plastics, to recycled plastics, to reclaimed product resulting from scrap within the plant.

Auxiliary equipment plays a vital and increasingly more automated role in receiving, conveying, conditioning, and delivering raw materials to molding machines. Robots become important in maximizing efficiency in post-mold parts handling and in processors' efforts to integrate additional value-added services, such as assembly, printed or painted graphics, packaging, and other functions. Granulators help reclaim valuable raw material from scrap parts, sprues and runners. Finally, heat transfer equipment, including mold-temperature control units and chillers, are indispensable when it comes to process optimization and cycle-time reduction.

7.1 Bulk Material Handling

Plastic materials usually are delivered to the molding plants in one of several ways: in large, pallet-mounted cardboard boxes called gaylords; in oversized reinforced plastic bags; in stacks of paper bags on pallets; or, extremely large quantities, by railcar or truckload, see Fig. 7.1. Pallets and large bags are usually brought into the plant by conventional forklift, but the massive quantities delivered by truck or rail, need to be unloaded and efficiently stored on-site. This requires specialized equipment. Specifically:

- *Cycling Loaders*—Railcar or truckload unloaders, called cycling loaders, are equipped with large vacuum pumps, large diameter lines, and a large integral hopper. In operation, these loaders "cycle" between a vacuum and pressure, regulated by a level switch in the hopper. The vacuum cycle draws resin from the railcar or truck until the hopper is filled to the appropriate level. Then, the level switch reverses the airflow cycle to drive the contents of the hopper to one or more outside silos. This cycling continues until the railcar or truck is empty.
- *Silos*—Bulk material storage silos are available in three basic types. *Bolted* silos consist of a series of rolled, epoxy-coated steel plates that are assembled on site,

Figure 7.1 Plastic material usually arrives at a plastics plant by bulk transport

equipped as needed with decks, nozzles, doors, ladders, and other equipment. They are inexpensive to buy and ship—even overseas—but are more expensive to assemble. *Welded* silos are built at a factory, transported to the site, and erected with a crane. While relatively expensive to transport, these can make storage capacity available very quickly. The third type, *spiral* silos, are constructed of aluminum, so they never need painting and the risk of material contamination from rust or flaking paint is eliminated. They are relatively inexpensive to ship and are quickly assembled on site using automatic equipment.

7.2 Pneumatic Conveying Systems

Although bulk shipments of resins are generally delivered to processors via heavy land transport, these same resins usually are handled with pneumatic conveyors at the plant level. These conveyors typically consist of a system of pipes, manifolds, valves, and controls that derive their motive force from a high-velocity stream of air. Typically, this air stream moves as a result of a differential in vacuum pressure that is created at the head of the system by a vacuum pump. In smaller, self-contained conveying systems, the vacuum may be created by a motor or by a fast-moving stream of compressed air that is pulled past a venturi. When granular, pelletized, or powdered materials are exposed to this vacuum pressure differential, they are drawn into the pneumatic conveyor, suspended in the air stream, and moved to their destination.

There are many advantages to pneumatic conveyance in a plastics processing plant: Contamination is held to a minimum. Waste and spillage are minimized, resulting in a cleaner plant and lower housekeeping costs. Personnel accidents, from lifting, or moving resins, or from slips and falls on spilled resin can be virtually eliminated.

While conveyors can be designed and optimized in many different ways, virtually all have many operating features and requirements in common. Each must create and sustain certain types of velocity required to move pellets through the system:

- *Minimum surface velocity*—the velocity required to move a specific material through the system without damage.
- *Settling velocity*—the velocity required to keep materials suspended and flowing in the air stream
- *Drop-out velocity*—the velocity required to lift pellets in vertical segments of the system

All plastic materials have certain characteristics which govern the amount of power or velocity required to move them properly. The specific gravity of a material is related to the amount of air that will be required to lift and move particles or pellets of it through the system. The bulk density of materials (in powders, granules, or pellets) is an important determinant of the size of pipes and receivers in the system. Materials with bulk densities between 400 and 880 kg/m^3 (25 and 55 lbs/ft^3) are relatively easy to convey, while heavier materials may require more power. Lighter materials have no effect on power requirements, but may require additional system or storage space.

Other material characteristics can create special challenges or problems in the system, see Table 7.1.

Table 7.1 Conveying Solutions for Specific Material Characteristics

Characteristic	Material behavior	Solution
Friability	Tendency to break into fine particles	Filtration
High cohesiveness	Tendency to clump together	Avoid pneumatic conveying
High erosiveness	More difficult to move, with a tendency to scratch and scuff the system	Reduce air velocity Utilize special system materials Increase power
Low melting point	Tendency to form stringy "angel hairs" that stick to bends of piping system	Avoid velocity in excess of recommendations
Acidity	A tendency to corrode system elements	Use corrosion resistant system materials
Aerated/De-aerated	Aerated materials flow freely in air; de-aerated materials tend to pile up and block airflow	Avoid moving de-aerated materials through system

A variety of factors, including the size of the plant, influence the processor's decision whether to install a centralized vacuum conveyor system or utilize one or more self-contained loaders, which will be discussed shortly.

High-volume, long distance material conveying requires the use of the most powerful type of vacuum pump: a positive displacement vacuum pump. Available in a range of sizes, positive displacement pumps employ a powerful rotating lobe blower, protected by a vacuum relief valve, and are factory set to draw 30 cm (12 in.) Hg, see Fig. 7.2. These pumps operate on three-phase power and incorporate a filter to protect the pump's inner workings. Due to their power, they can be quite noisy, and so often require a sound enclosure if located near working area. Central dust collection systems are installed just upstream from these large pumps to minimize the amount of residual particles, dust, or other contaminants that are generated by the conveying process.

Peripheral vacuum pumps are ideal for small to medium-sized resin-conveying applications. They are capable of moving free-flowing materials up to 61 m (200 feet) at rates of up to 1,360 kg (3,000 lbs) per hour or of supplying vacuum power to several smaller loaders at the same time. If dust generation is not a major problem, it is possible to substitute a filter canister for a central dust-collection system. If frequent loading cycles are expected, the pump may be equipped with a sequencing valve that allows the pump to run continuously and eliminate excessive starting and stopping.

Regardless of size, central conveying systems have a similar physical layout. The pump is at one end, preceded by a filter or dust collection system. At the other end is one or, more likely, many or other material receivers. These can include surge bins and other storage containers, hoppers that serve individual molding machines, or hoppers located on conditioning equipment such as dryers, blenders or feeders. Operation of the entire system is directed through an integrated control system.

Figure 7.2 Positive-displacement vacuum pumps and filters provide the motive force in central material handling systems

7.3 Receivers and Loaders

Vacuum material receivers provide the end point for materials drawn through a vacuum conveying system, see Fig. 7.3. Normally located near processing machines, they consist of a body, lid, and filter and are equipped with inlet and outlet connections. These connections are linked via tubes to the main vacuum line. Receivers often sit on top of some type of hopper, enabling pellets or other materials to be delivered direct to a point of use such as a blender, dryer, or processing machine.

As materials are drawn by vacuum through the main line, they are diverted to a single receiver through control system inputs that isolate the receiver by closing off a sequence of valves. This diverts the vacuum-driven flow to the receiver where the material strikes the filter, loses velocity, and falls into the receiver. The vacuum air, no longer carrying the material, passes through the filter and returns through the dust collector to the vacuum pump. Ordinarily, the level of material in the receiver is monitored by a level switch, which signals the central system to stop loading the receiver when it is full.

When the vacuum cycle is completed and the material is enclosed in the receiver, a flapper valve opens and allows the material to fall from the vacuum receiver into the hopper or the machine throat. The flapper then closes, ensuring that vacuum can be held through the receiver on the next conveying cycle.

As noted, multiple receivers usually operate off a single vacuum source. Each is equipped with vacuum sequencing valves, which are controlled at a receiver-mounted terminal box, with electronic connections to the master control panel for the entire system. Other control functions associated with receivers include material-line valve controls (for common material line isolation), internal demand-level switches, ratio valves to enable virgin/regrind mixes, and direct or gravity feeds.

Figure 7.3 Vacuum receivers, like these above a blending station, separate plastic pellets from the air that conveys them through the tubing of a central vacuum system

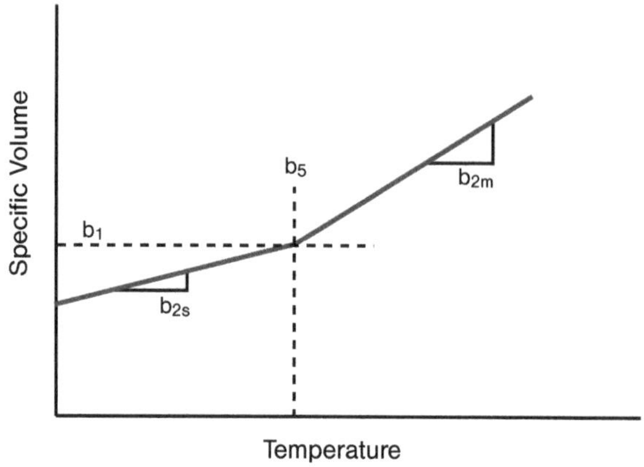

Figure 7.4 Loading system schematic

For many years, vacuum receivers looked very much like the cone-shaped hoppers that are common on injection molding machines. A lid was added to accommodate material inlet and vacuum outlet connections or, in the case of self-contained loaders, a vacuum motor would be mounted to the lid. More recently, however, manufacturers have been re-envisioning the vacuum receiver. With the trend toward lean production techniques, such as fast job and mold changes and lean inventory and management, direct feed "hoppers" appeared. These units usually have a cylindrical body, often made of clear glass, mounted on the feed throat of the molding machine. Only a small amount of material would be held at any given moment so they are much easier to clean out during a material change. Also, when loading dried material, residence times are reduced, minimizing the possibility of moisture regain in the

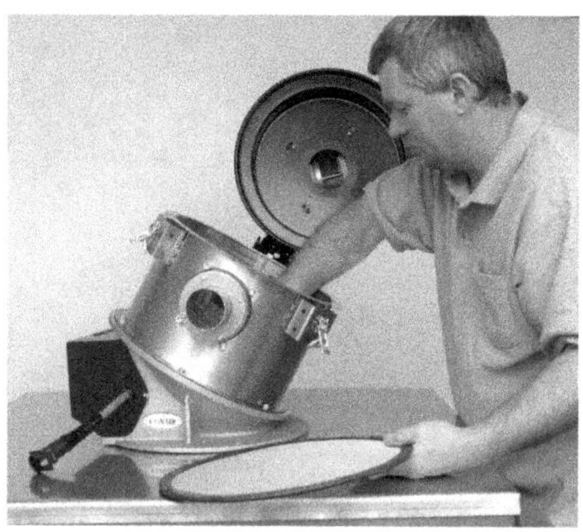

Figure 7.5 Today's newer vacuum receivers and loaders are angled to allow easier access for cleaning and maintenance

receiver. When supported by appropriate automation and a well-engineered pneumatic conveying system, direct feed provides a continuous and reliable supply of resin while eliminating the time required to clean out and manage excess resin in the machine hopper when it is time for a new job to run.

More recently, the market has seen the introduction of straight-through "tube loaders" and angled receivers (see Fig. 7.5). Both are substantially easier to manufacture, which keeps costs down, and the angled receivers make it much easier for operators to reach inside for cleaning or other maintenance.

7.3.1 Self-Contained Loaders

As an alternative to the dedicated vacuum pump, piping, and controls involved in a central conveying system, molders can employ one or more self-contained vacuum loaders, see Fig. 7.6. These can be used either as a supplement to a larger central system, or as a substitute material handling system. The function of self-contained loaders is identical to that of a central vacuum system, but these loaders operate locally rather than centrally.

Self-contained loaders are available in a range of sizes. They provide a versatile, portable conveying capability suited to moving pelletized or regrind materials to hoppers, bins, or processing machines. The smallest self-contained loaders are designed for low volume conveying to small machine hoppers, drying hoppers, or additive bins on feeders or blenders.

Larger models provide a larger body, motor, and filter and move larger volumes of material. While basic loaders are available in standard or built-to-order configurations, others are far more modular and flexible in design, enabling a wide range of configurations:

Figure 7.6 Self-contained loaders have a built in vacuum motor (or venturi vacuum generator). In this photo, the motor is in the black housing on top of the receiver

- Ratio loading capability (for virgin/regrind mixing)
- Closed-cycle air purging between different materials
- Automatic, post-load filter blowback and cleaning
- Gravity or air-powered discharge (for materials with poor flow)
- Direct-feed viewing chambers for machine throat loading

And, thanks to an ever-expanding array of controls, these loaders offer easy conveyance of both virgin and regrind materials, remote operation, and quick-connecting components to simplify disassembly, cleaning, and reassembly.

7.4 Material Handling Controls

There are three basic approaches to controlling central material handling systems. The original approach, referred to as central I/O (input/output), has each controlled device (usually loaders) wired discretely back to a main control panel, see Fig. 7.7. This configuration requires the lowest possible equipment investment, but the complicated wiring increases installation costs. Still, these control systems remain viable for smaller operations where the equipment layout is not likely to change.

Larger plants tend to install control networks utilizing "remote" or "distributed" I/O configurations. Remote I/O configurations have I/O hubs strategically located around the plant to support nearby wired material-handling equipment. Then these remote hubs are wired back to a central control panel. In distributed configurations, each loader has its own I/O modules, which are linked to a common communications bus to both main and remote operator interface panels. To monitor or manage materials, personnel can utilize operator interface panels, obtain system information, or

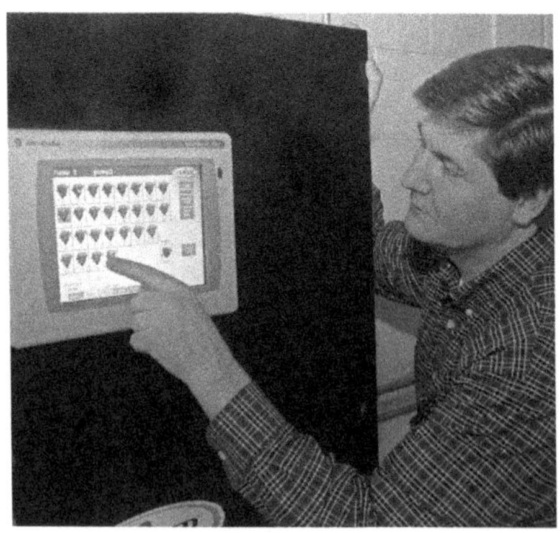

Figure 7.7 From a central control panel, molders can govern the operation of dozens of discrete loaders in a material handling system

Figure 7.8 Newer centralized control systems allow users to interface with other connected equipment. This control screen makes it possible to change settings on any blender in the plant from a central location

issue commands to the PLC and material handling system via a local network. These larger systems are readily scalable, able to handle simultaneous operator inputs, and capable of handling over 140 loaders, plus pumps and related equipment.

Advanced controls are capable of handling not only the most complex material handling requirements (with ratio loading, purging, fill-sensing, multi-source/multi-destination, proofing, reverse regrind convey, and other high-end functions), but can also interface to other auxiliary equipment including dryers, blenders, and heat-transfer equipment. These latter integrated systems don't actually control all the connected equipment (they are governed by their internal controls) but they do provide a convenient, central interface for production personnel to monitor and adjust settings on a wide variety of equipment. The system can even provide some alarm details so personnel across the plant can be alerted to investigate and correct equipment problems.

7.5 Feeders and Blenders

In addition to virgin resin, injection molders must manage a variety of other materials that are added to and blended with the virgin material, prior to molding. These additives include colorants and additives, as well as much larger volumes of regrind or recycled materials. To accomplish these jobs, processors depend on feeders and blenders. Feeders meter out precise quantities of minor ingredients, color concentrates, or additives and deliver them into a gravity-fed column of virgin or regrind material that may fall directly into the machine feed throat or into a multi-material blender.

Until relatively recently, most feeders have been volumetric. Volumetric feeders deliver additives through an auger screw, with the feed rate calculated based on the volume of additive passing through at the turn rate of the screw. Calibration is accomplished by cycling the feeder at a chosen speed, catching and weighing a series of feed samples, then comparing the sample weights to ensure that the auger is dispensing consistently.

Weight-based, or gravimetric, feeding is a newer and far simpler method (Fig. 7.9). Because of its historically higher cost, this technology has primarily been used only in high-volume, high-precision applications—in the pharmaceutical industry, for instance. However, lower-cost gravimetric feeders have more recently become available and this is now the preferred feeding method for plastics processors. Instead of relying on a fixed-volume auger to deliver a consistent weight, a gravimetric feeder continually measures the weight of additive using a load cell and digital controls. The load cell continually registers the loss-in-weight of the minor-ingredient hopper, then signals the feeder control. By comparing the delivery setting (desired g/min) in its control with the actual changes in hopper weight during the same period, a gravimetric feeder is essentially self-calibrating and self-regulating. A minor error in one dispense can be corrected for in the next dispense.

Thanks to gravimetric feeding, processors not only are realizing greater consistency and quality in their finished products, but also are saving money. In the past, the limitations of volumetric feeders led many processors to spend too much time with calibration, or to increase the level of additives beyond recommended levels as a means of problem solving. Today, gravimetric control eliminates these problems.

Figure 7.9 Gravimetric (loss-in-weight) additive feeders are now preferred because they are more accurate and easier to use than volumetric feeders

Figure 7.10 This gravimetric blender has storage for 6 different material ingredients, which are metered in precise proportion into a weigh bin before being mixed together and directed to the molding machine feed throat

Blenders automatically combine different materials, such as virgin or regrind resin, with colorants or other additives. To accomplish this combined feeding and metering function, blenders typically incorporate several large material feed hoppers with as one or more smaller feeders (Fig. 7.10). It is increasingly common to see blenders equipped with automatic loading mechanisms, with ingredient and additive hoppers fed automatically by local loaders or central conveying systems.

Like other process equipment, blenders are available with a wide range of throughputs, hopper capacities, feeder capacities, and control options. Today, the material savings possible through precise gravimetric controls make these blenders cost-effective in nearly every application, while volumetric blenders are gradually phased out.

While continuous blenders continue to be used in certain applications, the vast majority of blending applications in the typical molding plant are handled by gravimetric batch blenders. Batch blending involves delivering ingredients, one at a time in the proper proportion, into the blender's collection chamber. Because the ingredients are layered into the batch, thorough mixing is necessary to achieve a homogenous blend.

Blending can be performed at any location in the plant or at the press. If blending is performed at a central location, the blended material can be stored in a surge bin or other retainer for subsequent delivery to a dryer or machine hopper or feed throat. At the machine location, blenders can be located press-side or machine mounted to provide a direct feed.

7.6 Dryers

Many plastics, especially "engineering plastics" such as polyamide (nylon), polycarbonate (PC), acrylonitrile butadiene styrene (ABS), and fluoropolymer materials, are hygroscopic. This means the molecules of these polymers will absorb moisture until equilibrium is reached with the surrounding air. If they are processed while they contain excess moisture, the critical properties of these polymers can degrade, often with costly consequences:

- Processing is affected as moisture turns to steam in the barrel.
- Bubbles, sink marks, and surface irregularities cause cosmetic problems.
- Melt viscosity, molecular weight, and mechanical strength can degrade, causing potentially serious performance problems.

In fact, the producers of engineering resins report that the vast majority of customer complaints about resin performance are rooted in improper (or incomplete) drying. Because hygroscopic polymers bond to water molecules at a molecular level, it can be difficult to properly dry them. For this reason, dryer design and control technology continues to evolve as the industry seeks the most reliable and energy efficient drying solutions.

Figure 7.11 The desiccant dryer (right) supplies hot dry air to the drying hopper (upper left) to dry hygroscopic materials

Within a dryer, proper drying is governed by four fundamental factors:

- *Drying Temperature:* Tests conducted by polymer makers have set temperature requirements for proper drying. Heating a polymer to a specific temperature (i.e., nylon to 71 °C or 160 °F) dramatically reduces the strong molecular bond between water and polymer molecules. Such heating makes it possible, over a period of time, for the water molecules to be drawn away by a stream of dry (low dewpoint) air.
- *Dewpoint:* Before flowing into a drying hopper, air is first dehumidified to a very low residual moisture content. Today's most advanced dryers dehumidify air using a molecular sieve desiccant to an industry standard of –40 °C or –40 °F. Then, the dry air is heated to reduce its relative humidity, resulting in air with a very low vapor pressure. Within the drying hopper, the moisture molecules within the polymer pellets, freed of their bonds by heating, migrate toward the low-vapor-pressure air as it continually flows through.
- *Residence Time:* It takes time for the heated air to raise the temperature of the pellets and for water molecules to migrate to the pellet surface. Resin suppliers specify a "residence time" for how long each material must spend at the proper drying temperature and dewpoint.
- *Airflow:* Hot dry air is the medium that delivers heat, attracts moisture molecules from the polymers, and carries them away. Airflow must be sufficient to ensure consistent heating throughout the mass of the resin pellets as they flow through, ensuring that each pellet is dried for the specified time at the proper temperature.

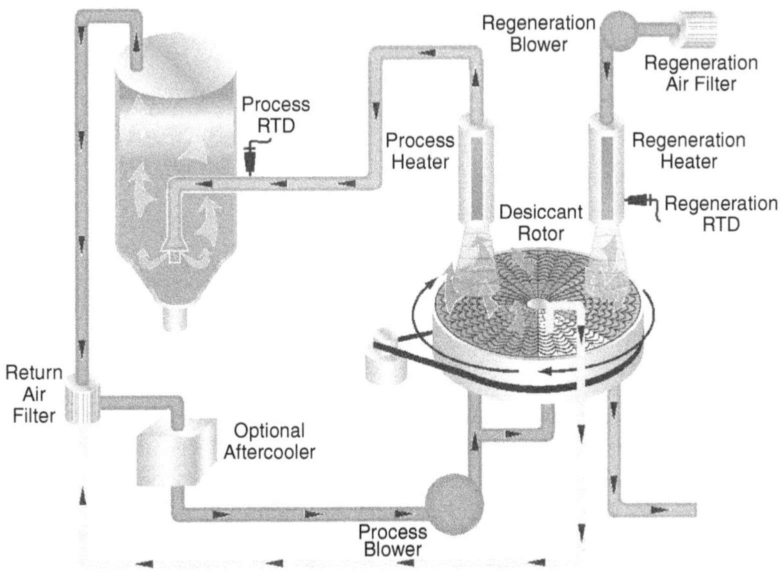

Figure 7.12 Schematic shows airflow in a desiccant wheel-type dryer. The wheel rotates continually so that desiccant passes through on-stream, regeneration and cooling

The overall design of dryers is important because any issues that affect these four factors can cause insufficient drying. Ideally, for example, all pellets should flow at an even rate through all parts of the hopper—resulting in a consistent mass flow and similar residence time for each pellet. Similarly, all pellets should be exposed to a consistently even airflow, held at a consistent speed and temperature in all areas of the dryer—top to bottom, center to sides. This can be accomplished with a well-designed, insulated hopper.

Dryer sizing is based on the throughput range (kg/h or lbs/h × h residence time) that a processor requires for the material being processed. This rule of thumb estimates the required hopper size in kilograms or pounds. In addition, however, processors should evaluate the overall efficiency and performance of various dryer technologies under their most demanding environmental conditions. For example, will the dryer perform as well in both dry winter conditions and humid summer conditions? How do dryers with comparable throughput compare in terms of energy consumption and total cost of ownership?

Table 7.2 Comparing Dryer Technologies

	Principle of drying	Normal dewpoint of drying air	Advantages
Twin tower	Desiccant/hot air	–40 °C (–40 °F)	Proven technology
Desiccant rotor	Desiccant/hot air	–40 °C (–40 °F)	Rock-steady dewpoint; lower energy consumption; fewer parts/easy maintenance
Vacuum (low-pressure)	Differential pressur (vacuum)	–18 °C (0 °F)	Low cost; shorter drying cycles
Compressed air	Compressed air (opt. membrane filter)	–18 °C (–40 °C with opt. membrane filter)	Low-cost; easy to maintain; lightweight unit suit mounting on small machines.

7.7 Integrated Drying and Conditioning Systems

Among the latest innovations for plastics auxiliary equipment are integrated conditioning and material handling systems. Built from modular core components—drying hoppers, desiccant dryers, vacuum pumps, and loaders—these units are pre-assembled and delivered with material-handling controls, conveying pipes and valves, and other components to ensure precise resin drying, blending, and delivery (Fig. 7.13). Compared to the traditional approach of press-mounted or press-side units, these integrated systems—built to match processor needs—can offer a host of capital cost, operating, energy, space, labor, and material cost savings in new or replacement applications.

Figure 7.13 Integrated material handling and conditioning systems streamline drying and delivery of engineering materials to injection molding machines. Dryer and drying hoppers are seen at right, distribution manifold in lower left and vacuum pumps and filters upper left

Using this approach, processors can maximize the utilization and efficiency of all dryers, ensure consistent drying of all materials at a central point, and eliminate press-side material drying and loading that can lead to safety concerns or material contamination. Material changeover is faster. Material manifolds enable multiple materials to pass through the same line, while integral purge valves ensure dispensing and delivery of precise amounts and eliminate the risk that small amounts of material could be left in non-conditioned spaces in the line.

The electronics, ductwork, and controls required for moving drying and conveying air to and from these integrated units are fully integrated to save space, enable isolation of an individual unit for maintenance, or enable addition of additional sleds of equipment to the same main ducting. Advanced airflow balancing ensures that each operating dryer receives the required cubic feet per minute (CFM) of conditioned airflow, while non-operating ducting or dryers are isolated.

7.8 Mold and Process Temperature Control

In an injection molding plant, temperature control centers on two things: mold temperature control and process equipment cooling. Mold temperature control is important because mold temperatures affect each step of the injection molding process: how a polymer behaves when it is injected, how smoothly it flows through a mold, and how rapidly and consistently it "freezes" into a part that can be removed from the mold. Typically, commodity plastics are injected at lower mold temperatures (for example, as low as 21 °C (70 °F) to 27 °C (80 °F)), while engineering plastics typically require higher initial mold temperatures (50 °C (120 °F) and up). However, as soon as a hot plastic is injected into the mold, the original mold temperature spikes upward. This temperature must be restored through cooling before the cycle can continue.

Mold temperatures are controlled through coolant loops. These consist of pump-driven plumbing, tanks, and valves that circulate a coolant (water or heat-transfer oil) through a mold, where it pulls out excess heat, then through a cooling device (a water-filled cooling tower or powered chiller) where the heat is extracted so the coolant can be circulated once again. Cooling towers, which can maintain a continuous flow of coolant at approximately 27 °C (80 °F) year-round, are generally located outdoors. A single large tower may function as a central cooling source, or a series of smaller portable towers may be used (Fig. 7.14). Chillers, which utilize refrigerant loops to drive coolant temperatures down to 2 to 7 °C (35 to 45 °F) year-round, may also be employed, either as large central or smaller portable units. In some large, centralized systems, both cooling devices are used, with cooling towers placed in advance of chillers to accelerate cooling and reduce overall energy consumption.

The cooling portion of the typical molding process accounts for up to 80% of the total cycle time. Therefore, reducing the time required for mold heat removal can reduce cycle time and increase productivity. At the same time, enhanced cooling can reduce warping and shrinkage of molded parts, resulting in higher quality. The latest improvements in heat-removal efficiency and technology are embodied in the latest generation of mold temperature control units, or TCUs (Fig. 7.15). These TCUs provide:

- Fine-tuning of coolant temperature to optimize performance for a variety of molds and materials
- Dual-zone coolant temperature control to match heat buildup in different segments of the mold
- Booster pump capability to increase coolant pressure and excite fluid turbulence, which is associated with increased heat removal.

Figure 7.14 Portable (machine-side) chillers are available in air- (left) and water-cooled models

Figure 7.15 Temperature control units circulate water or oil to precisely heat and cool injection molds

A typical injection-molding plant's heat-transfer system usually will be required to cool process equipment in addition to controlling the temperature of the mold. Systems requiring cooling can include machine hydraulic systems, valves and motors, as well as heated components such as the machine barrel and feed throat. While process cooling loops are plumbed separately from the mold temperature control loops, they often depend on the same cooling device (tower or chiller) for heat removal. Table 7.3 provides a list of factors involved in selecting a cooling system.

Table 7.3 Factors Involved in Selecting a Cooling System

System configuration	Central, portable, or combination
System type	Cooling tower, water-cooled chiller, air-cooled chiller
Mold cooling requirements (tons of cooling capacity)	Process type (injection molding, etc.); resin type(s); processing rate (kg/hour or lbs/hour), and number of machines
Process equipment cooling requirements	Hydraulics (0.1 ton per horsepower of hydraulic system); machine barrels (1.0 ton per inch of screw diameter); feed throats (0.5 ton per machine)

7.9 Automation

Robotic automation has proven to be a great investment for many plastics processors seeking ways to reduce labor inputs and costs. First used to automate the most repetitive jobs by captive and long-run molders, automation continues to expand its reach for both long- and short-run processors. It adds value by: increasing process

Figure 7.16 The simplest robotic device used in injection molding, a sprue picker pulls small parts or runner out of a mold and swivels to drop finished parts into boxes or deposit scrap into a granulator

speed and productivity; eliminating operator exposure to hot or hazardous operations; supporting greater part consistency, cosmetic appearance, and quality; and reducing maintenance and housekeeping costs.

The simplest robots are sprue pickers (Fig. 7.16), developed to perform reliable and complete separation of parts and runners, a basic, but tiring chore for human workers. Mounted above the molding cavity, these arm-mounted pickers simply extend downward, grasp the sprue, and then move upward or rotate outward to deliver the sprue/parts to a designated location—a bin, conveyor, etc.—a simple in/out operation. In addition, the sprue picker also helps to protect the mold by ensuring complete removal of the runner before the next injection cycle.

Beam or traverse robots (Fig. 7.17) are the next step up, providing a more complex range of motion and functions. These robots combine the in/out operation of a sprue picker with traverse motion along a beam, as well as up and down motion to place parts. Beam robots make it easy to run consistent, high-speed molding cycles, with placement of parts in one location (i.xe., a box), and runners in another (i.e., a granulator). Controls allow programming of insert loading functions, as well as complex part stacking and palletizing routines.

Articulated-arm robots represent a third level of automation, with the ability to move in up to six axes of motion (Fig. 7.18). While these robots demand a substantial programming and financial investment, they are far more versatile and flexible in operation, easier to install (alignment with the centerline of the machine is not

Figure 7.17 Today's 3-axis, servo-driven beam robots are capable of complex automation tasks before and after the part is molded

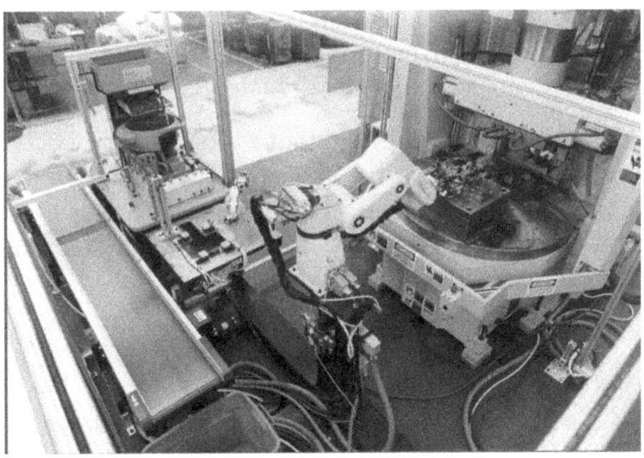

Figure 7.18 This 6-axis articulate robot picks up metal inserts from feeder at upper left, places them in a vertical-clamp injection mold (right), removes finished parts from the mold, positions parts for a trimmed and then places finished part on a conveyer

required), require fewer and less complex end-of-arm tooling, as well as less maintenance.

As molders seek more and more ways to improve process efficiency, automation's role is expanding at the press and in more, and more complex, downstream operations. At the start of a mold cycle, robots, supported by vibratory feeders or parts magazines, can load inserts such as nuts, bolts, pins, sensors, clips, or fasteners into mold cavities. With special tooling, they can also load overmolding films or substrates.

Downstream, using integrated automation controls, processors now automate other important operations: Parts removed from a mold can be placed directly into boxes or separation tubes, or stacked onto conveyors or pallets. Parts can be lifted, turned, or placed onto fixtures for secondary operations including degation, milling, cutting, or drilling. Or, in the most sophisticated operations, parts can be manipulated to complete gluing and assembly, or finishing operations including hot-stamping, painting, and labeling.

7.10 Granulators

Granulators help processors derive maximum value from every pound of resin by cutting post-mold scrap into pellet-like granules that can be re-melted and processed (Fig. 7.19). Scrap typically consists of post-mold sprues or runners, reject or excess parts, and purged or out-of spec material. To profitably reprocess scrap, it is essential that granulators provide material with a uniform particle size and shape. Uniformity is essential to ensure part quality: the scrap or re-grind material must convey and flow smoothly, mix easily with virgin material and additives, and melt at a rate and temperature matching that of virgin material. While granulators vary widely in size, horsepower, and options, much of their work is done by three main parts:

- First, there is a motor-driven rotor. Whether "open" or "closed," rotors are equipped with cutting knives, typically in a staggered arrangement.
- As the rotor spins, its knives drive scrap against an opposing set of stationary "bed" knives that are mounted on the walls of the granulation chamber. The clearance between the rotor and stationary knives is adjustable.

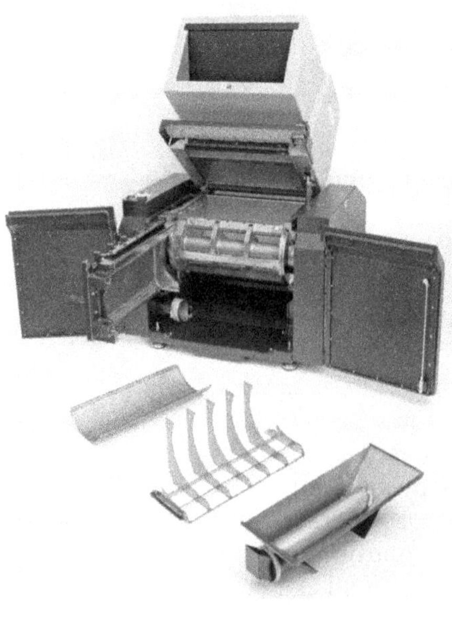

Figure 7.19 This press-side granulator has been opened to show interior components. In foreground are (L-R) screen, screen cradle and granulate evacuation catch-bin

- Third, there is a material screen, located at the bottom of the granulation chamber, that regulates regrind size by allowing only granules smaller than a certain size to fall through to a storage hopper. Smaller screen size equates with greater granule uniformity.

Granulators vary in size, based on whether they will be used next to the machine, between several machines, or used centrally to handle scrap generated throughout the entire plant. Surrounding the basic rotor/knife and chamber, granulators offer different infeed, hopper, and outfeed options. Infeeds, which should be sized to accept the largest parts being ground, provide either a "straight" or "tangential" feed path. Straight, drop-in feed designs are best for smaller or heavier-walled scrap because they orient the scrap on top of the rotor, enabling the knives to nibble away at the edges of the part instead of taking a big "bite" that could slow or stall the rotor (Fig. 20A). Tangential feed chambers (Fig. 20B) have rotors that are positioned at an offset from the feed opening so that bottles, TV cabinets and other large, bulky parts meet the rotor knives on the downward portion of their cutting circle. This allows the knives to take a bigger "bite" when the part first enters the chamber and draw the scrap into the cutting area.

Granulators should be optimized based on the type of material being ground (hard or soft, brittle or tough, filled or unfilled); the size/type of sprues, runners, or scrap parts; and the temperature of parts being granulated. These issues will help determine the appropriate rotor design, horsepower required, soundproofing/noise protection, and means of regrind removal. Tough, brittle engineering materials, for instance, shatter on impact, so they are better processed by closed rotors running at relatively low speeds. Softer, energy absorbing materials, and heat-sensitive materials, are more often ground with high-speed open rotors. Low-speed rotors are generally quieter and subject to less wear when filled materials are being processed.

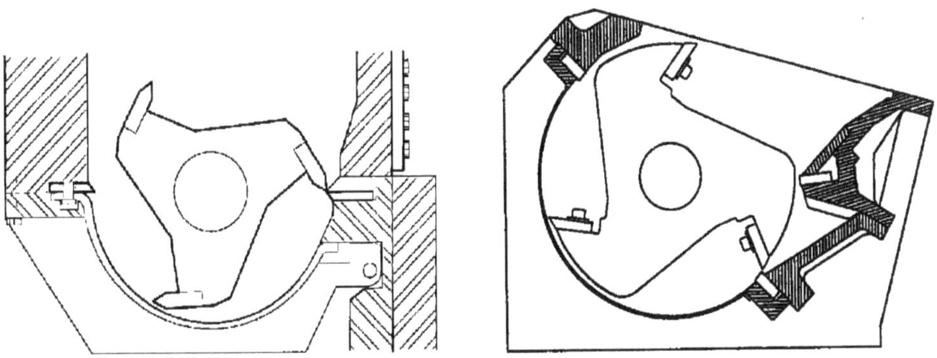

(A) (B)

Figure 7.20 Granulators with conventional (straight) feedthroat (A) are preferred for thicker scrap, while a tangential feedthroat (B) directs bulky scrap so that the rotor contacts it on the downstroke, taking a more aggressive bite

Throughput is an important consideration for large central grinders, which will see constant use, or for press-side granulators that may need to serve additional machines. These needs will determine motor horsepower, rotor design, number and design of rotating and bed knives, rotor speed, screen size and means of regrind removal. Generally, it is best to over-specify size and horsepower. The feed hopper and cutting chamber should be sized to handle the largest possible part without pre-cutting. If a large volume of material is being processed, automatic evacuation of regrind will improve grinder efficiency and reduce dependence on manual labor.

Feeding the granulator is another consideration. For manual feeding, the primary concerns will be operator ergonomics and safety. Hoppers open to the front of the unit for easy feeding and are designed to eliminate flyback and prevent operators from reaching the cutting chamber when the unit is operating. Robot-fed hoppers are open at the top so that parts can be dropped straight down. Some hoppers have powered infeeds that draw in sheet scrap, or augers to force-feed sprues and runners into the cutting chamber. Large central conveyors may also be configured to accept parts from belt conveyors. A host of special-purpose hoppers allow efficient feeding of pipe, profiles, film, sheet, or thermoforming skeletons.

With so many options available, it is wise to avoid "general purpose" granulators and invest in equipment that is customized to meet the material and material handling needs, while optimizing energy consumption, minimizing maintenance, and limiting noise.

8 Statistical Process Control

C. Rauwendaal

What is Statistical Process Control (SPC)? SPC simply means the use of statistical methods to monitor, analyze, and control a process, particularly process variation. SPC can be applied to injection molding, extrusion, blow molding, and essentially any other process. Statistics deals with the collection, organization, analysis, and interpretation of numerical data.

The goals of SPC are to improve and ensure quality, and, thus, reduce process cost due to waste as a result of rejects. Quality is determined by how much the characteristic of a product (i.e., the outside diameter of a pipe) differs from the target value. The closer the product characteristic is to the target value, the higher the quality.

It should be realized that SPC cannot solve all quality and production problems. It is not a cure-all! It will not correct a poor product design or poor employee training. Nor will it correct an inefficient process or worn-out machines and tooling; however, it can help detect all of these types of problems and identify what corrective action is necessary to solve the problem.

One of the most important tools in SPC is the control chart. The most common control chart shows how the average of a set of measurements and the range of the measurements vary with time. An example of a control chart is shown in Fig. 8.1. The average of a set of measurements is often called the *mean* and is represented by the symbol $\bar{x}$, pronounced x bar and sometimes written as x-bar. The range is the difference between the largest and smallest measurement in a group of data, it is represented by the symbol R. The chart shown in Fig. 8.1 is called an $\bar{x}$ and R chart or the Shewhart control chart. Control charts can be used for controlling processes, for identifying problems, and for spotting trends before they become problems. Control charts will be discussed in more detail later.

8.1 Statistical Process Control

8.1.1 Implementing Statistical Process Control

One of the key issues in implementing SPC is training. Training has to occur at all levels: production workers, supervisors, and management. SPC gives production workers the tools to monitor their efforts and to improve the product being made, enhancing both pride in workmanship and job satisfaction. Whereas SPC gives

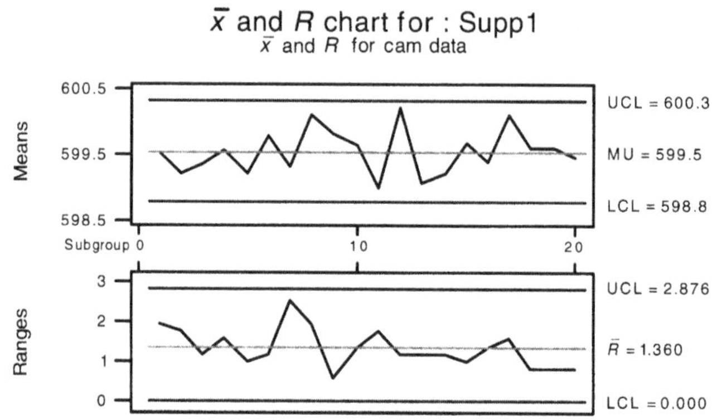

Figure 8.1 Example of an x-bar and R chart.

workers the proof that their work meets quality standards, it also gives both workers and supervisors the key to unlock the source of quality problems.

Management needs to learn SPC so that they can interpret SPC information in monitoring the output of their business. An added benefit is that statistical quality data can be a valuable product-marketing tool. The SPC data gives management the proof of the quality of their product. As mentioned earlier, many companies require proof of quality from their suppliers of everything from raw material to semi-finished and finished goods. A properly documented SPC program provides proof of quality required to obtain those contracts. The process of implementing SPC requires that several steps be taken.

The first step is the creation of an environment that allows and encourages problem solving. This step is often the most difficult one. SPC allows accurate detection of problems; however, detection of a problem is only part of the process of problem elimination. The next part of the process is recognition and acknowledgment of the problem by the appropriate people. Creation of an atmosphere that allows problem solving requires enlightened and capable supervisors and managers. This is an absolute necessity for efficient implementation of SPC. People should be encouraged to uncover and report problems. People reporting problems should be praised and rewarded, they should not be treated as troublemakers. The real troublemakers are the people ignoring problems, not the people reporting them!

The second step is training in SPC and process technology. In order to apply SPC successfully, the people involved with the process should have a basic familiarity with SPC concepts. As discussed earlier, training has to occur at all levels; however, training should cover statistical process control as well as process technology. It is very difficult to apply SPC to injection molding if there is not a good understanding of the injection molding process. The same is true for other processes, such as extrusion, blow molding, compounding, and the like.

The third step is to determine the key process problems and variables. Various methods can be used to identify and prioritize problems in SPC, including brainstorming sessions, fishbone (cause and effect) diagrams, Pareto charts, histograms, and scatter diagrams. These methods will be discussed in detail in the next sections. It is important to direct efforts to implement SPC to the most significant problems where an improvement in quality and productivity can be made and measured.

The fourth step is to make sure the measurement system is capable of measuring the level of variation that is occurring in the process. In any measurement system there is some level of measurement error. It is important to ascertain that the measurement error is small relative to the variation that we are trying to measure.

The fifth step is to use SPC to bring the process into control with its present set of conditions. These conditions can typically be grouped under the following main categories:

- man (personnel)
- machine
- material
- methods
- measurement
- milieu (environment)

The sixth step is to determine process capability. Once the process has been brought into control, it has to be determined whether the characteristics of the individual products meet the requirements of each characteristic. These requirements are often stated as engineering limits or product specifications. An example is a molded cover with a nominal thickness of 0.020 in, with an allowable upper limit of 0.021 in and a lower limit of 0.019 in. In this case the upper specification limit (USL) is 0.021 in, and the lower specification limit (LSL) is 0.019 in. *Specification* is often abbreviated as *spec.*

If the actual product variability is less than the specification width, the process to make the product is considered *capable* (see Fig. 8.2).

If the product variability is larger than the specification width, the process is considered not capable (see Fig. 8.3). In this case nonconforming products will be produced, which will result in rework or rejects. Various measures of process capability will be discussed in a later chapter.

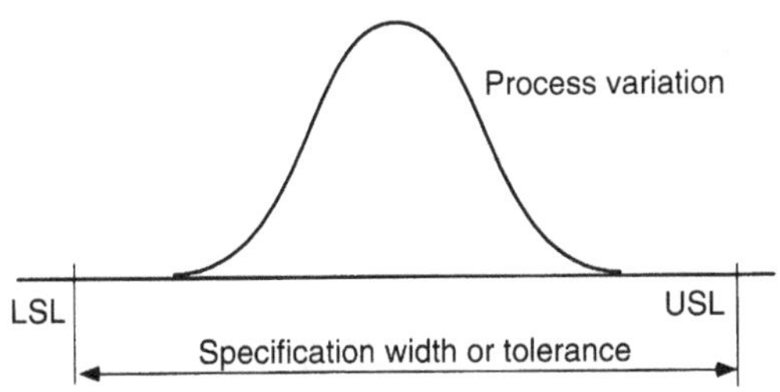

Figure 8.2 Example of a capable process.

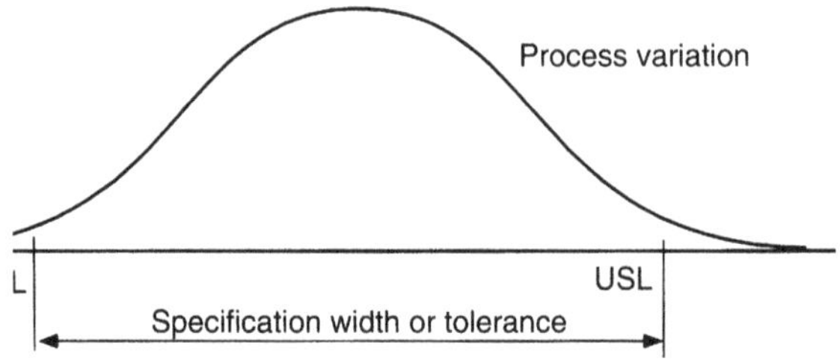

Figure 8.3 Example of a process that is not capable.

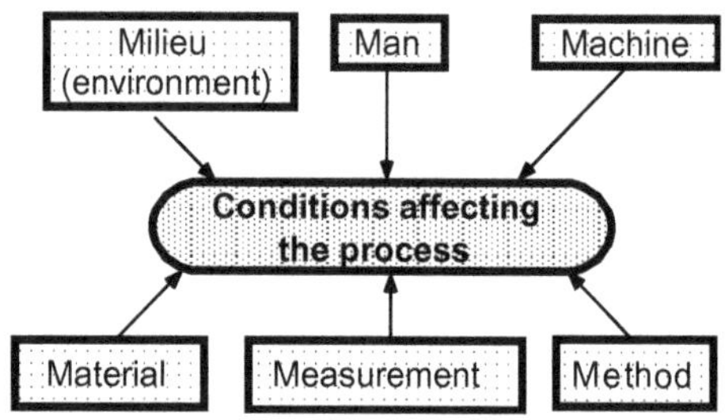

Figure 8.4 Main conditions affecting a process.

If the process is not capable, some corrective action is necessary. Step seven is to implement a plan of process improvement. The most desirable corrective action is generally to improve the process. This will typically involve a change in one of the conditions affecting the process. These are, as discussed earlier: man, machine, material, methods, measurement, and milieu (see Fig. 8.4).

After a change has been made, we move back to step three and repeat the process until the process becomes capable. In other words, until the products meet requirements consistently. Next, we have to determine whether the products meet requirements for all important product characteristics (e.g., dimensions, color, surface finish, impact strength, tensile strength, etc.). If the requirements for one characteristic are not met, then we have to take the next step. The eighth step is to identify the next important problem. We then move back to step two and repeat the various steps until all requirements can be met consistently.

8.1.2 Basic Statistical Concepts

Causes of Variability

All processes are subject to variability. Shewhart (1) distinguished two basic causes of variability: *common causes* and *assignable causes*. A common cause is a source of variation that is inherent or natural to the process. They are also called *random causes*, *chance causes*, and *natural causes*. Some examples of common causes in injection molding are ambient temperature, relative humidity, tooling adjustment, and barrel temperature. An assignable cause is a source of variation that is not inherent to the process, but which has an identifiable reason. They are also called *special causes*, *sporadic causes*, *chaotic causes*, and *unnatural causes*. Examples of assignable causes in injection molding are barrel heater burnout, unusual screw and barrel wear, sticking nozzle valve, and contamination in raw material.

The effect of a common cause is typically slight. No major part of the total variation can be traced to a single common cause. Common causes influence all data in a similar manner. They are stable and the collective pattern of common causes is predictable.

The effect of an assignable cause can be strong. Often a major part of the total variation can be attributed to a single assignable cause. Assignable causes influence some or all data in a dissimilar manner. They are not stable and the effect of assignable causes is unpredictable.

If only common causes of variation are present, the output of a process forms a distribution pattern that is stable over time and predictable. If assignable causes of variation are present, the output of a process will not be stable over time and is not predictable. When we want to reduce process variability we usually focus first on variability due to assignable causes. There are a number of good reasons for this. Assignable causes often have a strong effect on total variability and are usually easier to eliminate.

Basic Statistical Terms

When we collect data on a process, we pull *samples* from the process. A *sample* is one or several individual pieces or measurements collected for analysis. Samples are usually pulled in small groups called *subgroups*. Subgroups that are collected in a way that little variation can be expected between parts within the group, such as consecutive parts off an injection molding machine, are considered *rational subgroups*. They are called *rational* because the way the subgroup has been chosen makes a certain amount of sense. Subgroups typically contain 5 to 10 individual pieces or measurements.

Subgroups, of course, are part of the total number of parts that are produced. All of the parts we manufacture make up a *population* or *universe*. We try to determine the variation in the population by analyzing the variation of data in subgroups.

Mean, Median, and Mode

In most processes we try to manufacture parts to a certain target dimension. This target dimension is called the *nominal*. Because of the natural variation in processes, the actual dimensions will not be exactly at nominal. Some will be higher and some will be lower. It is hoped that all dimensions will be close to the nominal. This tendency to group around a certain dimension is called *central tendency*. The most common measure of central tendency is the *mean* or *average*. It is the sum of the observations divided by the number of observations and is usually represented by the symbol $\bar{x}$. If we have a number of n observations with values $x_1, x_2, x_3, \ldots x_n$, the mean is calculated as follows:

$$\bar{x} = \frac{x_1 + x_2 + x_3 + \ldots x_n}{n} \tag{8.1}$$

For example, we have five measurements of the thickness of molded caps:

$x_1 = 0.0203''$, $x_2 = 0.0208''$, $x_3 = 0.0197''$, $x_4 = 0.0198''$, and $x_5 = 0.0199''$

For this example the mean becomes:

$$\bar{x} = \frac{0.0203'' + 0.0208'' + 0.0197'' + 0.0198 + 0.0199''}{5} = 0.02010'$$

$\bar{x}$ is often termed the first moment of the distribution; it is analogous to the center of gravity. Another measure of central tendency is the *median*. It is the middle measurement with the numbers arranged in order of size. It is usually represented by $\tilde{x}$ (x-tilde), sometimes by the symbol *Mi*. For an even set of numbers, it is the average of the two middle numbers. If we arrange the numbers in the previous example in order of size we get:

0.0197″
0.0198″
0.0199″ ← median
0.0203″
0.0208″

The median is 0.199″ in this example.

A third measure of central tendency is the *mode*. It is the most frequently occurring value in a set of data. It can be represented by the symbol $\hat{x}$ (x-caret).

Range, Variance, and Standard Deviation

Mean, median, and mode are measures of central tendency. In addition to the central tendency, we also need to know about the spread of the data. Another term used for spread is *dispersion*. Commonly used measures for spread or dispersion are range, variance, and the standard deviation. The *range* is the difference between the largest and smallest measurement. The symbol for range is R; it is calculated from:

$$R = x_{max} - x_{min} \tag{8.2}$$

Variance is the mean of the squared deviations from the arithmetic mean. It is a kind of average that reflects the distance of individual measurements from the mean. The symbol for variance is v; it is calculated from:

$$v = \frac{(x_1 - \bar{x})^2 + (x_2 - \bar{x})^2 + (x_3 - \bar{x})^2 + \ldots (x_n - \bar{x})^2}{n - 1} \tag{8.3}$$

where n is the number of measurements.

Variance is also called the *second moment of the distribution*. It is analogous to the moment of inertia. We will calculate the variance for the previous example with the measurements written as whole numbers:

$$x_1 = 203, x_2 = 208, x_3 = 197, x_4 = 198, \text{ and } x_5 = 199$$

The mean was determined to be $\bar{x} = 201$. The variance now becomes:

$$v = \frac{(203 - 201)^2 + (208 - 201)^2 + (197 - 201)^2 + (198 - 201)^2 + (199 - 201)^2}{4} = 20.5$$

The *standard deviation* is simply the square root of the variance. The symbol for standard deviation is s; it is calculated as follows:

$$s = \sqrt{\frac{(x_1 - \bar{x})^2 + (x_2 - \bar{x})^2 + (x_3 - \bar{x})^2 + \ldots (x_n - \bar{x})^2}{n - 1}} \tag{8.4}$$

Of course, we can also write the much shorter expression:

$$s = \sqrt{v} \tag{8.5}$$

Thus, the standard deviation for the example is:

$$s = \sqrt{20.5} = 4.53$$

Another measure of dispersion that is used in some cases is the coefficient of variation or CV. The CV is the standard deviation divided by the mean:

$$CV = s\bar{x} \tag{8.6}$$

The coefficient of variation is frequently used in the analysis of mixing. The inverse of the coefficient is the signal-to-noise ratio, which is used in the Taguchi method.

The variance and standard deviation calculated so far are measures of the spread of the data in a sample. When we determine the variance and standard deviation of the entire population, we substitute the population mean μ for the sample mean $\bar{x}$. The population mean is usually represented by the Greek symbol μ (mu). The population standard deviation is typically represented by Greek symbol σ (sigma), whereas the sample standard deviation by symbol s.

If the population has a manageable size, such as 80 or 125, the population mean and standard deviation can be determined. When the population is very large,

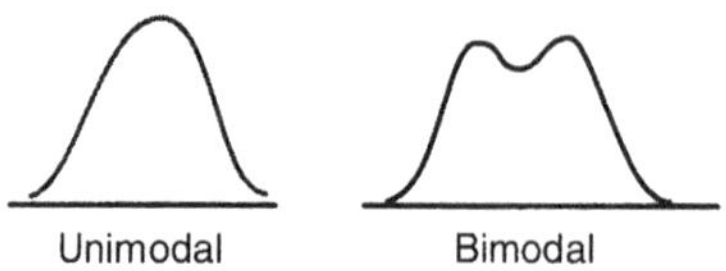

Figure 8.5 A unimodal and a bimodal distribution.

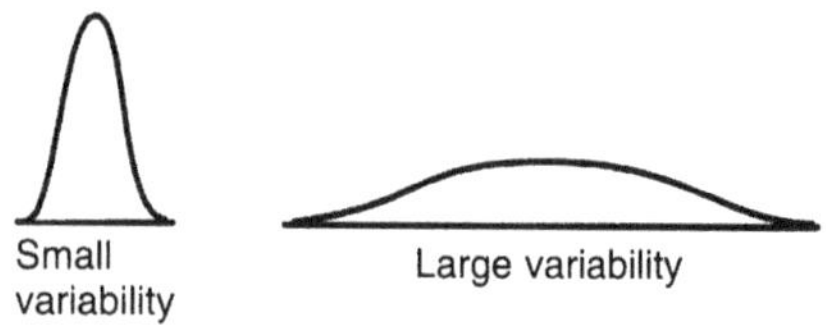

Figure 8.6 A narrow and wide distribution.

however, such as 450,000, it is often not practical to determine the mean and standard deviation of the population. In this case, the population mean is often approximated by the mean of the averages of several subgroups.

Variance has the property of additivity. This means that the total variance is equal to the sum of the variances of its parts. Standard deviation does not have the property of additivity. By analyzing the variances of different factors, it is possible to identify how these factors contribute to the total variance. This is known as the *analysis of variances* (ANOVA). It is an important tool in statistical process analysis.

Characteristics of a Frequency Distribution

Distributions can be described by their modality, variability, degree of skew, and degree of kurtosis. Some distributions have more than one point of concentration. These are said to be *multimodal* distributions. A distribution with a single point of concentration is called *unimodal*. Figure 8.5 shows a unimodal distribution on the left-hand side and a bimodal distribution on the right-hand side.

If one distribution is narrow and another broad, both using the same horizontal scale, it means that the first has less variability than the second. This is illustrated in Fig. 8.6.

A third characteristic is the symmetry of its variation; is it symmetrical or lopsided. If it is lopsided, the distribution is said to be skewed. Figure 8.7 shows a distribution with positive skew on the left and a negatively skewed distribution on the right.

A distribution that is symmetrical has no skewness. The fourth characteristic deals with the relative concentration of data at the center and along the tails of the distri-

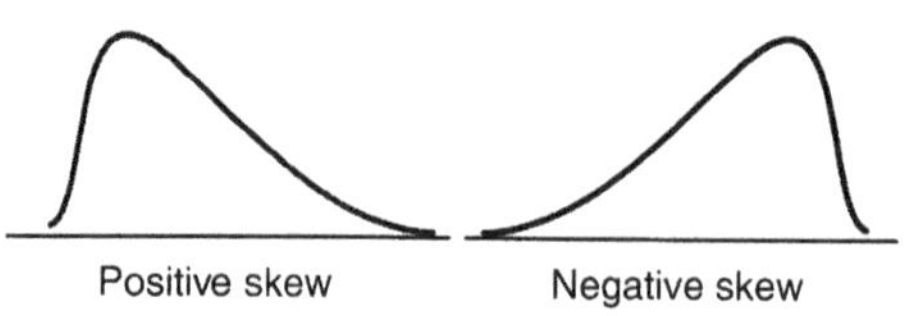

Figure 8.7 Illustrations of positive and negative skew.

Figure 8.8 Illustration of large and normal kurtosis.

bution. This characteristic is called *kurtosis*. If a distribution has a relatively high concentration in the middle and out on the tails, but relatively little in between, it has large kurtosis. If it is relatively flat in the middle and has thin tails, it has little kurtosis. This is illustrated in Fig. 8.8.

Different Distribution Patterns

Data can be distributed in a variety of different patterns. Some of the more important ones are:

1. The binomial distribution
2. The log normal distribution
3. The normal distribution

The binomial distribution is used when the result of an inspection can only have one of two possible outcomes, such as good or bad. A product characteristic that can only be described by one of two possible conditions is called an *attribute*. This is as opposed to a *variable*, which is a product characteristic that can be described by one of many possible values. Attributes and variables will be covered in more detail later. The binomial distribution is useful in SPC when we deal with product attributes. When we are dealing with large samples, typically 50 or more, then the binomial distribution is very closely approximated by the normal distribution. This is true as long as the less probable of the two possible outcomes occurs at least four or five times in each sample.

The log normal distribution is a special type of normal distribution. Log is short for logarithm. We talk about a log normal distribution when a frequency curve of the logarithm of each of the individual points forms normal distribution curve. The log normal distribution is often used in the analysis of particle size distribution of solid materials because many of them follow it closely.

The normal distribution is probably the most important one in SPC. It is also called Gaussian distribution. The distribution curve is bell shaped (see Fig. 8.9).

It has the following characteristics:

- It is symmetrical about the mean; therefore, the mean, median, and mode are all equal.
- It slopes downward on both sides to infinity.
- 68.25% of all measurements lie between $\mu + 1\sigma$ and $\mu - 1\sigma$ (see Fig. 8.9).

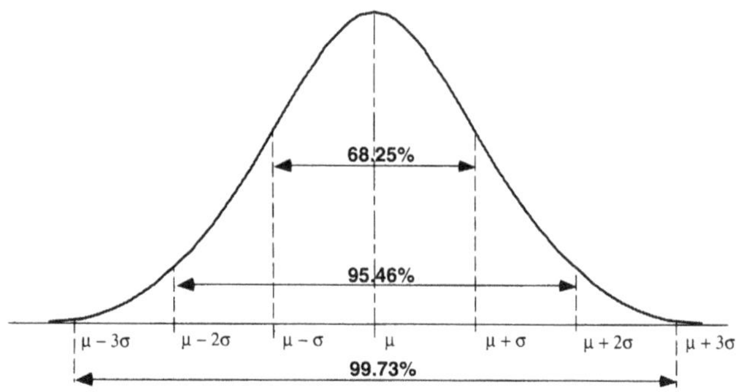

Figure 8.9 The normal or Gaussian distribution curve.

- 95.46% of all measurements lie between $\mu + 2\sigma$ and $\mu - 2\sigma$.
- 99.73% of all measurements lie between $\mu + 3\sigma$ and $\mu - 3\sigma$.

The Central Limit Theorem

In using the normal distribution to determine how many measurements can be expected to be within certain values, we have to be sure that the data follow a normal distribution. The central limit theorem developed by Shewhart (1) is very important in this respect. It states that sample averages (or x-bars) will follow a normal distribution as long as only common cause variations are present. This is true even if the individual measurements do not follow a normal distribution.

The standard deviation of the averages of subgroups, $\sigma_{\bar{x}}$, will be smaller than the standard deviation of the individual measurements, σ_x (see Fig. 8.10). The two standard deviations are related by the formula:

$$\sigma_{\bar{x}} = \sigma_x / \sqrt{n} \tag{8.7}$$

where n is the number of data in the subgroup.

The charting of averages is an important element of control charts. The central limit theorem is an important principle upon which many control charts are based.

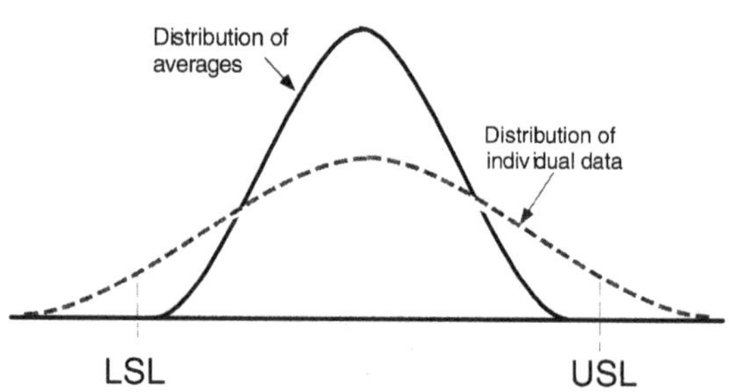

Figure 8.10 Comparison of the distribution of averages and the distribution of individual data.

8.2 Control Charts

8.2.1 Introduction

As discussed earlier control charts are one of the principal tools in statistical process control. Variations in a process can occur as a result of common causes and assignable causes. *Common-cause variation* is inherent to any process that is stable over time. Assignable cause variation results from significant and identifiable changes in the process, such as a different screw in the plasticating unit, a change in the polymer drying conditions, a new operator, and a new lot of raw material. In most cases, reducing variation by eliminating assignable causes is more practical than by eliminating common causes. Process control charts are the best tool for identifying and eliminating assignable cause variation.

Control charts show how process data evolve over time and allow determination of natural process variability. Three important uses of control charts are:

- They can identify assignable causes when they occur
- They can determine whether improvement action really reduces process variation
- They can be used to determine the true process capability

8.2.2 Control Charts for Variables Data

Control charts can be used for both variables and attributes data. Variables control charts are more sensitive to changes in the measured values and, therefore, are better for process control.

x & Rm The individual measurement, x, and moving range, Rm, chart is used when process control is based on individual readings. This can happen when the measurements are expensive. The moving range chart is sometimes combined with a moving average chart.

$\bar{x}$ & R The mean, $\bar{x}$, and range, R, chart is the most common control chart. It will be discussed later in more detail.

$\tilde{x}$ & R The median, $\tilde{x}$, and range, R, chart is similar to the $\bar{x}$ & R chart. The median is the middle value where data are arranged according to size. The median chart is sometimes used by itself without the range chart.

$\tilde{x}$ & x The median, $\tilde{x}$, and individual measurement, x, chart is a special chart used for family processes. This is especially useful for injection molding operations with multicavity molds. It allows separation of global and local factors; this is discussed in more detail later in the section on special SPC techniques in injection molding.

$\bar{x}$ & s The mean, $\bar{x}$, and sigma, s, chart is a more accurate indication of process variability than the $\bar{x}$ and R chart, particularly with a larger-size subgroup. The standard deviation, s, is more difficult to calculate than the range, R; however, this is immaterial when the data is processed by a computer.

CuSum A cumulative sum (CuSum) chart shows the cumulative sum of each $\bar{x}$ value minus the nominal value. This chart is more sensitive to a sustained shift from the nominal.

The *X*-Bar and *R* Chart

Because the $\bar{x}$ and R chart is so commonly used in SPC, we will treat this chart in depth. The methods of usage and analysis for other charts are similar to the $\bar{x}$ and R chart; therefore, what we learn about the analysis and interpretation of $\bar{x}$ and R charts can also be applied to other charts.

The creation of an $\bar{x}$ and R chart requires several steps. We will go through this process step by step.

Step 1 Select the subgroup size and how often samples are taken. A common subgroup size is $n = 5$, with n equal to the number of measurements in the subgroup.

Step 2 Record the raw data on the control chart. The data in each subgroup is listed in columns. If the number of subgroups is k, then the total number of measurements is n times k. Proper determination of control limits requires at least 25 subgroups ($k \geq 25$).

Step 3 Calculate the sum Σx, the mean, $\bar{x}$, and the range, R, for each subgroup using the following expressions:

$$\sum_{i=1}^{i=n} x_i = x_1 + x_2 + x_3 + \ldots x_n \tag{8.8}$$

$$\bar{x} = \sum x_i / n \tag{8.9}$$

$$R = x_{max} - x_{min} \tag{8.10}$$

Σx_i is a simplified notation for $\sum_{i=1}^{i=n} x_i$, which is a shorthand notation for the sum of the x-values from x_1 all the way through x_n.

Step 4 Calculate the central lines for the x-bar and R chart.

$$\bar{\bar{x}} = \frac{(\bar{x}_1 + \bar{x}_2 + \bar{x}_3 + \ldots \bar{x}_k)}{k} \tag{8.11}$$

$$\bar{R} = \frac{(R_1 + R_2 + R_3 + \ldots R_k)}{k} \tag{8.12}$$

where $\overline{R}$ = the average range

$\overline{\overline{x}}$ = the average of all samples or the "grand average"

Step 5 Calculate the control limits for the x-bar and R chart. First, the upper control limit for the x-bar chart:

$$\text{UCL x-bar} = \overline{\overline{x}} + A_2\overline{R} \tag{8.13}$$

The lower control limit for the x-bar chart is:

$$\text{LCL x-bar} = \overline{\overline{x}} - A_2\overline{R} \tag{8.14}$$

The upper control limit for the range chart is:

$$\text{UCL range} = D_4\overline{R} \tag{8.15}$$

The lower control limit for the range chart is:

$$\text{LCL range} = D_3\overline{R} \tag{8.16}$$

UCL is the upper control limit and LCL is the lower control limit. The constants A_2, d_2, D_4, and D_3 can be taken from Table 8.1; these factors only

Table 8.1 Factors for Determination of Control Limits

Subgroup size n	Factor A_2	Constant d_2	Factor D_3	Factor D_4
2	1.880	1.128	0	3.267
3	1.023	1.693	0	2.574
4	0.729	2.059	0	2.282
5	0.577	2.326	0	2.114
6	0.483	2.534	0	2.004
7	0.419	2.704	0.076	1.924
8	0.373	2.847	0.136	1.864
9	0.337	2.970	0.184	1.816
10	0.308	3.078	0.223	1.777
11	0.285	3.173	0.256	1.744
12	0.266	3.258	0.283	1.717
13	0.249	3.336	0.307	1.693
14	0.235	3.407	0.328	1.672
15	0.223	3.472	0.347	1.653
16	0.212	3.532	0.363	1.637
17	0.203	3.588	0.378	1.622
18	0.194	3.640	0.391	1.608
19	0.187	3.689	0.403	1.597
20	0.180	3.735	0.415	1.585

depend on the size of the subgroup. Factor D_3 is zero for $n = 2$ to $n = 6$; this means that the lower control limit for the range chart is zero for $n = 2$ to $n = 6$. Also, the central line of the range chart will not be centered between the control limits for subgroups smaller than $n = 7$. For subgroups of 7 and larger D_3 is nonzero, and thus the LCL is also non-zero and the central line will be centered between the control limits.

Step 6 Plot the sample averages and the sample ranges. Use the following guidelines:
- The grand average, $\bar{\bar{x}}$, should be near the center of the $\bar{x}$ chart.
- The UCL for the x-bar chart should be at about two thirds from the $\bar{\bar{x}}$ to the top of the scale.
- The UCL and LCL have to be equidistant from $\bar{\bar{x}}$.
- The scale of the R chart should extend from 0 to about one and one third times the UCL range.
- The points should be plotted as solid dots.
- Connect the points for easier pattern identification.

Step 7 Draw the central lines and control limits on the charts. The central lines should be solid; the control limits dashed. The lines should be labeled with their numerical value.

The data can be tabulated and plotted on regular paper or special chart forms can be used to make organization of the data easier. The control limits of the $\bar{x}$ chart ($\pm A_2\bar{R}$) represent the $\pm 3\sigma_{\bar{x}}$ limits, where $\sigma_{\bar{x}}$ is the standard deviation of the averages. Because we have a relationship between $\sigma_{\bar{x}}$ and σ_x (standard deviation of the individual data) we can express σ_x as a function of A_2 and $\bar{R}$. This expression is:

$$\sigma_x = \frac{A_2\sqrt{n}}{3}\bar{R} \tag{8.17}$$

Another expression that is commonly used is:

$$\sigma_x = \frac{\bar{R}}{d_2} \tag{8.18}$$

where d_2 is shown in Table 8.1.

From Eqs. (8.10 and 8.11) it is clear that $d_2 = 3/(A_2\sqrt{n})$. Thus, if d_2 is not known, it can be easily determined from A_2 and n.

An example of a tabulated data of the thickness of molded panels is shown in Table 8.2; the thickness is expressed in millimeters times 100.

The first five rows show the data of all the subgroups. The sixth row shows the averages of each subgroup; the bottom row shows the range of each subgroup. This information will be used to construct the x-bar and R chart for this data.

Table 8.2 Tabulation of Thickness Data

Subgroup 1	Subgr. 2	Subgr. 3	Subgr. 4	Subgr. 5	Subgr. 6	Subgr. 7	Subgr. 8
$x_1 = 138$	150	164	132	119	144	144	140
$x_2 = 146$	158	140	125	147	149	163	135
$x_3 = 161$	138	126	147	153	157	154	150
$x_4 = 168$	173	142	148	142	136	165	145
$x_5 = 146$	135	145	176	156	152	135	128
$\bar{x} = 151.8$	150.8	143.4	145.6	143.4	147.6	152.2	139.6
$R = 30$	38	38	51	37	21	30	22

For this example:

$$\begin{aligned}\bar{\bar{x}} &= 146.8,\\ \bar{R} &= 33.375,\\ \text{UCL}(x) &= 166.06,\\ \text{LCL}(x) &= 127.54,\\ \text{UCL}(R) &= 70.55, \text{ and}\\ \text{LCL}(R) &= 0.\end{aligned}$$

The $\bar{x}$ & R charts are shown in Fig. 8.11. Now that we know how to construct $\bar{x}$ and R charts, the next important step is the analysis and interpretation of $\bar{x}$ and R charts.

Interpretation of *X*-Bar and *R* Charts

The most common feature of a process showing stability is the absence of any recognizable pattern. The three main characteristics of a stable process are:

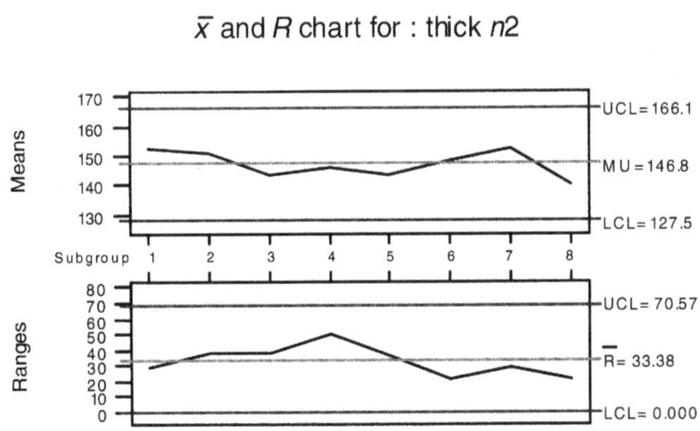

Figure 8.11 Example of *x*-bar and *R* control chart.

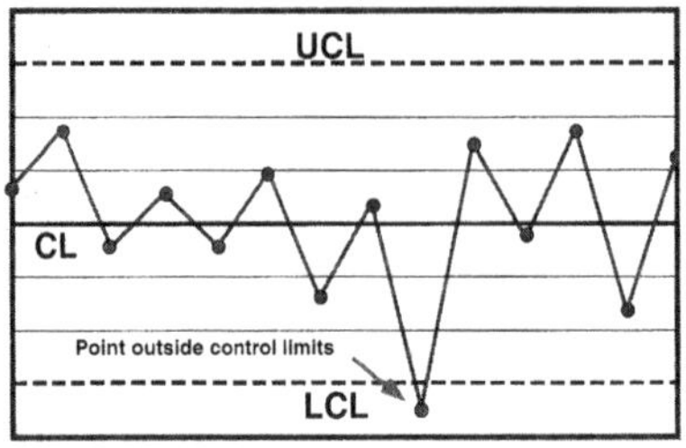

Figure 8.12 Illustration of a freak.

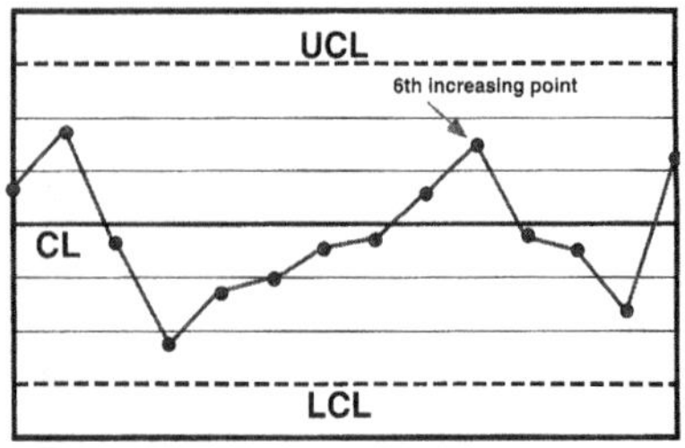

Figure 8.13 Illustration of a trend.

- Most points are near the center line
- Some points are spread out and approach the control limits
- No points beyond the control limits.

Characteristics of an unstable process influenced by assignable causes are:

- Freaks. One or more points outside the control limits (see Fig. 8.12).
- Trend. Six consecutive increases or decreases (see Fig. 8.13).
- Cycle. A repetitive pattern (see Fig. 8.14).
- Run. Seven or more points in a row above or below the center line (see Fig. 8.15).
- Zone rules:
  1. Two of three successive points fall beyond 2 standard deviations from the mean (see Fig. 8.16, right).
  2. Four of five successive points fall beyond 1 standard deviation from the mean (see Fig. 8.16, left).
- Shift in level. A sudden change in central tendency of the data (see Fig. 8.17).
- Stratification. Fifteen or more points very close to the center line (see Fig. 8.18).
- Clusters. The grouping of points in one area of the chart (see Fig. 8.19).
- Mixture. Identified by an absence of points near the center line (see Fig. 8.20).

To identify assignable causes it is best to focus first on the *R* chart. The *R* chart is more sensitive to changes in uniformity and consistency. Any change in the process

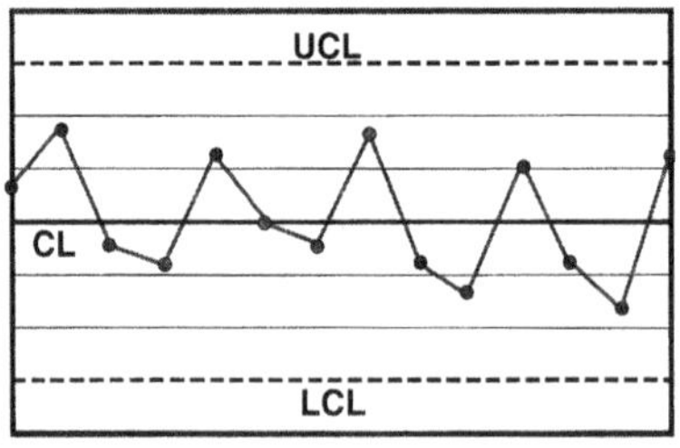

Figure 8.14 Illustration of repetitive pattern.

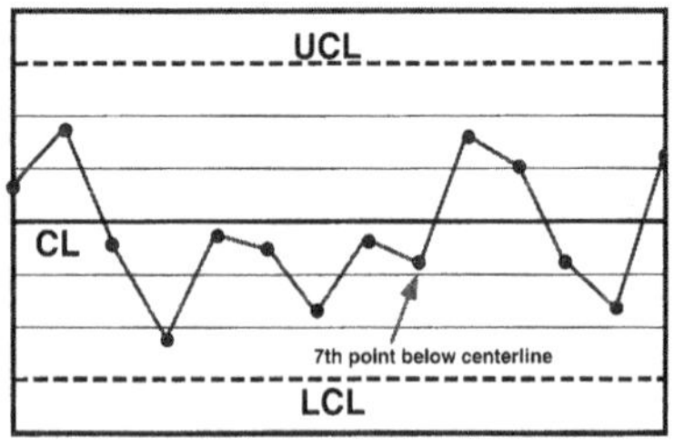

Figure 8.15 Illustration of a run.

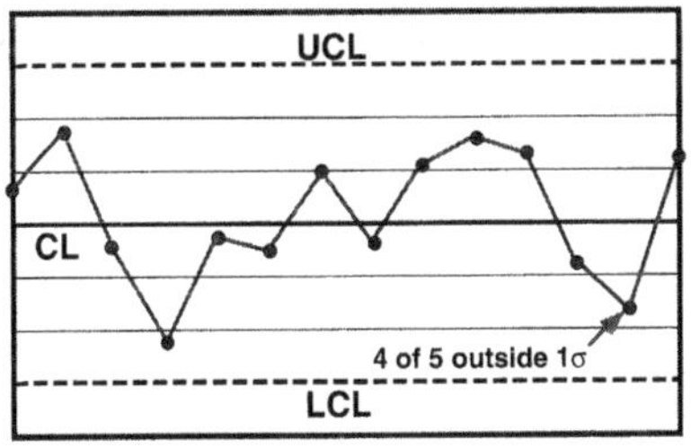

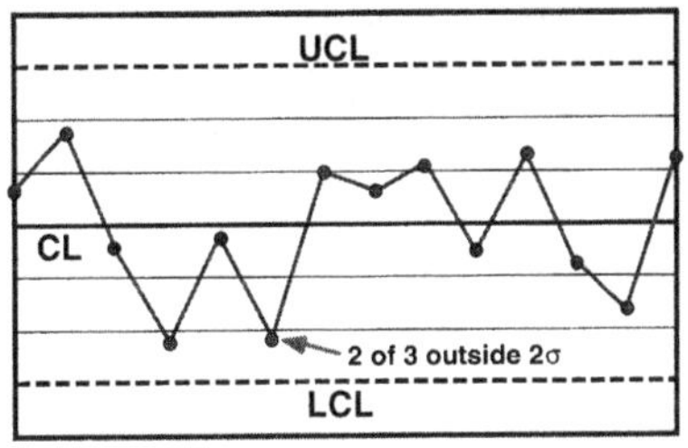

Figure 8.16 Illustrations of zones rules, four of five outside 1σ (left), two of three outside 2σ (right).

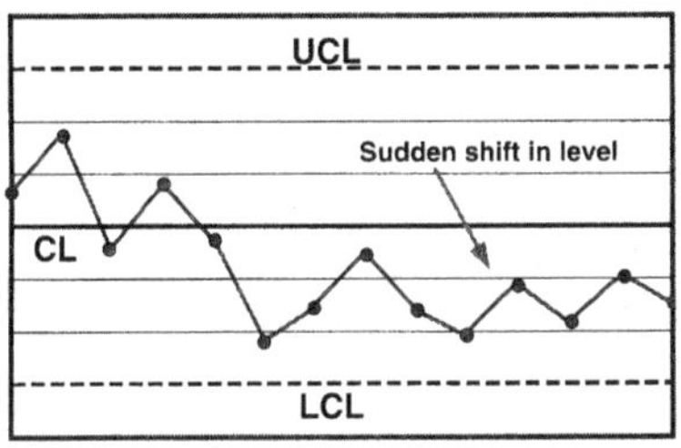

Figure 8.17 Example of shift in level.

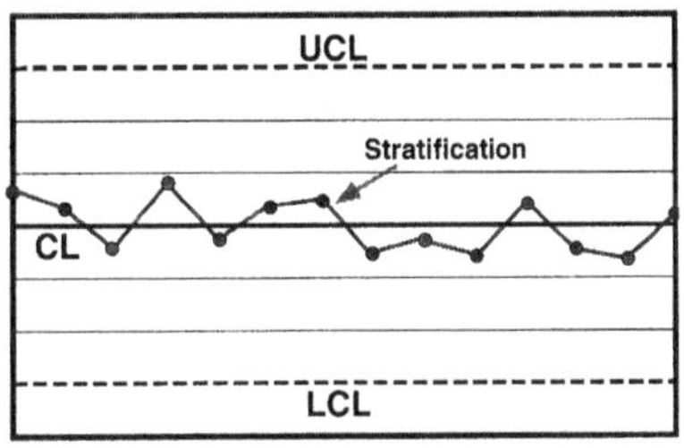

Figure 8.18 Example of stratification.

will tend to shift the R values upward. Because R charts are more sensitive to changes in the process the effect of process improvement will be noticed first in the R chart. The $\bar{x}$ chart should be analyzed after the R chart is stable. The $\bar{x}$ chart can be misleading when the R chart is unstable. When both the $\bar{x}$ and R charts are stable, the process is said to be in control.

SPC gives us various tools to identify assignable causes of variation; however, the fact that common causes of variation are inherent to the process does not mean that they cannot be reduced or eliminated. SPC unfortunately does not help us much in this area. To understand the common causes of variation in a process we have to understand the process technology and use other statistical tools, such as design of experiments or DOE. With a proper understanding of the process and with the use

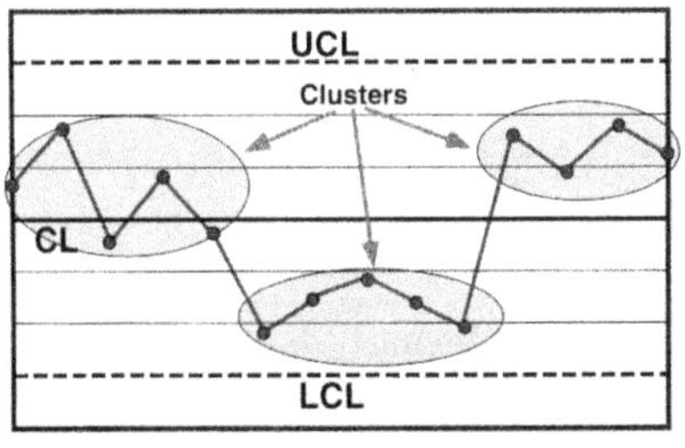

Figure 8.19 Example of clusters.

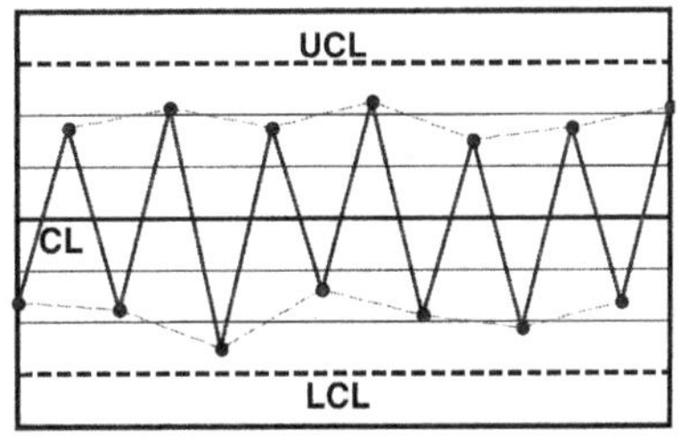

Figure 8.20 Example of mixture.

of DOE the effect of both assignable and common causes of variation can be analyzed and reduced to a minimum.

8.2.3 Control Charts for Attributes Data

Attributes data is a count of the number of nonconformities on a unit (number of defects per part) or the number of nonconforming units (defective parts). The following definitions are often used with attributes data:

Nonconformity (Defect)

Any aspect or fault that causes an item (production unit or product) not to conform to the specified criteria.

c = number of occurrences of nonconformities in a sample
$u = c/n$ = number of occurrences of nonconformities per production unit
n = number of production units in the sample

Nonconforming Item (Defective)

An item that does not meet measurement specifications or which contains one or more nonconformities (defects)

np = number of nonconforming items in a sample of n items
$p = np/n$ = proportion of nonconforming items in a sample

There are four types of control charts for attributes. Control charts for the count of nonconforming units are p-charts and np-charts. For p-charts the sample size varies; for np-charts the sample size is constant. Control charts for the count of individual nonconformities or defects on a product are c-charts and u-charts. For u-charts the sample size varies; for c-charts the sample size is constant. This is shown in Table 8.3.

It should be noted that the calculation of the three σ control limits for attributes is different from that of variables. For variables it is based on the normal (Gaussian)

Table 8.3 Control Charts for Attributes Data

Sample or subgroup size	What is counted	Type of control chart
variable (>50)	defective items	p-chart
constant (>50)	defective items	np-chart
constant	defects on an item	c-chart
variable	defects on an item	u-chart

distribution. For *p*-charts and *np*-charts it is based on the binomial distribution. For *c*-charts and *u*-charts it is based on the Poisson distribution. More information on control charts for attributes data can be found in Ref. [1,5].

8.3 Process Capability and Special SPC Tools for Molding

8.3.1 Introduction

Process capability is a measure of how capable a process is of making parts that are within specifications. Figure 8.21 shows a capable process where the spread of the process data is narrower than the tolerance or specification width.

In other words, all the process data are within the USL and LSL. Before the process capability is determined, one has to make sure that the process is in control. This is done by making sure that no assignable causes are affecting the process. Thus, if assignable causes are acting on the process, these have to be eliminated first. In determining process capability we need to determine the spread of the individual process data (x), not of the averages ($\bar{x}$). We therefore cannot simply use the control limits from an $\bar{x}$ chart because the control limits describe the $6\sigma_{\bar{x}}$ spread of $\bar{x}$. As discussed earlier, the standard deviation of the individual data, $6\sigma_x$, is related to the standard deviation of the averages $\sigma_{\bar{x}}$ by:

$$\sigma_x = \sigma_{\bar{x}} \sqrt{n} \tag{8.19}$$

Where n is the number of data from which $\bar{x}$ is calculated (the size of the subgroup). The spread of the individual data is always larger than the spread of the averages (see Fig. 8.22). Process capability is determined to find out whether a process can meet specifications and how many parts can be expected to be out of tolerance. Capability studies can also be useful in the following situations:

- Evaluation of new equipment purchases
- Equipment selection for production of certain parts
- Setting specifications
- Costing out contracts

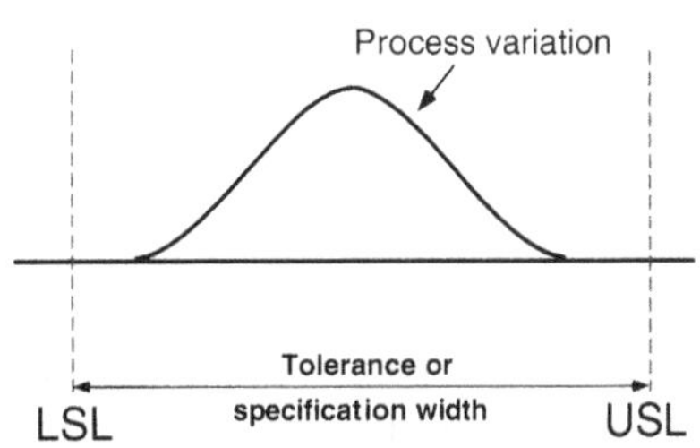

Figure 8.21 Illustration of a capable process.

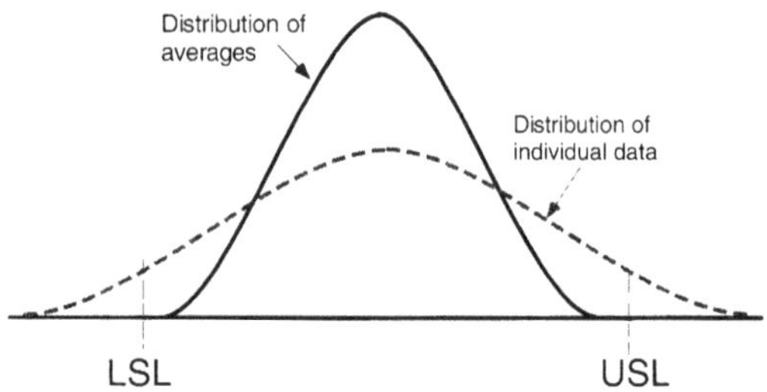

Figure 8.22 Distribution of the averages and the individual data.

The steps involved in a capability study are:

1. Collect process data and make sure measurement system is capable
2. Plot the data on control charts
3. Determine the control limits
4. Make sure the process is in control
5. Determine the process capability
6. If the process is not capable, improve the process and return to Step 1.

Before process capability is determined, one has to make sure that the process is in control and that the individual process data is normally distributed. It should be noted that a process can be in control and yet have its individual process data not normally distributed. This happens when the process is inherently nonnormal. Certain measurements will always result in nonnormal distributions. Examples are roundness, flatness, and so on; these measurements have a natural barrier at zero. A perfect measurement is zero; values less than zero are not possible. Standard capability indexes are not valid for such nonnormal distributions.

8.3.2 Capability Indexes

The most commonly used capability indexes are the process potential index, *CP*, and the capability index, *CpK*. The process potential index is the ratio of the tolerance to six times the process standard deviation of the *individual* data, σ_x:

$$CP = \frac{TOLERANCE}{6\sigma_x} \tag{8.20}$$

The tolerance is the difference between the USL and the LSL. Written differently:

$$\text{TOLERANCE} = \text{USL} - \text{LSL} \tag{8.20a}$$

The standard deviation of the individual data, σ_x, is often determined from $\bar{x}$ control limits of the $\bar{x}$ and R chart; this involves dividing the average range by a constant d_2. In some cases, the σ_x is determined from the sample standard deviation, such as when we use an $\bar{x}$ and s chart. When the σ_x is determined from the sample standard deviation, the ratio of tolerance to $6\sigma_x$ is also referred to as the performance index or *Pp*. There is unfortunately a lack of uniformity in the nomenclature used for process capability indexes. As a result, one should always verify exactly what definition of the capability index is used.

In a normal distribution 99.73% of the data will be within a $6\sigma_x$ spread. Thus, when $CP = 1.0$, it can be expected that 99.73% of the parts will be within the tolerance (assuming that the data is centered around the target value). The higher the value of *CP*, the more parts will fall within the specification limits. For example, when $CP = 1.33$, it can be expected that 99.994% of the parts will be within tolerance. In other words, only 6 out of 100,000 parts would be expected to be out of tolerance. When $CP = 1.0$, we would expect 270 out of 100,000 parts to be out of tolerance. Thus, a small change in *CP* can translate in a large change in the number of rejects.

CP is only a measure of spread of the process data. It does not take into account whether or not the data is centered around the target value. If the data is not centered around the target value, using the *CP* can result in a false sense of security. In this case, the *CpK* is more useful because it takes both the spread and the center of the data into account.

The expression for *CpK* is:

$$CpK = \frac{USL - MEAN}{3\sigma_x} \tag{8.21a}$$

$$CpK = \frac{MEAN - LSL}{3\sigma_x} \tag{8.21b}$$

The *CpK* is the lowest of these two values. If the mean is larger than the target value, Eq. (8.21a) will yield the lowest *CpK*. If the mean is smaller than the target value, Eq. (8.21b) will give the lowest *CpK*. The value of the *CpK* from Eq. (8.21a) is sometimes referred to as the upper capability index (*CPU*); conversely; the value from Eq. (8.21b) is referred to as lower capability index (*CPL*).

If the data are centered around the target value, the CP = *CpK*. When the data are not centered around the target value, however, the *CpK* will always be smaller than the CP. Figure 8.23 shows examples of *CP* and *CpK* values for four different distributions.

Many companies require a *CpK* index of 1.33 to 2.0 for the products from their suppliers. A *CpK* of 1.33 means that six or fewer parts per 100,000 are expected to be out of specifications. A CpK of 2.0 means that 2 or fewer parts per billion (ppb) are expected to be out of specification. A *CpK* of 2.0 clearly represents an almost negligible chance of rejects.

Another capability index is sometimes used: the *CR* index or capability ratio. The *CR* index is simply the inverse of *CP*:

$$CR = \frac{6\sigma_x}{TOLERANCE} \tag{8.22}$$

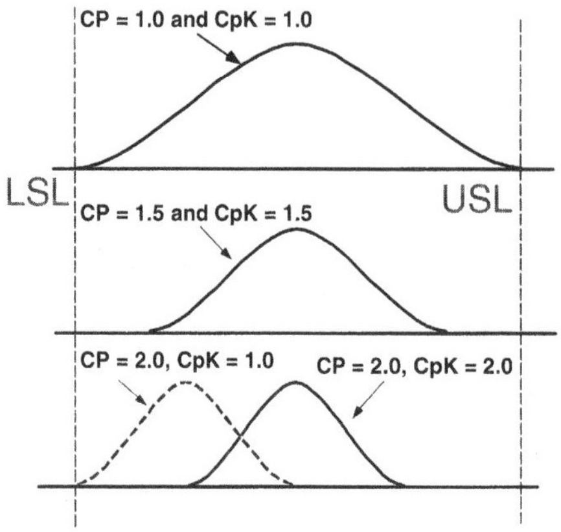

Figure 8.23 CP and *CpK* for different distributions.

The smaller the *CR* value the better. The *CP* index is used more often than the CR index, because it is easier to compare to *CpK*. Sometimes both the lowest and largest *CpK* values are used from Eq. 8.21. The larger of the two *CpK* values is also referred to as the *Zmax*/3 index. Using the same nomenclature, the smaller *CpK* index can also be called the *Zmin*/3 index.

Process Capability Shortcut with Precontrol

It is clear that determining process capability using conventional SPC is a rather time-consuming process. In the precontrol method of SPC, the process capability is determined simply by taking five consecutive units from the process. In precontrol the tolerance is divided into two zones: the middle zone is called the *green zone* (with width of half tolerance); the zones between the green zone and the specification limits are the *yellow zones*. The regions outside the specification limits are the *red zones* (see Fig. 8.24).

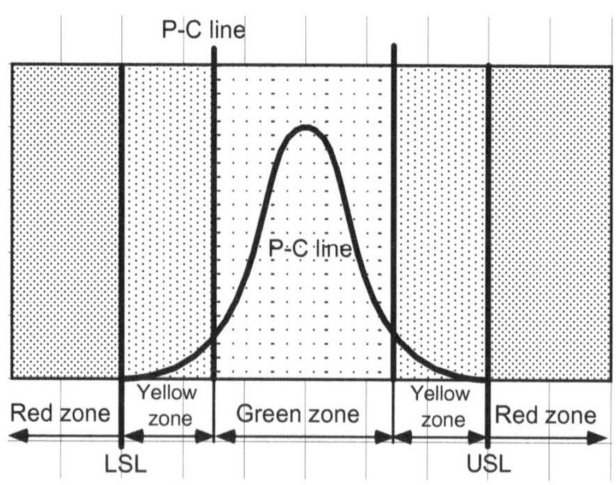

Figure 8.24 The three color zones in precontrol (shown with stippling).

If five consecutive units fall within the green zone, the process is considered capable. It can be shown that the capability index is at least 1.33. In conventional SPC, it takes at least 50 units from the process before the process capability can be determined.

8.3.3 Use of Computers

One of the most convenient ways of doing SPC is by using personal computers. There is a great multitude of good software commercially available that makes doing SPC much easier and quicker. An example of a quick overview of a substantial amount of SPC information is shown in Fig. 8.25.

The sixpack feature shows the x-bar and R chart, the actual data of the last 20 subgroups, the capability histogram of the data, the normal probability plot, and a capability plot showing how the process variation compares with the tolerance. Once the data is entered into a spreadsheet format, generating the plots shown in Fig. 8.25 takes only a few seconds. More detailed information on capability analysis is shown in Fig. 8.26.

Figure 8.26 shows the Cp, CPU, CPL, CpK, USL, LSL, k, n, mean, mean + 3s, mean − 3s, s, and so on; Fig. 8.26 was also generated using Minitab, a commercial software package from Minitab, Inc. When having to determine process capability with nonnormal data, a good option is to use a SPC software package that can handle nonnormal data. In fact, this would probably be the most expedient and accurate way to handle the nonnormal data analysis.

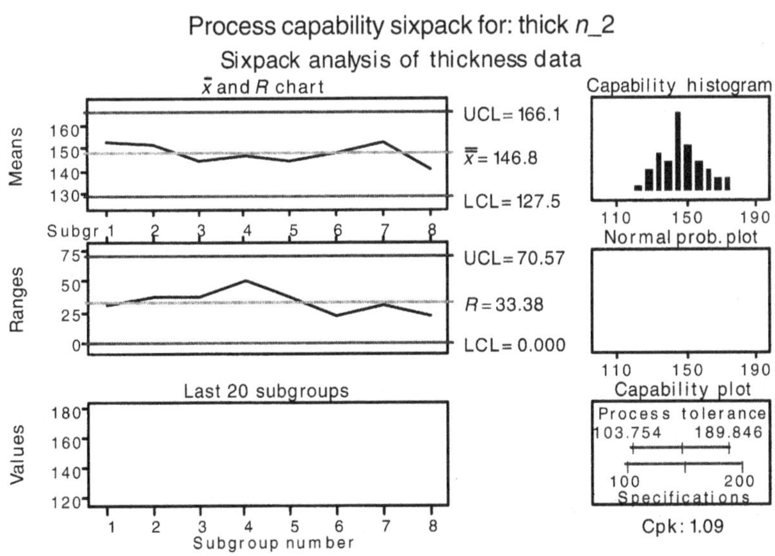

Figure 8.25 The sixpack option in capability analysis in Minitab™.

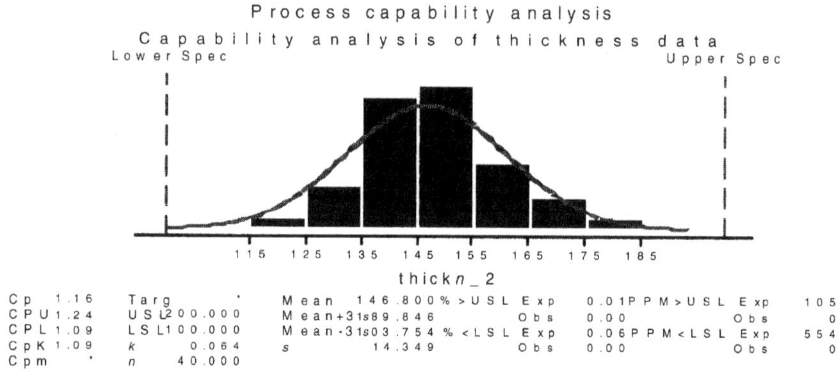

Figure 8.26 Process capability information on the thickness data.

8.3.4 Special SPC Techniques for Injection Molding

Because injection molding poses some special problems in SPC, as opposed to a continuous process like extrusion, some special techniques have been developed especially for injection molding process. These will be discussed next.

Family Processes

A *family process* consists of several statistically independent processes that are affected by common factors; these processes are also referred to as "multistream" processes. An example of a family process is injection molding using a multicavity mold. The filling and cooling in some cavities may be affected by factors not acting on the other cavities. If an operator samples five parts from a 32-cavity mold, the probability of a cavity not being included in the sample is 83.3%. If the samples are taken on an hourly basis, production may continue for a full shift or even a day without sampling one of the cavities. Thus, nonconforming parts may go undetected for a substantial period of time.

When a faulty cavity is part of a sample that leads to an out-of-control point, a common tendency is to adjust factors that affect all the cavities. These are the "global" factors. If the faulty cavity is not included in the next sample, the process may seem to have been properly adjusted. In actuality, the operator erroneously changed a process that was in statistical control.

Median/Individual Measurement Control Charts

One solution to this problem was proposed by Grant and Leavenworth in their book, *Statistical Process Control* (2). They proposed combining the results of median charts with individual measurements charts. The latter track local variations, whereas the former monitor global variation. One advantage of median charts is that they require no calculations by the operator.

In a median chart, each sample or observation consists of a unit from each family member. Remember that the median, $\tilde{x}$, is the point that divides the values of the individual measurements in half. The frequency of sampling is dependent on the process; initially, the sampling should be frequent enough to profile the process. If the process is stable then the sampling frequency can be reduced. Determination of the average median is similar to that of the grand average of a $\bar{x}$-chart:

$$\bar{\tilde{x}} = \frac{\tilde{x}_1 + \tilde{x}_2 + \tilde{x}_3 + \ldots \tilde{x}_k}{k} \tag{8.23}$$

where k = the number of subgroups.

The upper control limit is determined by:

$$UCL_{\tilde{x}} = \bar{\tilde{x}} + \tilde{A}_2\bar{R} \tag{8.24}$$

The lower control limit is determined by:

$$LCL_{\tilde{x}} = \bar{\tilde{x}} - \tilde{A}_2\bar{R} \tag{8.25}$$

Factor $\tilde{A}_2$ is given in Table 8.4. Note that the value of $\tilde{A}_2$ is different from the value A_2 for x-bar and R charts discussed earlier. The control limits for the individual measurement, x, control limits are determined from:

$$UIL_x = \bar{\tilde{x}} + E_2\bar{R} \tag{8.26}$$

The lower control limit is determined from:

$$LIL_x = \bar{\tilde{x}} - E_2\bar{R} \tag{8.27}$$

The factors $\tilde{A}_2$ and E_2 are given in Table 8.4.

Table 8.4 Factors with the Median ($\tilde{A}_2$) and Individual Measurement (E_2) Chart

n	$\tilde{A}_2$	E_2
3	1.187	1.772
4	0.796	1.457
5	0.691	1.290
6	0.548	1.184
7	0.508	1.109
8	0.433	1.054
9	0.412	1.010
10	0.362	0.975

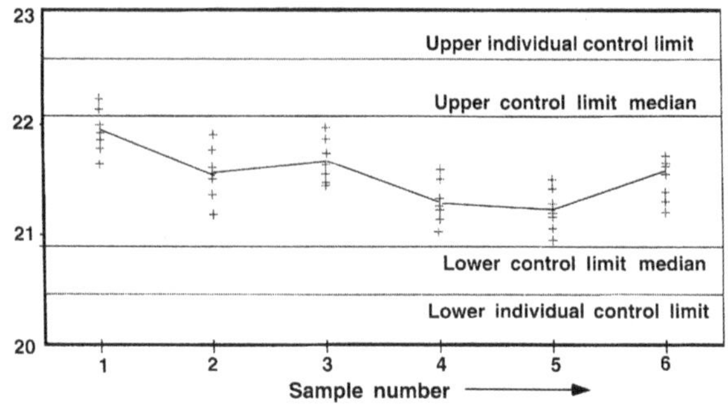

Figure 8.27 An example of a median/individual chart.

The median/individual chart shows the individuals as points and the median as a moving line (see Fig. 8.27). The measurements from one sample are shown along one vertical line; in this figure the measurements are shown as plus marks (+).

The median/individual (M/I) charts for family processes are easy to use and interpret. The analysis is more efficient because the effect of global and local causes of variation can be differentiated. The overall process variation can be reduced by centering the individuals.

If the median is within its control limits and one or more individuals exceed their limits, the problem lies with the offending individuals. An example of such a situation is shown in Fig. 8.28.

In this case, global factors should not be adjusted because it is local factors that cause the problem. Local factors could be factors like the temperature in a particular region of the mold, the size of the runner leading to one or more cavities, the size of the gate, and so on.

If the median exceeds its limits, a change in global factors would be appropriate. An example of the median going out of control is shown in Fig. 8.29.

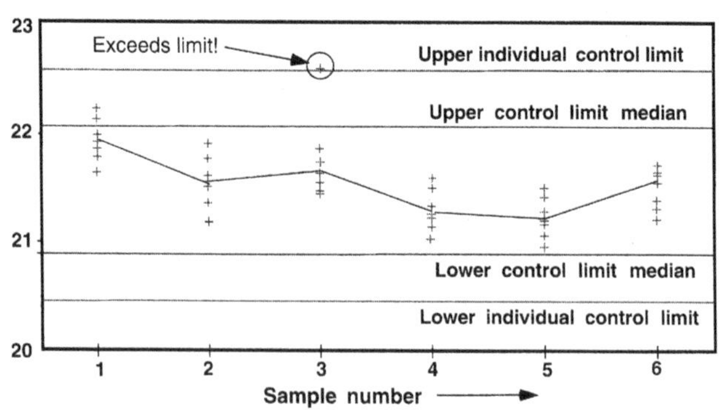

Figure 8.28 An example of an M/I chart with out-of-limit individual.

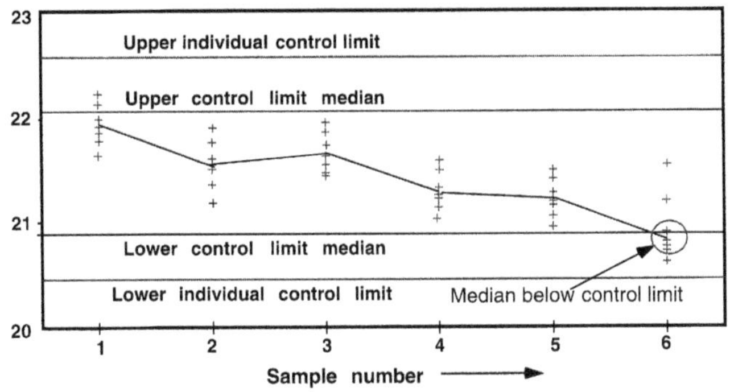

Figure 8.29 An example of an M/I chart with median out-of-control limits.

Van der Veen and Holst (3) describe an application of the M/I charting technique to the molding of a sprayer body in an 8-cavity water-cooled injection mold. The median chart indicated global shifts in the process. These were associated with changes in of color of the primary material. The individual's chart showed that cavity number 6 was consistently lower than all others. It was found that the cooling water temperature for cavity 6 was lower than it was for the other cavities, leading to faster cooling and less-efficient packing. This problem could be eliminated by adjusting the water flow. This charting technique is in some SPC software packages. Using a computer to generate control charts is certainly to be preferred over manual control charting. It is faster and reduces human errors, allowing people to be involved in more productive activities.

Group Charting

The problem of conventional control charting of multistream (family) processes was discussed by Bajaria and Skog (4). One method of dealing with multicavity injection

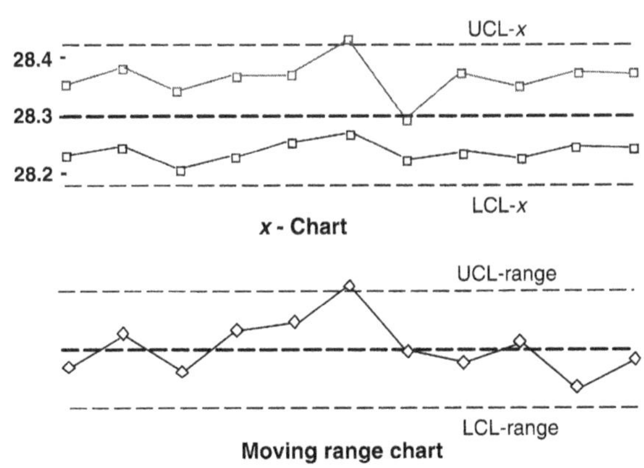

Figure 8.30 Example of a group chart.

molding processes is to use x and Rm charts (individual measurement and moving range) for each individual stream. This method is recommended for cases where the CpK of the cavities is equal to or greater than 3 ($CpK \geq 3$). When $CpK < 3$, the x-bar and R chart (average and range) is recommended. This approach, however, tends to be complicated and time consuming.

An alternative and more expedient approach is to use the group chart. In this chart the highest and lowest values are plotted on the x-chart and the largest moving range on the R-chart. An example of a group chart is shown in Fig. 8.30.

The process is running well as long as the high x-value is below the upper action limit and the low x-value above the lower action limit. The moving range chart shows the maximum range from any stream to its own previous value, making it sensitive to changes in any stream regardless of constant differences between streams. The group chart is most useful for day-to-day monitoring; however, a tabular report can be useful to identify consistently high or low stream. For instance, from a tabular report it may become obvious that cavity 3 is running consistently high. If we use an M/I chart, we can denote the individual points with the cavity number. In this case, it will be immediately obvious from the chart when one of the cavities is running consistently low or high. Identifying the low and high X values with the cavity number will do the same in the group chart. This method is preferred because it is easier to identify problems from a chart than from tabular data.

References

1. Shewhart, W., *Economic Control of Quality of Product* (1931), Van Nostrand Reinhold, New York.
2. Grant, E. I., Leavenworth, R. S., *Statistical Process Control*, 5th ed. (1980), McGraw-Hill, New York.
3. van der Veen, J., Holst, P., *Median/Individual Measurements Control Charting and Analysis for Family Processes* (1993), Northwest Analytical, Inc.
4. Bajaria, H., Skog, F., *Quality* (1994), December.
5. Rauwendaal, C.J., Statistical Process Control in Injection Molding and Extrusion (2000), Hanser, Munich.

9 Special Injection Molding Processes

L.-S. Turng*

Injection molding is one of the most versatile and important operations for mass production of complex plastic parts. The injection molded parts typically have excellent dimensional tolerance and require almost no finishing and/or assembly operations. In addition to thermoplastics and thermosets, the process is also being extended to such materials as fibers, ceramics, and powdered metals, with polymers as binders. Among all the polymer-processing methods, injection molding accounts for 32% by weight of all the polymeric material processed [1]. Nevertheless, new variations and emerging innovations of conventional injection molding have been continuously developed to extend the applicability, capability, flexibility, productivity, and profitability of this process further. To be more specific, these special and emerging injection molding processes introduce additional design freedom, new application areas, unique geometrical features, unprecedented part strength, sustainable economic benefits, improved material properties and part quality, and so on, that cannot be accomplished by the conventional injection molding process.

This chapter is intended to provide readers with a general introduction of these special injection molding processes with emphases on process description, relevant advantages and drawbacks, applicable materials, as well as existing and/or potential applications. References listed at the end of this chapter provide detailed information for more in-depth studies. With this information, readers will be able to evaluate the technical merits and applicability of the relevant processes in order to determine the most suitable production method. It is also hoped that a collective presentation of these various special molding processes, which descend from the same origin and yet mature with diversified creativity, will spark innovative ideas that lead to further improvement or new inventions.

In addition, thanks to a well-focused research effort conducted at many research and educational institutions (see, e.g., [2–20]), a solid scientific foundation for injection molding and related special processes discussed in this chapter has been established. Based on the resulting findings and theoretical principles, computer-aided engineering (CAE) tools have been developed and are now widely used in the industry. As a result, the design and manufacturing of injection-molded parts have literally been transformed from a "black art" to a well-developed technology for many manufacturing industries. These CAE tools help the engineer gain process insight, pinpoint blind spots and the problems usually overlooked, and contribute to the development and acceptance of many special injection molding processes discussed in this chapter.

*With a contribution from Mauricio Degreiff and Nelson Castaño on rubber injection molding.

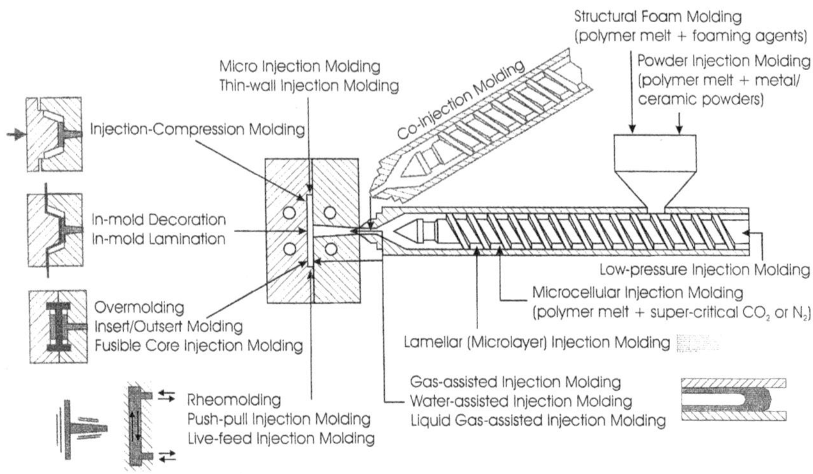

Figure 9.1 Special injection molding processes for thermoplastics.

It is very difficult to cover all special injection molding processes, not to mention those new processes that are being developed and field-tested. Furthermore, due to the diversified nature of these special injection-molding processes, there is no unique method to categorize them. As a preliminary attempt, Table 9.1 classifies the various processes based on the specific techniques employed by the process or the unique characteristics of the process. Figure 9.1 illustrates some of the characteristics of those special injection molding processes for thermoplastics. It should be noted that, for a special purpose or application, a new or viable special injection molding process could employ multiple specific techniques listed in Table 9.1 (e.g., gas-assisted powder injection molding, multicomponent powder injection molding, gas-assisted push-pull injection molding, coinjection molding with microcellular plastics, etc.). Detailed description of these processes can be found in the following sections dedicated to each individual process.

9.1 Coinjection (Sandwich) Molding

Coinjection molding (sometimes called "sandwich molding") comprises sequential and/or concurrent injection of a "skin" material and a dissimilar but compatible "core" material into a cavity. This process produces parts that have a sandwich structure, with the core material embedded between the layers of the skin material. This innovative process offers the inherent flexibility of using the optimal properties of each material to reduce the material cost, injection pressure, clamping tonnage, and residual stresses to modify the property of the molded part, and/or to achieve particular engineering effects.

Table 9.1 Categorization of Special Injection Molding Processes

1. Incorporation of additional material(s) or component(s) into the molded part
    a. Adding or injecting additional plastics
        i. Coinjection molding
        ii. Multi-component injection molding (overmolding)
        iii. Lamellar (microlayer) injection molding
    b. Injection around (or within) metal components
        i. Insert/outsert molding
        ii. Fusible core (lost core) injection molding
    c. Injecting gas into the polymer melt
        i. Gas-assisted injection molding
    d. Injecting liquid or water into polymer melt
        i. Liquid gas-assisted injection molding
        ii. Water-assisted injection molding
    e. Injecting gas into the metal (or ceramic) powder-polymer mixture
        i. Gas-assisted powder injection molding
    f. Incorporating reinforced fiber mats inside the cavity
        i. Resin transfer molding
        ii. Structural reaction injection molding
    g. Incorporating film, foil, fabric, or laminate to be back-molded by polymer melt
        i. In-mold decoration and in-mold lamination
        ii. Low-pressure injection molding
2. Melt formulation
    a. Mixing polymer melt with super-critical fluids
        i. Microcellular injection molding
    b. Mixing polymer melt with chemical or physical blowing agents
        i. Structural foam injection molding
    c. Mixing polymer melt with metal or ceramic powders
        i. Metal/ceramic powder injection molding
    d. Mixing prepolymer (monomers or reactants) prior injection
        i. Reaction injection molding
        ii. Structural reaction injection molding
        iii. Resin transfer molding
        iv. Thermoset injection molding
3. Melt manipulation
    a. Providing vibration and oscillation to the melt during processing
        i. Multi live-feed injection molding
        ii. Push–pull injection molding
        iii. Rheomolding
        iv. Vibration gas(-assisted) injection molding
    b. Using screw speed and back pressure to control melt temperature
        i. Low-pressure injection molding
4. Mold movement
    a. Applying compression with mold closing movement
        i. Injection-compression molding
5. Special part or geometry features
    a. Producing parts with miniature dimensions or relatively thin sections
        i. Micro-injection molding
        ii. Thin-wall molding

9.1.1 Process Description

Coinjection is one of the two-component or multi-component injection molding processes available today (see Table 9.2). Unlike other multi-component molding processes, however, the coinjection molding process is characterized by its ability to encapsulate an inner core material with an outer skin material completely. The process mechanics rely on the sequential and/or concurrent injection of two different materials through the same gate(s). Figure 9.2 illustrates the typical sequences of the co-injection molding process using the "one-channel technique" and the resulting flow of skin and core materials inside the cavity. This is accomplished with the use of a machine that has two separate, individually controllable injection units and a common injection nozzle block with a switching head. The principle of this type of process is relatively simple: Two dissimilar polymer melts from their injection units are injected one after the other into a mold cavity. In particular, in stage (a) of Fig. 9.2, a short shot of skin material (shown in black) is injected into the mold. Due to the flow behavior of the polymer melts and the solidification of skin material, a frozen layer of polymer starts to grow from the colder mold walls. The polymer flowing in the center of the cavity remains molten. As the core material is injected, it flows within the frozen skin layers, pushing the molten skin material at the hot core to the extremities of the cavity, as shown in stage (b) of Fig. 9.2. Because of the fountain-flow effect at the advancing melt front (an outward trajectory of fluid particles from the central region to the bounding walls), the skin material at the melt front will show up at the region adjacent to the mold walls. This process continues until the cavity is nearly filled, with skin material appearing on the surface and the end of the part as

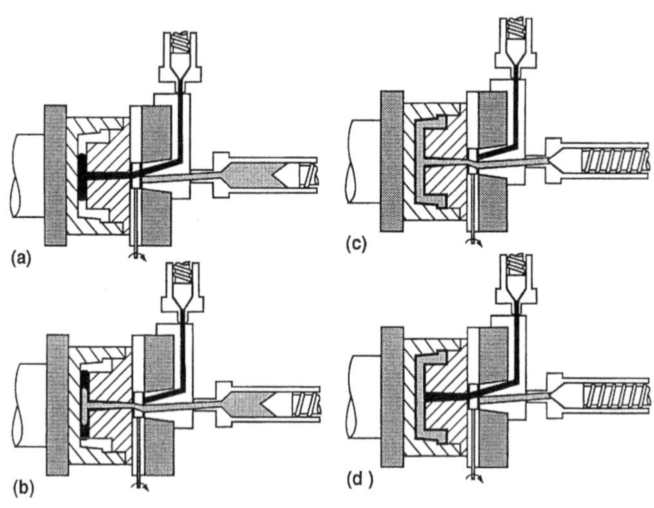

Figure 9.2 Sequential coinjection molding process [18] (Adapted from Ref. [21]. Reprinted by permission of the ASME International).

Table 9.2 Various Multicomponent Injection Molding Processes [29]

Brief description	Machine/Mold requirements	Materials requirements	Typical applications
		Coinjection Molding	
Two melts, injected sequentially or concurrently, flow into the cavity and separately form skin (material A) and core (material B) of the laminated part structure	Coinjection machine with special nozzle for mixture-free, concentric combining of the melt; alternative: three-way valve, staggered injection, material B penetrates within material A	Medium compatibility and melt viscosity requirement (skin material should have lower viscosity than the core material) Materials: PET, PA (reinforced and unreinforced), ABS (foamed and solid)	Headlight reflectors (core fiber reinforced, skin unreinforced to give good surface), Garden furniture (core foamed to reduce weight, skin solid to give good surface) Spectacle frame (core carbon fiber reinforced to generate high tensile strength, skin transparaent to give good surface)
		Two-Color Injection Molding	
A premolded component is placed in a second larger cavity and then encapsulated with the same, different colored material	With separately molded insert: standard molds and machines; More efficient set-up: tilting or transfer molds and two-color injection molding machine with two separate injection units	Low compatibility requirement; Materials: mostly ABS, PC, PA12, PA blends	Keyboard keys and push buttons for computers, telephones, and toys
		Overmolding (Multi-Component Injection Molding, Injection Welding)	
A premolded component is placed in a second larger cavity and then encapsulated with a different, compatible thermoplastic	Same as above; It is advantageous to use "core-back" molds (with retractable slides) that permit overmolding of a hot insert using the same mold and a molding machine with two separate injection units	Maximum compatibility requirement; Materials: frequently PA12 or PA 12 copolymers, also PA/PU pairs	Hand-held power tools grips, damping elements in ski shoe shells, elastic inserts in safty and sport goggles, sight glasses in technical equipment
		Bi-Injection Molding	
Two melts of different thermoplastics, guided separately (two runner systems, at least two gates) meet in the cavity	Machine with two separately controllable injection units; the melt front advancements determine the weld line location, high control quality needed	Medium compatibility requirement; greater number of pairs possible than overmolding	Flexible functional elements in an otherwise stiff part; positioning of core in thickenings by separate gates

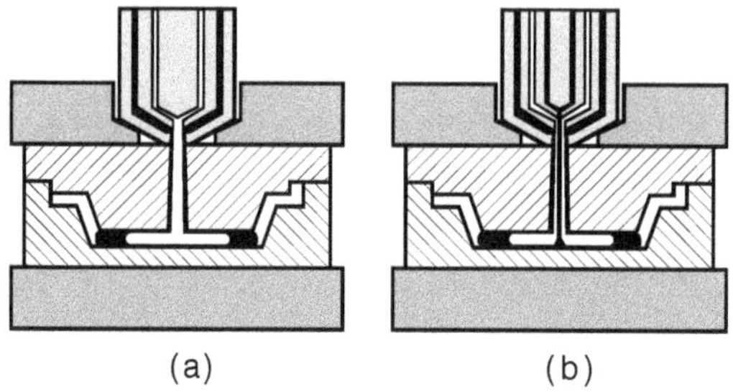

Figure 9.3 (a) Two-channel and (b) three-channel techniques [22].

shown in Fig. 9.2(c). Finally, a small additional amount of skin material is injected again to purge the core material away from the sprue so that it will not appear on the part surface in the next shot [cf. Fig. 9.2(d)].

When there is not enough skin material injected prior to the injection of core material, the skin material may sometimes eventually be depleted during the filling process and the core material will show up on portions of the surface and the end of the part that is last filled. Such "core surfacing" or "core breakthrough" is generally undesirable, although it may depend on the design requirement and final application.

There are other variations to the sequential (namely, skin–core–skin, or A–B–A) coinjection molding process described earlier. In particular, one can start to inject the core material while the skin material is being injected (i.e., A–AB–B–A). That is, a majority of skin material is injected into a cavity, followed by a combination of both skin and core materials flowing into the same cavity, and then followed by the balance of the core material to fill the cavity. Again, an additional small amount of skin injection will "cap" the end of the sequence, as described previously. In addition to the one-channel technique configuration, two- and three-channel techniques (cf. Fig. 9.3) have been developed that use nozzles with concentric flow channels to allow simultaneous injection of skin and core materials [22]. More recently, a new version of coinjection molding process that employs multi-gate coinjection hot-runner system has become available. Such a system moves the joining of skin and core materials into the mold, as shown in Fig. 9.4. In particular, this hot-runner system has separate flow channels for the skin and core materials. The two flow streams are joined at each hot runner coinjection nozzle. In addition to all the benefits associated with conventional hot-runner molding, this system allows an optimum ratio of skin and core in multi-cavity or single-cavity molds [23].

9.1.2 Process Advantages

Coinjection molding offers a number of cost and quality advantages, as well as design flexibilities and environmental friendliness as described later.

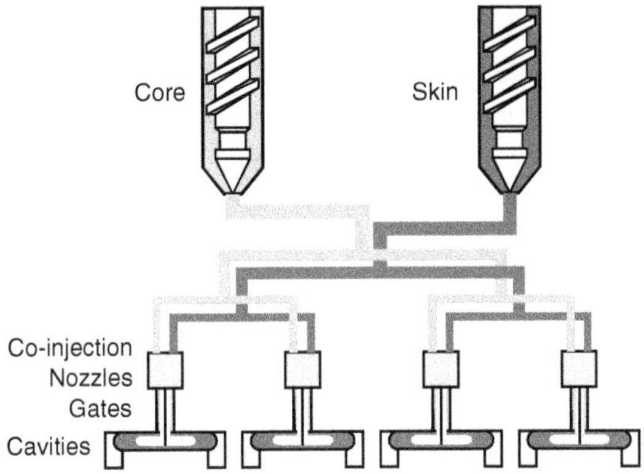

Figure 9.4 Multi-gate coinjection hot runner system with separate flow channels for skin and core materials, which join at each hot runner nozzle [23].

Material Cost Reduction and Recycling

High-performance and exotic engineering materials can be expensive but necessary for some applications. Coinjection provides the opportunity to reduce the cost of the product by utilizing lower-cost materials wherever the high-performance material is not necessary, perhaps in the core. That is, it permits the use of low-cost or recycled plastics as the core material, invisibly sandwiched within thin, decorative, expensive skin surfaces typically made of virgin plastic material. An example of this is a gear wheel (cf. Fig. 9.5) whose outer skin consists of a fluoropolymer or carbon fiber filled polyamide (nylon). As the recycling of postservice plastics becomes necessary by law in many countries, coinjection molding offers a cost-effective manufacturing technique to consume 100% recycled materials in high content.

Figure 9.5 Gear wheel (101.6 mm in diameter) with PTFE-filled white polyamide (nylon) 66 skin and glass-filled solid black polyamide 66 core made by coinjection molding [21].

Quality Surface with Foamed Core Material

In the case of thick-wall products, coinjection is preferred to conventional structural foam because of its superior cosmetic surfaces. Structural foam parts are often sanded, primed, base painted, and texture coated, all of which is expensive. A solid skin combined with a foamed core provides the advantages of structural foam, such as reduced part weight, low molded-in stresses, straight sink-free parts, and design freedom, yet without the objectionable elephant-skin surface defects. Coinjection molding with foamed material also features an excellent weight-to-strength ratio and produces better performance than gas-assisted injection molding for sensitive/fragile polymer materials. For thin-wall parts, such as food packaging and bottles, coinjection also offers additional benefit in terms of physical and mechanical properties of the part and cost saving with foamed core.

Modification of the Part Quality and Property

With coinjection, one can obtain a combination of properties by joining different materials in one part, which is not available in a single resin. For instance, an elastomeric skin over a rigid core will provide a structure with soft touch. Another example includes a combination of a brittle material with a high-impact-resistant material, which provides excellent material properties. In applications where the performance of the components demands the use of reinforced materials, coinjection offers a solution that combines the aesthetic and property attributes of an unreinforced skin material with the benefits of a highly reinforced core material. Additional performance and cost improvements can be made by combining the conductive plastic with a more impact-resistant and less-expensive grade of plastic through co-injection molding. Such an application includes using either a skin or core polymer filled with a conductive material (e.g., aluminum flakes, carbon black, or nickel-coated graphite fibers) to provide the molded part (e.g., computer housings) with electromagnetic shielding (EMI) properties and grounding characteristics (cf. Fig. 9.6) [24].

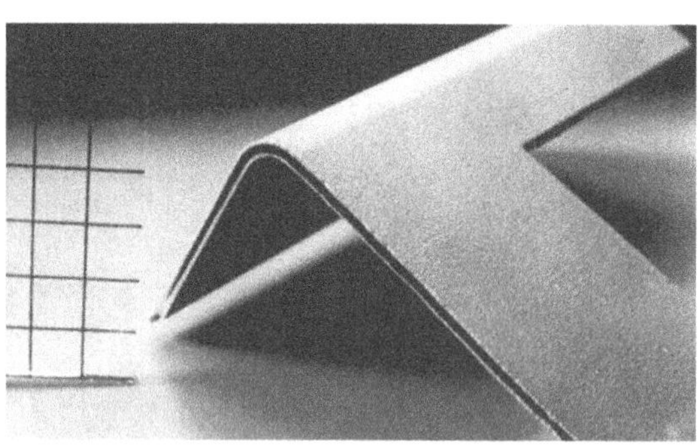

Figure 9.6 Section of a molding housing with an outer skin of ABS and a core of the same resin filled with 35 wt% of electrically conductive carbon black [24].

9.1.3 Process Disadvantages

Despite all the potential benefits of coinjection molding, the process has been slow to gain widespread acceptance for several reasons. First of all, the coinjection machine usually costs 50 to 100% higher than standard injection molding equipment [25]. This high investment cost offsets the benefits of developing unique processing techniques, improving part quality, and permitting the use of recycled materials. In addition, the development for a coinjection mold takes longer time than a conventional injection mold. This is also true for process set up, as the process requires additional control parameters for timing and controlling the injecting core material.

9.1.4 Applicable Materials

Coinjection molding can be employed for a wide variety of materials. Although most of the materials used are thermoplastic, there are some promising developments with using thermosetting materials, which are coinjected with thermoplastic materials. Because two materials are used in coinjection molding processing, the flow behavior (Fig. 9.2) and the compatibility of material properties are very important. In considering the material selection, the most important properties are viscosity difference and the adhesion between the skin and core material. Because the core material needs to penetrate the skin material in front of it, it is desirable to have a skin material with a viscosity lower than that of the core material. Using low-viscosity material in the core may cause the core flow front to travel too fast relative to the skin, which results in undesirable core surfacing. Experimental studies of coinjection molding have been conducted to examine the effect of relative viscosity ratio of skin and core materials on their spatial distribution within the part [26].

Because the materials are laminated together in the part, an effective adhesion of skin and core material is desirable for optimum performance. Table 9.3 provides some basic guidelines on a wide range of material combinations [27]. It should be noted that this information should only be used as a guide or benchmark. The actual performance must be determined by the application because the molding conditions and the operation/service conditions will influence the final performance. The other material property that needs to be of concern is material shrinkage. The rule of thumb is that materials with similar molding shrinkage should be paired in order to reduce stresses in the joining layers.

9.1.5 Typical Applications

Coinjection molding offers a technically and economically viable solution for a wide range of commercial applications in the emerging markets, which include automotive, business machines, packaging, electronic components, leisure, agriculture, and

Table 9.3 Material Adhesion Characteristics [25]

	ABS	ASA	CA	EVA	PA 6	PA 66	PC	PE-HD	PE-LD	PMMA	POM	PP	PPO mod.	PS-GP	PS-HI	PBTP	TPU	PVC-W	SAN	TPR	PETP	PVAC	PPSU	Blend PC-PBTP	Blend PC-ABS
ABS	■	■	■				■	X	X	■		X	X	X	X	■	■	■	■		■	░		■	■
ASA	■	■	■	■			■	X	X	■		X	X	X	X	■	■	■	■			░		■	■
CA	■	■	■	░				X	X			X	X	X	X	■	■	■	■						
EVA		■	░	■				■	■			■		■											
PA 6					■	■		░	░			░		X	X		■								
PA 66					■	■	░	░	░			░		X	X		■								
PC	■	■					■	X	X			X		X	X	■	■	■	■		■		■	■	■
PE-HD	X	X	X	■	░	░	X	■	■	░	░	X		X	X	X	X	░	X					X	X
PE-LD	X	X	X	■	░	░	X	■	■	░	░	■		X	X	X	X		X					X	X
PMMA	■	■						░	░	■		░		X	X			■	■						
POM								░	░		■	░		X	X										
PP	X	X	X	■	░	░	X	X	■	░	░	■		X	X	X	X	░	X	■				X	X
PPO mod.	X	X	X										■	■	■	X	X	X	X					X	X
PS-GP	X	X	X	■	X	X	X	X	X	X	X	X	■	■	■	X	X	░	X					X	X
PS-HI	X	X	X		X	X	X	X	X	X	X	X	■	■	■	X	X	░	X					X	X
PBTP	■	■	■				■	X	X			X	X	X	X	■	■	■	■						
TPU	■	■	■		■	■	■	X	X			X	X	X	X	■	■	■	■						
PVC-W	■	■	■	X			■	░		■		░		░	░	■	■	■	■						■
SAN	■	■	■				■	X	X	■		X	X	X	X	■	■	■	■		■	░		■	■
TPR												■								■					
PETP	■						■												■		■			■	■
PVAC	░	░																				■			
PPSU							■																■		
Blend PC-PBTP	■	■					■	X	X			X	X	X	X				■		■			■	
Blend PC-ABS	■	■					■	X	X			X	X	X	X			■	■		■				■

(From *Injection Molding Alternatives—A Guide for Designers and Product Engineers* (1998), Avery, J. A., Hanser Publishers, Munich, p.107, Table 5.4.1)
The symbols indicate: ■ = good adhesion; ░ = poor adhesion; X = no adhesion.
From *Innovation in Polymer Processing: Molding*. Stevenson, J. F. (Ed.) (1996) Hanser, Munich, pp. 338–339, Table 9.2.
Courtesy of Battenfeld Gmbh, Meinerzhagen, Germany.
Standard plastic material abbreviations: ABS, acrylonitrile–butadiene–styrene; ASA, acrylic styrene acrylonitrile; CA, cellulose acetate; EVA, ethyl vinyl alcohol; PA 6 or nylon 6, polyamide 6; PA 66—nylon 66, polyamide 66; PC, polycarbonate; PE—HD, polyethylene—high-density; PE—LD, polyethylene—low density; PMMA, polymethylmethacrylate; POM, polyoxymethylene [(poly) acetal]; PP, polyoxymethylene [(poly) acetal]; PP, polypropylene; PPO Mod., polyphenylene ether, modified; PS-GP, polystyrene—general purpose; PS-HI or HIPS, polystyrene, high-impact; PBTB or PBT, polybutylene terephthalate; TPU, thermoplastic polyurethane; PVC-W, polyvinyl chloride; SAN, styrene acrylonitrile; TPR, thermoplastic rubber; PETP or PET, polyethylene terephthalate; PVAC, polyvinyl alcohol; PVC, polyvinyl chloride; PPSU, polyphylene sulfide; Blend PC-PBTP, blend, polycarbonate–polybutylene–terephthalate; Blend PC-ABS, blend, polycarbonate–acrylonitrile–butadiene–styrene.

"soft touch" products. Example applications include canoe paddles, toilet seats and cisterns, computer housings, copier parts, cash register covers, television escutcheons, audio cabinets, circuitry and electronics enclosures, garden chairs, boxes and containers, shoes and soles, paint brush handles, metal hand-rims on wheelchairs, thin-wall containers and beverage bottles, automotive parts such as exterior mirror housings and interior door handles and knobs, components for high-end ovens, audio speaker housings, and music center mainframes.

Cost-reduction opportunities and the desire and mandated requirements to use recycled materials will drive more applications to coinjection in new and existing markets. More refined technology and a broader experience base among designers, processors and original equipment manufacturers (OEMs) will also expand the sporadic current use of the process. Useful design guidelines for coinjection molding and other multicomponent injection molding processes can be found in Ref. [28].

9.2 Fusible (Lost, Soluble) Core Injection Molding

The fusible (lost, soluble) core injection molding process produces complicated, hollow components with complex and smooth internal geometry in a single molding operation. This process is a form of insert molding in which plastic is injected around a temporary core of low melting-point material, such as tin-bismuth alloy, wax, or a thermoplastic. After molding, the core will be physically melted (or chemically dissolved), leaving its outer geometry as the internal shape of the plastic part. This process reduces the number of components required to make a final assembly or substitutes plastic for metal castings to boost performance (e.g., corrosion resistance) while saving weight, machining, and cost.

9.2.1 Process Description

Different techniques are available to produce single-piece components featuring complex, smooth internal geometry and a high dimensional stability, which cannot be obtained through the conventional injection molding process [1,30,31]:

- Fusible core technique
- Soluble core technique
- Salt core technique

All of these techniques employ the same principle: the production of injection molding with a lost core that gives the internal contour of the molded part. Among these lost core processes, fusible core technique is the most energy-intensive method. Nevertheless, this drawback is offset by the low core losses, smoother internal surface

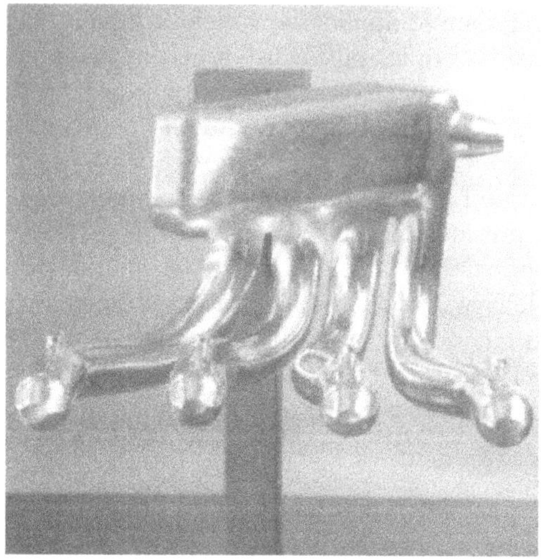

Figure 9.7 A eutectic bismuth-tin (BiSn 138) alloy core for an intake manifold [30]. Source: BASF. (From *Innovation in Polymer Processing—Molding*, Stevenson, J. F., (Ed.) (1996), Hanser, Munich, p. 159, Fig. 4.7.)

requiring low finishing cost, faster heat dissipation by using a stronger and highly conductive metal core. Fusible core injection molding basically comprises the following steps:

1. One or more core pieces are prefabricated at a separate station from the injection molding machine (cf. Fig. 9.7).
2. If more than one core piece is needed, the core assembly will take place in order to make a single-unit core.
3. The core then goes through a preheating process to reach temperature uniformity throughout the core before it is inserted into the mold (cf. Fig. 9.8).
4. Plastic is injected over the core in the injection machine mold, flowing evenly over the core without solidifying or melting the core.
5. After molding, the plastic part is taken with the core to a "melting" station, where the core is melted (or dissolved) out of the hollow plastic component (cf. Fig. 9.9).
6. The plastic part is cleaned with water and/or a cleaning solution and then taken to the next step for postmolding operations while the core material is recycled.
7. These postmolding operations include additional drilling or surfacing. In the end, a single-unit, hollow plastic part is obtained.

To minimize the core loss due to oxidation and energy consumption, a number of eutectic alloys have been found to be suitable for fusible core injection molding. These eutectic alloys exhibit stable dimensional stability (with low expansion or shrinkage), a sharp melting point, and a rapid phase transition. It is apparent that the

Figure 9.8 A eutectic bismuth-tin (BiSn 138) alloy core secured in the mold for an intake manifold [30]. Source: BASF. (From *Innovation in Polymer Processing—Molding*, Stevenson, J. F., (Ed.) (1996), Hanser, Munich, p. 164, Fig. 4.12.)

melting temperature of the fusible alloy should be lower than that of the overmolding plastic so that it can be melted out in the melting bath. Nevertheless, due to the high thermal diffusivity as well as its massive thermal inertia and latent heat of the alloy, the surface temperature of the core remains below its melting point during molding. This phenomenon is akin to the relatively small temperature rise at the mold-wall in conventional injection molding where the high thermal conductivity of the metal mold rapidly removes the heat from the molten plastic.

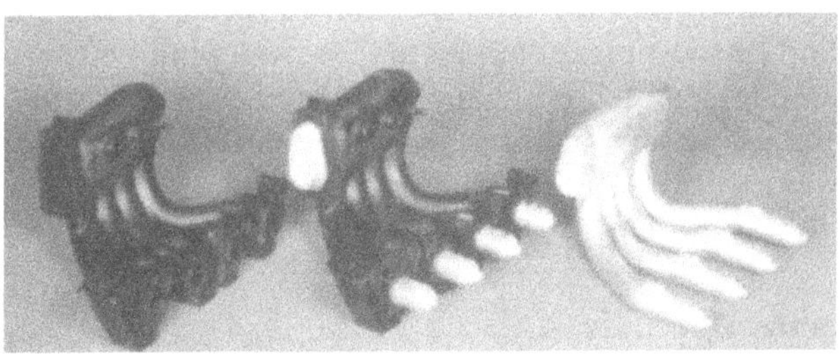

Figure 9.9 (Left) An intake manifold made from glass-fiber-reinforced polyamide 66 by fusible core technique; (center) core overmolded with plastic; (right) fusible core cast from bismuth-tin alloy [30]. Source: BASF. (From *Innovation in Polymer Processing—Molding*, Stevenson, J. F., (Ed.) (1996), Hanser, Munich, p. 155, Fig. 4.3.)

9.2.2 Process Advantages

The primary advantage of fusible core injection molding is the ability to produce single-piece plastic parts with highly complex, smooth internal shapes without a large number of secondary operations for assembly. Compared with options of aluminum casting and machining, cost savings with fusible core injection molding are claimed to be up to 45% and weight savings can be as high as 75% [25]. Plastic injection molding tools also have a much longer life than metal casting tools due to the absence of chemical corrosion and heavy wear. Additional advantages include:

- greater freedom in design
- high surface quality with no zones weakened by welding or joining
- dimensionally accurate internal and external contours and high structural integrity
- low labor intensity with only moderate levels of secondary operations (trimming, drilling, finishing, etc.)
- easier mold design
- little scrap in the overall process
- integration of inserts.

9.2.3 Process Disadvantages

The main disadvantage of fusible core injection molding is cost. First, it requires a significant capital investment, ranging from $6 to 8 million [25]. The fusible core molding unit also requires a very large injection molding machine, casting machine and equipment, a "melt-out" station, and robotic handling devices for the typically heavy plastic molding with core. There are also significant operating expenses, which come from core casting and electricity usage resulting from more machines being used as well as from core casting and melting. Another disadvantage is mold and machine development for casting and injection molding, which take time due to complex design and prototyping requirements. Because of this cost factor, the fusible core injection molding process for nylon air intake manifolds has been tempered by a shift to less-capital-intensive methods such as standard injection molding, where part halves are welded (twin-shell welding) or mechanically fastened [32].

9.2.4 Applicable Materials

Fusible core injection molding is generally employed for engineering thermoplastics, primarily glass fiber reinforced polyamide (PA) 6 and polyamide 66. Other materials used are glass fiber reinforced polyphenylene sulfide (PPS) and polyaryletherketone (PAEK); glass reinforced polypropylene (PP); and glass reinforced

Table 9.4 Various Fusible Core Injection Molding Applications [25]

Automotive applications	Nonautomotive applications
Air-intake manifolds	Bearing housings
Air-intake manifold seat flanges	Thermostat valves
Fuel rails	Water valves
Power steering inlet pipe adapters	Water pump impellers and housings
Turbo charger housings	Coffee machine and showerhead feed pipes
Thermostat housings	Plumbing fixtures
	Tennis racquets
	Canoe paddles
	Bicycle wheels
	Wheelchair wheels
	Helicopter components

polyoxymethylene (POM) [25]. Low-profile thermosets have reportedly been used in fusible core injection molding, which require a core with a higher melting point [30]. A number of factors must be considered in selecting the plastic materials. For example, the plastic component must be compatible with the core alloy and be able to withstand melting bath temperatures, which ranges from 100 to 180°C, in addition to typical considerations of the operation environment and mechanical property requirements.

9.2.5 Typical Applications

The fusible core injection molding has so far been used primarily for automotive air intake manifolds (see Fig. 9.9). Improved processing engineering and greater design freedom, however, will open up innovative and interesting applications for years to come. Other typical applications of fusible core injection molding are listed in Table 9.4.

9.3 Gas-Assisted Injection Molding

The gas-assisted injection molding process consists of a partial or nearly full injection of polymer melt into the mold cavity, followed by injection of inner gas (typically nitrogen) into the core of the polymer melt through the nozzle, sprue, runner, or directly into the cavity. The compressed gas takes the path of the least resistance flowing toward the melt front where the pressure is lowest. As a result, the gas penetrates and hollows out a network of predesigned, thick-sectioned gas channels, displacing molten polymer at the hot core to fill and pack out the entire cavity. The essentially inviscid gas ideally transmits the gas pressure effectively as it penetrates to the extremities of the part, thereby requiring only a relatively low gas pressure to produce sufficient packing and a fairly uniform pressure distribution throughout the cavity. As a result, this process is capable of producing lightweight, rigid parts that are free of sink marks and have less tendency to warp.

9.3.1 Process Description

Gas-assisted injection molding starts with a partial or nearly full injection of polymer melt, as in conventional injection molding, followed by an injection of compressed gas. The compressed gas is usually nitrogen because of its availability, low cost, and inertness with the polymer melt. During the gas-injection stage, gas penetrates the hot core of the thickest sections where the material remains molten and fluid, displacing it to fill the extremities of the cavity and pack out the mold.

In the so-called gas-pressure control process, the compressed gas is injected with a regulated gas pressure profile (constant, ramp, or step). In the "gas-volume control" process, gas is initially metered into a compression cylinder at preset volume and pressure; then, it is injected under pressure generated from reducing the gas volume by movement of the plunger. Conventional injection molding machine with precise shot volume control can be adapted for gas-assisted injection molding with add-on conversion equipment, a gas source, and a control device for gas injection, as shown in Fig. 9.10. Gas-assisted injection molding, however, requires a different approach to product, tool, and process design due to the need for control of additional gas injection and the layout and sizing of gas channels to guide the gas penetration in a desirable fashion. As an illustration, Fig. 9.11 shows the schematic of gas injecting through

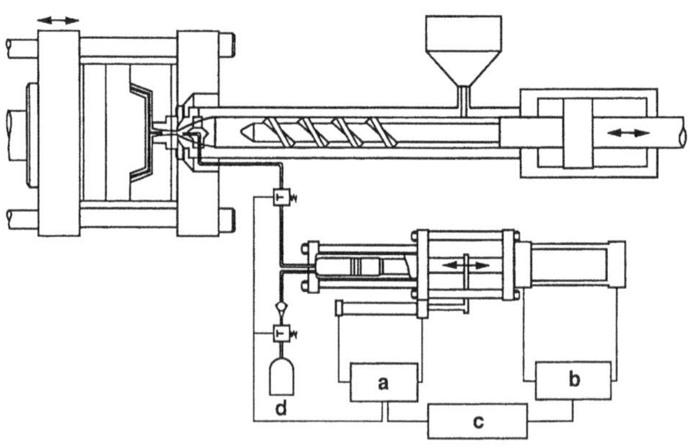

Figure 9.10 Schematic of a typical injection molding machine adapted for gas-assisted molding with add-on gas-compression cylinder and accessory equipment. (a) Electrical system, (b) hydraulic system, (c) control panel, (d) gas cylinder [34]. (From *Innovation in Polymer Processing—Molding*, Stevenson, J. F., (Ed.) (1996), Hanser, Munich, p. 46, Fig. 2.1.)

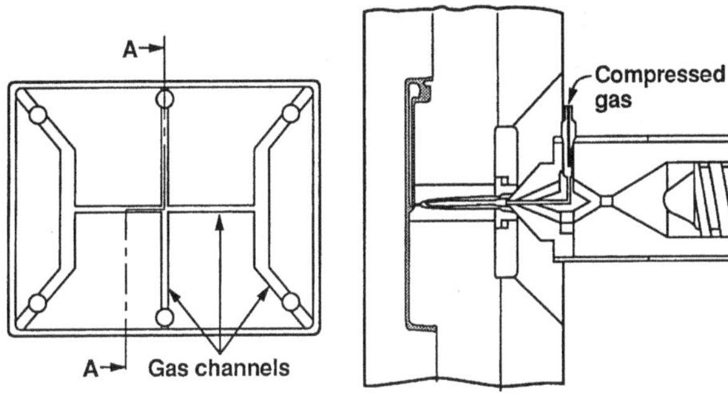

Figure 9.11 Schematic showing injection of gas through the nozzle and the gas distribution in the rib of the molding with a network of thick-sectioned gas channels [34]. (From *Innovation in Polymer Processing—Molding*, Stevenson, J. F., (Ed.) (1996), Hanser, Munich, p. 46, Fig. 2.2.)

the nozzle and the gas distribution in the rib of the molding with a network of thick-sectioned gas channels.

Process Cycle

As illustrated in Fig. 9.12, the gas-assisted injection molding process begins with a resin-injection stage (between Points 1 and 2), which is the same as the conventional

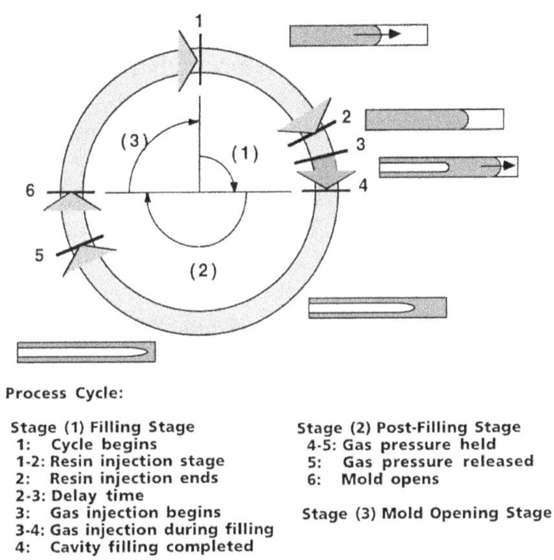

Figure 9.12 Process cycle of gas-assisted injection molding. Note that resin-injection time is between Points 1 and 2, the timer starts gas injection at Point 3, the fill time is between Points 1 and 4, and gas-injection time is from Points 3 to Point 5 [34]. (From *Innovation in Polymer Processing—Molding*, Stevenson, J. F., (Ed.) (1996), Hanser Publishers, Munich, p. 48, Fig. 2.3.)

injection molding process. Before the gas injection is triggered at Point 3, there is an optional delay time during which the polymer at the thin section cools so that the incoming gas can only core out the designated thick sections that serve as the gas channels. The gas injection can sometimes take place before the end of the resin injection to avoid hesitation marks at the melt front location during the gas delay time. It can be done, however, only when the gas-injection point is different from the polymer entrance; otherwise, simultaneous injection of polymer and gas at the same entrance will result in an undesirable, corrugated polymer skin thickness. The gas-injection time is the duration over which gas pressure is imposed at the gas entrance(s) (front Point 3 through Point 5). Because gas is injected to help fill and pack out the mold, the gas-injection stage actually consists of both gas-injection-filling and postfilling phases. The timer for gas injection and the gas-injection time are two important processing parameters that strongly influence the gas penetration. As in conventional injection molding, the part should be allowed to cool sufficiently inside the mold before the gas pressure is released at Point 5. The polymer skin thickness (also called residual wall thickness) over the cored-out gas channels ideally permits a significant reduction of cooling time. After that, the mold opens at Point 6 for the part to be ejected.

Although the total gas-injection time takes a significant portion of the entire cycle, the actual elapsed time from the introduction of the gas to the instant when the whole cavity is filled is very short, as illustrated in the process cycle clock in Fig. 9.12. Nevertheless, this relatively short gas-injection-filling stage is crucial to the success of the molded part because the various molding problems associated with this process, such as air trap, gas blow-through, short shot, gas permeation into thin section, uneven gas penetration, and the like, typically take place in this stage. The gas penetration created by displacing the polymer melt during the gas-injection-filling stage is often referred to as the *primary gas penetration*. This is typically the stage when the majority of gas penetration is determined (assuming a short shot of resin is injected), which impacts the quality of the final products. After the cavity is filled, a certain level of gas pressure is maintained to pack out the molded part. During this postfilling stage (or the holding stage), the so-called secondary gas penetration continues to compensate for the material shrinkage, primarily along the gas channels or thick sections. When the molded part cools down and becomes rigid enough inside the mold, the gas pressure is released prior to the ejection of the part. The essentially inviscid gas ideally transmits the gas pressure effectively as it penetrates to the extremities of the part, resulting in a uniform pressure distribution throughout. This process is consequently capable of producing lightweight, rigid parts that are free of sink marks and have less tendency to warp. This process provides tremendous flexibility in the design of plastic parts. A comprehensive review of the technology and applications can be found in Refs. [33,34] and the cited references.

Special Gas-Assisted Injection Molding Processes

Because of the versatile and promising capabilities of this process, some alternative gas-assisted injection molding processes have been developed and become commercially available. For example, instead of using the compressed nitrogen, the "liquid gas-assist process" injects a proprietary liquid into the melt stream. This liquid is converted to a gas in a compressed state with the heat of the polymer melt [35]. After the filling and packing of the cavity, the gas is absorbed as the part cools down, thereby eliminating the need to vent the gas pressure. In the so-called water-assisted injection molding process, water, which does not evaporate during displacement of the melt, is injected after resin injection. Compared with "conventional" gas-assisted injection molding, water facilitates superior cooling effect and thus a shorter cooling time, as well as thinner residual wall thickness and larger component diameters [36]. The "partial frame process" injects compressed gas into the strategically selected thick sections to form small voids of 1 to 2 mm (0.04 to 0.08 in) in diameter. Such a process can be employed to reduce sink marks and residual stresses [37] over the thick sections. Finally, the "external gas molding process" is based on injecting gas in localized, sealed locations (typically on the ejector side) between the plastic material and the mold wall [38,39]. The gas pressure maintains contact of the cooling plastic part with the opposite mold wall while providing a uniform gas pressure on the part surface supplementing or substituting for conventional holding pressure. This kind of process is suitable for parts with one visible surface where demands on surface finish are high and conventional use of gas channels is not feasible.

9.3.2 Process Advantages

The characteristics of gas-assisted injection molding lead to both the various advantages and the inherent difficulties associated with the design and processing. Recall that gas normally takes the path of least resistance as it flows toward the polymer melt front while hollowing out the thickest sections. To facilitate the gas penetration in a desirable pattern, the molded part typically has a network of built-in gas channels that are thicker than the main wall section. The gas ideally penetrates and hollows out only the network of gas channels by displacing the molten polymer melt at the hot core to fill the entire cavity.

Figure 9.13 schematically illustrates the mechanism of how the polymer melt and gas interact in the cavity. It is well known that in the conventional injection molding process, the pressure required to advance the polymer increases with the amount of polymer injected (or, equivalently, the flow distance). Figure 9.13(a) schematically depicts the rise in pressure during filling of the mold cavity without gas injection. It should be noted that the gap-wise averaged melt velocity is proportional to the magnitude of the pressure gradient and the melt fluidity. As the flow length of the polymer melt increases, therefore, the inlet pressure has to increase to maintain a certain pressure gradient if the flow rate is to be kept constant.

On the other hand, the pressure requirement with gas-assisted injection molding is the same as the conventional process during the resin-injection stage. Upon introduction of gas into the cavity, the gas starts to displace the viscous polymer melt, pushing it to fill the extremities of the cavity. Because the gas is essentially inviscid,

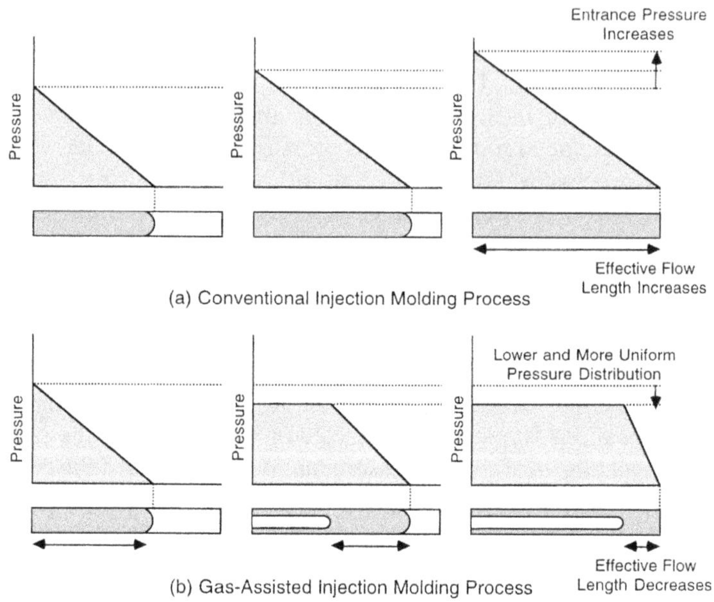

Figure 9.13 (a) The evolution of pressure distribution for the conventional injection molding process. (b) The evolution of pressure distribution for the gas-assisted injection molding process. Note the lower but more uniform pressure distribution in case of gas-assisted injection molding [34]. (From *Innovation in Polymer Processing—Molding*, Stevenson, J. F., (Ed.) (1996), Hanser Publishers, Munich, p. 49, Fig. 2.4.)

it can effectively transmit the gas pressure, without a significant pressure drop, to the advancing gas–melt interface [cf. Fig. 9.13(b)]. As the gas advances toward the melt front, therefore, the pressure required to keep the melt ahead of the gas moving at the same velocity decreases, because the effective flow length decreases. The gas pressure required to fill the mold cavity can consequently be lower than the required entrance melt pressure for the conventional injection molding process. Further, the resulting pressure distribution is more uniform in a gas-injected part, which induces less residual stresses as the polymer cools down during the post-filling stage.

The gas-assisted injection molded part can therefore be produced with a lower gas pressure requirement (which normally leads to lower clamping tonnage) and has less tendency to warp. Because the part is internally pressurized with gas during the packing stage, all shrinkage is taken up on the inside of the part, which eliminated the sink marks on the surface of the part. Other advantages of gas-assisted injection molding include:

- Increased part rigidity due to enlarged cross-sections at hollowed gas-channels, especially for large structural parts (also called open channel flow parts)
- Reduced material consumption, especially for rodlike parts (also called contained channel gas flow parts)

- Reduced cooling time since the thick sections are cored out
- Greater design flexibility to incorporate both thick and thin sections in the same part for part consolidation

9.3.3 Process Disadvantages

Because gas-assisted injection molding involves dynamic interaction of two dramatically dissimilar materials flowing within typically complex cavities, the product, tool, and process designs are quite complicated. Previous experience with the conventional injection molding process is no longer sufficient to deal with this process, especially with injecting gas and controlling its penetration pattern. There will therefore, be a learning curve before one can fully benefit from this process. The design process, more specifically, has to be carried out with the process physics in mind in order to avoid problems associated with the dynamic nature of this process. In addition, the start up investment of gas-assisted injection molding includes both the cost of additional gas injection unit and source of nitrogen as well as the cost of licensing if a patented process is being used.

Typical molding problems associated with gas-assisted injection molding include:

- air trapping
- gas permeation into thin sections
- uneven gas penetration
- gas blow-through
- surface defects
- short shot
- surface blisters

9.3.4 Applicable Materials

Most of the thermoplastic materials, technically speaking, can be used for gas-assisted injection molding. Gas-assisted injection molding has been extended to thermosetting polyurethane [40] and powder injection molding [41]. Note that the material selection should be based on requirements of application performance, such as stiffness, chemical resistance, and material strength at the operating temperature. It has also been reported that the reproducibility of the polymer skin thickness is normally poor for polymers with a strong shear-thinning viscosity [42].

9.3.5 Typical Applications

Typical applications for the gas-assisted injection molding process can be classified into three categories, or some combination of them:

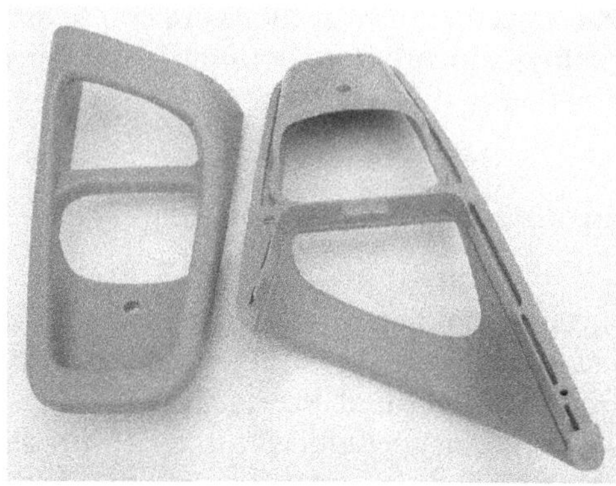

(a)

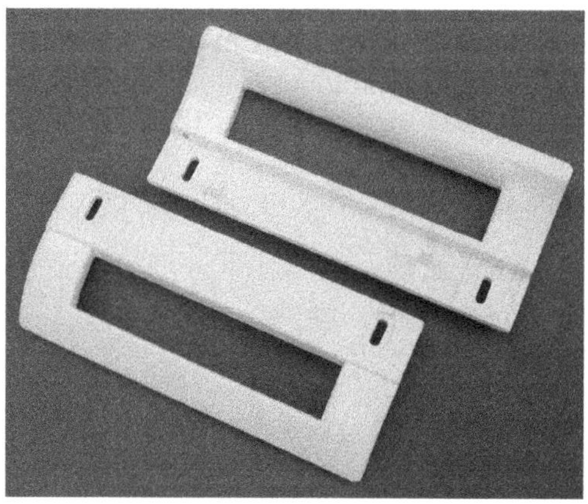

(b)

Figure 9.14 Gas-assisted door handles. [Parts courtesy of Battenfeld of America (a), West Warwick, Rhode Island, USA and Cinpres Limited (b), Tamworth, England.]

- Tube- and rodlike parts, where the process is used primarily for saving material, reducing the cycle time by coring out the part, and incorporating the hollowed section with product function. Examples are clothes hangers, grab handles (cf. Fig. 9.14), chair armrests, shower heads, and water faucet spouts.
- Large, sheetlike, structural parts with a built-in gas-channel network, where the process is used primarily for reducing part warpage and clamp tonnage as well as to enhance rigidity and surface quality. Examples are automotive panels, business machine housings, outdoor furniture (cf. Fig. 9.15), and satellite dishes.
- Complex parts consisting of both thin and thick sections, where the process is used primarily for decreasing manufacturing cost by consolidating several assembled parts into one single design. Examples are automotive door modules (Fig. 9.16), television cabinets, computer printer housing bezels, and automotive parts.

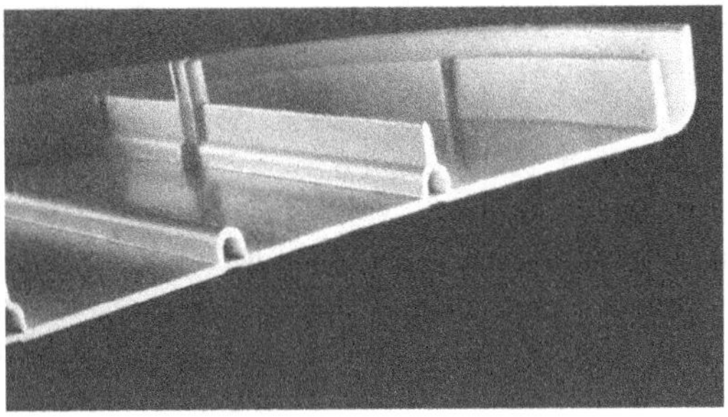

Figure 9.15 Cross-section of a garden tabletop. (Courtesy of Steiner Company, Austria.)

General design guidelines aimed at providing basic know-how in design and manufacturing of parts using gas-assisted injection molding can be found in the literature [34]. In addition, computer simulation for gas-assisted injection molding has also become available to help engineers gain process insights and make rational design decisions.

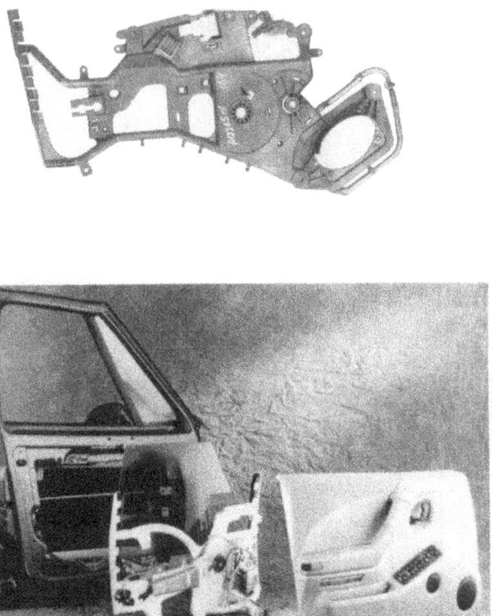

Figure 9.16 Automotive door modules (Courtesy of Delphi Interior & Lighting Systems) [25]. (From *Injection Molding Alternatives—A Guide for Designers and Product Engineers*, Avery, J. A., (1998), Hanser, Munich, p. 97, Fig. 5.3.18 and p. 98, Fig. 5.3.19.)

9.4 Injection-Compression Molding

Injection-compression molding (ICM) is an extension of conventional injection molding by incorporating a mold compression action to compact the polymer material for producing parts with dimensional stability and surface accuracy. In this process, the mold cavity has an enlarged cross-section initially, which allows polymer melt to proceed readily to the extremities of the cavity under relatively low pressure. At some time during or after filling, the mold cavity thickness is reduced by a mold closing movement, which forces the melt to fill and pack out the entire cavity. This mold compression action results in a more uniform pressure distribution across the cavity, leading to more homogenous physical properties and less shrinkage, warpage, and molded-in stresses than are possible with conventional injection molding.

9.4.1 Process Description

ICM is basically the same as conventional injection molding in the initial filling stage. After a preset amount of polymer melt is fed into an open cavity, a mold compression action is engaged and continues to the end of the molding process. Figure 9.17 illustrates both the initial injection stage and the subsequent compression stage. The compression can also take place when the polymer is being injected. Based on the process variation, ICM can be further classified into the following three categories [25]:

- Two-stage sequential ICM
- Simultaneous ICM
- Selective ICM

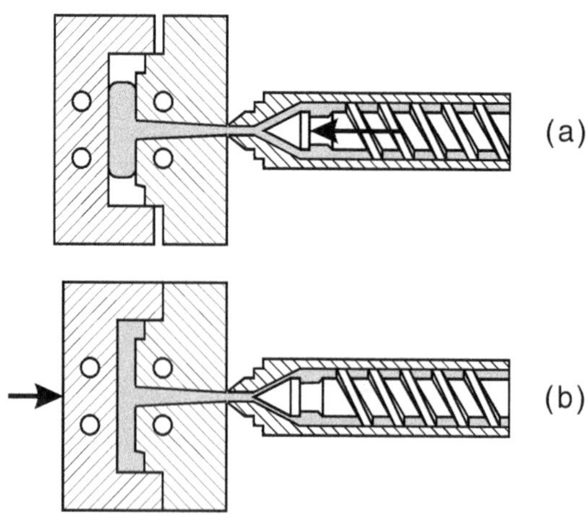

Figure 9.17 Typical injection-compression molding process sequence.

As the simplest type of these three techniques, the two-stage sequential ICM consists of separate injection stage and compression stage. During the injection stage, resin is injected into a cavity whose thickness is initially oversized by 0.5 to 10 mm greater than the nominal thickness. At the end of the resin injection, compression stage begins during which the mold cavity thickness is reduced to its final value. The mold compression action forces the resin to fill the rest of the cavity and, after the cavity is completely filled, provides packing to compensate for the material shrinkage due to cooling. A potential drawback associated with the two-stage sequential ICM is the "hesitation" or "witness" mark resulting from flow stagnation during injection-compression transition.

To avoid this surface defect and to facilitate continuous flow of the polymer melt, the simultaneous ICM activates mold compression while resin is being injected. The third technique, the selective ICM method, starts with the cross-section at the final nominal value. During the injection stage, melt pressure drives the mold back toward the cylinder, which is mounted to an unpressurized movable core. Based on cavity pressure or time, the compression stage is activated by pressurizing the cylinder to force the part-forming surface on the movable core to compress the melt [25].

A typical injection-molding machine with precise shot-volume control can be adapted for ICM; however, an additional control module is required for the mold compression stage. In addition, there are some requirements and suggestions for the injection-compression process and the molds [43]:

- A vertical flash face design is necessary to prevent uncontrolled leakage of the melt into the mold partition surface.
- A needle valve nozzle must be used to seal the cavity mechanically and to keep the compression delay time (from end of injection to beginning of compression) as low as possible to avoid hesitation marks.
- A mechanical stop is required in the hot runner system to ensure a precise metering of the melt to be injected into the mold. This way, it is possible to achieve a constant shot weight.

9.4.2 Process Advantages

The primary advantage of ICM is the ability to produce dimensionally stable, relatively stress-free parts, at a low pressure, clamp tonnage (typically 20 to 50% lower), and reduced cycle time. Recall that in conventional injection molding, a high injection and packing pressure has to be applied at the machine nozzle to produce sufficient pressure level at the extremities of the cavity to drive the flow and pack out the material. For thin-walled applications, such as compact disks, there is typically a significant pressure variation across the part due to high flow resistance. Such a high-pressure variation results in nonuniform packing and volumetric shrinkage within the part, leading to molded-in residual stresses and severe part warpage. With ICM the packing pressure is now applied in the thickness direction (as opposed to crossing

Table 9.5 Comparisons of Different Processes for Production of Thin-Walled Parts

Technology	Advantages	Drawbacks
Stretch-forming and vacuum thermoforming	Simple molds Low investment	Waste due to edge trim Wall thickness determined by the process
Dipping in solutions	Simple molds Low investment	Suitable for few plastics. Solvent vapors may be released Wall thickness determined by the process
Injection-compression molding	Any wall thickness possible Suitable for many plastics Short cycle times Good automation potential	High machine and mold costs Economic for high-volume products

the part dimension) for majority (if not all) of the part. As a result, ICM permits a lower and yet much more uniform pack/hold pressure distribution that can effectively pack out the mold and minimize molded-in residual stresses and part warpage.

Table 9.5 compares the ICM with other two processes commonly used for the production of thin-walled parts.

9.4.3 Process Disadvantages

The process disadvantages of ICM can be summarized as follows:

- The molds for this process are relatively expensive and subject to high wear during the compression stage.
- Additional investment is required for the injection molding machine; namely, the control module for the compression stage.
- This technology is only profitable for high-volume products, such as compact disks, or products that require minimum molded-in residual stresses, such as optical lenses.

9.4.4 Applicable Materials

For thin-wall applications, difficult-to-flow materials, such as polycarbonate and polyetherimide, have been molded to 0.5 mm [25]. On the other hand, high melt-flow-index polycarbonates are the most suitable materials for compact disks. In addition, most of the lenses are produced with polycarbonate due to its excellent optical properties. Other materials used in ICM include: acrylic, polyethylene, PPE/PA blends, and polypropylene, as well as thermoplastic rubber and most thermosetting materials [25].

9.4.5 Typical Applications

The ICM is the most suitable technology for the production of high-quality and cost-effective CD-Audio/ROMs. The ICM is also an appropriate technology for the production of many types of optical lenses. Figure 9.18 shows some of the typical ICM applications [44]. There has been a renewed interest in ICM for molding of thin-walled parts and for in-mold lamination (see Sec. 9.5).

9.4.6 Computer Simulation for Injection-Compression Molding

Several computer programs have been developed to simulate the ICM process (see, e.g., Refs. [45–47]). They use a Hele-Shaw fluid flow or 2.5-diameter-model to predict the melt front advancement and the distribution of pressure, temperature, and flow velocity dynamically during the injection melt filling, compression melt filling, and postfilling stages of the entire process. Based on the simulated results, it was found that ICM both shows a significant effect on reducing part shrinkage and provides much more uniform shrinkage within the molded part. The results have also shown that birefringence becomes smaller as the melt temperature increases, and as the closing velocity of the mold decreases. The flow rate and the mold temperature do not affect birefringence significantly. As far as the density distribution is concerned, the mold-wall temperature affects this parameter, especially near the wall. The flow rate, melt temperature, and mold closing velocity have insignificant effects on the density distribution [46].

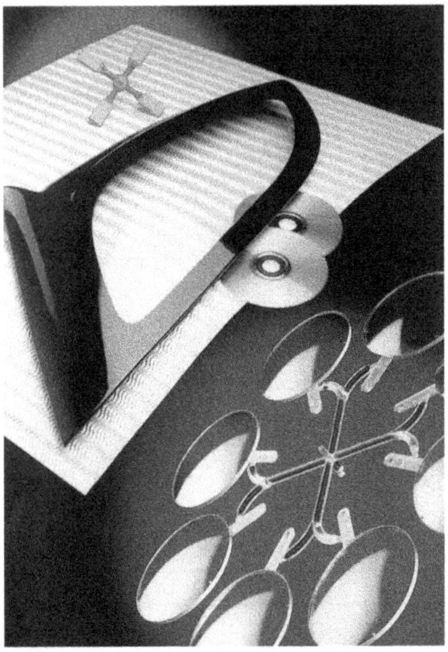

Figure 9.18 Typical injection-compression molding applications [44].

9.5 In-Mold Decoration and In-Mold Lamination

In-mold decoration (IMD) comprises insertion of a film or foil into the cavity, followed by injection of polymer melts on the inner side of the insert to produce a part with the final finish defined by the decorative film/foil. On the other hand, the in-mold lamination process resembles IMD except that a multilayered textile laminate is used. These two processes provide a cost-effective way to enhance and/or modify the appearance of product for marking, coding, product differentiation, and model change without costly retooling. The laminated component can have desirable attributes (e.g., fabric or plastic skin finish with soft-touch) or properties (e.g., electromagnetic interference, EMI, or radiofrequency interference, RFI, shielding).

9.5.1 Process Description

IMD

In the IMD process, a predecorated carrier laminated onto film stock from the roll is pulled through the mold and positioned precisely between the mold halves. The film stock may be decorated by printing methods (e.g., silk-screen or hot stamping) prior to molding [48]. During the molding stage, the polymer melt contacts the film and fuses with it so that the decoration can be lifted off from the carrier film and strongly attach to the surface of the molding. An injection molding machine with a typical IMD set up is shown in Fig. 9.19.

In one of the IMD techniques, the so-called paintless film molding (PFM) or laminate painting process, a three-layer coextrusion film with pigment incorporated into

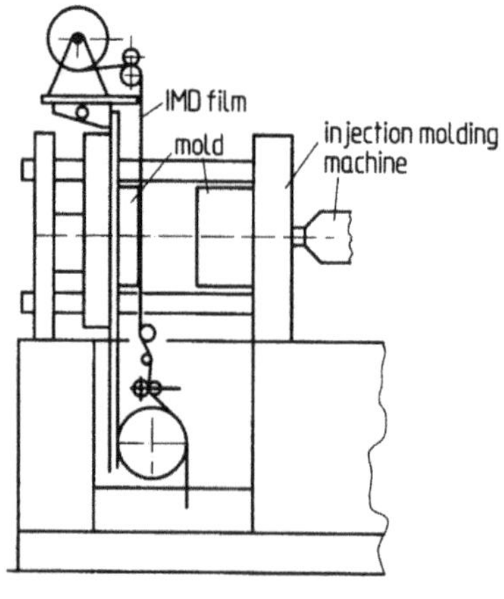

Figure 9.19 In-mold decorating setup with a foil-feeding device built into the injection molding machine [49]. (From *Injection Molding—An Introduction*, Pötsch, G. and Michaeli, W. (1995), Hanser, Munich, p. 175, Fig. 8.6.)

layers of clear-coat cap layers and core layer is first thermoformed into the shape of the finished part and then inserted into the cavity and overmolded with thermoplastics to produce a final part [50]. With this technique, it is possible to obtain a high-quality, extremely smooth paint finish on thermoplastic exterior body claddings and moldings ready for assembly without subsequent spray painting or finishing. Unlike in-mold coating and mold-in-color, this process provides high-gloss metallic and non-metallic finishes. It provides unique patterns and designs that are not feasible with paint. The paint laminate finish provides superior weatherability, acid etch resistance, and a safe worker environment because it is virtually pollution-free.

In addition to injection molding, IMD can be used with a variety of other processes, such as structural foam injection molding, ICM, compression molding, blow molding, thermoforming, resin transfer molding, and rotational molding [25].

In-Mold Lamination Process

Instead of using a thin film/foil as does IMD, the in-mold lamination process employs a multilayered textile laminate positioned in the parting plane to be overmolded by the polymer melt on the inner side. The decorative laminate can be placed in the mold as a cut sheet, pulled from the roll with needle gripper, or, by means of a clamping frame method. By means of a thermoforming operation, the clamping frame method allows defined predeformation of the decorative laminate during the mold-closing operation. In-mold lamination is also known as “laminate insert molding” [51] or “fabric molding” [52] for manufacturing automotive instrument panels and interior panels, respectively.

For in-mold lamination, the outer, visible layer of the decorative laminate can be made of polyester, PA, PP, polyvinyl chloride (PVC), or acrylonitrile-butadiene-styrene (ABS) film, cotton textile (woven, knitted, tufted, or looped fabrics), or leather. This outer layer typically comes with a variety of features to create an appearance or feel. In general, low surface texture is preferred because the ironing effect typically occurs with high pile or fiber loops. To provide the product with a soft-touch effect, there is typically an intermediate layer of polyurethane (PU), PP, PVC, or polyethersulfone (PES) foam between the top layer and the liner layer. Underneath the foam layer is the liner layer, which is used to stabilize the visible layer against shear and displacement, prevent the penetration of polymer melt into the intermediate layer, and provide thermal insulation against the polymer melt. This liner layer can be woven, knitted, or a nonwoven fabric. The typical structure of the in-mold laminated parts is shown in Table 9.6.

To avoid damage or undesirable folding of the laminate during molding, low injection pressure and low temperature are desirable. This makes low-pressure injection molding (see Sec. 9.8), ICM (see Sec. 9.4), compression molding, and cascade injection molding with sequential valve-gate opening and closing suitable candidates for in-mold lamination. Note that the latter process also eliminates the common problem associated with the weld lines, which are more evident with the IMD process.

Table 9.6 Typical Structure of the In-Mold Laminated Parts [53,54]

Structure	Typical dimension	Materials	Function
Coating (outer) material	Foil: 0.25 to 1 mm Textile: 0.25 to 4 mm	Foil: PVC, ABS, TPO Textile: PET, PA, PP	Decoration Tactile properties
Foam layer	1 to 3 mm	PU, PVC, PP	Soft touch Thermal insulation
Liner	0.5 to 1 mm	PET, PP	Thermal insulation, stabilizing coating material, increased tear resistance improved adhesion
Base (carrier, substrate) plastic material	2 to 3 mm	PP, PP-talc reinforced, ABS, ABS/PC blend	Define shape, give rigidity to part

9.5.2 Process Advantages

When compared with conventional surface decoration methods, such as painting, metallizing, hot stamping, PVC film laminating, and various painting methods, IMD and in-mold lamination offer a wide range of advantages, such as [55]:

- The process is a cost-effective way of surface decoration that replaces the traditional multistaged lamination process with a single molding cycle. The potential savings can amount up to 15 to 25%.
- The process is environmentally friendly with no volatile solvents released during adhesive lamination, and no posttreatment is needed.
- The decoration exhibits strong adhesion with the molding, as well as high surface-wear resistance and good chemical resistance.
- Use of the regrind is possible if permitted by the application and requirements.
- Three-dimensional decoration is possible within the limitations of the deformability of the film/foil and laminate.

9.5.3 Process Disadvantages

The disadvantages of IMD and in-mold lamination processes are additional equipment cost as well as extra steps for handling, die-cutting, performing, and placement of the decorative film/foil and laminate into the mold. Although conventional injection molding machines can be used with nonsensitive decorative materials, special machine equipment specifically made for these processes produces the best results and savings. Other disadvantages include longer cycle time (due to the insulating effect of the decorative layer), part warpage that results from unbalanced cooling, part rejection associated with damage, creasing, folding, shade changes, overstretching, and weld-line marks of the decorative layer.

9.5.4 Mold Design and Processing Considerations

To ensure the quality of the molded parts with decorative laminate, special considerations on designs of part and mold, selection of coating and base (carrier) materials, and set up of process conditions have to be taken into account, as we will discuss in the following sections.

Mold Design Considerations

Although the existing tools can often be adopted for IMD and in-mold lamination, it is generally recommended to design the tool specifically for these processes. In addition to the conventional design rules, such as appropriate radii and taper angle, the following design considerations have to be taken into account:

- For IMD, the film/foil must be carried into the mold between well-separated dowel pins, and run close against (but without rubbing) the mold surface.
- The design of the mold and the way the decorative laminate is fixed in the mold should ensure that the material is not overstretched during mold closing.
- Complex surfaces may result in problems caused by air entrapment or stretching of the film and laminate.
- The weld lines and sink marks associated with ribs become more evident, especially with thin films and textiles due to accumulation of decoration material at weld lines.
- The injection of polymer melt must be carried out in such a way that it will not adversely displace or deform the decorative laminate, resulting in creasing of the decoration, or damaging the surface structure of the film/foil or the three-dimensional structure of the textile.
- Because the decorative laminate typically sits on the moving platen, the part must be ejected from the sprue side to avoid leaving ejector pin mark on the decorated side, which faces the moving half of the mold.
- For IMD, the film/foil must be wide enough to cover all the cavities if a multi-cavity system is used.
- When decoration is three-dimensional, the extensibility of the IMD carrier film must not be exceeded.
- Venting should be provided between the film/foil and the moving mold half to prevent entrapped air and thus burning during the molding.
- Care must be taken to ensure that no melt reaches the display side of the decorative laminate, either through it or around the edges of the blank.
- The construction of the mold should allow automatic insertion of the decorative laminate and removal of the overmolded part.

Processing Considerations

The process conditions have to be changed slightly when running with IMD or in-mold lamination. For example, to avoid undesirable displacement of the decorative laminate, especially as melt first contacts the insert, the initial injection speed should be low. In addition, the injection rate should be set in such a way that the required injection pressure is at minimum. Note that for injection molding, if the injection pressure required to fill a cavity is plotted against the fill time, a U-shape curve typically results, with the minimum value of the required injection pressure occurring at an intermediate fill time. The curve is U-shaped because, on the one hand, a short fill time involves a high melt velocity and thus requires a higher injection pressure to fill the cavity. On the other hand, the injected polymer cools more with a prolonged fill time. This leads to a higher melt viscosity and thus requires a higher injection pressure to fill the mold. The curve shape of injection pressure versus fill time strongly depends on the material used, as well as on the cavity geometry and mold design. If the required pressure exceeds what can be tolerated by the decoration due to the flow-length/wall thickness ratios, special molding processes that permit low-pressure molding should be considered, as mentioned earlier [56].

For IMD or in-mold lamination, mold-wall temperature control is very important, because the decorative laminate is sensitive to the temperature. The melt temperature would normally exceed the maximum temperature that the decorative laminate can withstand. That is the reason why an additional backing layer is needed, which provides the insulation against the melt while reducing the possibility of melt breakthrough and folding. Meanwhile, because of the insulating effect of the decorative laminate, the mold-wall temperature on the decorated side should be set lower than the other side to promote balanced cooling and avoid part warpage.

9.5.5 Applicable Materials

As far as the base material is concerned, the processes are feasible with virtually all thermoplastics [57]. The largest volume resin used in in-mold lamination is PP, mostly for automotive applications. A wide range of other materials, including ABS, ABS/PC blends, PS, modified PPO, polyesters, PBT, PA 6, PA 66, and PE, have also been used successfully.

On the other hand, decorative laminate must have good thermal stability and resilience due to their exposure to high injection temperature and pressure. Color change that occurrs at the outer surface of the decoration typically results from thermal damage. In addition, the extensibility (stretching ability) of the decoration laminate is crucial for its applicability. Poor extensibility leads to tearing of the decoration laminate, whereas excessive extensibility results in overstretching that causes difference in surface brightness or show-through of the base material. Finally, compatibility and adhesion of the laminate with the base (carrier) resin are important considerations. The adhesion of the laminate to the part can be done through using a heat-activated adhesive layer or by melt penetration into a fabric backing to form a mechanical bond.

Because the recyclability of the laminated composite is becoming an important issue, the use of a single polyolefin-based system for the decorative, intermediate, and base layers seems to be a viable and cost-effective approach [58]. Moreover, due to the special characteristics of the processes, use of regrind as the base material is feasible. For example, recycled PP has been in favor by the automotive industry. It should be pointed out, however, that the presence of the metallic film in the laminate makes the recycling of decorated components more difficult.

9.5.6 Typical Applications

The IMD and in-mold lamination processes are suitable for many situations by providing integrated, single-step surface decoration and greater design freedom. For example, in-mold decoration has been used to produce rooftops, bumper fascia, and exterior mirror housings, as well as automotive lenses (cf. Fig. 9.20) and body-side molding with mold-in colors to help automakers eliminate costs and environmental concerns associated with painting. On the other hand, in-mold lamination has been used widely for automotive interior panels and other application areas. For example, Figs. 9.21 and 9.22 show the automotive interior panels and a shell chair featuring textile surfaces, respectively. These parts were made using a low-pressure in-mold lamination technique [35]. Finally, Table 9.7 lists some of the typical applications for

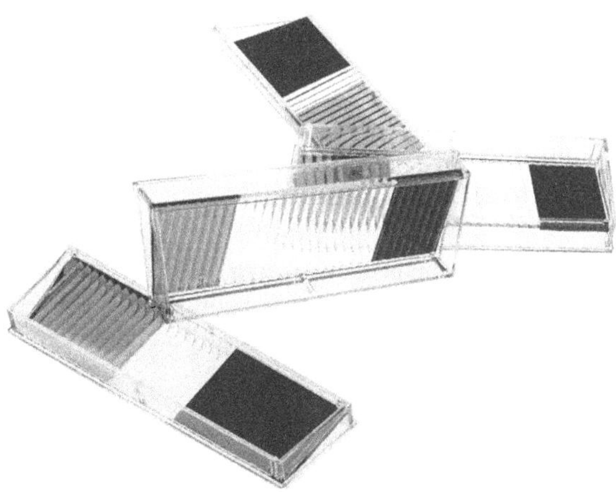

Figure 9.20 Automotive lenses with transparent film made by low-pressure in-mold decoration process. (Courtesy of Hettinga Technologies, Inc., Des Moines, Iowa, USA.)

Figure 9.21 Automotive interior panels molded with in-mold lamination using low-pressure injection molding. (Courtesy of Hettinga Technologies, Inc., Des Moines, Iowa, USA.)

Figure 9.22 A shell chair featuring a textile of pattern/strips has been successfully molded with in-mold lamination using a low-pressure injection molding technique called inverted force molding, IFM [35]. (Courtesy of Hettinga Technologies, Inc., Des Moines, Iowa, USA.)

Table 9.7 Typical Applications for In-Mold Decoration and In-Mold Lamination [25,55]

Application area	Molded parts
Automotive industry	Column cladding panels, consoles, door panels, headliners, heater/air conditioning controls, pillars, seat back panel, ventilation grilles, wheel covers, glove boxes, horn knobs, scale for tachometers, emblems
Household appliances	Appliance control panels (e.g., washing machine cover plates, front of microwave oven, toaster housing, etc.)
Telecommunication devices	Key pads and membrane switches for cellular phones
Radio industry	Cassette packs, covers for cassette players, video front plates
Sport equipment	Hockey sticks, water skis
Cosmetic industry	Caps for containers, powder-compact lids

various industrial sectors, which used to be manufactured by a labor-intensive and costly adhesive lamination process.

9.6 Insert and Outsert Molding

Insert molding is a process by which components consisting of metal or other materials are pre-placed in the cavity and then incorporated into a part as it is being molded. On the other hand, in the outsert molding process, the polymer melt is injected onto a metal frame (usually sheet metal) with holes punched out, at which the functional elements formed by the plastics are kept and anchored. For moderate to large production quantity, considerable savings in assembly and finishing can be realized by insert or outsert molding, especially with automated robot or machine-loading/unloading systems for insertion.

9.6.1 Insert Molding Process Description

For some applications, especially those in automotive, electric and electronics, medical, and fiberoptics, it is desirable to join plastics with metal parts. This can be done by the insert or outsert molding process. Insert molding is very similar to multicomponent injection molding (or overmolding, see Sec. 9.11) and in-mold decoration (see Sec. 9.5) in the way that an insert is pre-placed in the mold to be subsequently over-molded by the polymer melt. Note that for multicomponent injection molding, the insert is typically a premolded plastic component, whereas a film, foil, or fabric with desirable printing or color is used for IMD. Due to the similar insert approach, insert molding sometimes has been used to refer to these two processes. This section will focus on insert molding that incorporates inserts other than prefabricated plastic component, film, foil, or fabric.

For insert molding, polymer melt is injected into the mold where inserts have been preloaded. These insert components, such as fasteners, pins, rings, studs, terminals, thread inserts, and metallic strips that provide electric conductivity, are to be incorporated into the final molded parts. Figure 9.23 shows two of the typical parts made by insert molding. These inserts need to be loaded in the mold manually or by robot before polymer melt is injected into the cavity. The insert components are fixed in the mold by an undercut and/or by magnetic force [49]. The use of vertical insert molding machines with molds on a rotary table and automation support systems are quite common because they facilitate the loading and positioning of inserts. In addition, it can increase the production rate by concurrently loading the inserts in one mold and injecting the polymer in the other. Automation of insert loading also leads to faster cycle times, improved quality with precision, and enhanced production efficiencies.

Figure 9.23 A metal thread overmolded with plastic cap for better grip and an electrical coil component with molded-in metallic pins. (Parts courtesy of Engineering Industries, Inc. Madison, and Quantum Devices, Barneveld, Wisconsin, USA.)

One of the prospective developments of insert molding is to compete with transfer molding for encapsulation of integrated circuit (IC) chips. Such a new application is made possible by development of new series of epoxy molding compounds (EMC) that can stand extended residence time inside the barrel unit of an injection molding machine. An article on machine development on insert molding and this new IC packaging application can be found in Ref. [59].

9.6.2 Outsert Molding Process Description

The concept of outsert molding has been developed by exchanging the roles of metals with that of plastics in insert molding. The most common application is to inject polymer melt onto a metal frame (e.g., a sheet metal) preplaced and clamped in the cavity. The metal frame comes with punched-out holes where the plastic anchors and forms multiple functional elements, such as axles, bearings, bosses, guides, pins, spacers, shafts, snap connections, spiral and tongue springs, turning knobs, longitudinal and rotary movable elements. The possibility of having movable components on the sheet metal (e.g., a deflectable spring leaf, a sliding element movable in the longitudinal direction, or rotary turning knobs) result in considerable savings compared with metal components. Such a composite building method combines necessary load-bearing strength from the metal frame with design freedom from the material and processing advantages of thermoplastics. In the meantime, the economic savings may amount up to 40% compared with conventional manufacturing methods, which individual prefabricated elements of plastics or other materials have to fit with the sheet metal [60].

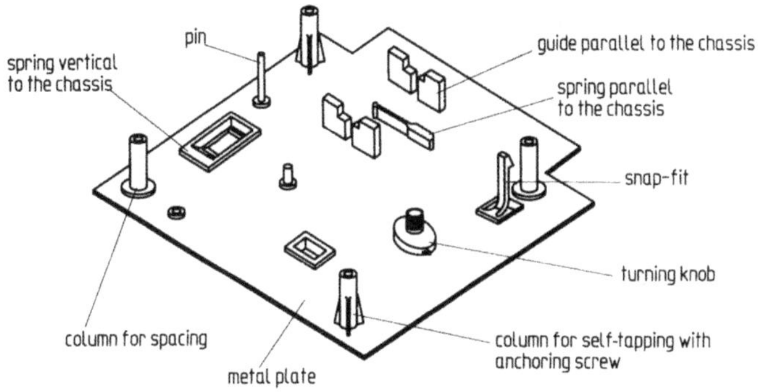

Figure 9.24 Various elements fabricated by outsert molding. (From *Injection Molding—An Introduction* (1995), Pötsch G. and Michaeli, W., Hanser, Munich, p. 181, Fig. 8.14.)

In outsert molding, the plastic elements to be molded onto the metal frame are connected by a distributor spider with restricted gating or, if the distance is too small, by channels so that they can be molded with a single shot. These channels are either detached after removal of the article from the mold or remain on the metal frame, provided that they do not interfere with the product functions. Due to a larger thermal expansion coefficient of the plastics as compared with that of metals, care should be taken so that the shrinkage of these connecting channels does not deform the metal frame or cause the plastic elements to break. With such a multicavity system, dozens of functional elements can be incorporated on the metal frame base. As an illustration, Figs. 9.24 and 9.25 show the various elements fabricated by outsert

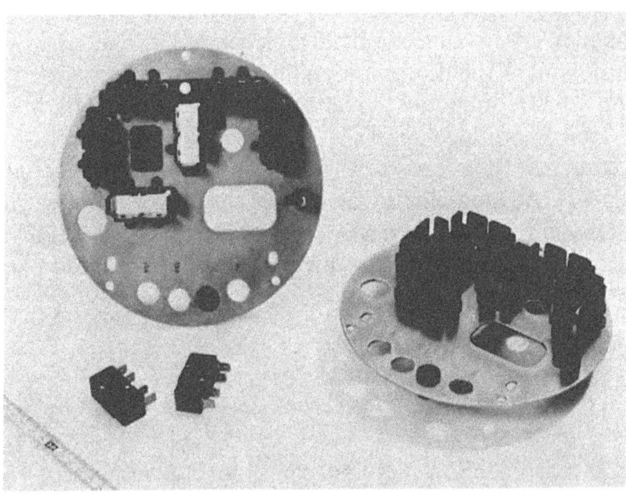

Figure 9.25 Building elements of a tube tester made with outsert molding [60].

molding and the building elements of a tube tester made with outsert molding, respectively.

In terms of the metal frame base, the sheet metals are usually 1- to 2-mm thick with thickness tolerance being smaller than 0.1 mm. Burrs at the edges of punched holes should be removed. Prior to injection molding, the sheet metal must be degreased and flattened. Two drilling holes with a distance larger-than-half of the sheet metal dimension are used for accurate positioning of the sheet metal inside the cavity. Expansion of the sheet metal should be taken into account as the metal touches the hot mold. It is recommended that one drilling hole be carried out as an oblong hole.

9.7 Lamellar (Microlayer) Injection Molding

Lamellar (microlayer) injection molding uses a feed-block and layer multipliers that combine melt streams from dual injection cylinders to produce shot volume in distinct microlayer morphological structure to be directly injection molded. The resulting parts that combine different resins in a layered structure exhibit enhanced properties, such as the gas barrier property, dimensional stability, heat resistance, and optical clarity, compared with parts made of single-phase resin or conventional polymer blends.

9.7.1 Process Description

Lamellar injection molding technology has been available since early 1990s [61]. It has been developed to create a mechanically micron-scale, layered morphology blend of two or more polymers for direct injection molding. The production of multimicrolayers is done via simultaneous injection of two components at predetermined rates in a feed-block to produce an initial, sandwiched layer structure (i.e., A–B–A). This resultant streams then passes through a series of layer multipliers that repeatedly subdivide and stack the layers to increase layer number and reduce individual layer thickness (cf. Figs. 9.26–9.28). At the end, the lamellar morphology is mechanically retained and remained in the final injection molded parts, as shown in Fig. 9.29 [62].

9.7.2 Process Advantages

Lamellar (microlayer) injection molding combines advantages of coextrusion and injection molding by coupling simultaneous injection with layer multiplication to produce a finely subdivided microlayer melt stream into complex multicavity molds

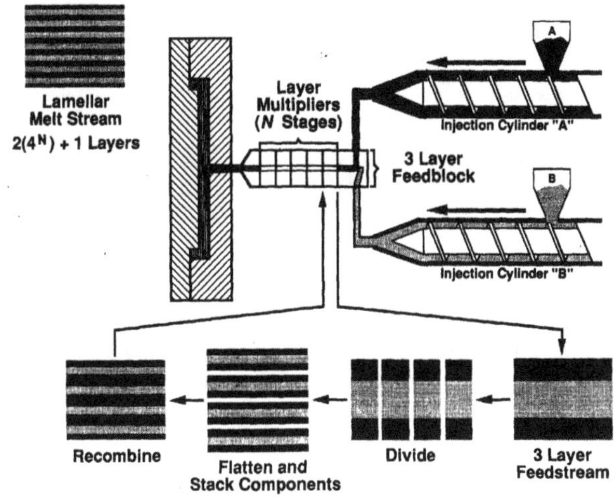

Figure 9.26 Schematic of lamellar injection molding showing the creation of micron-scale, layered morphology blend through the layer multipliers (see Figs. 9.27 and 9.28). (From *Innovation in Polymer Processing—Molding*, Stevenson, J. F., (Ed.) (1996), Hanser, Munich, p. 297, Fig. 8.2.)

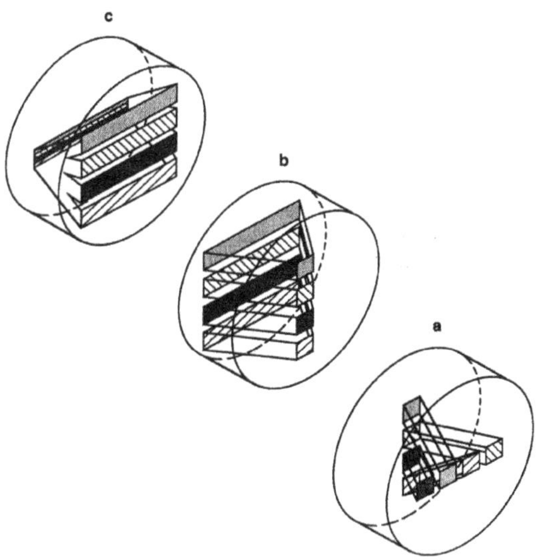

Figure 9.27 Three-piece layer multiplier assembly: (a) subdivision and layer rearrangement, (b) spreading of individual substreams to reestablish the original width, (c) recombination of individual substreams in a contraction zone that reestablishes the original cross-section. (From *Innovation in Polymer Processing—Molding*, Stevenson, J. F., (Ed.) (1996), Hanser, Munich, p. 300, Fig. 8.4.)

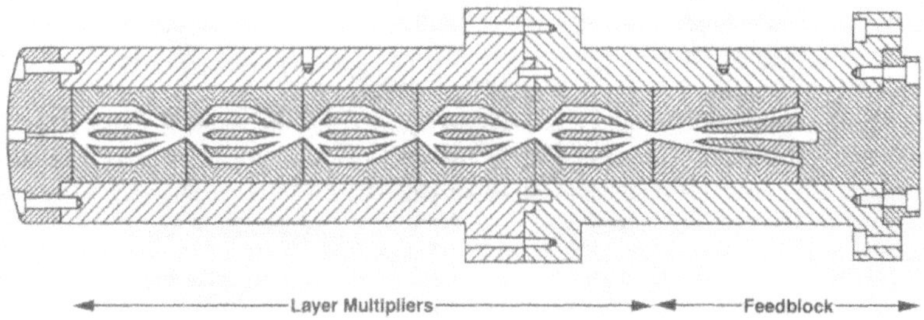

Figure 9.28 Three-layer feed-block and layer multiplication assembly contained in a tubular sleeve. (From *Innovation in Polymer Processing—Molding*, Stevenson, J. F., (Ed.) (1996), Hanser, Munich, p. 300, Fig. 8.5.)

without being restricted to circularly symmetric parts or metering individual mold cavities. In terms of producing continuous lamellar structure, it provides a superior alternative to conventional polymer melt blending technique for which the resulting morphology strongly depends on the material properties of the constituent materials and the processing parameters. It has been reported that one or two orders of magnitude reductions in oxygen permeability can be realized with lamellar injection molding relative to monolayer PET and conventional blends [62]. In addition, a 200-fold reduction in hydrocarbon permeability with HDPE-polyamide (nylon) 6 micro-layer structure compared with HDPE has been achieved. Furthermore, significant enhancement in chemical and temperature resistance, dimensional stability, and optical clarity were also made possible through lamellar injection molding.

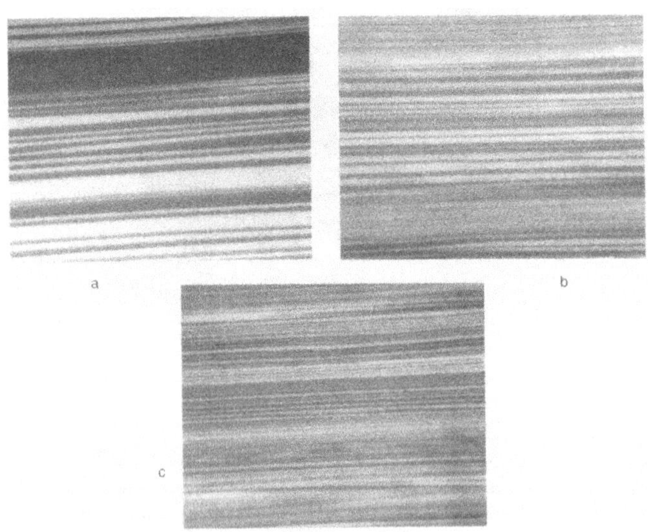

Figure 9.29 Micrograph of multilayer cross-sectional structure produced by the lamellar injection molding process. (From *Innovation in Polymer Processing—Molding*, Stevenson, J. F., (Ed.) (1996), Hanser, Munich, p. 306, Fig. 8.12.)

9.7.3 Process Disadvantages

The disadvantages of lamellar injection molding are the initial equipment cost and learning curve to acquire process know-how, which are typical of adopting any new technology. Major challenges of this process include determination and control of proper lamellar structure for the end application, effects of fountain flow behavior at the melt front (cf. Chap. 3), and high shear rate during injection molding on the break up and coalescence of lamellar structure, adhesion of incompatible polymers, and recycling of process scraps and postconsumer parts.

9.7.4 Applicable Materials

Lamellar injection molding has been employed with diverse combinations of polymers, especially with multiphase systems that combine commodity or engineering resins with a small quantity (2 to 20%) of barrier resin. Mutually reinforced composite properties can also be attained with proper selection of materials and morphology. Potential materials for lamellar injection molding applications are listed in Table 9.8.

Table 9.8 Potential Materials for Lamellar Injection Molding Applications [62]

Material combinations	Target benefits
PC-PET	Clarity, barrier, and chemical resistance
PC-PBT	Temperature and solvent resistance
Polyolefin-adhesive-EVOH	Gas and solvent barrier
PET-PEN	Barrier and temperature resistance and clarity
Polyolefin-adhesive-polyamide	Gas and solvent barrier
Transparent impact modified polystyrene-polyamide	Gas barrier and clarity
ABS-thermoplastic polyurethane	Solvent resistance
PC-thermoplastic polyurethane	Solvent resistance, temperature resistance, and optical clarity
Combinations of filled and unfilled polymers (e.g., PP-filled PP, PC-filled PC, PC-filled ABS, ABS-filled PC)	Preferred balance of properties, coefficient of linear thermal expansion, toughness, and temperature resistance
Combinations of brittle and ductile polymers	Mutually reinforced mechanical properties
Combinations of engineering thermoplastics with thermotropic liquid crystalline polymers (T-LCP)	Reinforcement with T-LCP
Brominated PC copolymers with ABS	Ignition resistance
Combinations of virgin and post-consumer recycled resins	Material property modification and cost savings

9.7.5 Typical Applications

Lamellar injection molding is believed to be advantageous over other competing processes in diverse application areas. In particular, potential application areas for lamellar injection molding technology include: (1) structural parts that require temperature resistance, chemical resistance, and/or dimensional stability, (2) houseware and durable goods that require optical clarity and/or temperature and solvent resistance, (3) health-care devices and components, (4) food and chemical packaging products that require barrier resistance, and (5) automotive components such as fuel and hydraulic systems [62].

9.8 Low-Pressure Injection Molding

Low-pressure injection molding (LPIM) is essentially an optimized extension of conventional injection molding. This process differs from traditional (high-pressure) injection molding in that it integrates a series of innovative practices that keep the injection pressure (and, consequentially, the clamp force) at the lower limits. The major benefits of LPIM include significant reduction of the clamp force tonnage requirement, less-expensive molds and presses, and lower molded-in stress in the molded parts. It also facilitates reduction of manufacturing cost by incorporating decorative film and/or textile into the molded parts.

9.8.1 Process Description

LPIM can be achieved by properly programming the screw rotation speed and the plasticating back pressure to control the melt temperature profile of the shot volume as well as by precisely profiling the injection speed and pressure to maintain a generally slow and controlled melt front velocity inside the tool. It also uses a generous gate size, a novel type of film gate, and/or valve gates that open and close sequentially based on volume of material injected to reduce the flow length while eliminating the weld lines. The packing stage is eliminated after complete injection of an exact shot volume with no shot cushion. These practices have been implemented into a number of processes by Hettinga [35]. Notable LPIM processes from these innovations are the Inverted Force Molding (IFM®), which maintains a constant melt front speed during injection, and the low-stress Thermoplastic Solid Molding (TSM™) for molding solid thermoplastics or molding with delicate films and textiles. By employing advanced electronics and hydraulics married with sophisticated software control, the process can maintain a solid, fully plasticized melt front with low pressure. The following sub-sections describe the characteristics of low-pressure injection molding process.

Control of Injection Temperature Profile

In LPIM, it has been indicated that nearly all of the heat required for mold filling comes from plastication as a result of the gentle injection speed and generous gate size [63]. It is different from conventional injection molding for which a substantial portion of the heat is generated by viscous heating. The control of shot volume temperature profile is done by manipulating the screw rotation speed (RPM) and the back pressure at the screw tip during the plastication stage. Note that the effective L/D ratio of the reciprocating screw decreases (by a length of generally 1 to 5 diameters) as the screw retracts in preparation for the shot volume. To compensate for the changing plastication length of the screw as it rotates and moves backward, an electrohydraulic device is employed to generally increase the back pressure at the screw tips to enhance the mixing and shearing (cf. Fig. 9.30). The screw RPM speed also follows a ramps-up-and-down profile to work with the back pressure for building up the desirable temperature profile within the shot volume (cf. Fig. 9.30).

Control of Injection Pressure and Injection Rate

The injection pressure profile is set in such a way that it delivers a controlled injection speed during filling. As illustrated in Fig. 9.30, the injection rate starts slowly to assure an even flow into the mold. Once the melt enters the cavity, the rate is set so that the melt front travels at the same speed throughout the injection stage. The injection pressure profile is generally in the shape of an inverted U (cf. Fig. 9.30), which

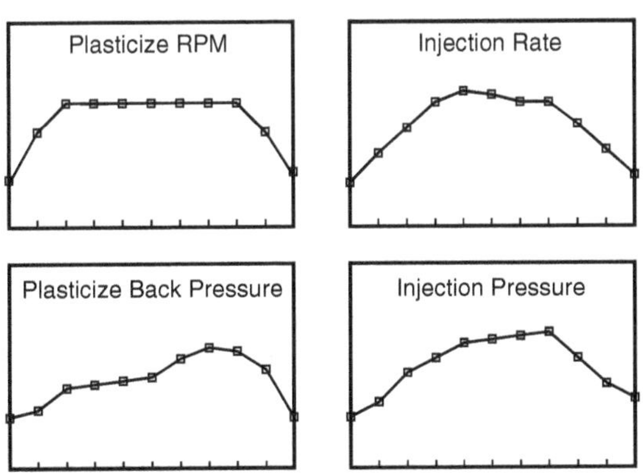

Figure 9.30 Typical profiles for screw RPM speed, back pressure, injection rate, and injection pressure for low-pressure injection molding.

reduces the rapid build up in the clamp force at the end of the cycle commonly seen in conventional injection molding [64].

Because the material solidifies almost immediately as it comes into contact with the cold mold wall, the velocity at the melt front determines the flow-induced stress and the degree of molecular and fiber orientation in regions adjacent to the part surface. Variable orientation within the part, as a result of changing velocity at the melt front during filling, leads to differential shrinkage and, thus, part warpage. It is desirable, therefore, to maintain a constant velocity at the melt front to generate uniform molecular and fiber orientation throughout the part.

Accurate Shot Size Control with No Cushion and Packing Stage

LPIM does not employ distinct high-pressure packing and holding stages as does conventional injection molding. The exact shot size with no cushion is injected into the mold in a carefully controlled, even fashion. Setting the correct injection volume is nevertheless important because a very minimum of packing is necessary for filling the cavity safely. Because the gate does not freeze off after the mold cavity fills completely, some compensation for thermal shrinkage is possible so there is no need for high-pressure packing and holding. It is conceivable that voids or sink marks will occur at thicker sections, ribs, and the like, due to the fact that compensation for the polymer volumetric shrinkage is not complete. For molding with approximately constant wall thickness, however, LPIM is thoroughly suitable [57]. In addition, it is claimed that the LPIM generally allows material to be molded at lower temperature. This results in comparable cycle time, even if the cavity filling is slower.

Multistation Injection Molding Machine

The lower pressure and clamp force requirements permit the use of less expensive tools. Because the process requires only one quarter to one third as much clamp tonnage for molding most parts, aluminum rather than steel can be used for tooling. In addition, the control hardware and software allow a single injector to serve multiple mold stations with different parts of various size, shape, weight, and configuration. As an example, the six-station, LPIM machine shown in Fig. 9.31 permits different parts to be molded consecutively without retuning the machine.

Multi-Point Volumetric Injection Control™

Multi-Point Volumetric Injection Control™ optimizes mold filing of large parts and highly viscous materials by sequentially opening and closing valve gates at strategically selected locations. The timing of gate opening and closing is controlled by volume injected instead of using standard time-based controls as the other cascading (or sequential valve-gate control) injection. The advantage of timing-by-volume approach is that the switching process of the hot runners are always connected to the current stage of filling, regardless of the speed changes required by the process. As an illustration example, Fig. 9.32 shows the mold filling in a bumper fascia. At first,

Figure 9.31 Multistation injection molding machine. (From *Innovation in Polymer Processing—Molding*, Stevenson, J. F., (Ed.) (1996) Hanser Publishers, Munich, p. 204, Fig. 5.7. Courtesy of Hettinga Technologies, Inc., Des Moines, Iowa, USA.)

resin is injected through the center gate. After the resin flows past the two downstream gates, it triggers the opening of those two gates, followed by closing of the first gate shortly. The process continues until the entire cavity is filled. It is conceivable that the use of multiple gates reduces the flow length, thereby, reducing the injection pressure requirement [65]. In addition, the sequential opening and closing of the gates eliminate the formation of weld lines. If necessary, the melt injection speed profile can be adjusted from gate to gate. For situations where packing pressure has to be used, all the valve gates can be opened again at the end of the injection stage.

Laminate Molding with Textiles, Foils, or Films

In-mold lamination or simply laminate molding is also known as composite or back molding. This process involves molding of plastics over layers of textile, film, or other materials preplaced inside the mold to form a laminated structure (see Sec. 9.5 for detailed descriptions on in-mold lamination). The inclusion of textile or film provides

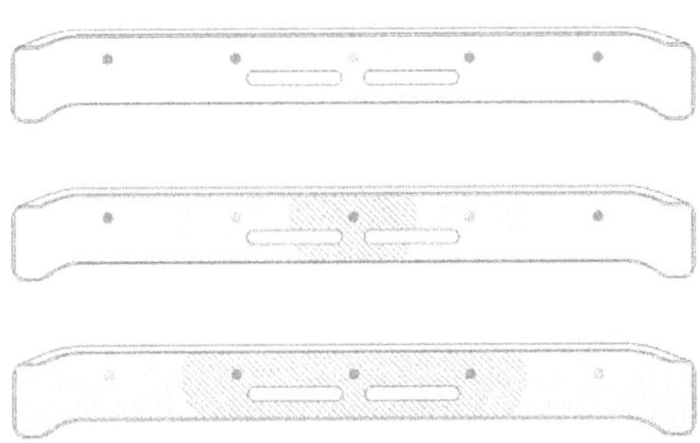

Figure 9.32 A bumper fascia being filled by the Multi-Point Volumetric Injection Control^{TM} or MPVIC Control. (Courtesy of Hettinga Technologies, Inc., Des Moines, Iowa, USA.)

an insulation layer that allows plastics melt to be injected at a slower rate to avoid the damage or shift of the textile or film. Laminate molding is one of the most successful applications with low-pressure molding because it eliminates the need for multiple-layer fabrics and special glue backings on textiles. A study has shown that the one-step laminate process offers weight reduction (12%) and cost saving (64%) compared with the traditional method of gluing the fabric to the molded part [63].

9.8.2 Process Advantages

The LPIM generally offers the following process advantages:

- Reduced tonnage requirements and thus lower cost molds
- Lower stresses in the molded parts and improved material properties
- Molded-in fabric or film capabilities
- Large part capability and material savings of 5 to 8%
- Lower melt temperatures and shorter overall cycle times
- Lower gate and cavity pressures
- Reduced tooling costs and overall manufacturing costs

9.8.3 Process Disadvantages

Because the various practices and techniques employed by low-pressure injection molding favorably reduces the injection pressure requirement, this process does not introduce any adverse process disadvantages for its intended applications.

9.8.4 Applicable Materials

LPIM is well suited to processing a broad range of materials, such as thermoplastics, thermosets, polymer alloys and blends, filled materials, recycled thermoplastics, and even rubber. In particular, polypropylene is used extensively with the LPIM due to the lower cost and improved physical and mechanical properties. There are a broad range of compatible textiles and films that can be molded with polypropylene using in-mold lamination. For fiber-filled materials, the generous gate size reduces the possibility of fiber breakage.

9.8.5 Typical Applications

In addition to the in-mold lamination components (cf. Figs. 9.21 and 9.22), the application of LPIM includes a variety of automotive components ranging from dashboards, consoles, interior panels (cf. Fig. 9.33), kick panels, and visors to glove boxes, lenses (cf. Fig. 9.20), headliners, bumpers, and trims. Other nonautomotive applica-

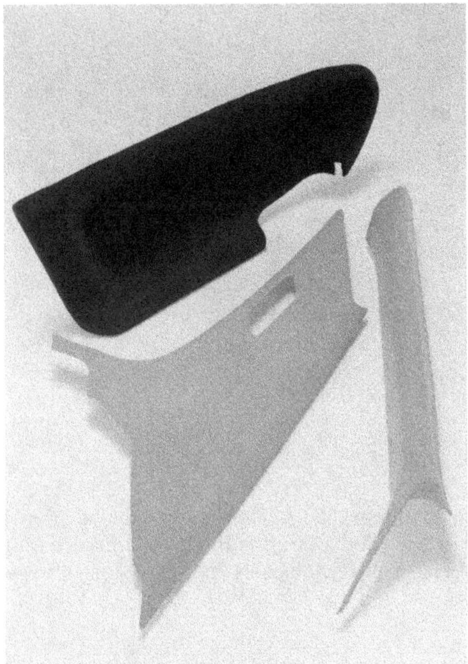

Figure 9.33 Automotive interior panels that eliminate heavy backed, expensive textile. (Courtesy of Hettinga Technologies, Inc., Des Moines, Iowa, USA.)

tions are material-handling containers (cf. Fig. 9.34), shipping pallets, business machine housings (cf. Fig. 9.35), lavatories (cf. Fig. 9.36), and building panels. Potential large applications include truck cabs and boat hulls.

9.9 Microinjection Molding

Microinjection molding (also called micromolding) produces parts that have overall dimension, functional features, or tolerance requirements that are expressed in terms of milli- or even micrometers. Due to the miniature characteristics of the molded parts, it requires a special molding machine and auxiliary equipment to perform tasks such as shot volume control, evacuation of mold (vacuum), injection, ejection, inspection, separation, handling, deposition, orientation, and packaging of molded parts. Special techniques are also being used to make the mold inserts and cavities.

9.9.1 Process Description

The demand for miniature injection molded parts and the equipment and processing capability to produce them with desirable precision began around 1985 and has been

Figure 9.34 An assortment of thin-walled kitchen-wares molded from high-density linear polyethylene, HDLPE. Each part was produced using the Inverted Force Molding (IFM) process on a 275-ton press. (Courtesy of Hettinga Technologies, Inc., Des Moines, Iowa, USA.)

growing ever since [66]. Among the various micromolding processes (see, e.g., [67,68]), injection molding possesses the advantages of having a wealth of experiences available in conventional plastics technology, standardized process sequences, and a high level of automation and short cycle times [69].

Figure 9.35 Business machine housings molded with Thermoplastics Cellular Molding, (TCM®) process. (Courtesy of Hettinga Technologies, Inc., Des Moines, Iowa, USA.)

Figure 9.36 Lavatory sink molded with Inverted Force Molding (IFM), which permits high gloss decorative films to be molded across the entire surface without damage or distortion of color on deep draws of film. (Courtesy of Hettinga Technologies, Inc., Des Moines, Iowa, USA.)

Microinjection Molding Classifications

Although there is no clear way to define microinjection molding, applications of this process can be broadly categorized into three types of products or components [70]:

- Microinjection molded parts (micromolding) that weigh a few milligrams to a fraction of a gram, and possibly have dimensions on the micrometer (μm) scale [e.g., micro-gearwheels, micro operating pins (cf. Fig. 9.37)]
- Injection molded parts of conventional size but exhibit microstructured regions or functional features (e.g., compact disc with data pits, optical lenses with micro-surface features, and wafer for making microgearwheels with Plastic Wafer Technology) [72] (cf. Fig. 9.38)
- Microprecision parts that can have any dimensions, but have tolerances on the micrometer scale (e.g., connectors for optical fiber technology)

Machine Requirements

Due to the previously mentioned characteristics, these applications all require special attention to the machine equipment, processing, making of molds, and the like.

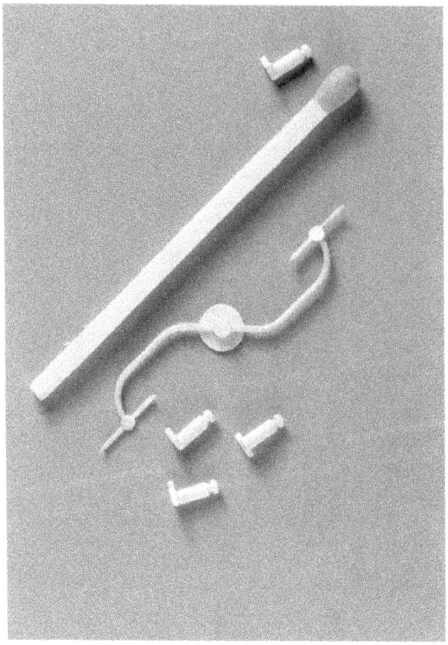

Figure 9.37 An microinjection molded operating pin by HB-Plastic GmbH, Korneuburg, Austria [71].

Although modern conventional injection molding machines can achieve impressive results, these machines must be adapted to meet the special requirements of microinjection molded parts such as [69]:

- Small plasticating units with screw diameters ranging from 12 to 18 mm and a shorter screw length with L/D ratio around 15 to avoid material degradation from prolonged residence (dwell) time
- Precise shot volume control and desirable injection rate

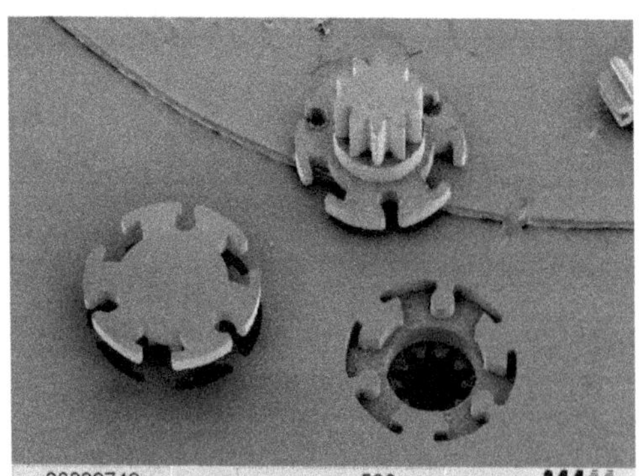

Figure 9.38 Microgearwheels manufactured by means of Plastic Wafer Technology [72].

- Repeatable control such as fill to pack switchover based on screw position or cavity pressure (preferred)
- Capability to raise the mold-wall temperature to such a level (sometimes slightly above the melting temperature of the polymer) to avoid premature solidification of ultrathin sections
- Mold evacuation if the wall thickness of the microinjection molded part is down to 5 μm, which is the same order as the dimension of vent for air to escape
- Shutoff nozzle to avoid drooling from the nozzle due to the high processing temperature of the melt
- Precise alignment and gentle mold-closing and -opening speeds to avoid deformation of delicate microinjection molded parts
- Special handling technique to remove and molded parts for inspection and packaging
- Possible local clean room enclosure or laminar flow boxes to avoid contamination of microinjection molded parts

It is apparent that with diminishing part volume and shot size, the conventional injection molding machine is no longer an economically viable solution for microinjection molding applications. As a result, many "micromolding machines" have been developed and have become commercially available (see, e.g., [66]). In addition to the various features described earlier, these machines are sometimes equipped with separate metering and injection ram (pistons) and screw design. These features are targeted to meter the shot volume accurately and to eliminate problems associated with material degradation resulting from by-pass and dead corners with the conventional injection screw. Because the size and weight of the microinjection molded parts differ significantly from the conventional parts, certain steps have to be taken to ensure proper part ejection. For example, a vision system can be installed in the molding machine to confirm ejection of the miniature parts. In addition, part removal can be performed by suction pads, which keep parts separate and oriented for quality control and packaging, or by electrostatic charging or blowing out. Traditional quality control methods, such as measuring the part weight, becomes impractical with microinjection molding. New quality control techniques employ a video inspection system for sorting acceptable and unacceptable parts. Because microinjection molded parts are usually part of the assembly, they are packaged in an oriented way ready for assembly.

Mold Making (Tooling)

Traditional methods of tooling, such as various machining methods and electrical discharge machining (EDM), can be used for making the mold for microinjection molded parts; however, they have quickly reached their limitations with decreasing dimensions of mold inserts and cavities. Existing technologies in the field of microelectronics have been employed to fabricate mold inserts and cavities for microinjection molding. One common practice is the LIGA technique, which is an acronym derived from the German words for deep x-ray lithography, electroforming (electroplating), and injection molding replication [73,74]. Other processes include microcutting, ultraprecision machining, laser machining, and micro-EDM technologies [69].

Molding Process

Microinjection molded parts typically have features on the order of a few micrometers to millimeters, and aspect ratios between 1 and 100. Several requirements are needed in order to reproduce these features with high fidelity. To be specific, the mold needs a special heating and cooling system. A few systems have been developed to provide the desirable dynamic mold temperature control. For example, the "variotherm" process control employs two oil circuits at different temperatures that heat up and cool the mold at filling and cooling stages, respectively [75]. On the other hand, the induction heating technique generates a peak mold temperature using an inductive system prior to injection [76]. Moreover, satisfactory results have been reported using electric cartridge heaters to control the mold temperature [77]. The use of adapted mold sensors, high-precision mold guides, mold evacuation system, integrated runner pickers and part removal robots, automatic gate-cutting systems, and mold cleaning systems that activate during each cycle are all vital to controlling the manufacturing processes and efficient handling and packaging of microinjection molded parts.

To ensure proper cavity filling, high injection speeds and pressures, maximum permitted melt temperatures, and elevated mold-wall temperature controls are needed. Large runner and sprues are used to have a shot size large enough to control the process and switchover reliably while moving polymer melt through the machine system to avoid degrading the material.

9.9.2 Process Advantages

The microinjection molding is intended for producing miniature parts. It does not compete with other macroinjection molding processes.

9.9.3 Process Disadvantages

Because ultrasmall parts do not weigh enough to register on a machine control system, some system uses oversized runners so that the machine can accurately set and monitor part production. Under these conditions, the runners can constitute up to 90% of the total shot weight. Waste is considerable because the material in the runners cannot be recycled most of the time for microinjection molding application. Finally, due to the typically high surface-to-volume ratios of the parts, the mold has to be heated to temperatures greater than the melting temperature during injection to prevent premature solidification, which leads to prolonged cycle time.

9.9.4 Applicable Materials

Almost any material suitable for macroscopic molding can be micro-molded. Materials reported for microinjection molding include POM, polycarbonate (PC), polymethylmethacrylate (PMMA), PA, liquid crystalline polymers (LCP), polyetherimide, and silicone rubber. Reaction injection molding has also been applied using material on the basis of acrylate, amides, and silicones [67].

9.9.5 Typical Applications

There is a rapidly growing demand for microinjection molded parts in such sectors as optical telecommunication, computer data storage, medical technology, biotechnology, sensors and actuators, microoptics, electronics, and consumer products, as well as equipment making and mechanical engineering. Examples of microinjection molded parts include watch and camera components, automotive crash, acceleration, and distance sensors, read–write heads of hard discs and CD drives, medical sensors, micropumps, small bobbins, high-precision gears, pulley, and helixes, optical fiber switches and connectors, micromotors, surgical instruments, and telecommunication components [66,78].

9.10 Microcellular Molding

Microcellular molding (also commercially known as MuCell process) blends "supercritical" gas (usually nitrogen or carbon dioxide) with polymer melt in the machine barrel to create a single-phase solution. During the molding process, the gas forms highly uniform microscale cells (bubbles) of 0.1 to 10 μm in diameter, and the internal pressure arising from the foaming eliminates the need of packing pressure while improving the dimensional stability. The original rationale is to reduce the amount of plastic used without sacrificing the mechanical properties. Because the gas fills the interstitial sites between polymer molecules, it effectively reduces the viscosity and glass transition temperature of polymer melt; therefore, the part can be injection molded with lower temperatures and pressures, which leads to significant reduction of clamp tonnage requirement and cycle time.

9.10.1 Process Description

Microcellular plastics (MCPs) are single-phase, polymer-gas solutions by dissolving or saturating a polymer in a supercritical fluid (CO_2 or N_2) and then precipitating it by triggering polymer nucleation by adjusting process conditions, such as pressure

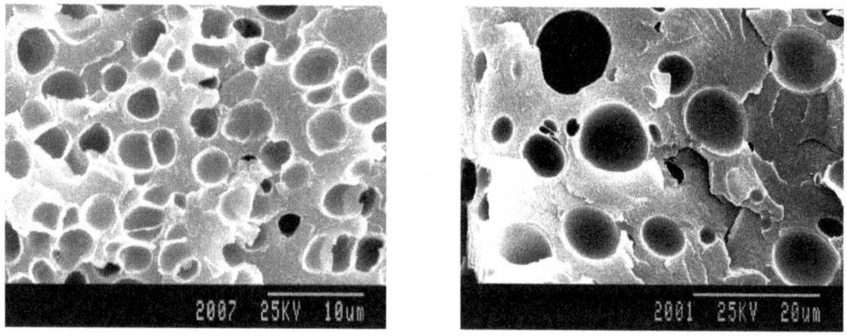

Figure 9.39 Micrographs of microcellular polystyrenes prepared under different conditions. (Photos courtesy of Institute of Chemistry, The Chinese Academy of Science.)

and temperature. The resulting MCPs are plastic foams with cell diameter sizes from 0.1 to 10 μm (cf. Fig. 9.39) and cell densities from 10^9 to 10^{15} cells per cubic centimeter. MCPs were initially conceived by Suh and co-workers at MIT as a means of reducing material consumption in mass-produced plastic parts [79]. The underlying rationale is to create enough voids smaller than the pre-existing flaw in polymers so that the amount of plastic used could be reduced without compromising the mechanical properties. Even though realizing a part weight reduction of 5 to 95% by replacing plastics with gas, the microcells also serve as crack arrestors by blunting crack tips, thereby, greatly enhancing part toughness [79]. When properly prepared, microcellular polystyrene (PS) has five times the impact strength of its unfoamed counterpart [80,81]. The fatigue life of microcellular polycarbonate (PC) with a relative foam density of 0.97 is four times that of its solid counterpart [82]. Furthermore, because the gas fills the interstitial sites between polymer molecules, it effectively reduces the viscosity [83,84] and the glass transition temperature of the polymer melt [79,85–87]. The material, therefore, can be processed at much lower pressures and temperatures. Table 9.9 lists the comparison of some of the properties between microcellular plastics and its solid unfoamed counterpart.

In addition to typical molding procedures, microcellular molding process involves four distinctive steps:

- Gas dissolution—atmospheric gas (N_2 and CO_2) is injected into the machine barrel to form a polymer-gas solution for processing.
- Nucleation—a large number of nucleation sites (orders of magnitude higher than conventional foaming processes) are formed by rapid and substantial pressure drop.
- Cell growth—cell growth is controlled by the processing conditions such as melt pressure and temperature.
- Shaping—the shaping of the part takes place inside the mold.

Although a number of fluids (or gases) can be dissolved into the polymer to form MCPs, many of them are either corrosive, less cost effective, or hazardous. Two of common gases used in microcellular molding process are N_2 and CO_2. The solubility of CO_2 is higher than that of N_2; however, N_2 tends to provide finer cell structure and gives better surface finish. To enhance the diffusion rate of N_2 and CO_2 into polymer, mechanical blending and shearing of the gas with the polymer is used to increase the contact surface area and reduce the striation thickness. In addition, the solubility of gas is further improved by bringing it to a "supercritical" state. At this state (the so-called supercritical fluid, SCF, state), gas behaves like a fluid so that it can be metered precisely into polymer melt, and yet it has the high diffusion rate of a gas for even distribution and dissolving in the melt. The introduction of supercritical gas into polymer melt is provided by a free-standing SCF conditioning system, which uses an air pump to bring the gas to the supercritical fluid state. The fluidlike gas is metered into the injection barrel by a solenoid metering valve, which is controlled by the machine controller. A specially designed screw is used to create the gas-polymer solution for both the microcellular molding process as well as for the solid plastics injection molding.

The solubility of gas increases with increasing pressure and decreasing temperature. Prior to injection, the melt is kept under sufficient pressure with higher screw back pressure to keep the gas in solution. In addition, the process requires a certain injection speed to keep the gas in solution before gas foaming starts in the mold [88]. Packing is not desirable because it impedes the foaming of dissolved gas. Furthermore, the internal pressure of growing cells takes the place of the holding pressure.

During injection molding, the formation of microcells depends on the processing conditions such as melt pressure and temperature. The size of the cells is generally inversely proportional to the cell density, both of which are determined by the thermodynamics of cell nucleation and growth and the amount of gas dissolved in the

Table 9.9 Property Comparisons of Microcellular Plastics with Unfoamed Solid Plastics

Properties	Unformed Plastics	Microcellular Plastics (MCPs)
Specific density ratio ($\rho/\rho_{unfoamed}$)	1.0	0.05–0.95
	100.0	35 (PS)*1
Glass transition temperature Tg (°C)	145.0	65 (PC/CO_2)*2
	90.7	50.8 ($PMMA/CO_2$)*3
Viscosity ratio ($\eta/\eta_{unfoamed}$)	1.0	~0.20*4
		~0.01–0.001*5
Tensile strength ratio ($T/T_{unfoamed}$)	1.0	~1.7 ($HDPE/iPP/CO_2$)*6
Elongation ratio at peak ($E/E_{unfoamed}$)	1.0	~1.1 ($HDPE/iPP/CO_2$)*6
Impact strength ratio ($I/I_{unfoamed}$)	1.0	~5.0*7
Fatigue life ratio ($F/F_{unfoamed}$)	1.0	~4.0 (PC/CO_2)*8

*1 PS with 10% CO_2 (Ref. [85]) *2 CO_2/PC at 9 Mpa (Ref. [86]) *3 CO_2/PMMA at 15.2 Mpa (Ref. [87]) *4 PS with 4% CO_2 (Ref. [84]) *5 Under suitable conditions (Ref. [83]) *6 Ref. [80] *7 Ref. [80,81] *8 At a specific density of 0.97 (Ref. [82]).

polymer. In principle, the larger the amount of gas dissolved in a polymer, the greater the cell density and the smaller of the cell size. The typical cell diameter ranges from 0.1 to 10 μm (as opposed to 250 μm or more with the conventional structural foam molding process). The cell size is so small that the cells cannot be seen by the naked eye; therefore, the molded parts can resemble solid plastics with acceptable physical appearance. The major difference between MCPs and conventional structural (macrocellular) foamed plastics lie in the method to create the formed structure, the size and distribution of the cells, the part-wall thickness limitation, and the physical properties and appearance of the final products. The microcellular molding process uses gas in its supercritical state instead of chemical blowing agents to create evenly distributed and uniformly sized microscopic cells throughout thermoplastic polymers.

9.10.2 Process Advantages

The microcellular molding process is claimed to offer the following processing improvements including potential cost savings [89]:

- Controlled product weight reduction of up to 95% depending on the material used
- Use of SCFs as blowing agents lowers the viscosity of the melt, which results in a substantial decrease in processing temperatures (as much as 78°C), 30 to 50% reduction in hydraulic injection pressure, and 30% or more reduction in clamp tonnage
- Lower melt viscosity and clamp tonnage permit molders to increase the number of cavities or mold larger parts without increasing clamp tonnage
- Faster cycle time due to reduced melt temperature (with lower glass transition temperature), improved mold filling (with lower viscosity), the endothermic reaction of cell nucleation and growth that accelerates cooling, recovery of material's normal glass transition temperature and quick nitrification after gas diffuses out from the MCPs, and the reduction in or elimination of holding pressure and time
- Elimination of sink marks, warpage, and molded-in stresses
- Significant retention of desired mechanical properties, such as high tensile strength, low permanent set, and impact strength, despite a sizable reduction in material content
- Ability of producing parts with thickness as small as 0.5 mm

In addition, molders can run both the microcellular molding process and conventional injection molding on microcellular-molding-capable machines. The microcellular molding process does not require additional chemical blowing (CBAs), hydrocarbon-based physical blowing agents, reactive components, or nucleating agents.

9.10.3 Process Disadvantages

Despite these developments and emerging applications, major challenges in processing MCPs lie in continuous generation of MCPs at an acceptable rate for mass production and control of the state of thermodynamic instability (via temperature and pressure variation) to create fine and uniform microcells throughout the part. In addition, the process requires changes in the machinery components (about 10 to 15% of the machine cost) and licenses. Due to the presence of microcells, it may have limited applications with parts that require clarity. In addition, swirled part surface has been reported that affects the cosmetic appearance. Finally, parts molded by the microcellular molding process needs to be stabilized for the gas inside the cells to diffuse out and its pressure equalize with the atmospheric pressure.

9.10.4 Applicable Materials

Most thermoplastics are suitable for the microcellular molding process, including amorphous and crystalline resins. To date, filled and unfilled materials such as polyamide (PA or nylon), PC, polypropylene (PP), PS, POM, polyethylene (PE), polycarbonate/acrylonitrile butadiene styrene (PC/ABS), high-temperature sulfones, polyetherimides (PEI), as well as polybutylene terephthalates (PBT), Polyetheretherketone (PEEK), and thermoplastic elastomers (TPEs) have been molded.

9.10.5 Typical Applications

For injection molding applications, MCPs will have potential in housing and construction, sporting goods, automotive, electrical and electronic products, electronic encapsulation, and chemical and biochemical applications [79]. Microcellular molding also has a potential in in-mold decoration and lamination due to its low pressure and temperature characteristics that reduce the damage to the film or fabric overlay during molding. Specific application examples include fuse boxes, medical handles, air intake manifold gaskets, automotive trims, and plastic housings.

9.11 Multicomponent Injection Molding (Overmolding)

Multicomponent injection molding (also called *overmolding*) is a versatile and increasingly popular injection molding process that provides increased design flexibility for making multicolor or multifunctional products at reduced cost. By adopting multiple mold designs and shot transfer strategy, this process consists of injecting a polymer over another molded plastic insert to marry the best features of different materials while reducing or eliminating postmolding assembly, bonding, or welding operations. Special machine equipment with multiple injection units, a rotating mold base, and/or moving cores is available to complete the overmolding in one cycle.

9.11.1 Process Description

Multicomponent injection molding is a special process used in the plastic processing industry, and is synonymous with overmolding, multishot injection molding, and the "in-mold assembly" process. This process, however, is different from coinjection molding, which also incorporates two different materials into a single molded part. To be specific, coinjection molding involves sequential and/or concurrent injection of two dissimilar but compatible materials into a cavity to produce parts that have a sandwich structure, with the core material embedded between the layers of the skin material (cf. Sec. 9.1). On the other hand, for multicomponent injection molding, different polymer melts are injected at different stages of the process using different cavities or cavity geometry. In particular, a plastic insert is first molded and then transferred to a different cavity to be overmolded by the second polymer filling inside a cavity defined by the surfaces of the insert and the tool. The adhesion between the two different materials can be mechanical bonding, thermal bonding, or chemical bonding [90].

Multicomponent injection molding dated back several decades ago as the multicolor molding of typewriter keys to produce permanent characters on the keys. This process has since advanced to allow consistent, cost-effective production of multicolor or multifunctional products in a variety of innovative and commonly used methods. The decision of choosing an appropriate molding technique depends on the production volume, quality requirements, and the molder's capabilities and preference. For example, without any additional equipment investment, one can use two separate molds and the conventional injection molding machine for producing multicomponent injection molded parts. In this approach, the insert is first molded and then transferred to a second mold, where it is overmolded with a second polymer. The disadvantage of this approach, however, is that it involves additional steps to transfer and load the prefabricated insert into the second mold. Nevertheless, the loading and pick up of the insert and molded system can be accelerated and precisely controlled by using robots or automated machine systems.

Another commonly used method of multicomponent injection molding employs a rotating mold and multiple injection units, as shown in Fig. 9.40 [90]. Once the insert is molded, a hydraulic or electricservo drive rotates the core and the part by 180 degrees (or 120 degrees for a three-shot part) allowing alternating polymers to be injected. This is the fastest and most common method because two or more parts can be molded every cycle. Utilities for the rotating mold (i.e., cooling water, compressed air, or special heating) are connected through a central rotary union. Rotational capability can be built into the mold or the machine platen. If there is a family of

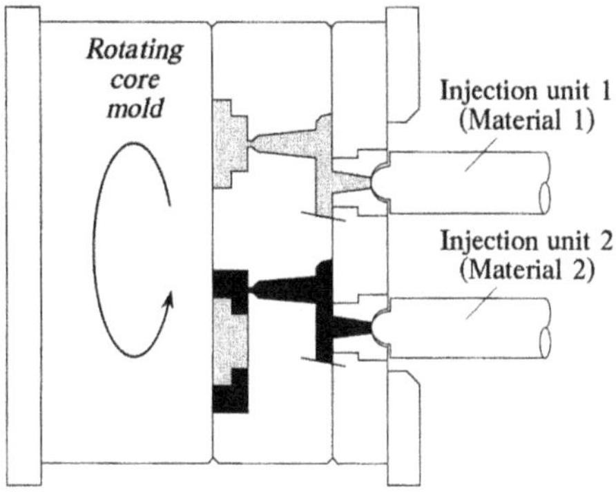

Figure 9.40 The rotating mold that is used to produce multi-component injection molded parts in an automatic fashion. (From *Plastic Part Design for Injection Molding—An Introduction*, Malloy, R. A., (1994), Hanser, Munich, p. 396, Fig. 6.82.)

parts that requires rotational transfers, it is more economical to have the rotational capability built into the machine (and pay for it once) than it is to buy a family of molds with the capability in each.

Another variation of multicomponent injection molding involves automatically expanding the original cavity geometry using retractable (movable) cores or slides while the insert is still in the mold. This process is called *core-pull* or *core-back*, as shown in Fig. 9.41. To be specific, the core retracts after the insert has solidified to create open volume to be filled by the second material within the same mold.

9.11.2 Process Advantages

Multicomponent injection molding offers an innovative approach that combines several components inside the mold, thereby, eliminating the need of secondary assembly and reducing the overall manufacturing cost. This process allows a new

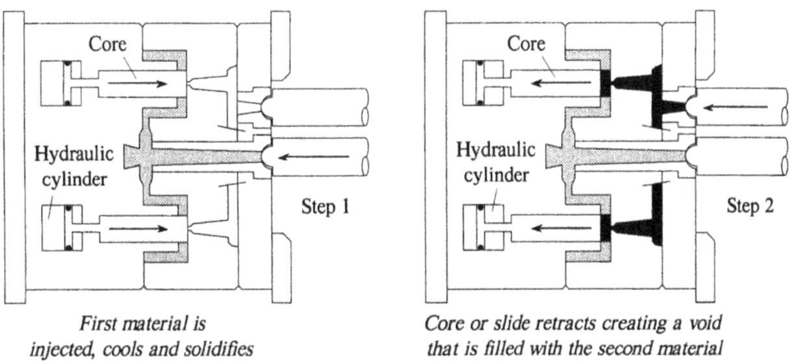

Figure 9.41 Multi-component injection molding using "core pull" or "core-back" technique. (From *Plastic Part Design for Injection Molding—An Introduction*, Malloy, R. A., (1994), Hanser, Munich, p. 396, Fig. 6.83.)

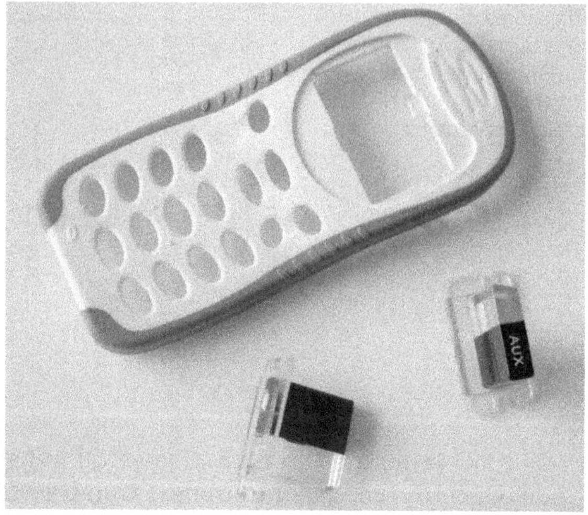

Figure 9.42 (Top) TPE (thermoplastic elastomer) overmolded on ABS/PC for a mobile phone to provide soft touch feeling and drop protection. (Bottom) Black color ABS with transparent acrylic, used on audio push button for light on transmitting. (Parts courtesy of Yomura Company Ltd., Taiwan.)

degree of freedom in industrial design and improves part esthetics, value, quality, and functions (cf. Fig. 9.42). Furthermore, the capability of multicomponent production with a single machine eliminates handling between processes, resulting in savings in part handling, in losses during assembly, and in the amount of floor space required [25].

9.11.3 Process Disadvantages

Process disadvantage is the investment for a more complex (rotating) mold and a special injection molding machine with multiple injection units and a special control system. The processes also require transferring of the insert to different mold or relatively complex tools and processing operations. It should be noted, however, that cost barriers to multicomponent injection molding have fallen as a result of molding equipment being modularized by machine manufacturers [91].

9.11.4 Applicable Materials

Material selection is vital for multicomponent injection molding. A thorough analysis needs to be conducted to determine material compatibility, chemical- and wear-resistance, environmental performance, and other program-specific requirements. Various material combinations will result in widely different levels of adhesion between the base material and subsequent overmolded materials. It is possible to achieve bonds that range from no-bond to true chemical bonds, where the materials interact at molecular levels and yield superb bond strength, along with chemical and environmental durability. Because multicomponent injection molding involves bonding of different materials into a single consolidated part, adhesion between pairing materials is important. Factors that influence adhesion include compatibility, process temperature, surface contact area and texture, molding sequence, and the design of mechanical interlock systems [25]. One of the most popular applications is the overmolding of a flexible thermoplastics elastomer (TPE) onto a rigid substrate to create the soft-touch feel and improved handling in a finished product. Table 9.3 provides basic guidelines on material adhesion characteristics for a wide range of material combinations; however, it is highly recommended to check with the resin supplier to confirm the material selection and compatibility.

9.11.5 Typical Applications

Complex parts in a variety of applications, from electrical, consumer, industrial, to automotive, are being molded with multicomponent injection molding. Figure 9.43

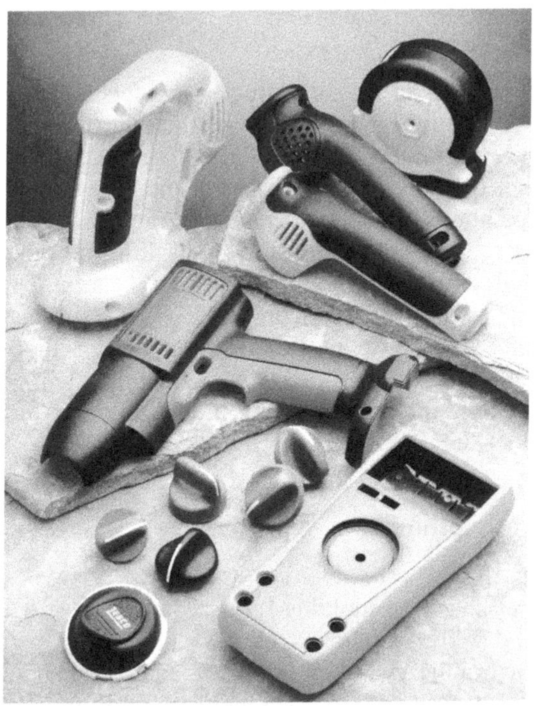

Figure 9.43 Examples of multi-component injection molded parts by Phillips Plastics Corporation, Hudson, Wisconsin, and Ingersoll Rand, Rockford, Illinois, USA. (Reprinted by permission of the Appliance Manufacturer Magazine.)

Table 9.10 Typical Multi-Component Injection Molding Applications [25]

Automotive	Nonautomotive
Lock housing	Toothbrush
HVAC vent grille	Tool handles
Airbag connectors	Articulating toys
Door rub strip and seal	Appliance knobs
Air duct adjustment wheel	Cosmetic compact
Air duct flap (no adhesion)	Hinge (no-adhesive)
Multi-colored taillight lenses	Electric shaver housing
	TV/VCR remote control unit
	Electric toothbrush cam assembly

shows a number of multi-component injection molded parts and products with typical applications listed in Table 9.10.

9.12 Multiple Live-Feed Injection Molding

The multiple live-feed injection molding process is a technique that applies macroscopic shears at melt–solid interfaces to control the microstructure of materials for producing high-integrity products. The oscillating force exerted at multiple polymer entrances comes from the action of pair(s) of pistons moving in the channels of a special processing head between the barrel and the mold. The melt oscillation keeps the material in the gates molten, whereas layers of different molecular or fiber orientation are being built up in the molding due to solidification. This process provides a means of making parts that are free from voids, cracks, sink marks, and weld-line defects. It also permits tailored orientation of the fibers for fiber-reinforced materials to obtain strengths greatly beyond what conventional injection molding can achieve. Multiple live-feed injection molding is also called *shear controlled orientation injection molding* (SCORIM) [92].

9.12.1 Process Description

The multiple live-feed injection molding requires a special processing head that is mounted onto the end of the injection screw barrel and adapter rings. This head has two double-acting hydraulic cylinders—one for each branch of the split molten stream (cf. Fig. 9.44). The head allows each melt stream to fill a chamber connecting the machine nozzle with the runners and gates. Each chamber has independently controlled pressures that may be alternately applied and removed to each branch of

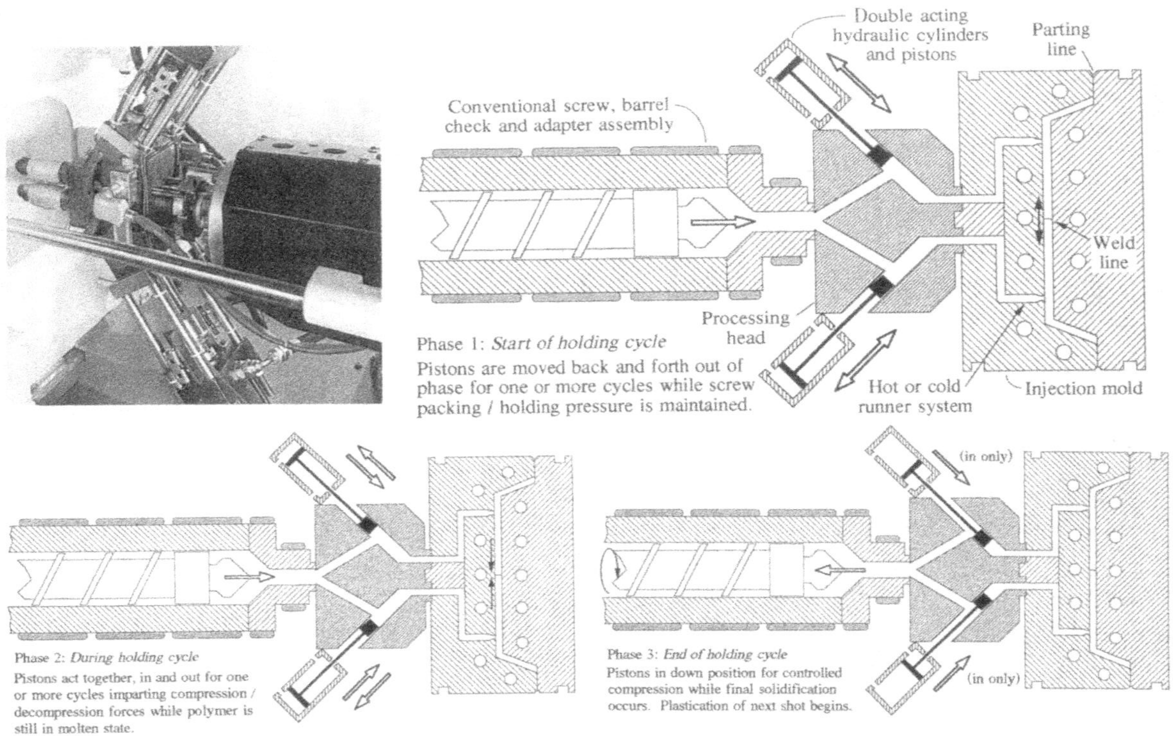

Figure 9.44 Multiple live-feed injection molding provides three different pressure modes that dynamically move and pack the polymer melt inside the mold. (From *Plastic Part Design for Injection Molding—An Introduction*, Malloy, R. A., (1994), Hanser, Munich, p. 60, Fig. 250.)

the split stream. The pistons provide programmed driving forces that dynamically move the molten plastic in the mold. Up to four pistons can be used in each processing head. To fit into existing injection molding machines, this processing head requires approximately 35 cm (14 inches) of space between the barrel of the machine and the mold. It also needs a hydraulic system to activate the hydraulic cylinders and pistons.

During the molding, the mold is first filled as it is in the conventional process. To be specific, the molten polymer is injected from the barrel into the mold through one or both piston channels (cf. Fig. 9.44). The pistons are initially withdrawn from the channel in order to permit the flow of the melt. Once the material starts to cool from the outside and inward, the pistons are activated in the programmed sequence. At first, the piston actions develop the fluctuating melt pressures (with pairing pistons oscillating 180 degrees out of phase) that move and shear the melt in the cavity and in the gates. The pistons then oscillate in phase, providing compression and decompression force to the melt. Meanwhile, new material is introduced to compensate for shrinkage and voids. At the last stage, the pairing pistons compress melt with equal constant pressure, as illustrated in Fig. 9.44.

The programmed sequences of forces drive the "live" melt in the cavity to eliminate or minimize physical imperfections in the part and change the morphology by aligning molecules and reinforced fibers in the preferential direction. This reportedly results in highly efficient packing and enhanced physical properties and performance of the molded product [92,93]. For example, fibers can be aligned circumferentially if a ring is molded, thus greatly enhancing the hoop strength. The desired directions of fiber alignment are decided during component design, and the mold will then be gated accordingly. Figures 9.45 and 9.46 show the micrographs of the resulting fiber orientation using SCORIM process and the comparison of fiber orientation at the weld line using both conventional and the SCORIM processes, respectively.

The multiple live-feed injection molding technology has been extended to two injection units and two double-line-feed devices. By sequencing and timing of the pairing pistons, a multiplayer laminated structure can be formed [94]. Moreover, the multiple live-feed injection molding technology has been employed in conjunction with "Bright Surface Molding (BSM)" for removing surface weld lines from highly reflective aluminum pigmented polypropylene injection molded parts [95].

9.12.2 Process Advantages

Multiple live-feed injection molding introduces programmed sequence of melt movement to create uniform distribution of polymer and/or desirable alignment of reinforcing fibers throughout the mold (cf. Figs. 9.45 and 9.46). Primary benefits of this process, therefore, are greater dimensional stability, part stiffness, and impact strength while eliminating product defects such as voids, cracks, sink marks, and weld-lines. Other benefits include additional design flexibility for thinner walls, fewer ribs and supports, tighter tolerances, and more cost-effective materials. As an example, Table 9.11

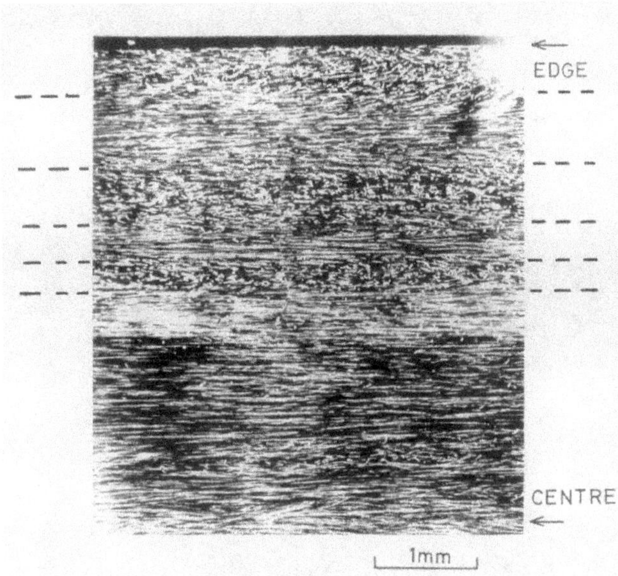

Figure 9.45 The multiple live-feed injection molding process permits controlled orientation of reinforcing fibers in the molded part. This micrograph shows the formation of shear-oriented layers and increased fiber orientation due to changing direction of polymer flow. (Courtesy of Cinpres Limited, Tamworth, England.)

Table 9.11 Comparison of Mechanical Properties of Molded Parts Using Conventional Injection Molding and SCORIM Processes [93]

Material	Weight (% of HA*)	Conventional injection molding			Multiple live-feed injection molding		
		Modulus at 1% strain (GPa)	Tensile strength (MPa)	Strain at break, ε_r (%)	Modulus at 1% strain (GPa)	Tensile strength (MPa)	Strain at break, ε_r (%)
HMWPE	0	1.20 ± 0.08	25.0 ± 0.5	13.0 ± 0.2	4.54 ± 0.61	83.6 ± 5.5	20.8 ± 4.2
HMWPE + HA	50/30	4.03 ± 0.12	39.2 ± 0.7	5.9 ± 0.4	7.45 ± 0.21	73.5 ± 4.1	18.7 ± 3.1
SEVA-C	0	1.81 ± 0.16	35.6 ± 3.2	5.3 ± 0.5	2.97 ± 0.23	41.8 ± 0.8	7.1 ± 2.0
SEVA-C + HA	50/30	5.24 ± 0.34	35.7 ± 5.6	1.8 ± 0.3	7.09 ± 0.21	42.8 ± 3.1	0.5 ± 0.1
S/CA	0	3.24 ± 0.31	70.2 ± 4.8	5.5 ± 1.2	5.81 ± 0.16	98.7 ± 6.8	8.6 ± 2.0
S/CA + HA	30/30	4.94 ± 0.36	60.4 ± 0.4	4.1 ± 0.3	8.63 ± 0.12	65.4 ± 5.3	3.1 ± 0.2

* HA (Hydroxylapatite) percentage for achieving maximum stiffness in conventional and multiple live-feed injection molding, respectively.

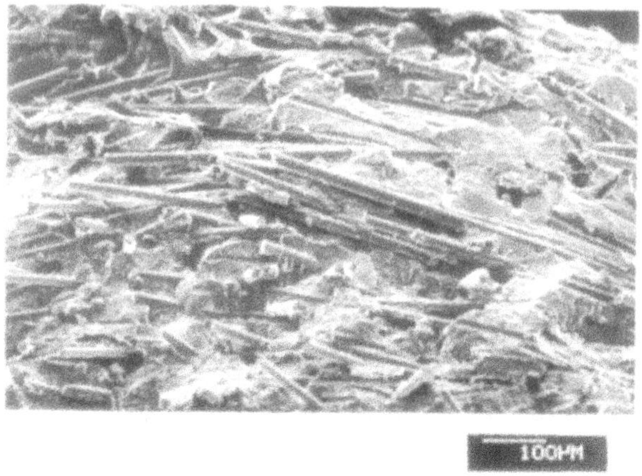

A

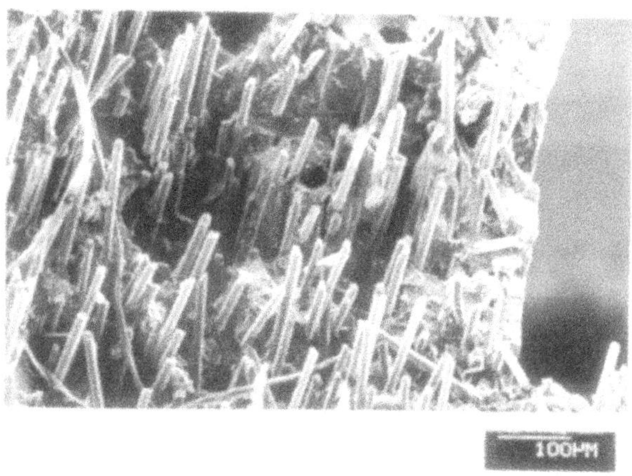

B

Figure 9.46 The fracture failure surfaces at the point of a weld line for a conventional sample (top) and a SCORIM sample (bottom). Note the change of fiber orientation at the weld line formation point. (Courtesy of Cinpres Limited, Tamworth, England.)

lists the comparison of mechanical properties of parts using conventional injection molding and SCORIM processes.

9.12.3 Process Disadvantages

As with several other special injection molding processes, the disadvantages of multiple live-feed injection molding is the requirement of the additional processing units, its packing pistons, plus the determination of pairing gates and associated process control parameters for oscillation of melt.

9.12.4 Applicable Materials

The process can be used for a variety of unreinforced and reinforced thermoplastics, thermosets, and liquid crystalline polymers (LCPs), including polyolefins, POM, PA, polyesters, PEEK, polyether sulfone (PES), and polyphenylene oxide (PPO).

9.12.5 Typical Applications

This process has been used to produce orthopedic devices such as prostheses, medical implants like bone fixation and fixation plates [93], and products that permit application of oscillating force at multiple gates to improve part integrity and strength.

9.13 Push–Pull Injection Molding

The push–pull injection molding process is a melt oscillation technique that creates orientation effects in melts containing fibers or LCPs. In this process, the material is pushed in and out of the mold between two injection units during injection and solidification stages. This process results in molecular and fiber orientation along the direction of the movement. It also minimizes effect of weld lines by dispersing them throughout the part and eliminates voids, cracks, and microporosities in large cross-section molding.

9.13.1 Process Description

The push–pull injection molding process is very similar to multiple live-feed injection molding (see Sec. 9.12). It was originally unveiled by Klockner Ferromatik Desma at the K'89 show [96]. As shown in Fig. 9.47, the push–pull injection molding system includes two injection systems and a two-gate mold [90]. The master injection unit injects the material into the cavity through one gate. It overflows the cavity and the extra material is pushed into the secondary injection unit, where the screw recoils 10 to 15 mm to make room for the material [97]. This process is then reversed with material being pulled back to the master injection unit, and the cycle repeats.

As the material flows back and forth through the mold, molecular orientation is continuously created and subsequently locked in as the material solidifies from the outer layers toward the hot core. By keeping the molten polymer in laminar motion during solidification, the molded parts acquire an oriented structure throughout the volume. If the mold is complex and the melt has to flow around obstacles, the motion will create better mixing in the "eddy current" area behind the obstacle and reduce the weakening effect of the weld lines.

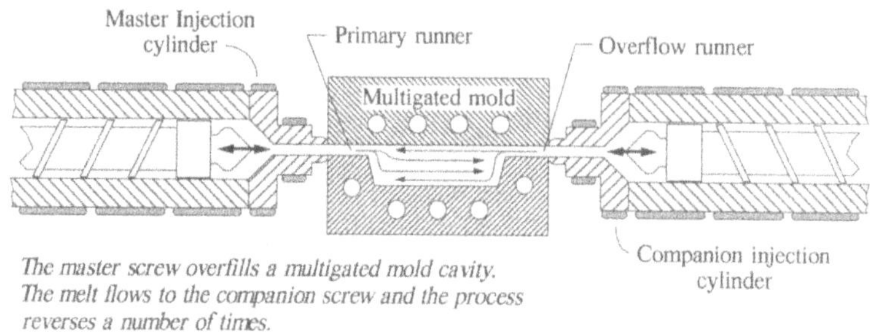

Figure 9.47 Principle of push–pull injection molding. (From *Plastic Part Design for Injection Molding—An Introduction*, Malloy, R. A., (1994), Hanser, Munich, p. 61, Fig. 2.51.)

Although 10 repetitions of push–pull cycles are standard in the production of the molded part, the push–pull sequence reportedly can be repeated as many as 40 times. The ram (screw) speeds for both injection units can be adjusted to generate different pressures inside the cavity. The injection cycle of push–pull injection molding is apparently, considerably longer than is that of conventional injection molding. Because the material is solidifying as the push–pull cycles take place, however, the filling and packing stages are virtually merged into one single stage.

A modification of the previously mentioned system with two injection units and four gates has been developed to produce materials with isotropic properties. The flow is basically activated in orthogonal fashion with two pairs of opposite gates opened and closed alternatively. Other developments of this process include a degassing step to prevent material degradation, variation of cycle time and sequence to compensate for cooling gradients, and using nitrogen to create hollowed parts similar to gas-assisted injection molding [96,98]. As examples, Fig. 9.48 shows a car stabilizer component using a hybrid push–pull and gas-assisted injection process. The material is a partial aromatic polyamide (polyarylamide) with 50% short glass fiber content. Six push–pull strokes were applied before gas injection. Figure 9.49 shows the parallel cut of a plastic component made by the hybrid push–pull and gas-assisted injection process.

9.13.2 Process Advantages and Disadvantages

The process advantages and disadvantages of push–pull injection molding are similar to those of the multiple live-feed injection molding (see Sec. 9.12).

9.13.3 Applicable Materials

Glass fiber reinforced PAs are materials well suited for push–pull injection molding. This process increases their stiffness by 30% along the direction of the melt flow.

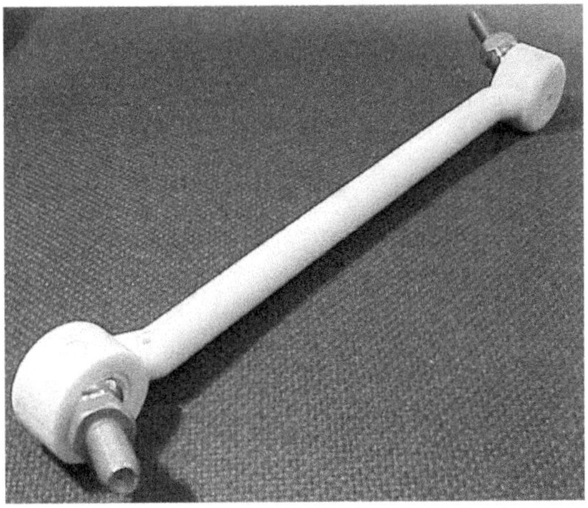

Figure 9.48 A car stabilizer component using a hybrid push–pull and gas-assisted injection process. The material is a partial aromatic polyamide (polyarylamide) with 50% short glass fiber content. (Photo courtesy of StructoForm GmbH, Aachen, Germany.)

Push–pull injection molding has been applied to LCPs and glass fiber–filled LCPs [98]. Increased orientation of the LCPs has boosted an increase of tensile strength of 150% and toughness increase of 250% over the material supplier's literature [25]. In another application, push–pull injection molding can also eliminate the mechanical anisotropy of LCPs by making them orthotropic [99]. This process has also been applied to several high-performance thermoplastics, such as PES, PPS, PPA, PEK, and PEAK.

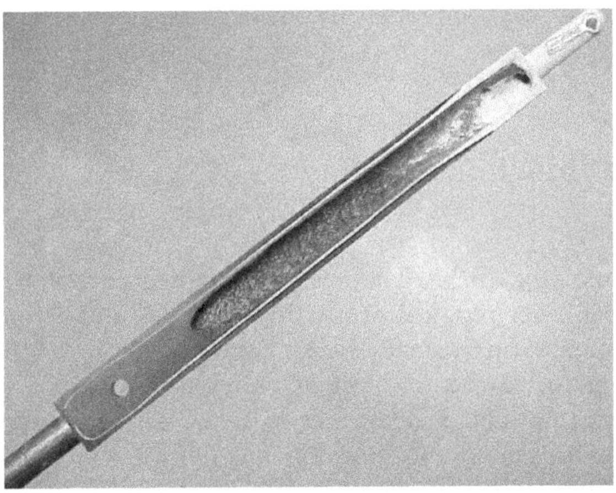

Figure 9.49 A test bar using a hybrid push–pull and gas-assisted injection process. The material is a liquid crystal polymer with 30% short glass fiber content. The material is partially colored to show the push–pull layers. (Photo courtesy of StructoForm GmbH, Aachen, Germany.)

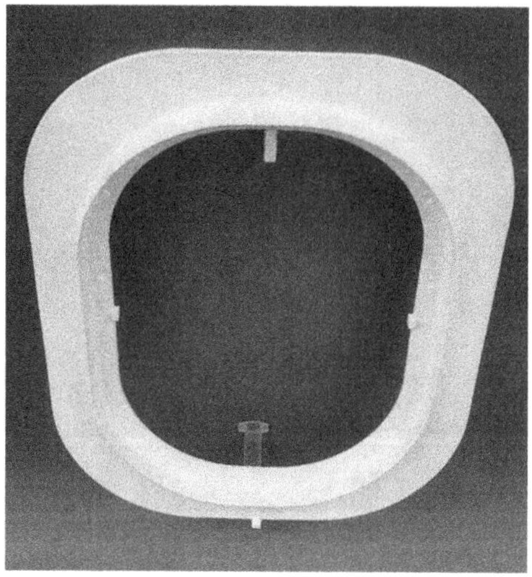

Figure 9.50 Airbus window frame made from LCP Vectra A using push–pull injection molding. (Photo courtesy of StructoForm GmbH, Aachen, Germany, with contributions from Klöckner Ferromatik, Germany, TU Delft, The Netherlands, IKV Aachen, Germany, Aerospatiale, France, and Geiger Technik, Germany.)

9.13.4 Typical Applications

Possible applications of push–pull injection molding are automotive parts, as well as aerospace and electronics components. For example, Fig. 9.50 shows an Airbus window frame using push–pull injection molding. The push–pull injection molding process is also said to have potential in powder injection molding of metals and ceramics (see Sec. 9.14).

9.14 Powder Injection Molding

Powder injection molding (PIM) combines the shaping advantage of the injection molding with the superior physical properties of metals and ceramics. It is rapidly emerging as a manufacturing technology for ceramics and powdered metals, where high volume, high performance, low cost, and complex shape are required. In these processes, a custom formulated mixture of metal or ceramic powder and polymer binder is injected into a mold and form into the desired shape (the "green part"), which is akin to conventional injection molding and high-pressure die casting. After the molding process, the binder is removed and the resulting "brown part" undergoes a sintering process to create metallurgical bonds between the powder particles imparting the necessary mechanical and physical properties to the final part.

9.14.1 Process Description

Powdered metal injection molding and ceramic injection molding are special injection molding processes used to produce highly complex metal and ceramic parts that otherwise would require extensive finish machining or assembly operations [100–104]. Figure 9.51 illustrates the basic steps involved in PIM: mixing for forming feedstock, molding, debinding (binder extraction), and sintering (densification). The last two steps can be combined into a single thermal cycle.

Mixing for Forming Feedstock

Feedstock is the mixture of powder and binder in forms of pellets and granules. Its formation involves mixing selected ceramic raw materials or powdered metal alloys and polymeric binders to form a moldable formulation. Powders are small particles that usually have sizes between 0.1 and 20 μm with a nearly spherical shape to ensure desirable densification. In principle, any metal that can be produced as a powder can be processed using this method. One exception is aluminum, which forms an oxide layer on the surface, thus preventing sintering. Commonly used particles are stainless steels, steels, tool steels, alumina, iron, silicates, zirconia, and silicon nitride. Binder is usually based on thermoplastics such as wax or polyethylene, but cellulose, gels, silanes, water, and various inorganic substances are also in use. The binder usually consists of two or three components: thermoplastics and additives for lubrication, viscosity control, wetting, and debinding [100].

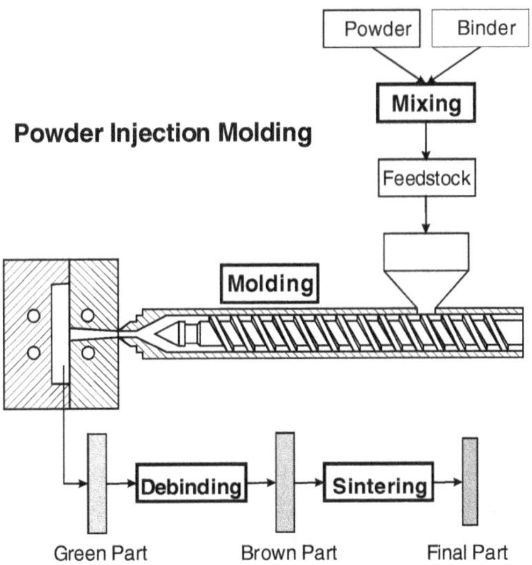

Figure 9.51 Basic steps involved in powder injection molding.

A typical feedstock is composed of approximately 60 vol.% powders and 40 vol.% binders. Twin-screw or high-shear cam action extruders are often used to mix the materials at elevated temperatures (100 to 200°C). Upon exiting the extruder, the mixture is chopped into pellets and granules to be used for injection molding. There are five factors that determine the attributes of the feedstock: powder characteristics, binder composition, powder–binder ratio, mixing method, and pelletization technique. An ideal feedstock is one that is easy to mold and that easily attains the final product dimensions.

Molding

The tooling for powder injection molding is similar to that used in conventional polymer injection molding except that molding is oversized to account for sintering shrinkage. During the molding stage, feedstock pellets are heated to temperatures of 150 to 200°C and injected into the mold at an injection pressure range of 35 to 140 MPa (5000 to 20,000 psi) and with cycle times of 30 to 60 seconds, depending on the part dimension and actual molding formulation employed [101]. The mold can be either heated to control the viscosity of the mixture or be kept at room temperature to facilitate cooling. At the end of the cycle, the molded components (so-called green part) have a strong crayonlike consistency and are suitable for autoejection from the mold via push-out pins or stripper rings. Note that, depending on the binder system used, the molded part could be very brittle; therefore, for critical parts it is desirable to remove the parts from the mold using a handling system.

Debinding (Binder Extraction)

The main objective of debinding is to remove the binder in the shortest time with the least impact on the compact. This step of the process is the most critical step in the powder injection molding process. Debinding is typically accomplished by one of the three general procedures: thermal binder removal method, catalytic binder removal agents, or solvent extraction method [100]. The debinding process, which removes as much as 30 to 98% of the binder, can be a long process. It typically takes many hours or even days, depending on part thickness and powder grain size. After the debinding stage, the resulting compact is called the "brown part."

Sintering (Densification)

The last step in the process is the sintering, which is a thermal treatment for bonding the particles into a solid mass. Furnaces are used at this stage that permit sintering at temperatures close to the material melting temperature under a controlled atmosphere or in a vacuum. Inside the furnace, sintering can be performed in air for ceramics or in a reduced atmosphere for metals to prevent oxidation or to reduce any existing oxides. A simple hydrogen atmosphere is sufficient for many metals. The atmosphere must contain corresponding carbon compounds to achieve the appropriate carbon content in the steel for steels in which carbon is an essential element.

During the sintering process, intergranular pores within the molded part are removed, resulting in significant shrinkage (typically on the order of 10 to 20%). Meanwhile, the finished part achieves a density of around 97%, at which the mechanical properties differ only slightly, if at all, from those parts manufactured by alternative methods.

9.14.2 Process Advantages

The process advantages of PIM can be summarized in the following:

- Reduced manufacturing cost savings of 20 to 40% over conventional metal and ceramic processing methods [101]
- A net-shape process technology with good dimensional tolerance control and with little or no secondary machining required
- Capability of producing complex shape and geometric features
- High production rates through the use of injection molding and/or automated production
- Parts with mechanical properties nearly equivalent to wrought materials

9.14.3 Process Disadvantages

The major process disadvantages of powder injection molding are:

- Constraints of part size and thickness
- Control of volumetric shrinkage and thickness uniformity

Although there is no technical limit to the maximum size of part that could be produced, two economic considerations restrict the sizes and part thickness. First, the larger the part, the greater the proportion of the overall cost that is attributable to the raw material, which is costly. Second, the thicker the section, the longer the debinding time and, thus, the higher the cost of that part of the process. At present, the limiting thickness is about 30 mm.

The "green part" contains a high-volume percentage of binder (as much as 50%), which leads to significant shrinkage during sintering. It is a major requirement of the sintering process that this shrinkage be controlled. Variations in wall thickness will cause variations in shrinkage during sintering, which makes dimensional control difficult. In addition, uniform wall thickness is critical in order to avoid distortion, internal stresses, voids, cracking, and sink marks.

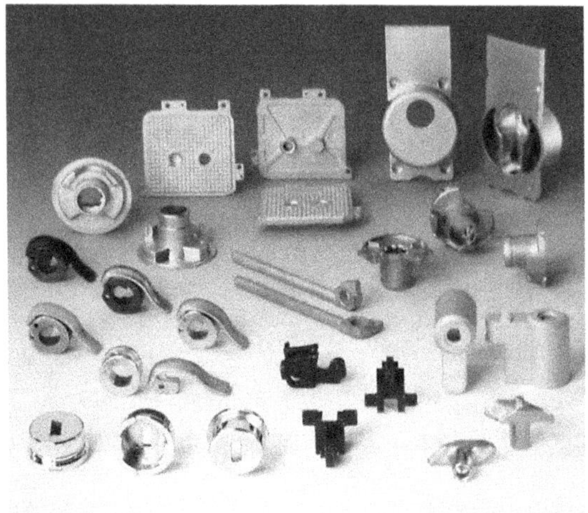

Figure 9.52 A range of powder injection molding components [104].

9.14.4 Typical Applications

PIM is applicable to a variety of automotive, consumer, electronic, computer peripherals, medical, industrial, military, and aerospace components. Automotive applications include turbocharger, brake, and ignition components, as well as oxygen sensors. Figure 9.52 shows a range of PIM components [104].

9.15 Reaction Injection Molding

Reaction injection molding (RIM) involves mixing of two reacting liquids in a mixing head before injecting the low-viscosity mixture into mold cavities at relatively high injection speeds. The liquids react in the mold to form a cross-linked solid part. The short cycle times, low injection pressures, and clamping forces, coupled with superior part strength and heat and chemical resistance of the molded part make RIM well suited for the rapid production of large, complex parts, such as automotive bumper covers and body panels.

9.15.1 Process Description

Reaction injection molding (RIM) is a process for rapid production of complex parts directly from monomers or oligomers. Unlike thermoplastic injection molding, the shaping of solid RIM parts is through polymerization (cross-linking or phase separation) in the mold rather than solidification of the polymer melts. RIM is also different from thermoset injection molding in that the polymerization in RIM is activated via chemical mixing rather than thermally activated by the warm mold. During RIM process, the two liquid reactants (e.g., polyol and an isocyanate, which were the precursors for polyurethanes) are metered in the correction proportion into a mixing chamber where the streams impinge at a high velocity and start to polymerize prior to being injected into the mold (cf. Fig. 9.53). Due to the low-viscosity of the reactants, the injection pressures are typically very low even though the injection speed is fairly high. Because of the fast reaction rate, the final parts can be demolded in typically less than one minute. Table 9.12 provides a brief comparison between RIM and thermoplastic injection molding.

There are a number of RIM variants. For example, in the so-called reinforced reaction injection molding (RRIM) process, fillers, such as short glass fibers or glass flakes, have been used to enhance the stiffness, maintain dimensional stability, and reduce material cost of the part. As another modification of RIM, structural reaction injection molding (SRIM) is used to produce composite parts by impregnating a reinforcing glass fibermat (preform) preplaced inside the mold with the curing resin. On the other hand, resin transfer molding (RTM) is very similar to SRIM in that it also employs reinforcing glass fibermat to produce composite parts; however, the resins used in RTM are formulated to react slower, and the reaction is thermally activated as it is in thermoset injection molding.

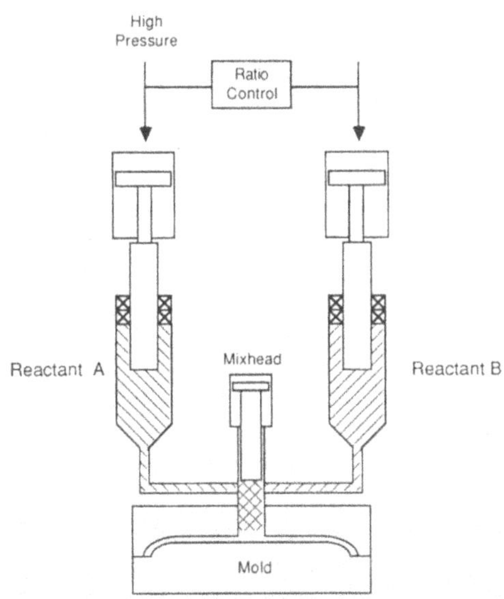

Figure 9.53 Schematic diagram of a reaction injection molding machine. (From *RIM Fundamentals of Reaction Injection Molding—Introduction*, Macosko, C. W., (1988), Hanser, p. 2, Fig. 1.1.)

Table 9.12 Comparison between RIM and Thermoplastic Injection Molding [25]

	RIM	Thermoplastic injection molding
Material processing temperature (°C)	40–60	150–370
Material viscosity (Pa-s)	0.1–1.0	10^2–10^5
Injection pressure (bar)	100–200	800–1000
Clamp force requirement per m² (metric ton)	45	1800–2700

9.15.2 Process Advantages

The process advantages of RIM stem from the low pressure, temperature, and clamp force requirements that make RIM suitable for producing large, complex parts with lower-cost molds. In addition, the capital investment on molding equipment for RIM is lower compared with that of injection molding machines. Finally, RIM parts generally possess greater mechanical and heat-resistant properties due to the resulting cross-linking structure.

9.15.3 Process Disadvantages

The mold and process designs for RIM become generally more complex because of the chemical reaction during processing. For example, slow filling may cause premature gelling, which results in short shot, whereas fast filling may induce turbulent flow, creating internal porosity. Improper control of mold-wall temperature and/or inadequate part thickness will either give rise to moldability problem or cause scorching of the materials. Moreover, the low viscosity of the material tends to cause flash that requires trimming. Another disadvantage of RIM is that the reaction with isocyanate requires special environmental precaution due to health issues. Finally, like many other thermosetting materials, the recycling of RIM parts is not as easy as that of thermoplastics.

9.15.4 Applicable Materials

Polyurethane materials (rigid, foamed, or elastomeric) have traditionally been synonymous with RIM as they and ureaurethanes account for more than 95% of RIM production. Alternative RIM materials include nylon (NYRIM), dicyclopentadiene (DCPD-RIM), acrylamate/acrylesterol, epoxies, unsaturated polyesters, phenolics, thermoset resins, and modified polyisocyanurate.

9.15.5 Typical Applications

Due to the low-viscosity of the reacting liquid mixtures, RIM is generally used to produce large, complex parts, especially for exterior and interior automotive components such as bumper beams, bumper fascias, and door panels. Other automotive applications include fenders, exterior trims, arm rests, steering wheels, and window gaskets. Nonautomotive applications range from furniture, business machine housings, medical and industrial enclosures, agricultural and construction parts, to appliances and recreational equipment.

9.16 Resin Transfer Molding and Structural RIM

Resin transfer molding (RTM) and SRIM are two similar liquid composite molding (LCM) processes that are well suited to the manufacture of medium-to-large, complex, lightweight, and high-performance composite components primarily for aerospace and automotive industries. In these processes, a reinforcement fiber mat (preform) is preplaced in a closed mold to be impregnated by a low-viscosity, reactive liquid resin in a transfer or injection process. These two processes differ in such areas as the resins used, mixing and injection set up, mold requirement, cycle time, fiber volume fraction, and suitable production volume.

9.16.1 Process Description

LCM processes such as resin transfer molding (RTM) and structural reaction injection molding (SRIM) are recognized as the most feasible and structurally efficient approach to mass produce lightweight, high-strength, low-cost structural composite components. In general, these processes consist of preparation of a reinforcing fiber mat (known as preform), preplacement of the dry preform in a closed mold, premixing and injection or transfer molding of reactive liquid resin (cf. Fig. 9.54), impregnation of the fiber mat by the curing resin, and removal of the cured, finished component from the mold (demolding).

The preform is the assembly of dry (unimpregnated) reinforcement media that is preshaped (via, e.g., thermoforming) and assembled with urethane-formed cores, if necessary, into a three-dimensional skeleton of the actual part. Both of the RTM and SRIM processes make use of a wide variety of reinforcing media, such as woven and nonwoven fiber products, die-cut continuous strand mats (CSM), random mats consisting of continuous or chopped fibers laid randomly by a binder adhesive, knit braiding or two- and three-dimensionally braided products, or hybrid preform made of layers of different types of media. The selection of the preform architecture depends on the desired structural performance, processability, long-term durability, and cost.

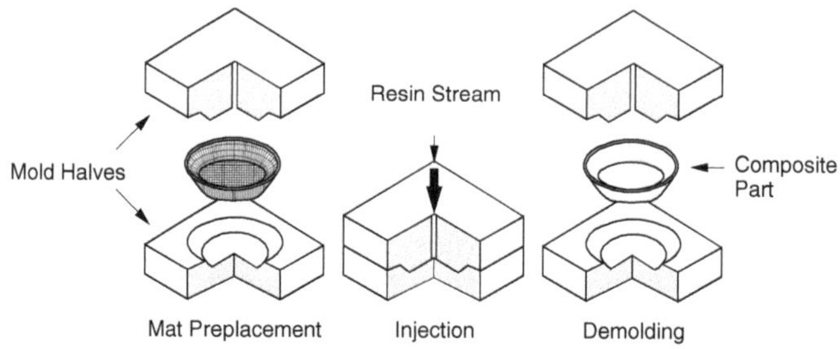

Figure 9.54 Schematic of the resin transfer molding (RTM) and structural reaction injection molding (SRIM) processes [105].

The main difference between RTM and SRIM stems from the reaction activation mechanism of the resins used [106]. This leads to different filling and cycle times, mixing and injection set-up, volume fraction and construction of the reinforcement, mold requirements, and suitable production volume. To be more specific, the chemical systems used in reaction molding processes can roughly be divided into two types: thermally activated and mixing activated. The resins used in RTM fall into the thermally activated category, whereas the resins used in SRIM are mixing activated. In the RTM process, a static mixer is used to produce the mixture at a typical mix ratios of 100 to 1 by volume. These thermally activated resin systems do not react appreciably at the initial resin storage temperature. They reply on heated mold wall to accelerate the chemical reaction. As a result, the filling times for RTM can be as long as 15 minutes, and the cycle time is on the order of an hour or longer, depending on the resin and application. Given the slower injection and reaction rates with RTM, high-volume fraction and more complex preform can be used to improve the part strength and performance further. In addition, the low viscosity of the resin and the slow injection rate result in low injection-pressure and clamp-force requirements. As a result, RTM allows the use of so-called soft tool (e.g., wood-backing epoxy mold or aluminum mold). Because of the long cycle time, RTM is generally limited to a low-volume production (i.e., less than 10,000 parts).

On the other hand, SRIM derives its name from the RIM process (cf. Sec. 9.15), from which its resin chemistry and injection techniques are adapted. That is, the chemical reaction is activated by impingement mixing of two highly reactive components in a special mixing head under high pressure. Upon mixing, the mixture is subsequently injected into the mold at a lower pressure. The resin starts to cure as it impregnates the preform and forms the matrix of the composite. Due to the fast reaction rate and rapid build up of viscosity from curing, the cavity has to be filled within a few seconds. In addition, the cycle time is as short as 1 minute. Flow distances for typical SRIM applications are therefore limited to 0.6 to 0.9 m from the inlet gate. Furthermore, the volume fraction and construction of the reinforcement have to be selected carefully to facilitate fast, complete filling before gelation occurs. Because of the high injection rate and short cycle time, SRIM generally uses steel tool and is suitable for medium- to high-volume production (10,000 to 100,000 parts).

9.16.2 Process Advantages

Prepreg/autoclave process has traditionally been the major manufacturing technique for producing lightweight composite components in the aerospace industry. This process however, is slow, expensive, and labor intensive. In addition, common manufacturing processes in automotive industry, such as thermoplastic injection molding and compression molding of SMC, can only incorporate low-volume, short fiber reinforcement. Thus, they cannot produce high-strength products needed for structural applications. RTM and SRIM, therefore, offer viable options for producing lightweight, high-performance structural components for both aerospace and automotive industries.

In addition, low injection pressure is another process advantage of both RTM and SRIM. Although the injection pressure varies with the permeability of the fiber mat, part geometry, and the injection rate, typical injection pressure varies from 70 to 140 kPa for low injection rate and reinforcement content (10 to 20%) to 700 to 1400 kPa for rapid mold filling and high reinforcement content (30 to 50%) [105]. Furthermore, the RTM and SRIM processes employ closed mold, which reduces or eliminates the emission of hazardous vapor. Other advantages include more repeatable part thickness and minimal trimming and de-flashing of the final part.

9.16.3 Process Disadvantages

The RTM and SRIM processes are more suitable for producing large, structural components with limited areas of intricate details. Tiny details, such as grooves or slots for assembly, as well as ribbed structures or bosses are not easy to mold in due to the difficulty involved in preform preparation and preplacement of preform for these features [105]. In addition, it is challenging to provide a Class A surface with standard resins and structural reinforcement materials, unless gel coats are used.

9.16.4 Applicable Materials

Of the several resin systems employed by RTM, polyesters are the mostly used resins because of the low cost. Other resins used in RTM include epoxies, vinyl esters, acrylic/polyester hybrid, acrylamate resin family, and methymethacrylate vinyl esters. On the other hand, common resins used for SRIM include urethane, acrylamate, and dicyclopentadiene.

9.16.5 Typical Applications

Typical applications for RTM and SRIM include automotive components (grille opening panels, hoods, deck lids, doors, and structural cross members), sport equip-

Table 9.13 Common RTM Applications [25]

Automotive	Others
Doors	Building doors
Panels	Cash dispensers
Roof panels	Chemical pumps
Spoilers	Electrical covers
Truck cabs	Instrument cases
Wind deflectors	Lifeboat covers
Aerospace	Machine covers
Aircraft engine blades	Manhole covers
Aircraft interior panels	Marine hatches
Aircraft wing bases	Medical equipment
Sport Equipment	Plumbing components
Bicycle frames	Railway body panels
Bicycle handlebars	Railway seats
Boat hulls	Roof tiles
Canoe paddles	Sinks
	Storage tanks
	Traffic light boxes

Table 9.14 Common SRIM Applications [25]

Automotive and Truck	Other
Bumper beams	Copier heat shields
Door panels	Office chair shells
Instrument panels	Satellite antenna disks
Landau roofs	
Noise shields	
Rear window decks	
Spare tire covers	

ment, marine, medical, construction, and corrosion resistance (for chemical industry). Table 9.13 lists the various RTM applications and Table 9.14 those for SRIM [25].

9.17 Rheomolding

Rheomolding is one of the melt-flow oscillation techniques that employs melt vibration (typically at low frequencies) to reduce the viscosity during filling. It also helps to alter the molecular orientation, thereby improving the mechanical properties, stiffness, strength, and clarity of the molded parts without resorting to processing aids such as thinning and nucleating agents.

(Note: in the area of die-casting process with metals, Rheomolding is sometimes referred to as a special process that employs semi-solid alloys as the material charge. The semi-solid alloys are generated by mechanically agitating solidifying liquid metals or by means of vibration.)

9.17.1 Process Description

Melt-flow oscillation or melt vibration can be applied in various forms to modify the mechanical, rheological, and/or optical properties of the molded materials [107]. For instance, melt vibration using mechanical shaking/vibration or ultrasonic vibration devices can homogenize, thus increasing the density of the molded material. In addition, vibration reduces the melt viscosity and changes the relaxation kinetics, thereby influencing diffusion and rate-sensitive processes such as nucleation and growth of crystals, blending, and orientation. Furthermore, vibration can also generate heat locally by internal friction, resulting in fusion at weld lines, reduction of surface stresses, orientation birefringence, and frictional coefficient at the wall as well as increase in the throughput. Other melt-flow oscillation or melt-vibration technologies, such as multiple live feed injection molding, push–pull injection molding, and vibration gas injection molding, are discussed in Secs. 9.12, 9.13, and 9.20, respectively. Practical applications of the principles of melt vibration have been implemented in injection molding, extrusion, compression molding, and thermoforming processes.

The underlying principle of Rheomolding is based on the fact that the material rheology is a function of vibration frequency and amplitude in addition to temperature and pressure [108,109], a phenomenon manifested itself as the more well-known shear-thinning behavior of polymer melts. It has been reported that low-frequency and high-amplitude oscillations, coupled with traditional cooling treatments, affect plastic materials in much the same way as does faster cooling [108]. This control of material's cooling rate can be used to manipulate the material orientation (for amorphous materials) and morphology (or semi-crystalline materials) for better material properties.

Rheomolding is produced by the reciprocating action of one or more melt accumulator pistons adjacent to the melt-flow path. As illustrated in Fig. 9.55, an add-on actuator is inserted between the barrel of the injection machine and the mold to provide the melt vibration. With such a configuration, the melt plastication and the injection vibration occur separately. In an alternative design, the mold is modified to incorporate small piston(s) to apply vibration directly to where it is needed, such as weld lines or other critical areas (cf. Fig. 9.56). In the Rheomolding processes, both hydrostatic pressure and shear force are oscillated during cycle time to modify the rheological properties, orientation, and crystallization kinetics of the polymer melts. The frequency, amplitude, and phase shift of the vibration modes are designed based on the material used and the objectives to impart quality and enhance performance.

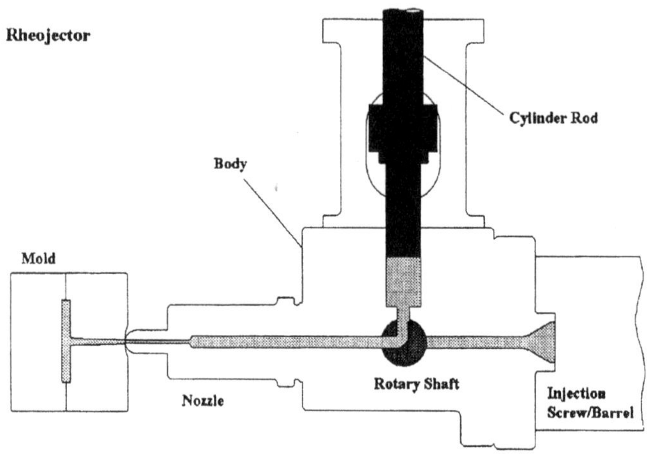

Figure 9.55 An add-on actuator inserted between the injection barrel and the mold applies the melt vibration to the molded material [107]. (Reprinted by permission of the Society of Plastics Engineers.)

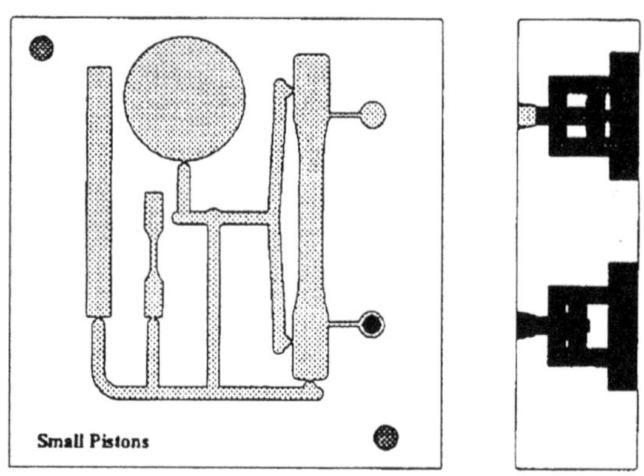

Figure 9.56 An ASTM mold equipped with a set of two pistons to vibrate the melt during filling and packing [107]. (Reprinted by permission of the Society of Plastics Engineers.)

9.17.2 Process Advantages

As with the objectives of other melt-flow oscillation techniques, Rheomolding offers the ability to modify the rheological and orientational states of polymer melts during processing, thus influencing the resulting mechanical (such as tensile strength, modulus, and impact) and optical properties (such as clarity) of the molded parts. It has been used to eliminate or strengthen weld lines and enhance the properties (by as much as 65 to 120%) [108]. Parts with thinner walls (which uses less material) can therefore be made or less expensive material can be used if the Rheomolding process enhances a specific property. Other process benefits include improved weld-line strength, reduced internal stresses and part warpage, potentially controllable crystallization kinetics, better surface finish, and lower friction coefficient at the wall surfaces.

9.17.3 Process Disadvantages

Similar to many other emerging processes, the know-how of the process has yet to be established to help fully utilize the potential benefits mentioned earlier. In addition, the required hardware and licensing fee adds additional costs to the production. Furthermore, it needs longer development time to design or build in the gas-vibration system and to identify the optimal processing conditions. In addition, there is a limitation to the melt temperature as warm material tends to relax and loses the orientation.

9.18 Structural Foam Injection Molding

Structural foam injection molding is an extension of injection molding for producing parts with cellular (or foam) core sandwiched by solid external skins. This process is suitable for large, thick parts that have enough strength-to-weight ratio to be used in load- or bending-bearing in their end-use applications. Despite the surface swirling pattern, which can be eliminated with special molding techniques or the postmolding finishing operations, this process permits molding of large parts at low pressure and with no sink marks and warpage problems. Structural foam parts can be produced with physical blowing agents like nitrogen gas or chemical blowing agents.

9.18.1 Process Description

Structural foam injection molding produces parts consisting of a solid external skin surrounding an inner cellular (or foam) core (see Fig. 9.57). There are several variants of this process. The most common one is the low-pressure structural foam injection molding. Other additional processes are gas counter pressure, high pressure, co-injection, and expanding mold techniques. As shown in Fig. 9.58, the low-pressure structural foam molding process involves a short-shot injection accompanied by expansion of melt during molding. The melt expansion is created from gas (typically nitrogen) dissolved in the polymer melt prior to injection. It can also be assisted by gas released

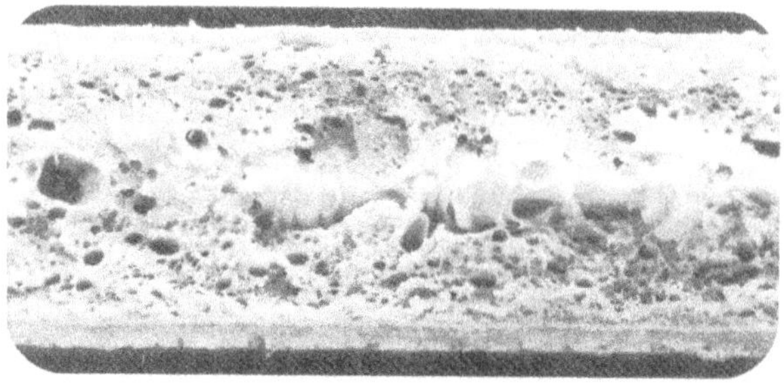

Figure 9.57 Cross-section of a typical structural foam molded part showing the integral skin layers and the foam core. (From *Plastic Part Design for Injection Molding—An Introduction* (1994), Malloy, R. A., Hanser, Munich, p. 115, Fig. 2.109.)

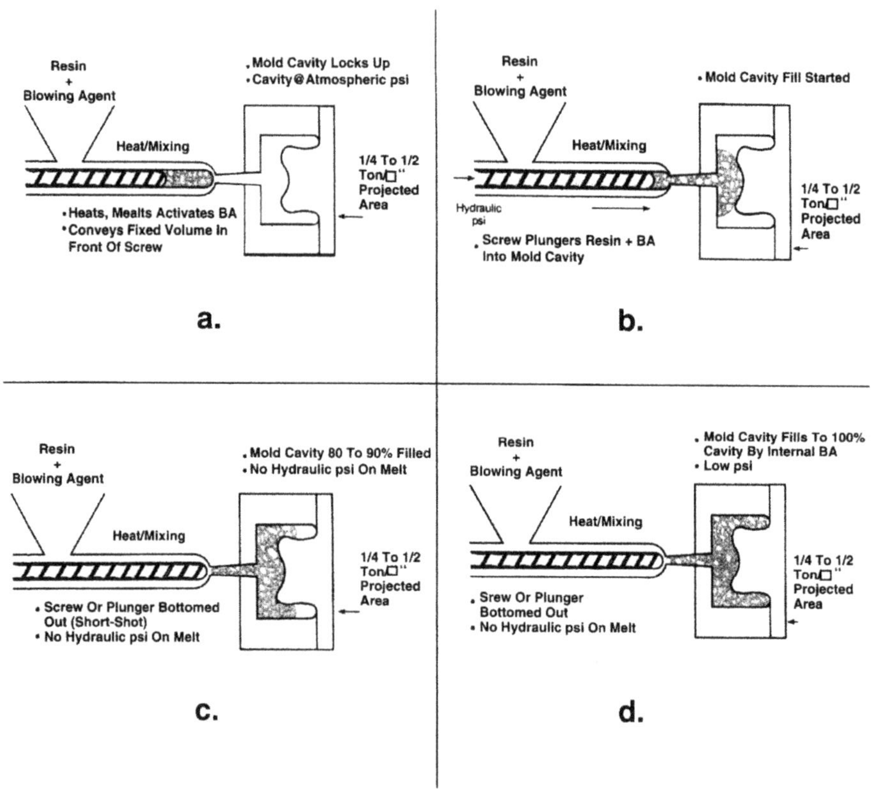

Figure 9.58 Various stages for the structural foam injection molding process. (From *Injection Molding Alternatives—A Guide for Designers and Product Engineers* (1998), Malloy, R. A., Hanser, Munich, p. 63, Fig. 5.5.2.)

from the chemical blowing agents in a resin-compatible carrier blended with the polymer pellets. The melt expansion from the foaming process generates an internal gas pressure of 21 to 34 Bars (2.1 to 3.4 MPa or 300 to 500 psi), which is sufficient to drive the polymer melt continuously to fill the extremities of the cavity. Note that this process requires only 10% of the pressure normally required to fill and pack out the mold with the conventional injection molding process. In addition, structural foam injection molding realizes part weight reduction by replacing plastics with gas and eliminates the typical shrinkage associated with thick parts with the melt expansion. The foam structure provides insulation effect so that the part cools down at a slower rate. Due to the low injection pressure and slow cooling rate (which allows material relaxation), the process produces parts with very low molded-in residual stresses.

As indicated earlier, both nitrogen and chemical blowing agents can be added to the polymer melt to create foam structure. To add nitrogen into the polymer melt and to hold the polymer-gas mixture under pressure until it is injected into the mold, it requires special structural-foam molding equipment with an accumulator. When the external pressure of the melt is relieved as the material enters the mold cavity, the gas begins to emerge from the solution. At the end of a short-shot injection, the pressure decreases rapidly as the material relaxes. Thus, gas expands immediately at this point, pushing the material to fill the rest of the mold. On the other hand, the chemical blowing agents generate gas (nitrogen, hydrogen, or carbon dioxide) by either thermal decomposition or chemical reaction in the melt. Selection of chemical blowing agents depends on the polymer material used, cost, convenience, safety, cell structure, the nature and quantity of gas generated, and the exothermic (heat releasing) or endothermic (heat absorbing) nature of the reaction. If a blowing agent is exothermic, it increases the material temperature and the gas pressure. On the other hand, the endothermic blowing agent helps to cool down the cell structure and reduce internal gas pressure, thereby, allowing the part to be removed in a shorter time. After foaming, the gas inside the core of the part must diffuse to the atmosphere to release the internal pressure [110].

A gas counterpressure technique has been developed and employed to produce parts with smooth, thicker, and nonporous skin surfaces, and more uniform cell density in the core. In this process, the mold is pressurized with an inert gas to create a pressure level of 14 to 34 Bars (1.4 to 3.4 MPa or 200 to 500 psi) during filling. The pressure in the mold must be high enough to prevent the melt from foaming. To maintain the pressure level, the mold must be sealed with O-ring. It may also be necessary to seal ejector pins, sprue bushing, and other moving slides or cores. The control of a sufficient short-shot weight and a critical timing for the counterpressure to be released are also critical. After the injection, the gas is vented and mold is depressurized to allow the expansion of polymer. Because of the counterpressure, the flow length of the material may be reduced by 10 to 20%, and the density reduction is limited to 8 to 10%, as compared with a level of 10 to 15% with the low-pressure structural foam injection molding [25]. Gas counter pressure using a hot runner mold has been reported in Ref. [111] to mold a polystyrene part with surface smoothness substantially better than that of the parts molded under the same processing conditions using low-pressure structural injection molding.

Similar to conventional injection molding, the part performance strongly depends

Table 9.15 Differences between Design Criteria for Conventional Injection Molding and Structural Foam Injection Molding

Criterion	Conventional injection molding	Structural foam injection molding
Wall thickness	Maximum 5 mm No abrupt changes in thickness	Up to 20 mm possible Optimum 4 to 8 mm Abrupt changes in thickness possible, therefore, greater design freedom
Rib thickness	Two thirds of the wall thickness	Any thickness
Surface feature	Smooth or depending on mold	Irregular and structured
Heavy section	Must be avoided because of sink marks	Possible. But longer cooling time is required to prevent "postblow" (local swelling at thick sections due to material expansion after ejection)

on the material, processing conditions, and the part and mold designs. The size and distribution of foam cells depend on the material, the molding technique, and the processing conditions. They have direct influence on the physical properties of the part. Because of the pressure and temperature fields, a variation of cell size exists within the foamed part with large voids concentrating at the center of the part and areas away from the gate, and gradually decreasing cell size toward the solid part surface. In addition, the surface roughness was found to increase progressively in the direction of melt flow from the cold sprue toward the end of the part [110]. The processing conditions found to enhance the surface smoothness of the part were a higher melt temperature, higher mold temperatures, higher short-shot weight, higher injection speed, and lower blowing agent concentration.

Due to the unique volume expansion feature of structural foam injection molding, the design guidelines for this process are different from those of conventional injection molding. Table 9.15 lists the differences in design criteria [112].

9.18.2 Process Advantages

Structural foam injection molding was introduced on a commercial basis in early 1960s [110]. The advantages of this process are:

- Large, complex parts can be produced with low clamp tonnage (one order of magnitude lower) compared with conventional injection molding using solid thermoplastics.
- The process also produces parts with very low mold-in residual stresses, resulting in more dimensionally stable and flatter parts.
- Even though the increased part thickness enhances part rigidity and load-bearing capability, weight reduction as much as 10 or 15% can be realized.

9.18.3 Process Disadvantages

The major drawback of structural foam injection molding is the swirling pattern on part surface. The swirl results when the gas from the blowing agent breaks through the melt front and is trapped between the mold surface and the polymer. To improve surface appearance and quality, secondary operations, such as sanding, priming, and painting, are required. These postmolding operations result in higher part cost. It should be noted, however, that parts with substantially improved surface quality have successfully been produced using gas counterpressure structural foam injection molding and co-injection structural foam, as well as expanding, high-pressure methods [110].

Because of the reduction of thermal conductivity due to the foam structure, the cooling time of structural foam parts is longer than that of the solid plastic part. If the part is removed from the mold before the material is sufficiently cooled, the internal gas pressure will cause blister on the part surface, especially at the thick sections. Moreover, if the part is painted before the internal gas pressure reaches the ambient, atmospheric pressure, blisters will also occur.

It should also be noted that the selection of blowing agents should avoid the possible material degradation from the by-products of the blowing agents during their decomposition. Finally, due to the characteristic cellular structure, the mechanical properties of structural foamed parts can be significantly lower than the solid parts, even though thicker foamed parts are more rigid than are their solid counterpart on a weight-to-weight basis.

9.18.4 Applicable Materials

Table 9.16 lists the variety of materials used for structural foam injection molding. Among these materials, the most common are low-density and high-density polyethylenes for their low cost, ease of processing, chemical resistance, and low-temperature impact strength. Polypropylene is an attractive material for its increased stiffness and chemical resistance, and its reinforced grades exhibit increased temperature resistance. High-impact polystyrene and modified polyphenylene ether (m-PPE), modified polyphenylene oxide (m-PPO), and polycarbonate are commonly found in high-performance applications. The versatility inherent in polyurethane

Table 9.16 Materials Used in Structural Foam Injection Molding [25]

Acrylonitrile-butadiene-styrene (ABS)	Polycarbonate
Polyamide 66 (Nylon 66)	Polypropylene
Polyetherimide	Polystyrene
Polyethylene (Low- and high-density)	High-impact polystyrene
Modified polyphenylene ether (m-PPE)	Polybutylene terephthalate
Modified polyphenylene oxide (m-PPO)	Polyurethan

Figure 9.59 Pallets that provide all standard sizes with 13,600 kg (30,000 lb) static load capacity, four-way entry for forklift and two-way entry for floor jacks. Material used is high-density polyethylene [114]. (Reprinted by the permission of the Society of Plastics Industry.)

Figure 9.60 Gas chromatograph main frame that supports a 18 kg (40 lb) transformer and withstands temperatures up to 71°C (160°F) from power supply. Material used is modified polyphenylene oxide [114]. (Reprinted by the permission of the Society of Plastics Industry.)

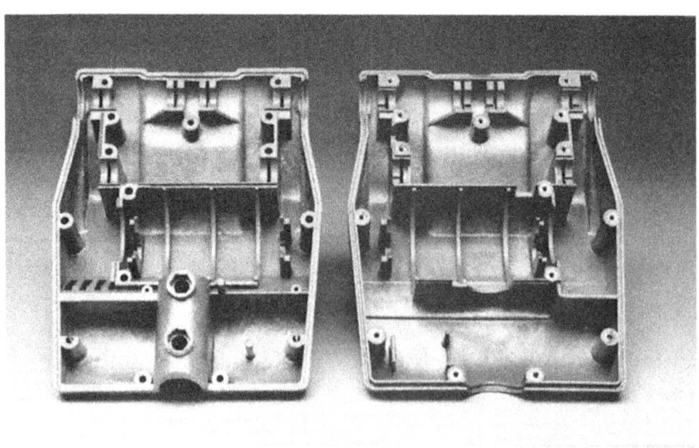

Figure 9.61 Weeder/cultivator that supports motor and gear train and provides chemical resistance to garden chemicals. Material used is 5% glass-filled polycarbonate [114]. (Reprinted by the permission of the Society of Plastics Industry.)

Figure 9.62 Office and home furniture that have structural integrity plus ability to hold tolerances with durable, abrasion-resistant, and colorable finish. Material used is RIM molded rigid polyurethane [114]. (Reprinted by the permission of the Society of Plastics Industry.)

Figure 9.63 A telecommunications housing produced with structural foam injection molding. The material used is foamed polycarbonate. (From *Injection Molding Alternatives—A Guide for Designers and Product Engineers* (1998), Avery, J. A., Hanser Publishers, Munich, p. 77, Fig. 5.2.16.)

chemistry and the advantages of the RIM process (cf. Sec. 9.15) makes the utilization of polyurethane structural foams possible in many applications [113].

9.18.5 Typical Applications

Because structural foam injection molding enables molding of large, complex parts with rigidity, dimensional stability, and load-bearing capability, it finds many suitable applications in material handling, business machines, automotive components, and medical analysis equipment. Figures 9.59 to 9.63 show a number of structural foam molding applications [114].

9.19 Thin-Wall Molding

Thin-wall (injection) molding is a high-speed, high-pressure injection molding process for producing parts with a nominal wall thickness less than 1.2 mm or flow-length-to-wall-thickness ratios ranging from 100:1 to 150:1 or more. Given the thickness restriction and extreme processing conditions, thin-wall molding has a smaller processing window. Nevertheless, this process becomes increasingly important due to the economic advantages of using thin walls and the unprecedented growth of portable electronic and telecommunication devices that require thinner, smaller, and lighter housings.

9.19.1 Process Description

The rapid growth in the telecommunication and portable electronics markets has created a strong demand for thinner wall-section applications and technology. These hand-held and portable devices, such as cellular phones (see Fig. 9.64), PDAs, and notebook computers, demand light-weighted plastic shells that are much smaller and thinner (less than 1.2 mm), yet still provide the same mechanical strength as conventional parts. Other emerging applications for thin-wall molding are in the manufacturing of medical, optical, and electronic parts or parts that have microscale features. Because decreasing wall thickness realizes part-weight reduction, material savings, and dramatic reduction of cooling time, the thin-wall molding concept is also gaining popularity with cost-conscious end users for large applications like computer monitor housings and automotive instrument panels and fascias.

The term *thin-wall* is relative. Conventional plastic parts are typically 2- to 4-mm thick. Thin-wall designs are called *advanced* when thicknesses range from 1.2 to 2 mm, and *leading-edge* when the dimension is less than 1.2 mm. Another definition of thin-wall molding is based on the flow-length-to-wall-thickness ratios. Typical ratios for these thin-wall applications range from 100:1 to 150:1 or more. Regardless of the definitions, thin-wall parts have a more restrictive flow path compared with conventional injection molded parts [cf. Fig. 9.65(a)]. As a result, they freeze off quickly during molding. To combat this situation, molders often attempt to increase the melt temperature by as much as 38 to 65°C beyond recommended range [115]. Another alternative to avoid premature freeze-off is to inject the material at an order-

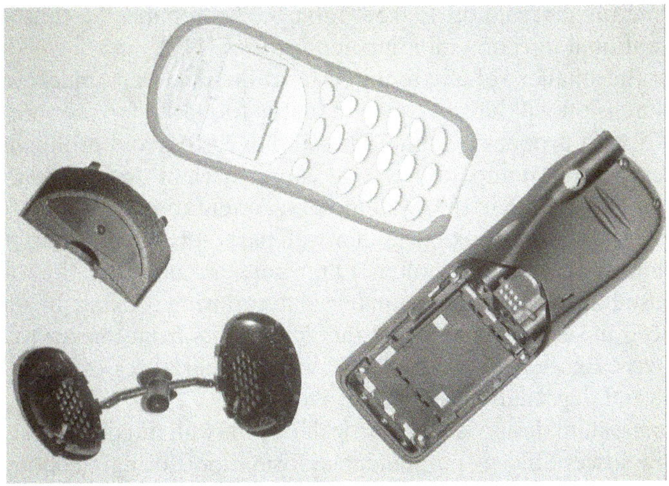

Figure 9.64 Plastic housings and components of cellular phones (Parts courtesy of Flambeau Micro, Sun Prairie, Wisconsin, USA). The cellular phone on the top was overmolded with thermoplastic elastomer to provide soft touch feeling and drop protection. (Part courtesy of Yomura Company Ltd., Taiwan.)

of-magnitude–higher injection speed (500 to 1400 mm/s ram speed). It therefore requires very high injection pressures (2400 to 3000 Bars) to achieve fast injection and adequate packing. Given the thickness restriction and extreme processing conditions, thin-wall molding has a smaller processing window as shown in Fig. 9.65(b). Other commonly used techniques to mold thin-wall parts include use of hot runners, multiple gates, and/or dynamic valve-gate control (e.g., sequentially opening and closing valve gates to relay the injection while avoiding weld lines). An optimized ram-speed profile and optimal fill time also help to reduce the pressure requirement [116]. In comparison to a standard injection molding fill time of around 2 s, thin-wall

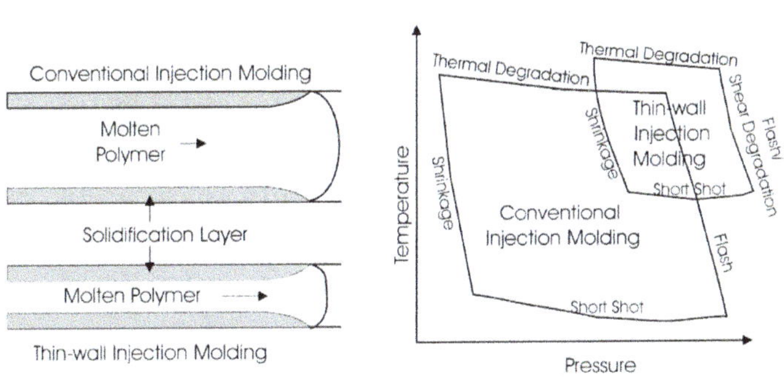

(a) (b)

Figure 9.65 Comparisons of (a) flow cross sections and (b) processing window between conventional injection molding and thin-wall injection molding.

molding fill time ranges from 0.1 to 0.5 s. Table 9.17 compares the thin-wall molding with the conventional injection molding applications [117].

Because of the smaller volume required from the thin walls, machines with standard barrel dimension will have a capacity that is too large for the shot volume. As a result, barrels with proper size should be employed to avoid prolonged residence time and excessive melt temperatures that cause thermal degradation. Due to the high velocity and shear rate in thin-wall molding, orientation occurs more readily. To help minimize anisotropic shrinkage in thin-wall parts, it is important to pack the part adequately while the core is still molten. Large gates, greater than the wall thickness, are generally used to ensure sufficient material flow during packing. In addition, gates should be placed in such a manner that the flow occurs from thicker to thinner sections of the cavity. Excessive residence time, melt temperatures, or shear, all of which can cause material degradation, should be avoided.

The ejection system design is more critical for thin-wall molding. First, thin-walled parts are more susceptible to permanent deformation during ejection. Second, to increase the part rigidity, thin-walled parts require reinforcing ribs, which requires higher ejection forces and complicates the ejection system design. Third, the part tends to adhere to the mold wall due to high injection and packing pressures. Finally, if underdesigned, the ejector pins may also buckle or even break during operation, interrupting production and incurring high maintenance costs.

Thin-walled parts for electronic or telecommunication devices should be designed with styling lines and curved surfaces to boost stiffness and enhance part aesthetics. Impact strategies involve using unreinforced plastic housings to absorb the load, using filled thermoplastics to transfer it, or using overmolding process to incorporate a protective thermoplastic elastomer layer (cf. Fig. 9.64). For all cases, the internal components should be fastened snugly, and the design should avoid stress concentration and sharp notches.

9.19.2 Process Advantages

Thin-wall molding realizes part-weight reduction, material saving, and dramatic reduction of cooling time. They also fulfill customers' demand, such as design requirements, cost reduction, and disposal concerns. In particular, conventional injection molding cycle time typically ranges between 30 and 60 s per cycle. Thin-wall molding, however, can run at a cycle time of 6 to 20 s, which cuts the cycle time by two thirds. This reduction in cycle time results in significant manufacturing cost reductions.

9.19.3 Process Disadvantages

Thin-wall molding is more technically challenging than conventional injection molding due to restrictive flow path, excessive processing conditions, and a smaller

Table 9.17 Comparison of Thin-Wall Molding with Conventional Injection Molding [117]

Key factors	Conventional applications	Thin-wall applications	
Wall thickness	2.0–3.0 mm (0.08–0.12 in)	1.2–2.0 mm (0.05–0.08 in)	<1.2 mm (<0.05 in)
Machinery	Standard	High-end	Customized high-end equipment
Injection pressure	620–970 Bars (9000–14,000 psi)	1100–1380 Bars (16,000–20,000 psi)	1380–2410 Bars (20,000–35,000 psi)
Hydraulic system	Standard	Standard	Accumulators on injection and clamp units, servo valves
Control system	Standard	Closed loop, microprocessor control over: injection speed, hold pressure, decompression speed, screw RPM, back pressure, all temperatures (including feed throat and oil temperature)	Microprocessor controlled with the following resolutions: speed—1 mm/s, pressure—1 Bar, position—0.1 mm, time—0.01 s, rotation—1 RPM, clamp force—0.1 ton, temperature —1°C
Screw design	Compression ratio: 2.0:1 to 2.5:1, L/D = 20:1 to 24:1, flights 5/10/5: Nitriding not suggested	Compression ratio: 2.0:1 to 2.5:1, L/D = 20:1 to 24:1, flights 5/10/5: Nitriding not typically used	Compression ratio: 2.0:1 to 2.5:1, L/D = 20:1 to 24:1, flights 5/10/5: Nitriding not typically used
Fill time	>2 s	1–2 s	0.1–1 s
Cycle time	40–60 s	20–40 s	6–20 s
Drying	Dew point of −29°C to −40°C; hoppers sizes for material throughput	Dew point of −29°C to −40°C; hoppers sizes for material throughput	Dew point of −29°C to −40°C; hoppers sizes for material throughput
Tooling	Standard	Better venting, heavier construction, more ejector pins, better polish	Extreme venting, very heavy molds, mold interlocks, precise mold surface preparation, extensive ejection features, mold costs 30–40% higher vs. standard

processing window. It also requires rugged, costly tools, and perhaps an upgraded or customized molding machine for high-speed, high-pressure injection. Finally, the high shear rate level that results from the high-speed injection causes shear-induced degradation, as observed by the reduction of failure strain in tensile test results [118].

9.19.4 Applicable Materials

Most of the thermoplastics can be used for thin-wall applications; However, engineering resins, such as PC, acrylonitrile-butadiene-styrene (ABS), PC/ABS blends, and polyamide 6, are more commonly used for thin-wall molding than are commodity resins. This is probably because, as the wall thickness reduces, better physical properties of engineering resins are needed to maintain certain part strength.

9.19.5 Typical Applications

Thin-wall molding is more popular in portable communication and computing equipment, which demand plastic shells that are much thinner yet still provide the same mechanical strength as do conventional parts. Typical applications include cellular phones, pagers, notebook computer housings, medical devices, car stereo faces, minimally invasive surgical devices, electronic connectors, interior and exterior panels for automobiles, and optical storage media with pits on the thin plastic substrate (0.6 mm for DVD).

9.20 Vibration Gas Injection Molding

Vibration gas injection molding (Vibrogaim or vibrated gas-assist molding) is one of the melt-flow oscillation techniques that uses pressurized and vibrated gas to manipulate the polymer melt during processing to modify its flow behavior and to improve the mechanical and optical properties of the molded parts.

9.20.1 Process Description

Vibration gas injection molding is among the latest melt-flow oscillation techniques that provide melt manipulation capabilities during molding [119,120]. Other melt-vibrating technologies, such as multiple live feed injection molding, push–pull injection molding, and rheomolding, have been discussed in Secs. 9.12, 9.13, and 9.17, respectively. Vibration gas injection molding uses the pressurized gas (typically nitrogen) as the tool or as a "gas spring" for delivering and controlling the vibration from the generating devices to the plastic melt. Gas vibration can be applied at various stages and locations during processing and vibrated from subsonic, low frequency (1 to 30 Hertz) all the way up to ultrasonic range (15,000 to 20,000 Hertz). Whereas high-frequency vibrations fuse weld lines, accelerates the stress relaxation of plastics, and alter the rate of crystallization and crystal growth, low-frequency vibrations help change molecular orientation in a desirable way. In addition, resonance, which is a condition at which the entire system vibrates at its natural frequency, extracts the energy from the vibration and produces physical changes in the targeted system. One of the unique features of vibration gas injection molding is its ability to combine different modes of various frequencies and amplitudes to take advantages of their distinctive effects on the plastic melt.

The pressurized gas can be introduced directly into the mold from the back of the cavity, through the injection nozzle, or via channels or shaped chambers at specific locations in the mold near the runners and gates. Gas vibration is generated by the vibration sources, such as pneumatic, electrical, or mechanical transducers. Other hardware required by vibration gas injection molding includes a gas mixing unit, gas pressuring/injection units, and a computer control system for monitoring and maintaining resonant condition.

The unique advantage of this process is the ability to alter the morphology of the plastics, thereby modifying the mechanical properties, such as tensile strength, modulus, and impact. For example, it reportedly boosted the crystallinity in polypropylene by 25%, which resulted in improvement in low-temperature impact and clarity [119]. Detailed descriptions on the process benefits and drawbacks can be found in the various sections that deal with other melt-flow oscillation techniques.

9.21 Water-Assisted Injection Molding

Water-assisted injection molding was developed in the late 1990's as an alternative to gas-assisted injection molding and its variants, such as the liquid gas-assisted process (see Section 9.3) [121, 122]. It is also called water-assisted injection technology (WIT). Although water-assisted injection molding is very similar to gas-assisted injection molding, this process offers shorter cycle times, reduced wall thicknesses, smoother inner surfaces, and greater part and cored-out diameters compared to its gas counterparts. It also achieves material savings for the production of large cross-sectioned, tube- and rod-like, and/or fluid-carrying parts such as handles, arm rests, roof rails, and media ducts and pipes (Fig. 9.66).

9.21.1 Process Description

Akin to gas-assisted injection molding, water-assisted injection molding consists of a partial or nearly full injection of polymer melt into the mold cavity, followed by injection of water using one or more hydraulic pumps into the core of the polymer melt. At the interface of the water front and the melt, a thin layer of cold polymer is formed, thus acting as a piston or plow that displaces molten polymer at the hot core of thick-sectioned or tube- and rod-like parts to fill the rest of the cavity. For certain

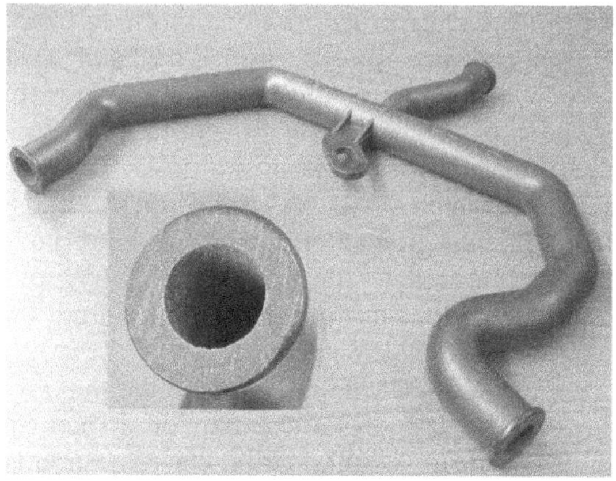

Figure 9.66 A water-assisted injection molded automotive coolant tube using a 30% glass-filled polyamide (nylon). The wall thickness and inner surface are shown in the inset.

applications, overflow wells or flash cavities are used and later trimmed off, allowing production of entirely hollowed components like the automotive coolant tube shown in Figure 9.66. After the cavity is volumetrically filled, water holding pressure is maintained to pack out the entire cavity. Typical water pressures range from 200 to 300 bars but can go much higher. The volumetric flow rate of water can vary from 27 L/min to 130 L/min. Water temperature can run as high as 60 °C with no detrimental effects on part properties while lower water temperatures may run the risk of freezing the melt around the water injector or nozzle. After the material is sufficiently cooled and solidified, the water either stays in the part and is later drained off outside of the mold, or it is blown out of the part by compressed air which also helps dry the part [122]. As long as the water injection molding takes place rapidly, the hydrolytic degradation of hydrophilic polymers (e.g., polyamides) is not a concern [121]. Typically, the water is recirculated at a regulated temperature using a storage tank. A purification step is needed to avoid possible contamination from the resin, fillers, or additives.

9.21.2 Process Advantages

The main advantage of water-assisted injection molding is the shorter cycle times [122]. Due to better cooling effect and higher heat capacity of water, the cycle time for water-assisted injection molding is much shorter than that of gas-assisted injection molding. Since less amount of polymer melt has to be cooled as a result of the water penetration and cored out along the thick-sectioned channels, the cycle time for water-assisted injection molding can also be shorter than that of conventional injection molding. Unlike gas, water is incompressible and thus is capable of pressurizing the cavity effectively and evenly and requires less energy to produce the same results as gas-assisted injection molding. Other process advantages include reduced material consumption, design freedom to incorporate hollowed sections, smooth inner surface that is desirable for fluid flow within the part, elimination of sink marks, and better dimensional stability.

9.21.3 Process Disadvantages

The disadvantages of water-assisted injection molding include an additional operation to remove the water from the molded part; jetting or spray of water; formation of voids if water injection and holding pressures are too low; partial shaping with stream, bubbles, turbulence, chambers, pores, and fingering (water permeation into thin sections, [123]) resulting from errors in processing or design; and rust-proof requirements of the tool. It also requires potential licensing fees and precise control of additional process parameters, such as delay time for water injection, water volumetric flow rate, and water pressure and temperature, all of which affect the water penetration and residual wall thickness. Like gas-assisted injection molding, employing water-assisted injection molding for a multi-cavity system is challenging due to the complexity of process control and the competition of water penetration into each individual cavity.

9.21.4 Applicable Materials

Water-assisted injection molding has been successfully applied to a number of semi-crystalline polymers such as polyamides (PA), polyethylenes (PE), polypropylenes (PP), polybutylene terephthalates (PBT), and polyacetals or polyoxymethylenes (POM), as well as amorphous materials such as polycarbonates (PC), acrylonitrile-butadiene-styrene (ABS), PC/ABS, and high impact polystyrene (HIPS). In addition, thermoplastic copolyester elastomers (copolyester TPEs) are being used with water-assisted injection molding. Some amorphous materials have been found to be prone to formation of voids, especially with low water pressures.

9.22 Rubber Injection*

9.22.1 Rubber Molding Processes

Compression Molding

During compression molding, a piece of preformed material is placed directly in the mold cavity and compressed under hydraulic clamp pressure. At the completion of the required cure cycle, the hydraulic clamp is released and the part is stripped

* Contributed by Mauricio DeGreiff and Nelson Castaño.

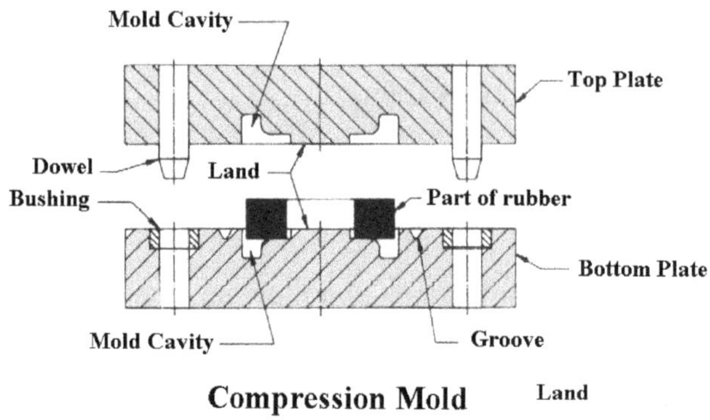

Figure 9.67 Schematic of compression molding for rubbers.

from the mold (cf. Fig. 9.67). Compression molding is usually done in hydraulically clamped presses acting through a ram.

As the mold cavities are open when the preform is introduced, compression molding usually possesses the least tolerance control and most flash. Exact weight is required in the preforming stage to ensure that additional variation is minimized. In addition, stock flow and air venting must be carefully addressed. Along with the preform shape, careful selection of the press closing rate and the bump cycle are required. The press bumping is a short pressure release after initial clamping to "burp" the mold.

As little shear is evident in the compression process, temperatures are obtained only by platen contact with the mold. Preheating of the rubber is occasionally used to speed the curing process. The dimensional requirements of the part most often determine whether or not to use compression methods.

Transfer Molding

Transfer molding is a precision molding technique. As shown in Fig. 9.68, the preformed stock flows from a transfer pot, usually above the mold cavities, to the parts below through sprues. The transfer ram and pot can be part of the mold or can be part of the press. All transfer molding is done with a closed cavity; therefore, less fin-

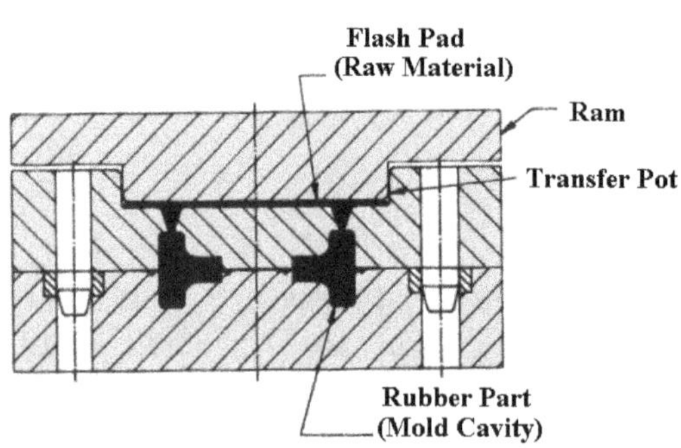

Figure 9.68 Schematic of transfer molding for rubbers.

ishing is required. In addition, more precision is obtained. Transfer molding, however, is also a low shear process and little self-heating of the compound takes place. As a result, cure times are not much shorter than compression molding.

Injection Molding

Injection molding is a versatile, high-precision process. The cost to produce a part is often the lowest, although the press and molds are the most expensive. The injection process employs a closed mold, with the stock delivered under high shear from an injection barrel (cf. Fig. 9.69). Several advantages result from this high shear. First, the rubber is at elevated temperatures as it enters the cavity, allowing for short cure times. This is most particularly advantageous in thick cross-section parts. Second, the viscosity of the rubber compound is lowered, allowing for considerable thermoplastic flow.

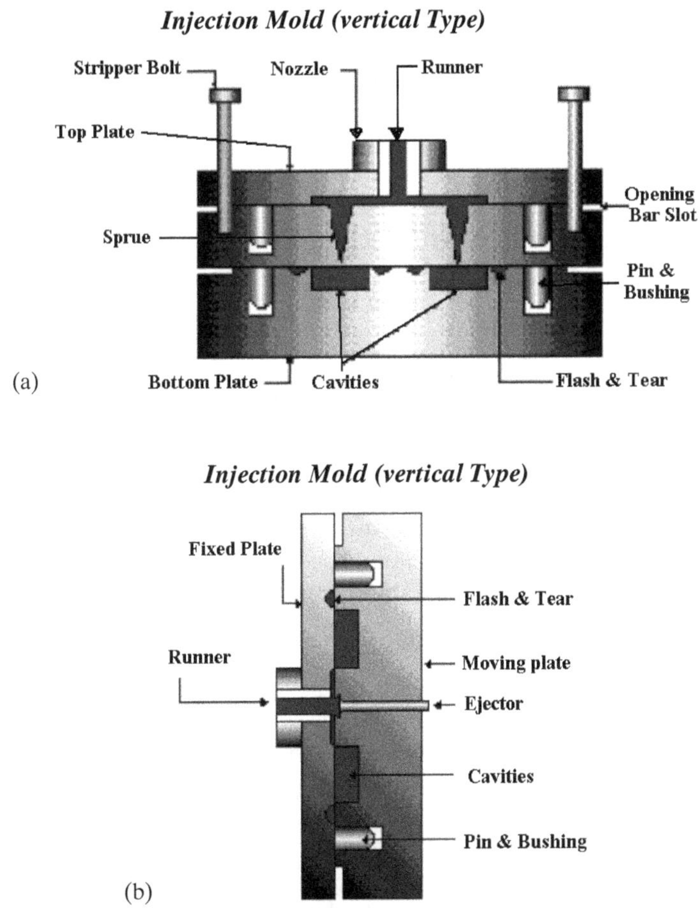

Figure 9.69 Schematic of injection molding for rubbers.

Injection presses can be vertical or horizontal. They can also use either a ram or a screw to masticate and deliver the uncured compound to the mold. The reciprocating screw masticates stock better and maintains even heat, but shot sizes are limited. The ram can deliver large quantities of stock to the mold, frequently at the expense of uniform mastication and heat. Hybrid processes have thus been developed, with a screw/ram combination commonly used. In these cases, the screw is used only for stock mastication, whereas the ram is used in injection.

Referring to the following figures, the uncured stock is delivered through an injection nozzle into a sprue bushing. Rubber flows through the runner system into the parts via sprues. It is most important that the flow of stock and heat history of the compound be as uniform as possible. Symmetrical or "balanced" runner systems are therefore employed.

As the mold is closed and the rubber heated during injection, tight tolerance control is achieved with minimal flash. If finishing is required, it is usually adequate to employ a bulk cryogenic tumbling technique.

A negative factor with injection molding is the large material waste generated through the runner system. "Cold runner systems" have been designed to address this deficiency. On the other hand, injection molding has many advantages. Tooling for this process, however, is most expensive because high-precision, hardened tools are often required for the high volumes encountered.

9.22.2 Curing Systems in Rubber Injection Process

Curing Systems in Common Use

A manufacturer transferring from a compression to an injection molding process may fairly safely carry out the first trials without modifyirig the compound, relying on adjustments of barrel temperature to obtain reasonable operating conditions.

Some typical formulations for NR and NBR polymers are listed in Tables 9.18 and 9.19, together with curing systems selected to offer a range of processing and cure requirernents. These are based either on MBTS, sulphenamides, or Sulfasan R because of the need for a certain minimum of scorch safety in the compounds.

Efficient Vulcanizing Systems

Efficient vulcanizing (EV) systems are defined as those where a high proportion of sulphur is used for a cross-linking purpose. These systems have two main advantages over conventional systems, giving vulcanizates with reduced reversion and better aging characteristics. In addition to these advantages, EV systems based on dithiodimorpholine (DTM) are very versatile, enabling a wide range of scorch times, cure rates, and states of cure to be chosen at will.

It is particularly important for injection molding of thick sections to avoid reversion. EV systems give the complete answer to this problem. The conventional system (sulphur/MBTS/DPG) shows reversion immediately after the maximum modulus is reached, whereas the EV system (DTM/MBTS/TMTD) shows no reversion even

Table 9.18 Typical Formulations for NR Polymers

Black NR formulations					
Natural Rubber	70				
Whole tyre reclaim	60				
Carbon black	75				
Zinc oxide	40				
Stearic acid	5				
Paraffin wax	2				
Antiozonant	1				
Curing systems	**A**	**B**	**C**	**D**	**E**
Sulphur	2.5	2.5	2.5	—	—
Dithiodimorpholine—DTM	—	—	—	1.4	1.2
Sulphenamide	1.2	1.2	—	1.4	1.2
MOR	—	—	1.2	—	—
Thiuram	0.3	—	—	0.2	0.5

A, B, and C are suitable for thin-section products and are in ascending order of scorch time. D and E are efficient vulcanizing systems suitable for thick sections. They give much-reduced reversion and improved ageing resistance.

after three times the optimum cure time. EV systems can be developed to give equivalent cure properties with much improved aging as compared with a conventional cure, even when antioxidants are omitted.

Conclusions

Accelerator systems for injection molding should be chosen to give adequate scorch time, fast cure without reversion, and appropriate product properties. When molding

Table 9.19 Typical Formulations for NBR Polymers

Nitrile rubber formulations (NBR)				
Nitrile	100			
Carbon black	80			
Dioctyl phthalate	5			
Zinc oxide	5			
Stearic acid	1			
Curing systems	**A**	**B**	**C**	**D**
Sulphur	1.5	1.5	0.5	—
TMTD	0.5	—	3.0	—
MBTS	1.0	1.5	—	3.0

A and B are conventional curing systems that may be adequate where aging resistance is not a particular problem. C is a low sulphur system giving much-improved aging, but its scorch time is usually sufficient only for ram type injection. D combines excellent aging with a scorch time long enough for most applications.

thick sections from polymers that revert (e.g. NR), EV systems should be used to minimize reversion. Combinations of a sulphenamide, DTM, and TMTD are ideal, and the ratios can be varied to meet precise machine operating conditions and product requirements. Conventional sulphur/accclerator systems can be used where reversion is not a problem, and the following accelerators will give the best cure rates for each scorch time requirement.

Conventional sulphur/accclerator systems can be used where reversion is not a problem, and the accelerators, such as MOR, TBBS, and CBS/TMTD (in the order of decreasing scorch time), will give the best cure rates for each scorch time requirement. Accelerator loadings may be increased to give improved product properties or to counter the effect of oil addition.

9.23 Liquid Silicone Rubber* of Injection Molding

Over the past 25 years, injection molding of liquid silicone rubber (LSR) has evolved the thermosetting nature of the material, liquid silicone injection molding requires special treatment, such as intensive distributive mixing, while maintaining a cool material temperature before the material is pushed into the heated cavity and vul-

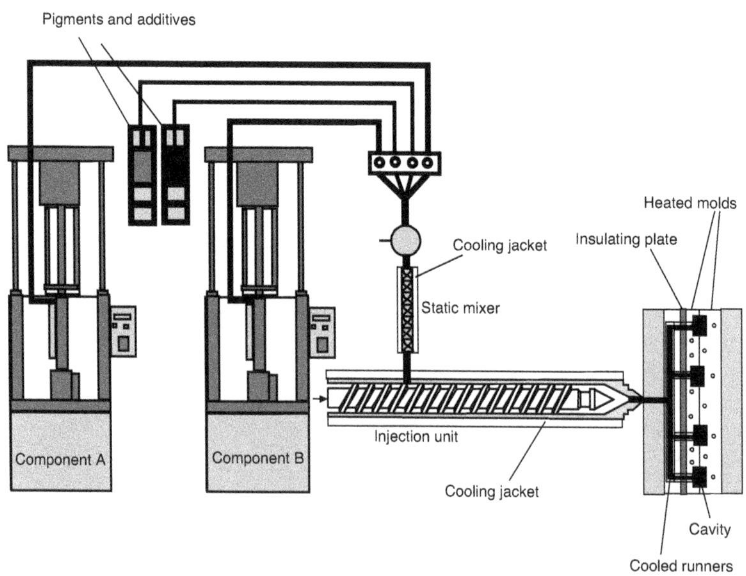

Figure 9.70 Schematic diagram of a liquid silicone rubber injection molding system.

* Contributed by Tim A. Osswald

canized. Figure 9.70 schematically depicts an LSR injection molding process. Liquid silicone rubbers are supplied in barrels or hobbocks. Because of their low viscosity, these rubbers can be pumped through pipelines and tubes to the vulcanization equipment. The two components (labeled A and B in the figure) are pumped through a static mixer by a metering pump. One of the components contains the catalyst, which is typically platinum based. A coloring paste, as well as other additives, can also be added before the material enters the static mixer section. In the static mixer, the components are well mixed before being transferred to the cooled metering section of the injection molding machine.

The static mixer renders a homogeneous material that yields products that are not only consistent throughout the part, but also from part to part. This is in contrast to solid silicone rubber materials that are purchased premixed and partially vulcanized. As a result, hard silicone rubbers are processed by transfer molding and result in less material consistency and control, leading to higher part variability. Additionally, solid silicone rubber materials are processed at higher temperatures and require longer vulcanization times.

From the metering section of the injection molding machine, the compound is pushed through cooled sprue and runner systems into a heated cavity where the vulcanization takes place. The cold runner and general cooling result in no loss of material in the feed lines. A schematic of a liquid silicone rubber mold with cooled runners is depicted in Fig. 9.71. The cooling allows production of LSR parts with nearly zero material waste, eliminating trimming operations and yielding significant savings in material cost.

Silicone rubber is a family of thermoset elastomers that have a backbone of alternating silicone and oxygen atoms and methyl or vinyl side groups. They constitute about 30% of the silicone family, making them the largest group of that family. Silicone rubbers maintain their mechanical properties over a wide range of temperatures and the presence of methyl-groups in silicone rubbers makes these materials extremely hydrophobic. Typical applications for liquid silicone rubber are products that require high precision, such as seals and sealing membranes; electric connectors;

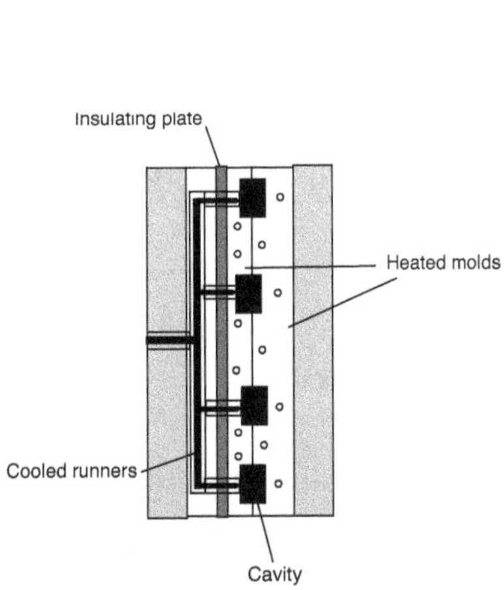

Figure 9.71 Schematic diagram of a liquid silicone rubber injection mold.

Figure 9.72 Photographs of injection molded liquid silicone rubber parts.

(a) (b)

Figure 9.73 Photographs of (a) an injection molded liquid silicone rubber single wire seal, and (b) a peripheral seal. These parts were molded with an oil bleeding formulation, a silicone rubber-silicone oil blend for a continuous lubrication effect. (Courtesy Simtec Silicone Parts, LLC, Madison, WI, USA).

multi-pin connectors; infant products where smooth surfaces are desired, such as bottle nipples; medical applications; and kitchen goods, such as baking pans and spatulas. Figures 9.72 and 9.73 present some examples of liquid silicone injection molded parts.

Acknowledgment

The author would like to express his appreciation to Professor Tim A. Osswald and Dr. Paul Gramann for reviewing this chapter and for their useful suggestions. In addition, he wants to acknowledge the students who took the special-topic course on

"Fundamental of Injection Molding" in the Fall of 2000 for their contribution to this chapter. In particular, he likes to thank Cory Celestino, Juan Hernandez, Chanyut Kolitawong, Maria D. P. Noriega E., Michael Peic, Jeff Stauffacher, Andreas Winardi, Sam Woodford, and Mingjun Yuan for their help.

References

1. Rosato, D. V., Rosato, D. V., *Injection Molding Handbook: The Complete Molding Operation Technology, Performance, Economics*, Second Ed. (1995), Chapman and Hall, New York.
2. Wang, K. K., et al., *Computer-Aided Injection Molding Systems, Progress Reports Nos. 1–21* (1975–2000) Cornell Injection Molding Program (CIMP), Cornell University, Ithaca, New York.
3. Kamal, M. R., Kenig, S., *Polym. Eng. Sci.* (1972) 12, 294, 302.
4. Berger, J. L., Gogos, C. G., *Polym. Eng. Sci.* (1973) 13, 102.
5. Tadmor, Z., *J. Applied Polym. Sci.* (1974) 18, 1752.
6. Hieber, C. A., Shen, S. F., *J. Non-Newtonian Fluid Mechanics* (1980) 7, 1.
7. Wang, V. W., Hieber, C. A., Wang, K. K., *J. Polym. Eng.* (1986) 7(1), 21.
8. Isayev, A. I. (Ed.), *Injection and Compression Molding Fundamentals* (1987) Marcel Dekker, New York.
9. Coyle, D. J., Blake, J. W., Macosko, C. W., *AICHE Journal* (1987) 33(7), 1168.
10. Manzione, L. T. (Ed.), *Applications of Computer Aided Engineering in Injection Molding* (1987) Hanser, Munich.
11. Lee, L. J., Wu, C. H., Chiu, H. T., Nakamura, S., *International Polymer Processing* (1998), 13(4), 389.
12. Tucker, C. L., *Fundamentals of Computer Modeling for Polymer Processing* (1989) Hanser, Munich.
13. Advani, S. G. and Beckermann, C. (Ed.), *Heat and Mass Transfer in Solidification Processing* (1991) HTD-Vol. 175/MD-Vol. 25, ASME, New York.
14. Castro, J., Conover, S., Wilkes, C., Han, K., Chiu, H. T., Lee, L. J., *Polymer Composites* (1997) 18(5), 585.
15. Wang, H. P., Turng, L. S., Marchal, J. M. (Eds.), *CAE and Intelligent Processing of Polymeric Materials* (1997), MD-Vol. 79, ASME, New York.
16. Turng, L. S., Wang, H. P., Ramani, K., Benard, A. (Eds.), *CAE and Related Innovations for Polymer Processing* (2000), MD-Vol. 90, ASME, New York.
17. Wang, J. T., In *CAE and Related Innovations for Polymer Processing*, Turng, L. S., Wang, H. P., Ramani, K., and Benard, A. (Eds.), (2000), ASME, New York, pp. 257–271.
18. Turng, L. S., Wang, V. W., Wang, K. K., *J. Eng. Mat. Technol.* (1993) 115, 48.
19. Turng, L. S., *Adv. Polymer Technol.* (1995) 14(1), 1.
20. Chen, S. C., Chen, Y. C., Cheng, N. T., Huang, M. S., *J. Reinforced Plastics and Composites* (1999), 18(8), 724.
21. Escales, E., *German Plastics* (1970) 60, 16–19. Translated from *Kunststoffe* (1970) 60, 847.
22. Eckardt, H., Davies, S., *Plastics and Rubber International* (1979) 4(2), 72.
23. Kirkland, C., *Injection Molding Magazine* (May 1996), 85.
24. Eckardt, H., *German Plastics* (1985) 75(3), 10–15. Translated from *Kunststoffe* (1985) 75(3), 145.
25. Avery, J. A., *Injection Molding Alternatives—A Guide for Designers and Product Engineers* (1998), Hanser, Munich.
26. Young, S. S., White, J. L., Clark, E. S., Oyanagi, Y., *Polym. Eng. Sci.*, (1980) 20(12), 798.
27. Avery, J. A., In *Innovation in Polymer Processing—Molding*, Stevenson, J. F. (Ed.) (1996), Hanser, Munich, 331–365.
28. Eckardt, H., Paper presented at the Fourteenth Annual Structural Foam Conference and Parts Competition, Boston, MA, April (1986).

29. Hagen, R., *German Plastics* (1989) 79(1), 27–30. Translated from *Kunststoffe* (1989) 79(1), 72.
30. Hauck, C., Schneiders, A., In *Innovation in Polymer Processing—Molding*, Stevenson, J. F. (Ed.) (1996), Hanser, Munich, 151–191.
31. Schmachtenberg, E., Polifke, M., *Kunststoffe Plast Europe* (1996) 86(3), 16. Translated from *Kunststoffe* (1996) 86(3), 322.
32. Gabriele, M. C., *Mod. Plast.* (April 1999), 70.
33. Eckardt, H., In *Innovation in Polymer Processing—Molding*, Stevenson, J. F. (Ed.) (1996) Hanser, Munich, 1.
34. Turng, L. S., In *Innovation in Polymer Processing—Molding*, Stevenson, J. F., (Ed.) (1996) Hanser, Munich, 43.
35. Hettinga, S., Product Brochure, Hettinga Technologies, Inc. Des Moines, IA, Web site address: *http://www.hettingatechnology.com/*
36. Michaeli, W., Brunswick, A., Kujat, C., *Kunststoffe Plast Europe*, (2000) 90(8), 25. Translated from *Kunststoffe* (2000) 90(8), 67.
37. Moore, S., *Mod. Plast.*, January (1993) 26.
38. Pearson, T., Product Brochure, Gas Injection Limited, Cheshire, U.K.
39. Moore, S., *Mod. Plast.*, November (1999) 28.
40. Kleba, I., Haberstroh, E., *SPE ANTEC Tech. Papers* (2000), 559.
41. Hopmann, C., Michaeli, W., *SPE ANTEC Tech. Papers* (1999), 417.
42. Potente, H., Hansen, M., *Int. Polym. Processing* (1993), 8(4), 345.
43. Dubois, J. H., Pribble, W. I., *Plastics Mold Engineering Handbook* (1987) Van Nostrand Reinhold Company, New York.
44. Bürkle, E., Wohlrab, W., *Kunststoffe* (1999) 89(9), 64.
45. Wang, J. T., In *CAE and Intelligent Processing of Polymeric Materials*, Wang, H. P. et al. (Eds.) (1997), ASME, New York, pp. 83–96.
46. Kim, I. H., Park, S. J., Chung, S. T., Kwon, T. H., *Polym. Eng. Sci.* (1999), 39(10), 1930, 1943.
47. Chen, S. C., Chen, Y. C., Peng, H. S., *J. of Applied Polym. Sci.* (2000), 75(13), 1640.
48. *Engineering Design Database Design Guide* (1991) GE Plastics, Pittsfield, MA, USA.
49. Pötsch, G., Michaeli, W., *Injection Molding—An Introduction*, (1995) Hanser, Munich.
50. Gabriele, M. C., *Mod. Plast.* April (1999), 76, 70.
51. Grande, J. A., *Mod. Plast.* September (1998) 75, 35.
52. Snyder, M. R., *Mod. Plast.* October (1999) 76, 54.
53. Jaeger, A., Fischbach, G., *Kunststoffe German Plastics* (1991) 81(10), 21. Translated from *Kunststoffe* (1991) 81(10), 869.
54. Bürkle, E., Rehm, G., Eyerer, P., *Kunststoffe Plast Europe* March (1996) 86, 7. Translated from *Kunststoffe* (1996) 86(3), 298.
55. Böcklein, M., Eckardt, H., *Kunststoffe German Plastics* (1986) 76(11), 17. Translated from *Kunststoffe* (1986) 76(11), 1028.
56. Steinbichler, G., *Kunststoffe Plast Europe* March (1995) 85, 18. Translated from *Kunststoffe* (1995) 85(3), 337.
57. Anders, S., Littek, W., Schneider, W., *Kunststoffe German Plastics* (1990) 80(9), 20. Translated from *Kunststoffe* (1990) 80(9), 997.
58. Shah, S., Kakarala, N., *Plastics Engineering* September (2000) 56, 51.
59. Snyder, M. R., *Mod. Plast.* November (1998) 75(11), 71.
60. Anonymous, *Kunststoffe German Plastics* (1978) 68(11), 4. Translated from *Kunststoffe* (1978) 68(7), 394.
61. Schrenk, W. J., U.S. Patent 5 202 074 (1993).
62. Barger, M. A., Schrenk, W. J., *In Innovation in Polymer Processing—Molding*, Stevenson, J. F. (Ed.) (1996) Hanser, Munich, pp. 293–330.
63. Hettinga, S., *In Innovation in Polymer Processing—Molding*, Stevenson, J. F. (Ed.) (1996) Hanser, Munich, 193.
64. Turng, L. S., Chiang, H. H., Stevenson, J. F., *Plastics Engineering* October (1995) 33.
65. Turng, L. S., Chiang, H. H., Stevenson, J. F., *SPE ANTEC Tech. Papers* (1995) 668.
66. Snyder, M. R., *Mod. Plast.*, January (1999) 76(1), 85.

67. Ruprecht, R., Bacher, W., Hausselt, J. H., Piotter, V., *Proc. SPIE Symp. Micromachining and Microfabrication Process Technology* (1995) 2639, 146.
68. Weber, L., et al. *Proc. SPIE Symp. Micromachining and Microfabrication Process Technology II* (1996) 2879, 156.
69. Weber, L., Ehrfeld, W., *Kunststoffe Plast Europe* (1998), 88(10), 60–63. Translated from *Kunststoffe* (1998) 88(10), 1791.
70. Kukla, C., Loibl, H., Detter, H., Hannenheim, W., *Kunststoffe Plast Europe* (1998) 88(9), 6–7. Translated from *Kunststoffe* (1998) 88(9), 1331.
71. Seidler, D., Zelenka, R., *Kunststoffe Plast Europe* (1998), 88(9), 7. Translated from *Kunststoffe* (1998) 88, 1338.
72. Weber, L., Ehrfeld, W., *Kunststoffe Plast Europe* (1999) 89(10), 64. Translated from *Kunststoffe* (1999) 89(10), 192.
73. Becker, E. W., et al. *Microelectron. Eng.* (1986) 4, 35.
74. Ehrfeld, W., et al., *J. Vac. Sci. Technol. B* (1988) 6, 178.
75. Michaeli, W., Rogalla, A., *Kunststoffe für die Mikrosystemtechnik* (1997) 6, 50.
76. Schinköthe, W., Walther, T., *Kunststoffe Plast Europe* (2000) 90(5), 17–19. Translated from *Kunststoffe* (2000) 90(5), 62.
77. Holzhauer, M., Zippmann, V., *Kunststoffe Plast Europe* (2000) 90(9), 18. Translated from *Kunststoffe* (2000) 90(9), 64.
78. Mapleston, P., *Mod. Plast.*, July (1997) 74, 41.
79. Suh, N. P., "Microcellular Plastics," In *Innovation in Polymer Processing—Molding*, Stevenson, J. F. (Ed.) (1996) Hanser Publishers, Munich, 93–149.
80. Park, C. B., Doroudiani, S., Kortschot, M. T., *Polymer Engineering and Science* (1998) 38(7), 1205.
81. Shimbo, M., Baldwin, D. F., Suh, N. P., *Polymer Engineering and Science* (1995) 35(17), 1387.
82. Seeler, K. A., Kumar, V., *Journal of Reinforced Plastics and Composites* (1993) 12, 359.
83. Gulari, E., Manke, C. W., Paper presented at the *219th ACS National Meeting, I&EC*—106, San Francisco, CA, March 26–30 (2000).
84. Royer, J. R., Gay, Y. J., Desimone, J. M., Khan, S. A., *Journal of Polymer Science, Part B: Polymer Physics* (2000) 38, 3168.
85. Kwag, C., Manke, C. W., Gulari, E., *Journal of Polymer Science, Part B: Polymer Physics* (1999) 37, 2771.
86. Zhang, Z., Handa, Y. P., *J. Polym. Sci. Part B: Polym. Phys.* (1998) 36, 977.
87. Handa, Y. P., Kruus, P., O'Neill, M., *J. Polym. Sci., Part B: Polym. Phys.* (1996) 34, 2635.
88. Knights, M., *Plast. Tech.* (Sept. 2000) 46, 40.
89. Trexel, Inc. Web site, *http://www.trexel.com/*.
90. Malloy, R. A., *Plastic Part Design for Injection Molding—An Introduction* (1994), Hanser Publishers, Munich.
91. Mapleston, P., *Mod. Plast.* April (2000), 46–51.
92. Product brochure, *The SCORIM Process*, SCORTEC, Inc., Gulph Mills, PA.
93. Reis, R. L., Cunha, A. M., Bevis, M. J., *Mod. Plast.* May (1999) 76(5), 73.
94. Hills, K. Paper presented at Molding '95. March (1995), New Orleans, LA.
95. Rawson, K. W., Allan, P. S., Bevis, M. J., *Polym. Eng. Sci.* (1999) 39(1), 177.
96. Becker, H., Malterdingen, G. F., Müller, U., *Kunststoffe German Plast* (1993) 83(3), 3. Translated from *Kunststoffe* (1993) 83(3), 165.
97. Theberge, J., *Plast. Eng.* February (1991) 47, 27.
98. Becker, H., Ekenhorst, D., Fischer, N., Paper presented at the *Int. Symp. on Advanced Materials for Lightweight Structures*, March (1994), *ESTEC*, Noordwijk.
99. Harada, T., *Kogyo Zairyo (Engineering Materials)* (1991), 39(15), 65 (in Japanese).
100. German, R. M., Bose, A., *Injection Molding of Metals and Ceramics* (1997), Metal Powder Industries Federation, Princeton, NJ.
101. Ballard, C., Zedalis, M., *SPE ANTEC Tech. Papers* (1998), 358.
102. Bayer, M., *Kunststoffe Plast Europe* (2000), 90(7), 17. Translated from *Kunststoffe* (2000) 90(7), 46.

103. Schumacher, C., *Kunststoffe Plast Europe* (1999), 89(9), 23. Translated from *Kunststoffe* (1999) 89(9), 70–72.
104. Petzoldt, F., *Kunststoffe Plast Europe* (2000), 90(7), 14–16. Translated from *Kunststoffe* (2000) 90(7), 40–43.
105. Johnson, C. F., *Engineering Plastics*, Vol. 2, Engineered Materials Handbook, ASM International (1988) 344.
106. González-Romero, V. M., Macosko, C. W., *Polym. Eng. Sci.* (1990) 30(3), 142.
107. Ibar, J. P., *Plast. Eng. Sci.*, (1998) 38(1), 1–20.
108. Ibar, J. P., *Mod. Plast.* January (1995) 72(1).
109. Ibar, J. P., U.S. Patent 5,605,707 (1995).
110. Brewer, G. W., In *Engineered Materials Handbook* (1998) Vol. 2, ASM International, 508.
111. Wu, J. S., Lee, M. J., *Plastics, Rubber and Composites Processing and Applications* (1994) 21(3), 163.
112. Tewald, A., *Kunststoffe Plast Europe*, July (1998) 88(7), 13. Translated from *Kunststoffe* (1998) 88(7), 970.
113. Macosko, C. W., *RIM Fundamentals of Reaction Injection Molding—Introduction* (1988), Hanser, Munich.
114. Farrah, M., Hanson, D., Wendle, B., (Eds.), *Structural Foam* (1984), Structural Foam Division, Society of Plastics Industry, Washington, DC, USA.
115. Fassett, J., *Plastics Engineering*, December (1995) 35–37.
116. Stevenson, J. F., Dubin, A., *J. Injection Molding Technology*, December (1999) 3(4), 181.
117. GE Plastics, Thin-wall Technology Guide (1998).
118. Wang, H. P., Ramaswamy, S., Dris, I., Perry, E. M., Gao, D., In *CAE and Related Innovations for Polymer Processing*, Turng, L. S., Wang, H. P., Ramani, K., Benard, A., (Eds.) (2000), MD-Vol. 79, AMSE, New York, 193.
119. Ibar, J. P., *SPE ANTEC Tech. Papers* (1999), 552.
120. Ibar, J. P., U.S. Patent 5,705,201 (1998).
121. Michaeli, W., Brunswick, A., Gruber, M., *Kunststoffe Plast Europe* (1999), 89(4), 20. Translated from *Kunststoffe* (1999) 89(4), 84.
122. Michaeli, W., Brunswick, A., Pohl, T., Kunststoffe Plast Europe (1999), 89(9), 18. Translated from *Kunststoffe* (1999) 89(9), 56.
123. Liu, S.-J., Lin, S.-P., *Adv. Polymer Technol.* (2006) 25(2), 98.

10 Part Design

J. Beaumont

10.1 The Design Process

The successful design of a plastic part is one of the greater challenges and financial risks an engineer or a company has to face. Before a plastic part can be truly evaluated, a mold must be custom designed and built that can easily cost tens to hundreds of thousands of dollars. This can be several million times the selling price of the part being designed. Furthermore, the process of designing, building a mold, and molding the first plastic part can easily take 20 weeks. Finally, after molding the parts in the new mold, the parts must be evaluated for size, shape, and mechanical properties. It is rare that these first parts are to the required specifications. The next stage is typically a long, costly process of trying to produce parts to specification. This commonly includes changes to the part design and/or the plastic material, resulting in changes to mold design and process.

The problem with successfully producing plastic parts arises from the complex nature of plastic materials and how their characteristics can be significantly influenced by the part design, tooling, and molding process. This problem is accentuated by the fact that a majority of plastic parts are designed with only a minimum consideration of the molding requirements, and a minimum appreciation of the significant impact of these requirements on the molded part. As a result, it is common that a mold is built and then modified numerous times in order to produce the required product.

After years of numbing experience, many companies have adopted a process of rushing through the part design, and then to have the mold designed and built in order to shoot parts just to find out what is wrong and begin the debugging process. This process continues today despite the availability of new computer aided engineering (CAE) technologies, such as injection molding simulations and structural analysis, which can reduce many of the risks. The perception is that seeing a mold being built is making progress, whereas up-front design engineering increases the time to produce a product. The additional time to debug the design of the mold with the lost potential of an optimized product is considered part of the business. One of the reasons for the limited acceptance of up-front engineering is that it is less tangible and more abstract. It is difficult to determine what should be done, how much up-front engineering is required, what tools are required, how well the tools work, how efficiently they are being used, and how to measure their actual contribution to the final product.

Each product presents a different degree of challenge. A judgment must be made as to what level of up-front engineering should be applied versus the risk. The more time invested in engineering, the lower the risk involved in the project. However, there is a diminishing return on the use of the up-front time. If left alone, an engineer could spend a lifetime trying to perfect the part design. The product may never make it to market if the engineer is expected to design a product guaranteed to succeed.

The successful design of a plastic part requires numerous activities. Some of these are common to that of any new product or design. A product concept is developed by someone with the intention of making a profit. To get from the concept phase to the money-making phase, a number of tasks are required of numerous different individuals. Some of these tasks must be performed sequentially, whereas others are best performed concurrently. The progression of sequential tasks is not always the same. For example, some material requirements dictate a design, whereas some designs dictate the material. The process of selecting a material may come before, after, or concurrently with the part design. Some tasks need more attention in some projects than others; regardless, the paramount factors on which the success of the project depends are to identify clearly the objective of the project, to plan, and then to communicate these objectives to all parties involved.

The design process presented in the following provides a list of tasks in the order that they might be conducted. This sequence may vary from project to project. Following data collection, and the development of a project plan, many of the tasks are best performed concurrently to minimize problems with product performance and production.

1. Data collection and product specifications
   - Collect information on the product and customer
   - Start a folder for information as soon as it comes in
   - Develop product specifications based on the collected data
   - Determine a beginning and end date for the project
   - Once all data has been collected, product specification written, and timetable established, have your customer review and sign off
   - Do *not* begin designing a product until all parties agree on the product specifications and timetable
2. Project plan
   - A project plan should be made in the early stages of the project. A preliminary plan may be made before *data collection* and then be altered and developed based on any newfound information. It can be expected that the plan will be altered throughout the project, but it provides an important guideline to accomplish your design and time objectives.
   - The project plan may be very basic or quite elaborate, depending on the complexity of the project. At a minimum a project plan should:
     - Establish a *timetable* for project *tasks*
     - Allocate *resources*. Resources can include personnel, equipment, technology, money, and so on

- Without a project plan you significantly increase your risk of failure to meet your obligations

It is critical to meet both design objectives as well as time objectives.

- Establish project deadline
- Identify resources (including personnel)
- Identify tasks
- Assign resources to tasks
- Develop schedule for resources and tasks (check for overlap)
- Determine the critical path
- Order long delivery items or information first

3. Preliminary Design
   - Develop sketches of a variety of design options
   - Be creative. Attempt to isolate yourself from previous designs in order not to inhibit creativity and originality. If a design team exists, develop initial concepts individually.
   - After initial concept designs, review with team members and against existing products. Channel and merge ideas progressively until an optimum design is established.
   - Challenge your design and the design of team members. Do not become overly attached to a given design that might blind you from new and better ideas
   - Review viability of the various options that consider manufacturability, cost, and risk
   - Can components in the design be merged, or can value be added, because of inherent material or process capabilities of plastics and the injection molding process? (This is commonly referred to as "Value Engineering".) Speculate on alternate approaches to the design.
   - Isolate weaknesses and assumptions that might compromise the final product. Review these and further develop them.
   - Anticipate part requirements (structural, environment, cosmetic, etc.)
   - Refine sketches and develop mock-ups (these may be done with cardboard, foam, CAD, etc.)
   - Do not be limited by things you do not know. Seek assistance.
   - Never design too much at a given session. Break it into as many discrete pieces as possible, avoiding fatigue, but preserving continuity.
   - Perform preliminary analyses where appropriate. This should be simplified work that might include hand calculations, use of spreadsheets, and mold filling analysis.
4. Material Selection: An initial material selection must be made before the detailed design can begin. In some case the further design of the part might dictate that

the material be changed. Material selection is presented in more detail later in this chapter.

5. Develop Detailed Design
   - *Function*: Gather the design team and interested parties into a small group and narrow concepts, addressing functions, into a single design. Make critical decisions.
   - Apply design guidelines as specific to the selected plastic materials and processes.
   - *Structural*: Perform appropriate structural analyses through calculations or structural analysis software. Apply appropriate safety factors.
   - *Process*: Review process effects as related to the design. This might relate to wall thicknesses, tooling requirements, and so on.
   - Develop detailed part drawings suitable for production or prototyping.
   - Simple solutions are always preferred.
6. Testing/Prototyping
   - It may be appropriate to test or prototype your design, or components of your design, before committing to production. Prototypes can serve various functions, including evaluation of fit, form, function, and structural considerations. For structural and fit evaluation it is often required that the part be produced from the same material and by the same process that will be used in production. Other than for structure and fit, evaluation may often be serviced with alternate materials and may be produced through means other than those to be used for the production product. Techniques that are common include:
     - Machining of materials (wood, plastic)
     - Stereolithography or alternate rapid prototyping methods
     - Casting of urethane parts from silicone molds produced with a master
       - Masters can be machined or formed through stereolithography or an alternate rapid prototyping method
     - CAD and simulation software
   - With plastic products there may be particular concern with the final size, shape, and mechanical properties due to the influence of the process. This has classically been addressed by producing parts that employ prototype tooling. Caution must be maintained because differences between the prototype tool and the production tool will create variations in the prototype part. One must clearly ascertain the objectives of a prototype mold and assess what is required of the tool to achieve these objectives. Variations in mold steel, cooling, gating location, number of cavities, and the like, will create differences between the parts molded in the prototype versus the production mold.
     - Many advancements in CAE and material characterization techniques are improving the ability to evaluate a part without physical prototyping.
7. Review design and revise through Steps 4 to 6
   - Review all assumptions
   - Review and critique any simulation work closely
   - Try to view design with a new critical eye
   - Avoid design momentum—do not stick with a design that is considered flawed

8. Commit to the design and develop project plan to bring to production
   - Avoid running out of resource
   - Avoid simultaneous efforts and over commitment
   - Establish critical path, flags, and goals—both final and intermediate

10.2 The Four Building Blocks of Plastics Part Design

The successful development of injection molded plastic parts is significantly handicapped by the complex interaction between plastics materials, part design, mold design, and processing. These four building blocks each affect how the part is designed, and how it will behave once it is produced. Each of these four factors individually present a significant challenge. The task of considering their interrelation in the successful development of a plastic part can seem overwhelming, but new plastic party are developed every day.

Due to the complexity of developing an injection molded part, it is best that the development be performed with the cooperative effort of individuals with expertise in each of the four major areas. By involving a process engineer and a mold designer during the early stages of part design, many manufacturing problems might be avoided. Technologies such as structural analysis and injection molding simulation programs can also act as a significant aid in product development. With the use of injection molding simulation, a skilled analyst can evaluate many of the manufacturability issues as the design is evolving. However, without a good understanding of this technology, a designer can develop false confidence. It should be recognized that these programs provide information that requires skilled interpretation.

The following sections discuss each of the four key building blocks of developing a plastic product and some of the complexity that they contribute.

10.2.1 Material

More than 100,000 variations of plastic materials are available. This creates significant challenges for a designer to sort through and determine which provides the combination of optimum performance and manufacturability at a minimum cost. In addition to screening material requirements based on part application, one must consider whether the material can be molded into the geometry of a physical part design. The physical part design must satisfy the mechanical requirements of the part, and have the ability to be produced in, and ejected from, a mold.

The environments in which the part is to live must be considered when selecting a material for a given application. These requirements are often factors that cannot be influenced by the physical design of the plastic part. Material properties that cannot be affected practically by design include chemical resistance, UV resistance, colorability, transparency, HDT, and low shrink. For a given product application these would be characterized as material requirements and would be a high priority in the selection process. Material properties such as modulus, impact strength, and tensile strength are mechanical properties and have a lower priority when selecting a material. Mechanical properties of a product are affected by both the product design and the material; therefore, selecting a material based on mechanical properties might limit the opportunities when designing a new part.

In addition, the characteristics of plastic materials are much more complex than most other materials. Few of their properties are constant, and often only limited information on them is available. These nonconstant properties include fundamental characteristics such as modulus, yield strength, and viscosity. These can be affected by strain, strain rate, temperature, humidity, time, and process. Figure 10.1 illustrates the effect of temperature and strain rate on the stress–strain relationship of a plastic material. If a designer requires a material's modulus with which to design, the conditions under which the data was derived must be considered as to its relevance to the particular application. The standard published modulus for a material derived using ASTM test procedures has limited practical use in designing an under-hood automotive part. The ASTM procedure specifies that the plastic specimen be prepared at room temperature and at a relative humidity of 50%. The conditions under the hood of an automobile are significantly different from this.

Figure 10.2 contrasts the stress–strain relationship of a metal versus a plastic. Note the lack of a significant linear region from which the modulus of the plastic material can be derived. When designing with metal, it is relatively easy to derive a modulus because of its strong linear region. In addition, the metal is generally used in low strain applications where the modulus is constant. With plastics, the modulus is nonlinear except at very small strains. This creates a problem because plastics are commonly used in applications that experience relatively high strains. The result is that the modulus is actually changing as the part is being distorted. Figure 10.3 illustrates the existance of six different moduli derived for a given plastic dependent on deflection. This indicates that as a component is deflecting, its modulus is constantly changing as the deflection progresses (see Fig. 10.4).

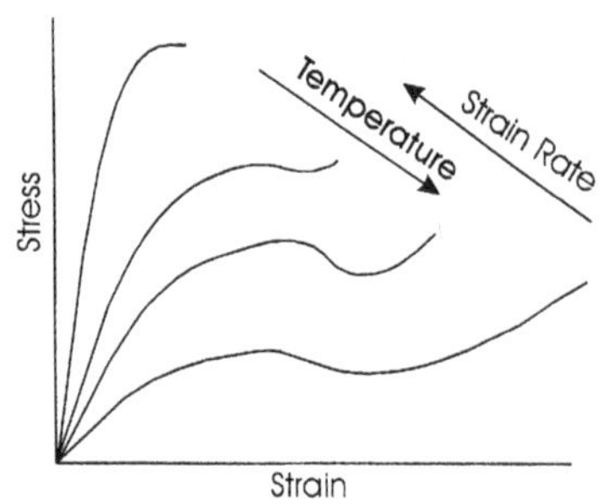

Figure 10.1 The effects of temperature and strain rate on the stress-strain relationship of a plastic material.

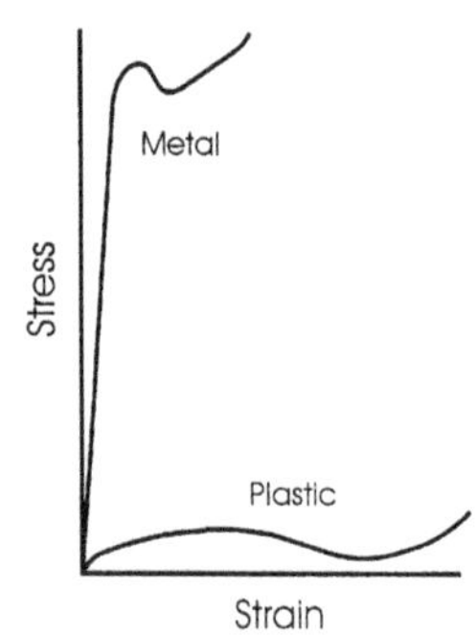

Figure 10.2 Comparison of a typical stress-strain curve between a metal and plastic.

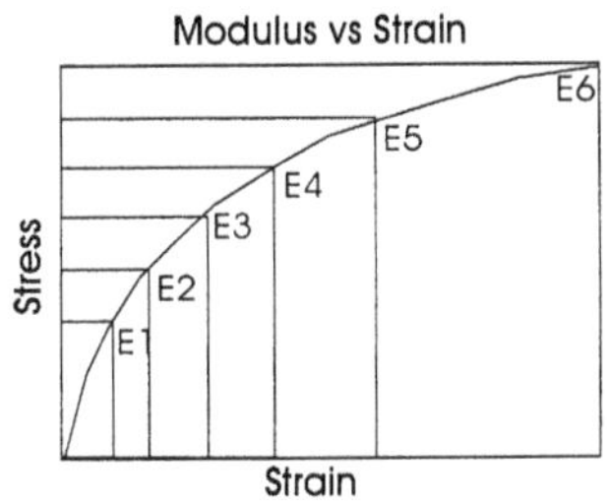

Figure 10.3 Six different moduli derived for the same plastic.

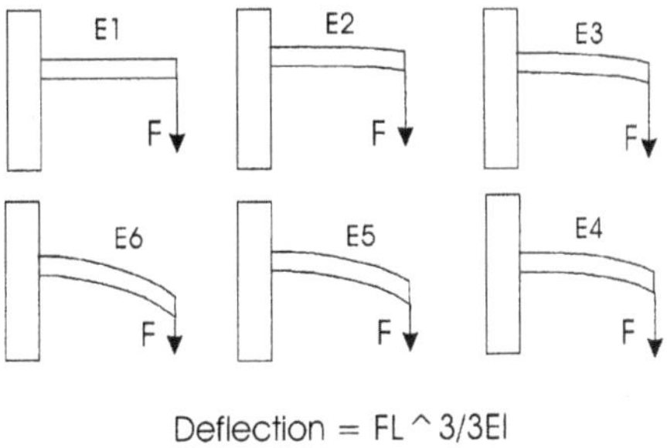

Figure 10.4 Modulus is constantly changing as the deflection progresses.

Plastic material properties also change with time. A part under load may initially experience very little strain. With time, under the same load, the part will continue to distort. This is commonly referred to as "creep."

Properties of plastics are also significantly affected by process. Factors such as cooling rate, flow direction, and flow type can all influence the orientation of the polymer molecule and any asymmetric fillers or reinforcements within the polymer. The orientation will create anisotropic properties in a molded part. ASTM tests do not specify how a specimen is to be prepared. The result is that most test specimens are molded at ideal conditions with the most favorable orientation. A designer must again be cautious of using this data and question its relationship to the part being designed.

Material Selection

With the wide variety of plastic materials, it is important to develop some methodology in identifying the optimum material for a given application.

1. Develop lists of the materials specifications. The first list should be *material requirements*. Requirements would include material properties that cannot be effected by varying the physical design of the part. These properties might include:
   - chemical resistance
   - colorability
   - clarity
   - thermal resistance

   The second list should be *material preferences*. Preferences are material properties that appear to fit the application, but which can be affected by part design. These are commonly mechanical properties. Though the part may require good rigidity, this can also be influenced by the design of the part; therefore, it is not mandatory that the material be a certain modulus for the part to be rigid. Other material preferences might include:
   - impact strength
   - tensile strength
   - modulus

2. Material selection should consider:
   - Mechanical properties
   - Ability to survive its environment
   - Wear
   - How the part is to be produced
   - Specialty properties/requirements (optical, lubricity, etc.)
   - Cost. In structural applications, cost should consider rigidity and strength of the material rather than just cost per volume. (Material cost issues are presented later in this chapter)
3. List the materials that satisfy your requirements and have some of your preferences. Next, rank these materials by cost and availability. Finally, rate each material for *processability*.
4. If your material selection is still not clear, pick the one with which your company is most familiar.
5. If a structural, or injection molding, analysis is to be performed, then the appropriate material data should be collected. In some cases the availability of this data may be considered in your material selection process, i.e., if you cannot analyze it, your design capabilities will be limited.

Material Cost Considerations

After identifying candidate materials that satisfy requirements and preference, it is im portant to consider their cost relative to the parts performance. Material costs are generally provided in a cost per weight relationship ($/pound). Cost per weight does not always provide a good indicator as to the material's actual cost relative to the part application requirements. The effect of material density, strength, and rigidity on cost much be evaluated. The part may require high strength, which would indicate that there is a need for a high strength material. However, part strength can be affected by the structural design of the part. By establishing a required modulus too early, final material selection is unnecessarily limited and part design is potentially compromised.

Consider a structural part where both HDPE and a glass-filled nylon satisfy the material requirements. The cost per pound of the HDPE is considerably less than that of the nylon. Because the nylon has a much higher modulus, the part can be designed with a much thinner wall than if the HDPE is used. The thinner wall will reduce the amount of material used and the cooling time required. Both of these will reduce the cost of the molded part. Therefore, it is not clear which of these two materials is less expensive to use with the part and thus must be evaluated.

At a minimum, material costs should be contrasted based on volume rather than weight. In a structural application, the relative required wall thickness requirements of the candidate materials can also be contrasted. If the strength of a part is being considered, it should be realized that a part made from a Nylon could have a thinner wall than if it were produced from an HDPE. This will reduce the amount of material required and molding cycle time, thereby possibly reducing total part cost despite the higher cost per pound for the Nylon.

Although contrasting the structure of two parts made with different materials is best evaluated using structural analysis programs, the following method provides a simplified means of determining equivalent cost of plastic materials based on structural requirements. Candidate materials can be evaluated based on their relative strength, as indicated by their *yield strength*, or for their rigidity, as indicated by their relative *moduli.*

Contrasting the Cost of Plastic Materials Based on Strength This process provides a means to contrast the cost of plastic materials based on their ability to withstand stress.

1. Determine the primary type of stress the part will experience (tensile, compressive, flexural, shear, torsion).
2. Find the corresponding *yield stress* values of your candidate materials.
3. Determine the *yield strength ratio* (σ^*) of the candidate materials utilizing one of the materials as a reference material. In this example, there are two candidate materials: A" and "B."

$$\textit{Yield Stress of Material "A"} = 10{,}000\,psi \quad \text{(Reference Material)}$$

$$\textit{Yield Stress of Material "B"} = 7000\,psi$$

$$\sigma^* \ (\textit{Yield Strength Ratio of "B" relative to "A"}) = \frac{7000}{10{,}000} = 0.7$$

4. Realizing that material "B" has only 70% of the strength of material "A," an equivalent geometry can be derived that would give a structure the same strength, whether it was made from material "A" or "B." To accomplish this, a formula is derived from the stress formula for the particular loading case for which your part will experience in application. In addition, the designer must consider which geometrical dimension in the part is considered a variable. For example, in a beam that will be exposed to a flexural load, is the width of the beam (wall thickness) or its height a variable?

   In a flexural loading case, the following standard stress formula would be used. M = Max Moment, c = distance from neutral axis to surface of maximum stress and I = Moment of Inertia.

$$\sigma = \frac{Mc}{I}$$

   In a tensile or compressive loading case, the following simple relationship can be used. σ = Stress, F = Force and A = Area.

$$\sigma = \frac{F}{A}$$

Example 10.1

Find the equivalent geometry for a *flexural load case* where part height (h) is the variable (i.e., what height is required of a part made from material "B" so that it has the same strength of a part made from material "A"). As we are only looking for equivalence, one does not need to know the actual load conditions, nor the actual part dimensions. Regardless of the part's actual shape, the materials are contrasted assuming the structure is a simple beam with a height (h) and a width (w).

$$\sigma = \frac{Mc}{I} = \frac{M.5d}{wh^3/12} = \frac{6M}{wh^2} \qquad d = \sqrt{\frac{6M}{w\sigma}}$$

where M = bending moment
c = distance from the neutral axis to outer surface
I = moment of Inertia

Because we are looking to contrast candidate materials under identical structural loads, and part width is not a variable, then

$$\frac{6M}{w}$$

is the same for either material and can be canceled from the equation; therefore, finding the equivalent height can be reduced to

$$h^* = \sqrt{\frac{1}{\sigma^*}}$$

Find the *equivalent height* (h^*) of material "B" relative to material "A"

$$h^* = \sqrt{\frac{1}{\sigma^*}} = \sqrt{\frac{1}{.7}} = 1.195\,h$$

This indicates that a part that is produced from material having a yield stress of 7000 psi must be 1.195 time taller than a part being produced from the reference material that has a yield stress of 10,000 psi.

5. Find the current cost/volume ($\$/in^3$) of the material from the material suppliers. This will most likely require that cost be converted from cost per weight ($/pound) to cost per volume ($\$/in^3$).
6. Find the *equivalent cost/volume* of the candidate materials based on their relative heights and their relative costs per volume.

$$\text{Equivalent Cost/Volume} = (\$/in^3)(h^*)$$

Contrasting the Cost of Plastic Materials Based on Rigidity The following method provides a means to contrast the cost of plastic materials based on their ability to resist deflection.

Example 10.2

1. Determine the primary type of load the part will experience in application (flexural, tensile, compression, shear).
2. Find the corresponding *modulus* values for the candidate materials.
3. Determine the *modulus ratio* (E^*) of the candidate materials using one of the materials as a reference. In this example, there are two candidate materials: "A" and "B."

$$\textit{Modulus of material "A"} = 350{,}000\,\text{psi} \quad \text{(Reference Material)}$$

$$\textit{Modulus of material "B"} = 210{,}000\,\text{psi}$$

$$\textit{Modulus Ratio of Material "B" relative to "A"} = E^* = \frac{210{,}000}{355{,}000} = 0.592$$

4. Realizing that material "B" has only 59.2% of the rigidity of material "A," an equivalent geometry can be derived that would give a structure with the same rigidity, whether it was made from material "A" or "B." To accomplish this, a formula is derived from the deflection formula for the particular loading case for which your part will experience in application. Again the designer must consider which geometrical dimension in the part is considered a variable (height or width).

   In a flexural loading case, any standard beam equation under a flexural load can be reduced to:

$$y = \frac{1}{EI} = \frac{1}{E\left(\frac{wh^3}{12}\right)}$$

   Again, because we are looking to find the relative height (h^*) for a part made of material "B," which will provide the same rigidity as when produced from material "A," y and w will cancel out and be reduced to:

$$h^* = 3\sqrt{\frac{1}{E^*}}$$

5. Find the cost/volume ($/in^3)
6. Find the "equivalent cost/volume" based on performance

$$= (\$/\text{in}^3)(h^*)$$

Sample Problems

Three materials have been identified that satisfy the requirements of a part to be designed. These materials have the following properties. The part will be under a flexural load during application.

Material	Cost ($/lb)	Flexural Modulus (psi)	Flexural Yield Strength (psi)	Density (g/cm^3)
A	0.60	200,000	8,000	1.25
B	0.52	105,000	7,500	1.05
C	0.75	260,000	10,000	0.98

** **Given** the following calculations for a rectangular beam under a flexural load.

$$\sigma = \frac{FLc}{I} \quad y = \frac{FL^3}{3EI} \quad I = \frac{wh^3}{12}$$

Problem 10.1

You are selecting a material for a part which will be under a flexural load during application. The height (h) of the part can be changed (is the variable). Determine which of these materials is the most cost effective based on rigidity.

1.1. Determine the Cost per Volume of the materials ($/in^3)
1.2. Determine the ratio of moduli of the three materials (use material "A" as the reference material)
1.3. Derive the formula for the equivalent geometry
1.4. Find the equivalent geometry (height)
1.5. Find the equivalent cost

Answers Problem 10.1

1.1. "A" = $0.0271 "B" = $0.0197 "C" = $0.0266
1.2. "A" = 1 "B" = 0.525 "C" = 1.3
1.3. $h^* = \sqrt[3]{\frac{1}{E^*}}$
1.4. "A" = 1 "B" = 1.2396 "C" = 0.9163
1.5. "A" = 0.0271 "B" = 0.0244 "C" = 0.0243

In this example, material "C," which is the most expensive material based on weight, is the least expensive material to use in application.

Problem 10.2

Determine which of the preceding materials is the most cost effective based on strength.

1.1. Determine the ratio of the yield strength of the three materials (use material "A" as the reference material)
1.2. Derive the formula for the equivalent geometry
1.3. Find the equivalent geometry (height)
1.4. Find the equivalent cost

Answers Problem 10.2

1.1. "A" = 1 "B" = 0.9375 "C" = 1.25

1.2. $h^* = \sqrt{\dfrac{1}{\sigma^*}}$

1.3. "A" = 1 "B" = 1.0328 "C" = 0.8944

1.4. "A" = 0.0271 "B" = 0.0204 "C" = 0.0237

In this case, material "B" is the least expensive material in application and in cost per pound. However its cost advantage over material "C" has been decreased.

Problem 10.3
Derive the formula for equivalent geometry if the primary load is flexural and the height of your part is fixed (therefore, only width w can be changed).

Answer Problem 10.3

- For Sample Problem 10.1 (solving for rigidity)

$$w^* = \frac{1}{E^*}$$

- For Sample Problem 10.2 (solving for strength)

$$w^* = \frac{1}{\sigma^*}$$

Problem 10.4
Derive the formula for equivalent geometry if the primary load in application will be compressive and you are concerned with the strength of the part.

Answer Problem 10.4

$$A^* = \frac{1}{\sigma^*}$$

Material Selection Effect on Cycle Time

To refine material cost impact on a molded part further, the effect of wall thickness on cycle time must be considered. Cycle time consists of fill, packing, cooling, and mold open (ejection) time. A thinner wall will normally reduce the filling, packing, and cooling time of a molded part.

Consider a structural case of a plastic shelf. The rigidity of the shelf is to be gained primarily from a series of parallel ribs spanning the length of the shelf. The height of the ribs are fixed at 1.5 in. Rib width (w) is considered the only variable. The two candidate materials that satisfy the requirements, and are being considered, for this application are HDPE and nylon. The cost of the nylon is more than three times that of the HDPE when contrasted based on dollars per pound ($1.50/lb vs. $0.43/lb). If you consider the relative modulus of the materials, the equivalent width of the ribs if produced from the Nylon is only 31% than if produced from HDPE. If you consider density difference the equivalent cost of the Nylon is now only 1.28 times that of the HDPE. Even though the Nylon still appears to be more expensive than the HDPE, the difference is significantly less than the original 3.49 cost ratio.

For full evaluation, however, the effect on production cost should be considered. A structural analysis determines that a 0.200-in rib thickness is required if the HDPE is to be used. Because the nylon part can be only 31% of this width to get the equivalent rigidity, the nylon part's wall thickness would be only 0.062 in thick. Cooling time for the 0.200-in-thick HDPE part is calculated to be approximately 26 seconds, whereas the Nylon is expected to cool in less than 3 seconds. Fill, pack, cooling, and mold open time are estimated to be 50 seconds for the HDPE and 13 seconds for the nylon part. Given a single cavity mold with production cost of $50/hour, one direct laborer at $10/hour and 5% spoilage, the production cost for the HDPE part is $ 0.88/part, whereas the Nylon part is only $0.23/part. Thus, the production cost for the HDPE part is nearly four times that of the Nylon part.

Relative cooling times of molded parts, as affected by material and wall thickness, can be approximated using the following formulas. Cooling time (t) is calculated based on wall thickness (h), thermal diffusivity (α), melt temperature (T_m), mold wall temperature (T_w), and ejection temperature (T_e). For solid round sections, as in bosses and runners, the radius (R) is used.

Two formulas are given for calculating approximate cooling time of both a flat plate and a round section. The first (t_c) calculates the time for the centerline of the plastic part to reach ejection temperature. The second (t_a) calculates the time for the average of the cross-section of the plastic part to reach ejection temperature.

Flat Plate

$$t_c = \frac{h^2}{\alpha\pi^2} \ln\left[\frac{4}{\pi}\left(\frac{T_m - T_w}{T_e - T_w}\right)\right]$$

Note: T_e = the desired temperature of the centerline of the part at ejection.

$$t_a = \frac{h^2}{\alpha\pi^2} \ln\left[\frac{8}{\pi^2}\left(\frac{T_m - T_w}{T_e - T_w}\right)\right]$$

Note: T_e = the desired temperature for the average part temperature at ejection.

Cylinder

$$t_c = 0.173 \frac{R^2}{\alpha} \ln\left[1.6023\left(\frac{T_m - T_w}{T_e - T_w}\right)\right]$$

Note: T_e = the desired temperature of the centerline of the part at ejection.

$$t_a = 0.173 \frac{R^2}{\alpha} \ln\left[0.6916\left(\frac{T_m - T_w}{T_e - T_w}\right)\right]$$

Note: T_e = the desired temperature for the average part temperature at ejection.

It is nearly impossible to predict an actual cooling time of a molded part accurately. Cooling time is only partly related to the time it takes for the plastic to reach a given temperature. Uniformity of cooling, melt flow patterns in a part, and material will affect shrinkage and the development of residual stresses that will tend to warp a part. Cooling time will also affect these and therefore becomes a process variable, which is difficult to determine until the part is molded. In addition, the shape of the part can have a significant effect. A flat part will have less structure to resist residual stresses and therefore might have to be constrained in a mold longer than a cylindrical part, which has a better structure to resist warpage. The longer cooling time will both lower the part wall temperature and thereby increase its rigidity at the time of ejection and increase the amount of time that the mold can act as a fixture, helping to control the parts final size and shape. Because these factors are not considered in the preceding calculations for cooling time, they should be used with caution when relying on them to estimate cycle times.

10.2.2 Product Design

The designer must design a product to satisfy its functional, structural, aesthetic, cost, and manufacturing requirements. These variables are often at odds with each other, and judgments must be made in balancing the conflicting requirements. An example might be a rib that is required for added structure or stability to reduce the potential for warpage. This rib, however, can create a sink on the primary show surface due to the high localized shrinkage resulting at the intersection of the rib and the primary wall. Widely accepted guidelines indicate that the rib should be thinned relative to the primary wall to reduce the large volume of material causing this localized shrinkage [see Fig. 10.5(a)].

A thinned rib now presents a restricted flow channel that may result in mold filling problems. The height of the rib, therefore, becomes limited, which requires that multiple shorter, thinned ribs be used to replace a single rib. Regardless of these approaches, the thinned ribs will shrink differently than the thicker primary wall. This

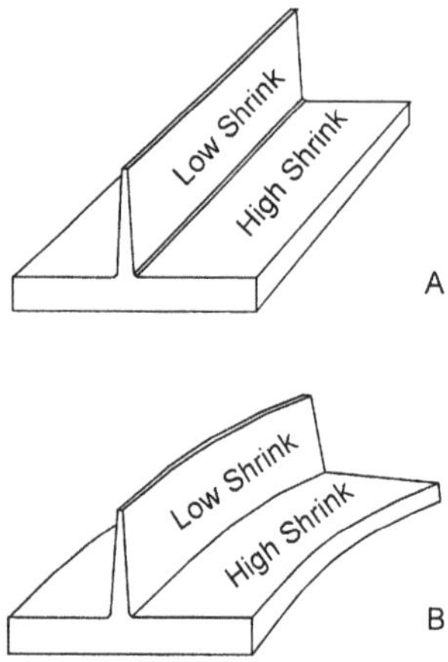

Figure 10.5 (A) Thinned rib to reduce localized shrinkage. (B) Warpage caused from variation in shrinkage between thin and thick regions.

variation in shrinkage will result in residual stress and often warp the part [Fig. 10.5(b)].

One of the significant advantages of plastic parts is their ability to be molded to incorporate a multitude of features that might otherwise require machining and assembly of multiple parts. The expectations of the plastic part and the challenge to the designer to satisfy the multiple functions will present further challenges. The need to combine features while not overly complicating the tooling requirements presents a compounding challenge.

Radiator end tanks are a good example. These parts include an oversized flange (thicker than the primary wall) for mounting and sealing the tank to the radiator, a number of coolant hose connections, and a multitude of miscellaneous mounting brackets, widgets, and thingamajigs. These designs are produced for nearly every automotive company; however, the nature of the design puts a significant burden on the mold designer and molder. Despite years of experience with these tanks, a molder still must go through numerous revisions on nearly every new tank in order to reduce warpage on these parts to an acceptable level. Part and mold design modifications, which might reduce the molding problem, may often be discarded because they are a departure from accepted historical methods. The short-term cost, time, and risk of developing new approaches often distract from the potential long-term gains of pursuing them.

The oversized flanges (see Fig. 10.6), typical of these parts, create many of the actual molding and warp problems, as well as create analysis problems for today's injection molding simulation programs. The thin shell–type mesh used by these programs does not represent the low width-to-thickness aspect ratio of the flange very well. Even though the programs are used regularly in the design of new molds for automotive end tanks, these modeling limitations inhibit their full use and require skilled consideration by the analyst.

For structural applications a designer will make use of either standard design formulas for calculating deflections and stress, or, for more demanding designs, use finite

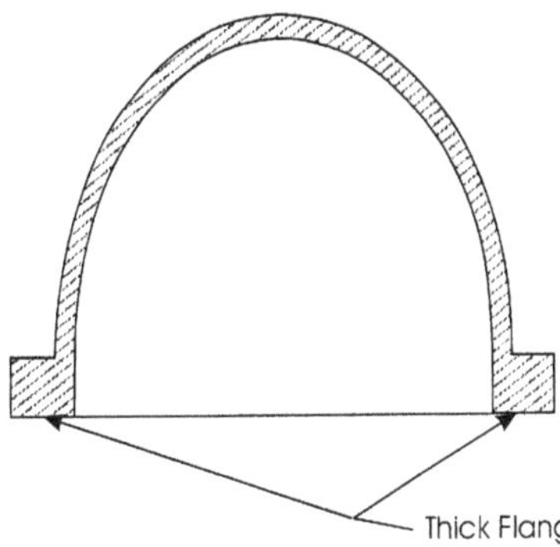

Figure 10.6 Cross section illustrating common flange/primary wall relationship responsible for molding and warpage problems common in production.

element structural analysis. However, these methods are significantly challenged due to the complex characteristics of plastic materials. Even if a structural analysis program has the sophistication to solve the nonlinear problems required, it is doubtful that the material data will exist for the given material that is to be analyzed.

In addition, there is the unique requirement of the plastic part designer to consider the mechanical properties of the material as well as its melt characteristics. The materials melt characteristics will reflect on the manufacturability of the part. Again, the materials properties during mold filling are expensive and difficult to capture. The viscosity, like the modulus, is nonlinear and will continually change during mold filling. The flow channel is constantly changing as plastic freezes when it contacts the relatively cold channel walls, restricting the flow channel.

It is commonly desirable to prototype a part before committing to production. Again, for an injection molded plastic part, this is not a simple procedure. Although parts may be machined or built by many different methods to produce a prototype, these machined parts only provide limited information. They do not represent the actual mechanical properties or the problems with acquiring the desired size and shape once they are injection molded. They do not predict the mold filling problems that might be expected during injection molding.

Prototype molds might be built, but the information gained from this is limited. The molding process and resultant plastic parts characteristics are affected by differences that result from molds built of alternate materials and have different runners, gating, cooling, and processes.

10.2.3 Mold Design and Machining

Molds for the production of plastic parts must be custom designed and built. The construction of these custom-designed molds requires highly skilled craftsmen to operate a wide range of precision machining equipment. The injection mold must replicate the geometry and surface finish requirements of the part. One of the primary challenges in the design of a mold is to assure that when it opens, all features of the part are relieved so that it can be ejected. One mold designer told me that 70% of the parts he receives could not be ejected from a mold as designed. This is somewhat of an interpretive point in that some mold designers will tell you they can design a mold to eject nearly any part as long as cost of the tooling and productivity are not a concern. Ejection may require combinations of crafted parting lines and a multitude of movable components that result in a tool that is essentially a fairly complex machine.

However, the challenge of designing a part goes much further than simply ejecting the part. Many other aspects of mold and part design have to be considered:

- Machinability of mold components
- Sized cavity dimensions to account for part shrinkage
- Adequate and uniform cooling
- Venting of gases
- Product surface finish
- Incorporate tolerances
- Include delicate inserts
- Accommodate delivery of melt
- Automatic separation of runner and part
- Built to withstand millions of cyclic internal loads from injection pressure (up to 30,000 psi)
- Built to withstand external clamp pressures (up to 7000 tons)

The most fundamental requirements when designing a mold are that the cavity can be filled with the specified plastic, be robust enough to accommodate the internal and external forces, and be built so the molded part can be ejected from the mold. The requirements for ejection and the forces are mechanical and are more clearly accommodated. Formal structural analysis is rarely performed on the mold designs; rather, the designer relies on skill, intuition, and a conservative approach to structural requirements. A conservative approach to the design of the melt delivery system and positioning of the gates is usually taken because the filling of the cavity, and its effect on the part, is less predictable and more abstract. In return, the conservative approach will result in oversized runners and poor positioning of the gates. This conservative approach to addressing the fundamental requirement of mold filling is often in contrast to productivity and product quality issues.

The mold design illustrated in Fig. 10.7 is used to produce a cap that has both an inside and outside thread. Due to this thread arrangement the mold includes an internal unscrewing mechanism, side action, and a stripper mechanism. In this case a simple-looking plastic part weighing approximately 10 g requires a mold weighing in at approximately 300 kg. Despite the eloquence of this mold design and the applaudable engineering that went into it, this mold is limited to producing only two parts per cycle and will result in a weld and gas trap that must be vented through some stationary pin or plug that requires regular maintenance.

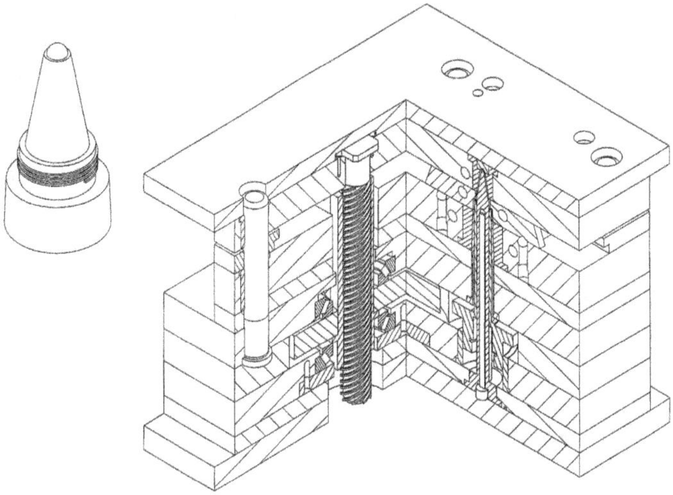

Figure 10.7 Mold including an internal unscrewing mechanism, side action, and a stripper mechanism used to produce a cap with both an internal and external thread (Drawing by Tom Garman).

10.2.4 Process

After the material has been selected, the part designed, and the mold built, an attempt to produce a part can be made. At this time, it can be determined what the product will look like, how it will perform, and how inexpensively it can be produced. This is often the beginning of an iterative cycle, which may include alterations in material, part design, and the mold before achieving a part within specifications.

The size, shape, and mechanical properties of an injection molded plastic part are significantly affected by the molding process. The complex shrinkage of plastic materials is probably the most dominate, underlying problem associated with successful design and manufacture. During the injection molding process, plastic materials go through a volumetric change of as much as 35%, as they are cooled from a molten to a solid state. Adding to the challenge, flow of plastic through the mold induces molecular orientation that introduces a directional component to this shrinkage. The direction and magnitude of the shrinkage varies significantly from part to part, and even within a given part. The direction and magnitude of the resultant shrinkages affects the final size, shape, and mechanical properties of the finished product. It becomes imperative to understand these shrinkage phenomena in order to develop design and molding strategies that will minimize their potential negative effect.

As discussed previously, thermoplastics are comprised of long, organic molecular chains that are made up of numerous repeating carbon based units (monomers). The atoms within these chains are held together by relatively strong covalent bonds. The individual polymer chains are held together by relatively weak electrostatic forces. These weaker secondary forces have only a fraction the strength of the primary bonds that hold together the individual atoms within the polymer chain.

During the injection molding process, a polymer mass is heated to a point at which it melts. In this molten phase, the material can flow and is forced into a cold mold where it will take the shape of the cavity. Following cooling, the material can be removed from the mold, and it will retain the general shape of the cavity.

Shrinkage of thermoplastics can be characterized into two broad classifications, volumetric and linearized. Volumetric shrinkage can be attributed to thermal contraction that will occur in all polymers and crystallization that occurs in semicrystalline polymers. With the absence of any external forces, this shrinkage will be isotropic. Linearized shrinkage is developed from shear and extensional forces acting on a polymer during mold filling and packing. These forces develop orientation in both the polymer and the fillers, which may be contained in the polymer. This orientation can vary in direction and magnitude and has a direct effect on how the polymer shrinks. The following presents the phenomena of volumetric shrinkage resulting from thermal contraction and crystallization and linearized shrinkage as developed by flow in the mold.

Volumetric Shrinkage

Thermal contraction and expansion is a well-known and understood phenomenon, which occurs with most all known matter. With few exceptions, a material will expand when heated and contract when cooled.

When relatively low levels of heat are introduced to a polymer, this external energy increases molecular motion and weakens the cohesive bond energies of both the primary atomic and secondary molecular bonds. The result is an increase in the specific volume of the polymer mass as the atoms and molecules move apart. The amount of expansion will be proportional to the increase in heat input to the body. As the secondary bonds are weaker, the introduction of heat will have a more dramatic effect on them. During subsequent cooling, if no external forces are applied, the polymer would experience uniform orthotropic contraction (shrinkage).

With continued heating, the secondary bonds will continue to weaken until motion between polymer chains become relatively free and the polymer begins to exhibit distinct fluid like characteristics. Through examination of a P-V-T (pressure-volume-temperature) graph, the relationship between pressure, specific volume, and temperature can be evaluated. Figure 10.8 illustrates a characteristic P-V-T relationship of an amorphous material. By following the shrinkage of the material at 0 MPa, one can see that the material shrinks more than 10% as it is cooled from its process temperature of 250°C to room temperature, 23°C. The inflection in the curve indicates the glass transition temperature for this material.

As there is a significant amount of free space as a plastic is heated, applied pressure will result in compression of the material. The effect of this compression at various temperatures can also be determined by examination of Fig. 10.8, which shows the PVT characteristic of a polystyrene. At the process temperature of 250°C, the material can be seen to compress from a specific volume of approximately 1.06 cm^3/g at atmospheric pressure to 0.95 at 160 MPa. These pressures correspond to the potential range experienced during the injection molding cycle on a typical injection molding machine. The combination of thermal contraction and pres-

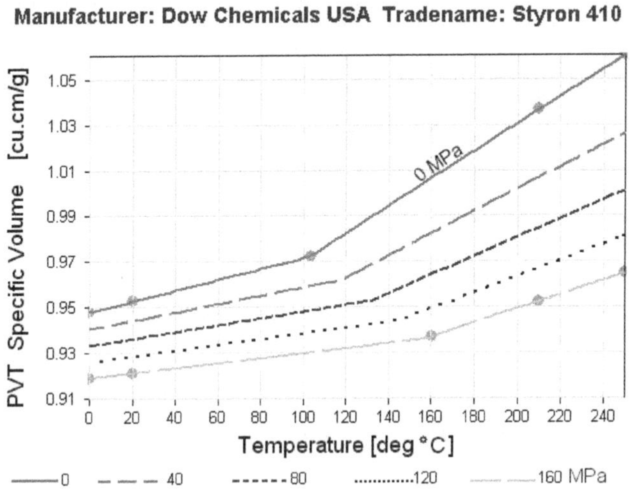

Figure 10.8 Characteristic PVT relationship of an amorphous material.

sure yields potential variances of 14% in volume during molding in this particular case.

Volumetric Shrinkage of Semicrystalline Polymers

Ordered polymers, such as HDPE, can form crystals in which the polymer chains fold back upon themselves resulting in densely packed parallel chains. The effect of this can be seen in Fig. 10.9, which is a characteristic P-V-T graph of a semicrystalline polymer. At the elevated temperatures where the polymer is molten, its structure is

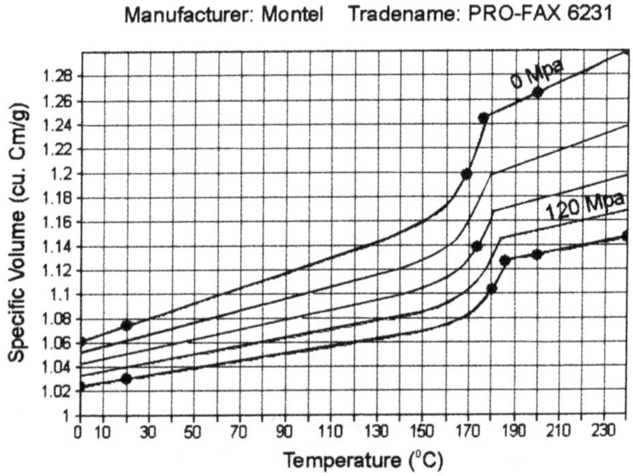

Figure 10.9 Characteristic PVT relationship of a semicrystalline material.

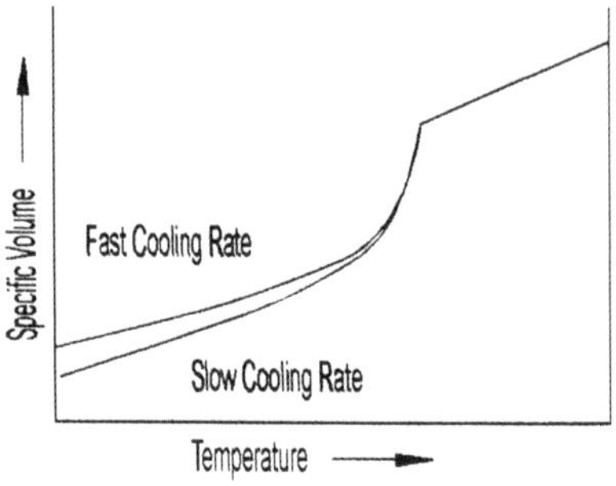

Figure 10.10 Illustration of the relationship of cooling rate and shrinkage of a semicrystalline material at atmospheric pressure.

amorphous. As the material is cooled, initially there is a gradual linear reduction in specific volume. At the materials crystallization temperature, there is a sudden decrease in specific volume as the polymer transforms from a molten amorphous to a solid semicrystalline structure. Once the material has solidified, the reduction in specific volume with change in temperature again becomes more gradual and linear. In this case, the volumetric change is nearly double the amorphous material under similar conditions.

The formation of the crystallization structure requires the molecules to organize themselves from the random state that exists while molten. As this physical structuring of molecules requires freedom of movement during the time that the crystals are forming, the rate at which the polymer is cooled will effect how much of this structuring occurs (i.e., what percentage of the polymer will be crystalline versus amorphous). It can also affect the size of the forming crystals. As the crystals are denser than the amorphous regions of the polymer, it can be expected that the rate of cooling will effect how much the polymer will shrink. Therefore, unlike amorphous materials, the density of a semicrystalline material will be affected by cooling rate. Figure 10.10 illustrates the relationship of cooling rate and shrinkage of a semicrystalline material at atmospheric pressure. The differences in specific volume at room temperature result from changing the cooling rate of the molten semicrystalline polymer. Also note the abrupt change that occurs in specific volume between its melt temperature and room temperature. This is the region in which the polymers crystals are forming during cooling (disassociating during heating).

Linear/Directional Shrinkage

During the injection molding process the polymer is subjected to thermal energy and plasticated in the injection barrel. The molten plastic is then forced under high pressure into a cold mold. The resultant shear field, acting on the expanded polymer mass, results in the molecules becoming oriented in the direction of the principle strain. The degree of orientation is a function of the applied shear stresses that are commonly over 100,000 pascals. This orientation, or ordering of the molecules, results in a relatively high energy state which in effect reduces its entropy. Given time, the molten mass would lose its orientation and return to a random state. This restoring force may be described as entropy elasticity. As the dominant orientation is normally in the direction of flow, this returning to a random state, or elastic recovery, would effectively reduce the length of the mass in the direction of the original applied stress (i.e., increase the shrinkage in the direction of flow versus transverse to flow).

In response to economic goals, a mold is kept relatively cold to minimize the time it takes to form a part, and thereby its apparent cost. This results in the outermost layers of the laminar flow front freezing nearly instantly as they contact the cold steel wall of the cavity. This freezing action takes place so quickly that the rapid reduction in free volume restricts the expected effects of entropy elasticity. Immediately under this frozen layer, the material is insulated from the steel wall and therefore cools more slowly. This allows time for recovery to take place, which effectively results in the expected linear or directional shrinkages resulting from flow. These variations in linear shrinkages are continuous through the entire cross-section of any given region, or element, of the part and are set up during the filling, packing, and compensation phases of the molding cycle. As these laminates with variable shrinkages are mechanically linked, a stress will develop between them.

It is apparent that if unconstrained by adjacent laminates within any given cross-section of a molded plastic part, the material would shrink differently within each discrete laminate. The orientation in the outermost laminates will be higher as a result of the stronger shear field. Figure 10.11 shows the results of a flow analysis providing information on Shear Rate distribution through a cross section during filling. These outer laminates are also the first to cool and thereby lock in the orientation.

During the compensation phase of the mold filling cycle, the shear field will move more toward the center as the outer laminates quickly freeze and thereby locking in the high orientation. However, despite the increase in the shear field near the center, the rate of cooling is much slower. This allows the material to relax (disorient). The result is that the net shrinkage is a balance of the stain between each of the laminates.

These differences in strains also result in a stress between laminates. Studies have been conducted, which have looked at the stress distribution through the cross section of plastic parts [1]. These studies have evaluated stresses by studying the birefringence of a specimen through its cross-section. The results have shown that the outer

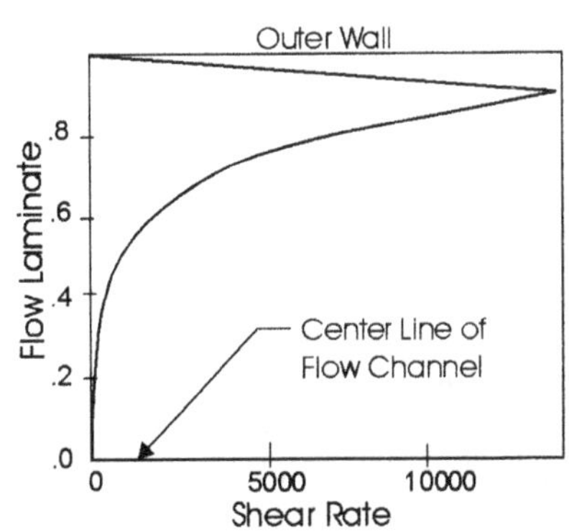

Figure 10.11 Shear rate distribution through a cross-section of the melt flow during filling. Orientation is highest in the outer laminates.

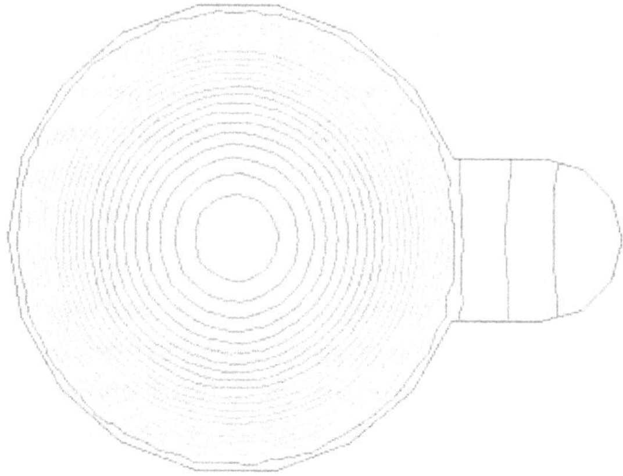

Figure 10.12 Filling pattern during injection molding. Each line represents the melt front advancement at equal increments of time.

laminates are under a compressive stress while the center laminates are under a tensile stress. This results from the conflicting strains between the two regions.

Linearized shrinkages, resulting from flow, are complicated further by changes in flow direction and flow velocity that often occur through the cross-section during the filling and compensation phases.

Figure 10.12 is the filling analysis results from a simple part that is gated in the center of the large circular region. The geometry is used to describe the complex flow conditions that can be established, even in this simple shape, and how this affects shrinkage. Each of the lines is referred to as isochrones and represent the advancement of the melt front at equal increments of time. Close spacing represents slow movement; further spacing represents rapid movement. The part is filled from a single gate at a constant injection rate. As the melt emerges from the gate the velocity of the flow front decreases as it diverges in a radial pattern through the increasing volume of the part. Upon reaching the edges of the large, circular main body of the part, the flow becomes limited to filling the appendage on the right side. At this moment all flow entering the cavity is redirected towards this last region to fill. This results in a sudden surge in the flow front velocity into the appendage. This velocity increase between the gate and the appendage causes an increase in shear stress in this region. The high shear stress is shown in Fig. 10.13 as the lighter colored region. Additionally the sudden increase in velocity results in a pressure spike. Figure 10.14 illustrates the resultant pressure spike, which in this case nearly triples the filling pressure during the last 15% of fill time. With the changing velocities and stresses, there is a resultant increase in the degree of orientation taking places in the flow path between the gate and through the appendage. This, in turn, will result in variation in the magnitude of the directional shrinkage throughout the part. Therefore we have variations not only in the direction of the shrinkage, but also in its magnitude.

Shrinkage is complicated even further by changes in flow direction that occur under the frozen layer as the part is filled. Figure 10.15 is a plot that provides infor-

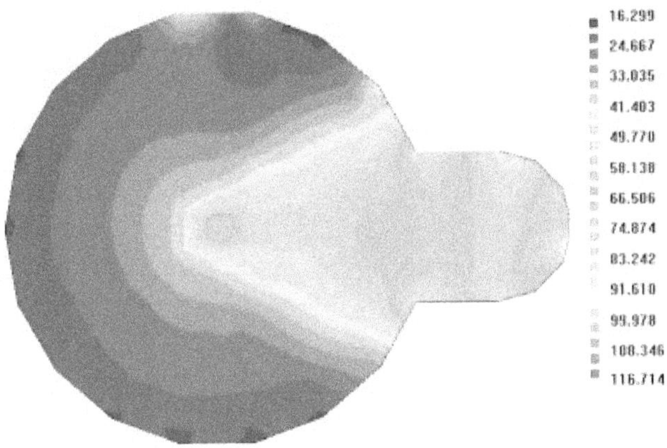

Figure 10.13 Shear stress distribution through the part as a result of melt front velocity.

mation on the direction of melt flow at the instant of fill. This is the direction of flow under the frozen plastic layer that was formed during the original mold filling. When contrasted to the original filling pattern, Fig. 10.12, one can interpret where flow direction, and resultant orientation, will be different between the outer laminates and the inner laminates of the molded part. The orientation of the outer laminates will be controlled primarily by the original filling pattern, while the orientation of the inner laminates will be primarily controlled by the later flow conditions under the frozen layer. One can see that in some regions of the part the flow direction remains constant throughout filling while others have changed significantly under the frozen layer. Therefore, it can be expected that not only does linear shrinkage vary 360

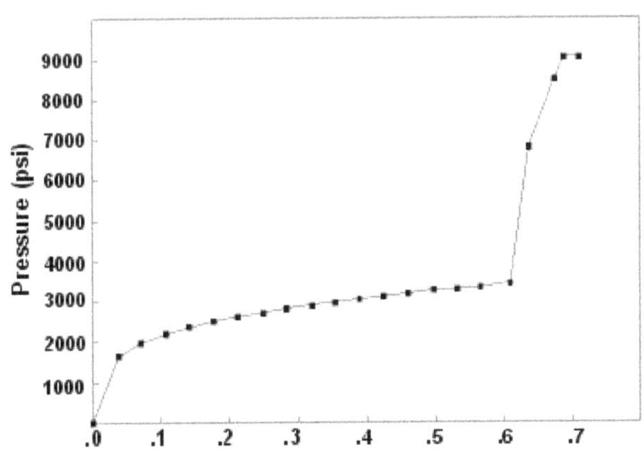

Figure 10.14 Pressure to fill the cavity nearly triples at the same instant the melt front velocity increases.

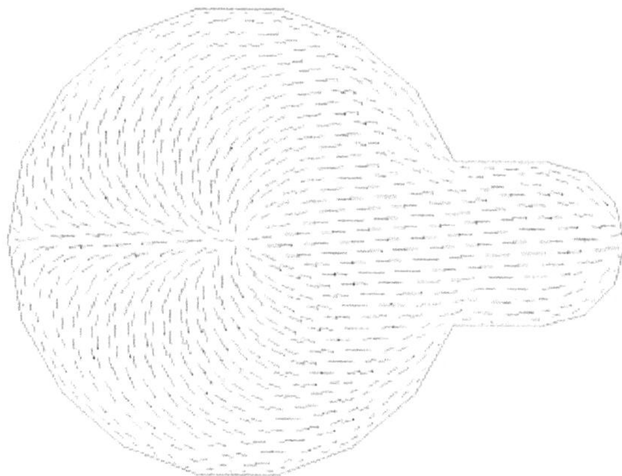

Figure 10.15 Plot providing information on the direction or angle of melt flow at the instant of fill. This is the direction of flow under the frozen plastic layer that was formed during the original mold filling.

degrees globally throughout this part, but within any given cross section the shrinkage direction and magnitude can be different.

Shrinkage of Fiber Filled Materials

The inclusion of fibers (typically glass or carbon) can have a significant effect on the linear shrinkage of a polymer during injection molding. The asymmetric ridged nature of the fibers restricts shrinkage of the polymer matrix in the direction of the fibers orientation. Therefore, if the fibers become oriented in the direction of flow, then the material will be restricted in its shrinkage in that direction. The result is the highest shrinkage will take place in the transverse to flow direction rather than the common, high flow direction shrinkage. In cases where there is a high degree of orientation of fibers, the relative shrinkage of flow and transverse to flow can be quite pronounced.

Figure 10.16 is the results of a study by B. G. Jaros [2] on the anisotropic shrinkage developed in parts molded of a 30% glass fiber filled PBT. Shrinkage in three different geometries were contrasted. The three geometries created variations in flow type. The first geometry created a near ideal linear flow. The test specimen was 50-mm wide, 200-mm long, and 2-mm thick. The part was fed from one end by a specially designed fan gate so that the flow would be linear across the entire length of the part. Under these conditions the resultant shear field will act on the glass fibers, as with the polymer chains, and orient themselves along the length of the sample. Note that the average shrinkage, labeled LINEAR, in the flow direction is 0.18% whereas the transverse to flow shrinkage is 1.7%. This is a variation of more than 9:1 (transverse to flow: flow).

Glass fiber orientations can be significantly affected not only by the direction of flow, but by whether the flow is dominated by shear or extensional flow effects. In the case above, flow within the cavity is almost exclusively shear dominated. Virtually no extensional flow exists. This can be contrasted to a center gated disk. In the case of the disk, the melt enters into the center and expands outward as it fills the cavity. As the melt diverges through the cavity, its flow is not only radial from the gate, but in fact, has a component which acts perpendicular to this direction which is referred to as extensional flow. The condition is analogous to the surface of an expanding balloon where the surface is stretched perpendicular to the direction of its expansion. This extensional flow will act on the glass fiber and result in orientation in the transverse to flow direction.

The net orientation of the glass fiber will depend on whether the dominant force acting on any given fiber results from shear flow or extensional flow. In the outermost laminates adjacent to the cavity wall, it is expected that shear stresses will be the dominant factor. Midway through the cross-section of the flow channel the shear field is nonexistent. As a result the tensile forces of the expanding flow front will be dominant and orient the fibers perpendicular to flow. The cross over as to which is dominant will depend on process conditions and location within a mold. In Jaros's study, radial flow effects on shrinkage was evaluated using a 101.6-mm radiused half disk. The same 30% glass fiber filled PBT was run in this mold with the same thickness as with his linear flow cavity. This radial flow shrinkage data is summarized in Fig. 10.16. The average flow direction shrinkage was found to be 0.33% and the

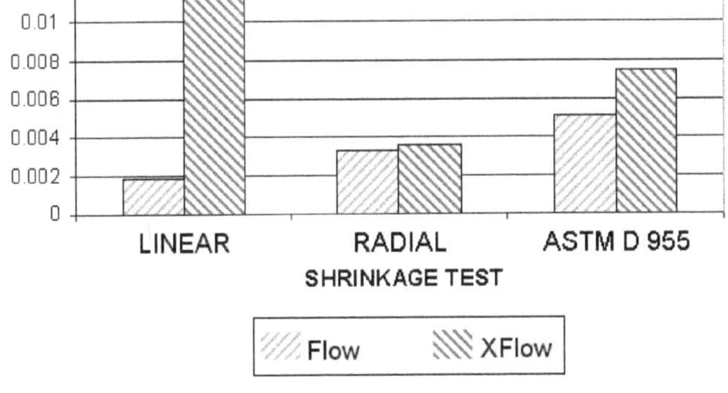

Figure 10.16 Results of anisotropic shrinkage study developed by B. G. Jaros [2] in parts molded of a 30% glass fiber filled PBT.

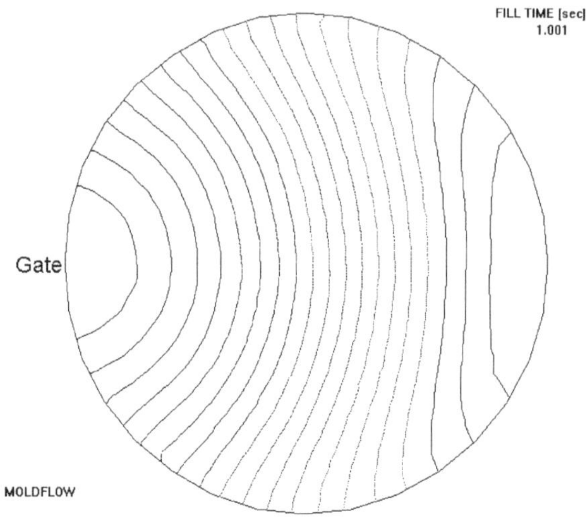

Figure 10.17 Filling pattern of 102 mm diameter test plaque. Initial melt front is radial as it diverges from the gate.

average transverse to flow was 0.36%. This is a variation of only 1.09:1 (transverse to flow: flow) versus the 9:1 ratio found in the pure linear flow part.

Finally the shrinkage results from the linear and radial geometries that Jaros molded are contrasted to the data provided for the same material by the material supplier. The material supplier's flow and cross flow shrinkage data was generated from a standard 102 mm diameter test plaque. The sample is edge gated from a narrow tab gate.

As the flow enters the standard test plaque cavity, its flow is initially radial as it diverges from the gate (see Fig. 10.17). This initial extensional flow establishes cross-flow orientation of glass fibers in the midstream of the cavity. As the flow front progresses, its direction begins to become linear through the remainder of the part. However, in the absence of an offsetting shear force in the midstream, the cross-flow orientation will be maintained through filling of the rest of the part. This results in complex variations in orientation not only through a given cross-section of the part but globally across the part. The result is that the standard test method provides data showing a difference of only less than 1:1.4 in shrinkages (flow: cross flow). This is in contrast to Jaros' data, which was based on pure linear and pure radial flow data.

Determining Mold Cavity Dimensions

As discussed previously, shrinkage of plastics will vary both in magnitude and direction throughout a molded part. Therefore determining the required cavity dimensions to achieve the specified size of a plastic part can be extremely elusive. As a result, it is common that a cavity has to be modified numerous times to achieve the specified part size. This is an extremely time consuming and costly task which can far outweigh the cost of the initial tool.

Most mold cavities are sized based on data collected through standardized test methods such as ASTM D955. This data must be used with caution, as the source of this data virtually never represents the geometry or process of your molded part. It is common that only a single shrinkage value may be known for a material. This value is often used regardless of flow length, cavity thicknesses, or process. Based on Jaros' study, published data for a 30% Glass Filled PBT developed using ASTM D955 states that shrinkage ranges from 0.4% to 0.8% depending on flow direction and process. Jaros' test data on the same material found significantly different shrinkages. Jaros' samples were molded using a wide range of processing conditions with two different cavities creating either pure linear or pure radial flow conditions. Depending on flow type and process, Jaros found that shrinkage with the same material varied from 0.15% to 2.04%.

Correctly dimensioning mold cavities is a common problem in the plastics molding industry. Mold shrinkage analysis software is attempting to remedy this situation. However, most mold builders find it difficult to justify the cost of performing a shrinkage simulation. Additional problems include the limited availability of materials characterized for shrinkage, the time to conduct the simulation and the potential for error in the simulation. Further developments and new technology are providing solutions to these limitations.

The most common method of sizing cavities today remains using hand calculations based on published shrinkage data provided by the material supplier. The following formula is used when calculating required cavity dimensions from a fixed shrinkage constant.

$$L_M = \frac{L_P}{1-\alpha_L}$$

Where L_M is the length of the mold cavity, L_P is the target length of the molded plastics part and α_L is the shrinkage coefficient for the given material.

Sample Problem: Given a molded part that is to be 610-mm long. The shrinkage coefficient (α_L) for the given material is 0.021 mm/mm. The required mold cavity dimension is found by:

$$L_M = \frac{L_P}{1-\alpha_L} = \frac{610\,\text{mm}}{1-0.021\,\text{mm}/\text{mm}} = 623.08$$

Shrinkage coefficients for some common polymers are shown in Table 10.1. It should be realized that these values can vary dependent on process, geometry, material grade, material supplier and types and amount of additives.

Shrinkage during the injection molding process can also have a positive contribution. An example is in the ejection of parts from the mold. Shrinkage of the part onto a core controls which side the part will adhere to as the mold opens. This allows for the design of an ejection system for positive removal.

Warpage in Plastic Parts

Warpage of plastic parts results from variations in shrinkage. These variations create stresses that can overcome the mechanical properties of the molded plastic part and

Table 10.1 Sample Shrinkage Values Used for Neat Injection Molded Thermoplastics

Name of material	Shrinkage (mm/mm)
ABS	0.004–0.006
Reinforced ABS	0.001–0.003
ABS/PVC alloy	0.004–0.006
Acetal	0.020
Acrylic	0.002–0.006
Modified acrylic	0.002–0.006
Cellulose acetate	0.005–0.008
Cellulose acetate butyrate	0.003–0.006
Teflon TFE	0.030–0.070
Teflon FEP	0.040–0.060
Nylon Type 6	0.010–0.015
Nylon Type 66	0.015–0.020
Reinforced nylon	0.002–0.005
Phenylene oxide-based resin	0.005–0.007
Polycarbonate	0.005–0.007
Reinforced polycarbonate	0.003–0.005
Polyaryl ether	0.003–0.007
Polyesters (PTB)	0.004–0.008
Reinforced polyesters	0.0025–0.0045
Polyether sulfone	0.0015–0.003
Polyallomer	0.001–0.002
Polyethylene Type I	0.030
Polyethylene Type II	0.030
Polyethylene Type III	0.015–0.040
Polyethylene Type IV	0.015–0.040
Reinforced polyethylene Type IV	0.003–0.005
EVA copolymer	0.010
Polyphenylene sulfide	0.001–0.004
Polypropylene	0.010–0.020
Reinforced polypropylene	0.003–0.005
Polypropylene copolymers	0.010–0.020
Polystyrene general purpose	0.004–0.006
Polystyrene impact	0.004–0.006
SAN copolymers	0.003–0.007
Reinforced SAN copolymer	0.001–0.003
Polysulfone	0.007
Reinforced polysulfone	0.001–0.003
PVC rigid	0.004–0.006

cause it to distort. The ability of the plastic part to resist these residual stresses depends on the rigidity of the plastic material at the time of ejection and the rigidity of the parts shape. If a part is ejected from the mold while it is still relatively hot, the modulus of the material will be lower and therefore more prone to warping. Holding the part in the mold longer will lower the temperature of the plastic and increase the modulus of the plastic material. The shape of the plastic part can also have a significant effect on whether the part will warp. A flat part has little structure and is much more susceptible to warpage than a part like a cup, which has good structure. Ribs are sometimes added to a flat part to add rigidity and thereby resist warpage. However, the ribs themselves can often create variations in shrinkage and have an opposite effect, thereby inducing warpage.

Mold and molding factors that affect warpage by creating variations in shrinkage include:

- Direction and magnitude of orientation which are affected by:
  - Fill pattern indicates direction of orientation on the outer laminates of a molded part.
  - Transient flow (effects the direction of orientation in the midplane of a molded part.
  - High shear stress results in high orientation. Regions of high shear stress are commonly near gates and the end of fill where melt may accelerate into restricted regions.
- Uniformity of cooling. Variations in cooling can occur from side to side or region to region within a plastic part. Side to side variations will create a bending conditions with the part warping toward the hot side. Regional variations in cooling can create an unstable buckling condition.
- Pressure variations across a part. It is normal that pressures will be highest near the gate. Given a part with a constant wall thickness, it can be expected that volumetric shrinkage near the gate will be less than regions away from the gate. In a center gate circular part, like a disk, this would tend to cause the part to warp in a bowl shape.

Residual Stresses in Plastic Parts

Variations in shrinkage in a part will create residual stresses. If these stresses can overcome the rigidity of the part, it will warp. However, regardless if the part warps or not, stresses will remain in the part. Over time, these stresses may cause premature failure through a multitude of factors. The same mold and molding factors identified as contributors to causing warpage will create residual stresses. The residual stresses developed from variations in shrinkage can add to the stresses developed from external loading. Additionally, environmental factors such as chemicals or temperature can accelerate the effects of residual stresses, causing unexpected premature part failure.

Mechanical Properties

The mechanical properties of a molded part can be effected by both flow induced orientation and the existence of weld lines created when two melt fronts combine. Orientation creates anisotropic mechanical properties which are much like the effect of the grain structure in a piece of wood. Under flexural loading, a part will be stronger when loaded perpendicular to orientation. Under tensile loading, the part will be stronger when load parallel to orientation. Like orientation induced anisotropic shrinkage, these variations in mechanical properties are most severe with fiber reinforced materials.

Weld lines are formed when two melt fronts join. Molecular orientation is reduced across a weld line, so there will generally be a reduction in mechanical strength along the junction. A butt weld, where the flow meets straight on, will be weaker than a meld line, where the two flow fronts meet at an angle and merge together. A fiber reinforced material will have a significant reduction in strength at a weld. Generally the higher the fiber content the higher the percentage loss in strength. A transient flow can develop across a weld and effect its strength. This is a condition in which a flow continues through the weld, under the outer frozen layers, after the melt fronts initially meet. The continued flow causes molecular and fiber orientation across the weld and can have dramatic effects in increasing weld strength. This can be induced by special tooling or processes. In some cases this transient flow can be created by strategic gating within the mold.

10.3 Part Design Guidelines for Injection Molded Plastic Parts

General guidelines for designing plastic parts have evolved over the years and are focused at issues related to manufacturability. These can include consideration of material shrinkage, part ejection, cooling, and mold filling; however, many of these guidelines contradict each other. A guideline developed to address one problem can create a new, and sometimes more severe, problem. It is important to understand the logic behind the guidelines and the potential consequence of utilizing them. With this understanding a designer can make informed decisions regarding a design and anticipate the consequences of including a specific feature in the design. This may cause the designer to seek an alternative, improved approach to achieve the products objectives.

The most troublesome issue related to successfully developing a new plastic part is probably anticipating how it will shrink and warp after molding. Many part design guidelines attempt to address these issues, but fall far short. Shrinkage of plastic parts will basically vary from material to material and within a given material. This makes it difficult to design a mold and process that will produce a part to the desired size. In addition, variations in shrinkage within a given part develops residual stresses that act to warp the part. The stresses that do not warp the part will reside in the part and potentially cause delayed dimensional and structural problems. The more the shrinkage variations within a part, the more potential there is for warpage and residual stresses.

High-shrink, and fiber-filled, materials create more potential for shrinkage related problems than low-shrink and nonfiber-filled materials. A high-shrink material has a higher potential for shrinkage variation than a low-shrink material. Unfilled, or neat, amorphous materials are relatively low-shrink materials, most having linear shrinkages of less than 0.010 mm/mm. Neat semi-crystalline materials will normally have high shrink values ranging from 0.010 mm/mm to more than 0.050 mm/mm. These shrinkage values will vary for each material. They will also vary for each material dependent on part design, mold design, and process.

Polymer additives, such as rigid fillers and reinforcements, will reduce overall shrinkage. If the additive has no real aspect ratio, it will generally reduce shrinkage-related problems. Fillers with an aspect ratio, such as a fiber where the length is longer than its diameter, will reduce overall shrinkage but create significant variations in directional shrinkages. These directional shrinkages develop from orientation established during mold filling and compensation phases. The result is that fiber-filled materials will tend to increase a materials desire to warp.

One of the primary objectives for the designer of injection molded plastic parts is to maintain a uniform wall thickness in the part. All designs should be approached with this objective in mind. A uniform wall will minimize injection molding problems, particularly those related to shrinkage. It is common that a plastic part be designed such that it includes relatively thick features attached to thinner wall sections. This is a major source of many part design problems. A designer should make every effort to eliminate or core out any thick features in a plastic part.

Most of the following design guidelines are based on how to combine various design features of a part with the primary wall. This will include how to add features such as ribs, bosses, and holes. The addition of ribs that are classically thinner than the primary wall is of particular interest, as they break the most fundamental design objective of maintaining a uniform wall thickness.

Before applying any design feature, or guideline, that creates variations in a parts wall thickness, the designer should ask:

1. What is the reason for modifying the thickness of this feature relative to the primary wall?
2. Does the reason for the variation have significance in this part?
3. Is there an alternate approach that will avoid this variation?

4. If the feature creating a variation in wall thickness must be used, what are the potential problems created by this variation and what can be done to address these problems?

After answering these questions, the designer must judge the benefits, or needs, of varying the wall thickness against the potential problems that will be created. This chapter is intended to help a designer make these judgments. It should be understood, however, that *there are no fixed rules in designing plastic parts*. The following guidelines are suggestions to minimize problems, but they must always be judged individually for a particular case.

10.3.1 Designing the Primary Wall

Injection molded plastic parts are generally designed around the use of relatively thin walls. Unless the injection molded plastic part is foamed, or is produced with gas assisted injection molding, its walls are normally less than 5 mm (0.2 in). When determining the thickness of the primary wall one must consider the structural, functional, and aesthetic issues related to the wall, as well as its impact on manufacturability. Manufacturability issues include consideration of the injection pressure required to fill the mold cavity, cooling time, and the influence on ejection from the mold. These will require consideration of available injection pressure and injection rates, mold rigidity, ejection techniques, and melt delivery means.

Maintain a Constant Wall Thickness

Maintaining a constant wall thickness should be the primary objective of the product designer. Each region in a part that has a different thickness will want to shrink differently. These variations in shrinkage will complicate achieving the desired size of the molded part and be major contributors to warpage and residual stresses. Variations in wall thickness also affect the mold filling and packing, or compensation, phases of the molding cycle. Variations in the wall thickness can also result in irregular, and difficult to predict, filling patterns. This can further complicate orientation-induced shrinkage as well as cause problems with venting, gas traps, and weld lines. In addition, a thick region fed through a relatively thin region can result in sinks and voids in the thicker region. A thin region, such as a rib, attached to a thicker region can result in hesitation effects during mold filling that could potentially cause no fills.

If variations in wall thickness cannot be avoided, then try to keep the variation to a minimum and provide a gradual transition rather than a sudden change (see Fig. 10.18). Be conscious of the negative effects of having these variations that would include variations in shrinkage, warpage, residual stresses, and undesirable mold filling patterns.

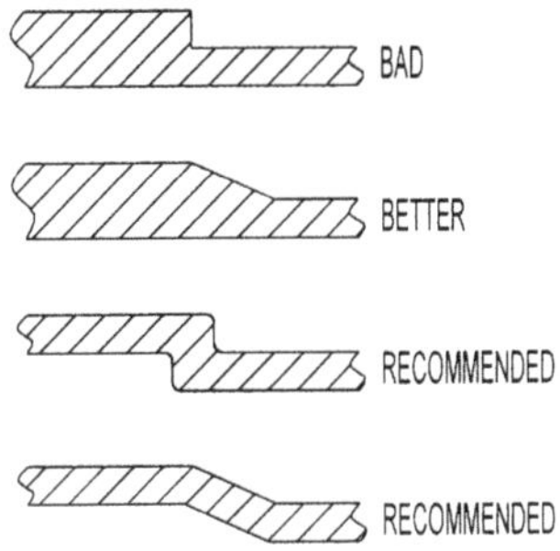

Figure 10.18 Provide as gradual a change as possible (or eliminate the change if possible) at the intersection of a thick/thin region.

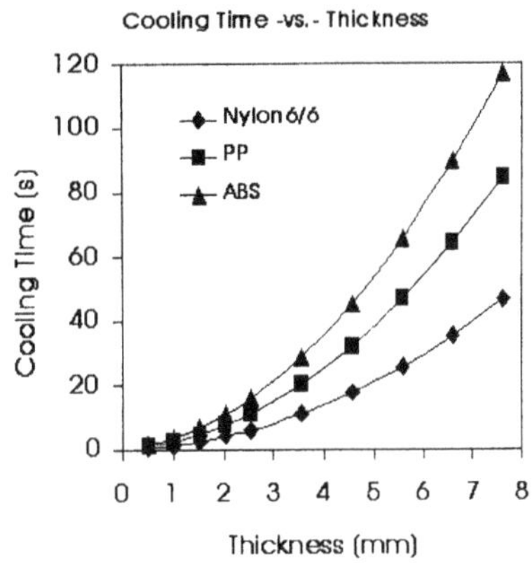

Figure 10.19 Exponential increase of cooling time with respect to wall thickness for three common plastic materials.

Minimize Wall Thickness

For injection molding, thinner walls will lead to more cost-effective production through the use of less material and shorter molding cycles. The plastic material used and the cooling time can often be attributed to more than 70% of the part's cost. Reducing the amount of plastic material has an obvious direct proportional effect on reducing material costs. Reducing cooling time through a thinner wall has an even greater potential benefit, particularly where cooling time is a significant portion of the part cost. It can be seen from Fig. 10.19 that cooling time increases exponentially with an increase in wall thickness. Doubling wall thickness of an ABS part from 1.0 to 2.0 mm increases cooling time by approximately 2.6 times. Doubling the wall thickness again, from 2.0 to 4.0 mm, increases cooling time approximately 3.8 times.

The overall part thickness is generally determined by the amount and type of loading that the part is expected to withstand. Depending on the application of the part, other factors that should be considered include electrical, sound and thermal insulation, and gas permeability. Regardless of the particular requirements, it should be the objective of the designer to minimize the wall thickness while satisfying the part's functional requirements. The designer must be conscious of the relationship between wall thickness and pressure to fill in accomplishing this. Reducing wall thickness will exponentially increase pressure to fill. Decreasing the wall thickness by half can easily cause a threefold increase in the pressure to fill.

In applications where plastic parts are to replace metal parts it is common that the plastic part will be cored out relative to the original metal part its replacing. Metal parts are often made bulky and solid, as their formation normally requires machining, or removal of material from a solid chunk of steel. This is a subtractive process where increasing the amount of steel to be removed increases machining time and related costs. The more steel removed, therefore, the more expensive the product. In

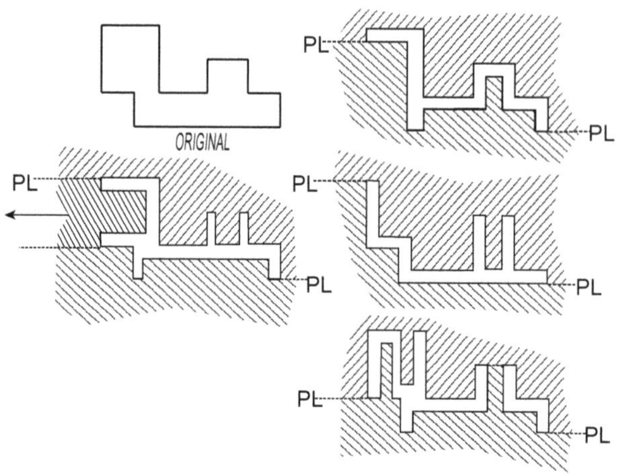

Figure 10.20 Design alternatives are shown for a part with thick regions. The optimum option will depend on the functional and structural requirements of the part. The molds parting line and ejection requirements must be considered.

contrast, injection molding of plastics parts is an additive process. Here, the opposite is true: the less the material, the lower the cost of the product.

Figure 10.20 illustrates four alternative designs for a part that might have been designed to be machined from steel. The three design cross-sections on the right would be easily produced with injection molding and are shown as cavities formed between two mold halves that would open vertically. Consider how the need to remove the molded part easily has influenced the design. The alternative design on the left bottom could also be easily produced, but it would require a side core. Without the side movement of the side core, the feature on the left would create an undercut that would trap the part in the mold.

Shape of the Wall

The shape of the wall can affect the rigidity of the part and the potential for warp. A domed or circular shape is much more structurally stable than a flat part. A flat part has little structure with which to resist warpage. When molding a flat part, therefore, one must be more concerned with establishing a flow pattern, during mold filling, and a process that will minimize stresses. It should also be expected that successfully molding parts, with shapes that are prone to warping, will be more sensitive to changes in process and are likely to result in higher spoilage.

The shape of the wall can also affect cooling time. A nonstructural shape will be more sensitive to warping. To resist warping, the part must be constrained in the mold longer while it cools further and becomes more rigid. The cooler, more rigid part will be able to better resist distortion from its residual stresses. Regions in a part formed by tall cores will increase the local thermal load, making cooling more difficult.

Features such as flanges, positioned along the perimeter of the part, can increase rigidity without adding thickness. These flanges should ideally be perpendicular to the primary wall to which they are attached, and be of the same thickness or slightly thinner. A thicker flange around the perimeter of a part, like a disk, will cause the center to buckle, forming a bowl-like shape.

Surface Finish

Polishing the core in the direction of draw (ejection) is normally recommended to assist in ejection when small draft angles are required. The exception is polyolefins and elastomers where a courser 400 grit polish is recommended.

10.3.2 Ribs, Gussets, and Bosses

Ribs, gussets, and bosses are features that are added to the primary wall for structural, assembly, or other functional reasons. When adding these features to the primary wall, the plastics designer must again consider manufacturability. This includes consideration of part ejection, venting of air during mold filling, and effects on mold filling and packing. The driving force behind most of the standard design guidelines for these features is the attempt to balance these manufacturability issues that are often conflicting.

Ribs

Ribs are primarily used to increase the rigidity of a part or a specific region of the part. To achieve the desired rigidity of a given part an excessive thickness might be needed in the primary wall. This would negatively affect part cost. The addition of ribs can often attain the required rigidity while maintaining the more desirable thinner wall.

From a manufacturing standpoint the ribs should be kept as short as possible and be drafted (i.e., tapered on either side) (see Fig. 10.21). Both of these requirements ease ejection problems. In addition, a short rib reduces mold filling problems and excessive thinning of the rib tip, resulting from the required draft. Several short ribs, therefore, are preferred to one tall rib.

The intersection of the rib to the primary wall creates an increased volume (region "A" in Fig. 10.21). This is compounded by the need to radius this intersection to minimize stresses. The increased volume at this intersection is generally filled by material flowing through the thinner primary wall (region "B" in Fig. 10.21). During the compensation, or packing, phase of the molding process, this thinner wall will freeze off first, thereby blocking compensating flow to the still shrinking region "A." This will cause sinks and/or voids to form at the intersection. In addition, the thicker region can act as an undesirable flow leader. To address these problems, ribs are normally thinned relative to the primary wall.

Figure 10.22 illustrates a recommended rib design. The base thickness of the rib should be 50 to 75% of the primary wall thickness, depending on a material's shrinkage characteristics. A rib thickness of 50% of the primary wall is generally recommended for high-shrink materials, and 75% for low-shrink materials. High-shrink materials are normally considered those with shrinkages of more than 0.015 mm/mm. Low-shrink materials are those with shrinkages less than 0.010 mm/mm. Because shrinkages can vary from less than 0.001 mm/mm to more than 0.05 mm/mm, the

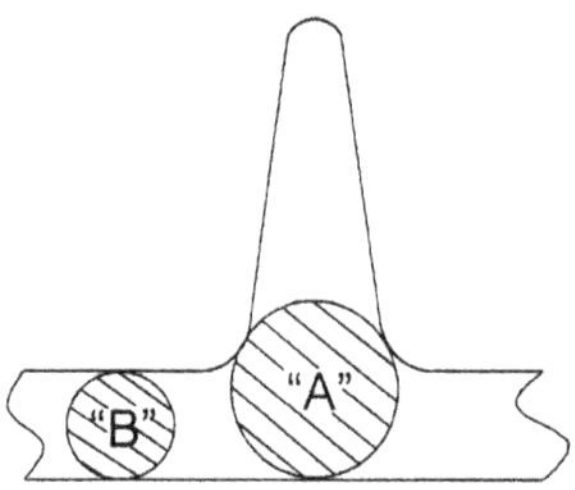

Figure 10.21 The required draft and radius used for a rib results in the large volume of material at region "A" relative to the adjoining primary wall "B." This will create packing problems for region "B."

designer must make judgments as to how to apply these guidelines. There is no set rule that will work in all cases.

The base of the rib should have a radius. The size of the radius should be 0.25 times the primary wall thickness with a minimum radius of 0.25 mm. The height of the rib should be no more than 2.5 to 5 times the primary wall thickness. Draft angles range from 1/2 degree to a maximum of 2 degrees, with 1/2 to 1 degree used most commonly. Excessive rib thickness and radii should be avoided because these may lead to sinks (or voids) and extended cycle times.

Figure 10.23 illustrates the potential problem created by the improper combinations of draft angle, height, and rib base thickness. The thin top edge can result in both a structural and a mold filling problem. Figure 10.24 illustrates the stress distribution through a rib under flexural loading. It can be seen that the greatest stress is at the top, or free, edge of the rib. Because this region experiences the highest stresses under load it is important that it is not thinned too much.

Design Example:

Given a rib to be attached to a 2.5-mm thick wall, molded of a high-shrink material. The rib should be dimensioned as follows:

- Thickness at base = 1.25 mm
- Height (Maximum) = 6.25 mm

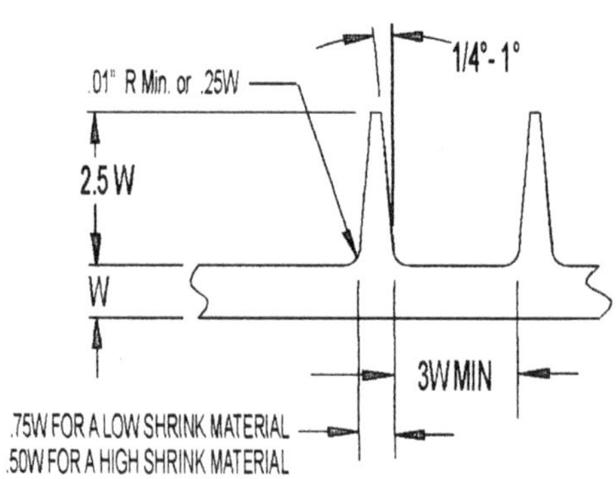

Figure 10.22 Guidelines for designing a rib with respect to the primary wall thickness of the part.

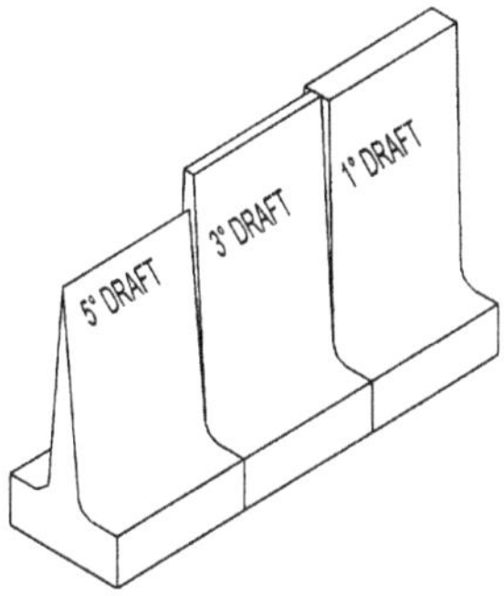

Figure 10.23 Excessive draft on a rib will restrict heights of ribs and result in excessively thin, fragile, free edges.

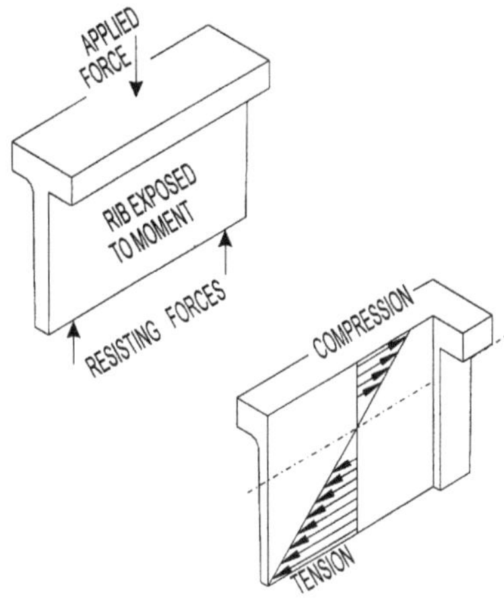

Figure 10.24 Illustration of the stress distribution through a rib under flexural loading. The greatest stress is on the thin, free edge of the rib.

- Radius at base = 0.625 mm
- Draft = 0.5^0/side

This results in the top, free edge, of the rib being 1.141-mm thick.

Gussets

Gussets are thin features, much like a rib, that are normally used to reinforce a local feature in a part. The reinforced feature could be a side wall, boss, or some other projection. Gussets are normally triangular in shape and should be designed using the same guidelines for thickness and radii as they are for a rib. In addition, they should be no more than four times the primary wall thickness in height and two times the primary wall thickness in width. Again, this is to address ejection and mold filling problems. Standard design guidelines for gussets are shown in Fig. 10.25. Here, two gussets are shown supporting a boss that is also connected to the primary wall with a single rib.

Design Caution

A major flaw in the previously presented design guidelines for ribs and gussets is that they create a variation in wall thickness that will often cause a part to warp. Ribs, in particular, can often span the entire length or width of a part. The thicker primary wall will take longer to cool and will shrink more than the thinner rib. As a result, the part will tend to warp away from the rib. Ribs added to flat plastic parts to add rigidity to minimize warpage can actually have the opposite effect and can cause warpage.

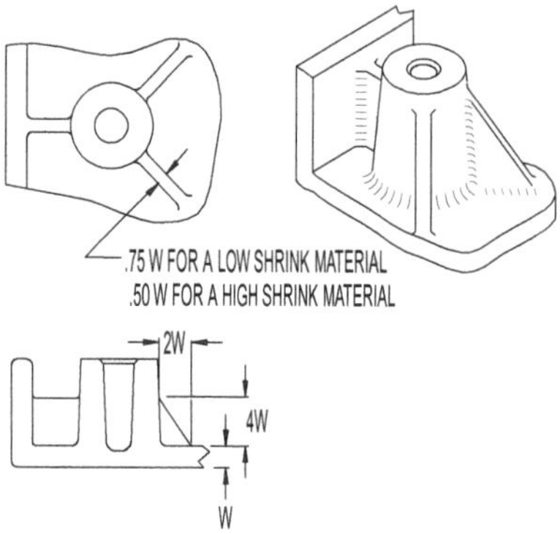

Figure 10.25 Standard guidelines for gusset design.

As a result a designer must be sensitive to application and aesthetic issues related to the part. If cosmetics are not a concern, reducing the rib thickness becomes less important. A resulting sink from a thicker rib will have virtually no negative structural consequences. This is particularly true when contrasted to the stresses that would develop between a primary wall and a thinner rib. Voids also may be tolerated in many parts at the intersection of a rib to the primary wall. During flexural loading the highest stresses are on the tip of the rib. Next would be the outermost surface of the primary wall opposite the rib. It is most likely that the part failure would be at the tip of the rib rather than due to a sink located in a relatively low stress location.

A further consideration should be the rigidity of the material. A sink is formed when the frozen material on the surface of the part is drawn in as the material below shrinks. A low-modulus material will easily be drawn in by the shrinking material. A high-modulus material that has been cooled will resist having the outer skin being pulled in. As a result, the high-modulus material is more likely to form a void and the visibility of a sink might be minimized. Design guidelines for heights of ribs and gussets are broken all the time with parts still being successfully molded. A designer must be aware of the potential problems created when deviating from the guidelines and should be prepared to address them.

10.3.3 Bosses

Bosses are normally either solid or hollow round features that project off the primary wall. They can be used for assembly with self-tapping screws, expansion inserts, force-fit plugs, drive pins, positioning, and other means. Bosses are often supported with ribs, adjoining walls, or gussets. Bosses can be free standing or connected to the side walls using ribs. A boss should not be attached directly to a side wall (see Fig. 10.26)

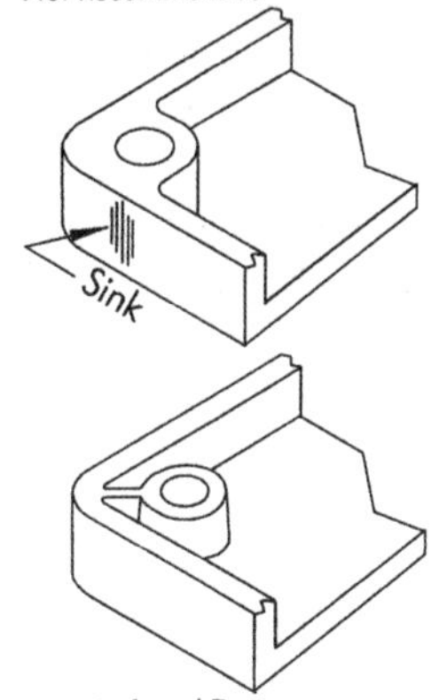

Figure 10.26 Comparison of a boss attached directly to the side wall of a part (not recommended) and the preferred design in which the boss is offset from the wall.

because the intersection with the wall will result in a thick region, causing sinks or voids. A hollow boss can result in weld lines on the outside show surface of the part because the material flows around the core, forming the hollow center. In addition, the resultant weld will commonly form a butt weld. This type of weld is particularly prone to structural failure if a hoop stress is developed from the insertion of a screw or stud used for assembly. The boss can be placed near the side wall and attached to it with a rib that will provide both rigidity and a means for venting (air can travel from the boss, across the rib, to the side wall, where venting is generally available at the natural parting line of the mold).

A standard recommended boss design is shown in Fig. 10.27. The inner diameter of a boss is normally determined by its function. When used with self-tapping screws,

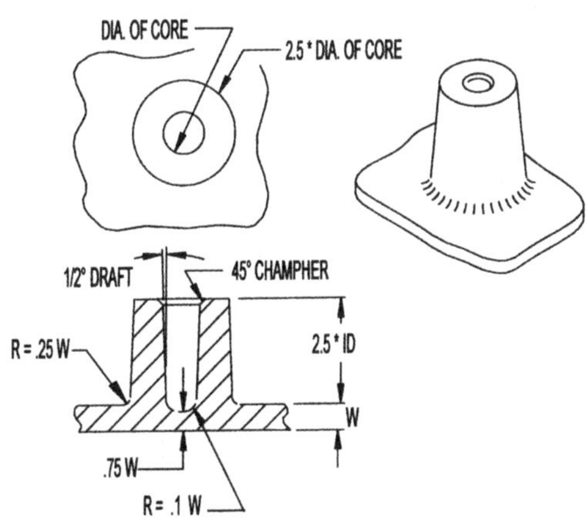

Figure 10.27 Standard recommended boss design. Though an inside draft of one half degree is suggested to assist with ejection, a zero degree draft is often preferred for functional purposes.

the inside diameter is commonly the pitch diameter of the screw. The outer diameter is approximately 2.5 times the inside diameter. This is a compromise between providing strength and minimizing the thickness of the boss. If a boss is too thick it could cause voids. As a self-tapping screw is inserted, it might break through to the void and thus reduce the pull-out strength. Outside diameters may vary slightly depending on screw design and the plastic material. Medium-to-low modulus materials can use a thread-forming screw. A variety of specially designed screws are available that will reduce the hoop stress on the boss. Very high modulus materials require a thread cutting screw because the strain from a thread forming screw would cause the boss to crack. There are a number of companies that provide specially designed thread forming and thread cutting screws for plastics. Each design has been developed for different material moduli and each company has its own design recommendations for inside and outside boss diameters. Some of these specialized screws are the Trilobe® from Continental Screw Company and the Hi-Lo from Elco® Industries.

The height of the boss should be no more than 2.5 times the inside diameter. This is to minimize the potential of deflection of the core forming the inside of the boss as well as filling, packing, and ejection problems. The intersection of the boss and the primary wall should have a radius that is 25% of the primary wall thickness, with a minimum radius of 0.25 mm. The intersection of the inner wall of the boss and the base should have a radius that is 10% of the primary wall thickness. The connecting wall thickness at the inside base of the boss should be 75% the primary wall thickness. A 45-degree chamfer should be placed on the inner top edge of the boss. A minimum of a 1/2-degree draft should be placed on the outside walls of the boss. Draft on the inside of the boss is also desirable, but it is often not used because it might compromise the assembly with a screw or a press fit. As a result, a stripper sleeve is generally required to assure ejection of the boss.

It should be expected that the relatively large cross-section at the intersection of the boss and the primary wall will normally result in a sink, or gloss variation, on the primary wall surface opposite the boss. An alternate design for reducing the material volume at the intersection of the boss to the nominal wall may be used (see Fig. 10.28). Even though this design reduces potential for sinks and voids, it will

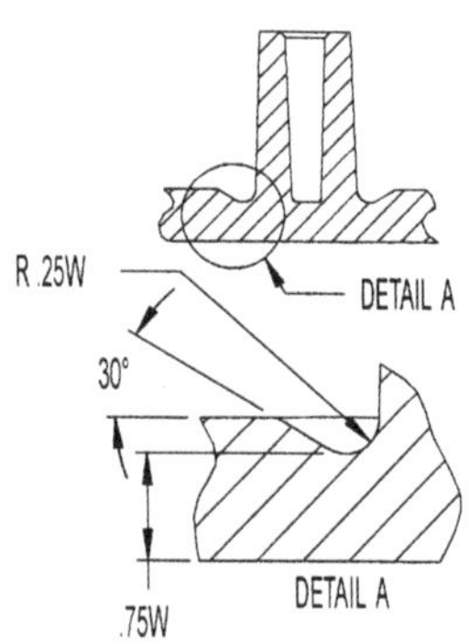

Figure 10.28 Alternate boss design for reducing the material volume at the intersection of the boss and the primary wall.

require an additional mold insert to form the depression that will increase the cost of the tooling.

10.3.4 Corners, Fillets, and Radii

Sharp corners should be avoided because most plastics are notch sensitive. This is particularly true with an inside corner that will act as a stress concentrator under load. Stress concentration will increase with the ratio of the corners fillet radius to the wall thickness. A preferred minimum radius-to-wall thickness (R/T) ratio would be 0.5. The sharper the radius, the more brittle the material will appear when loaded. In order to achieve this, the opposing wall surface must be radiused in order to avoid a thick region at the intersection.

Figure 10.29 illustrates the proper design of corners in a plastic part. This design provides for a constant wall thickness throughout the corner region. The inside radius should be a minimum of 0.5 (50%) the thickness of the primary wall. The outside radius should be the inside radius plus the wall thickness. This assures that the wall thickness is constant throughout the corner. (Note: The larger the inside radius the better.)

For example: Given a part having a corner with a primary wall thickness of 2.0 mm. The minimum inside radius of the corner should be 1.0 mm and the outside radius of the corner would be 3.0 mm.

This design maintains a uniform wall thickness in the corner. It is preferred that the inside radius be as large as possible. A larger radius can both improve the part structurally and reduce expected warpage developed from unbalanced cooling in corners. It may also improve filling in some circumstances, particularly with fiber-filled materials. In some cases, sharp corners are used for aesthetic purposes or to accommodate a mold machining issue. High loads and stresses should be avoided in these regions.

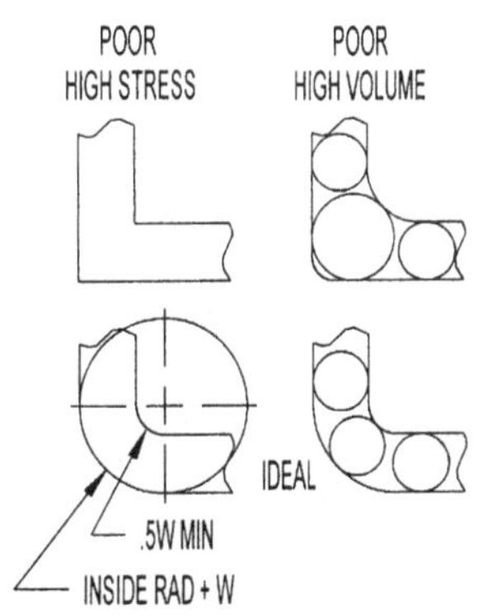

Figure 10.29 Proper design of corners in a plastic part.

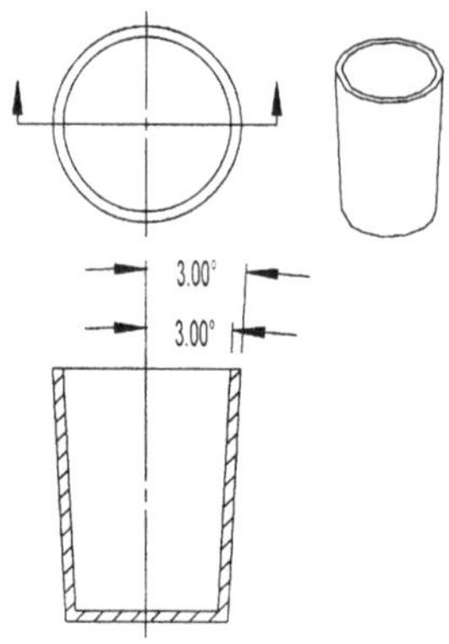

Figure 10.30 Draft on the inside and outside surfaces of a part should be equal and parallel.

10.3.5 Taper and Draft Angles

Drafts or tapers are angles put on vertical walls of an injection molded part to provide easier ejection from the mold. These angles generally range from 1/8 degree to several degrees, depending upon the material, anticipated ejection problems, and product design requirements. Draft on inside and outside surfaces of side walls should be equal and parallel (see Fig. 10.30). The greater the draft, the less potential for ejection problems.

10.3.6 Undercuts and Holes

Undercuts in a part can significantly increase the difficulty of ejecting molded parts from a mold. Examples of undercuts include a hole in the sidewall of a part, an internal snap ring which is often used in container caps, and internal and external threads. These require the use of special mold features such as side cores, split cavities, collapsible cores, unscrewing devices, and stripper plates. These features add to the initial cost of the mold, its maintenance, and molding cycle time. If possible, avoid holes or undercuts that would require these special mold requirements. If a hole is placed in a wall that is approximately perpendicular to the direction of the molding machine's platen movement, then a simple core may be used to form the hole and can easily be extracted during normal mold opening with the use of simple ejection; however, if the hole is created on the sidewall of a part, which is not perpendicular to the direction of the molding machine's platen movement, then the core forming the hole must generally be extracted by some special means. This side action would normally be performed by a side core, or split cavity, that would require some special means to move it perpendicular to the normal mold opening direction.

Some holes in side walls can be designed such that side action is not required. A stepped parting line can be used such that the hole is created while eliminating an undercut (see Fig. 10.31). A minimum shutoff angle of 5 degrees allows a clamp force

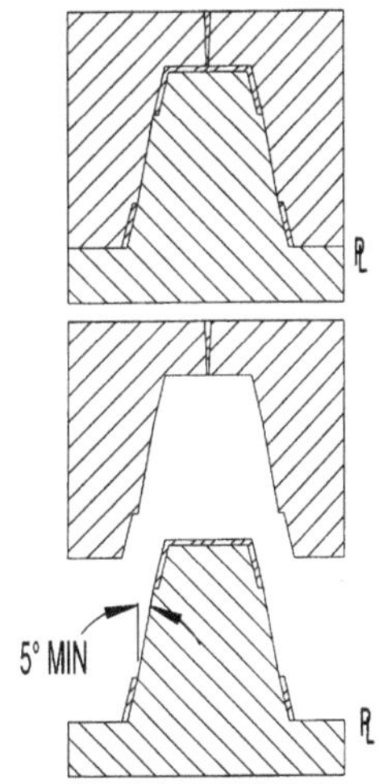

Figure 10.31 A stepped parting line can sometimes be used to create holes in the side walls of a part. This can eliminate the need for side action.

to be developed between the mating mold halves to prevent flashing and helps to assure that the two halves will not scrape against each other during mold opening and closing. This method simplifies ejection by eliminating the need for side cores/slides.

Another example of the use of the stepped parting lines for creating undercuts and holes is the part shown in Fig. 10.32. The vents on the top of this part are designed to allow airflow and prevent objects from falling through the gaps. The intermeshing mold features that create these vents are illustrated in Fig. 10.33. A similar approach is used to create the vents on the back of the part. These intermeshing shutoffs create holes on both the top and side of the part without creating undercuts, thereby eliminating the need for special side action in the mold.

Holes should provide sufficient spacing to minimize ejection problems and weaknesses created by weld lines. If weld line strength is a particular concern, the holes may be spotted during molding and drilled as a postmolding operation. If the holes

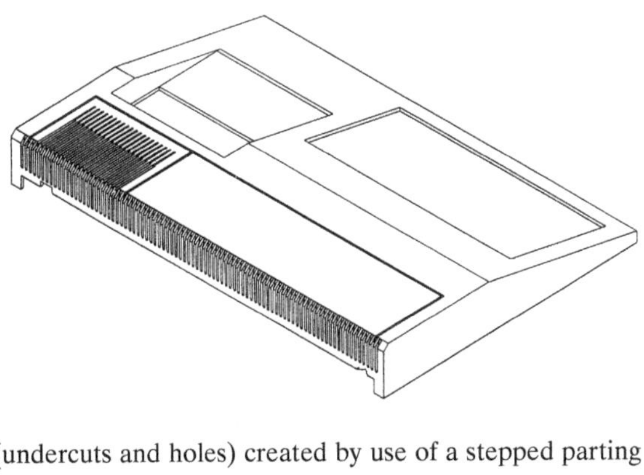

Figure 10.32 Vents (undercuts and holes) created by use of a stepped parting line.

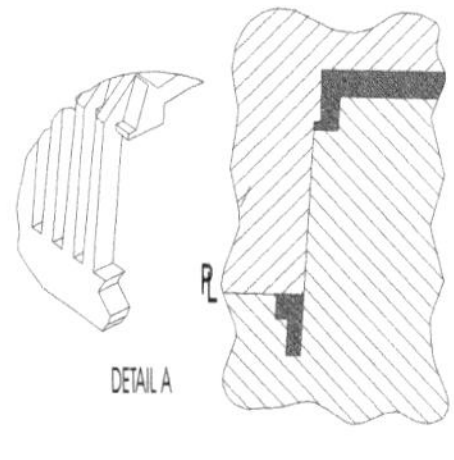

Figure 10.33 "Detail A" and the stepped parting line which creates the vents in Fig. 10.32.

are to be molded and will be structurally loaded, then the outer perimeter of the hole should be a minimum of one diameter from the outside edge of a side wall (see Fig. 10.34A). For holes that serve a nonstructural need the spacing between them can be as little as twice the wall thickness of the part (Fig. 10.34B).

Some undercuts, like the snap ring on many container caps, may be stripped off the core forming it. When designing these snap rings, the designer must consider the design of the snap ring, the adjoining wall, and the material properties. The snap ring must be designed such that the part can be stripped, or slid off the core forming it without the snap ring, or the part, being damaged. This requires that the snap ring be designed with a reduced lead angle. A lead angle of 45 degrees is preferred (see Fig. 10.35). In addition, the sidewall must be able to be deflected during ejection. This requires consideration of the depth of the undercut, the diameter of the core, and the materials stress or strain limits. Because the properties of the material are not normally available at the relatively high temperatures when a part is ejected from the mold, designing the maximum undercut that can be stripped from the mold is often

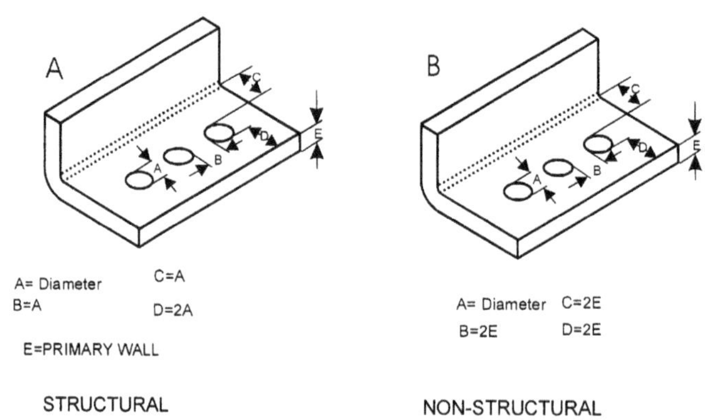

Figure 10.34 (A) Guideline for placement of holes in a part used for structural applications. (B) Guideline for placement of holes in a part used in nonstructural applications.

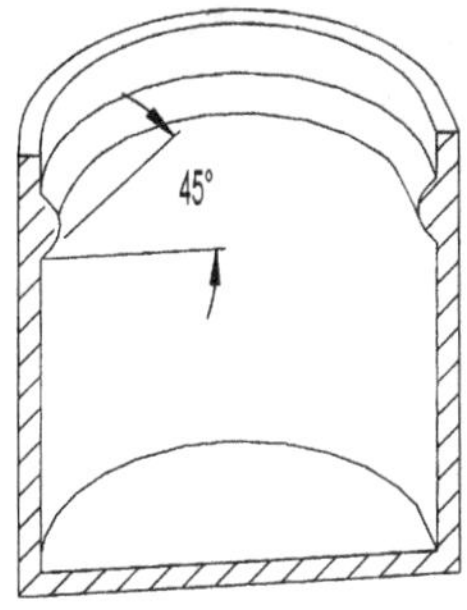

Figure 10.35 To facilitate stripping from a mold, a lead angle of 45 degrees is suggested.

based on the dynamic strain limits available for the room temperature material. This is generally a conservative design approach because the strain limits of most plastics are much higher at their ejection temperatures. The dynamic strain limit for a material is most representative of a materials characteristic under the rapid strain experienced by snap fits.

For example:

Determine the maximum height which can be used for a snap ring that can be used on a part which has an ID of 70 mm. The dynamic strain limit of the material is 3%.

$$\varepsilon = \frac{\Delta r}{r} \qquad \Delta r = \varepsilon \times r$$

$$\Delta r = 0.03 \times 35\,mm = 1.05\,mm$$

The inside radius is shown as r, Δr is the maximum allowable change in radius; that would be the maximum allowable undercut. ε is the dynamic strain limit for the material. It should be expected that a stripper plate or stripper sleeve would be required to strip the part from the core.

The preceding is a simplified approach to designing an undercut. Prototyping is often required for verifying the design.

10.3.7 Gating and Process Considerations

One must always be cautious of the melt flow pattern during mold filling relative to a hole. There will be a weld line any time there is a hole in a part unless a disk-type gate is used inside the hole. This weld line will reduce strength and create potential negative cosmetic effects. A designer should look to position the weld line strategically relative to these issues.

Whenever an undercut must be stripped from a core, be sure that there are no weld lines across the undercut. The high stress developed during ejection could cause the part to fail. The special core or cavity components, required to create these holes, or undercuts, are generally separate mold inserts that require movement. This required movement, and the fact that these parts are not integral with the primary mold plates, result in these mold components being less rigidly anchored; therefore, when these mold cores become relatively long they are more susceptible to undesirable deflection created from the high melt pressures developed during mold filling and pack stages. This deflection can affect mold filling, part geometry, and development of flash problems. When selecting a gate position in a cavity that includes these components, the resulting applied forces should be considered. It may be desirable to position the gates away from these features to reduce the pressure acting on them.

10.3.8 Cores

Molded parts may include features created by tall cores, such as the inside of a boss, or may be formed entirely from a tall core, such as a test tube or pail. These tall cores can present a number of potential problems. High pressures can deflect the core causing variations in wall thickness and disruption of the expected fill pattern. For this purpose, the part should be designed such that the maximum height of the core forming it is 2.5 times the core outer diameter. Even a core that is center gated, where the melt is expected to flow symmetrically down its sides, can deflect. This can be caused by slight changes in melt conditions (shear and temperature effects) developed in branching runners or other factors, such as non-uniform cooling, that create an imbalance in the mold filling. This unbalanced filling will continue into the packing stages. The filling and packing imbalance causes a pressure variation from one side of the core to the other. Even 300-mm diameter cores used for molding a pail can deflect from this imbalance. Even with today's 2.5-D mold filling simulations, the melt variation created in a branching runner or its effect on a deflecting core is not recognized.

Several designs have been used to help reduce core deflection. Figure 10.36A illustrates a design that incorporates a feature that has a self-centering effect. Here, if the melt leads over one side of the core and begins to deflect it, the flow path near the gate feeding this leading flow will be restricted while opening up the flow path on the opposing side. This local, bending-induced variation in wall thickness will potentially correct for the initial core deflection.

Figure 10.36B illustrates a second design where the base of a molded cup-shaped part is thickened relative to the side walls. If there is any imbalance developed out of the gate, the leading melt front will hesitate when it reaches the thinner side walls, allowing the slower flow fronts to catch up.

A part with a core gated on its side will be particularly sensitive to core deflection. In addition, side gating can result in severe weld lines or gas traps opposite the gate location. With edge gating, the creation of weld lines and gas traps can develop in even relatively short cores. A weld line might appear in a round core where the height is only 50% of the core's diameter. A gas trap during mold filling could appear in a core where the height is as little as 75% of the core diameter.

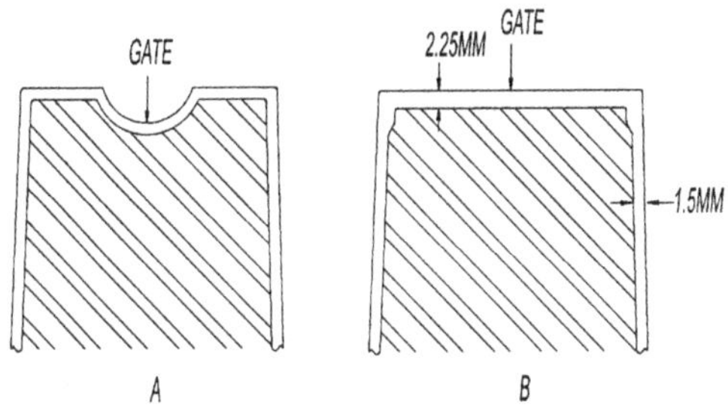

Figure 10.36 Two concepts of self-stabilizing cores.

10.3.9 Avoid Picture Frame Features

A design where the perimeter of the parts is thicker than the interior is referred to as a picture frame design. (Fig. 10.37). Picture frame designs can create numerous problems:

- Gating in the thicker perimeter will result in a racetrack effect, where the melt quickly follows the path of the thicker perimeter and moves slowly through the thinner center region during mold filling. This can result in a gas trap or a severe weld line in the thinner region.
- Gating in the thinner center will result in voids and sinks in the thicker perimeter.
- In either gating location, the thicker perimeter will shrink more than the center region, which will cause the part to warp in a bowl-, or oil can–, like shape.

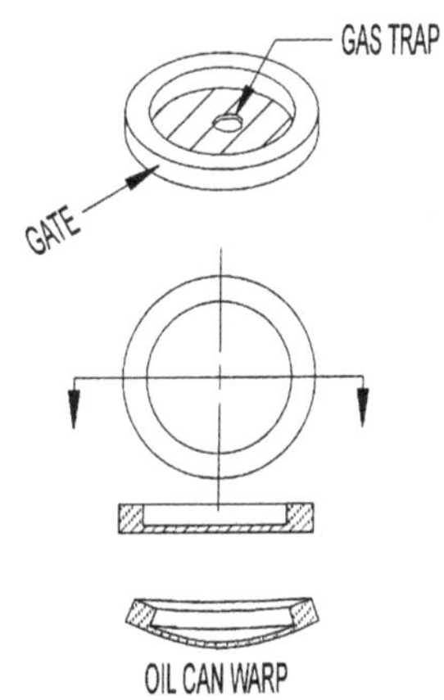

Figure 10.37 "Picture frame" part design. This type of design can result in numerous mold filling and warping problems.

10.3.10 Integral Hinges

Integral hinges are very thin features molded into a part that provide local flexibility. These are commonly used in plastic containers and directly connect a box and its

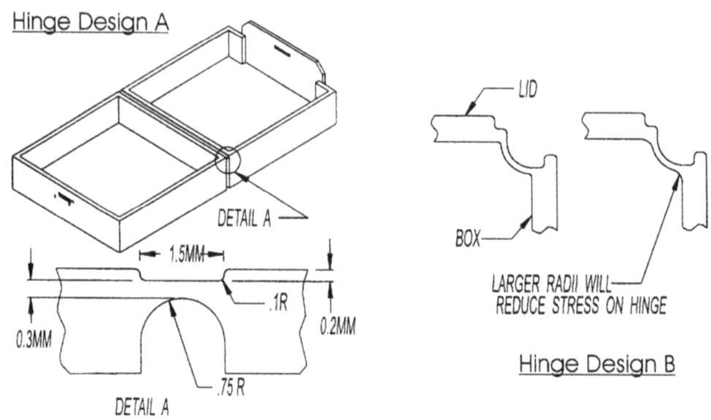

Figure 10.38 Common designs for integral hinges.

lid (Fig. 10.38). Although most commonly used with polypropylenes, integral hinges can be used with most of the more ductile plastic materials. Integral hinges made of polypropylene can last hundreds of thousands of cycles when correctly designed and molded. The strength and durability of the integral hinge comes from the high polymer orientation gained when material is forced to flow under high stresses across the thin feature. Poor flow or weld lines along the hinge will result in its premature failure.

There are many design variations for the integral hinge. They are generally very thin and short continuous strips joining the two parts sections to be hinged brackets (example: box and lid). Thickness is commonly only 0.3 mm with a width of less than 1.0 mm (see Fig. 10.38, Design A). These hinges can generally be bent in a 180+ degree arch and allow for a very small radiused closer at the hinge. A thicker noncontinuous hinge can also be used. These are more like straps and are longer to allow for deflection (see Fig. 10.38, Design B). These normally require a larger radius when closed.

Care must be taken when gating and molding the thinner integral hinge. With a container and lid combination, a gate is normally placed in the larger container. The gate should be placed away from the hinge such that it is the last place for the melt to reach. The melt should approach perpendicular to the hinge so that the flow front and the hinge are parallel. These strategies will help to avoid any hesitation of the melt front as it hits the restricted hinge. If the melt front reaches the hinge, while other regions are still being filled, it can hesitate and potentially freeze off.

An additional potential problem with integral hinges is packing of the cavity regions on the opposite side of the hinge from the gate. During packing, the material in the hinge will quickly freeze off while material in the thicker lid is still molten. Control of packing is lost and the lid can be expected to shrink differently than the container portion and may include undesirable sinks, voids, or loss of surface finish. If a gate is placed on both sides of the hinge to improve packing, extreme caution needs to be taken that the resulting weld line between the two gating locations does not occur at the hinge.

10.4 Sample Part Design

Given a simple manifold block to be produced from steel (see Fig. 10.39). The functional components are the central hole, which provides passage for low pressure air, and the four corner holes to position bolts for assembly. The only real structural requirement is that the part must withstand the compressive load of 70 pounds from each of the four assembly bolts. When produced from steel, the outer boundaries of the part would be machined from a solid block of steel (top figure) and then the five holes would be drilled (bottom figure). The process of machining steel is a subtractive process where it is desirable to minimize the amount of material to be removed in order to keep costs down. Coring out the unused material around the holes would simply add to the cost of machining the metal part. This additional step would only be done if there were some other requirements, such as weight reduction.

When designing an alternate plastic part design, the designer should no longer be treating the part as something that is to be machined. The designer should focus on the functional component of the part, which is in this case the five holes, and build the part as though it is an additive process. During this building stage, the designer should be addressing the functional requirements while keeping in mind the objective to maintain a minimum and constant wall thickness.

First, the five holes are created individually using thin features, which in this case are tubular structures (Fig. 10.40). The center hole has no real structural requirement and is kept relatively thin. The four mounting holes must withstand a compressive force (F) of 70 pounds each. Given the inside diameter (ID) of 0.25 in, the minimum outside diameter can be calculated. Applying a safety factor of 4:1, each boss must withstand a compressive force of 280 pounds. Given that the plastic material, which

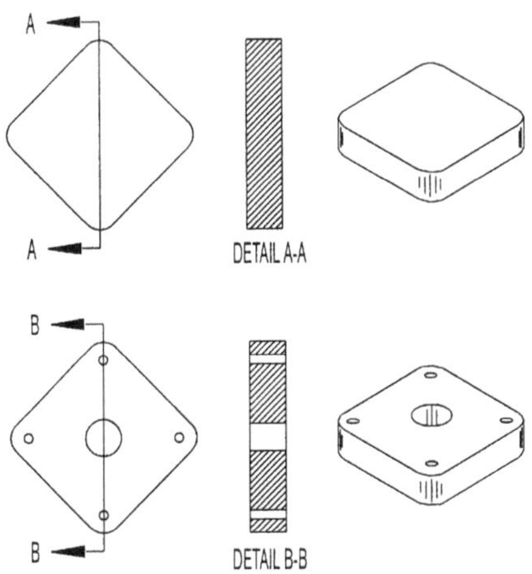

Figure 10.39 Manifold block machined out of steel.

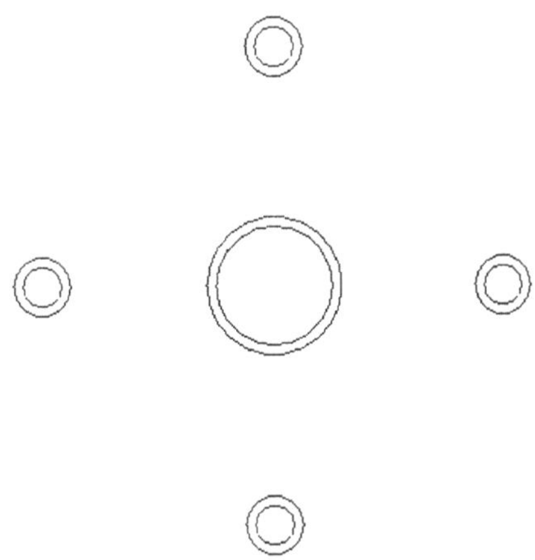

Figure 10.40 Functional components of manifold block in Fig. 10.39.

is to be used, has a compressive yield stress (σ) of 10,000 psi, the minimum outside diameter (OD) is found by:

$$\sigma = \frac{F}{A}$$

$$\sigma = \frac{F}{\left(\frac{\pi(OD)^2}{4} - \frac{\pi(ID)^2}{4}\right)}$$

$$OD = \sqrt{\frac{4\left(\frac{F}{\sigma} + \frac{\pi(ID)^2}{4}\right)}{\pi}}$$

$$\text{OD} = 0.313\,\text{in}$$

The minimum required wall thickness of boss is found to be 0.0313 in.

As the four mounting holes, or bosses, act to locate the center air passage, they must be connected to it. Figure 10.42 illustrates one approach in providing this connection. Here a single thin wall connects each of the mounting bosses to the center air passage. At this stage all of the functional components have been included and combined into a single part. If more stiffness is required, ribs, or gussets might be added along the connecting walls. These are kept short and placed on either side of the wall to maintain symmetry to minimize potential for warpage (Fig. 10.43).

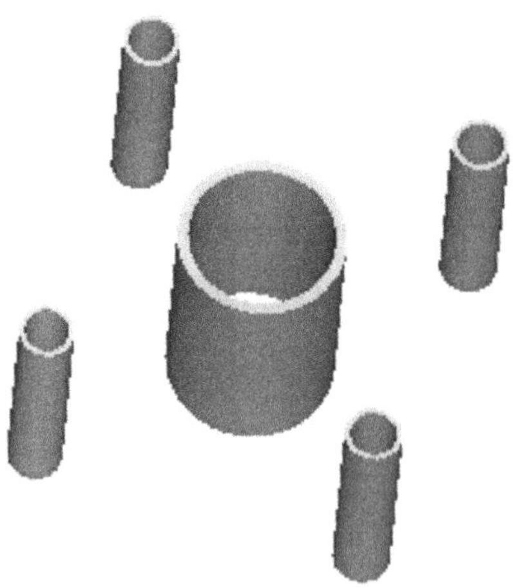

Figure 10.41 Plastic design applied to functional components.

Figure 10.42 Tying the functional components of the manifold together to form a functional part.

Wall thicknesses must now be determined. As the only real structural requirement is satisfied with the 0.0313-in thickness of the boss walls, all other walls could be the same or less. The 0.0313-in thickness, however, is already quite thin. An injection mold filling simulation might be required if the designer wants to consider using a wall this thin. Once a wall thickness has been found for which the part can be safely filled (including consideration of the runner and gating system), all wall sections including the flow passage, the four connecting walls, and the ribs that support these walls could be designed to be the same thickness. The potential is that sinks or voids will appear at the intersection of any of the wall junctions. The only concern might be structural because this is not a cosmetic part. To minimize the risk, four gates could be positioned on the interior of the air channel, opposite the location of the four walls connecting to the boss. This would feed melt directly into the relatively thick intersecting wall locations that would act as flow leaders. These are all directly linked in this design. The result is that during the packing stage of the molding cycle, melt can be continually fed to these regions to compensate for the shrinking material. The gate would actually be slightly offset from the connecting wall so that it would not jet during initial injection.

Radii of 25% of the wall are added at the intersection of all walls to reduce stress at these intersections. The radii will increase the volume of material at the intersections, increasing the potential for voids, but the impact is again minimized due to the location of the gate and the connection of all of these regions to improve packing.

A variation of this design would be to thin the connecting four wall sections and their associated supporting ribs. The concern of varying wall thickness causing residual stress and warpage should not be a concern in this part. Unlike a thinned rib placed on the side of a thicker wall, a bending moment is not created. In this part, the wall thickness variation might cause the heights of the center air passage and the four bosses to shrink in height more than the thinner connecting walls. This might create a residual stress, but due to their relative relationship and structure, would be unlikely to cause the part to warp. If the boss height shrinks by more than the connecting walls, then the stress during assembly may be diverted to the taller walls, which might engage connecting components. This could be addressed by reducing the height of the walls in the design stage. This would assure that the bosses take up the stress, and a close interface of the center air passage is maintained.

Further concerns with this design variation might include feeding material from the gate, through a thin wall to the thicker walls of the boss. This is somewhat alleviated by the fact that the melt flow channel along the connecting rib is created by the intersection of actually four walls (the top and bottom of the connecting wall and the two sides of the supporting ribs). This increases the channel size and should improve the ability to feed material to the boss during packing stage.

Throughout the design, ejection from the mold must be considered. With either of the preceding designs, the cavity could be positioned such that the parting line of the mold is along the center plane of the part, with equal portions of all components formed in both the stationary (cavity) half and the movable (core) half of the mold. The inside diameters could similarly be formed from cores projecting from either

Figure 10.43 Addition of ribs and draft to the molded plastic manifold for increased stiffness and ease of ejection.

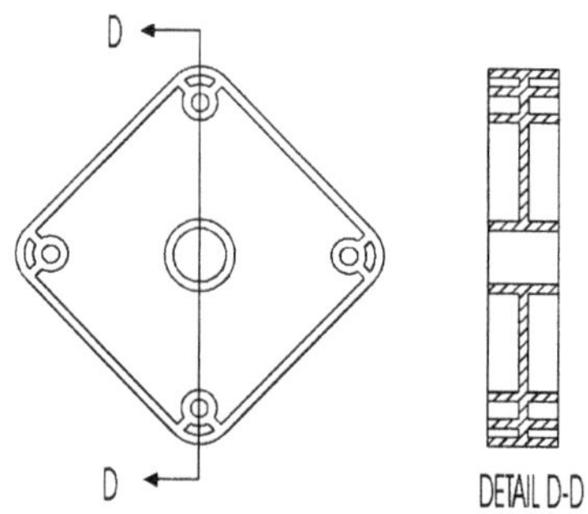

Figure 10.44 Variation of the manifold design that can be used for injection molding.

mold half. Drafts are added to facilitate ejection and are located on the inside and outside walls of the center flow channel, the four bosses and the four connecting walls. To assure that the part sticks to the ejection half of the mold during mold opening, the draft angle applied to this half of the mold should be less than the stationary half. Stripper sleeves should be used to eject the part from each of the five cores creating the inside diameter of the holes.

Give the development of this part to five different designers, and five different designs will develop. Another variation of this design is shown in Fig. 10.44. The functional components are again created with thin-walled features and then connected. In this design a frame is developed around the perimeter of the part. Supporting gussets, or ribs, connect the four bosses to the outside wall.

10.5 Estimating Part Costs

Due to the high cost of an injection mold, the cost of injection molded parts are significantly affected by production volume. It is rare that injection molding can be justified if production is expected to be less than 10,000 parts per year. Alternate processes such as vacuum forming, pressure forming or machining need to be evaluated for smaller production levels.

Part cost is a function of direct, packaging, shipping and indirect costs. Direct costs include materials, direct labor plus any dedicated mold or other tooling that is required to produce the parts. Indirect costs include the elements of the business that are required for its operation, but don't directly produce a product. These include the building expenses, equipment, and support staff. Support staff can include engineering, sales force, company management, maintenance personnel, etc.

Direct costs can be found as follows:

1. Determining the volume of the molded part and the material cost per volume.
2. Direct labor is based on knowing how many parts are produced per hour, the hourly rate associated with any direct labor (including overhead), and the number of laborers required for a particular job. If a job is to run fully automatic, there is still a fraction of a labor cost assigned as someone must be responsible for handling the molded parts, loading material, etc.
3. Dedicated tooling/equipment costs. This cost is very sensitive to production volume. In addition, this cost can be significantly influenced by the risk and business strategy of the company as effected by the duration over which investments are to be amortized.

Given a case where a part requires a mold that costs \$200,000. Of these parts 100,000 are to be produced per year over 5 years (total of 500,000 parts). The cost of the mold is commonly amortized over some period of time that is dependent on the company's business strategy.

The company could choose to spread the cost of the mold over the 5 years and 500,000 parts. This would result in a cost of approximately \$0.40/part (\$200,000/500,000 parts). If they take 5 years to pay for the mold, however, the \$200,000 used to purchase the mold is a long-term debt that must reflect an interest cost to them. Even if the company had the cash to pay for the mold, this is money lost to them which could have been providing capital gains through investments. Either way a cost for use of this money must be factored. The cost therefore will be something more than \$0.40/part and will be presented later.

As with any purchase that is spread over time, the advantage is that the short-term costs are minimized. By this means the cost of the product is also minimized. The further advantage of this is that the cost to the customer is lowered. Spreading the cost over an extended period of time puts the manufacturer at an increased risk. If the product does not sell and production is terminated after only a year, the manufacturer will only have recovered \$40,000 of the \$200,000 for the cost of the mold. The manufacturer will therefore lose \$160,000.

The alternative is to try to amortize the mold cost over one year. In this case the cost of each part during the first year would have to be \$2.00 versus only \$0.40. The disadvantage here is that the part will cost more than a company who elected to amortize parts through 5 years; however there are a number of benefits to this method:

1. If you succeed in selling your part for the \$2.00 during the first year, then your mold has been paid for. During the following 4 years of production you can either maintain your prices and make a hefty profit or drop your price to combat any competition.

2. If the product does not sell after the first year, the cost of the mold will have been recovered and a loss is not taken.
3. Since it only took one year to pay for the mold, the cost of money will be significantly less.

In determining the cost of a product, indirect costs are commonly factored into an hourly rate that is developed around the primary piece of manufacturing equipment to be used to produce the parts. In the case of an injection molded article, an hourly rate is assigned to each of the injection molding machines according to its cost. This is typically reflected in the size (tonnage rating) of the machine. These costs will vary significantly from company to company and even plant to plant. Example of costs for various injection molding machines are as follows.

100 Ton = \$30/hr
200 Ton = \$34/hr
300 Ton = \$38/hr

An example of calculating part cost follows. This example does not include the cost of packaging or shipping.

Number of cavities	= 16
Production cost	= 50 \$/hour
Direct labor cost	= 8 \$/hour
Annual production	= 5,000,000 parts/year
Number of year for amortizing tooling	= 3
Molding cycle	= 16 seconds
Total number of payments during the time that the mold will be amortized	= 36
Mold cost	= \$100,000
Material cost	= 2.0 \$/lb
Number of annual payments on the mold	= 12
Part weight	= 10 grams
Production Yield	= 95%

Material Cost per Part Determine material cost per part. Given that the material cost is given in dollars per pound (\$/lb), and part weight is given in grams(g), we must convert part weight from grams to pounds when determining material cost per part:

$$\text{Material cost per part} = \left(\frac{\text{PartWeight}}{454}\right) \cdot \text{Material Cost}$$

$$\text{Material cost per part} = \$0.0441$$

Production Cost First, determine the number of parts produced per hour.

$$\text{Parts per hour} = \frac{3600}{\text{Molding cycle}} \cdot \text{Number of cavities}$$

$$\text{Parts per hour} = 3600$$

Second, determine production cost for each part:

$$\text{Production cost per part} = \frac{\text{Production Cost} + \text{Direct labor}}{\dfrac{3600}{\text{Molding cycle}} \cdot \text{Number of cavities}}$$

$$\text{Production cost per part} = 0.0161$$

Mold Cost per Part First, determine the total cost of the mold (mold cost + interest)

$$\text{Total cost} = \text{Mold cost} \cdot \text{Total number of payments} \cdot \left[\frac{\left(\dfrac{\text{Interest rate}}{\text{Num annual pay}}\right) \cdot \left(1 + \dfrac{\text{Interest rate}}{\text{Num annual pay}}\right)^{\text{Total number of payments}}}{\left(1 + \dfrac{\text{Interest rate}}{\text{Num Annual pay}}\right)^{\text{Total number of payments}} - 1} \right]$$

$$\text{Total cost} = \$116{,}162$$

Second, determine the mold cost per part: Total cost of the mold divided by the total production over the years for which the total mold cost will be amortized.

$$\text{Tooling Cost per Part} = \frac{\text{Total Cost}}{\text{Annual Production} \cdot \text{Number of Years Amoritized}}$$

$$\text{Tooling Cost per Part} = 0.0077$$

Total Cost = (Material Cost + Production Cost + Mold Cost) × Yield

$$\text{Total part cost} = (\text{Material cost per part} + \text{Production cost per part} + \text{Tooling cost per part}) \cdot \text{Yield}$$

$$\text{Total Part Cost} = \$0.0645$$

The preceding example applies the 5% spoilage to the total cost. In some cases a different spoilage might be applied to material cost, production cost, and materials.

References

1. Wagner, A. H. Yu, J. S., and Kalyon, D. M., Orientation and Residual Stress Distribution in Injection Molded Engineering Plastics, Proceedings of the Society of Plastics Engineering 47th Annual Technical Conference, SPE, Brookfield Center, CT, 1988, 303–307.
2. Jaros, B. G., Shrinkage of Glass-Filled Parts as Developed by Radial Flow, Proceedings of the Society of Plastics Engineering 5th Annual Technical Conference, SPE, Brookfield Center, CT, 1991, 2646–2648.

11 Simulation in Injection Molding

B. A. Davis, A. C. Rios, V. Yang

11.1 Introduction

At this date, nearly everyone interested in injection molding has either read of the "marvel" of flow analysis for injection molding, or used the technology to aid in the design of injection molded products. It is commonplace for new designs to be numerically simulated or even optimized via computer simulation before the first piece of steel is cut for the mold. Rather than expend time and money building a series of prototype tools to fine-tune a new product, computer simulation has dramatically affected the way new parts are designed. Designers and process engineers have effectively leveraged the power of simulation to shorten the design process and to eliminate processing problems that once caused reductions in quality and profitability.

Computer simulations offer the tremendous advantage of enabling designers and engineers to consider virtually any molding option without incurring the expense associated with material waste and machine time. The ability to "try out" new designs or concepts on the computer gives the engineer the opportunity to eliminate problems before beginning production. Additionally, the engineer can quickly and easily determine the sensitivity of specific molding parameters on the quality and production of the final product. This computer-aided engineering (CAE) *virtual* design offers tremendous flexibility to the engineer to determine the effects of different gating scena–rios, different runner designs, geometric features, and different molding and processing conditions on the moldability and quality of the final product. Furthermore, these CAE tools allow these virtual designs to be expediently completed in a matter of days or even hours rather than the weeks associated with experimental trial and error.

Computer-aided engineering in injection molding is not a panacea, however. There are certainly limitations to how and when computer simulation should be used to analyze the injection molding process. As with any computer simulation, there are a certain number of assumptions and simplifications that have been built into the computer program. Some of which allow the program to solve complicated analyses in a reasonable amount of time, and others are generalizations of material behavior to allow the numerical method to solve complicated coupled field equations. Others are not associated with difficulties in creating a suitable numerical algorithm, but rather with problems in carrying out the experimentation needed to characterize a material response or processing effect. Unfortunately, most users are not completely aware of the assumptions and simplifications and more importantly not familiar with their implications on the validity of the solution. With the power of modern injection molding CAE software, it is easy to get caught up with the "pretty picture" mentality and to lose sight of the link with the actual real-life molding process.

Computer-aided engineering of injection molding must be tempered with an understanding of the actual molding process as well as the assumptions and simplifications embedded in the simulation program. The implications of these assumptions on the validity of the solution and how to interpret the results is one area where experience is invaluable. However, having a basic understanding of the underlying theory and numerical methods being used by the computer program gives the user greater confidence in applying the software to their particular problem.

With the tremendous pressures on molders to cut cycle times, improve quality, reduce lead-time, and simultaneously reduce costs, CAE tools must be used to maintain competitiveness in a globally aggressive market. However, these tools must be used with a certain understanding of their theory and limitations to ensure accurate use. This chapter serves to introduce this very topic in the field of computer-aided engineering in injection molding. The history of CAE in injection molding is introduced along with the underlying theory and assumptions for simulation. A brief synopsis of available numerical methods is also given to educate the reader on the ability and accuracy of each method. This chapter then illustrates a number of practical examples and case studies across a variety of injection molding topics.

11.2 History

While simulating the injection molding process is commonplace today, it was in the not to distant past that simulation was the exception rather than the norm. Early "simulations" for injection molding were based on "lay-flat" approaches. These early methods required the geometry to be "unfolded" and described by several one-dimensional flow paths. This type of analysis was capable of providing velocity and temperature distributions across the flow direction as well as the prediction of short-shots and pressure gradients. However, the use of such a method required gross simplifications of the actual part geometry as well as an experienced user to make the appropriate geometric simplifications. Although these early simulations were not particularly computationally intensive, they did not gain widespread acceptance due, in part, to the difficulties in determining the lay-flat equivalent.

The roots of modern computer simulation in injection molding can be traced back to early work by Wang and Hieber [1] at Cornell University. During the mid 1970s a tremendous amount of research and development was done in order to understand the complex phenomena that occur during the injection molding of thermoplastic material. In particular, the formation of CIMP (Cornell Injection Molding Program) in 1974 served as an example of how University/Industry/Government collaboration can be organized to solve a technologically challenging problem. CIMP was organized to develop a fundamental scientific basis so that injection molding part and process design could be accomplished through rational scientific laws rather than only on experience.

Although the early work of CIMP served as the basis for the modern day simulation programs, today's numerical methods can be specifically traced back to early work by Heiber and Shen [2]. Heiber and Shen's considerable breakthrough was the finite element method as applied to injection mold filling. Though at the time not commercially practical, their method paved the way for ensuing commercial software in injection molding. Numerous other researchers have contributed to the body of research by incorporating new models and methods to compute mold filling, packing, orientation, residual stress, shrinkage, and warpage. The precursory work at CIMP spawned a number of additional worldwide research institutions to advance the technology of CAE in injection molding.

One of the major difficulties with simulating the injection molding process is the transient free surface associated with the flow front. The material constantly changes shape as it flows and deforms inside the cavity, making it necessary to redefine the geometry of the fluid domain of interest after each successive time step. Redefining the finite element mesh or finite difference grid was one of the largest obstacles to producing a feasible computer simulation method.

Wang, Hieber, and Wang [3] implemented a mesh editing procedure or dynamic mesh editor into an injection molding simulation. After carefully choosing a time step and advancing the flow fronts, the user was required to fill the gap between the old and the updated melt fronts with new finite elements. This procedure not only required extensive user interaction, but also made the mesh sizing dependent on the size of the chosen time step.

An alternative technique by Lee, Folgar, and Tucker [4] used a finite element mesh to represent the initial charge during compression molding. The same mesh was used after each time step to fit the material flow fronts through the use of a special finite element elasticity calculation. This procedure is analogous to drawing the original mesh on a sheet of rubber and then stretching it to conform to the shape of the flow front at any time step. Although an improvement upon the initial dynamic mesh editor technique, this method suffered from problems with multiple charges, inserts, or cases where the initial charge differs significantly from the final part shape.

Tadmor, Broyer, and Gutfinger [5, 6] had previously used a spatial finite difference formulation to solve two dimensional flow problems in complex geometrical configurations. Using a Hele-Shaw [7] formulation to simulate the flow, their method was applicable to flows in narrow gaps of variable thickness such as those encountered in injection molding. This technique is known as the flow analysis network (FAN) and works well for both Newtonian and non-Newtonian fluids. The method uses an Eulerian grid of cells that covers the flow cavity. Each cell is assigned a factor that is related to the amount of fill that ranges from 0 for an empty cell, to 1 for a completely filled cell. Localized mass balances are made around each cell and the resulting system of linear algebraic equations are solved to compute the flow field. Although this method is well suited to many problems, it requires relatively fine grids especially if curved boundaries are present in the geometry.

More recently Wang et al. [8] and Osswald [9] modified the flow analysis network to model the non-Newtonian non-isothermal flow inside thin three-dimensional cavities using finite elements. The technique, which is commonly called the control volume approach (CVA), requires that the three-dimensional molding surface be divided into flat shell finite elements. The cells, or control volumes, are generated by connecting the element centroid with the element mid-sides. When applying the mass balance to each cell, the resulting equations are identical to those arising from a Galerkin method for finite elements. This allows the use of a standard finite element assembling technique when generating the set of linear algebraic equations.

By this stage of history, a sufficient body of research had been developed to warrant commercialization of injection molding simulation software. Commercially available simulation had been available as early as the mid 1970s in the form of time-shared computer simulation using a lay-flat approach. However, it was not until the mid 1980s that commercial finite element based simulation tools were available for injection molding. By the late 1980s a number of different companies, such as Moldflow, C-Mold, Moldex3D (previously known as Moldex), and Simcon, offered simulation tools for CAE in injection molding. The commercial software packages offered by these companies began as cumbersome command line interpreters and slowly developed into the full-featured graphical user interfaces that today's users have become accustomed to. In addition to the commercially available packages, a number of research codes were developed by universities and independent companies. However, many of these independent codes have been replaced by the above-mentioned commercial codes because of a lack in continuous improvement and innovation.

By the late 20th century, many attempts have been devoted to extend the Hele-Shaw assumption, or the so-called two-and-one-half-dimensional ($2^1/_2$-D) analysis model, into the true three-dimensional analysis model. As was for the $2^1/_2$-D model, the three-dimensional simulation struggled to accurately calculate the polymer-air interface of complex topology and its evolution over time. Rider [10] derived a reconstruction scheme of second-order accuracy both in space and time and extended it to the unstructured mesh. Later Gao [11] successfully combined this scheme with the finite element method to simulate the three-dimensional mold filling during casting processes. The FEM has been conventionally adopted to simulate polymer processing because of its flexibility in modeling arbitrary boundary shapes. However, it tends to produce a large, sparse matrix, generally with high condition numbers, and consequently becomes very inefficient and memory demanding with a three-dimensional large-scale system. In light of these issues, Chang and Yang [12] developed an efficient and robust three-dimensional finite volume method (FVM) for its advantage over FEM in computation time and numerical stability. This important work opened the door for three-dimensional commercial codes of injection molding simulation.

The commercial development of software has not, however, subdued the research and development into injection molding. On the contrary, the fundamental research necessary to further the method into new advances in injection molding has benefited from the success of commercialization. New models and methods to compute mold filling, packing, post fill shrinkage, warpage, and residual stresses have been developed since the inception of commercial CAE in injection molding. The methods have also been expanded and complimented to enable processes such as gas assist, injection/compression, co-injection, insert molding, or multiple-component molding to be simulated as well. The continuing requirements on molders to lower costs, increase quality and performance, and decrease lead time will ensure continuing advancements in CAE for injection molding both in the form research and development and as new commercialized software tools.

11.3 Governing Equations

To be able to predict and model complex polymer flows that occur during injection molding, one must first have a basic understanding of the mathematics that govern flow. Regardless of the complexity of the flow, it must satisfy certain physical laws. These laws can be expressed in mathematical terms as the conservation of mass, conservation of momentum, and conservation of energy as shown in Eqs. 11.1-11.3. In addition to the three governing equations there may also be one or more constitutive equations that describe material properties as previously introduced in Chapter 3. Since these equations may also be coupled together, e.g., temperature and shear rate dependent viscosity, the solutions can become even more complex. The goal of the modeler is to take a physical problem, use these mathematical equations and solve them to predict the flow phenomena. Although analytic solutions to the conservation equations for some simple geometries are available, when more complex two-dimensional problems or three-dimensional analyses are required, numerical methods are necessary.

$$U_{i,i} = 0 \tag{11.1}$$

$$\rho(U_{i,t} + U_j U_{i,j}) = \sigma_{ji,j} + \rho g_i \tag{11.2}$$

$$\rho C_v (T_{,t} + U_i T_{,i}) = -q_{i,i} - T[P_{,T}]_v U_{i,i} + \tau_{ij} U_{i,j} + \dot{S} \tag{11.3}$$

While the specific details involved with numerically solving these conservation equations are beyond the scope of this chapter, the underlying models used for molding analysis must be presented to educate the user to assumptions and limitations of the methods.

11.3.1 Flow Models

In general, the majority of computer simulations of the injection molding process are based on the 2½-D finite element/control volume method. While the concept of using finite elements to simulation fluid flow is not itself novel, the application to thin walled structures in injection molding requires some additional information. By and

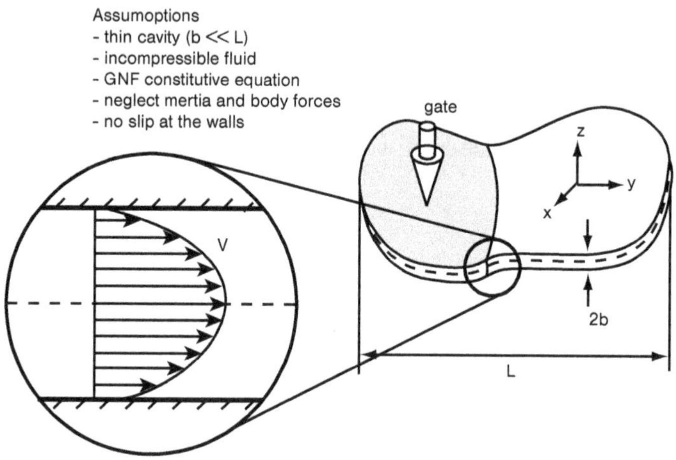

Figure 11.1 General Hele-Shaw flow geometry, relevant assumptions, and typical velocity profile through the thickness.

large, the most prominent model used for flow in these thin cavities is based upon a Hele-Shaw [7] formulation.

As with many modern computer algorithms, the major underlying theory of flow in injection molding is based on a century old work in classical mathematics. This theory was originally published in 1899 by a naval engineer named Henry S. Hele-Shaw. Primarily, Hele-Shaw was interested in discovering the physics behind flow in narrow gaps. Although decades ahead of the advent of injection molding, his work provided the necessary foundation for flow analysis using the finite element method.

In essence, Hele-Shaw flow can be applied to a generalized Newtonian fluid that may be shear thinning and temperature dependent. As shown in Fig. 11.1, the principle of Hele-Shaw flow is based on the thin cavity assumption. In this case, the thickness across the flow gap is assumed to be much smaller than along the main flow direction. Starting with the assumptions of incompressibility, generalized Newtonian fluid behavior, negligible inertia, and no slip boundary conditions at the walls the thin cavity enforces conservation of mass and conservation of momentum on an integrated basis through the thickness direction. This formulation reduces the governing equation for flow to the LaPlace equation, shown in Eq. 11.4, where P denotes pressure, x and y are planar directions, and S is the flow conductance. Furthermore, the flow conductance may be expressed as shown in Eq. 11.5. Here, the viscosity η varies across the thickness of the gap as well as spatially in the x and y directions. As shown in Eq. 11.5, the viscosity may also depend on the strain rate, $\dot{\gamma}$, and temperature, T, as well.

$$\frac{\partial}{\partial x}\left(S\frac{\partial P}{\partial x}\right)+\frac{\partial}{\partial y}\left(S\frac{\partial P}{\partial y}\right)=0 \tag{11.4}$$

$$S(x,y)=\int_0^b \frac{(z')^2}{\eta(\dot{\gamma},T,z')}dz' \tag{11.5}$$

All modern 2½-D mold filling simulation programs for injection molding rely on this general Hele-Shaw methodology to compute the mold filling pattern. While Hele-Shaw flow is valid for much of the flow in injection molding, it does have certain limitations that must be considered. The Hele-Shaw flow is valid distant from the gate and from the active flow front. In these regions, the flow is governed by a boundary layer problem in which the layer is approximately equal to the thickness of the gap.

At the flow front, the flow field is considerably more complex because of the existence of the fountain flow effect. Here, the progression of the flow front in combination with the no-slip boundary condition at the walls, leads to a breakdown in the Hele-Shaw theory. In this case, flow is no longer parabolic in nature but begins to flow outward from the center of the gap to the walls. This phenomenon is illustrated schematically in Fig. 11.2. Notice that near the flow front the velocity profile maintains a zero slip boundary condition at the mold surface. However, because the flow front advances there is a discrepancy for the material located adjacent to the mold surface. This causes a non-zero flow component across the gap and leads to the fountain flow effect. As the figure represents, the melt that flows inside the cavity freezes once it contacts the cool mold walls. The additional melt that subsequently flows into the cavity enters between the frozen layers and forces the melt skin at the flow front to stretch and unroll onto the cool mold walls.

Another breakdown of the Hele-Shaw theory is when thick sections are considered. While most injection-molded parts are relatively thin in nature, there are some classes that can have relatively thick local regions. Because one of the founding assumptions of Hele-Shaw is the thin section, error increases as the aspect ratio (b/L) of the geometry increases. From order-of-magnitude analyses, it can be shown that errors in continuity and shear stresses are roughly of the order $\vartheta(b/L)$. Therefore, as the thickness increases the Hele-Shaw model deviates more from actual flow phenomenon occurring during injection mold filling.

Another limitation of Hele-Shaw theory is when the flow resistance from lateral walls becomes significant. The deviation may be significant when a grill-type of structure exists or when the thickness-to-width ratios of the flow channel become small.

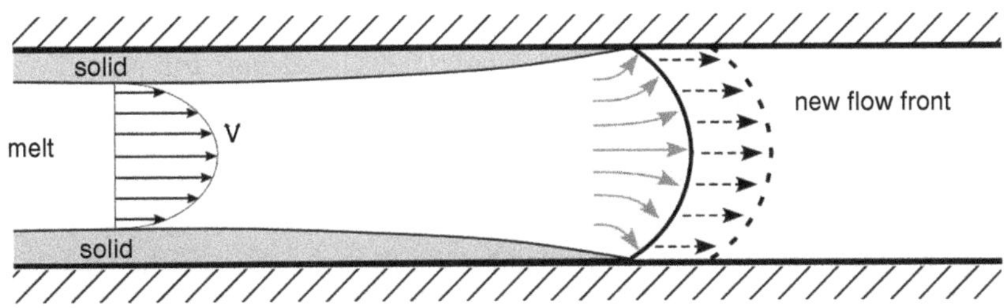

Figure 11.2 Fountain flow effect in injection molding.

As opposed to the 2½-D flow model, the governing equations (11.1), (11.2), and (11.3) are adopted in the true 3-D simulation without further assumption and mathematical simplification. Therefore, the above-mentioned limitations of Hele-Shaw theory are eliminated in 3-D simulations.

11.3.2 Orientation Models

During processing the molecules, fillers, and fibers are oriented in the flow and greatly affect the properties of the final part. When thermoplastic materials are injection molded, the complicated flow phenomena create variations in the orientation throughout the part. This variation arises both on a bulk basis where orientations will vary from point-to-point in the geometry as well as variations through the thickness of the part. The polymer molecules, themselves, are stretched and because of the intermolecular entanglement are not able to relax sufficiently before the part cools and solidifies. Although the molecular orientation of polymers is important to many mechanical properties, it is currently beyond the scope of CAE in injection molding to permit an in-depth analysis of this type of orientation. However, when fillers and fibers are considered, there is a sufficient body of research to reasonably predict their orientation during injection molding.

Critical research conducted by Folgar and Tucker [10–11] modeled the motion of a single fiber in a concentrated suspension. Their approach, based on the Jeffery Model, added the interaction that occurs between fibers. Folgar and Tucker derived the governing equation for fiber orientation in planar flows as Eq. 11.6.

$$\dot{\phi} = \frac{-C_I \dot{\gamma}}{\Psi}\frac{\partial \Psi}{\partial \phi} - \cos\phi \sin\phi \frac{\partial v_x}{\partial x} - \sin^2\phi \frac{\partial v_x}{\partial y} + \cos^2\phi \frac{\partial v_y}{\partial x} + \sin\phi \cos\phi \frac{\partial v_y}{\partial y} \quad (11.6)$$

In general, the physical model originally derived by Folgar and Tucker relates the orientation distribution function, Ψ, of the fiber field to the magnitude of the strain rate, $\dot{\gamma}$ the derivatives of velocity, v_x and v_y, and the angle, ϕ, at which the fiber is oriented with respect to the flow field. The model also includes a fiber-to-fiber interaction coefficient, C_I, which varies between 0 for a fiber without interaction with its neighbors, and 1, for a closely packed bed of fibers. Their fiber orientation model has been shown to hold for both thermoplastic and thermoset materials.

Advani and Tucker [12–14] developed a more efficient method to represent the fiber orientation through the use of orientation tensors. This technique dramatically reduced the computational requirements when solving the fiber orientation problems using the Folgar-Tucker method. Instead of representing the orientation of a fiber by a single angle, ϕ, Advani and Tucker proposed the use of the components of a unit vector along the fiber axis and various second order tensors. This compact notation fully describes the orientation state, a_{ij}, at any point in space and can be written as Eq. 11.7.

$$\frac{Da_{ij}}{Dt} = -\frac{1}{2}(\omega_{ik}a_{kj} - a_{ik}\omega_{kj}) + \frac{\lambda}{2}(\dot{\gamma}_{ik}a_{kj} + a_{ik}\dot{\gamma}_{kj} - 2\dot{\gamma}_{kl}a_{ijkl}) + 2C_I\dot{\gamma}(\delta_{ij} - \alpha a_{ij}) \quad (11.7)$$

Here, δ_{ij}, defines the unit tensor while ω_{ij} and $\dot{\gamma}_{ij}$ are the vorticity and the rate of deformation tensors, respectively. The tensor form of the governing equation also contains the magnitude of the strain rate, $\dot{\gamma}$, the interaction coefficient, C_I, and α which equals 3 for a 3-D orientation and 2 for a 2-D orientation field. The equation also makes used of a fourth order tensor, a_{ijkl}. This tensor must be approximated by an appropriate closure model [14-16] to make Eq. 11.7 solvable. To obtain the orientation state of the fibers during mold filling simulations, Eq. 11.7 may then be solved using a traditional finite element/control volume mold filling approach.

As indicated in Eq. 11.7, fiber orientation is three-dimensional and affected by the vorticity and rate of deformation of the flow field. Therefore, the fiber orientation prediction in a $2^1/_2$-D flow model can be less satisfactory because of a lack of velocity component in the wall thickness direction, especially in areas around gates, corners, melt front, variable wall thickness, etc. The error in fiber orientation predictions can lead to errors in anisotropic mechanical properties and warpage prediction. The true 3-D simulation, on the other hand, can provide much more reliable and accurate fiber orientation-related calculations.

11.3.3 Heat Transfer Models

Injection molding, by nature, is a fully non-isothermal process. This means that at any stage of the injection molding process, temperature is non-uniform and furthermore depends greatly upon time and processing conditions. From the melting of the raw material to the injection and cooling of the final part, complex heat transfer mechanisms are at work. In general, during the molding process energy is transferred via conduction, convection, and dissipation.

Most simulation packages assume that the injection temperature is uniform and constant. Clearly, this is not the case in actual injection molding where fluctuations in temperature can occur axially along the shot as well as non-homogeneous temperatures through the thickness of the shot. The uniform injection temperature assumption is not a result of an inability of the computer simulation, however, but in difficulty in measuring the actual injection shot temperature. Therefore, the uniform injection temperature is reasonable for most calculations.

In order to solve the heat transfer during injection molding, the governing equation for energy must be solved. The conservation of energy, Eq. 11.3, mathematically imposes that the energy and heat transferred through the polymer must balance with the amount of heat generated and consumed by the sample. This classic equation balances the heat transfer by conduction, convection and dissipation with the heat storage capacity of the polymer itself.

As the material in injected through the gate, the higher shear rates cause an internal heating of the polymer molecules. This heat generation is commonly referred to as viscous dissipation. The molecular friction of polymer chains rubbing against one another can create a rather significant temperature rise in the melt. Fortunately, this effect is readily accounted for by modern simulation techniques.

Additional heat transfer occurs due to the polymer flow in the form of convection. This form of energy transfer is governed by temperature differences in the flow direction. The fluid motion transfers energy along its flow path and thus *convects* heat during mold filling. A prime example of convection during injection molding is the effect where the hot melt from the gate is transferred downstream to the material flowing in the mold.

The last major mode of energy transfer during the injection molding cycle arises because of conduction. While conduction occurs in all directions of the polymer, the thin nature of injection molded parts causes conduction through the thickness direction to dominate. Therefore, simulation programs will typically only consider conduction across the thin gap. This simplification is quite valid when the low thermal conductivity of polymers is considered. Because the cool mold wall is significantly cooler than the hot melt stream, conduction is far greater into the mold than along the flow direction.

One confounding effect that arises when molding semi-crystalline polymers is the heat of crystallization or heat of fusion. During the process of crystalline structure formation, a certain amount of energy must be conducted out of the material before the cooling process can continue. This heat of fusion is reflected in the shape of the enthalpy-temperature curve that can be obtained from standard DSC tests as shown in Fig. 11.3. Fortunately, there are well-tested models and methods to account for the heat of fusion during numerical calculations.

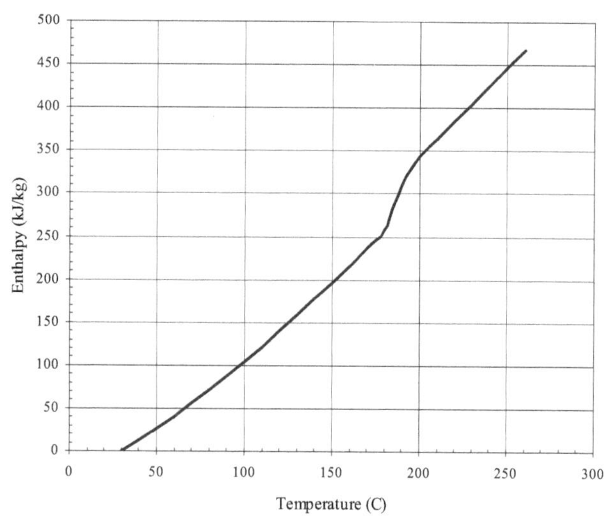

Figure 11.3 Specific enthalpy-temperature (DSC) curve for a PBT material.

11.3.4 Constitutive Equations

To be able to solve the governing equations for injection molding, several additional constitutive equations are required. As discussed in Chapter 3, a constitutive equation is one that describes the material behavior as a function of processing and applied conditions. An example of a constitutive equation is the model used for viscosity.

The shear rate and temperature dependence of viscosity is well documented for polymeric materials. Chapter 3 illustrates several graphs and examples of viscosity models. However, the Cross-WLF viscosity model is oftentimes preferred for numerical studies. This model is shown in Eq. 11.8,

$$\eta(T,\dot{\gamma},p) = \frac{\eta_o(T,p)}{1+\left(\frac{\eta_o\dot{\gamma}}{\tau^*}\right)^{1-n}} \tag{11.8}$$

where the zero shear viscosity is based on the WLF functional form as shown in Eqs. 11.9–11.11.

$$\eta_o(T,p) = D_1 \exp\left[-\frac{A_1(T-T^*)}{A_2+(T-T^*)}\right] \tag{11.9}$$

$$T^* = D_2 + D_3 p \tag{11.10}$$

$$A_2 = \bar{A}_2 + D_3 p \tag{11.11}$$

Like the Carreau-WLF equation introduced in Chapter 3, the Cross-WLF model does an effective job of describing the low shear rate Newtonian plateau as well as the shear thinning region of viscosity. The WLF shift for the Cross model also permits accurate characterization of the temperature dependence of viscosity.

In order to describe the change of volume and shrinkage during the injection molding process a constitutive model for the pressure-volume-temperature interrelation for the polymer must be described. The density or its reciprocal, the specific volume, is a commonly used property of polymers. The specific volume of a polymer is dependent upon its temperature as well as the pressure of the material itself. The common method of illustrating this dependence is plotted on a pvT diagram. Because the injection molding process has varying pressures and temperatures throughout the cycle, an accurate model to predict the behavior is required. This experimentally obtained data is used to obtain information about the compressibility and volumetric expansion of polymeric materials. If the data are obtained under equilibrium, they are fundamental thermodynamic properties of the material. The data are seen to reflect transitions as the material moves from one physical state to another.

The pvT behavior of an amorphous thermoplastic at atmospheric pressure can be summarized as shown in Chapter 3. The kink in the curve shown in Fig. 11.4 characterizes the glass-transition temperature, which is a function of pressure. The slopes

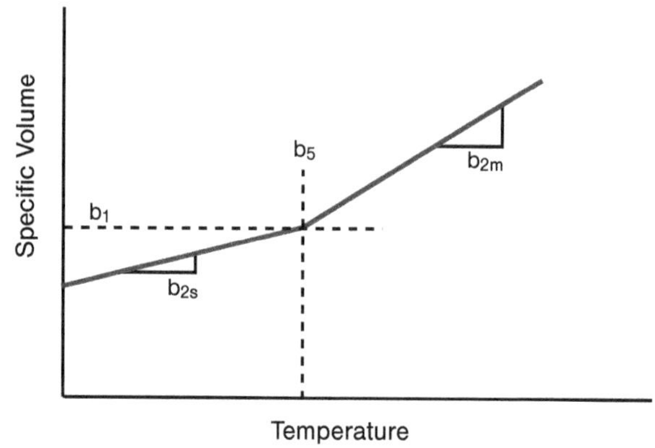

Figure 11.4 Schematic of a pvT diagram for an amorphous polymer.

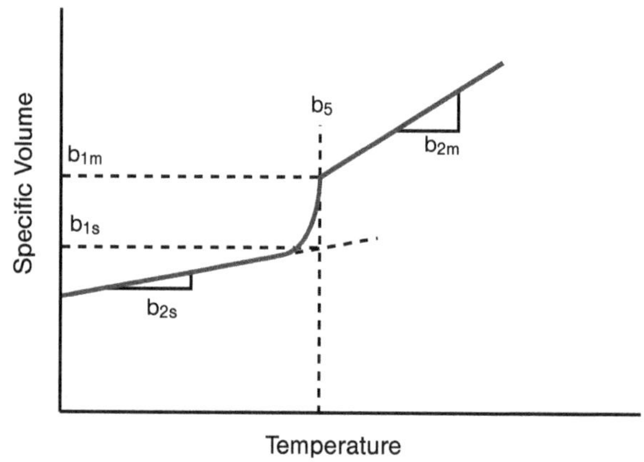

Figure 11.5 Schematic of a pvT diagram for a semi-crystalline polymer.

of the T curve, b_{2m} and b_{2s}, denote the bulk thermal-expansion coefficients in the liquid and solid phases, respectively. On the other hand, a schematic diagram for the pvT behavior of a semi-crystalline material under atmospheric pressure shows an abrupt transition in the T curve that is associated with the crystallization temperature, as shown in Fig. 11.5. For many commercially available programs a modified Tait equation [17] is used to empirically fit the properties. A thirteen-constant dual domain Tait equation is shown in Eqs. 11.12–11.18.

$$V(T,p) = V_o(T)\left[1 - C\ln\left(1 + \frac{p}{B(T)}\right)\right] + V_t(T,p) \tag{11.12}$$

where for $T > T_t(p)$ one uses

$$V_o = b_{1m} + b_{2m}\overline{T} \tag{11.13}$$

$$B(T) = b_{3m} \exp(-b_{4m}\overline{T}) \tag{11.14}$$

$$V_t(T, p) = 0 \tag{11.15}$$

and for $T<T_t(p)$ one uses

$$V_o = b_{1s} + b_{2s}\overline{T} \tag{11.16}$$

$$B(T) = b_{3s} + b_{2s}\overline{T} \tag{11.17}$$

$$V_t(T, p) = b_7 \exp(b_8 T - b_9 p) \tag{11.18}$$

Here, the value of C = 0.0894 is taken as a universal constant and the following relationships are used to fully describe the polymer response.

$$\overline{T} \equiv T - b_5 \tag{11.19}$$

$$T_t(p) = b_5 + b_6 p \tag{11.20}$$

In both the viscosity as well as shrinkage, temperature plays an extremely important role. Furthermore, during the computation of temperatures, the material thermal properties also depend upon the current temperature. Properties such as the thermal conductivity and the specific heat change over the range of processing temperatures as well as upon the phase of the polymer. As the polymer melt cools through the glass transition temperature the conductivity and specific heat will change discontinuously. Likewise, for semi-crystalline materials, the latent heat of fusion of the material must be removed from the melt before solidification. These polymer properties can be included in numerical analyses either by simple tabulation of the data or by relatively simple material constitutive models.

11.4 Numerical Methods

Four basic classes of numerical techniques exist that are commonly used to solve fluid flow problems: the finite difference method (FDM), the finite element method (FEM), the boundary element method (BEM), and the finite volume method (FVM). Each of these methods has its own advantages and disadvantages and, therefore, one may be preferred for a certain type of problem. Although it is not the intended purpose of this chapter to provide a detailed derivation of the numerical methods used, it is necessary to provide a general description of the methods. Each of the numerical methods is described and some basic considerations are discussed as they apply to solving the injection molding process.

11.4.1 Finite Difference Method

The finite difference method started gaining prevalence in the 1930s for use in hand calculations and is the simplest to use and understand. Figure 11.6a shows the grid that must be constructed to represent the geometry of a 2-D domain. Once the grid is created, the governing differential equations are rewritten in discretized form and then applied at each nodal point. The resulting system of algebraic equations can then be solved by standard Gaussian elimination or by more elaborate numerical algorithms. Because of the simplicity of the method, it can be implemented in a wide variety of problems. Since the method discretizes the governing equations at the start of the analysis, it is fairly easy to model non-linear problems and those that couple flow and heat transfer together. The finite difference method is also relatively straightforward to program for computer simulations and can provide quick computation times. While the simple nature of the finite difference method allows for easy implementation, this simplicity also yields certain disadvantages. The first point to make about FDM is that it is best suited to problems that have relatively simple geometries. Even though more complex geometries can be modeled with special differential operators or coordinate transformations, the other methods presented in this section often lend themselves to be more efficient than FDM. Furthermore, since discretization of the governing equations occurs at the start of the analysis, more error is introduced than in other methods. Therefore, FDM can have problems with obtaining a convergent solution for non-linear problems.

As applied to the injection molding problem, the finite difference method is not particularly well suited. Because FDM prefers relatively simple geometries, the generation of FDM grids for complex injection molded articles is cumbersome. The finite difference method is, however, well suited to solving the heat transfer through the thickness direction of thin walled injection molded parts. Additionally, there are some rudimentary calculations that can be accomplished with FDM to estimate moldability of injection molded parts.

11.4.2 Finite Element Method

In contrast to FDM the finite element method is a relatively new technique used for solving polymer processing problems. Popularized in the 1960s, along with the advent of digital computers, the finite element method has become the basis for many commercial flow packages. Like FDM, FEM is a domain method in which the entire geometry to be modeled must be discretized into nodes and elements. The mesh shown in Fig. 11.6b represents the discretization required for FEM to model a simple two-dimensional geometry. Although several different methods are available to obtain the final equations, the Galerkin method of weighted residuals is normally preferred. Once the mesh has been created, the governing differential equations are then expressed in integral form and numerically integrated to obtain an algebraic system of equations. Because of the nature of the finite element method, it is capable of mod-

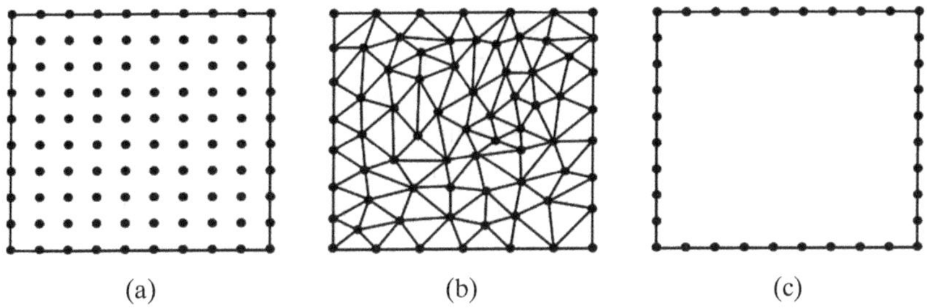

Figure 11.6 Discretization of a 2-D domain for (a) finite difference method, (b) finite element method, and (c) boundary element method.

eling much more complex geometries than FDM. It can also provide quite accurate solutions to the field variables, such as fluid velocities or pressures, for a wide variety of problems that include non-linear flows. Like finite difference, however, the finite element method does contain a certain amount of intrinsic error even before numerical errors are included.

Nonetheless, the finite element method has proven well suited to solving injection molding problems. By taking advantage of the Hele-Shaw theory, complex three-dimensional problems can be solved efficiently by using shell finite elements. Furthermore, the control volume approach is well suited to computing the advancement of the flow fronts during mold filling, and the same finite element mesh can be used to compute additional packing, cooling, and warpage of the injection molded part. Overall, most commercial injection molding packages utilize the finite element method as the principle numerical technique.

11.4.3 Boundary Element Method

In contrast to both the finite difference and finite element methods, the boundary element method only requires that the boundary of the geometry be discretized. As shown in Fig. 11.6c, a two-dimensional geometry only requires a discretization of the curve that makes up the boundary of the part. BEM also gained prevalence around the same time as FEM, but because of the relatively complex mathematics involved with BEM, it has been relatively slow to gain acceptance compared to FEM. The boundary element method begins with a different form of the governing equations that are expressed in terms of domain integrals. These integrals are then manipulated by Green-Gauss transformations until they are reduced to boundary integrals. The integrals are then numerically evaluated to yield an algebraic system of equations. Up to the point of evaluating the integrals, no approximations have been made in the governing equations. Therefore, the boundary element method unlike FDM or FEM, introduces error into the solution only through numerical sources. Another advantage of BEM is that the accuracy of higher order derivatives is excellent.

However, the boundary element method is not particularly well suited to solving flows during the injection molding process. Because most injection molded parts are relatively thin, the reduction of dimensionality gained by BEM actually complicates the solution compared to shell simplification enjoyed by FEM. Additionally, the presence of highly non-linear equations and coupled flow with heat transfer make BEM not well suited. These disadvantages considered, however, BEM has been used successfully in computing mold temperatures during the injection molding process.

11.4.4 Finite Volume Method

Similar to FEM, the whole computation domain is subdivided into a finite number of non-overlapping control volumes or "cells" of arbitrary topology in the FVM. The transport variables are usually stored at the centroid of the cells, and linear interpolation is used to approximate the values of the variables at cell-face centers. The transport equations are then integrated over each cell and applied with the Gauss divergence theorem whenever possible to obtain an integral formulation for each variable. This approach can be applied to Eqs. 11.2 and 11.3 to obtain equations for velocity components and temperature. However, some special means are needed to compute the pressure field, as no explicit equation exists for the pressure field. Further detail can be found in [12].

Unlike FEM, the sparse matrix of the resulting algebraic equations of FVM is always diagonally-dominated. It, therefore, can be efficiently solved with iterative. The efficiency of direct solver is very good for small matrices, but very demanding in time and memory if the matrices are large. On the contrary, the iterative solvers generally require much less memory and are more efficient than the direct solvers for large matrices. Therefore, FVM is generally superior to FEM in the three-dimensional simulation.

One more advantage of FVM over FEM is its versatility in mesh support. It's very straightforward in FVM formulations to treat various mesh types, such as tetrahedral, hexagonal, prismatic, pyramid, or polyhedralas "cells" of different topology. The basic FVM formulation remains unchanged. Although FEM can also be tailored to adopt different mesh types by using different shape functions, it generally leads to additional numerical accuracy and stability issues.

This advantage makes it very easy for FVM to model the complex geometries in injection molding. For example, the tetrahedral mesh is excellent for describing complex geometries, but very limited for thin wall areas. Applying a tetrahedral mesh in thin wall regions generally leads to meshes with only one layer of elements across the thickness. It's, therefore, not enough to capture the drastic variation of temperature and melt viscosity across the thickness. Mesh refinement will improve this issue. However, isotropic refinement of tetrahedral meshes leads to an unacceptably large number of elements, and anisotropic refinement leads to very poor mesh quality that deteriorates solution accuracy. On the other hand, the thin area can be modeled with prismatic or hexagonal meshes in FVM. The number of elements across the thickness can also be optimized to balance computation time and accuracy.

11.5 Simplified Calculations

Simple calculations are those intended to require very little user setup effort but give useful information early in the design stage of injection molded parts. Typically, such calculations involve substantial simplifications to the geometry and/or material behavior to provide quick calculations. However, newer techniques allow more detailed analysis of realistic geometries with a good degree of accuracy. Ultimately, the results from such analyses are able to give trends and indications of molding problems early in the design stage.

11.5.1 Finite Difference Based Calculations

There are a large number of finite difference based calculations that can be made on injection molded articles. Some of the earliest calculations on record are those based on the FAN method used by Tadmor [5] and others. The FAN method provides a means by which the mold filling of rather complicated geometries can be solved via a lay-flat approach. By unwrapping complicated geometries and making the appropriate simplifications, useful insight into the molding process can be gained from such methods. However, one of the major downfalls of such a method is the difficulty in appropriately unwrapping the geometry.

Other approaches rely on dramatically simplifying the geometry to a one dimensional flow field, but fully accounting for all material behavior during filling. One specific approach used by Buchmann [18] simplified the analysis to investigate flows in spiral and radial molds. His treatment included a one-dimensional finite difference calculation that accounted for non-isothermal non-Newtonian polymer flow with

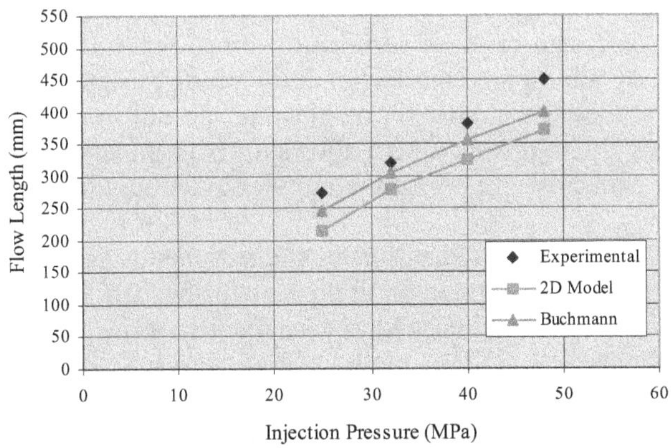

Figure 11.7 Flow length versus gate pressure for polystyrene in a spiral mold at a constant flow rate of Q = 3.68 cc/sec.

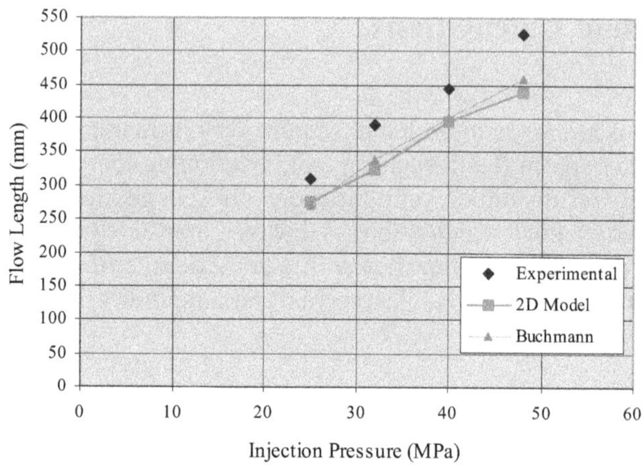

Figure 11.8 Flow length versus gate pressure for polystyrene in a spiral mold at a constant flow rate of Q = 7.4 cc/sec.

convection, dissipation, and solidification. The results from this analysis provided decent accuracy when compared with experimental values for flow length as shown in Figs. 11.7-8. The main purpose of such a simulation, however, was to allow different materials to be rapidly compared and to develop a processing window for these materials.

11.5.2 Midplane Model-Based Calculations

A substantial improvement over these rudimentary finite difference simulations is the advent of FEM/FDM hybrid approach of midplane-based analysis. As long as a midplane model can be successfully extracted from the given CAD geometry, it can be analyzed very efficiently and accurately. More importantly, it is very easy to perform a series of design revision verifications once you have a midplane model available. Extracting a midplane model generally requires significant user intervention, especially for those geometric topologies that do not have midplane counterparts. Nevertheless, the midplane model allows users to understand and control the analysis accuracy explicitly and perform computer-aided molding analysis and optimization very easily.

The FEM/FDM hybrid approach divides the midplane surface into a number of planar elements, usually triangular, to approximate the part geometry. The corresponding wall thickness is then assigned to each element as its attribute. One-dimensional beam elements are used to model the runner system. The corresponding diameter is then assigned to each beam. If the runner cross section is not circular, a proper shape factor needs to be assigned accordingly. To capture the temperature and shear rate variation across a wall or in the radial direction, FDM discretization is employed.

This method has been widely used in many commercial codes since late 1980s, and has been extended to simulate packing, cooling, and warpage. However, its warpage prediction is sometimes not very reliable because of the limitation of the midplane approach.

11.5.3 Solid Model-Based Calculations

A relatively new approach by Yu and Thomas has received patent status [19] and serves to directly link analysis with the role of designer. Their solid model-based approach utilizes a specialized technique to perform a molding analysis "directly" on a solid model rather than on a mid-plane representation of the geometry. Thus, by promoting molding analysis during the part design stage a more effective product can be created. The solid model based calculations take advantage of a common feature of CAD design software, namely stereolithography.

Stereolithography (SLA) is a "rapid-prototyping" process that produces a physical, three-dimensional object directly from a 3-D CAD model file. A common stereolithography machine uses a computer-controlled laser to *solidify* a photocurable resin. In essence, the machine builds up a 3-D representation of the model in a layer-by-layer approach. More accurately SLA is really a "rapid-modeling" technique since the objects generated from existing photopolymers do not have the physical, thermal, or mechanical properties required of end-use production material. Stereolithography techniques embedded within solid modeling software generate a triangular grid across all surfaces of the solid model. This three-dimensional faceted representation of the geometry fully defines the features and volume of the part to be analyzed.

The resulting "STL" file exported by most solid modelers consists of a set of triangular "elements" on surfaces and features of the geometry. However, unlike a structured finite element mesh, these stereolithography grids often contain severely distorted triangles. While this distortion is uneventful for SLA machines, they pose serious problems for numerical techniques used to solve the injection molding process. For this reason, commercial solid model based calculations typically utilize mesh improvement algorithms to improve the aspect ratios of the triangular elements.

The resulting solid model based approach utilizes a Hele-Shaw model for the flow during injection mold filling. Like the finite element based advanced calculations described in section 11.6, their method uses a finite element/control volume approach to solve the non-isothermal non-Newtonian mold filling problem. However, unlike these advanced calculations, this method does not require a mid-plane model to be created.

Rather than creating a mid-plane model, the solid model based calculations consider all the surfaces of the geometry. Based on the improved mesh from the stereolithography file, the method implements a system of pairings and synchronizations between opposing surfaces in the geometry. The method strives to match features from one surface with those from the matching surface of the geometry. Through the use of element normals, the distance between matching surfaces is applied as a thickness to each triangular element.

The most simplistic view of such a technique can be visualized by considering the molding of a simple flat plate. The 3-D representation of the flat plate will have both an upper surface and a lower surface. The solids based stereolithography method strives to make pairings between elements that lie on the upper surface with the corresponding ones on the lower surface. Figure 11.9 illustrates a 2-D slice from a typical ribbed geometry encountered in solid model based calculations. By carefully analyzing the element normals and adjacent elements, one can obtain synchronization between upper and lower surfaces of specific features. These synchronized pairs are then linked by both boundary conditions and advancement of flow fronts.

In Fig. 11.9, it is relatively straightforward to construct the pairings in planar sections, but when curvatures from fillets or complex features are considered the synchronization is not possible. Therefore, in solid model based calculations, certain built-in assumptions are used for those elements without well defined synchronized pairs. In these cases, the thickness of the element is often taken as an average of its adjacent elements. For example, the nodal pairings of A-a and B-b in the planar section are clearly evident. However when nodes such as G are considered, there is no distinct node to form the synchronized pair. Therefore for nodes like G, the thickness property is taken to be the average of its adjacent nodes.

The resulting element structure has thickness assigned to each element as well as special synchronization pairings to be used for the molding analysis. The particular thickness assigned corresponds to the whole gap thickness between the pairing surfaces so that the mesh volume is approximately double that of the actual geometry. Accordingly, the injection flow rate imposed by the analysis is approximately double that of the geometry. The nodal pairings are used to synchronize the values for pressure and the position of the flow front. During a standard analysis, an injection gate is placed on one of the surfaces of the geometry. If the node at which the gate is located is one of a synchronized pair, then a complimentary gate is placed on the other node in the pair. The analysis then uses a Hele-Shaw type solver to compute

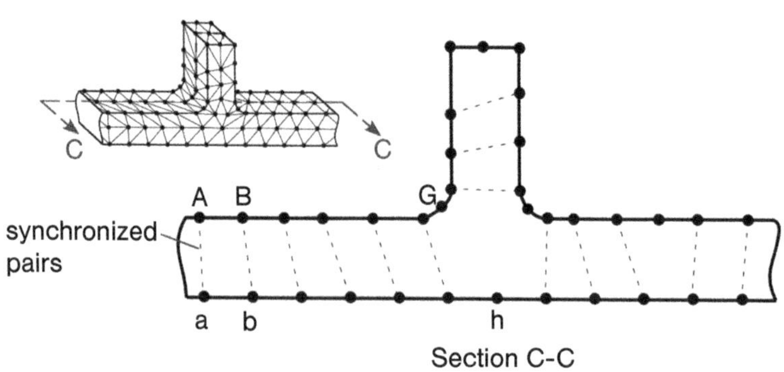

Figure 11.9 Solid model–based synchronization of surfaces.

the flow in the mesh on each side of the object. The key point in this analysis is to obtain similar flow fronts on each side of the matched surfaces. Therefore, the synchronization pairs are used to match the boundary conditions, pressures, temperatures, and flow state from each side of the mesh.

Results from a solid model based analysis can include direct quantities such as injection pressures, time to fill, location of knitlines, and temperatures. Subjective measures like confidence of fill, air entrapment, cooling quality and surface quality can also be computed and displayed. An additional feature of this type of analysis is the ability to give recommendations as to the optimal regions to place the gate. More recent advancements [20] have begun to incorporate warpage analyses into the solid model based calculations. However, the readers should be aware that because the solid model–based calculation is derived from the midplane model-based calculation, it inherits the same limitation as the midplane calculation.

Figure 11.10 shows a validation experiment where the mold filling of a part was simulated using a solid model based software and compared two injection molded

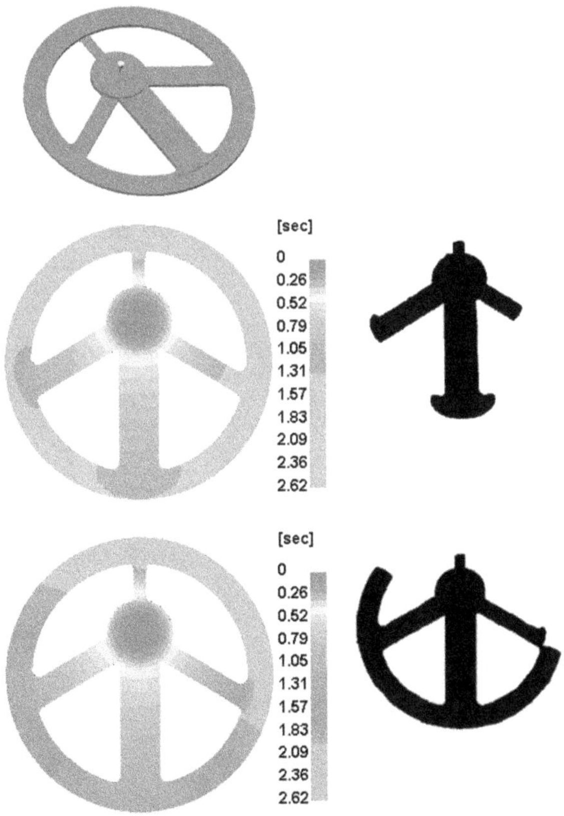

Figure 11.10 Mold filling of universal peace symbol used to validate simulation (Courtesy of Moldflow Corporation)

short shots. Figure 11.10, from top to bottom, shows the part geometry indicating a center gated system (arrow in center of part) and two mold filling snapshots at two subsequent filling times. The validation demonstrated excellent agreement between simulated and experimental results.

The overall benefit from the solid model based calculations is the tight integration that can be achieved between standard CAD and processing. By incorporating these analysis techniques directly in the design stage, the engineers and designers have useful insight into how geometric features will affect the processing of the actual part. Furthermore, this insight gives the designer the opportunity to fine-tune the design to ameliorate problems before beginning production. An example is shown in Fig. 11.11 where regions with possible part cooling problems are directly indicated by the software. Here the software highlights potential problem areas with red contours and green color to reveal regions of effective cooling quality. This information helps the designer to concentrate efforts in optimizing problem regions.

A complementary tool to the solid model based calculations is mold-based analyses. This method still utilizes the solid model based calculations but now expands the analysis to include such molding features as runner and gate design. The goal of these mold-based calculations is to create and optimize the mold design early during the preliminary stages of part design. By rapidly assessing the effects of gates, runners, sprues, and cavity layouts, the molding engineer can avoid potential problems. The effects of distribution systems in single cavity or multi-cavity molds as well as family molds can be rapidly evaluated to determine the effects on clamping force, shot size, and cycle times. Quick analysis of the polymer flows through the distribution system of runners, sprues, and gates provides the ability to create balanced runner systems and measure the effect of different gate designs on the final product. Typically embedded within such calculations is also the ability to estimate costs and waste for

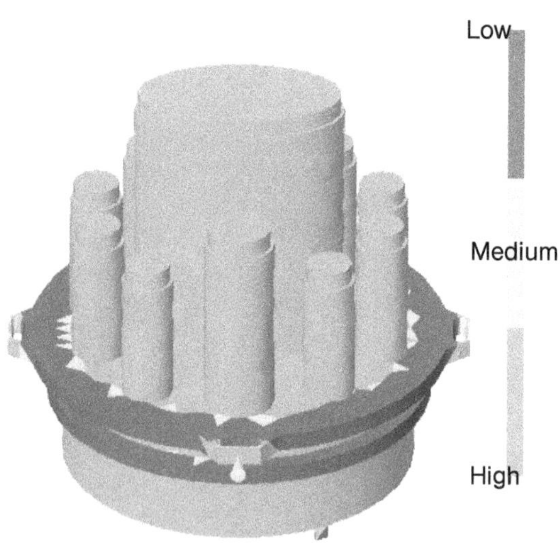

Figure 11.11 Cooling quality of a fiber optic joint housing (Courtesy of Moldflow Corporation)

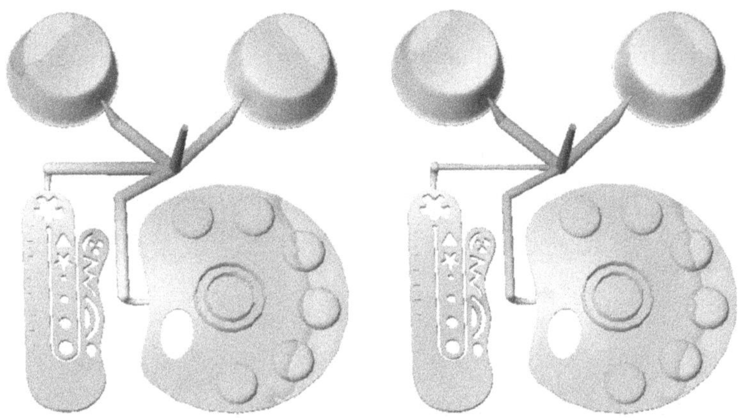

(a) (b)

Figure 11.12 Runner balancing of a Hasbro painting set; (a) before and (b) after balancing. (Courtesy of Moldflow Corporation)

the tooling under consideration. These mold-based calculations in conjunction with the solid model based calculations compliment each other and can serve to further integrate CAD and CAE for injection molding.

An example of these capabilities is shown in Fig. 11.12 where the mold for a children's painting kit is balanced to ensure even filling of four cavities. The software automatically and simultaneously balanced the flow and the overall runner volume is minimized to reduce scrap material.

11.5.4 True 3-D Calculations

The fully three-dimensional approach was first attempted in late 1990s. However, it was not ready for commercial application until the first complete suite of true 3-D programs were introduced by CoreTech System Co., Ltd., in 2002. Thanks to the advancement in both numerical simulation algorithms and computer hardware, true 3-D calculations are now becoming much more popular and are widely recognized as the state-of-the-art simulation technology for injection molding.

The primary advantages of true 3-D simulation over the conventional midplane-based or the above-mentioned solid model–based calculations are:

1. A three-dimensional solid model can be directly adopted for simulation without any model simplifications. The molten material is allowed to flow in all directions. This is not only very suitable for thick parts, but it is also especially useful for those parts with comparable thickness and flow length, such as connectors, lens, gears, etc.

2. Full 3-D flow and heat transfer in all directions can be considered, such as fountain flow, heat transfer through lateral walls, flow resistance from lateral walls, flow imbalance between runners, etc.
3. Full 3-D Navier Stokes solvers employ inertia and gravity effects, which are generally neglected in midplane-based calculations.

The viscosity, pvT, and thermodynamic models remain unchanged in true 3-D calculations. However, all components of velocity and heat transfer are included without further assumptions.

11.6 Advanced Calculations

Advanced calculations are those that are typically based on the finite element method and require substantially more user interaction than the solid model based calculations. Accordingly, the advanced calculations give a higher degree of confidence when analyzing parts and can include more effects from material behavior and processing conditions. While solid model based calculations give trends and insight into processing for the designer, the advanced calculations are used to ascertain detailed information during the injection molding process. While a detailed derivation [27] of the numerical methods used to solve advanced calculations is beyond the scope of this chapter, a basic understanding of key concepts is useful.

Advanced calculations require the generation of a mid-plane model of the part in order for analysis to occur. This can require substantial user interaction and knowledge of how to "best" mid-plane a complicated geometry. This mid-plane model is a shell element representation of the full three dimensional geometry. The shell elements are located on the mid-plane of all geometric surfaces and carry the prescribed thickness of the geometry. An advanced calculation relies on a combined finite element/control volume method to solve the mold filling during injection molding. The method will typically utilize a non-Newtonian non-isothermal Hele-Shaw formulation to solve the governing equations for flow. Most advanced calculations utilize a finite difference method to compute the temperature profile through the thickness of the elements and to solve for layered results in the analyses.

The principle benefit of advanced calculations is associated with the fine detail solutions that can be obtained for mold filling, pressures, packing, temperature, orientation, residual stress, shrinkage, and warpage. The computations and functionality of advanced calculations are based on firm scientific basis of the process. They can readily handle thin cavity, three-dimensional geometries with either cold or hot runner systems. The method is also capable of simulating mildly compressible material properties during Hele-Shaw flow as well as asymmetric thermal boundary conditions between the mold halves. Many potential tribulations, such as weld lines, air entrapment, and short shots can be detected prior to tool build. The additional benefits from advanced calculations are that the gate location, melt delivery system, part thickness, and processing conditions can all be determined via numerical predictions rather than experimental trial-and-error.

Further advantages of advanced calculations involve the post fill analysis required for injection molding. Once the cavity has filled, the packing phase maintains sufficient pressure to overcome polymer volumetric shrinkage. Advanced calculations account for this phase of the molding cycle by introducing accurate volumetric shrinkage predictions from pvT data. Additionally, the complex thermal distributions throughout the geometry contribute to uneven shrinkage and warpage in the final molded part.

The cooling stage during the injection molding process is concerned with lowering the bulk temperature of the polymer and allowing it to solidify before demolding. However, to analyze this stage of the process more than the cavity must be considered. To adequately predict the temperature variations throughout the part, the mold cooling must be accounted for by numerical means. While it is possible to assume a constant mold temperature during cooling, this is not realistic of actual moldings. A more accurate method is to consider the location and size of cooling lines within the mold to determine the effects on part cooling. Modern simulation tools can account for the cooling in the mold via a combination of boundary element and finite element methods. These results can then be used to predict the shrinkage and warpage of the final part.

Overall, advanced calculations are used to ascertain detailed information on the injection molding process. Through accurate models and numerical methods, it is fully possible to compute the mold filling, location of knitlines, pressures, gas entrapment, layered effects, fiber orientation, temperatures, shrinkage, warpage, cycle times, cooling, and residual stress in the final part.

11.6.1 Commercial Codes

There are a number of commercial and research codes available to simulate the injection molding process. These software packages often have a variety of modules that can be implemented to simulate each phase of the injection molding process: mold filling, packing, cooling, and warpage. In essence, each simulation package is founded on the same numerical method, but each may have subtle variations in the manner in which material properties are account for. Far improved over the pioneering codes, these software packages contain full-featured graphical user interfaces and a multitude of features to aid in analyzing results. General features of such codes include the ability to solve for the mold filling, location of knitlines, gas entrapment, pressures, fiber orientation, temperatures, and shrinkage & warpage. Some pertinent results from commercial software are shown along with comparisons to existing experimental result

11.6.1.1 Mold Filling

With mold filling the advancement of the material inside the mold cavity can be studied. When analyzing an injection-molded part, mold filling is the first result generated by the CAE software. During mold filling the velocity vectors, pressure and temperature are computed over time and form the basis for post-fill analyses. Mold filling is used to predict cycle times, to ensure complete filling of the mold, and to predict knit lines and gas entrapment.

Figure 11.13 shows mold filling results for an upper door panel assembly. In this case a two cavity mold produces two parts at the same time with four injection gates on each one. Mold filling results are generally shown with color contours representing isochronous filling times. In Fig. 11.13 the color legend corresponds to the percentage of fill and varies from 0 to 100 percent.

Figure 11.14 shows another example of mold filling where a short shot of an electrical switch breaker box was simulated and compared to the molded short shot. This experiment was used to validate the simulated mold filling results to the real mold filling pattern. As seen in Fig. 11.14 the experimental and computed mold fillings are in excellent agreement.

11.6.1.2 Knit lines/Gas Entrapment

The mold filling analysis can be used to predict the formation of knit lines or weld lines and gas entrapment. These can cause weak points and surface finish problems that can lead to cracks and failure of the final part. A knit line occurs when distinct flow fronts meet each other. Gas entrapment occurs when two flow fronts meet around an unfilled area on the part. The unfilled region will contain gas that becomes entrapped by the surrounding material.

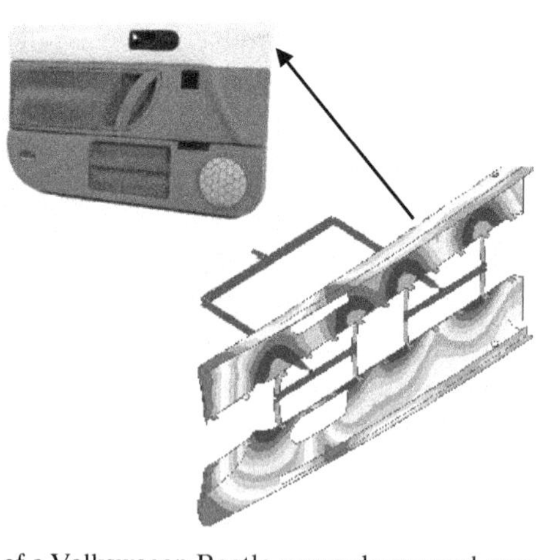

Figure 11.13 Mold filling of a Volkswagen Beetle upper door panel assembly (Courtesy of Simcon/IKV and SAL Automotive SAI GmbH

In Fig. 11.15 the mold filling analysis of an electrical switch breaker box shows the formation of a knit line where air is also entrapped. The knit line is recognized in the circled area in Fig. 11.15 where the two flow fronts meet together. Air is also trapped in the unfilled spot seen in the circled area. Through simulation, proper gating

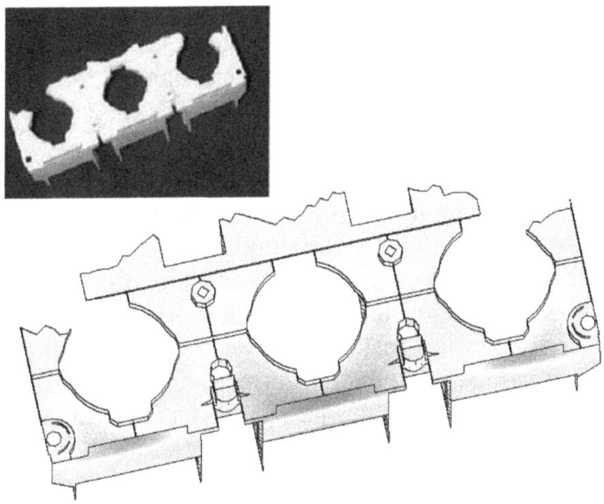

Figure 11.14 Mold filling of an electrical switch breaker box showing a short shot (Courtesy of Square D Company)

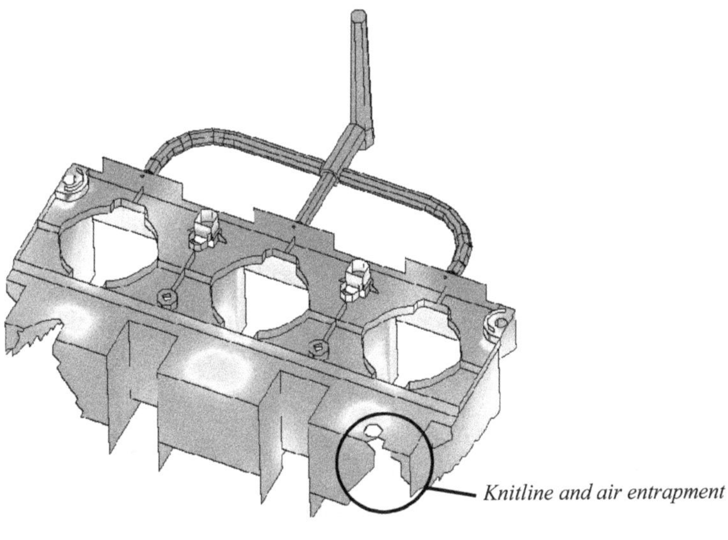

Figure 11.15 Mold filling of an electrical switch breaker box showing formation of a knitline and air entrapment (Courtesy of Square D Company)

and venting locations were determined for the breaker box to completely avoid air entrapment.

11.6.1.3 Orientation

Orientation of molecules, fibers or additives occurs because of the natural tendency for them to align in the direction of flow. Orientation leads to non-homogeneous material properties throughout the injection molded part. Predicting orientation is important to understanding and predicting the mechanical behavior of a part. The orientation results coupled with structural analysis software can be used to predict the mechanical response of the part to specific loads.

Figure 11.16 shows the fiber orientation in an injection molded rib [22]. The experimental study shows the results from an image analyzer used to highlight the fibers in a transversal cross-section of the rib. The simulation result in Fig. 11.16 shows computed fiber orientation where each vector is in the direction of the orientation and the vector length shows the magnitude of the orientation.

In Fig. 11.17 fiber orientation results were taken a step further and exported to structural analysis software where the flow-direction tensile modulus of the part at different locations was computed [29]. The part tested was an end gated strip made out of Polycarbonate. The graph in Fig. 11.17 shows tensile modulus for experiments and three different algorithms used to compute the fiber orientation. The details of this algorithm can be found in the reference and are beyond the scope of this chapter.

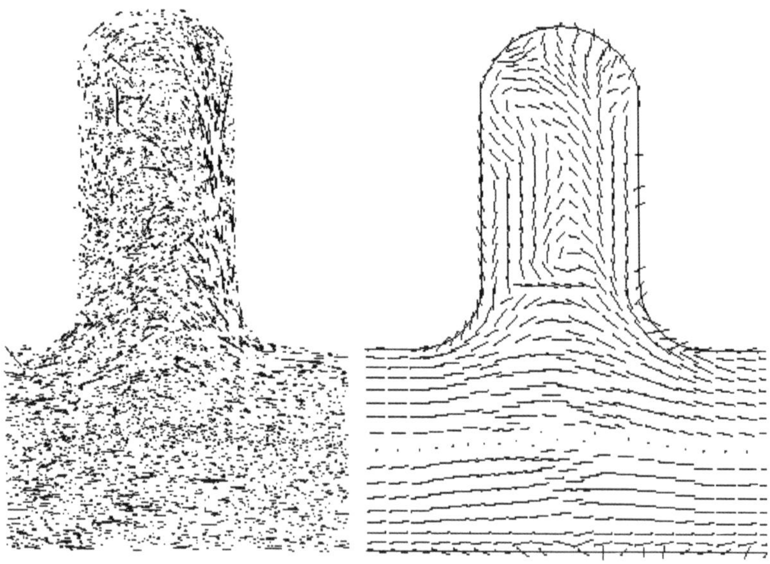

Figure 11.16 Experimental (left) and simulated (right) fiber orientation for an injection molded rib [Figs. 9–10 from Ref. 22]

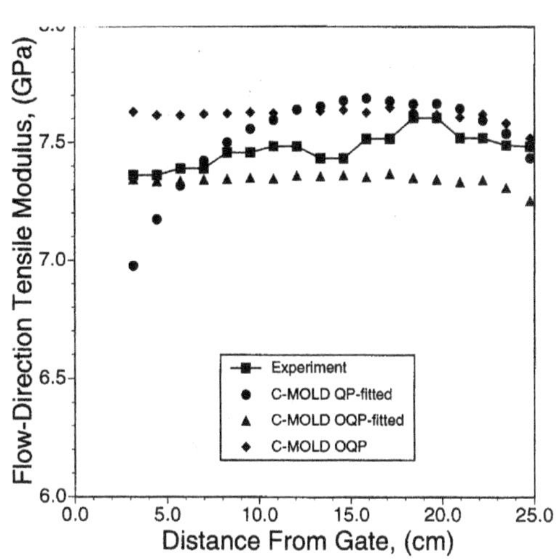

Figure 11.17 Experimental and simulated flow-direction tensile modulus along the centerline of an end-gated strip made from PC [Fig. 16 from Ref. 23]

11.6.1.4 Temperatures

The temperatures of the mold surface can also be simulated with CAE software. Besides the implications that mold temperature has in mold filling, variations in temperature may also lead to warpage of the part. The layout of mold cooling lines has a great effect on the temperature distribution of the mold surface. With the aid of CAE software it is possible to compute the effect of mold cooling lines on the mold temperature and therefore on the final part. Figure 11.18 shows the mold of an electrical switch breaker box with its cooling lines. The software was successfully used to optimize the location of cooling lines to obtain a uniform mold temperature and therefore minimizing warpage.

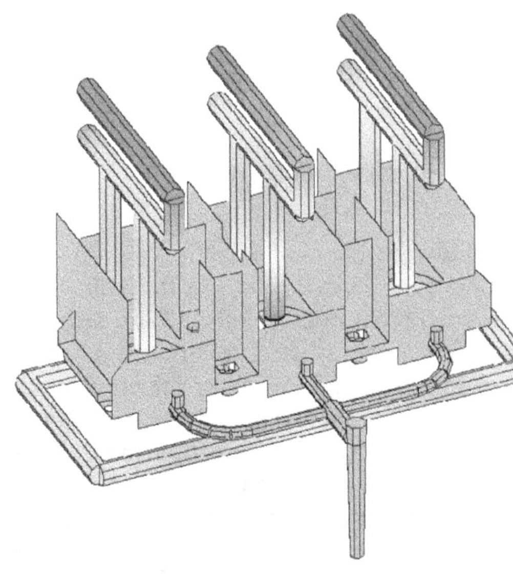

Figure 11.18 Cooling channel layout for an electrical switch breaker box (Courtesy of Square D Company)

Figure 11.19 Simulated shrinkage and warpage of a miniature palette (Courtesy of Simcon/IKV)

11.6.1.5 Shrinkage and Warpage

Shrinkage and warpage of an injection molded part is a consequence of orientation and temperature distributions. After the CAE software computes orientation and temperatures it is able to predict the final shrinkage and warpage of the part. The simulation software can be used to test the effect of different materials and processing conditions on the final shape of the part.

Figure 11.19 shows the simulated final shape of a miniature palette after removal from the mold. The simulated results are exaggerated five times to facilitate the analysis. The CAE software can be successfully used to prevent parts with high deformation and reduce the number of unacceptable parts.

11.6.2 Specialty Calculations

In addition to classical injection molding of thermoplastic materials, there are a number of specialized processes that can be analyzed by advanced calculations. While there are a great number of specialty processes that are used in the industry, two of the more commonly used are discussed here to show the benefit of simulation. Both the co-injection and gas assist injection processes are equivalent in the sense that each process contains two distinct materials that are injected into the same mold cavity. In the case of co-injection molding, the second material is usually another polymer. However, in the case of gas assisted injection molding, the second material is a vapor used to reduce weight and pressures.

11.6.2.1 Gas Assist

In the gas assisted injection molding (GAIM) process a short shot of molten material is injected into the mold. Then an inert gas such as nitrogen is injected into the melt to penetrate into the mold and push the molten material to finish filling the part. As the gas flows inside the mold it takes the path of less resistance, which is normally the mid-plane of the part, leaving a hollow cavity in the melt. The gas pressure is then maintained during the packing stage before it is vented out of the part.

The GAIM process offers many benefits as reduction of cycle times, reduction of material, molding pressure and clamping force. At the same time it is a more difficult to setup process which requires selecting gas injection location, gas channel size, and gas injection timing and pressure. If these variables are not properly selected it will lead to a non-uniform wall thickness throughout the part leading to structural failure and surface finish defects.

To be able to take full advantage of the GAIM process use of CAE software becomes extremely important. With GAIM CAE software the gas injection point, gas channels layout and size can be optimized to ensure proper filling and wall thickness of the part. Software tools allow the user to test scenarios where channel size, injection gates, gas injection time and pressure are varied to obtain the desired wall thickness and mold filling. Otherwise the optimization of the process requires many trial and error experiments that are costly and time consuming.

Figure 11.20 shows the simulated mold filling and gas penetration of a towel holder. Using a GAIM CAE software tool the gas injection point, pressure and timing were selected to provide a part with high quality finish and uniform wall thickness. The simulated wall thickness can be further exported to a structural analysis program to assess the mechanical behavior of the part.

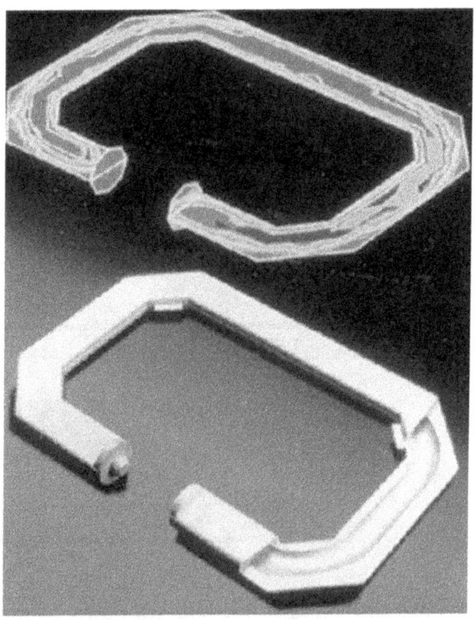

Figure 11.20 Mold filling of a Battenfeld towel holder (Courtesy of Simcon/IKV)

11.6.2.2 Co-Injection

Co-injection simulates the sequential injection of a skin polymer followed by subsequent injection of a different core polymer. Co-injection molding benefits directly from the fountain flow effect prominent in injection molding. As the material in the center of the cavity is pushed outwards to the mold walls at the flow front, the polymer cools and forms a solidified skin against the mold surface. This skin layer provides an insulating layer at each wall surface through which new polymer is transferred to the melt front.

A sequential co-injection process will typically have two or more barrels and one nozzle in the injection molding machine. The skin polymer is injected into the mold first and is followed by the core polymer. Optimally, the skin polymer is expected to solidify against the mold walls and the core polymer will fill the inner regions of the cavity. The end product is a sandwich-like structure, as shown in Fig. 11.21.

Simulation of the co-injection traces the spatial distribution of the skin and core polymers throughout the cavity during the filling process. These analyses can account for the differences in material properties and in processing temperatures of the skin and core polymers, as well as the mass, heat, and momentum interactions between them. Results from simulation can provide designers and engineers with the ability to qualitatively predict part performance, improve mold designs, and optimize process controls. In particular, simulation is an efficient tool for determining the best combination of skin and core polymers, and the most appropriate switch-over time.

11.6.2.3 Shear-Induced Imbalance Between Cavities

It is a common practice to use geometrically symmetric runner systems when designing a multi-cavity mold. The geometrically symmetric runner system makes the flow distances from sprue to each cavity equivalent to provide consistent conditions between cavities. However, one often observes imbalanced filling patterns in each cavity with such a "symmetric" runner system caused by viscous dissipation effects that lead to asymmetric temperature distributions inside the runner system. This problem has been investigated by numerous groups, and it has been concluded that the shear-induced imbalance in sprue and primary runners leads the imbalanced filling pattern in cavities [24]. The patented MeltFlipper® technologies manage the

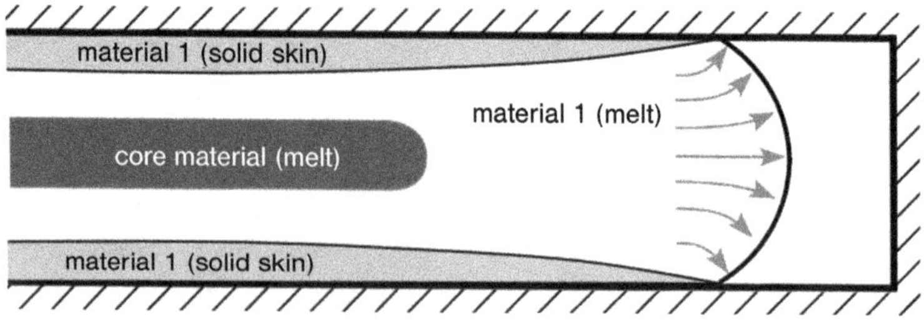

Figure 11.21 Sandwich layers from the co-injection molding process.

melt variations so that they are symmetrical relative to a downstream runner branch or cavity. The result is a true rheological balance within the mold filling, virtually eliminating the molding inconsistencies found in traditional runner designs. Thus, the improved balance can often provide for increasing the efficiencies and outputs from a multi-cavity mold.

The conventional midplane model–based calculation and its derivatives employ one-dimensional beam elements for runners. The temperature and shear rate profiles across radius are assumed always axisymmetric. Such a simplified computational model is, therefore, incapable of predicting the runner imbalance. Chien et al [25] developed a proprietary three-dimensional mesh algorithm to capture the temperature and shear rate variations in the runner system. By using a 3-D FVM flow solver, it successfully predicted the flow imbalance of a geometrically symmetric runner system and the improved balance with MeltFlipper®. Figure 11.22 shows both the experimental and simulation results of an eight-cavity mold with geometrically symmetric runner layout. The inner cavities are generally filled faster at normal filling speed because of shear-induced imbalance. Figure 11.23 shows that the high-sheared material developed in the primary runner (sprue effect neglected) is axisymmetric until it branches into the secondary runner, then onto the inner side of the secondary runner. The high-sheared material generally exhibits a lower viscosity because of shear thinning and viscous heating, so the inner cavities will be filled faster than the outer ones. The simulation results clearly help visualize the melt temperature development in the runner system. Figure 11.24 shows the experimental and simulation results with the MeltFlipper®. This melt rotation apparatus positions the high- and low-sheared material top to bottom, so it is symmetric side-to-side. By this means, the imbalance between cavities would be eliminated.

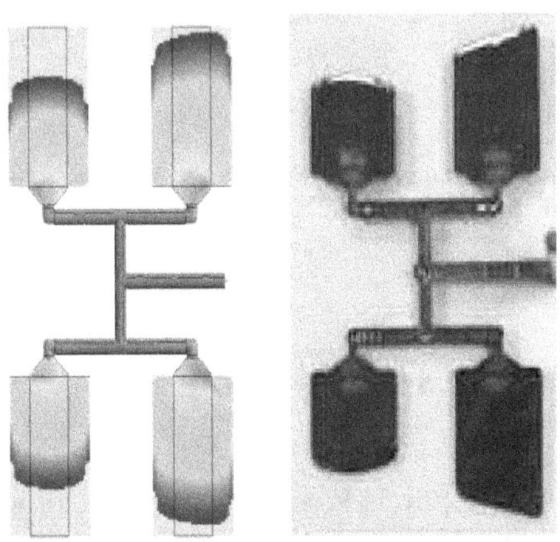

Figure 11.22 Experimental and simulation results of runner imbalance of a 8-cavity mold with geometrically symmetric runner. (Courtesy of Beaumont Technology Inc. and CoreTech System Co., Ltd)

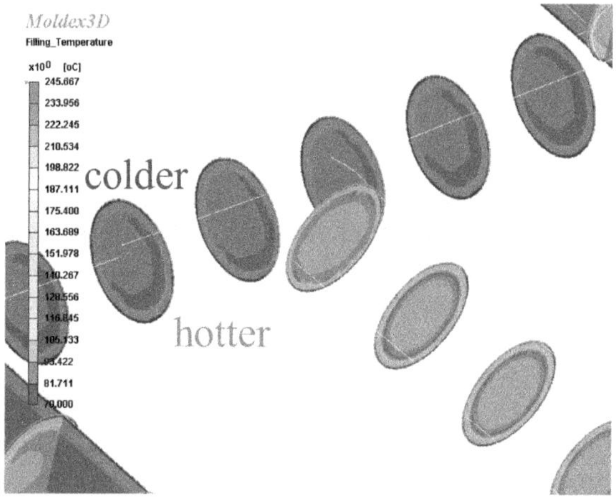

Figure 11.23 Three-dimensional simulation shows the axisymmetric high-temperature zone in the primary runner diverging into asymmetric temperature profile in the secondary runner.

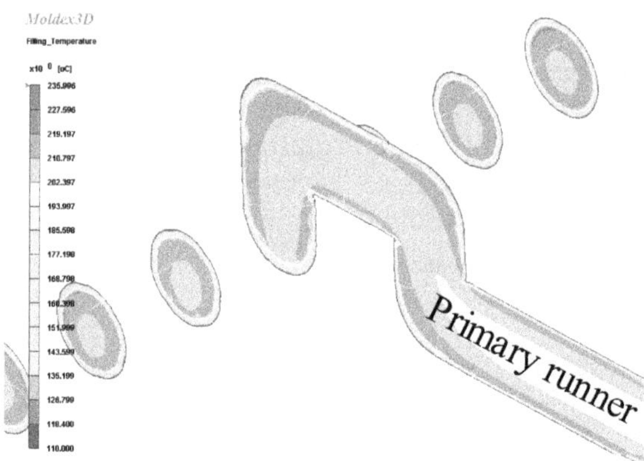

Figure 11.24 Three-dimensional simulation shows the melt rotation apparatus repositioning high sheared material to be symmetrical side-to-side.

11.7 Injection-Compression Molding

The injection-compression (IC) cycle is a hybrid molding process that incorporates both the features of injection and compression molding. Although the actual details of the process can be quite complex, from the standpoint of simulation it can be described as shown in Fig. 11.25.

Figure 11.25 shows the general nomenclature required to understand the injection-compression molding process. Initially, the tool is closed sufficiently such that

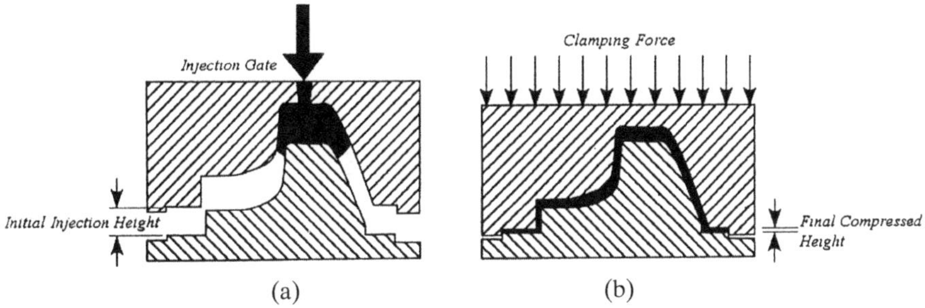

Figure 11.25 Schematic of the injection compression (IC) molding process.

the shear edges have engaged and the cavity is closed to prevent leakage. This state is illustrated in Fig. 11.25a. Here, only a portion of the tool is shown where the shear edges and support surfaces have been removed for clarity. At this point, the *Initial Injection Height* is defined as the total gap height between the mold halves. The polymer is then injected through the *Injection Gate* until the volume is sufficient to fill the finished part. Then the mold is closed to the Final Compressed Height as shown in Fig. 11.25b.

As evidenced from the prior discussions, there are several substantial differences between injection-compression cycle and the standard injection molding process. The first difference in the process is that the tool is now initially separated at a fixed gap to ensure that the shear edges are engaged to seal the mold cavity. The distance that the mold is backed-up from the final compressed height is referred to as the *Backup Height.* Using the notation from Fig. 11.25, the backup height can be formally defined by Eq. 11.21. Typical values for backup height in injection-compression molding range from 1–10 mm.

$$\textit{Backup Height} = \textit{Initial Injection Height} - \textit{Final Compressed Height} \quad (11.21)$$

The backup height can be highly instrumental in the formation/avoidance of knitlines. Unlike standard injection molding, the injection-compression molding cycle benefits from a compression phase. Standard injection molding is characterized by a radially expansive flow field. However, compression molding produces an equibiaxial flow regime that differs from the flow during injection. Accordingly, the flow field can be substantially altered by the amount of compression included in the cycle. Also, because the thickness of the cavity changes during the mold closing process, knitlines can be avoided through proper backup height selection

Modern simulation techniques can study the location, number, and size of gates as well as their individual flow rates. If the gate cross sectional area is too small, then problems with viscous heating or short shots can occur. Also, if fiber reinforced polymers are considered, a undersized gate can cause detrimental fiber damage during injection. For multiple gates or edge gates, there also exists the possibility for non-uniform injection rates. If the runner system is not properly designed, then individual gates or regions can experience different flow rates during injection. This can, subsequently, lead to knitlines or improper mold filling patterns.

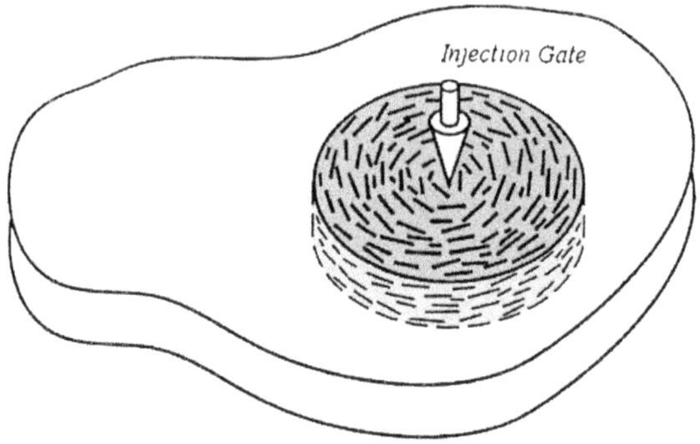

Figure 11.26 Fiber orientation field from radially expansive flow during injection molding.

There also exists, in injection-compression molding, an initial fiber orientation in the material after the injection phase. In standard compression molding processes, the initial charge material may be manufactured to have a prescribed initial fiber orientation field. However, in the injection-compression molding process, the flow during the injection phase tends to orient the fibers in the circumferential direction as shown in Fig. 11.26. Because the injection process produces a radially expansive flow outward from the gate, the fibers show a proclivity for orientation in the tangential direction. However, as obstacles in the mold cavity alter the flow field, the fiber orientation becomes more complex.

11.7.1 IC Molding of Thermoplastic Materials

Injection molding has become one of the most important polymer processing operations in the plastics industry because of its ability to efficiently produce near net-shape components with high dimensional accuracy. Thermoplastic polymers can be, however, intrinsically difficult to process due to their pvT and viscosity characteristics. In some applications, such as optical disks and high-precision molding, standard injection molding is unable to satisfy production requirements. Especially when residual stress limits are imposed on the final part, an alternative method of processing is often warranted.

A surrogate processing method is the injection-compression molding process. This process, which is sometimes referred to as stamping, coining or hybrid molding utilizes a discrete two-stage process. The IC molding process initially maintains the cavity with an enlarged volume during the injection phase. This ensures that lower pressures and stresses are generated during the injection cycle. At a stage during or after filling, the mold is physically forced together in a compression operation that brings the cavity to the final part thickness. During the compression phase, the resin is forced into the unfilled regions of the cavity by a more uniform compressive packing pressure across the cavity. This results in a more homogeneous part in terms of physical properties and molded-in stresses.

Historically, the in-mold design for injection-compression process has been developed on a trial-and-error basis. By relying heavily on experience and intuition this method has been limited and inefficient when applied to the molding of larger, more precise, and more complex parts. However, current CAE simulation tools can provide a viable alternative to experimental design. Possible molding problems can be detected early in the design phase and product development cycles can be shorted significantly.

11.7.2 IC Molding of Thermoset Materials

Although most of this chapter is written specifically towards the injection molding of thermoplastic materials, a sufficient body of research exists to enable thermoset materials to be simulated as well. As with thermoplastic materials, the injection-compression molding of thermoset materials is characterized by a distinct two-phase cycle. Initially, the material is injected into the separated cavity and then subsequently compressed to fill the entire part. However, unlike thermoplastics, thermoset materials are more often fiber-reinforced materials.

A typical bulk molding compound (BMC) or vinyl ester material used in the injection molding process will have glass loadings from 15–30% by volume. While the material behavior is considerably different for these thermoset materials, the mold filling can be accurately computed via a combined finite element/control volume approach. Current state of the art software [26–27] is capable of fully solving the mold filling, location of knitlines, molding pressures, clamping forces, fiber orientation, curing behavior, thermal history, and resulting shrinkage and warpage of the final part.

Figure 11.27 illustrates the mold filling pattern for an injection-compression molded valve cover. This filler-filled thermoset composite was selected for IC (integrated circuit) molding to enable higher levels of automation and more consistent product

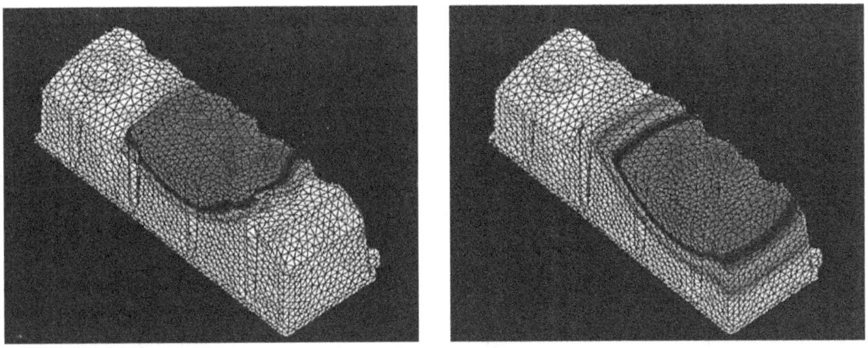

(a) (b)

Figure 11.27 Mold filling for an injection-compression molded valve cover for (a) a centrally located gate and (b) an offset gate.

quality. The figure shows the mold filling pattern for a centrally located gate. This gating scenario produces a knitline on the downstream side of the oil fill hole and leads to unacceptable material properties. By modifying the location of the injection point in conjunction with a change in the backup height, acceptable processing can be accomplished. However, the flow inside an IC package is often three-dimensional with material crossing the opening of leadframe. The true 3-D approach is, therefore, more accurate and reliable than the conventional $2^1/_2$-D approaches.

11.8 Molding Process Optimization

Most of the CAE tools and techniques introduced to this point have been concerned with analyzing and designing injection molded parts. However, for the molding engineer, the grand challenge problem is related with finding the optimal molding configuration. Confronted with requirements on part design, choices of material selection, limitations on processing conditions, and requirements on tooling choices, the molding engineer has a daunting task to optimize the process. Traditional methods utilize design heuristics and some advanced CAE simulation tools to discover problems before they occur, but to fully optimize the design and process is a formidable task.

11.8.1 Optimal Gating

Recent attempts have been made to automatically determine optimal gating configurations for injection molded parts. However, to determine *optimal* conditions, one must determine what parameters and measurements are critical process variables. Additionally, rigorous mathematical methods must be implemented to ensure repeatability of optimization routines and to guarantee convergent solutions.

A recent method proposed by Ye and Wang [36] compared various methods for optimizing gate location in injection molding. They discussed and compared a simulated annealing approach with a more efficient genetic algorithm to ascertain optimal gate location. Their technique was applicable to single and multi-gated parts and strived to balance flow paths while minimizing warpage across the entire part. While these were specific requirements to this precursory research, the generalized method is capable of optimizing any program variables.

11.8.2 Active Process Control

Obviously as simulation techniques improve and the results become more accurate, the next logical step is to attempt to use these simulated results in a more active role in process control. As injection molders realize, a consistent optimization strategy for machine operation and production control must be implemented to ensure quality. While advanced statistical process control (SPC) has been used to troubleshoot processes, a more proactive means would be to utilize CAE along with experimental feedback to aid in direct process control.

The basic idea of a "knowledge-based system" is that a body of information and experience can be captured and utilized to improve the injection molding process. By reducing the dependency on highly experienced personnel, the goal of active process control is to standardize the set-up and troubleshooting. Some researchers [31–32] have implemented these expert systems in preliminary feasibility studies. In essence, the molding conditions determined by the set-up phase are used to develop a robust processing window that produces acceptable parts. Automatic "design of experiments" (DOE) are completed and relationships between CAE results and molded parts are correlated. The outcome from such DOE produces a set of guidelines for processing variables and part quality. The optimization routines can then ascertain the most robust set of processing conditions to yield acceptable parts.

Furthermore when implemented in an active control loop, an expert system can utilize feedback from downstream quality measurements to modify machine set points to achieve optimal parts. These modification to set points can be conveyed back to the machine operator via control panels or, optimally, directly back to the injection molding machine itself. By incorporating the vast potential from CAE results along with physical results from the molding cycle, an expert system for the injection molding process is realizable.

11.9 Conclusions

Computer-aided engineering for injection molding has evolved substantially from its primitive origins. What was once a "black art" for designing molds and processes, has become firmly founded on fundamental science so that injection molding part and process design can now be accomplished through rational scientific laws rather than solely on experience. Continual improvements and advancements in software tools and computer hardware have made CAE an attainable tool for any size molding business. What began as a tool for only the largest mold shops has become mainstream software that even the smallest "mom-and-pop" molding shop can utilize and afford.

With increasing requirements to decrease costs, improve quality, and reduce cycle time computer aided engineering is paramount for profitability. However, blind reliance on simulation should not categorically replace engineering. Rather, CAE should continue to be used to solve processing problems as well as to help achieve the goal of designing "flawless" products. With current research centered towards improving the integration between design, production, and quality, the original goal of CAE in injection molding is as clear today as in its beginning.

Acknowledgements

The authors would like to express their gratitude to all those who contributed information and results to help make this chapter possible. Specifically, the figures and results provided by Towfiq Gangjee of Square D Company, Peter Foss of General Motors, Murali Anna-Reddy and Peter Kennedy of Moldflow Corporation, Paul Filz of Simcon Kunststofftechnische Software GmbH, and The Madison Group: PPRC are greatly appreciated.

References

1. Wang, K. K., Hieber, C. A., Proc. Of The Int. Conf. On Prod. Eng., Tokyo, Feb (1976)
2. Hieber, C. A. and Shen, S. F., *J. Non-Newtonian Fluid Mechanics*, (1980) 7, p. 1–32
3. Wang, V. W., C. A. Hieber and K. K. Wang, *ANTEC*, (1985), p. 826–829
4. Lee, C. C., F. Folgar and C. L. Tucker, *J. Eng. Ind.* (1984) 106, p.114–125
5. Tadmor, Z., E. Broyer and C. Gutfinger, *Polym. Eng. Sci.* (1974) 14, p. 660–665
6. Broyer, E., C. Gutfinger and Z. Tadmor, *Trans. Soc. Rheol.* (1975) 9, p. 423–444
7. Hele-Shaw, H. S., Proceed. Royal Inst. (1899) 16, p. 49–64
8. Wang, V. W., C. A. Hieber and K. K. Wang, "Applications of Computer Aided Engineering in Injection Molding," Ed. L. T. Manzione," Hanser Publishers, Munich, 1987.
9. Osswald, T. A., Ph.D. Thesis, Dept. of Mech. Eng., University of Illinois, (1987)
10. Rider, W. A., and D. B. Kothe, *J. Comput. Phys.* (1998) 141, p. 112–152
11. Gao, D. M., *Int. J. Numer. Meth. Fluids* (1999), 29, p. 877–895
12. Chang, R. Y. and W. H. Yang, *Int. J. Numer. Meth. Fluids* (2001) 37, p. 125
13. Folgar, F. P., Ph.D. Thesis, University of Illinois at Urbana-Champaign, (1983)
14. Folgar, F. P., and C. L. Tucker, *J. Reinf. Plast. Comp.*, (1984) 3, p. 98
15. Jackson, W. C., S. G. Advani and C. L. Tucker, *J. Comp. Mat.*, (1986) 20, p. 539
16. Advani, S. G., Ph.D. Thesis, University of Illinois at Urbana-Champaign, (1987)
17. Advani, S. G. and C. L. Tucker, *Polym. Comp*, (1990) 11, p. 164
18. Verleye, V., and F. Dupret, *Proc. ASME WAM*, New Orleans, (1993)
19. Cintra, J. S., and C. L. Tucker, *J. Rheol.* (1995)
20. Tait, P. G., Physics and Chemistry, Vol. 2. Part IV (1889)
21. Buchmann, M., M. S. Thesis, University of Wisconsin-Madison, (1993)
22. Yu, H. G., and R. Thomas, U.S. Patent 6096088 (2000)
23. Fan, Z., et al., ANTEC, (1999)
24. Beaumont, J. P., and J. H. Young, "Mold Filling Imbalances in Geometrically Balanced Runner Systems," J. Injection Molding Tech. (1997) 1, 3
25. Chien, C. C., C. C. Chiang, W. H.Yang, Vito Tsai, and David Hsu, ANTEC (2002)
26. Beaumont, J. P., Runner and Gating Design Handbook (2004) Hanser
27. Kennedy, P., Flow Analysis of Injection Molds (1995) Hanser, Munich
28. VerWeyst, B. E, C. L. Tucker III, P. H. Foss and J.F. O'Gara, International Polymer Processing, (1999) 14, 4, p. 409
29. VerWeyst, B. E, C. L. Tucker III and P.H. Foss, International Polymer Processing, (1997) 12, 3, p. 238
30. Christensen, S. K., et al., ANTEC, (1997) p. 782–789
31. Cadpress, The Madison Group (1996), Madison, WI
32. Hétu, J.-F., et al., Polym. Eng. Sci. (1998) 38, p. 223–236
33. Talwar, K, et al., ANTEC (1998)
34. Han. R. and Gupta, M, ANTEC (1999)
35. Capellmann, R. and Haberstroh, E., ANTEC (1999)
36. Ye, H. and Wang, K. K., ANTEC (1999)
37. Pantelelis, N. G., Kanarachos, A. E., and Procos, M. M., ANTEC (1999)
38. Speight, R. G. and Thomas, A. R., ANTEC (2000)

12 Process Troubleshooting

J. Wickmann*

Troubleshooting has always been a reactive process, where one or more processing conditions have moved out of adjustment. By definition a problem has to occur in order to be fixed. Part of troubleshooting is often that the problem is not fixed permanently. It often occurs again in a few hours, the next shift, or a few days or weeks. For many, such a process for solving problems works quite well; however, in the injection molding arena, processing, parts, materials, standards, and people are constantly changing, leading to new troubleshooting approaches, which result in an increase of process control and product quality. This chapter presents troubleshooting methods, covering the traditional cause and effect approach, as well as technically advanced techniques that include process monitoring, data collection, design of experiments, and statistical process control. Statistical process control was discussed in more detail in the Chap. 11.

12.1 Introduction to Troubleshooting

Traditional troubleshooting approaches require that a set-up sheet be completed after the injection molding process is developed and before going into production. The set-up sheet has been the result of trial and error and a small amount of sampling. Its goal is to identify the critical plastic conditions that affect the part characteristics and give guidelines for injection molding technicians. Troubleshooting guides provide some information on which variables to change for different problems, such as splay, part distortion, variation in dimensions, burning, and others. The following eight points are a checklist recommended for the early stages of process set up:

1. Check the machine and mold set up according to a set-up sheet.
2. Check the material dryer.
3. Check transfer point if possible.
4. Check cushion.
5. Check the mold steel temperatures.
6. Check material melt temperature.
7. Check the injection time.
8. Check the cycle consistency.

* With contributions from Treasa Springett and Raghu Vadlamudi.

The plastic injection molding industry has changed considerably in a very short period of time. This involves materials, machines, sensors, data collection and analysis, product design, part failure analysis, plastics additives, machine cycle times, and others. The automotive and medical markets coupled with the globally competitive economy has driven much of the injection molding technology and troubleshooting. Troubleshooting historically has been a reactive process, after the fact when the damage has already been done. Much like everything else the process of troubleshooting has had to change. Troubleshooting, whether it is for short or long production runs, must be proactive in order to be effective.

The injection molder would historically be given an idea or possibly a mold and be told to produce a given quantity of parts. It was the molder's responsibility to make the system work. The success of being able to run good parts depended on compounders, material suppliers, mold builders, machine operators, material handlers, designers, and the like. Each one of them had their own concept of how best to produce the part and had their own agenda for running their business. Each one of them essentially does what is most expedient for them. In the final analysis the molder has a solution to producing parts, but not the optimal solution. The only option left for the molder is sometimes the traditional form of troubleshooting, after the fact, to fix the system and run as many good parts as possible and cope with nonconforming parts. In many cases the damage was already done by some unsuspecting player, not intentionally, and not knowing the impact of their decisions.

Additional problems arise as a result of the molder having to do more than just molding and shipping parts. Packaging, assembly, painting, and a host of other post-molding operations cause additional problems. Many molders are small shops (fewer than 250 employees) and have not had access to costly solution resources available. There have been a number of cost-effective resources (books, periodicals, software, interactive training tools, design tools, and others) available only in the recent past.

The level and extent of product sophistication has changed. Customers have products that require thinner walls, reduced cycle times, greater chemical resistance, and improved mechanical, physical, and electrical properties that no one had previously associated with plastics. With requirements to meet three to four defects per million opportunities, the entire process must also mature. Injection molding has the ability to produce some very sophisticated parts using very sophisticated technology. In order to remain competitive one has to see the injection molded part, the injection molding process, and molding materials as essential components of a complex integrated system.

Data collection and analysis (quality, design of experiments, statistics, etc.) has moved from being a university topic of curiosity for a few to being absolutely critical for lean manufacturing in a global environment.

The process of benchmarking comes along with the data collection and analysis. What were the key design criteria, material specifications, product requirements, and injection molding processing variables last quarter? Where are those same variables today? The technology has allowed continuous process improvement to become a common part of our everyday vocabulary. Most of the information available indicates that the trends in technology, such as greater speed, better software, sensors, memory, and access, are going to continue to mature.

The ability to quantify sources of variability is an outgrowth of the benchmarking process. If one assumes a given distribution there will be assignable and unassignable sources of variation. If one assumes an initial spread of data of +/– 3 sigma, one begins logically to attack all major sources of variability and to reduce as many sources of variation as possible. Using 6 sigma three to four defects per million (ppm) opportunities becomes a very real goal and provides competitive opportunities and advantages.

The new materials, part design, metallurgy, screw design, nonreturn valve design, drier technology, mold design, mold temperature controllers, ejection systems, cooling line layout, machining systems, rapid prototyping, testing (mechanical, physical, electrical, and chemical) and associated equipment, sensor technology, control technology, artificial intelligence, mold filling software, and product development, coupled with globalization, all contribute toward a constant improvement of injection molding technology. This is coupled with the fact that the price of technology has dropped. The technology that is available has to be justified at one point in time by the production of somewhat costly engineered parts. The cost of the available technologies is now such that commodity items require sophisticated technology to maintain competitive position, to say nothing of quality.

Troubleshooting in the future is going to require, at a minimum, a core of long-term, highly skilled employees with a sophisticated knowledge base. This knowledge base may well be imbedded in the software and use varying amounts of artificial intelligence and fuzzy logic.

There are at least 33,000 plastic materials available with roughly 1500 new plastic materials being developed annually. Growth of compounding makes an infinite range of tailored resins possible. New markets continue to open up for plastics. New methods of molding/shaping plastics continue to develop. Some of these new processes directly compete with injection molding for market share of product and processing.

These factors all contribute toward the continuous change in troubleshooting. One's approach to troubleshooting must therefore also evolve. It has to encompass the whole injection molding process. It must be seen as a system that unites everything from design, materials selection, mold design, processing, and automation. A seamless integrated approach needs to be recognized. In order for lean manufacturing to take place successfull, troubleshooting must be more robust, accurate, precise, and timely.

Traditional troubleshooting has served a useful purpose [1]. Companies today have increasing amounts of information available to them to aid in decision making. Because competitors have access to the same data, the companies that transform data into knowledge and then organize and use that knowledge to their benefit have the advantage. An organization's ability to learn faster than its competitors may be its only sustainable advantage [1].

The molder is constantly being confronted with new product ideas, new production methods, new material combinations, and new processing technologies that can only be accommodated within complex and integrated production systems. There are eight* fundamental steps or stages involved in applying what is called the *breakthrough strategy* to achieve six sigma performance in a process, division, or company. The eight stages [1] are recognize, define, measure, analyze, improve, control, standardize, and integrate. These eight steps or stages should be applied anywhere on the sigma journey. The relative gains or improvements will be more dramatic at the three sigma level than at the five sigma level.

If you are currently producing parts at about the three sigma level you can expect roughly 66,810 defects per million opportunities. Does your current troubleshooting process benchmark; do you know where you stand? Six-sigma is producing at roughly 99.99966 percent error free or 3 to 4 defects per million opportunities. Traditional process troubleshooting will not get you, your product, or your process to six-sigma. When was the last time you did a gap analysis?

Table 12.1 represents the values of a Six Sigma organization [1]. There are 25 issues addressed on the left side. Each of those issues is intimately related to the process of troubleshooting. People issues and troubleshooting are dominant under the heading of Six Sigma.

Newer approaches to plastics manufacturing involve benchmarking. This powerful tool helps organizations compare their processes with those of their competitors; however, it is not a one-time event. Benchmarking that does not result in process improvements is a waste of time, money, and energy [1]. Quantitative benchmarking allows companies to evaluate performance in every division on a level playing field using standardized indeces of performance and capability and then compare performance both between divisions and with other companies considered "Best-in-Class."

The American Society for Quality (ASQ) Futures Team [2] identified eight key forces likely to have the greatest impact in the immediate future for service and manufacturing sectors. The eight key factors were:

- Partnering—Superior products and services will be delivered through partnering in all forms, including partnerships with competitors. Over time this will force greater cooperation and efficiencies within businesses.
- Learning systems—Education systems for improved transfer of knowledge and skills will better equip individuals and organizations to compete.
- Adaptability and speed of change—Adaptability and flexibility will be essential to compete and keep pace with the increasing velocity of change.
- Environmental sustainability—Environmental sustainability and accountability will be required to prevent the collapse of the global ecosystem.
- Globalization—Globalization will continue to shape the economic and social environment.
- Knowledge focus—Knowledge will be the prime factor in competition and the creation of wealth.
- Customization and differentiation—Customization (lot size of one) and differentiation (quality of experience) will determine superior products and services.

* Five to eight, depending on the author or consultant.

Table 12.1 The Values of a Six Sigma Organization

Issue	Classical Focus	Six Sigma Focus
Analytical perspective	Point estimate	Variability
Management	Cost and time	Quality and time
Manufacturability	Trial and error	Robust design
Variable search	One-factor-at-a-time	Design of Experiments
Process adjustment	Tweaking	Statistical process control
Problems	Fixing	Preventing
Problem solving	Expert based	System based
Analysis	Experience	Data
Focus	Product	Process
Behavior	Reactive	Proactive
Suppliers	Cost	Relative capability
Reasoning	Experience based	Statistically based
Outlook	Short-term	Long-term
Decision making	Intuition	Probability
Approach	Symptomatic	Problematic
Design	Performance	Productivity
Aim	Company	Customer
Organization	Authority	Learning
Training	Luxury	Necessity
Chain-of-command	Hierarchy	Empowered teams
Direction	Seat-of-pants	Benchmarking and metrics
Goal setting	Realistic perception	Reach out and stretch
People	Cost	Asset
Control	Centralized	Localized
Improvement	Automation	Optimization

- Shifting demographics—Shifting demographics (age and ethnicity) will continue to change societal values.

These eight keys apply to the injection molding of plastics as well as to any other global industry. How well your company survives and implements process troubleshooting into the future is a function of how you interpret and implement some of the preceding key factors. Those key factors revolving around people, process, and culture are going to be critical for effective troubleshooting.

The next generation of injection molding troubleshooting is going to require well-trained technicians and engineers. Using the most sophisticated sensors, knowledge-based systems, and injection presses available, they will strive to produce consistently higher-quality user-friendly parts. In some specialized applications, designs must offer robust and sophisticated concepts and include almost one-of-a-kind inexpensive components, subassemblies, or assemblies using tailor made materials.

12.2 Troubleshooting Guide

Correct process control is critical for making identical parts to tight tolerances and meeting numerous quality standards. The trends in process control will be discussed in more detail in the Sec. 12.3. This section presents a troubleshooting guide that is specifically geared toward processes that have a velocity controlled first stage or mold filling stage. Velocity control means the hydraulic or plastic pressure limit allowed on first stage is higher than the hydraulic or plastic pressure at the position of transfer from first to second stage. Time and hydraulic pressure modes of transfer are not acceptable. If you are filling and packing the part on first stage, you should not use the troubleshooting guide presented in this section. The objective of a velocity-controlled mold filling stage is to have a wide operating window to make good parts by having the machine adjust (on the current shot) for normal viscosity changes. For our discussion first stage is used to fill the part 99% full by volume. At the end of this stage the part is not fully packed and may appear short or contain sink marks. The last area to fill does not see much pressure and the flow front may still be visible if the press is stopped at the end of the first stage.

12.2.1 Troubleshooting Table

J. Bozzelli

This sections presents a troubleshooting table that lists possible injection molding or process defects. The table presents possible causes of the given defects and suggests possible remedies.

- **Air bubbles.** See *Bubbles.*

- **Black specks or flakes in the part.** Carbon or other contamination that affects cosmetics and part performance, can be any shape or size from chunks to fine particles or flakes. Can be from resin degradation or contamination. See also *Color Mixing*.

Possible causes	Possible remedies
Poor screw or nozzle design	Pull and clean the screw barrel. Check for carbon build up behind the flights in the metering and transition sections of the screw. Most general purpose screws have this problem due to dead space behind the flights. Molding acrylic, polycarbonate (PC), styrene acrylonitrile (SAN) for lens application with a general purpose screw is unwise. Specify a correctly designed screw with no dead spots and melt uniformity. Check nonreturn valve, nozzle end-cap and nozzle tip for abrupt changes in flow path. See "*Screw design*."
Excessive nozzle length	Make sure the nozzle length is as short as possible and it is PID temperature controlled.
Faulty or poorly designed hot runner flow channels	Check for open or incorrect thermocouple readings. Repair and replace any that are working on % or variacs. Redesign hot runner. Check watt density and location of heaters. Check location of thermocouples.
High melt temperature; hot spots in screw or barrel	Check melt temperature via the hot probe technique or an appropriate IR sensor. Few IR sensor work correctly. Adjust temperatures only if necessary. Check screw and barrel for hot spots. Are any zones overriding their temperature setting? If so raise that zone or the one before it. Lower back pressure. Slow screw speed RPM.
Contamination in virgin resin	Check virgin resin using a white pan, spread 1.5 kg of resin over a 1500 cm^2 area, and inspect for 5 minutes under appropriate lighting. If there are black specks note if they are on the surface or imbedded in the granules. Inform resin supplier if embedded.
Contamination in reprocessed or reground resin	Check screw design, material grinding and handling procedures. Discard or resell the material
Excessive, high temperatures or long residence times at melt temperatures	Use 25–65% of the barrel capacity. Using lower than 25% provides long residence time for the resin or additive to degrade. Go to smaller barrel if possible. Check heater band function, nozzle should be PID controlled, check screw for high shear dispersion mixing elements. Repair/remove.
Excessive fines	Remove fines before processing; a must for lens applications.
Special note:	It is rare that a purging compound will solve a black specking problem if the cause is screw design—80% of the time it will be wiser to pull the screw and clean it. Screws should have a high polish.
Vented barrel	On vented barrels often the vent is poorly designed with a dead space or hang-up area. This is a hard to get to area for mechanical cleaning, but it must be cleaned.

- **Black streaks.** See also *Black specks* and *Color mixing.*

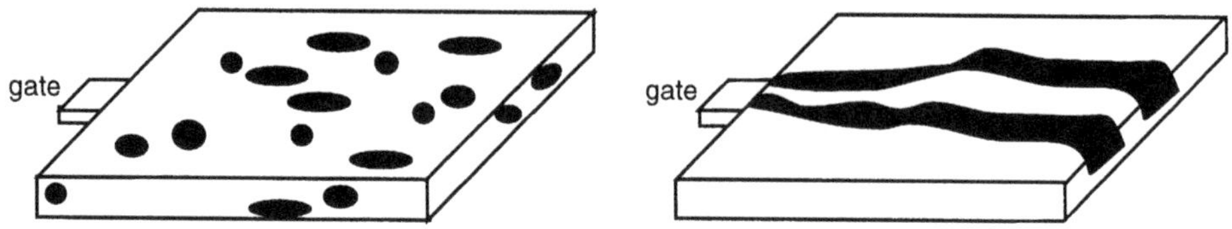

Black specks (left) and black streaks (right) [10].

Possible causes	Possible remedies
Sharp angle or corner at the gate	Polish gate area, radius sharp corners, change gate type, or enlarge gate
Hot runners	Check for proper temperature control, burnt out heating elements, open thermocouples. Check depth of probe relative to gate surface, account for thermal expansion. Check for hang up areas or dead spots.
Poor nozzle temperature control	Nozzle temperature is notoriously poorly controlled. Placement of the thermocouple should be one third the distance of the nozzle from its tip if he nozzle is more than 3″ (75 mm) long. For 3″ (75 mm) lengths and shorter it is OK to have the thermocouple imbedded into the hex on the nozzle body. For longer nozzle bodies the thermocouple should be underneath the heater band via a butterfly-type thermocouple. The nozzle should be PID temperature controlled, use of a % or variac is unacceptable. Do not allow the thermocouple to be attached via the screws on the clamp for the heater band. Clean out the nozzle, inspect for blockages.
Sharp angle on screw tip or broken nonreturn valve	Check entire flow path of plastic looking for burrs, sharp corners, grooves, and the like, in any of the plasticating components: screw, tip, check ring, flights, and so on.
Contamination	Check for foreign material resin.

- **Blisters on the part's surface.** See also *Delamination*. Thin film of plastic that bubbles up from the surface.

Possible causes	Possible remedies
Gas traveling across surface during fill or pack	Check for moisture, trapped air, or excessive volatiles. Also, check for excessive decompression or suck back
Trapped air due to inadequate venting	Check number of vents, check vent depth, compare with manufacturer's recommendations. Clean all parting line and core pin vents

Possible causes	Possible remedies
Trapped air due to flow pattern	Perform a short shot sequence changing transfer position or shot size to make various-sized short shots from 10 to 95% full parts. Note flow path for any back flow or trapped air in blind ribs. Jetting also traps air. Changing first stage pressure cannot do this short shot sequence correctly!
Trapped air due to excessive clamp tonnage	Reduce clamp tonnage, especially for small molds: Mold should take up 70% of the distance between the tie bars.
Trapped air due to decompression or suck back	Minimize decompression or screw suck back, especially with hot runners or hot sprues. Be careful to maintain proper nonreturn valve function.
Trapped air due to low L/D screw	Especially for general purpose screws with 18:1 L/D or lower. Raise back pressure to 1000 to 1500 psi melt pressure.
Degraded resin or additive package	Check melt temperature for proper range as recommended by resin supplier. Minimize residence time by shortening cycle, take it off cooling time first. Try virgin only, a new lot

- **Blooming.** This can be a solid powder, wax or liquid buildup on the surface of a part. An additive migrating to the surface, immediately or over a period of time, usually causes this. Difficult to eliminate with processing, usually requires a different additive or formulation. See also *Mold build up*, *Blush*, and *Surface finish*.

Possible causes	Possible remedies
Additive migrating to the surface of the part	Trial different lots or grade. Trial slower injection rate. Seek a different additive or formulation.
Inadequate venting	Clean vents. Check vents by bluing or plastigauge. Add more vents.
White powder on part or mold surface	If the resin is acetal it may be formaldehyde depositing on the mold or part. Dry the acetal before processing.
Melt temperature too high or low	Try higher and lower melt temperatures. Usually not much of a benefit here, wasted time.
See also *Mold build up*, *Surface finish*, and *Blush*	

- **Blush.** A haze or dullness on the surface, often seen as rings or half circles near the gate; this is usually worse with impact modified resins such as HIPS or ABS. One of the more difficult cosmetic issues to resolve. The problem is a different morphology of the impact modifier. This develops as shear rate changes at the flow front during fill. See also *Gate blush*, *Surface finish*, and *Gloss*.

Blush (*continued*)

Possible causes	Possible remedies
Improper temperature control of nozzle	PID temperature control nozzle and tip, make sure temperature is uniform over length of the nozzle. Minimize length of nozzle. %, or variac heating of the nozzle, is not acceptable. See *Black specs* for correct thermocouple placement. Trial higher and lower nozzle temperatures to see if blush changes.
Sharp corners at the gate	Break the corners, provide a minimal radius to reduce shear at the sharp corner
Improper gate type	Change gate type for less shear or sharp corners
Improper gate location	Change gate location
Size of the gate	Decrease land, increase area of the gate. This will change gate seal time!
Fast injections speeds	Change injection velocity. May need to profile injection; use no more than 2 to 3 velocities. Slow fill as plastic enters gate area then increase velocity. If possible use one slow velocity to allow for repeating in other machines.
Too high or low melt temperature	Target middle of melt range specified by resin producer, try 25°F (12°C) higher then same amount lower.
Too low or high a mold temperature	Try high end and low end of recommended mold temperature range suggested by resin supplier.
Trapped gas	If blush is near an area of non-fill or short, check venting. Pull a vacuum on the mold cavity during fill.
If just in a small area	Look for excessive ejector speed, part sticking remedies or hot spot on the mold surface. See *Sticking*.

- **Brittleness.** If parts crack easily or upon cooling; causes can be from over- or underpacking and often from molecular weight degradation or contamination.

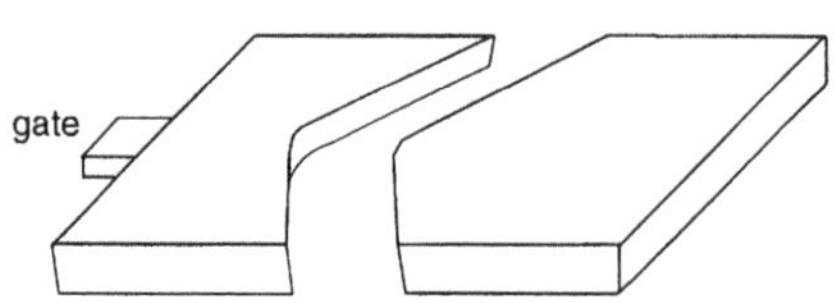

Degraded part tends to be brittle and break easily [10].

Possible causes	Possible remedies
Defining the brittle areas	First find out if the brittleness is throughout the part or localized in one area. If localized it may be a packing issue. If throughout the entire part look for molecular weight degradation via melt flow rate testing of the material before and after molding. If brittle at lower temperatures check t_g of material for proper resin selection.
Over or underpacking the gate	If brittleness is near the gate, do a gate seal analysis.* Check if brittleness changes with or without gate seal. Running without gate seal allows plastic to discharge and may cause underpacking or excessive residual tensile stress. Running with gate seal may overpack the gate and lead to brittleness due to overpacking, too many molecular chains packed too closely together, and they do not have space to get closer with cooling. This leads to retained compressive stress.
Molecular weight degradation due to hydrolysis	Check to see if this polymer can undergo hydrolysis. This is where minute amounts of water react with the polymer in the barrel of the machine. Water can act as a pair of scissors cutting the long polymer chain into short segments. This leads to lower melt viscosity. Therefore check moisture level. Check the melt pressure at transfer from first- to second-stage and see if it is lower than normal or "good" runs.
Molecular weight degradation due to excessive temperatures	Check melt temperature, via the hot probe method or appropriate IR temperature sensor. Adjust only if necessary. Check barrel heats and duty cycles. Check for temperature zone override. Check for proper screw and barrel wear and condition. Check for cracked, chipped, or restrictive nonreturn valve or check ring. Check residence time in the barrel. Add some colored granules to the empty feed throat and count how many shots it takes to see the color come out. Multiply the number of shots times the cycle time for the true residence time. Note also if it comes out in streaks or blended. If streaks occur there are dead spots in the barrel. See *Screw design*.
Too much degraded regrind or contamination	Check amount and quality of regrind. Check for contamination. Run 100% virgin and check properties.
Part not conditioned properly	Certain resins, especially nylons, need to regain moisture to achieve full physical properties. Have the parts been conditioned according resin manufactures specifications?
Too much retained orientation or stress	Check gate location and for orientation effects. Check part design for sharp corners, appropriate nominal wall, and its uniformity

* To determine gate seal start the experiments with a very long pack and hold or second-stage time that ensures time enough for the gate to seal. Continue molding keeping the cycle time constant, but reducing the pack and hold or second-stage time, 1 or 2 seconds, and adding the same amount of time to the cure time. By switching time from pack and hold to the cure time the cycle should remain constant. The gate seal time corresponds to the pack and hold time when the part weight starts to drop.

- **Bubbles.** Bubbles are either trapped gas or vacuum voids. Before proceeding you must determine which you have. *Test*: On a freshly molded part, gently and slowly heat the area containing the bubble. A hot air gun is best, a small lighter next, and a torch if you know what you are doing. As the resin softens, the bubble will expand or contract. If it expands it is trapped gas; if it contracts or pulls the side wall in it is a void.

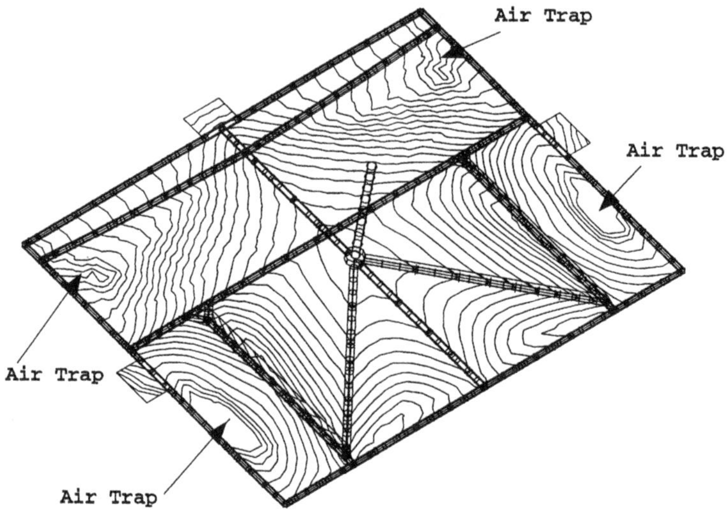

Air trap locations indicated by the computer-predicted melt-front advancements [10].

Possible causes	Possible remedies
Trapped gas	Progressively short shoot the mold using shot size or position transfer to make 10, 20, 40, 60, 70, 80, 95% full parts. Note if any show the plastic flow front coming around on itself. Is there a racetracking effect? This includes jetting. Note if ribs are covered before completely filling. This test cannot be done correctly if you reduce first-stage pressure or velocity. You must be velocity controlled for first stage. The concept is to find where the gas is coming from and eliminate its source. Vent the tool properly or use porous steel to eliminate gas traps. Jetting can cause gas, air, to be trapped. Check for moisture, steam is a gas. Change gate location. Pull a vacuum on the mold cavity during fill.

Possible causes	Possible remedies
Vacuum voids	Thin up the nominal wall in this area by adding steel to core out thick sections. Change gate location; fill thick to thin. Voids occur upon cooling in thick sections. You need to pack more plastic into the cavity with: (1) consistent cushion, make sure you are not bottoming out the screw, (2) higher second-stage pressures, (3) longer second-stage time, (4) very slow fill rates, (5) use counterpressure, (6) open the gate for longer gate seal times to allow more packing during second-stage, or (7) increase the runner diameter. Also, raise the mold temperature significantly and/or eject the part sooner. These will allow the outside walls to collapse more upon cooling. Reduce melt temperature.

- **Burns.** Black carbon soot, or white residue, glossy areas on a part. Burns are a result of dieseling, which can lead to pitting on the mold surface in that area. Due to this mold damaging dieseling do not continue molding until a remedy has been made, severe tool damage may occur. See also *Black streaks*.

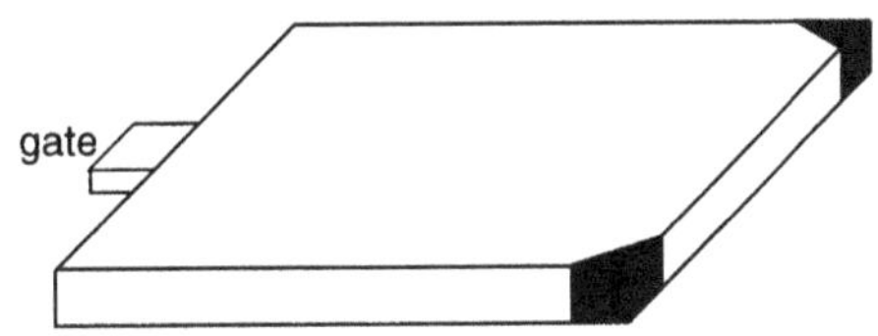

Burn marks [10].

Possible causes	Possible remedies
Trapped air or volatiles dieseling	Air trapped in a mold can be pressurized to thousands of pounds of pressure. Air with a little hydrocarbon as fuel will diesel when pressurized. Check for proper venting in this area. If there is a question of venting: Rub the suspect area with a hydrocarbon solvent and shoot a part; if carbon forms venting is not adequate. Glossy spots on parts are sometimes an indication of trapped gas.
Inadequate venting, plugged vents, hobbed over vents, undersized or not enough vents	The recommendation would normally be to slow down the injection rate to allow time for air to escape. This is unacceptable as this will add seconds to the cycle time, significantly decreasing profits. Including runners, 30% of the perimeter should be vented. Porous steel may be used to vent blind pockets. Use an ejector pin as a vent. Vacuum may also be applied to the mold to avoid trapping air and dieseling. Are the vents cut to the resin manufacture's specifications? Solve the venting issue, do not work around it. Dieseling will pit and erode the mold steel. The extra time, coupled with increased mold maintenance, will continually erode profits.

Burns (*continued*)

Possible causes	Possible remedies
Core pins not vented	Core pins must be vented as air is trapped as plastic flows around the core pin.
Excess volatiles	Resin may have a large percentage of mold release agents, lubes, or other additives that volatilize during injection filling. Try a lower melt temperature if allowed by the resin supplier. Try a different grade of resin with reduced additive concentration.
Excessive decompression	Decompression pulls air into the nozzle because nozzle contact force is off. Because nozzle contact force is "on" before injection, air is trapped in the nozzle and is pushed into the sprue, runner, and so on, possibly causing splay, or burns. This is especially important in hot sprues and runners. Reduce amount of decompression.

- **Burns at the gate.** See *Burns*; see also *Black streaks.*
- **Check ring.** See *Nonreturn valve.*
- **Cloudiness or haze in clear parts.**

Possible causes	Possible remedies
Excess additive or noncompatible additive package	Try a different lot, change material supplier.
Contamination	Clean barrel and screw thoroughly. Disassemble screw, barrel, and nonreturn valve. If there is carbon on the screw flights, see *Screw design*. Check resin for contamination in virgin lot or regrind or excessive fines
Nonuniform melt	Check melt quality by adding a small amount of compatible color granules. If they come out as streaks rather than uniform tint or pastel, the screw is not adequately mixing or providing melt uniformity
Incorrect mold temperature	Raise mold temperature. Lower mold temperature for amorphous PETG.
Moisture in resin	Check moisture content, function of dryer, residence time, dew point, and so on.
Worn or improper mold texture or surface	Check mold surface for deposits, plate-out. Check for appropriate polish.
Crazing	Check for stress in the area; for example stress whitening near ejector pins or places where the part sticks. Is there any contact with chemicals such as solvents or mold spray that attacks this type of plastic?

- **Color mixing.** Color streaks, marbling, dark streaks, and unmelted solids all fall into this category. The root cause in almost every case is poor screw design or, less commonly, a worn screw and/or barrel. The goal is to distributive mix the color concentrate (master batch) or liquid coloring agent *uniformly* into the base resin. Do not try to achieve dispersive mixing with injection molding screws. Please refer to Chapter 4 of this handbook for a full description of screws and their method of melting plastic. Others are available, but these two are the ones with which I am most familiar. It is not recommended to use powder colorant. It is cheaper, but the fine dust is detrimental to your lungs, machinery components, and good housekeeping.

Possible causes	Possible remedies
Back pressure incorrect	Increase back pressure to 1000–1500 psi. Inspect after 3 to 4 shots. Maximum back pressure to trial is ~5000 psi. Be careful using high backpressures for resins that tend to degrade such as acetal, or PVC. High backpressure cannot be used on vented barrels due to vent flooding. You will rarely have problems of color dispersion or melt uniformity with vented barrels and screws.
Melt temperature	Make sure the melt temperature is within the resin supplier's recommendation. Use the hot probe technique or appropriate IR sensor. Adjust if necessary.
Color concentrate carrier incompatible with base resin	Check the color concentrate carrier by calling the supplier and asking. It should be of the same resin family, usually only with higher melt flow, as the part. There is no such thing as a universal color carrier.
Incorrect colorant let down ratio	Weigh blend a few pounds to make sure there is the correct ratio of colorant to base resin. Too low or too high a concentration can cause problems. Do not use more than recommended as the cost of using extra concentrate will hurt profits and often part properties are compromised with too much colorant.
Melt uniformity	If possible, run natural resin and add a few colored granules into the feed throat. Wait until the color appears in parts. If the color comes out in streaks, then you do not have melt uniformity. If the color comes out as a uniform pastel (lighter shade), then you have melt uniformity. If you have a melt uniformity issue, specify a new screw; see "*Screw design*" below. Worn barrels and screws provide better mixing due to some back flow over screw flights.
Low L/D screw	Try higher back pressure as above. This will most likely help, but it will not alleviate the mixing problem. Check L/D if below 18:1. Forget processing fixes and use precolored resin or get a melt uniformity screw. The screw should not have mixing pins or a dispersive mixing element in it. Not recommended but a possible solution is to replace the existing nozzle with one fitted with mixing elements similar to a reversing helix. With mixing nozzles be careful of high backpressures. Also, some are difficult to purge. Pineapple mixers or dispersion disks are strongly discouraged because they are cheap but detrimental.

Color mixing (*continued*)

Possible causes	Possible remedies
Fast screw RPMs	Slow screw rotation speeds provide the best mixing. Use all but 2 to 3 seconds of the cooling or mold closed timer to get the screw back ready for the next shot.
Screw design	Standard general-purpose screws are known to provide unmelted solids within the melt. Do the melt uniformity test above. Melt uniformity screws are strongly recommended. These have a gentle mixing section near or within the metering section as well as filled radiused rear flight sections. Minimum recommended L/D is 20:1 for normal injection molding situations.
Miscellaneous	Use the maximum amount of regrind allowed. Regrind acts like additional coloring agent

• Color variation, instability, or color shift

Possible causes	Possible remedies
Melt temperature too high	Make sure melt temperature is within resin supplier's recommended range via hot probe or appropriate IR sensor. Adjust if necessary
Colorant thermally unstable	Run the colorant and base resin at the minimum temperature allowed by the resin supplier. Then run the colorant at the maximum allowed by the resin supplier. View identical parts in a light booth. If no light booth available, look at the identical parts at the same angle under florescent light, then sun light, and then incandescent light. Thermally stable colors will all look the same.
Incorrect coloring agent	Double-check the coloring agent for correct color, type, and base resin. Call the supplier if necessary.
Long residence time	Use 25 to 65% of the barrel capacity. Minimize cycle time to decrease residence time.
Virgin resin color shift or instability	Check color of different virgin lots; should be similar. Master batch or coloring agent cannot hold color tolerance if base resin shifts in color. Is natural resin thermally stable colorwise?
Gloss differential between parts	In an assembly, two mating parts often look like different colors due to gloss differences. Check gloss level for each part.
Surface finish of the mold	In an assembly, two mating parts often look like different colors due to surface finish differences. Check mold cavity for identical surface finish.
View angle different	In an assembly, two mating parts will often look different in color because the angle of viewing each part is different because of the assembly. Disassemble the components and look at them at the same angle.

Possible causes	Possible remedies
Colorant let down ratio not constant	Weigh blend a few pounds at the correct blend ratio and test for color accuracy. If this matches the color target, recalibrate feeder and make sure auger only turns during screw rotation. Auger feeders can get out of sync with screw rotation and it has been known to happen that operators incorrectly adjust feed rate.
% regrind not constant, or discolored	Check virgin only samples for color accuracy. Check level and quality or regrind.

- **Core pin bending or core shift.** Cores and core pins move due to non-uniform pressure on the core during packing.

Possible causes (if throughout the part)	Possible remedies
Nonuniform pressure on the core or core pin during first-stage or filling	Take second-stage pressure to minimum to make a 99% full shot. If minimum is more than 1000 psi plastic pressure take second-stage time to 0 seconds. Do not reduce first-stage pressure to do this test, you must be in velocity control for first-stage. Look at the part for evidence of core or core pin shifting. If the pin has shifted, try faster fill rates. Find the toolmaker and see if there is a way to lock or support the core during fill. Would a gate location change help? If with a 99% full short shot you do not have core shift, the problem occurs during second-stage. Contrary to popular opinion most core pins shift occurs during pack and hold.
Nonuniform pressure on the core or core pin during second-stage or pack, hold and cooling	Do a gate seal analysis (see p. 575, footnote). Inspect parts for core shift with and without gate seal. Set second-stage time from the best results of the gate seal study. Next, determine the best second-stage pressure by starting out with 5 shots with the lowest second-stage pressure that makes a full part. Inspect parts for core shift. Then produce 5 shots with 1000 psi higher melt pressure and check parts for more or less core shift. Repeat until second-stage pressure begins to flash the part or mold damage would result with that second-stage pressure. Choose pressure that provides least core shift.
Core or core pin shift due to mold misalignment	Check mold for proper alignment upon closing. Wax or silly putty will work. Check for thermal expansion mismatch. Correct as needed.

- **Cracking.** First determine if the cracking is localized at one area of the part or is it occurring throughout the part. See also *Brittleness* and *Sticking.*

Cracking (*continued*)

Possible causes (if throughout the part)	Possible remedies
Molecular weight degradation	Check part and resin for proper melt flow rate ASTM 1260; check for contamination, try a different lot, check regrind level and quality.
Improper resin	Check grade and type of resin
Improper type or amount of colorant	Check let down ratio and type of colorant carrier
Possible causes (if localized)	**Possible remedies**
Solvent, surfactant, or chemical attack	Inspect mold and part handling for possible contamination by oils, solvents, fingerprints, mold sprays, soaps, cleaners, etc. Remove if source is found.
Contamination	Check localized area for off color or foreign material. Check regrind for trace contaminates.
Exposure to radiation	Check the resin for UV, sunlight, or gamma radiation stability

- **Crazing.** Fine lines or small cracks usually confined to a small area.

Possible causes	Possible remedies
Solvent, surfactant, or chemical attack	Inspect mold, part handling, or assembly for possible contamination by oils, solvents, fingerprints, mold sprays, soaps, cleaners, etc. Remove if source is found.
Exposure to radiation	Check the resin for UV, sunlight, or gamma radiation stability
Part distorting upon ejection	Note location of craze, if near or at an ejector pin the part may be sticking in the mold; see *Sticking*

- **Cycle time too long**

Possible causes	Possible remedies
Thick or nonuniform nominal wall	Use minimum nominal wall and keep it uniform within the guidelines for good piece part design. Maximum variation for amorphous is 20 to 25%; for semicrystalline it is 10 to 15%. Thick is not stronger in plastic parts. Thinner with ribs will provide better performance, save plastic, and save cycle time. Teach these principles to your designer and clients.

Possible causes	Possible remedies
Slow filling	Increase injection rate, make sure first stage is velocity controlled filling 95 to 99% of the cavity.
Improper cooling	Note temperature in vs. out for each water line, maximum allowable difference is 4°F or 2°C. Make sure water flow is turbulent, Reynolds number >5000.
Robot movement too slow	Optimize or update robotics.
Mold movement too slow	Optimize mold opening speed and distance. Optimize mold closing and mold protection.
Long or excessive ejection strokes	Optimize speed of ejection, but not too fast to cause pin marks, if part sticks fix the problem do not compensate with extra ejection strokes.
Screw recovery too long	Start with the rear zone at the lowest temperature setting recommended by the manufacturer. Note recovery time. Raise rear zone temperature 15°F (7°C) and note recovery time. Repeat until rear zone is at maximum recommended temperature setting of resin manufacturer. Plot data and pick best temperature for minimum recovery time. Keep backpressure and rpm constant for this test.
Not enough ejector pin area	Add or enlarge ejector pins. Larger surface area will allow ejection earlier.
Excessive second stage time	Perform gate seal experiment to optimize second-stage time setting.

- **Dark streaks.** See *Black streaks*, *Color mixing*, and *Black specks.*

- **Deformed or pulled parts.** See *Sticking* and *Warp.*

- **Degradation.** See *Black specks.*

- **Delamination.** Can be on the surface or when a part breaks and you see layers at the break. See also *Blisters.*

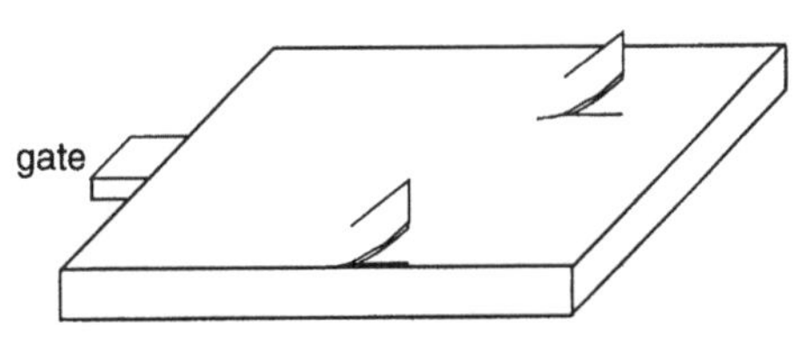

Delamination causes layer-wise peel-off on the surface of a molded part [10].

Delamination (*continued*)

Possible causes	Possible remedies
Contamination	Delamination is most often caused by incompatible resin contamination. Run only virgin, inspect virgin, try a new lot, change resin supplier, check color carrier for compatibility to base resin. Run natural resin without colorant. Check regrind.
High molecular orientation at the surface	Try a hotter mold, slow injection rate down for longer fill times. Review gate location.
Hot spot in the mold	Sometimes seen opposite a gate due to plastic impinging on the metal. The incoming polymer remelted the skin layer. Check cooling lines for appropriate Δ T (<4°F, 2°C) Reynold's number, >5000. Try lower mold temperature and slower injection speed.

- **Dimensional variations.** If this is the first trial of the mold, verify that the steel is dimensioned correctly, develop a stable process, and center the steel and the process to the center of the part specifications. Part size variations can be caused by one or a combination of the following: pressure gradients, improper measuring technique, different amounts of material in the mold, degree of crystallinity, cooling rate, amount of molecular chain orientation, and/or fiber orientation.

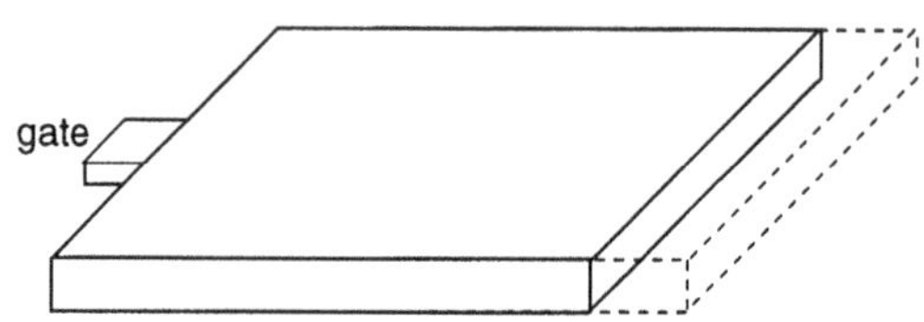

Dimensional variation is an unexpected change of part dimension [10].

Possible causes, parts too small	Possible remedies
First stage not consistent	Take second-stage pressure to minimum. If minimum is above 1000 psi plastic pressure take second-stage time to 0 seconds. Do not reduce first stage pressure to do this test, you must be in velocity control for first stage. Make parts 99% full and run 20 shots. Are they all consistent in size? If not, fix the nonreturn valve. If consistent go to next possibility.

Possible causes, parts too small	Possible remedies
Pressure gradient variations	Do a gate seal experiment. It is critical to know if you are using the correct amount of second-stage time. This time should not vary run to run. Do a gate seal experiment. It takes 10 to 20 minutes. Should be done on every mold before release to production. Running without gate seal allows plastic to discharge from the gate, and parts would be smaller. Running with gate seal, (longer second-stage times) packs the most plastic into a part and holds it in. Thus, larger parts may be possible. Careful not to overpressurize the gate. Also check resin viscosity and percentage of water if hygroscopic.
Too little second-stage pressure	Increase the amount of second-stage pressure, if flash occurs do not run with flash. Make sure machine switchover from first to second stage is less than 0.1 seconds.
Nonreturn valve malfunction or screw and barrel wear. Different amount of material in the mold	Check nonreturn valve function, check cushion repeatability. Try adding more decompression to help seat the check ring. Watch out for splay development with too much decompression. If nonreturn valve is leaking, replace with one that has a stepped angle for better sealing. If nonreturn valve checks out OK, then check barrel for roundness for the entire stroke distance. Repair if more than 0.003 inches clearance between barrel wall and flight. Check machines repeatability of second-stage pressure, it should be < +/– 5–10 psi.
Degree of crystallinity	If a semi-crystalline resin, check cooling rate by measuring mold temperature. A hot mold allows the plastic to cool slowly and yield higher crystallinity, which provides more shrinkage and a smaller part. Cooling must be identical run to run and throughout production. Check colorant for change in formulation: Colors affect crystallinity and therefore size of parts. Try the resin with a nucleating agent.
Too much post mold shrinkage	Not advised, but longer cooling time may hold the part to size. This is using the mold as a shrink fixture which is costly. Not recommended, but try cooling the part in water after ejection. Rapid cooling prevents some shrinkage.
Low melt temperature	Increase melt temperature. This may lower density and allow for more shrink.
Gate too small, gate seal occurs too quickly	If the gate is too small and gate seal occurs before the part is fully packed out, the part may be too small. Larger gate may be indicated. Obtain proper approval before modifying the tool. Adding a filler such as glass fibers can also reduce shrinkage. Again, obtain proper approval first.

Dimensional variations (*continued*)

Possible causes, parts too small	Possible remedies
Improper measurements	Check for identical conditioning of parts before measurements are taken. Have the parts been cooled the same and for the identical times after molding. Even laying down the part with the same side up consistently is important. Check measuring device for calibration.
Possible causes, parts too large	**Possible remedies**
Pressure gradient variations	Do a gate seal experiment. It is critical to know if you are using the correct amount of second-stage time. This time should not vary run to run. Should be done on every mold before release to production. Running with gate seal, longer second-stage times allow for keeping all the plastic in the mold and a more robust process. Try decreasing second-stage time to allow for gate unseal. Also, check resin viscosity and percentage of water if hygroscopic.
Too much second-stage pressure	Reduce second-stage pressure
Degree of crystallinity	If a semi-crystalline resin, check cooling rate by measuring mold temperature. A colder mold will prevent crystalline growth, thus making the part larger. Cooling must be identical run to run and throughout production. Check colorant for change in formulation: Colors affect crystallinity and therefore size of parts. If possible, try a resin with no nucleating agent to increase crystallinity for a smaller part
Not enough post mold shrinkage	Reduce cooling time to eject the part hotter to allow for more shrinkage.
Improper measurements	Check for identical conditioning of parts before measurements are taken. Have the parts been cooled the same and for the identical times after molding. Even laying down the part with the same side up consistently is important. Check measuring device for calibration

- **Discoloration.** See *Color instability*, *Color mixing*, and *Black specks.*

- **Ejector pin or push marks.** These are blemishes or part deformation near ejector pins, often stress whitening is seen. See also *Sticking.*

Possible causes	Possible remedies
Part sticking to the ejector side	See sticking in mold, especially gate seal analysis, mold issues, and cooling or mold temperature
Ejector velocity is to fast	Slow velocity of ejection
Not enough ejection pin surface area	Add ejector pins or use larger diameter ejector pins
Ejector plate cocking	Check the length of the knock out bars, they should be identical in length, maximum difference is 0.003 inches
Ejector pins not all the same length	Check ejector pins for proper length

- **Flash**

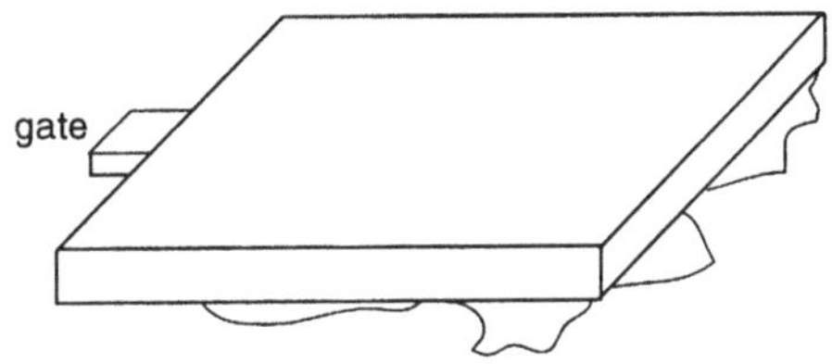

Flash [10].

In separating first stage (fill) from second stage (pack and hold) it is critical that first stage fills the part to 95 to 99% full. This troubleshooting procedure mandates that first stage ends before the cavity is full. Flash may be caused by a combination of the causes stated below.

Possible causes	Possible remedies
Parting line mismatch or mold damage	Take second-stage pressure to minimum. If minimum is above 1000 psi plastic pressure take second-stage time to 0 seconds. Do not reduce first-stage pressure to do this test, you must be in velocity control for first stage. If part is short, inspect part for flash. If flash is present, first clean mold surfaces and carefully inspect for any material on the surface or in the cavity that would prevent proper clamping at the parting line. Ensure there is no flash behind slides. Mold another shot under identical conditions. If flash is still present, there is a parting line mismatch or tool damage that must be repaired; get a moldmaker and blue one side of the parting line while the tool is in the press. Close clamp and note blue transfer. Plastigauge can be also be used. Repeat if necessary. If this is a thin-walled part and flash is present on short shots see the following possible causes.

Flash (*continued*)

Possible causes	Possible remedies
Too much plastic delivered to cavity during first stage	Take second-stage pressure to minimum. Do not reduce first-stage pressure to do this test, you must be under velocity control for first stage. If part is full increase the position cutoff or reduce shot size. Also, check machine response on switchover from first to second stage. If there is a sharp rise in pressure upon switchover, the machine needs repair. Switchover should occur with no rise or dip in hydraulic pressure as it transfers from first to second stage.
Clamp tonnage too low; cavity and runner pressure exceeds clamp force	If flash occurs during second stage or pack and hold, reduce second-stage pressure. If this makes an unacceptable part. Check machine level, level if necessary. Check tie bar stretch for each tie bar, all should be within 0.002 in maximum of each other. Check platen parallelism. Try mold in a larger clamp tonnage press.
Clamp pressure too high	If flash is concentrated in the center of the tool, lower the clamp tonnage. Small molds in large platens tend to cause platen warp. The mold should take up 70% of the distance between the tie bars.
Melt temperature too high	Take melt temperature via hot probe technique or appropriate IR sensor. If not within the resin manufactures specified range, adjust accordingly. Reduce residence time if degradation is possible. Checking melt flow rate (MFR) before and after molding may tell you if molecular weight changed. A 30% change in MFR is acceptable for unfilled resins; 40% is unacceptable. Higher for filled resins.
Viscosity too low	If resin is hygroscopic check moisture content. Check for correct resin. Try different lot. Check for degradation, as earlier. As a very last resort you can lower injection rate which will increase viscosity. This is *not* recommended.
Mold improperly supported	Check number, placement and length of support pillars in mold.
Sprue bushing to long	For nonsprue-gated parts, check length of sprue bushing under production conditions. Both thermal expansion and under load of nozzle contact force, sometimes this is 5 to 15 tons, the sprue bushing can become a mold opening force.

- **Flow lines.** See *Weldlines.*

- **Gate blush.** See also *Blush.*

Possible causes	Possible remedies
Gate design	Break any sharp corners on the gate. Change gate type.
Gate location	Change gate location.
Angle of impingement from gate	Change the angle at which the plastic enters the cavity.

Possible causes	Possible remedies
Injection velocity too fast	Slow injection rate to avoid blush.
Wrong mold temperature	Raise or lower the mold temperature significantly.
Wrong melt temperature	Check nozzle for proper size and temperature control. Nozzle and tip should be as short as possible. Temperature control should be PID, *not* a % or variac.

- **Gloss.** Too high or low a gloss level on the surface of a part. Gloss is how shiny a surface is. Usually measured as a Gardner Gloss at 60%. See also *Blush.*

Possible causes	Possible remedies
Mold surface finish	Clean and inspect mold surface finish under good lighting conditions. Make sure there is the correct finish and that it is clean of any mold buildup. See *Mold build up*.
Incorrect mold temperature	A hotter mold will generally provide higher gloss. Raise mold temperature to increase gloss, lower mold temperature to decrease gloss.
Pressure in the cavity	Check second-stage pressure. Is the part packed well enough to replicate the steel surface. Higher second-stage pressures generally provide higher gloss. Lower second-stage pressures provide lower gloss.
Injection velocity	Increase injection velocity for higher gloss, decrease injection velocity for lower gloss.
Resin type	Certain ABS and PC/ABS resins are made specifically for high or low gloss applications. For high gloss you want emulsion-type ABS; for low gloss use mass-type ABS.
Venting	If gloss is different at small areas, check renting.

- **Hot tip stringing**

Possible causes	Possible remedies
Residual melt pressure	Plastic is compressible, during screw rotation with 1000 to 1500 psi plastic pressure the plastic is compressed. Use screw decompress to relieve this pressure; be careful not to pull air into the nozzle.

Hot tip stringing (*continued*)

Possible causes	Possible remedies
Volatiles	Moisture, low molecular weight additives, or degraded resin can turn to gas when heated in the barrel or hot manifold. This gas formation can provide enough pressure to push molten plastic out the tip. Trapped air in the hot manifold or runner system will do the same. Check moisture content of the resin. Check for resin degradation, or excessive residence time.
Poor hot tip design or geometry	Check tip design and clearance with gate surface. Check thermocouple placement and heater energy distribution. Replace if incorrect.
Poor temperature control of the hot tip/high temperature	Check thermocouple placement in hot tip. If poor design, redesign for proper temperature control and cooling.

- **Jetting.** This appears as worm tracks on the part.

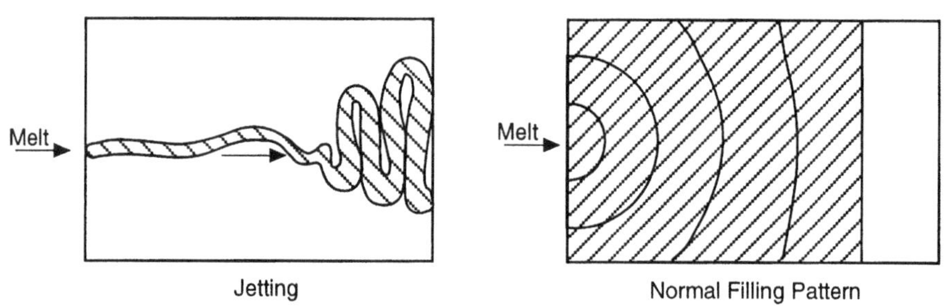

Jetting versus normal filling pattern [10].

Possible causes	Possible remedies
Gate location	The gate should not be in the center of the nominal wall. Plastic must have something to stick to or impinge upon as it enters the cavity. Extend runner to go beyond the gate. This allows a small amount of plastic to enter the gate, providing a normal flow front before full injection flow is started into the cavity.
Gate type	Break any sharp corners of the gate. Change gate type. Extend the runner to go beyond the gate. Increase the gate depth.

Possible causes	Possible remedies
No flow front formation	Provide an impingement pin 40 to 70% of the nominal wall thickness just opposite the gate. This pin may be withdrawn during second stage or pack and hold by a spring or hydraulic mechanism.
Fast injection rate	Slowing the injection rate is usually the first approach to solve jetting. It is the most expensive due to the increase in cycle time. Consistency and overall part quality often suffer.

- **Knitlines.** See *Weldlines.*

- **Mold build up or deposits.** Mold build up can be a solid (dark or white), liquid or wax accumulation on the mold or part. The cause is usually an additive in the plastic resin. Take a cotton swab and accumulate as much of the material as possible on the tip and send to a laboratory for analysis. See also *Surface finish.*

Possible causes	Possible remedies
Additive package; i.e. mold release, stabilizers, colorant(s), antioxidants, antistatic agents and flame retardant packages	Try natural resin with no colorant. Request a different grade with lower additive content, try a different suppliers resin.
Additive package temperature stability	Reduce melt temperature to lower end of acceptable range as stated by resin supplier. Minimize residence time in the barrel by shorter cycle times or using a small barrel. Shot size should be 25 to 65% of the barrel capacity.
Inadequate venting	Check vents for proper depth and number, add or repair as needed. Provide a vacuum to the cavity during fill.
Moisture	Check moisture level in resin. If high humidity environment, run a trial with dehumidifier blowing over mold surface.
Mold temperature	Check for high mold steel temperatures in the area of build up a hot or cold spot. Remedy any cooling issues.
Contamination	Check for contamination.
Infrequent mold cleaning	Some residue build up is to be expected after long runs of time. Molds must be cleaned on a routine basis.

Mold build up or deposits *(continued)*

Possible causes	Possible remedies
Injection rate too high	Fast injection rates can provide shear separation of low viscosity additives which bloom to the surface. Slow injection rate.
Mold steel—resin incompatibility	Certain metals and resins are incompatible, for example some flame retardant packages and PVC can form acid gases that will corrode mild steel, like P-20.
White powder on surface of mold or part	A flame retardant additive package sometimes blooms. If on acetal parts try drying the resin before processing.

- **Nonfill.** See *Shorts.*

- **Nonreturn valve.** A mechanical device that acts like a valve in a car engine. When closed it becomes a piston pushing the plastic forward through the nozzle. It often leaks or does not seal properly. It is critical that this valve holds a consistent cushion. Also known as *Check Ring.*

Possible causes	Possible remedies
Performance test	Turn first-stage pressure limit down to the normal second stage pack and hold pressure for this mold. Bring the screw back to 90% of the full shot and inject for 10 seconds with the previous part still in the mold, runner and sprue. Note if the screw drifts forward. It should not. Repeat at the 50% shot size and again at the 10% shot position. Any screw forward drift is an indication that the barrel or nonreturn valve are worn. Replace or repair; both barrel and valve must be inspected. There is no process technique to work around a broken or worn nonreturn valve.
Proper design	The seat and the sliding ring commonly have mating angles. This is not correct. Like a valve in a car it works better when it has a step angle to assure a more positive seat. *Few valves are built correctly.* Buy the step angle valves. It is also best not to have it rotate with the screw. Avoid ball checks unless dealing with very easy flowing material.

- **Nozzle drool or stringing.** See also *Hot tip stringing.*

Possible causes	Possible remedies
Volatiles	Moisture and, low molecular weight additives turn to gas when heated in the barrel. This gas formation can provide enough pressure to push molten plastic out the nozzle. Trapped air in the screw flights will do the same. Check moisture content of the resin. Increase back pressure. Make sure screw L/D is greater than 18:1.

Possible causes	Possible remedies
No decompression after screw rotate	Plastic is compressible. During screw rotation with 1000 to 1500 psi plastic pressure the plastic is compressed. Upon sprue break or mold open, plastic will be pushed out of the nozzle. Use screw decompress to relieve this pressure; be careful not to pull air into the nozzle.
Incorrect or Poor temperature control on the nozzle	Nozzle temperature is notoriously poorly controlled. Placement of the thermocouple should be one third the distance of the nozzle from its tip if the nozzle is more than 3 in (75 mm) long. For 3 in (75 mm) lengths and shorter it is OK to have the thermocouple imbedded into the hex on the nozzle body. For longer nozzle bodies the thermocouple should be underneath the heater band via a butterfly-type thermocouple. The nozzle should be PID temperature controlled, use of a % or variac is unacceptable. Do not allow the thermocouple to be attached via the screws on the clamp for the heater band.
Wrong nozzle length	Try a shorter nozzle.
Wrong nozzle tip	Avoid standard tips, try a reverse taper, nylon or a full taper, ABS tip.

- **Nozzle freeze or cold slug**

Possible causes	Possible remedies
Nozzle tip too cold	The sprue bushing is a heat sink for the nozzle tip and draws heat from the tip while in contact. Use sprue break if possible. Insulate the tip from the sprue bushing with high temperature insulation.
Poor temperature control of nozzle	Use PID temperature control. Ensure that the thermocouple is one third the distance of the nozzle from its tip if the nozzle is over 3 in (75 mm) long. For 3 in (75 mm) lengths and shorter it is OK to have the thermocouple imbedded into the hex on the nozzle body. For longer nozzle bodies the thermocouple should be underneath the heater band via a butterfly type thermocouple. Do not allow the thermocouple to be attached via the screws on the clamp for the heater band.

- **Nozzle stringing.** See *Nozzle drool.*

- **Odor.** Any smell or odor that is unusual. Odor due to off-gases of plastics may have a potential to be a health and safety issue. Get help, find the source, vent it, or remedy the cause, but do not continue to "live" with it. See also *Black specks.*

Odor (*continued*)

Possible causes	Possible remedies
Polymer degradation	When overheated or oversheared certain polymers break down to acid gases. PVC, Acetal (POM), and certain flame-retardant resins are the most common. Proper processing does not provide these odors. If detected check for excessive temperatures, high back pressure, worn or scored screw or barrel.
Contamination	Check base resin for foreign material. Contaminates may cause the odor themselves or cause the plastic to degrade. Check regrind: How clean is it? Try virgin resin to see if odor disappears. Omit coloring agent.

- **Orange Peel.** Surface cosmetic imperfections in gloss or smoothness, sometimes concentric lines called *record grooves* or *recording*. See also *Surface finish.*

Possible causes	Possible remedies
Mold build up or deposits	Check for residue or deposits on the mold/cavity surface. If there are mold deposits, see *Mold build up*
Mold surface finish	Check surface of cavity for proper polish or finish and whether it is clean. Repair and clean.
Slow filling	Increase injection rate, this decreases resin viscosity and allows more pressure to be transferred to the cavity if first- to second-stage switchover is <0.1 seconds. Ensure velocity control is not pressure limited.
Low cavity pressure	Increase second-stage pressure. Increase second-stage time, and, if possible, remove the same amount of time from the cooling or mold closed timer to keep cycle time constant.
Mold temperature	Increase mold temperature. Decrease mold temperature.
Melt temperature	Check melt temperature, adjust to within the manufacture's guidelines if temperature is outside limits. Try higher end and lower end of resin suppliers guidelines.
Uneven filling of a single cavity	Balance flow path with flow leaders if possible. Increase injection rate.
Unbalanced filling in multicavity molds	Adjust runner size to balance filling. Do not adjust gate size to balance filling because this will provide various gate seal times and vary part dimensions, weight, etc.

- **Pinking of the part.** It happens relatively rarely that parts will turn pink while in storage. The cause is usually carbon monoxide gas reacting with components of the plastic.

Possible causes	Possible remedies
Carbon monoxide	Minor amounts for carbon monoxide are known to discolor certain resins. Remove parts to open area and see if discoloration disappears. Exposing the part(s) to sunlight can accelerate this. If discoloration reverses, remove all gas fueled lifts, etc., from storage area. Improve storage area ventilation. Go to battery-operated fork lifts.

- **Pitting**

Possible causes	Possible remedies
Trapped gases dieseling	See *Burns*. If it is due to dieseling, do not run the mold, further damage will result.
Corrosion or chemical attack by the resin or additive on the steel	Check for resin compatibility with the steel of the mold. If acid gases are possible, a more chemically resistant surface may be required. A different steel or coating the existing surface should be specified.
Abrasive wear, erosion	Highly filled resins can pit and erode a mold's surface finish. Change gate location, coat cavity with a wear resistant finish. Rebuild tool with appropriate hardened steel.

- **Poor color mixing.** See *Color mixing.*

- **Racetracking, framing, or nonuniform flow front.** The flow front should be a continual half-circle fill from the gate.

Possible causes	Possible remedies
Nonuniform wall thickness	Thicker sections of part fill preferentially due to lower melt pressures required to fill. Plastic flow will accelerate in thicker sections and hesitate filling a thin section. This may allow the plastic to "racetrack" around the perimeter or section of a part and trap air or volatiles. Try faster injection rates, but it is unlikely that this will solve the problem because you are fighting a law of physics. Round the edge or taper the junction between the nominal wall change. The correct fix is to redesign with a uniform nominal wall.
Gate location	Gate into the thick area and provide flow leaders to the thin areas to provide uniform filling.
Hot surface or section in the mold	Allow the mold to sit idle until mold is at uniform temperature. Make and save first shot for 99% full. If flow path is different than later shots it is a tool-steel temperature, and cooling issue. Check mold for hot spots. Get uniform cooling.

- **Record groves, ripples, wave marks.** These are concentric groves or lines usually at the leading edge of flow. The flow front is hesitating, building up pressure, then moving a short distance and hesitating again. This is almost always related to lack of adequate pressure at the flow front or slowing of injection velocity.

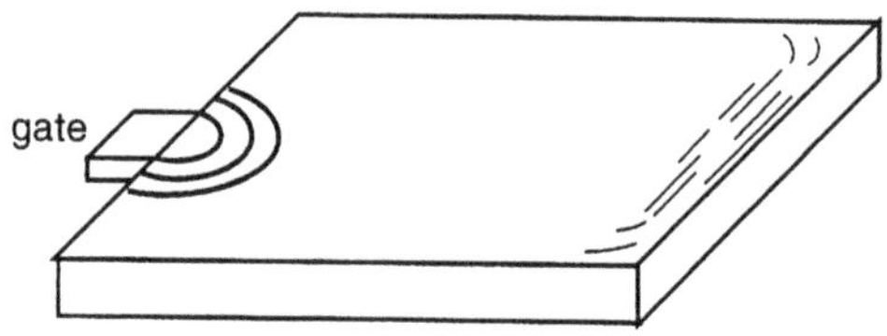

Ripples [10].

Possible causes	Possible remedies
Pressure limited first stage or lack of velocity control	Double check that the pressure at transfer is 200 to 400 hydraulic psi lower than the set first stage limit. For eletrics, use 2000 to 4000 psi. Make sure this is enough Δ P.
Incorrect position transfer	Take second-stage pressure to 1000 psi plastic pressure or, if the machine does not allow this, take second-stage time to zero. The part should be 95 to 99% full. Unless this is a thin-walled part, then the part should be full with only slight underpack near the gate. Adjust position transfer to provide appropriate fill volume.
Melt temperature too low	Check melt temperature via the hot probe technique or appropriate IR sensor. Make sure it is within the resin supplier's recommended range.
Poor first- to second-stage switchover response	Note response of hydraulic pressure at switchover. It should rise to the transfer point then drop rapidly to the set second-stage pressure. If hydraulic pressure drops much below set second-stage pressure then the flow front may be hesitating and building a high viscosity. Repair machine or valve.
Low pack rate or volume	Increase pack rate or volume of oil available for second-stage.
Low mold temperature	Increase mold temperature 20 to 30°F. Decrease cycle time. This will raise steel temperature in the mold.

- **Screw design**

Possible causes	Possible remedies
Standard general purpose screw	These are known to produce unmelt due to solids bed break up. They should be replaced with melt uniformity screws: Minimum L:D is 20:1. This is an industry problem: Most (99%) general purpose screws do not provide uniformly melted polymer coming out of the nozzle.
Screw and barrel metals	Recommend: bimetallic or hardened barrels and soft screws like stainless steel. Chemically resistant screw material is especially critical for clear resins. Screw should be polished with sharp edges on flights with the back of the flights filled with metal, or large radius to prevent dead spots and carbon build up. A modified barrier should transition the transition zone to the metering zone.
Barrier flights	Generally not recommended unless short and at the end of the transition zone or beginning of the metering section.
Vented barrels	Vented barrels though uncommon do have their purpose. They provide excellent melt uniformity and process resins that are not subject to hydrolysis more uniformly; unfortunately, design is poor. The two-stage screw must be designed with a continuous flight through the decompression section. The first stage should be cut such that it cannot overpump the second stage. Vented barrels require near zero backpressure to prevent vent flooding. This presents purging and residence time problems.

- **Screw recovery, slow recovery, screw slips, or does not feed.** The metering section of the screw pumps plastic forward, which pushes the screw back.

Possible causes	Possible remedies
Feed throat temperature	Run throat temperature at 110 to 120°F for most resins. For high-end engineering resins you may want to go higher. Do not run feed throat at 60 to 80°F. Feed throat should be PID temperature controlled.
Feed problems	Check size of granules and flow through hopper and feed throat. Ensure that material gravity feeds correctly when resin is being loaded into the hopper. Vacuum loading may interrupt normal gravity feeding.
Heavily carbonized or blocked flights	Standard general purpose screws are notorious for dead spots behind flights. These can have large carbon or other deposits that block plastic flow. Check screw for clean polished flights.
Worn screw and/or barrel	Worn screws and barrels will provide better mixing, but slow recovery rates as plastic flushes back over flights.

Screw recovery *(continued)*

Possible causes	Possible remedies
Moisture	Check moisture content of plastic; check feed throat for cracks, leaking water.
Granule size	Plastic granules should be uniform in size and shape. A wide range in granule size, fines, and small granules, along with large chunks of regrind, will cause feeding problems. This includes large and small pellets in virgin.
High back pressure	Target 1000 to 1500 psi melt back pressure. Try lower back pressure.
High RPM	Try lower screw rotate speeds, better melt uniformity, and mixing is obtained with slow screw speeds. Use all but ~2 seconds of the cooling time for plasticating. Do not lengthen the cycle.
Incorrect barrel temperature settings	Start by setting front and center zones to the center of the resin supplier's recommended range. Set rear zone at the minimum of the range. Back pressure set at 1000 melt psi. Average recovery time for 10 cycles. Repeat with rear zone 10°F higher until you have reached the rear zone setting at the maximum recommended by the resin supplier. Pick the temperature that gives you the minimum recovery time.
Poor screw design	See *Color mixing* and *Screw design.*

- **Screw slip.** See *Screw recovery.*
- **Shorts or short shots or nonfill.** Part is short or some section of the part, like a rib is not completely filled out.

Possible causes	Possible remedies
Consistent short shots	
Incorrect shot size	Take second-stage pressure to 1000 psi plastic pressure or, if the machine does not allow this, take second-stage time to zero. The part should be 95 to 99% full. Unless this is a thin-walled part, then the part should be full with only slight underpack near the gate. Adjust position transfer to provide appropriate fill volume. Do not use first-stage pressure limit to do this test or make appropriate volume on first stage.
Pressure limited first-stage or lack of velocity control	Double check that the pressure at transfer is 200 to 400 hydraulic psi lower than the set first stage limit. For electrics, use 2000 to 4000 psi. Make sure this is enough Δ P.
Injection rate	Increase injection rate to decrease viscosity.

Possible causes	Possible remedies
Nonreturn valve or barrel worn or broken	Check nonreturn valve and barrel: Are they in specification? If the nonreturn valve is OK, double-check barrel for wear and ovality. Repair or replace as needed. Note: Nonreturn valve should not have mating angle between seat and sliding ring. This should be a stepped angle for positive shut off.
Large pressure drop	Perform a short shot analysis for pressure loss. Note pressure at transfer for shots making: (1) 99% full part, (2) sprue, runner, and gate, (3) sprue and runner only, (4) purge full shot through the nozzle into the air. Use intensification ratio and calculate pressure drop for (1) nozzle, acceptable range 200 to 4000 psi (2) sprue and runner, acceptable range 200 to 5000 psi (3) gate acceptable range 200 to 5000 psi (4) part acceptable range 200 to 40,000 psi. Evaluate where largest pressure drop is and remedy. This will find if there are blockages and where.
Trapped gas or air	Progressively short shoot the mold using shot size or position transfer to make 10, 20, 40, 60, 70, 80, and 95% full parts. Note if any show the plastic flow front coming around on itself. Is there a racetrack effect? This includes jetting. Note if ribs are covered before completely filling. This test cannot be done correctly if you reduce first-stage pressure or velocity. You must be velocity controlled for first stage. The concept is to find where the gas is coming from and eliminate its source. Vent the tool properly or use porous steel to eliminate gas traps. Jetting can cause gas, air, to be trapped. Check for moisture, steam is a gas. Change gate location. Pull a vacuum on the mold cavity during fill.
Insufficient second-stage pressure	Make sure first stage is at the right shot volume. If OK, raise second stage pressure.
No cushion	Ensure adequate cushion to allow for transfer of packing or second-stage pressure.
Melt temperature	Verify melt is within the resin supplier's recommended range.
Mold temperature	Try higher mold temperatures and or faster cycle times.
Resin viscosity	Change resin to a higher melt flow rate. Be careful because properties may decrease due to lower molecular weight. Parts must be fully tested in the application for correct performance.
Long flow length	For every millimeter of flow there is a pressure drop. Long flow lengths have large pressure drops. Add gates or flow leaders. Last resort: increase nominal wall.
Thin nominal wall	Add flow leaders if possible. Increase nominal wall if all other avenues fail. Thicker nominal wall will reduce pressure loss.

Shorts or short shots or nonfill (*continued*)

Possible causes	Possible remedies
Intermittent Short Shots	
Nonreturn valve or barrel worn or broken	Check nonreturn valve and barrel: Are they in specification? If the nonreturn valve is OK, double check barrel for wear and ovality. Repair or replace as needed. Note: Nonreturn valve should not have mating angle between seat and sliding ring. This should be a stepped angle for positive shut off.
Cushion not holding	Note cushion repeatability: If varying by more than 0.200 in or 5 mm, check nonreturn valve, as earlier. Try a larger decompression stroke to help "set" the check valve. Be careful not to suck air into the nozzle and cause splay.
Contaminated material	Check gates and parts for foreign material. Check quality of regrind.
Melt temperature	Verify melt is within the resin supplier's recommended range.
Mold temperature	Try higher mold temperatures and or faster cycle times.
Unmelt	Look for unmelted granules in the part, color streaks, see *Screw design* and introduction to *Color mixing*. Provide uniformly melted material to the gate.
Cold slug	It is occasionally possible for a cold slug from the nozzle to go beyond the sucker pin and plug a gate. Check nozzle for cold slug formation. See *Nozzle drool.*
Insufficient second stage pressure	Raise second stage pressure.
Trapped gas	See earlier; see also *Bubbles.*

- **Shrinkage.** This is one of the biggest problems encountered in the polymer industry. ASTM Test Method 955 was never intended to be used to determine shrinkage for parts out of molds, yet it is what most people use. Its purpose was to compare "lot to lot" differences. In addition, the test specimen is 0.125 in (3.17 mm) thick. Shrinkage depends on nominal wall thickness, cooling rate, nucleation in semicrystalline resins, and pressure gradients in the cavity. See also *Short shots* and *Flash.*

Possible causes	Possible remedies
Too much shrinkage	
Insufficient plastic	Make sure cutoff position provides the right volume of plastic on first stage. Do a gate seal analysis. Add more second-stage time to achieve gate seal. Add more second-stage pressure. Check that you have adequate cushion.

Possible causes	Possible remedies
Degree of crystallinity	If a semi-crystalline resin, lower mold temperature to increase cooling rate. This will decrease crystallinity and decrease shrink. Look at a nucleated resin. Watch for postmold warp over weeks of time. Do a thermal cycle test on parts to show amount of internal stress and warp. Check all properties.
Low second-stage pressure	Raise second stage pressure, see also *Insufficient plastic.*
See *Short shots*	
Too little shrinkage	
Too much plastic	Make sure cutoff position provides the right volume of plastic on first stage. Do a gate seal analysis. Add more second-stage time to achieve gate seal. Add more second-stage pressure. Check that you have adequate cushion.
Degree of crystallinity	If a semi-crystalline resin, raise mold temperature to decrease cooling rate and increase crystallinity. This will make a smaller part. Check all properties.
High second-stage pressure	Lower second-stage pressure. Lower second-stage time to allow for gate unseal. See *Flash.*
See *Flash*	

- **Silvery streaks**. See *Splay.*
- **Sinks.** Sinks are depressions on the surface of a part that do not mimic the mold steel surface. Sinks and voids are signs of internal stress and are warning signs that the part may not perform as required. See also *Bubbles*, the void section.

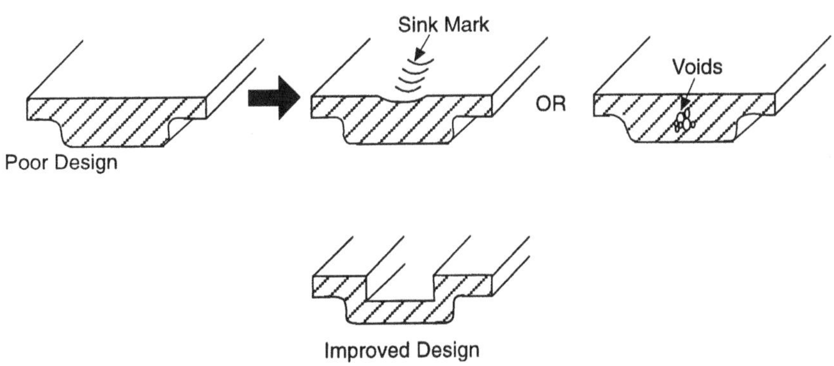

Sink marks and voids are created by material shrinkage without sufficient compensation [10].

Sinks (*continued*)

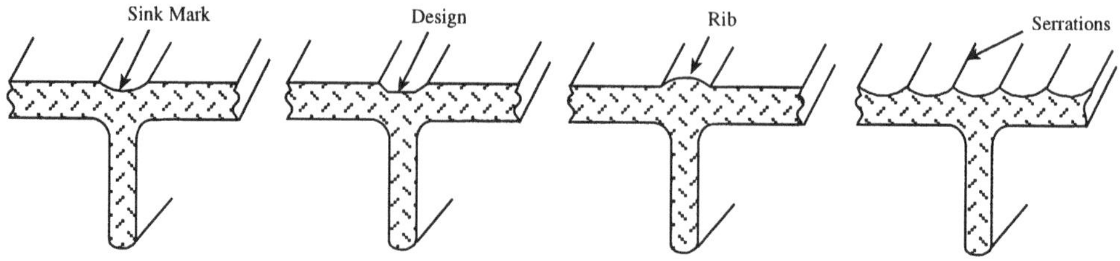

Sink marks can be masked by creating a design, rib, serrations [10].

Possible causes	Possible remedies
Insufficient plastic	These occur upon cooling in thick sections. You need to pack more plastic into the cavity with: (1) consistent cushion, make sure you are not bottoming out the screw, (2) higher second-stage pressures, (3) longer second-stage time, (4) very slow fill rates, (5) use counter pressure, (6) open the gate for longer gate seal times to allow more packing during second stage, or (7) increase the runner diameter. Lower the mold temperature significantly to freeze the outside wall of plastic, though this might produce voids internally. Reduce melt temperature. Thin up the nominal wall in this area by adding steel to core out thick sections. Change gate location; fill thick to thin.
Postmold cooling	Thick sections can remelt parts outside surface once ejected, allowing the surface to collapse. Trial cooling in water or between aluminum sheets rather that air.
Thick nominal walls	Core out to thin the nominal wall. Thicker is not stronger in plastic parts. Thick nominal walls should be redesigned with ribs if strength is needed. This will save plastic and cycle time.

- **Splay and silver streaks.** These are cosmetic blemishes on the part surfaces. Splay is usually caused by a gas, usually water (steam), but can be caused by unmelt, dirt, chips or flakes from three-plate molds, cold slugs, excessive volatile additives, air, and/or decompression. Remember: If the resin is subject to hydrolysis (reacts with water), you may need to discard samples with splay rather than regrind them. Water degrades molecular weight in some resins such as polycarbonate, nylon, and polyesters. Also called chicken tracks, tails, and hooks.

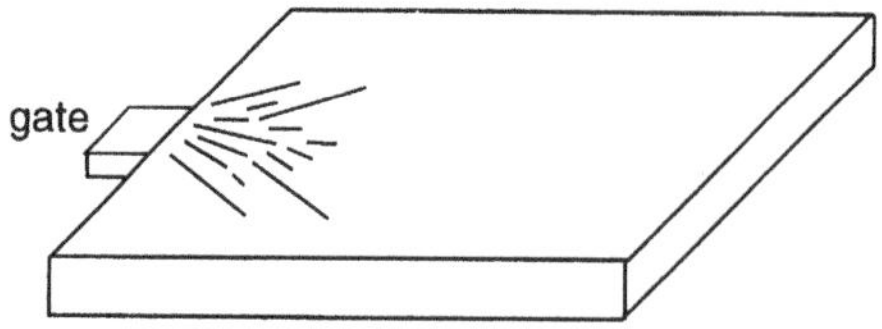

Silver streaks [10].

Possible causes	Possible remedies
Moisture in the plastic granules	Check moisture content with a moisture specific method (not weight loss): Karl Fisher or Computrac 3000. Weight loss does not tell you if it is moisture or volatiles. Check dryer.
Dryer not functioning properly	Check for the correct drying temperature set point. Remove the bottom 20 to 50 lbs. of material in the dead space of the cone of the hopper and put it on the top. Check for funnel flow: Hopper must drain via mass flow. This can be achieved by having the angle of the cone at the bottom of the hopper 60 degrees included. Check the temperature of the incoming air and the temperature of the air exiting the hopper: It should be within 20 to 40°F (10 to 20°C) of the incoming air. Check for fines clogging the filters. Check that the regeneration heaters are functioning. Check the temperature of the air going into the desiccants: It should be below 150°F (65°C). Check dew point of air before and after desiccants if possible. Dew point should be lower after desiccants. Dew point should be −25 to −40°F.
Condensed moisture on pellets	If resin has been brought inside from a cold climate, condensation can occur like condensation on glass. Even though the resin is not hygroscopic it still must be "dried." For condensation moisture, drying at 150°F (65°C) for 1 to 2 hours should be fine.
Moisture due to high humidity	In un-airconditioned shops during high-humidity days (summertime) there are times when nonhygroscopic resins pick up moisture or the air around them carries enough moisture that a dryer is necessary.
Moisture in the feed throat	Check feed throat temperature and look for cracks. Run feed throat at 110 to 130°F (40 to 55°C) via PID temperature control.
Moisture from mold	Check mold, particularly cores, for condensation. Raise mold temperature if condensation is present. Check for water leaks: from fittings dripping into the mold and from cracks or leaks in the mold. Repair all leaks!
Poor nozzle temperature control	Nozzle temperature is notoriously poorly controlled. Placement of the thermocouple should be one third the distance of the nozzle from its tip if the nozzle is more than 3 in (75 mm) long. For 3 in (75 mm) lengths and shorter it is OK to have the thermocouple imbedded into the hex on the nozzle body. For longer nozzle bodies the thermocouple should be underneath the heater band via a butterfly-type thermocouple. The nozzle should be PID temperature controlled, use of a % or variac is unacceptable. Do not allow the thermocouple to be attached via the screws on the clamp for the heater band.
Air trapped in melt from screw or decompression	Check L/D of screw: if less than 18:1, there is a good possibility that the screw is at fault. Raise backpressure, slow screw rpm, and extend cooling time. Raise rear zone temperature. If none of these work order

Splay and silver streaks (*continued*)

Possible causes	Possible remedies
	a melt uniformity screw. See *Screw design*. If you have decompression on, minimize or turn off decompression, especially when using hot runner or hot sprue molds. Too much decompression sucks air into the nozzle, which gets trapped in the flow stream because nozzle contact force is loaded before injection.
Chips or flakes from three-plate molds	Open mold and carefully inspect for plastic chips, flakes, or particles. These can upset the flow front and cause marks similar to splay.
Contaminants, dirt, or cold slugs	Solid particles of any kind can upset the flow front and cause marks similar to splay. Inspect plastic granules for contamination or dirt. Check for degradation on the screw.
Unmelted granules and gels	Solid particles of any kind can upset the flow front and cause marks similar to splay. Unmelt can be a cause due to poor screw design or short L/D. Unmelt is to be expected from general-purpose screws. Gels are high molecular weight or cross-linked resin that does not melt. Gels are rare and can only be remedied by the resin supplier. Try a different lot or grade of resin.
Volatiles due to degradation of polymer or additives	Due to high shear, hot spots in the barrel, flame retardant packages, degraded regrind, and the like, a polymer can produce off-gases that will mimic moisture and provide splay. Check temperature set points vs. actual melt temperature as measured with a hot probe.

- **Sprue sticking**

Possible causes	Possible remedies
Nozzle tip radius mismatch with sprue bushing	Check whether nozzle radius matches with sprue bushing by inserting a piece of cardboard and bring in nozzle contact. Look for perfect tip imprint in cardboard; any tears, cuts or mismatch: change tip. There are inexpensive gauges that have the 3/4- and 1/2-inch radii.
Nozzle tip orifice too large	Check orifice of sprue and tip. Tip orifice should be at least 0.030 in (0.75 mm) smaller in diameter.
Scratches or incorrect polish on sprue	Sprue should be draw polished to a #2 finish. Vapor honing is better for soft touch material. Circular polishing provides minute undercuts that get filled with plastic and "stick" the sprue. Look for scratches or gouges that form undercuts, even when small, these too will stick a sprue.
Sprue puller problems	Is the sprue puller large enough and the right design? Add undercuts, Z-puller, or more reverse taper.

Possible causes	Possible remedies
Sprue too soft, not frozen	If sprue is still soft at mold open, down size sprue or cool the sprue bushing. Try lower nozzle, then melt temperature. Last resort, add more cooling time.
Improper taper on sprue	Is taper 1/2 in (13 mm) per foot (305 mm). Increase taper.
Overpacked sprue	Do a gate seal study (see p. 575, footnote). If possible take off some second-stage time and add to cooling or mold closed timer.

- **Sticking in mold**

Possible causes	Possible remedies
Mold build up or deposits	Check for residue or deposits on the mold/cavity surface. If there are mold deposits, see *Mold build up*.
Mold surface finish	Check surface of cavity has proper polish or finish and is clean.
Slow filling	Increase injection rate: this decreases resin viscosity and allows more pressure to be transferred to the cavity if first- to second-stage switchover is less than 0.1 s. Ensure velocity control is not pressure limited.
Low cavity pressure	Increase second-stage pressure. Increase second-stage time, and, if possible, remove the same amount of time from the cooling or mold closed timer to keep cycle time constant.
Mold temperature	Increase mold temperature. Decrease mold temperature.
Melt temperature	Check melt temperature, adjust to within the manufactures guidelines if temperature is outside limits. Try higher end and lower end of resin suppliers guidelines.
Uneven filling of a single or multicavities	Balance flow path with flow leaders if possible. Increase injection rate.
Unbalanced filling in multicavity molds	Adjust runner size to balance filling. Do not adjust gate size to balance filling; this will provide various gate seal times and vary part dimensions, weight, and so on.
White powder build up on acetal parts	Formaldehyde is condensing on the mold or bloom to the part surface. Dry the acetal before processing.
Improper polish or undercuts causing part to stick to the "A" side of the mold upon mold opening	In manual mode, open the tool *slowly* and note any noises or cocking of the part. Before ejection, inspect part for deformation. If deformed before ejection, the part is sticking to the opposite side of the mold and deforming upon mold open. Draw polish or remove undercuts in areas where part sticks. Many soft touch polymers need a vapor honed surface to aid release. High polish forms a vacuum.

Sticking in mold (*continued*)

Possible causes	Possible remedies
Improper mold polish or undercuts causing part to stick to the "B" side of the mold upon ejection	Slowly eject the part and note if a corner or section hangs up or sticks or there is cocking of the part at this stage. Draw polish or remove undercuts if present. Many soft-touch polymers need a vapor-honed surface to aid release.
Incorrect second-stage time or improper gate seal time	Perform a gate seal analysis (see page 575, footnote). Determine if gate seal or unseal changes the problem.
Overpacking, not enough shrinkage	Reduce pack and hold or second-stage pressure. Make a few parts with shorter pack and hold time to achieve gate unseal. Add any time taken off second stage to the cooling, cure or mold closed time to keep cycle time the same. Also, try a longer cycle time.
Underpacking causing excessive shrinkage	If reducing pack and hold pressure makes it worse, the cause could be the part shrinking onto a mold detail or core. Increase pack and hold pressure, increase second-stage time, and take the identical number of seconds off cooling to keep cycle time the same. Also, try a shorter cycle.
Vacuum, high polish can cause a vacuum to form during molding that holds the part to the steel	Provide a vacuum break before mold opening or ejection.
Excessive ejection speed	Slow ejection velocity.
Parts handling or robot removal	Check end of arm tooling and movement of the part on the robot arm.
Contamination	Check regrind for contamination or excessive fines.
Plate out on the mold surface	Inspect mold surface for build up, clean appropriately; for example, with toilet paper and semi-chrome for nonoptical parts; for optical finishes, check with mold maker.
Incorrect mold finish	Too high a polish can cause a vacuum to hold the part to that side of the mold. Change polish or texture.
Crazing	Check for stress in the area. For example, stress whitening near ejector pins or places where the part sticks. Is there any contact with chemicals such as solvents or mold spray that attacks this type of plastic?
Mold temperature too cold or too hot	Raise and or lower mold temperature significantly as long as you do not cause mold damage. Check cooling lines for water flow rate in gallons/minute (GPM). You need a Reynold's number of 5000 or greater to achieve turbulent flow (optimum cooling). Make sure the difference in temperature between in and out of water lines is less than 4°F (2°C).

Possible causes	Possible remedies
Ejector plate cocking	Check for even length of knock out bars. Should be within 0.003 in of each other.
Mold deflection	Take dial gauge measurements on top and bottom of both sides of the mold. Repeat for both sides. Note mold deflection looking for deflections opposite each other.
Inadequate mold release in resin	Add mold release. Try same resin from different suppliers. To solve the problem avoid using mold sprays except for start-up situations.

- **Streaks.** See *Color mixing* and *Black specks.*

- **Surface finish.** See also: *Orange peel*, *Gloss*, and *Gate blush.*

- **Unmelted particles.** See *Screw design* and *Color mixing.*

- **Vent clogging.** See *Mold build up.*

- **Voids.** Here, you must absolutely sure it is a void and not a bubble of gas. You cannot tell by looking—it must be tested. The test is found under "*Bubbles.*" See *Bubble* and *Sink.*

- **Warp: Dimensional distortions in a part.** One of the hardest problems to resolve through processing is often due to the most common root cause of poor part design, or nonuniform nominal wall sections especially in semicrystalline resins. Warp is a result of retained compressive, tensile, orientation and or crystalline stresses. The trick is to figure which stress is ruling. After setting the production process, always run a thermal cycle study to see if the part is stable. That is, slowly heat the part to the maximum possible application temperature for a few hours, then cool it slowly to the minimum temperature it might see in use or application, normally around –20 to –40°F. Repeat this cycle for an appropriate number of times determined by the application. Does the part hold its shape and dimensions for proper performance in the application?

Possible causes	Possible remedies
Warp for amorphous Resins, low shrink	
Molecular orientation	Take second stage to less than 1000 psi plastic pressure and mold a 99% full part. See *Short shots* for this procedure. Inspect the 99% full part for warp. If the part warps during this experiment, then the warp is caused by either orientation or shrinkage. If there is

Warp (*continued*)

Possible causes	Possible remedies
	uniform nominal wall and no standing cores, the warp is most likely orientation. Try fast to slow injection rates. Raising mold temperature 20 to 30°F (10 to 15°C) may also help. If nonuniform nominal wall, find the designer or client and explain the importance uniform nominal wall. *Note*: With nonuniform nominal wall you are now battling a law of physics, difference in cooling rate for thick vs. thin, and you will continue to have problems even if you do find a cooling method to minimize the amount of warp! If there is no warp on the 99% full part, then the warp is due to packing pressure or pressure gradient.
Pressure or pressure gradient	Run a gate seal study (see p. 575, footnote). Note if there is a difference in amount of warp between samples with and without gate seal. If you have warp with gate seal, try lower second-stage pressure with gate seal. If this works, run with lower pressure and gate seal. If lower second-stage pressure does not provide a well packed part do a DOE to find what second-stage time and pressure provides the best compromise running with gate unseal. Running with gate unseal will allow some plastic to backflow out of the cavity to provide less of the pressure gradient from end of fill back to the gate. Different gate location may help.
Nonuniform cooling/nonuniform shrink	This can be due to either mold steel cooling or nonuniform wall sections. To determine whether it is from the mold allow plenty of time for the mold steel to come to a uniform temperature, usually 20 to 30 minutes. Use this time to check water routing and GPM to assure turbulent flow, Reynold's number >5000. Upon start up . . . save the first part. Compare first part warp to parts after 30 minutes of steady-state molding. If warp is less for the first part than others the warp is due to poor cooling in the mold. Fix the cooling in the mold by adding cooling lines, by increasing water flow, or by replacing existing steel with a metal that has higher thermal conductivity. Cores especially need optimum cooling and often require high thermal conductivity steel. If it is due to a thick section on the part, the part will need to be cored out. Using a higher thermal conductivity steel may help. Uniform nominal wall is the correct answer. If nothing can be done you can try an extended cycle time. This uses the mold as a shrink fixture but cuts profit margins.
Nonuniform wall thickness	Thicker sections stay hot longer and will shrink more than thinner sections. Taper the junction. Running different temperatures on the mold halves may help; however, care must be taken not to damage the mold. As earlier, the correct answer is uniform nominal wall.
Warp for semi-crystalline resins, high shrink	

Possible causes	Possible remedies
Molecular orientation	Same as above!
Pressure or pressure gradient	Same as above!
Nonuniform cooling/nonuniform shrink	Same as earlier, plus: Try a filled or nucleated version of the resin if available. Also, if there is a differential in shrinkage flow direction vs. cross-flow, then there is the possibility of using two fillers such as glass and mica to try and compensate.
Nonuniform wall thickness	Same as previously!
Crystallinity	If all the preceding fails, try (providing the mold will not be harmed) an amorphous resin such as General Purpose Polystyrene (GPPS) or High Impact Polystryene (HIPS), ABS, or Polycarbonate/ABS blend. If the amorphous parts do not shrink, you have proven that it is a temperature-cooling rate-crystallinity issue! Try a nucleated or filled version of this semicrystalline resin.

- **Weldlines.** This is rarely due to cold flow fronts. Polymer fountain flows with the flow front regenerating with each millimeter of flow distance. The issues are: few molecular chains cross this boundary, air is often trapped as the flow fronts meet due to the shape of the flow front, and shear migration to the flow front of the lights and volatiles (additives) in the resin. Also known as *Flowlines* and *Knitlines.*

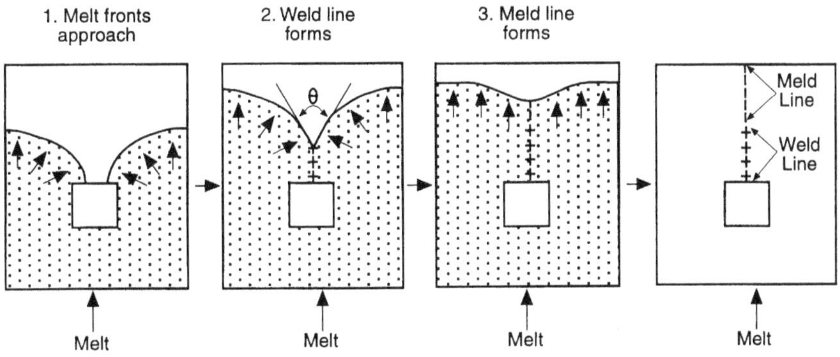

Weld and meld lines [10].

Weldlines (*continued*)

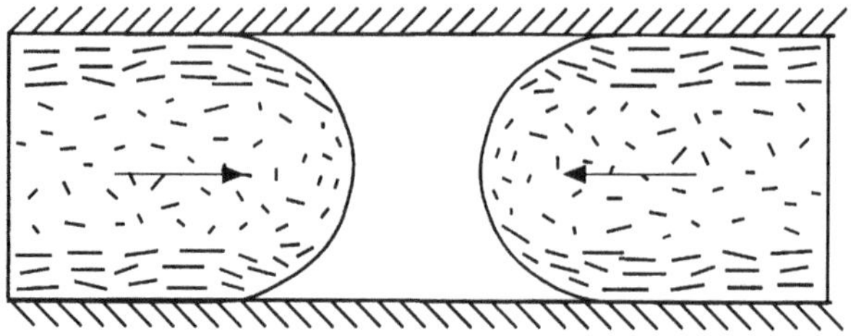

(a)
Fiber reinforced material; melt fronts approach.

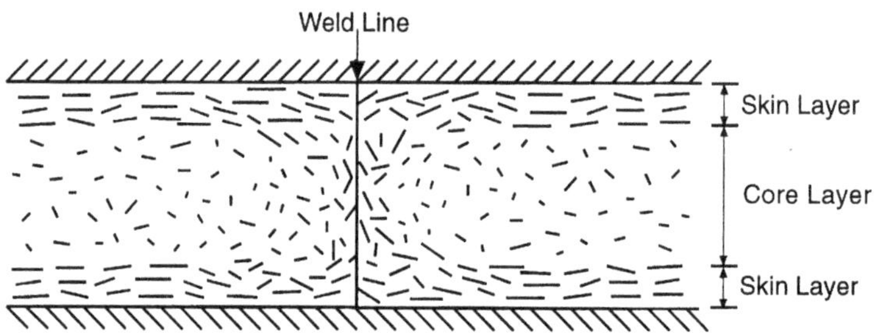

(b)
Weld line formation in fiber-reinforced material.

Fiber distribution parallel to the weld line leads to a weaker bond [10].

Possible causes	Possible remedies
No molecular chain entanglement	Increase injection rate. Raise the mold temperature. Try a higher melt flow rate resin, check properties and performance. Improve venting. Provide a flow channel at the weldline to aid venting and provide some movement at the weldline. This will help with some chain entanglement. Try resin without filler. Change gate location to put weldline in a nonstressed area. If multigated part, try blocking some gates; get permission first!
Trapped air	Improve venting, use porous steel. Provide a flow channel to aid venting. Vent core pins. If flow fronts are folding back onto themselves, try different flow rate to change fill pattern. Add flow leaders to direct flow path.
Low pressure at weldline	Increase second-stage pressure. Increase injection rate. Raise melt temperature as last resort.

- **Worm tracks.** See *Jetting*.

12.2.2 Troubleshooting on Injection Molding of Rubber

M. De Greiff

- Distortion

Distortion on removal from the mold is basically a problem of scorch or a combination of temperature and time effects. It may be caused by one of, or any combination of, the following:

High injection temperature (barrel temperature too high, screw speed too high, back pressure too high, injection pressure too high, nozzle runners or gates too small)
High mould temperature
Long injection time (injection pressure too low or nozzle, runners or gates too small)
Compound too scorchy
Compound viscosity too high

The first action to take is to check temperatures, pressures, injection times, mix viscosity, and scorch time. Put right if necessary. If still distorted, reduce mold temperature 5 to 10°C. If necessary, also reduce barrel temperature 5 to 10°C. Reduce injection time by using maximum injection pressure and if this is not successful try a wider nozzle.

Long-term action would be to Choose a safer acceleration system which will permit safe, scorch-free mold filling, higher injection temperatures, greater plasticity, and shorter mold filling times.

Gates and runners can be widened as a last resort but this step is irreversible.

- Delamination (Onion Skin, Orange Peeling)

A molding may be seen to be composed of distinct layers.

The fault is caused by rubber starting to cure while mold filling is still taking place. Rubber on the outside walls of a molding is stationary and cures while more rubber is pushed through the center to fill more remote parts of the cavity. The fault is mostly seen in long, thin moldings with a narrow flow path.

Causes can be any of those contributing to the distortion fault just described. The delamination fault may be a milder form of the first fault, but it is often associated with the flow pattern in the mold cavity.

The remedy is to fill the mold completely before cure starts. Flow patterns are almost inevitable, but if mold flow is complete before cure starts the high kinetic energy of the rubber molecules will allow some randomization of rubber chains before cure takes place.

The simplest remedy is to use maximum injection pressures for quick mold filling. If this does not give quick enough filling to solve the problem, then the mold temperature can be lowered to reduce the cure rate.

In the long term the compound must be made safer to allow quicker and safer mold filling, preferably at a higher injection temperature.

If these remedies fail, then the nozzle, the runners, and the gates can be widened to allow quicker mold filling.

No particular injection time can be recommended as correct because safety will depend on machine, mould and compound.

- Scorch in the Barrel

Scorch in the barrel is not as serious now as it was in the earlier days of injection molding when a barrel could be locked solid with cured rubber and the remedy was to dismantle the machine and extract the screw and rubber from the rear. Now, nearly all the machines used here have been shown to have the power to churn out crumbed rubber on removal of the nozzle.

This fault is caused by the barrel control fluid being too hot, the screw speed too high, the back pressure too high, the compound too scorchy, or the cure time too long for the safety of rubber in front of the ram.

The fault may be detected in its early stages by injection temperatures that are higher than normal. The reason is that more work is done by the screw and the ram when rubber is just beginning to cure in the shallow Mooney curve region before the scorch value has been reached. Detection is immediate if a machine has an ammeter across the screw motor.

Small fragments of cured rubber may appear in the moldings if this phase is not detected. Later stages of scorch are indicated by obvious groaning of the screw motor and by slow, erratic, noisy injection of partially cured rubber.

The remedy is to lower the barrel temperature 5 or 10°C and quickly use the screw motor to clear the barrel of partially cured rubber. Temperatures can be raised gradually to within about 3°C of the temperature at which scorch occurred with perhaps slightly slower screw speeds and lower back pressures. Starvation of feed to the screw can lead to excessive screw speed or too long a period of screw rotation, which may cause scorch in the barrel.

- Long Cure Cycles

This fault is more or less the reverse of the distortion problem. The cause may be due to either one or a combination of the following:

Too low injection temperature (barrel temperature too low, screw speed too low, back pressure too low, injection pressure too low, nozzle, runners or gates too wide)
Mold temperature too low
Compound too safe

The first action to do is to check temperatures, pressures, injection times, viscosity, and scorch.

Increase barrel temperature steadily to get the highest safe, scorch-free barrel conditions. Increase screw speed, back and injection pressure, and reduce nozzle diameter to get the highest safe injection temperatures. Operate a screw delay procedure so that hot rubber in front of the ram is not cooled off before it can be injected. Finally, increase mold temperature.

The last action to do is as before, but also to sharpen up the acceleration system and try using a more reinforcing filler that will increase the viscosity and give the screw action and the ram a chance of doing more work and producing more heat. Use of a more reinforcing filler will only help if higher screw speeds are possible and if there is spare injection pressure available. Care must be taken to remain within hardness specifications.

- Porosity

This fault consists of regular round bubbles and is usually a sign of undercure. Small amounts of water form steam and "blow" a soft, undercured molding when it is released from the press. Porosity can usually be remedied by increasing the cure time, but attention to factors that produce higher injection temperatures and use of a higher mould temperature can help to reduce porosity without spoiling the production rate.

Increasing the time of action of the injection pressure or the hold pressure may help to prevent porosity from developing, as long as it does not lead to backrinding.

The last term of action is to check the water content of rubber and fillers. The presence of some water is inevitable, but excessive or variable water content is worth controlling because it affects cure rate and physical properties. In addition, check the volatility of processing oils. Avoid volatile matter where possible. Reject unduly wet fillers. Draw off as much water vapor as possible by the extraction system at the internal mixer.

- Blisters in the Body of a Molding

Blisters are not usually round; rather, they are elongated and flat. They can be due to air trapped during former processing in a Banbury, or on a mill, or during the barrel phase of the injection molding process itself. Visual examination of the injection moulding feed stock should settle whether the air trapping occurred before reaching the injection machine. Entrainment of air by the screw can occur. Air can also be trapped by rubber folding over itself and enclosing air during mold filling.

If entrainment of air does occur in the barrel of the injection machine it may be expelled with characteristic, explosive violence during injection shots into the air.

The remedy is to maintain regular feeding of stock to the throat, to reduce screw speed, and/or increase back pressure. Reduction of barrel temperature would sometimes help by increasing viscosity but this would reduce productivity. The most useful remedy is to increase back pressure to allow trapped air to escape at the throat. Reduction of the injection pressure to increase the injection time and give the air a chance to get away can help to prevent air entrapment in a mold.

The last remedy may be to compound for a higher viscosity. Mould design including use of a vacuum will prevent trapping of air in a moulding.

- Air Trapping on the Surface of Moldings

Air can be trapped by rubber as it flows into a molding cavity. Air traps are nearly always formed in the same place in every successive molding, and the places of trapping are the extremities of mold flow or in corners where the flow of rubber in two or more streams cuts off air so that it cannot escape.

Short-term remedies are to reduce injection pressure or decrease the injection ram speed without reducing the injection pressure, thus increasing injection times and allowing air to get away at the same rate as rubber fills the mold.

Long-term remedies are concerned with venting the mold to get the air away as quickly as possible. Vents are placed at the site of the air trapping and made so that they do not cause undue flashing, thus causing trimming problems and spoiling the appearance of the molding. The application of a vacuum pump may be necessary in extreme cases.

Mold design to control the way in which the mold fills should prevent obstacles that obstruct expulsion of air and it should eliminate dead corners either in the cavity or the overflow groove. Sharp corners should also be avoided in sprue channels.

Compounding to increase viscosity can be helpful.

- Backgrinding

Strain or distortion at the gates of a molding is caused by thermal expansion of rubber while cure is taking place. It can be made worse by maintaining the injection, and/or hold pressures, for too long. It can also be exaggerated by high mold temperatures.

The remedy is to reduce the time of action of the hold pressure. Reduction of mold temperature or increase of the injection temperature will help by reducing the temperature difference between rubber and mold, thus reducing the thermal expansion.

A safer compound will allow necessary expansion to take place before cure sets in.

- Reversion

If reversion is the problem, then the first remedy is to lower the mold temperature and, if possible, to increase the injection temperature to maintain the production rate.

The long-term remedy is to examine the cure system carefully. Increase of accelerator level should help. This and a simultaneous reduction of the sulphur level is really the answer, but the proposed modifications should be checked by rheometer to be sure that equal or slightly greater cross-linking density is given by the modified cure system.

- Oxidation

Air trapping frequently leads to very severe local oxidation and stickiness in the neighbourhood of the air trap. This is sometimes referred to as reversion.

The remedy is to avoid air trapping by good mold design. A good antioxidant system backed up by use of thiuram disulphide accelerators helps to minimize the effects of very minor air traps.

12.2.3 Important Troubleshooting Considerations

T. Osswald

Understanding the flow characteristics of polymers during mold filling is critical when developing a strategy for setting up the first stage. When troubleshooting injection molding, it is important to consider that polymers are shear thinning*. A major aspect in maintaining a stable process is to achieve a constant flow rate into the mold cavity during filling. The first strategy for a stable process is to fill all cavities at the same time and to be consistent from shot to shot. In order for every cavity to experience the same flow and pressures, the runner system must be balanced. How a runner system is balanced correctly is discussed in Chapter 3 of this handbook. The ram speed or the filling time must be maintained constant from cycle to cycle. Since many machines do not have the capability of recording the ram speed, the filling time can be used as a control parameter.

Most hydraulic machines have some type of valve to control the flow of oil to the hydraulic ram. This valve can be manual, servo-valve, proportional, cartridge, and so on. These valves can be adjusted manually or electrically, closed-loop, or open-loop, to regulate the flow of oil to the ram cylinder. The pressure differential, Δp, needs to be set for these valves to function properly. The first-stage pressure must be set higher than the maximum pressure required to fill the mold cavity 99%. If the first-stage pressure on the pump's side of the hydraulic system nears the pressure in the ram's side, the speed of injection will decrease. This will inevitably lead to a short shot due to the multiplying effect of the polymer melt's shear thinning behavior.

Caution must be exercised when setting Δp. In order to achieve a constant injection speed, Δp must be sufficiently large. However, we must also protect the system from detrimental influences such as contaminants or unmolten pellets that can block one of the cavities in a multi-cavity system, which can lead to flash around the cavities that are overfilled. Here, a dangerous situation can arise that can result in tool damage. To avoid this type of problem, the ram must slow down in the last stage of mold filling. However, a well-built mold should be able to withstand a certain amount of excess pressure. Overpressurizing the mold cavity is likely to happen for a number of reasons throughout its life. The question is how much extra pressure is required to still have velocity controlled filling, without risking mold damage during its service life.

Methods to find the minimum Δp are available, but are not commonly known. One way to determine this pressure is to raise the hydraulic pressure limit in the first stage while molding short shots until the filling time stops decreasing. It should then be easy to determine how much Δp is required to fill the mold cavity 99%. The Δp is the difference between the pressure requirement of the hydraulic ram and the pressure set for the first stage. This Δp must be positive. The first-stage set point limit is a relief pressure. The next step is to set-up a trial with the most viscous lot shipped by the resin supplier.

* The shear thinning behavior of polymers is discussed in detail in Chapter 3 of this handbook.

It is important to point out that when talking about set pressures for the first-stage, we are discussing actual polymer melt pressures and not hydraulic pressures. It is also important to note that this troubleshooting guide and strategy assumes that the injection molding machine is in proper working condition. Machine problems must be fixed before continuing troubleshooting.

12.3 Technology and Process Troubleshooting

J. Wickmann*

Innovations in product development, new materials development, speed to market, and high-volume automated manufacturing has made the injection molding process a viable option. The injection molding process involves a molding machine, mold, and plastic material. With all these elements, injection molding processes, like any other manufacturing process, is continually faced with the issues of process consistency and part variation. The following discussion focuses on the technological innovations in the injection molding industry and how they affect the troubleshooting of the process.

Problem solving is often referred to as troubleshooting. Albert Einstein once said that a problem understood is half solved. Understanding the problem is not an easy task, especially when it comes to top-end precision injection molding processes, because there are so many variables to consider. Troubleshooting becomes a trial and error method when the troubleshooter does not exactly know which variable is affecting the part quality and to what extent. The troubleshooting process can become time consuming and costly.

As the process and parts become more sophisticated, troubleshooting has to be approached using a more scientific method. One must be able to distinguish between the material, machine conditions (settings), process outputs, and the part characteristics. Troubleshooting becomes easier when the process outputs can be measured and the troubleshooter is able to draw from a known relationship between the process inputs, outputs, and the part characteristics.

Troubleshooting needs to start when the customer identifies his requirements. Process troubleshooting can be done effectively when it is approached as a team effort. The troubleshooting team should consist of representatives from the following functions of an enterprise:

1. Customer
2. Design engineering
3. Project engineering
4. Mold design and building
5. Process engineering

* With contributions from Treasa Springett and Raghu Vadlamudi.

6. Manufacturing engineering
7. Quality engineering
8. Management
9. Customer service

The customers in this team brings in his requirements and experience with similar products. Together with Customer Service, the customer can specify the time lines. Design for manufacturability (DFM), *design for assembly* (DFA), and *design for disassembly* that can be achieved with the cross-functional working of all team members.

12.3.1 Technology Implications

Technological advances in the injection molding industry have made available design software, mold filling simulation software, finite element analysis (FEA) programs, and statistical techniques for the team members to contribute to the troubleshooting process. All of the preceding mentioned functions of an enterprise working together can lead to an efficient, robust process, designed to withstand the inherent variations found in the injection molding process.

Computer-aided engineering (CAE) software packages help design engineers develop optimum designs suited for injection molding in a relatively short period of time. The chapter on simulation in injection molding in this handbook discusses CAE in more detail.

Simulation software is available along with the design packages to help predict mold filling behavior in the mold. These simulation software packages, as well as FEA programs and various statistical tools, can identify areas of concern that should be investigated. Such areas might include wall thickness transitions that can pose problems during injection, hot spots in the mold cavity or core, pressure distribution in the cavity, and knit lines. Once the problematic areas are identified the design engineer can make changes in part geometry to improve material flow characteristics. A number of iterations can be done before cutting the cavity material. In an effort to minimize costly in-process troubleshooting it is essential to both economical and high-quality part production that a thorough detailed analysis be done proactively.

Commercially available software can be used as a tool for validating part and mold design prior to production. Validation of part and mold design involves sizing the feed system (sprue, runners, and gates) to the mold cavity to obtain uniform melt pressure distribution. It also involves designing an optimized cooling circuit to control hot spot development in the mold steel during processing. If this validation and analysis is not done up front, time and effort will be spent troubleshooting the process during valuable production hours. The key step to having effective troubleshooting in place is to transfer the information learned early during the development process to the production floor.

12.3.2 Injection Molding Process and Sensors

The three important elements in the injection molding process are machine, mold, and material. The four critical variables in the injection molding process are time, temperature, pressure, and injection speed. These four variables interrelate with each other. Due to changes in technology and better understanding of the injection molding process it has been found that cavity pressure is the most critical variable. Troubleshooting becomes easier when one can correlate the part quality to the process variables. This can be done during the process development stages using sensors and statistical techniques. The statistical techniques include design of experiments (DOE), Analysis of Variance (ANOVA), control charts, and capability studies. Some of these aspects are covered in the chapter on statistical process control in this handbook.

The first step toward achieving process control is to use an injection molding machine that can monitor the process parameters. In recent years, machine manufacturers, have developed sophisticated injection molding machines that can provide process parameter statistics. These statistics can help injection molding technicians to understand the process condition, and to react accordingly, as long as they are trained in using statistical process control systems.

It is very important for the process troubleshooter to understand how process variables will affect part quality and functional, dimensional, and aesthetic requirements. Numerous studies [3] have repeatedly shown that process variables such as cavity pressure, melt temperature, and mold temperature are the primary factors that affect the part characteristics.

The most important process variable is cavity pressure. The cavity pressure profile is considered as representative of the quality of the molded part. A typical cavity pressure profile is shown in Fig. 12.1. Machine conditions can be monitored using either peak cavity pressure or cavity pressure integral [3].

Any changes in the mold temperature, hold pressure, hold time, and injection speed will have an effect on cavity pressure. In thin-wall parts peak cavity pressure becomes a critical variable, whereas in thick-wall parts cavity pressure integral becomes a critical variable. Hold pressure and hold time affect the peak cavity pressure, whereas the injection time and mold temperature affect the cavity pressure integral. The part dimensions will vary with the change in cavity pressure integral. The cavity pressure integral changes with the fluctuations in mold temperature. As the mold temperature varies, so does the time for the cavity pressure to reach atmospheric pressure at which the part shrinkage starts.

The cavity pressure profile is unique to each mold and material combination. Depending on its composition each material has its own flow behavior inside the mold. The amorphous materials have a wide range of melt temperatures, whereas the crystalline materials have a sharp melting point. Figure 12.2 shows typical cavity pressure profiles for amorphous and crystalline materials.

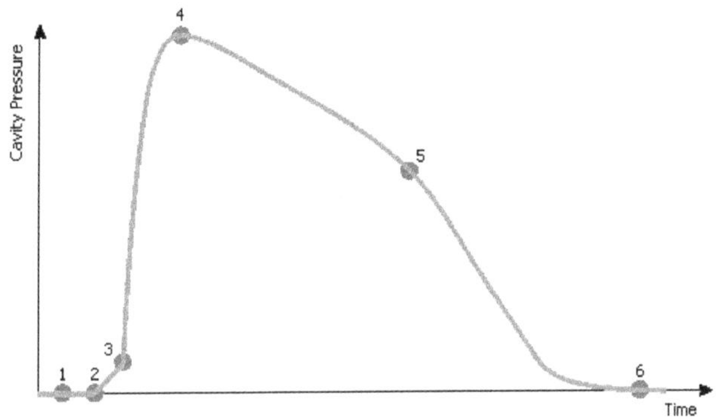

Figure 12.1 Typical cavity pressure profile. 1. start of injection; 1–2. injection of melt into cavity; 2. melt reaches sensor; 2–3. filling cavity; 3. volumetric filling; 3–5. compression of melt; 4. maximum cavity pressure; 5. gate sealing point; 6. atmospheric pressure = start of shrinkage. (Courtesy of Kistler Instruments Corporation).

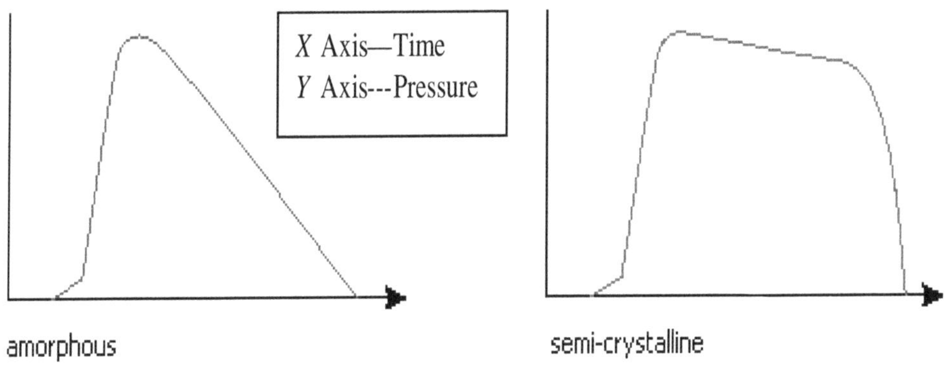

Figure 12.2 Typical cavity pressure profiles for amorphous and crystalline plastics (Courtesy of Kistler Instrument Corporation).

12.3.3 Pressure Sensors

The pressure sensors can be either direct contact– or behind-the-ejector-pin–type. Figures 12.3 and 12.4 show the illustrations for installation of both types of sensors. Behind-the-ejector-pin–type sensors will be used when there is no space for the direct contact–type pressure sensors. The pressure sensors located in the mold will provide more useful information than will the pressure sensors located in the machine's hydraulic cylinder. The pressure drop through the injection barrel, nozzle, sprue, runner, and gates provides cavity pressure information that is more beneficial.

The installation and location of these sensors is very critical. The location can be near the gate or at the end of fill. The sensor near the gate will give more information than will the sensor at the end of fill with regard to the material flow behavior inside the cavity. Another important guideline is to locate the pressure sensor in the thick-wall area. The advantage of locating the pressure sensor in the thick-wall area is that it is the last place the resin in the mold cavity solidifies. Even though process control can be achieved by using the sensor near the gate, sensors at the end of fill can be used efficiently to sort good and bad parts so that the defective parts will not be shipped to the customer. The sorting of good and bad parts can be done by connecting the cavity pressure monitoring system to a flip chute or a robot that can be used to sort parts. The peak cavity pressure or cavity pressure integral windows can be established during the process development stages. The pressure sensors, which are used in injection molding industry, are either strain gauge–type, piezoelectric, or ultrasonic. Strain gauge technology uses electromechanical transducers that convert force into voltage. The change in voltage produced can then be converted to digital read-out using data acquisition systems.

Piezoelectric sensors consist of "quartz (Silicon dioxide)" that develope an electric charge proportional to the force when force is applied [4]. This charge can then be converted to a voltage using a charge amplifier. This voltage signal can then be converted into a digital read-out. Figure 12.5 shows quartz cavity pressure transducers. Although the preceding two types are intrusive in nature, ultrasonic sensors are nonintrusive and have fast response times. These sensors can be attached to the outside surface of the mold. Research is being done, using ultrasonic pulse-echo techniques, to monitor the injection molding process [5].

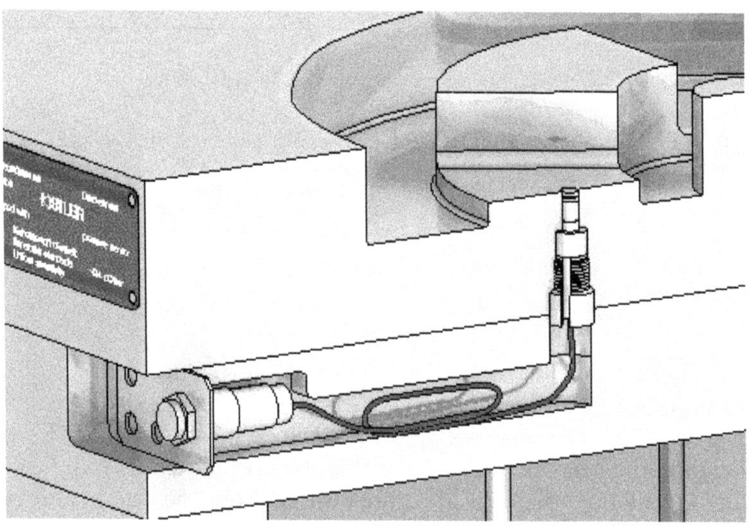

Figure 12.3 Flush-mounted pressure sensor (Courtesy of Kistler Instruments Corporation).

Piezoelectric and strain gauge transducers are widely used in the injection molding industry. Each piezoelectric sensor costs around $1000. The sensors are available in sizes as small as 1 mm. Installation of pressure sensors is easier if they are designed into the initial mold design. Depending on the number of cavities and the degree of process control needed, the number of transducers required will vary, as will the total cost.

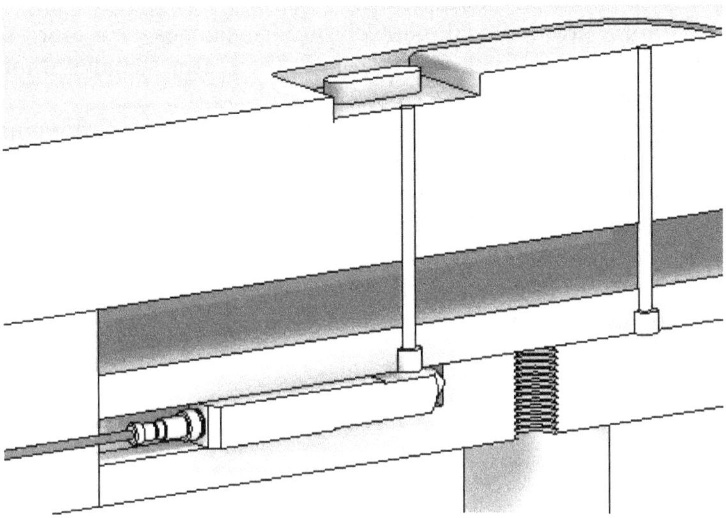

Figure 12.4 Behind the ejector pressure sensor (Courtesy of Kistler Instruments Corporation).

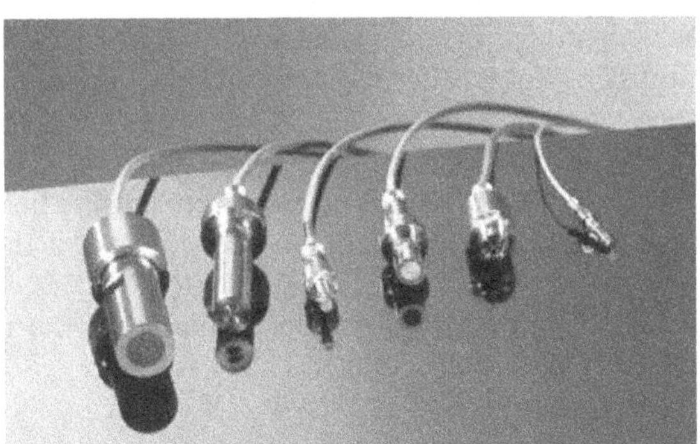

Figure 12.5 Quartz cavity pressure sensors (Courtesy of Kistler Instruments Corporation).

12.3.4 Temperature Sensors

Temperature is another important variable in the injection molding process. Research shows that melt temperature can greatly influence product quality. The standard practice in the injection molding industry to measure melt and/or mold temperature is to use temperature probes. Using temperature probes involves cycle interruption and completely depends upon the technique and the operator. Measuring melt temperature using probes involves shooting the melt with the nozzle pulled back and inserting melt temperature probe into the hot plastic. It becomes even more difficult to measure melt temperatures with melt probes when using smaller shot capacity machines.

The melt temperature can be measured and monitored using flush-mounted thermocouples. The better location for thermocouples to measure and monitor melt temperature is the molding machine barrel end cap [6]. Melt temperature can also be measured by using IR pyrometry. The IR transducer can be used to detect melt homogeneity [7]. The IR pyrometers have much faster response times than do conventional thermocouples. The response time of IR sensors is 1000 times faster than conventional thermocouples and is measured in milliseconds [8].

12.3.5 Process Monitoring

Process monitoring systems serve as tools in troubleshooting the injection molding process due to the fact that real-time data acquisition can be accomplished. The process monitoring system provides continuous feedback to the technicians and the operators to help in process improvement. These systems acquire signals from the machine, pressure sensors, and temperature sensors, and display them in a graphical manner. Another advantage of today's process monitoring systems is that they are equipped with statistical tools. These systems all have the capability to monitor up to roughly 60 process parameters.

The monitoring systems allow data collection and show the data graphically in a user-defined format. Data collection and process monitoring are beneficial only when procedures are in place for the action plans on how to react if there is an adverse trend in the data.

There are a number of monitoring systems available in the injection molding industry.

12.3.6 Automatic Troubleshooting

Traditional injection molding methods obtain volumetric filling of the cavity by switching over from injection pressure to hold pressure based on time, hydraulic pressure, screw position, or cavity pressure. The switchover is performed at a fixed point.

The disadvantage with this type of switchover is the need to re-establish the switchover value when there is a change in the process. Changes may be needed due to material lot-to-lot variation, injection pressure variation, temperature variation, or a combination of all these things. Any such change causes variations in material viscosity that in turn produces different cavity pressure profiles and different part characteristics.

Kistler Instruments Corporation has developed a microchip based on artificial intelligence (Automatic Quality Control System, AQCS Type 5881A) that can detect a change in the slope of the cavity pressure curve during the volumetric filling of the mold cavity just before the compression of melt. The switchover from injection pressure to holding pressure is done after detecting the volumetric filling of cavity to produce consistent parts.

The next generation of technology in this field will use artificial intelligence along with neural networks.

12.3.7 Design of Experiments

Design of experiments (DOE) is a very powerful and critical tool for improving the performance of a manufacturing process. It is also very useful in the development of new processes. The objective of DOE is to develop a process that can minimally be affected by ambient conditions and material lot variability, and to identify the critical variables of the process.

According to Montgomery [9], there are a few different DOE experimental strategies. The first one is the best-guess approach performed by experienced personnel who have a very good technical knowledge in the field. The drawback of this approach is taking another guess if and when the initial approach does not produce acceptable results. This approach might provide acceptable results but probably not the optimal results. The second experimental strategy is the one-factor-at-a-time approach. This approach involves studying the effect of one factor over its range on the process by keeping all other factors constant. This approach is very time consuming and does not give any information about the interaction between the factors. The interaction between the factors is present when an independent variable's effect on a response variable is depended upon another independent variable's level. The best approach is to use a multifactorial experiment. In this approach the factors are varied together. Factorial experiments allow the experimenter to estimate the effects of a factor at different levels of other factors. This approach also provides information about the interaction of various processing factors.

Adopting design of experiments to injection molding is a critical step to working in 6 sigma, International Organization For Standardization (ISO), and Food and Drug Administration (FDA) environments. It is very important and a requirement in the medical device industry to validate the process before the commencement of production. The validation of the injection molding process includes identification of critical variables and to show the capability of the process over short or long term. The critical variables in the process can be identified using design of experiments. Design of experiments can be used to develop a process that reduces the variability and improves capability. The data obtained from the DOE can be analyzed by using ANOVA to obtain signal-to-noise ratio. The objective should be to maximize the signal and minimize the noise.

The important elements of DOE are the independent variables (factors) and dependent variables (responses). In the injection molding process the independent variables are the process variables like time, temperature, and pressure. The dependent variables could be those defined by the customer. The process characterization should be done to reduce variation to satisfy customer's requirements and to simplify manufacturing.

12.3.8 Factorial Design of Experiments

The number of experiments in a factorial design of experiments is 2^n where n is the number of factors taken into consideration. Because there are many variables in the injection molding process, it is important to screen out unimportant factors to minimize the number of factors and experimental runs. For example, a 6 factor 2 level DOE would require 64 experimental runs. A DOE otherwise becomes a lengthy, costly, and laborious process that potentially produces no significant information.

It is standard practice in the industry to consider machine settings like barrel temperatures, cooling time, hydraulic pressures, and water temperature set points as principal factors. These conditions are related to part quality and characteristics only to some extent. With the pressure and temperature sensors that are available in the industry, it is feasible to measure the cavity pressure, melt temperature, and mold steel temperature. DOE will be more meaningful if melt and mold temperatures and cavity pressures are considered as the factors. Screw speed, back pressure, barrel temperatures, and water temperature set points are considered as background factors (secondary importance) that affect the previously mentioned plastic conditions. DOE software that is available in the industry allows the user to determine the main effects of factors on part quality and characteristics and also the interaction between the factors. DOE software geared specifically for injection molding process is also available.

The DOE is more effective when plastic conditions are considered as factors rather than the machine settings. It is critical to seek help from the people experienced in injection molding before designing an experiment. If the DOE is not set up properly the outcome of the experiment may not be useful. One has to be prepared to conduct DOE again by adding new factors and responses, if the factors chosen do not show any effect.

It is important to do preliminary work in designing an experiment. Information from material manufacturers and similar parts that were molded previously is of great importance to maximize the effectiveness of a DOE. A typical work sheet and a matrix for a DOE are shown in Tables 12.2 and 12.3.

Table 12.2 Design of Experiments Work Sheet

Objective: Define clearly the process, expectations, and actions to be taken

Factors:

Name	Units	Type	Maximum	Minimum

Responses:

Name	Units	Measurement Type	Maximum	Minimum

Comments:

Courtesy of Phillips Plastics Corporation.

Table 12.3 Design of Experiments Matrix

Run Number	Factor 1	Factor 2	Factor 3	Response 1	Response 2	Response 3	Response 4
1	Low	Low	Low				
2	Low	Low	High				
3	Low	High	Low				
4	Low	High	High				
5	High	Low	Low				
6	High	Low	High				
7	High	High	Low				
8	High	High	High				

Courtesy of Phillips Plastics Corporation.
High and Low are the levels of factors chosen for DOEs.
Response 1, 2, 3, 4 . . . can be part characteristics.

12.4 Conclusions

Process troubleshooting has to go beyond the nuts and bolts of cause and effect. Injection molding of top-end/precision parts has become a very sophisticated process. The trends are very clear and global competition will drive the industry. Domestic and global demand for top-end/precision parts will continue to increase. The issue of standardization (ISO, QS, FDA, etc.) is going to become more common. Defects per million opportunities will continue to be quantified and reduced, and 6 sigma and the cost of quality will be very common. High-quality, inexpensive technological applications for injection molding will continue to flood the global market. Everyone will continue to have access to high-technology injection molding. Demand for high skill levels will continue to increase. Many of us will work in cross-functional teams. People with knowledge and skills will be seen as assets. Design of experiments becomes commonplace and routine. Material specifications will become quantified and variability of many properties reduced. If materials variability is not reduced, machine sensors will compensate for it. Key Processing Characteristics (KPCs) will be identified, quantified, controlled, and reduced.

References

1. Harry, M., Schroeder, R., Six Sigma (2000), Doubleday Press, New York.
2. ASQNet\Foresight 2020, *http://www.asqnet.org/member/futures/key_forces.shtml.*
3. Manzione, L. T., *Applications of Computer Aided Engineering in Injection Molding* (1987), Hanser Publishers, Munich.
4. Bichler, M., Schwaig, *Germany Selecting and Evaluating Process Parameters for Quality Assurance*, Plast Europe (1994).

5. Wen, S.-S.L., Jen, C.-K., Nguyen, K. T., Derdouri, A., Simard, Y., *Recent Advances* in *Ultrasonic Monitoring of the Injection Molding Process*, (1999), Industrial Materials Institute, National Research Council Canada, Quebec.
6. Anonymous, When it comes to injection, molding can learn a lot from extrusion *Injection Molding Magazine* (1999), February.
7. Reddy, V. N., Schott, N. R., In-Situ Real Time Monitoring of The Transient Melt Temperatures In The Nozzle During Injection Molding Via IR Pyrometry, ANTEC '95.
8. Anonymous, Learning lessons from IR temperature measurement, *European Plastics News* (1995), March.
9. Montgomery, D. C., *Design And Analysis of Experiments* (1997), John Wiley & Sons, New York.
10. Moldflow Corp. (www.plasticszone.com)

13 Materials Troubleshooting

M. Sepe

A successful plastic product represents a combination of good part design, good design and construction of the mold used to produce the part, sound processing techniques, and proper material selection. These four disciplines are interconnected; therefore, any discussion of materials must factor in an awareness of the other three. In addressing the problems that materials can cause in a product, however, we can focus on some characteristics that are unique to materials.

The first of these is *composition*. In order for a part to perform as expected, it must be made of the specified material. Despite the fact that we tend to think of the raw material in terms of the polymer type (ABS, polypropylene, polycarbonate, etc.), we deal in the real world with compounds, not pure polymers. Any commercial molding material is a mixture of polymer, additives, and, in many cases, fillers and other modifiers. All of the prescribed ingredients must be present in the correct amounts. We have to start with the right material and we have to ensure that it remains free from contamination as it is converted into the finished part.

The second area of concern is *degradation*. This concern most often focuses on the polymer. Polymers are unique materials that derive their properties from the large size and extended chain structure of the molecules. Injection molding is a rigorous treatment of this molecular structure, and the possibility exists, albeit more in some polymers than in others, that the chain size will be altered significantly during processing. This modification usually takes the form of a reduction in the average size of the molecules, which in turn is related to the molecular weight of the polymer. For the part to be successful, the process must preserve the molecular weight of the raw material. The process must also prevent the excessive loss of the stabilizers that are added to the polymer to protect it from adverse chemical reactions such as oxidation.

Finally, the material selected must be right for the part. Successful material selection begins with an early and accurate assessment of application conditions. As self-evident as this sounds, the number of times when material selection is actually approached scientifically is shockingly small. As a result, a lot of the pain in product failure is self-inflicted during the material selection stage of the development process. This may come in the form of an inappropriate polymer selection, or it may manifest as an ignorance of some of the finer points of material selection, such as using a compound with a minimal level of a particular stabilizer or selecting a grade with an inappropriately low molecular weight.

This chapter will discuss the processes and tools that are used to diagnose and confirm the nature of material-related product failures. Because manufacturing is performed in increasingly far-flung regions of the world, and as material suppliers abandon their traditional role as the analytical adviser for the processor, timely analysis of these types of problems becomes a competitive necessity. At the same time, this is the real world. The focus must be on the practical and the cost effective. There are always several ways to approach a problem that involves the material. The answer often feeds back to one of the other three disciplines mentioned previously. This chapter will seek to provide a familiarity with the most important tools of material analysis from the problem-solving perspective, while at the same time touching on some of the more sophisticated techniques that are occasionally needed to get to a permanent solution.

13.1 Composition Problems

Performance problems that arise from composition fall into two major categories:

- The first is a slight variation in the intended composition that causes the properties to drift from those expected by the customer. A homopolymer polypropylene may have been substituted for a copolymer in an application where cold-temperature impact properties are important. A glass-fiber content may be lower than specified, causing a product to lack stiffness and strength. An application that requires a material with special antioxidant to withstand an aggressive end-use environment may be changed to a material containing a general-purpose stabilization package.
- The second type of composition problem is *contamination*. Here, the correct material becomes mixed with a second material to produce an unpredictable result.

Contamination is usually fairly easy to detect, even during processing. The foreign material may process at a significantly different temperature than the primary resin, leading to either an unexplainable increase in viscosity (polycarbonate gets into a batch of polystyrene), or an appearance of rapid degradation (acetal is mixed with nylon 6/6). There may be other appearance problems, such as a change in gloss or color. Because very few polymers are naturally compatible, there are often obvious signs of *delamination*, a phenomenon where the material peels into distinct layers. This behavior may be especially noticeable in high-stress regions of the flow path, such as near gates and weld lines. These types of problems are usually caught early because they disrupt production, but the reason for the problem may not be readily identifiable. More importantly, once the foreign material is identified the task of pinpointing the source may not be straightforward. With the increased use of recycled materials, contamination can find its way into the product stream in a variety of ways. As we will see, it pays to be analytical and not assume too much.

The difficulties that arise from substitution of one compound for another are much harder to identify until the product gets into the field. The problem will sometimes not be obvious until months of field service have elapsed when a chemical resistance or a long-term heat aging condition finally catches up with the part. For this reason, this type of problem is more expensive and up-front testing of the incoming material may be a cost-effective strategy; however, the tools of the trade are essentially the same in both types of situations. We will deal first with problems related to polymer composition, and then move to issues related to fillers and reinforcements. Finally, we will address some of the difficulties caused by additives.

13.1.1 Diagnostic Tools for the Polymer

There are three primary analytical tools that are well suited to the task of identifying the polymer or polymers in a compound. They often overlap and provide supplemental information. The first of these, and probably the one that is used most often, is infrared (IR) spectroscopy. IR is uniquely suited to analyzing polymeric materials because polymers are organic. The types of chemical bonds that are found in organic materials absorb radiation in the infrared region and therefore leave a very good fingerprint of their structure. Thirty years ago the business of analyzing the test results of an IR scan relied on the expertise of the analyst in picking out the distinguishing characteristics of the spectrum. Today, the use of the computer permits the incorporation of extensive libraries of known materials into the software for rapid comparison and matching.

In using any analytical technique it is important to understand its limitations. The use of computerized aids often fosters an attitude of blind faith in the result. Distinguishing between materials is only possible if the technique can identify differences. Figure 13.1 shows a comparison between the IR spectrum of general-purpose polystyrene and impact-modified polystyrene. Following the absorption peaks from left to right shows that there is great similarity between the two materials. This is due to the fact that both the polystyrene and the butadiene rubber that is used as the impact modifier are both hydrocarbon materials containing similar types of chemical bonds; however, the butadiene rubber has an important absorption that the polystyrene does not have. This occurs at a wavenumber of $965\,cm^{-1}$ and is a signature that the analyst or the computer can identify to distinguish between the two materials. If care is taken in preparing the samples, the relative intensity of this peak can be compared between different impact-modified materials to identify differences in the level of butadiene used in each compound.

Figure 13.2 shows an IR spectrum for an ABS material. This spectrum resembles those of the two materials in Fig. 13.1. This might be expected because ABS contains both polystyrene and butadiene rubber. ABS, however, also contains acrylonitrile, which is a component that imparts heat resistance and chemical resistance and extends the utility of the material beyond the range where impact polystyrene can

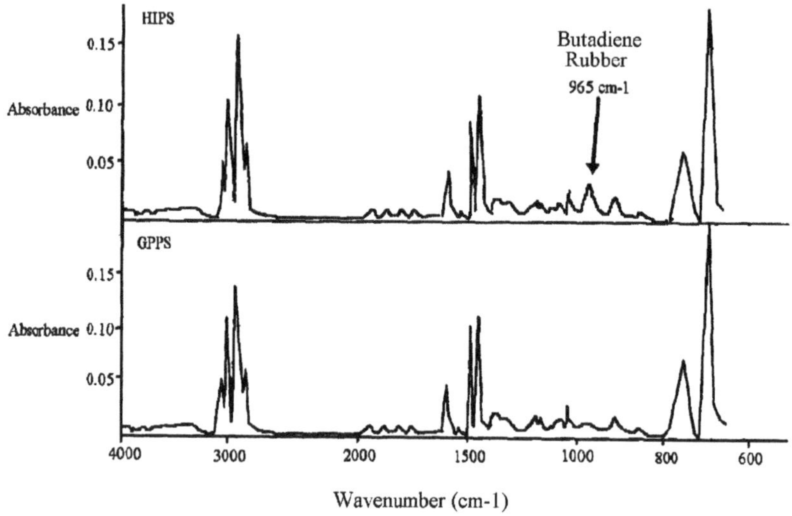

Figure 13.1 Comparison of high-impact and general-purpose polystyrene by IR spectroscopy.

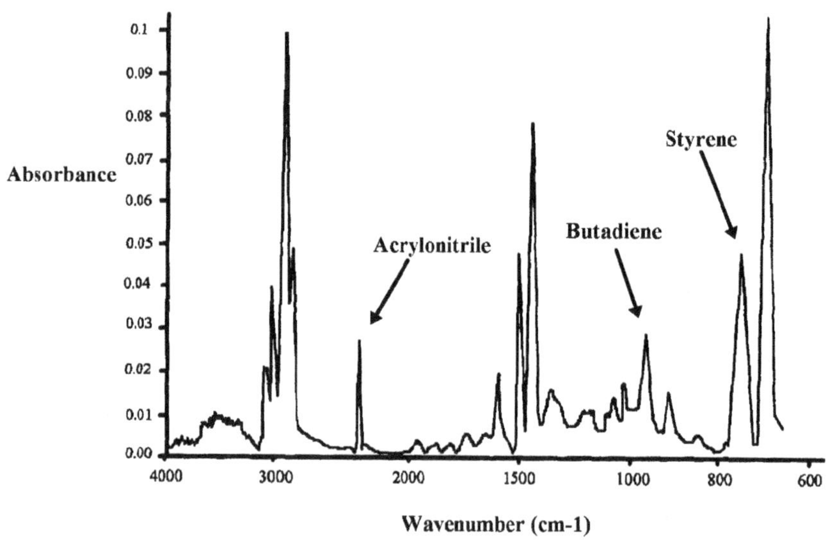

Figure 13.2 Infrared spectrum of ABS showing key peaks that identify each monomer.

operate. Acrylonitrile shares many of the same bonds that are found in styrene and butadiene, but it also contains a unique linkage known chemically as *a nitrile group*. This group possesses a very distinctive absorption at 2237 cm^{-1}. Once the absorptions unique to each of the three ingredients have been identified, it may be possible to compare the intensity of these absorptions to make quantitative judgments about the composition of a particular grade of material. Acrylonitrile, butadiene, and styrene can be combined in various concentrations to produce a broad range of property combinations. IR spectroscopy is the best technique for examining these differences by focusing on the key peaks shown in Fig. 13.2.

The intensity of the absorptions that we have been pointing out is dependent on several considerations. Some of these have to do with the inherent characteristics of the chemical bonds being examined. Some bonds are relatively weak infrared absorbers, whereas others are inherently strong absorbers. Other factors have to do with sample preparation. Sample thickness will have an influence on the intensity of the absorption peaks. The thicker the sample, the stronger the absorption peaks will be. The method of measurement is a factor. The two most common methods of evaluation are transmittance and absorbance. In transmittance, the peak intensity represents a ratio of the IR beam passing through the sample compared with a beam passing through a reference. In absorbance, the comparison represents the logarithm of this ratio. Absorbance is typically the preferred method for quantitative work; however, it is important even here that certain rules be observed. If an absorbance spectrum has a vertical axis that exceeds a value of 1.0, then the analyst is working in a region where the mathematics is unreliable.

For all of the reasons mentioned above, any strict quantitative interpretation of a spectrum such as the one in Fig. 13.2 should be treated with caution. It is perfectly legitimate to cite shifting ratios in peak heights as evidence of a change in composition. But when an analyst starts using the peak heights to calculate absolute percentages of acrylonitrile, butadiene, and styrene, they have extended the utility of the method beyond the point of sound judgment. Only with the use of several known standards can such absolute percentages be derived, and the availability of such standards is very limited. In any analytical method, beware of overinterpretation.

Because IR spectroscopy looks at the basic chemical linkages that hold the molecules together, it is a useful technique for distinguishing differences in essential chemistry. There can be differences, however, in materials that do not change the essential chemistry but significantly alter the performance of the material. Figure 13.3 shows an IR spectroscopic comparison of an acetal homopolymer and an acetal copolymer. A thorough examination of these two spectra shows no differences because the types of chemical bonds present in both compounds are the same; however, there are substantial differences in performance that arise from this difference in structure. The homopolymer has a greater degree of structural regularity, which gives rise to a higher degree of crystallinity. In the short term this results in higher strength and stiffness and greater heat resistance. All of these differences can be confirmed by examining a property sheet for commercial grades from each family. Even though both types of acetal have similar strengths and weaknesses in chemical resistance, the ingredients used to produce the copolymer give it a slightly higher

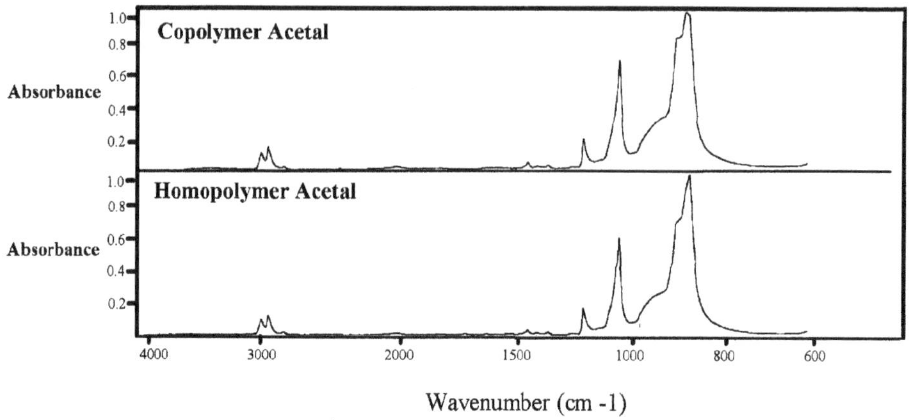

Figure 13.3 Comparison of acetal copolymer and homopolymer by IR spectroscopy.

degree of long-term stability. Extended periods of exposure to elevated temperatures in air or in water therefore favor the copolymer. In addition, moderately acidic and basic media that will degrade the homopolymer will have much less of an effect on the copolymer.

The infrared spectra in Fig. 13.3 were part of an initial study to compare good and bad parts from a food processing application; the mixing of large batches of barbecue sauce. The good parts were holding up well to the temperatures and the acidic chemistry of the sauce while the bad parts were eroding badly. A comparison by IR spectroscopy indicated that the materials were the same. While the labels on the graphs in Fig. 13.3 indicate that one is a homopolymer and one is a copolymer, this information was not known at the time the tests were done and was only discovered by falling back on an alternate technique.

That technique is called *differential scanning calorimetry* (DSC). DSC measures the heat flow into or out of a material as it changes temperature. All materials have a property called *heat capacity*. It is a measure of the energy input required to increase the temperature of that material by a fixed amount. Because the total energy is governed by the mass of the material and the size of the temperature change, it is convenient to normalize this property for both mass and temperature change. This normalization gives us the measurement of specific heat. *Specific heat* is defined as the amount of energy required to raise the temperature of 1 g of material 1°C. This property has become very important in the plastics industry as an input into simulation programs for mold filling and cooling. It is also important to processors because it governs the amount of work that the heater bands and the screw must do together to establish a homogeneous melt. In order to raise the temperature of a material, energy must be put into the material. DSC quantifies that input.

Transition temperatures such as the melting point offer an alternate tool in the task of material identification. Breaking the crystal structure of a material requires an additional energy input that causes the output from the DSC to deviate from the baseline established by the specific heat properties. Figure 13.4 shows a typical DSC result when a semi-crystalline polymer, in this case a polypropylene, is melted. As the crystals begin to melt, the energy input required to maintain the programmed rate of rise in temperature increases. This increase continues until the majority of the crystals have melted. The curve then returns rapidly to baseline and heating of the molten polymer continues. There is a great deal of information in this energy well between solid and liquid. The endpoints of the curve define the onset and endpoint of crystal melting. The peak temperature at the minimum point of the curve is typically considered to be the melting point, even though it is obvious that melting is a process that spans a relatively broad temperature range. In addition, the area under the curve is a relative indication of the concentration of crystals present in the polymer.

This brings us to a very important side discussion of polymer structure. We tend to categorize polymers in many different ways, but one of the most important of these categorizations has to do with whether or not a material is capable of forming crystals as it cools from the melt. Those materials that do not form crystals as they cool are known as *amorphous*; those that do are called *crystalline*, or, more appropriately, *semi-crystalline*. The polypropylene example in Fig. 13.4 is semi-crystalline. We know this because we can identify the melting event associated with the presence of those crystals. If the same test were performed on a polycarbonate or an ABS, we would

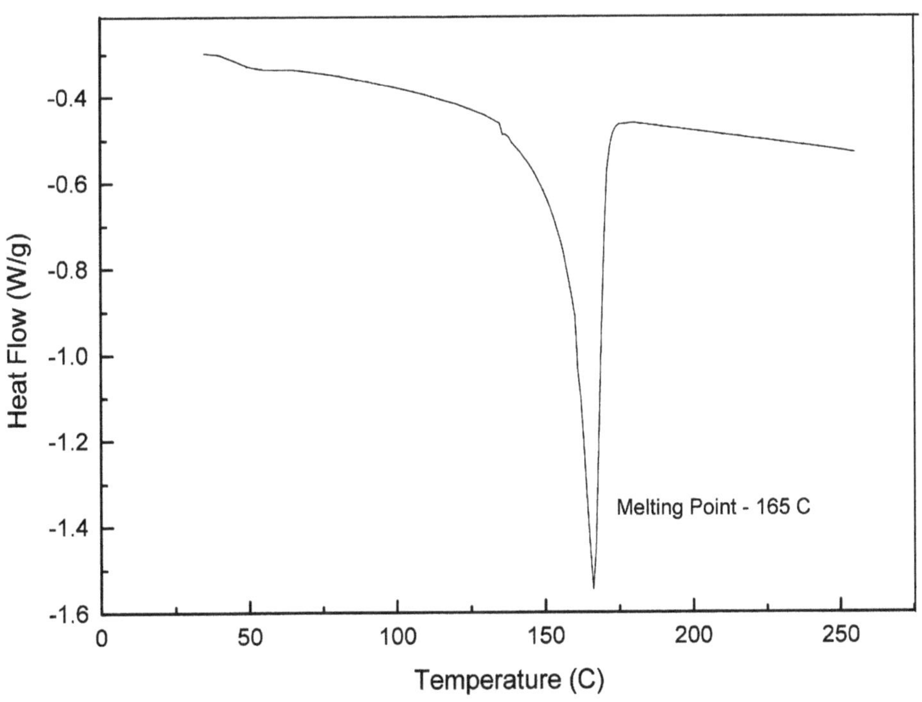

Figure 13.4 DSC heating scan for polypropylene homopolymer.

not detect such an event because there is no structure to melt in the traditional sense; however, even when a material is capable of forming crystals as it cools, this is not like ice, where it is assumed that virtually every water molecule takes up a predictable position in the crystal matrix as the water freezes. The complex shape and large size of the polymer molecules cause the crystallization process to be incomplete. Some regions of the polymer become well organized into crystals, but other regions remain amorphous. The relative degree of crystallized regions and amorphous regions is an extremely important determinant of final properties in the molded part. We therefore refer to a degree of crystallinity rather than talking in absolute terms of crystalline or amorphous structure. The area under the curve for the melting event is a relative indication of the degree of crystallinity, and the absolute measurement of this area is known as the *latent heat of fusion*.

How, then, does this relate to the problem with the blades in the barbecue sauce mixer? Remember that we mentioned that the structural regularity of the homopolymer was greater than that of the copolymer, and that one of the consequences of this difference was superior short-term heat resistance in the homopolymer. This superior resistance is reflected in a higher melting point for the homopolymer. Figure 13.5 shows a comparison by DSC of the good and bad parts from the application. Although the IR spectra indicated no difference, the DSC clearly shows both an offset of 10°C in the melting point as well as a quantitative difference in the heat of fusion. Because the homopolymer has a higher degree of crystallinity due to its more regular structure, it also has a higher heat of fusion. The acetal copolymer was suited

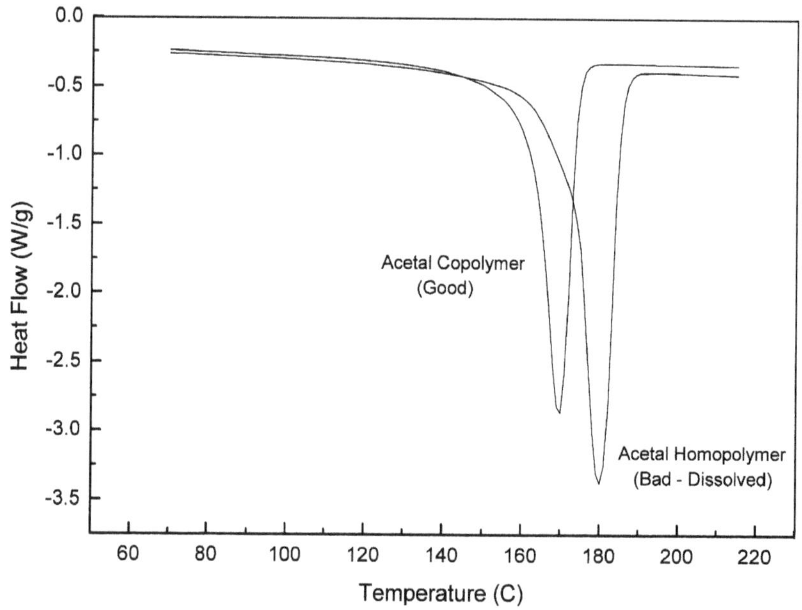

Figure 13.5 DSC comparison of acetal copolymer and homopolymer.

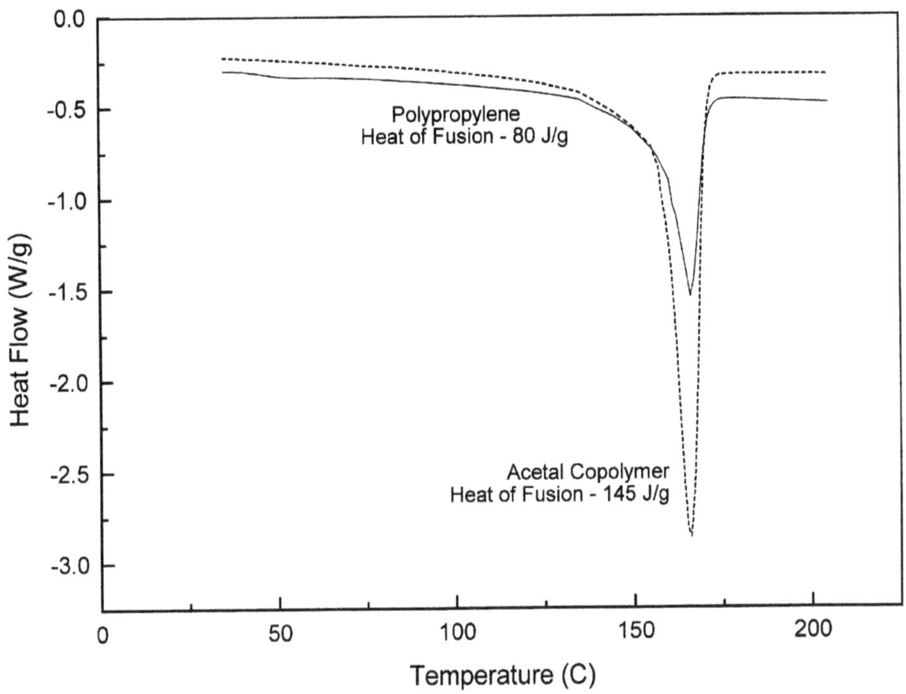

Figure 13.6 DSC first heat comparison of polypropylene and acetal copolymer.

for the high temperatures and acidic chemistry of the application; the homopolymer was not. What seemed like a minor substitution in materials was actually the difference between success and failure.

Other material differences that are better identified by DSC than by IR also involve variations in the structure of a polymer that is otherwise chemically the same. Examples include various densities of polyethylene, where heat of fusion and melting point rise with increasing density, but where the IR spectrum does not change. Different nylons are similarly more easily distinguished by DSC than by IR, as are PBT and PET polyester. Where melting points do not provide a unique fingerprint, details of the heat of fusion can frequently be used to make a determination. For example, acetal copolymer has a melting point between 165 and 168°C, whereas some polypropylene homopolymers cover a range from 164 to 170°C. This makes the determination of composition by melting point alone an ambiguous proposition. The heat of fusion in acetal copolymer, however, is significantly higher, as is shown in Fig. 13.6. Even at its most crystalline, the heat of fusion for a polypropylene will reach a maximum of 115 to 120 J/g; in this case it is only 80 J/g. Typical values are 140 to 150 J/g in acetal copolymer.

Controlled cooling of a material by DSC can reveal other distinguishing characteristics. The instrument used to produce the test results shown in this chapter employs a sign convention such that a downward movement from the baseline is

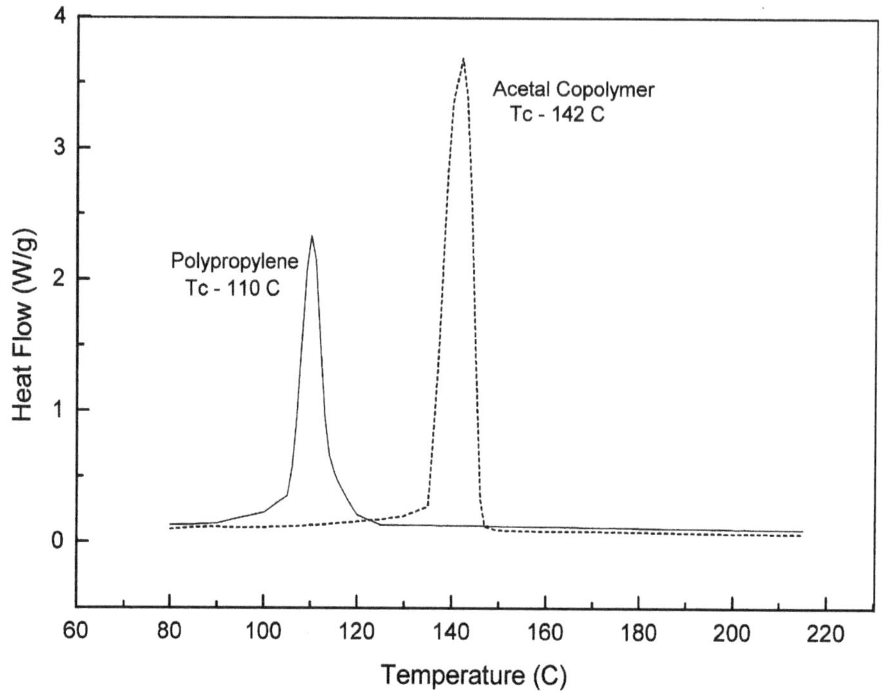

Figure 13.7 DSC cooling comparison of polypropylene and acetal copolymer.

endothermic and an upward movement is exothermic. Melting is an endothermic process; recrystallization of the material from the melt is an exothermic process. The amount of energy released during recrystallization will be close to that consumed during crystal melting; however, the recrystallization temperature also provides an additional fingerprint. Figure 13.7 compares the recrystallization process for the acetal copolymer and the polypropylene. This particular polypropylene exhibits a peak in the recrystallization process of 110°C although a survey of a variety of polypropylene materials shows a variation from 105 to 130°C, depending upon the type of material and the presence of various pigments, nucleating agents, and fillers. The acetal copolymer, however, consistently recrystallizes at a much higher temperature of 142°C. This difference alone is sufficient to distinguish between polypropylene and acetal copolymer.

Another example in similar melting points is the comparison of nylon 6 with PBT polyester. Both of these materials have melting points between 222 and 225°C, depending upon the exact grade. In addition, these polymers develop similar levels of crystallinity so that the heat of fusion comparison that served to distinguish between the polypropylene and the acetal does not help. Figure 13.8 illustrates a comparison between unfilled PBT and nylon 6. Although the recrystallization temperature of PBT polyester is usually 10 to 15°C higher than that of nylon 6, various additives can shift this temperature to a point where a clear distinction becomes doubtful. In such a case, IR spectroscopy is the logical method of making such a determination; however, thermal analysis offers an alternate technique that can also capitalize on differences in the polymer structure.

Thermogravimetric Analysis

This method is known as *thermogravimetric analysis* (TGA). At its basis, TGA operates on the principle that all organic materials will decompose if heated to a sufficiently high temperature. TGA monitors weight loss as a function of temperature or as a function of time at a constant temperature. This method is used to assess comparative thermal stability, monitor various reactions that may occur at elevated temperatures, and quantify inorganic content such as glass fiber in an otherwise organic material. It also has a little-known ability to distinguish between different types of polymer backbone structures if the test is run in a particular way.

Most analysts tend to run TGA tests in a single atmosphere, usually either nitrogen or air. Although this may provide a useful fingerprint, it misses a very important structural property. When a material is decomposed in an inert atmosphere like nitrogen, a process known as *pyrolysis*, it may form a char that is part of the original

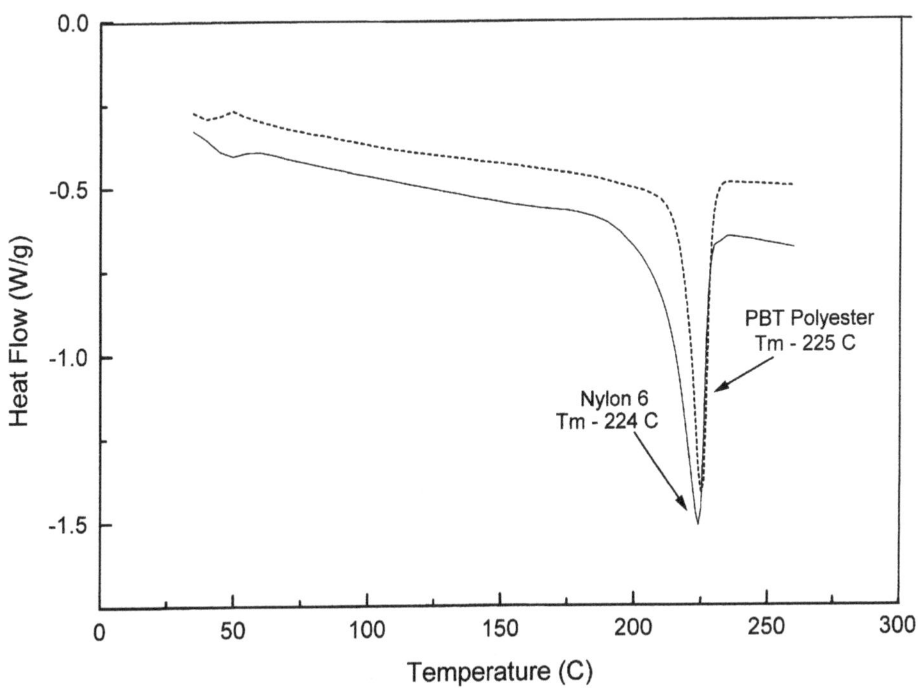

Figure 13.8 DSC first heat comparison of PBT polyester and nylon 6.

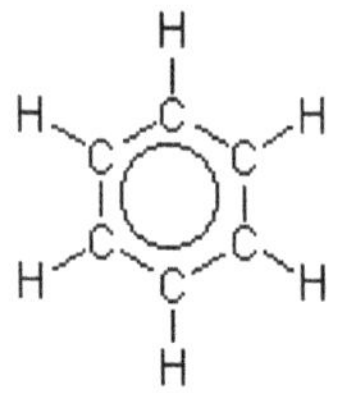

Figure 13.9 Chemical structure of a benzene ring.

organic structure but which is stable at elevated temperatures. The amount of char formed is related to the type of chemistry found in the backbone of the polymer chain. The primary determinant of char formation is something called an *aromatic ring*. In organic chemistry an aromatic ring is a very special type of structure that consists of six carbon atoms arranged in a hexagonal pattern. The simplest molecule that contains this ring is benzene, where all the members attached to the central carbon atoms are hydrogen atoms. Figure 13.9 shows this structure. When this ring is present in the backbone structure of a polymer it tends to stiffen the chain, which translates to higher strength, modulus, and heat resistance. It also constrains the chain so that certain types of motion are more difficult to achieve. This can result in reduced crystallinity, which is something we will discuss later.

The more aromatic rings we find in the polymer backbone, the more char will be formed during the pyrolysis of the material. Once this char has been formed, the best way to measure it is by introducing air or oxygen into the TGA. This causes a reaction that forms carbon dioxide and results in a weight loss associated with char formation. Figure 13.10 shows the repeating chemical structure for nylon 6 and PBT polyester. Note that the nylon has no aromatic rings, but the PBT does. Based on our discussion, we would expect that the PBT would form more char than the nylon. A TGA comparison of the two materials in Figure 13.11 shows that we would be correct. Nylon only produces about 1.7% char, whereas PBT produces approximately 7.7%. This is a distinguishing characteristic that can be used to distinguish between nylon

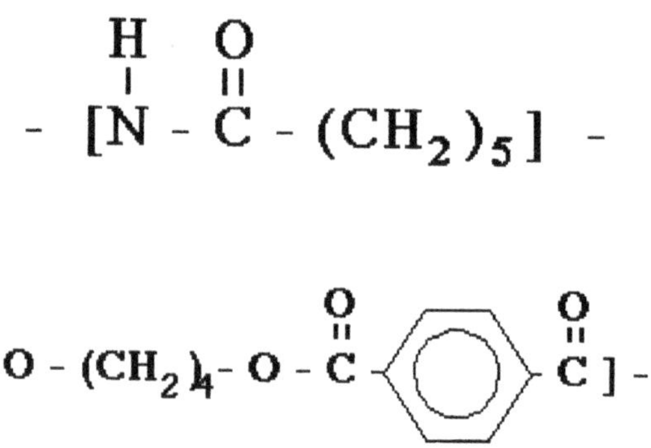

Figure 13.10 Structure of the repeating chemical units in PBT polyester and nylon 6.

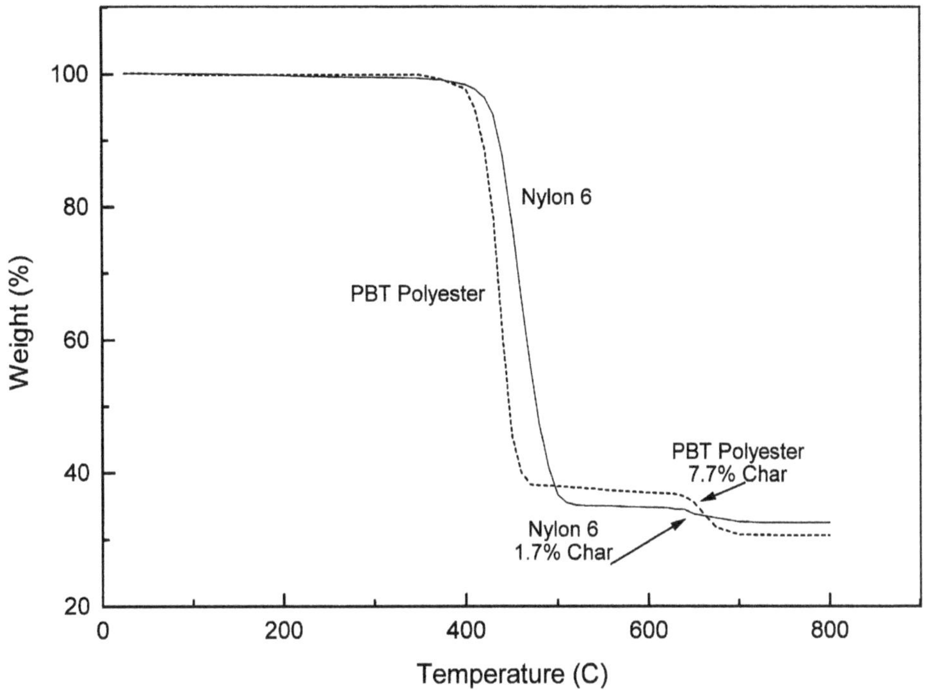

Figure 13.11 TGA comparison of PBT polyester and nylon 6 containing 33% glass fiber.

6 and PBT. It can also be seen that because it is less thermally stable than nylon 6, the PBT polymer begins to decompose at a lower temperature. This is another possible method for determining whether an unknown material with this melting point is made from nylon 6 or PBT polyester.

DSC Test Results

The primary point of this discussion is to emphasize the importance of using techniques that capitalize on differences between materials. Figure 13.12 shows DSC test results for ABS, high-impact polystyrene (HIPS), and general-purpose polystyrene (GPPS). Because these are all amorphous materials, the only thermal transition that we observe is the glass transition. These occur within a very narrow temperature region between 100 and 110°C. Without prior knowledge of the composition of each material, we would be hard pressed to say with any certainty which material was which; however, we have already seen that IR spectroscopy does a very capable job of determining the differences between the three compounds.

Finding various ingredients in a compound is a task that also must be tailored to the best analytical method. Figure 13.13 shows a DSC scan for two PPS materials compounded with PTFE and glass fiber. The glass fiber makes no obvious

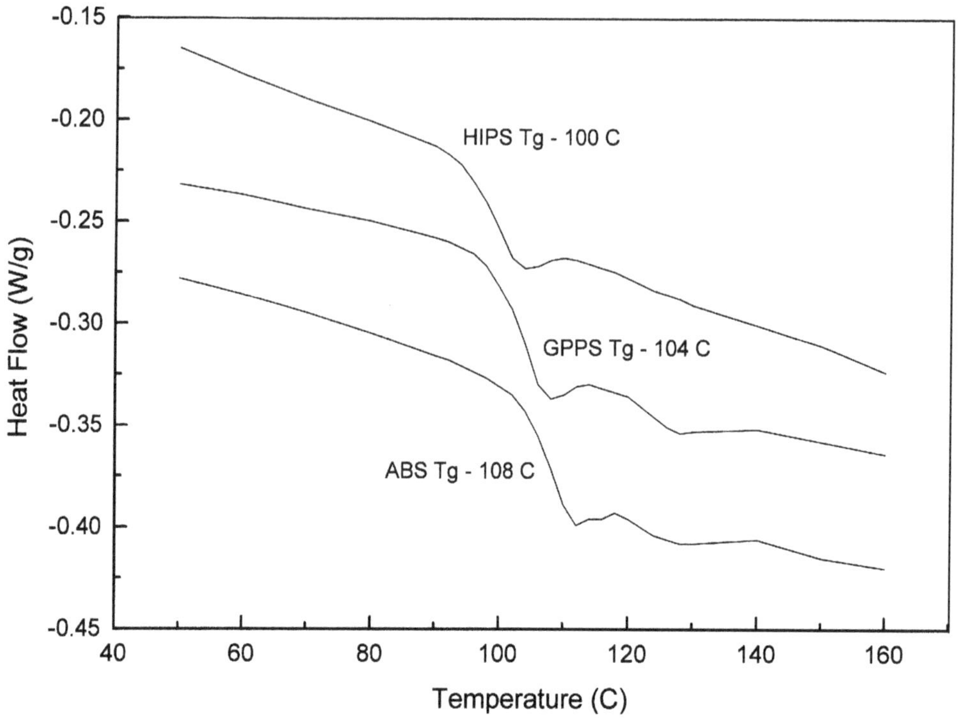

Figure 13.12 DSC measurement of glass transitions for ABS, high-impact polystyrene, and general-purpose polystyrene.

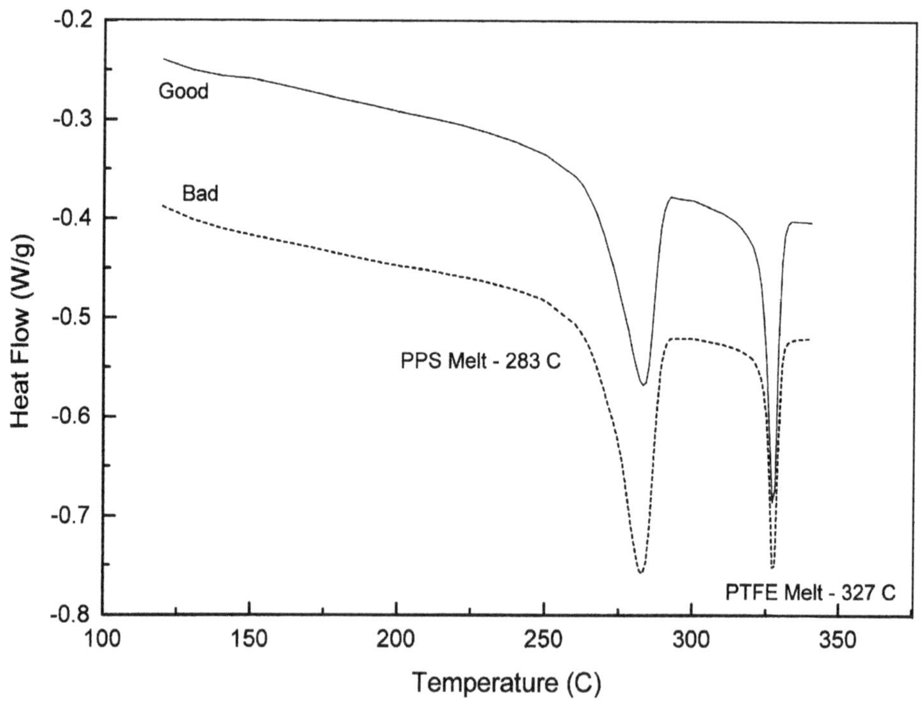

Figure 13.13 DSC first heat comparison of good and bad gear molded from a PPS material filled with PTFE.

contribution to the DSC scan because it has no transitions in the temperature region of interest. Both the PPS and the PTFE, however, are crystalline and therefore have melting points. The PPS melts just above 280°C and the PTFE has a characteristic sharp melting point at 327°C. This particular product was used to mold a part that acted as a bearing surface, and the PTFE was incorporated to reduce the coefficient of friction and improve wear resistance. It is possible to see without quantitation that the good part has a stronger melting endotherm associated with the PTFE, which is an indication that the intended composition of the material has changed. If the task of the analyst were simply to detect the PTFE, IR spectroscopy would be a valid method for performing this evaluation. A semi-quantitative finding like the one in this case, however, would be extremely difficult with IR.

Figure 13.14 shows the same comparison of the two materials as they cool from the melt. In making a comparison of relative peak intensities, the cooling phase or the second-heat phase is more reliable than first heat because the scan on first heat also includes the thermal history of the molding process. This can interfere with the inherent composition of the material. After the material has been melted in the DSC, that processing history has been erased and we can see the material in its inherent state.

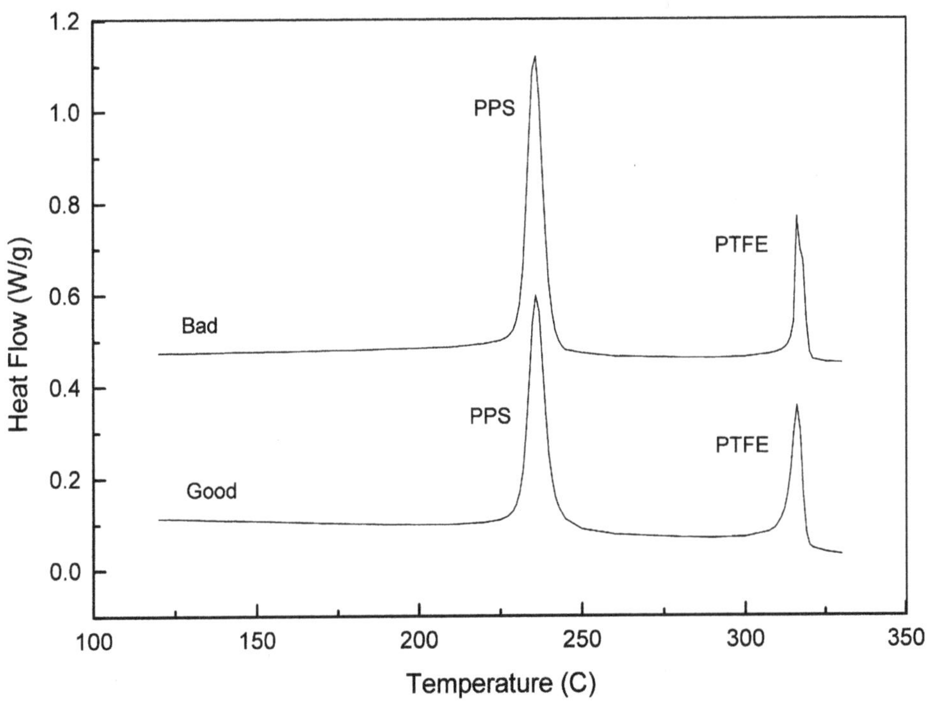

Figure 13.14 DSC cooling comparison of good and bad gear molded from PPS filled with PTFE. The higher relative intensity of the PTFE peak indicates a higher loading of PTFE in the good part.

The difference in relative concentration between the PPS and the PTFE in the two samples is even easier to detect on the cooling scan. We can integrate the curves to obtain a ratio that confirms what we are seeing visually. At this point there may be a temptation to assign absolute concentration values to the two components based on the strength of these transitions. This is dangerous for two reasons. First, because the material is glass-fiber reinforced, we must account for this glass fiber to ensure that some of the differences are not due to a variation in the filler content. It is also important to note that we have no assurance that a pure PPS and a pure PTFE have the same heat of reaction. In fact, the chances are very good that they do not. If we have access to pure PPS and pure PTFE samples, we can use the heat of recrystallization of those materials to assign quantitative values to the test results. These types of standards, however, with the desired purity and the assurance that they represent the same type of material found in the compound, are historically very hard to obtain. It is safer to state the relative concentrations and leave it to the material supplier to account for the differences.

If a quantitative assay is necessary, however, TGA can offer a solution, although not a straightforward one. Since TGA essentially measures the temperature of thermal decomposition, separation of a mixture depends upon the ability to distinguish between materials according to the way they degrade. Figures 13.15(a–c) show this principle for physical mixtures of acrylic and polycarbonate. In recycling, keeping these two materials apart is a big undertaking, particularly if there is a desire to maintain clarity in the recycled product. When they become mixed, a simple test is needed to determine the composition of the mixture so that it can be properly dispositioned as a lower performance product.

Figure 13.15(a) shows a TGA scan for a pure acrylic. Note that the temperature range of decomposition for the polymer is relatively low and that the material produces no char. In this plot we are also showing the weight loss rate of the decomposition as a means of highlighting the degradation process. Figure 13.15(b) shows the same result for a pure polycarbonate. Here, we can see that the polymer decomposition step occurs at a much higher temperature range; the polycarbonate does not begin to decompose until the temperature is above the endpoint of the acrylic decomposition. This provides at least the potential for full separation of the two components in a mixture. Note also that the polycarbonate produces a little more than 25% char, which is a typical result for unmodified polycarbonates. Figure 13.15(c) shows a result for a 60:40 mixture of acrylic and polycarbonate. Note that the separation of the two polymers is quite effective: The acrylic is almost completely gone before the polycarbonate begins to degrade. Even though the numbers appear to be unsatisfactory at first glance, when we add the char back into the polycarbonate weight loss we find that we are very close to the stated composition of the contaminated "blend."

This may leave the impression that quantitative separations of this type are possible all the time by TGA. It is unfortunately not that simple. Most polymers have degradation points that are similar, so extensive overlaps are more the rule than the exception. In our particular problem of PPS and PTFE, both materials are uncommonly stable, and they essentially degrade in a single step. There is hope for this

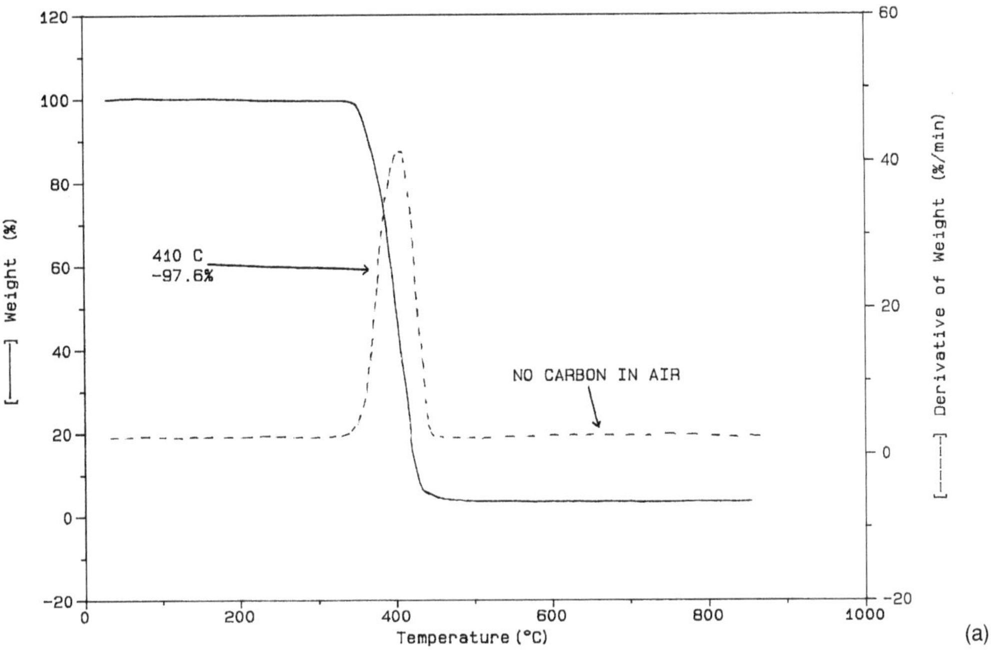

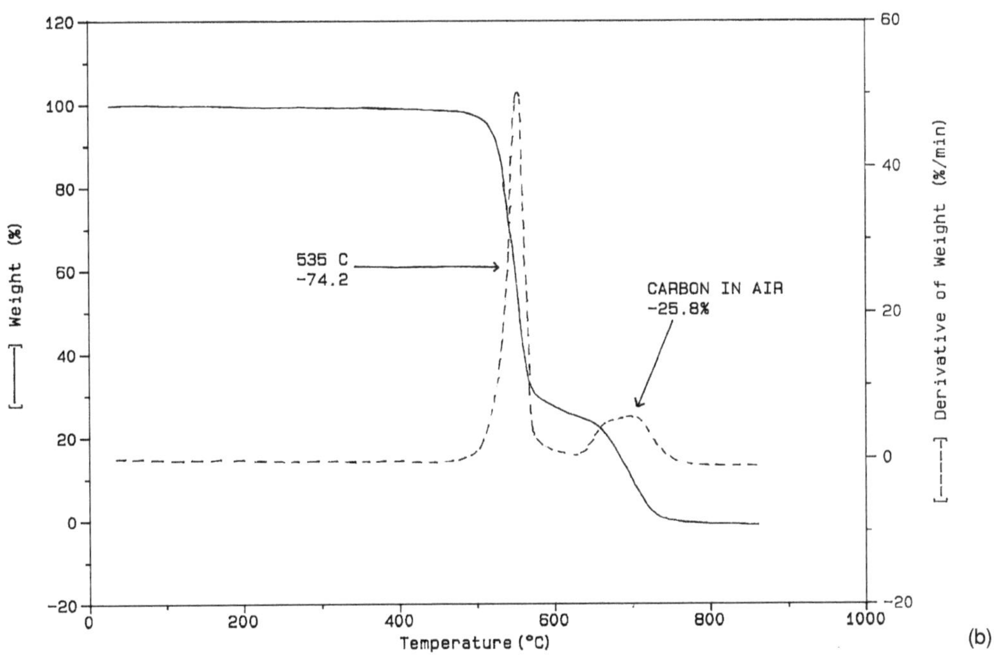

Figure 13.15 (a) TGA scan for a pure acrylic showing no by-products from pyrolysis. (b) TGA scan for a pure polycarbonate showing substantial char formation from pyrolysis.

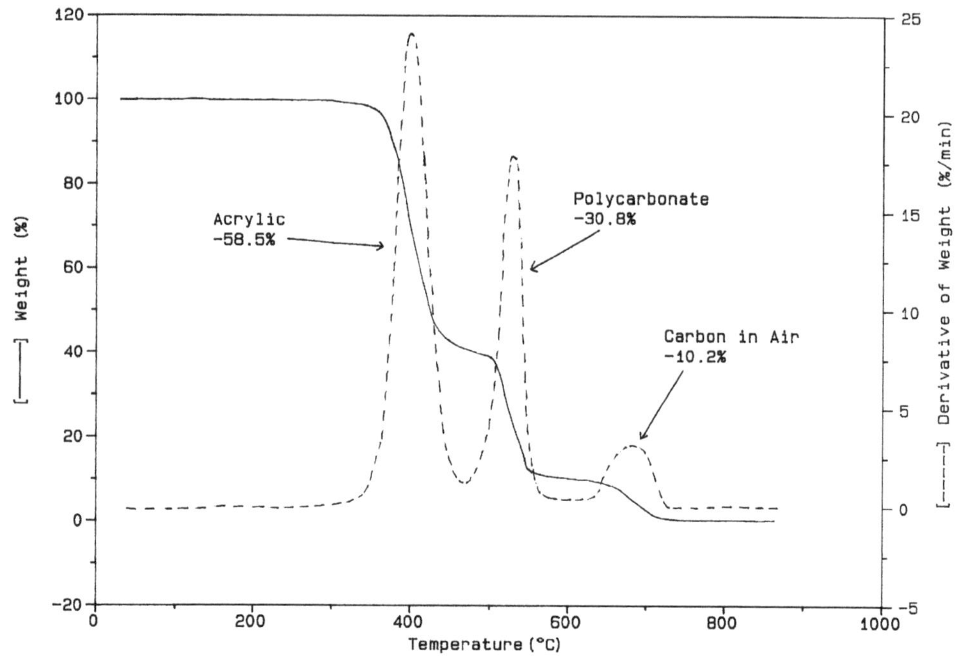

(c)

Figure 13.15 (c) TGA scan for 60/40 blend of acrylic and polycarbonate showing complete separation of the two polymers.

method, however, and it lies in the char formation. PPS is a highly aromatic compound, so much so that only 50% of the compound is lost when a pure PPS compound degrades in nitrogen. The remaining 50% remains until air is introduced and the char is removed through oxidation. PTFE, on the other hand, has no aromatic character and produces virtually no char. In addition, the TGA scan will give us a quantitative measurement of the other major component, the glass fiber. With a little math, and a knowledge of the behavior of the pure compounds, it is possible to use char formation to quantify the presence of the PPS and the PTFE. This is an excellent example of how all three of our primary techniques may be used, but the degree to which they can capture the problem varies significantly. Figures 13.16 and 13.17 show the result for the TGA tests of the good and the bad bearing material.

Char Content

An examination of the various weight loss steps shows that the fiber content was 32.25% in the good sample, leaving a resin content of 67.75%. The char given off by this material represented 27.12% of the compound. Remember that pure PPS gives off equal parts in nitrogen and in air when it decomposes. Thus, a char content of 27.12% gives a total PPS content of 54.24%. The remaining resin in the 67.75% can

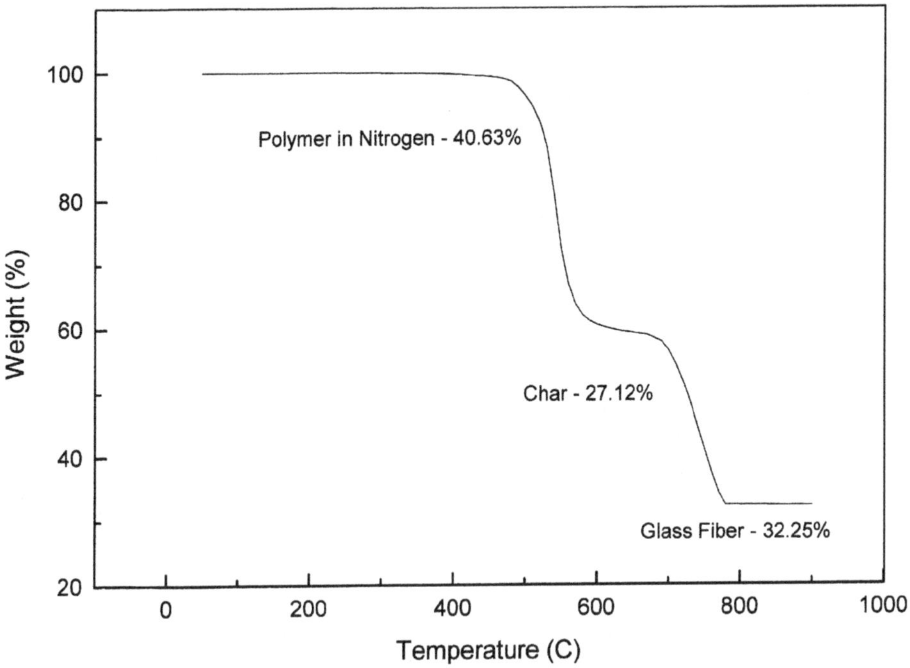

Figure 13.16 TGA scan for a good gear molded from PPS filled with PTFE.

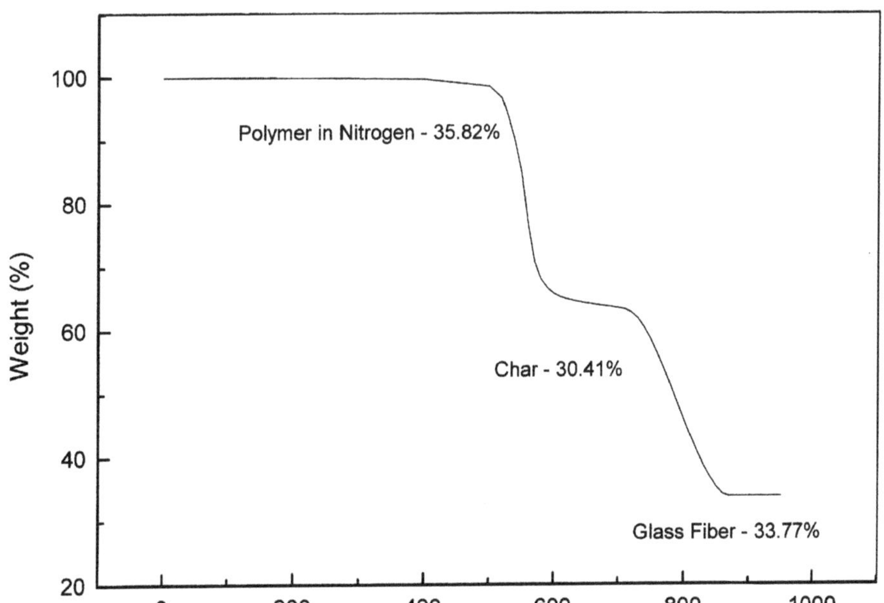

Figure 13.17 TGA scan for a poorly performing gear molded from PPS filled with PTFE.

be accounted for by the PTFE, giving a result of 13.51%. The advertised PTFE content in the material was 13%. A sample from a later part with wear problems had a glass-fiber content of 33.77%, leaving a resin content of 66.23%. Included in this was a char value of 30.41%, which translates to a PPS content of 60.82%. This leaves only 5.41% PTFE. An apparently small change in the TGA scan involved a reduction in the PTFE content by more than half of the intended amount. Two other batches of problem product also gave PTFE contents of 5 to 5.3%.

Other cases are less complicated. Another polymer that makes use of PTFE as a wear-resistant additive is acetal. It would seem reasonable to expect that we could at least verify the presence of the PTFE in the acetal by the same DSC technique used for the PPS product. There is a problem, however. Acetal polymers are among the least thermally stable materials commercially available. Anyone who has worked with the materials in manufacturing knows this, and also knows that the material, once overheated, rapidly decomposes to formaldehyde. Even in the gentle confines of a DSC instrument, with no shear and with nitrogen flowing through the instrument, acetal homopolymer will begin to degrade at about 275°C. Acetal copolymer begins to go near 290°C. Because the mode of decomposition is gas formation, the reaction is endothermic and very strong. It overwhelms the crystalline melting event of the PTFE and fouls the DSC cell with the degradation by-products.

There is a way out, however. The very difference in thermal stability that handicaps us with the DSC makes the TGA method extremely effective. There is hardly a polymer known to manufacturers that decomposes earlier than acetal, and hardly a polymer that degrades at a higher temperature than PTFE. This makes quantitative separation a simple proposition, as can be seen in Fig. 13.18. This is a test that is frequently performed to confirm that an unmodified acetal has not been substituted for the much more expensive PTFE-filled material. Notice also that there is no char formation to consider in the calculation. Both of these materials are free of aromatic character; therefore, they decompose completely at their respective temperatures and the calculation of absolute concentration is straightforward.

A final example of the use of these three techniques in attacking a problem will point out some subtle problems with calculations. One of the workhorse materials in the commodity resin category is a copolymer of polyethylene and vinyl acetate, known commercially as EVA (ethylene-vinyl acetate). The basis for these compounds is typically low-density polyethylene (LDPE). Pure LDPE has a fairly high degree of crystallinity and a melting point of approximately 110°C. The purpose of adding the vinyl acetate is to reduce the crystallinity of the material by breaking up the structural regularity inherent in polyethylene. As more vinyl acetate is added the material becomes softer and more pliable. It also increases in clarity. DSC is a useful method for observing the effect of the vinyl acetate on the crystallinity because the melting point and the heat of fusion both decline as the vinyl acetate content increases.

There are several routes to manufacturing a soft, clear, polyolefin-based material. First, there are other modifiers that can be used to achieve the same types of property modifications. Ethylene-acrylic acid (EAA) and ethylene-methacrylic acid

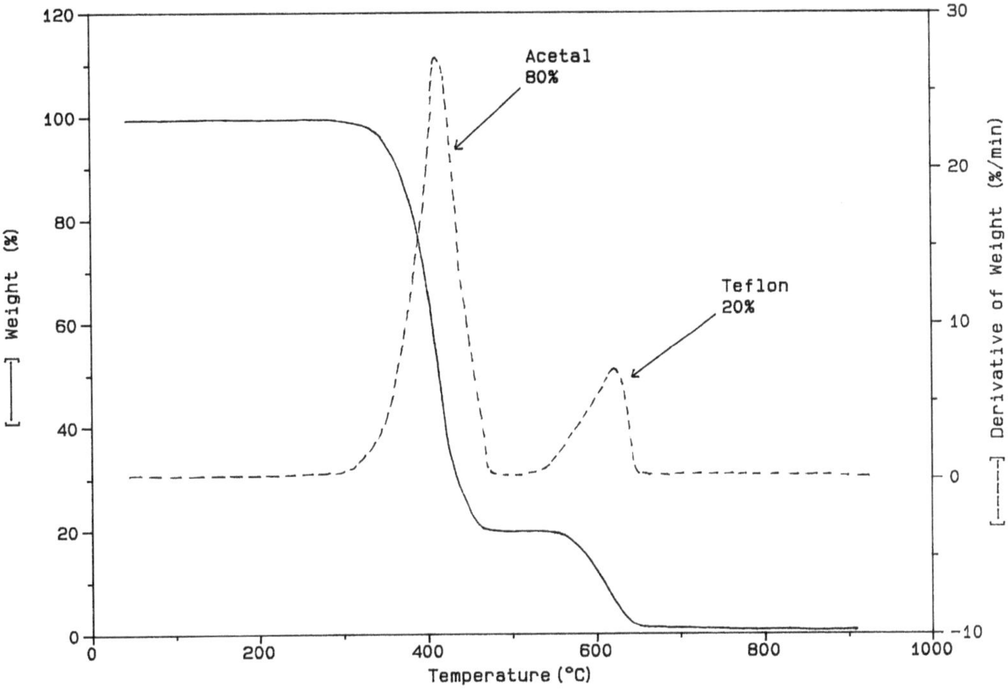

Figure 13.18 TGA scan of a Teflon-filled acetal showing separation of the two polymers.

(EMAC) copolymers will resemble EVA in appearance and properties. With the explosion in metallocene catalyst technology, it is now possible to achieve these same types of properties without resorting to such modifiers at all. Ethylene-octene copolymers that chemically look just like polyethylene but melt at temperatures typical of EVA copolymers are now common.

The initial task of identifying the nature of the modification relies on IR spectroscopy. The various modifiers will present different spectral results, and the metallocene copolymer will look like polyethylene, providing a sort of negative proof of composition. In addition, a metallocene material will achieve the same reduction in hardness without an increase in density; however, if the IR spectrum confirms the presence of vinyl acetate, the amount of added vinyl acetate is usually of interest. We have already mentioned that the melting behavior of the material is influenced by the vinyl acetate content. With increasing vinyl acetate levels, the melting point declines and the melting event begins to broaden into two distinct but overlapping processes. Figure 13.19 shows a comparison by DSC of a pure LDPE and two copolymers containing different levels of vinyl acetate.

It is tempting to use the relative strength of these two endotherms as a means of calculating the vinyl acetate content, just as we did with the PPS–PTFE alloy. The

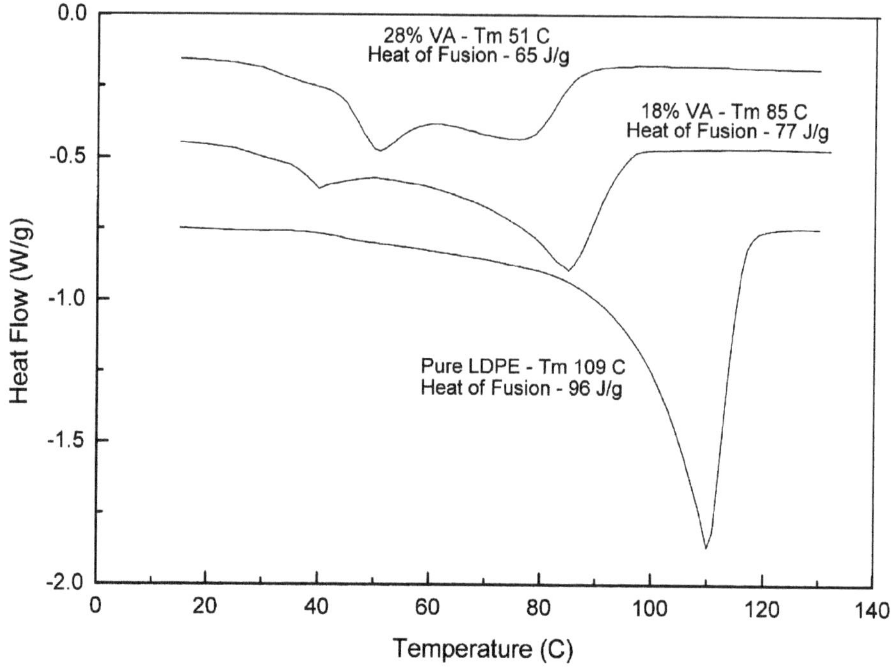

Figure 13.19 DSC comparison of low-density polyethylene and two analogs modified with different concentrations of vinyl acetate.

EVA is a copolymer, however,— it is a new compound, and not simply a physical mixture. This is an important distinction. In a mixture, the components maintain their properties and can be separated and analyzed based on those properties. A copolymer, however, is a new material with distinct properties that are not necessarily a straightforward combination of the original ingredients. A good way to think about this is to consider a mixture of salt and sand. It is possible to grind both the salt and the sand to a very small particle size and intimately mix the two components to produce a functional product. For example, sand provides traction on icy surfaces and salt melts that same ice. Combining the two may result in a useful product, but the salt and the sand would maintain their distinct properties. They would act the same way on the ice as they did when they were separate materials, and they could be physically separated with appropriate methods. Let us now consider the salt—sodium chloride. As its name suggests sodium chloride (NaCl) is made up of sodium and chlorine. It is not, however, a physical mixture. Sodium is a soft, shiny, and highly flammable metal; chlorine is a yellowish toxic gas. In combination they form an edible product. Sodium chloride has none of the properties of the two materials used to create it, and physical techniques will not be effective in separating the salt back into its elemental constituents. Trying to detect the presence of sodium and chlorine in the salt by measuring their respective melting points will be futile, so knowledge of the chemistry of the product is sometimes necessary to arrive at a sound analytical technique.

The EVA Copolymer

In the case of EVA copolymer, we can capitalize on a difference in thermal stability between vinyl acetate and polyethylene to perform the desired measurement. Figure 13.20 shows the results of a TGA on an EVA. The initial weight loss represents the decomposition of the vinyl acetate. If the material is heated slowly enough, full separation can be achieved between the vinyl acetate component and the polyethylene. Even after this separation has been achieved, it is still important to have some knowledge of the chemistry of decomposition in order to make this technique quantitative. The great temptation is to equate the weight loss with vinyl acetate content. If the vinyl acetate decomposed as vinyl acetate this would be correct and the determination would be straightforward, but it is not that simple. When copolymerized vinyl acetate decomposes, it gives off acetic acid. It is the acetic acid that is volatile and leaves the system. There is a 1:1 relationship between the number of molecules of vinyl acetate present and the number of molecules of acetic acid evolved. The molecular weight of vinyl acetate, however, is 86 g/mole, but the molecular weight of acetic acid is only 60 g/mol. The actual weight lost therefore only represents 60/86

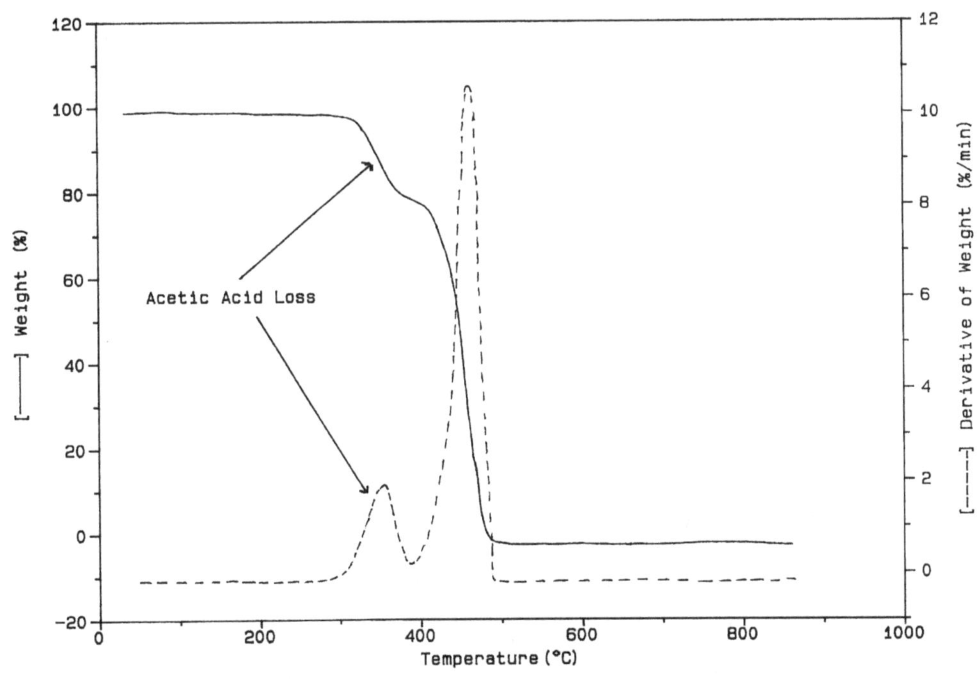

Figure 13.20 TGA scan of ethylene-vinyl acetate copolymer.

(or 70%) of the vinyl acetate. To obtain a correct value for the vinyl acetate content we must multiply the weight loss by 1.43.

13.1.2 Diagnostic Tools for Fillers and Reinforcements

Most fillers and reinforcements used in polymer systems are either partially or completely inorganic. There are exceptions, of course. We have already discussed the use of PTFE as an additive to reduce the coefficient of friction, and the commercial success of Kevlar, which is a type of nylon, has been extended to the world of moldable plastics. In addition, carbon can be used in a variety of forms, including conductive powders, carbon black, graphite, and fibers to modify the properties of plastic materials. Although carbon is organic it will not have a melting point or an infrared spectrum and therefore cannot be detected by at least two of the three methods discussed earlier.

The most popular reinforcement is glass fiber. Common fillers include talc, calcium carbonate, mica, wollastonite, and kaolin clay. With the exception of calcium carbonate, these are all wholly inorganic. The first task in dealing with a filled compound is determining the filler content. Most commercial compounds use a single type of filler or reinforcement; therefore, establishing the amount that is present is often sufficient. This is most easily accomplished with some type of ash test that is designed to decompose the polymer completely and leave the filler behind. In air, most polymers decompose completely at temperatures of less than 600°C, and the char that forms will decompose soon after. Inorganic systems can resist extremely high temperatures, so in most cases a heating routine that reaches 850 to 900°C will successfully remove all of the polymer and leave the filler behind. A high-temperature furnace called a muffle furnace has been traditionally used to "ash" 2 to 3 g samples of raw material or a molded part to establish filler content by simply weighing the sample before and after.

Filler Content Analysis

These furnaces unfortunately offer no good means of observing the weight loss process. The preferred method to accomplish this is TGA. Even though TGA instruments rely on smaller samples, they provide a picture of the entire decomposition process. We have already seen that this can be crucial in developing a profile of the polymer composition, but this can prevent mistakes even in the area of fillers, as we will show in a moment. At its simplest, a filler-content analysis involves decomposing the polymer until a constant mass is obtained. Figure 13.21 shows a test result for a 33% glass-fiber–reinforced nylon. The test result is quite straightforward. A small amount of the sample mass, usually less than 3%, is lost as moisture and volatile additives in the early part of the test. The polymer degrades over a fairly narrow temperature range. If the test is run in air, the polymer decomposition will be com-

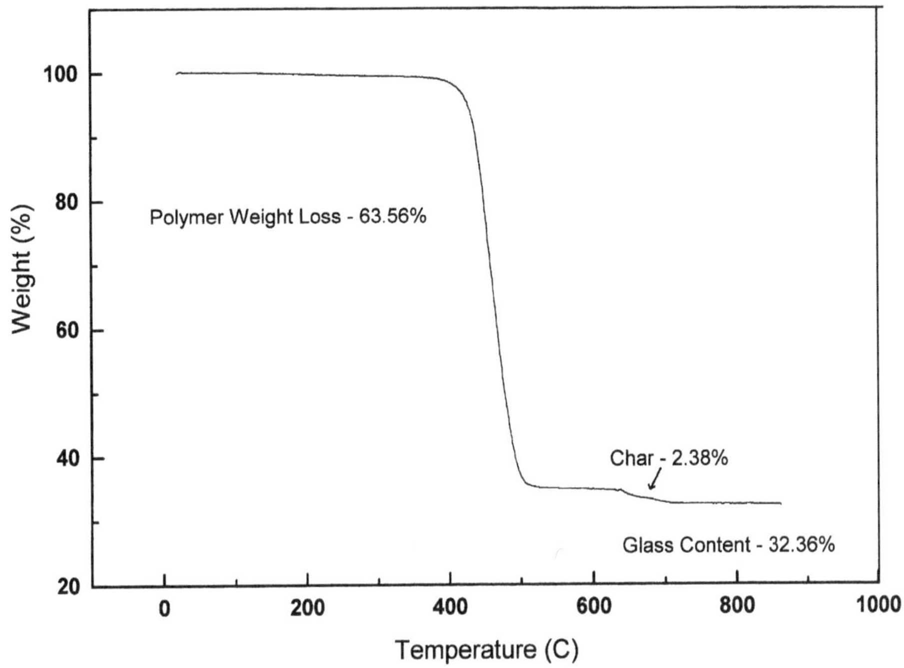

Figure 13.21 TGA scan for a nylon 6/6 filled with 33% glass fiber.

plete at this stage. If it is run in a relatively inert gas like nitrogen or argon, the polymer will leave a char that will then decompose once air is introduced. In this result, the char represents only 2% of the material and can be seen as a small weight loss at more than 600°C. No further events occur beyond this point. The remaining material represents the inorganic materials, in this case the glass fiber. The test confirms that the glass content is within the desired range. A physical examination of the residue, usually under some degree of magnification, will verify the filler as glass fiber. With proper equipment it is even possible to measure the fiber lengths and diameters, a property critical to the performance of the fiber. It may also be possible to distinguish various shapes of glass such as fiber, milled glass, and glass beads or hollow spheres.

Fiber content and condition are critical to the performance of a compound. A survey of property charts for a given family of materials will show that strength and modulus increase significantly as the amount of glass fiber in the compound increases. As a result, one of the fundamental concerns is filler content and type when a filled material fails to perform as expected. Figure 13.22 shows a TGA evaluation of a good part, a bad part, and associated raw material produced in a glass-fiber–reinforced polyetherimide. The raw material and the good part contain very close to 30% fiber, the amount prescribed in the raw material called for on the part drawing. The bad

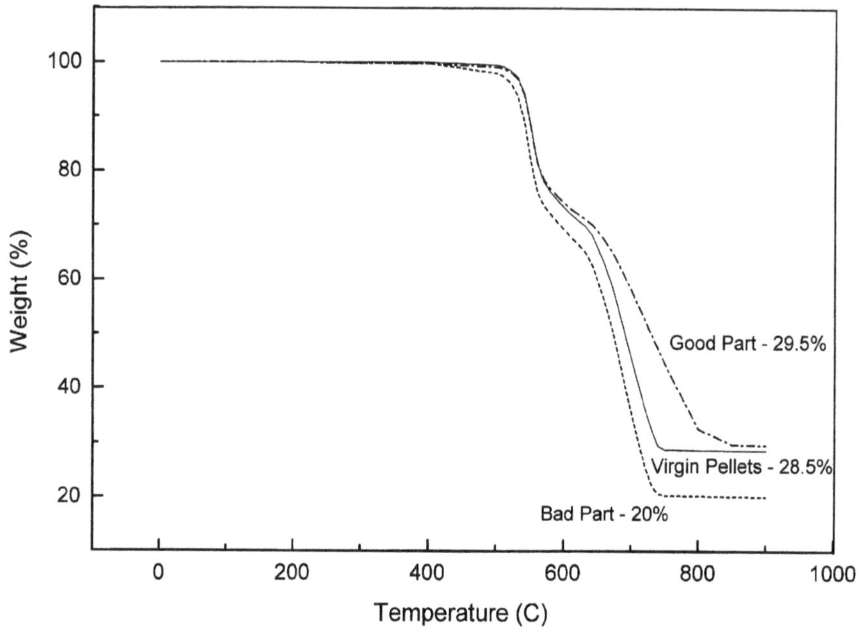

Figure 13.22 TGA scans on polyetherimide showing different levels of glass fiber.

part, which exhibited inadequate stiffness, was shown to contain only 20% fiber by TGA.

Mixed systems that employ a combination of glass fiber and a mineral are often used to reduce cost and solve problems related to fiber orientation such as warpage. Most mineral fillers, however, lack the high length-to-diameter ratio, called the *aspect ratio*, that makes glass fiber so useful. There is consequently a trade-off between the lower cost of the mineral filler and reduced performance. Keeping the mixture of glass fiber and mineral filler is therefore important to the maintenance of satisfactory properties. In this case a simple value of total filler content is insufficient, so an examination of the ash becomes very important. Figure 13.23 shows a comparison by TGA of material from two parts made from a nylon 6 containing a combination of 15% mineral filler and 25% glass fiber. One of the parts was performing properly, whereas the other was failing due to a lack of strength and modulus. The TGA showed no significant difference in overall filler content, but a microscopic examination of the residue showed that there was almost twice as much fiber in the good part. The 25:15 recipe had become reversed in favor of the mineral.

X-Ray Spectroscopy

This type of problem does not need to be left to a visual approximation. A very useful technique for determining the composition of inorganic and partially organic substances is known as energy dispersive x-ray spectroscopy, which is abbreviated as EDS, EDX, or even EDAX. More quantitative data can be obtained by an alternate technique known as x-ray fluorescence (XRF), but the advantage of EDS is that it is part of a scanning electron microscope (SEM). High-magnification images of the material can therefore be produced at the same time that the EDS analysis is being conducted. These methods capitalize on the fact that different elements and crystalline compounds respond in unique ways to being bombarded by x-rays. Using these techniques, an elemental map of a material can be produced for materials where the infrared spectrum will be inconclusive or will provide little or no information. Most inorganic fillers are based on silicates, which are a combination of silicon and oxygen; however, other elements will tend to be distinctive to a particular type of filler. For example, glass fiber will primarily be silicon and oxygen, but wollastonite will also contain a substantial amount of calcium, talc will contain magnesium, and kaolin clay will contain potassium and aluminum. In the case of the mineral–glass filler recipe gone awry, the mineral used was kaolin clay. A comparison of EDS tests run on the good and the bad filler showed very high levels of potassium and aluminum in the filler from the bad product, which confirms the problem with the ratio of mineral: glass fiber.

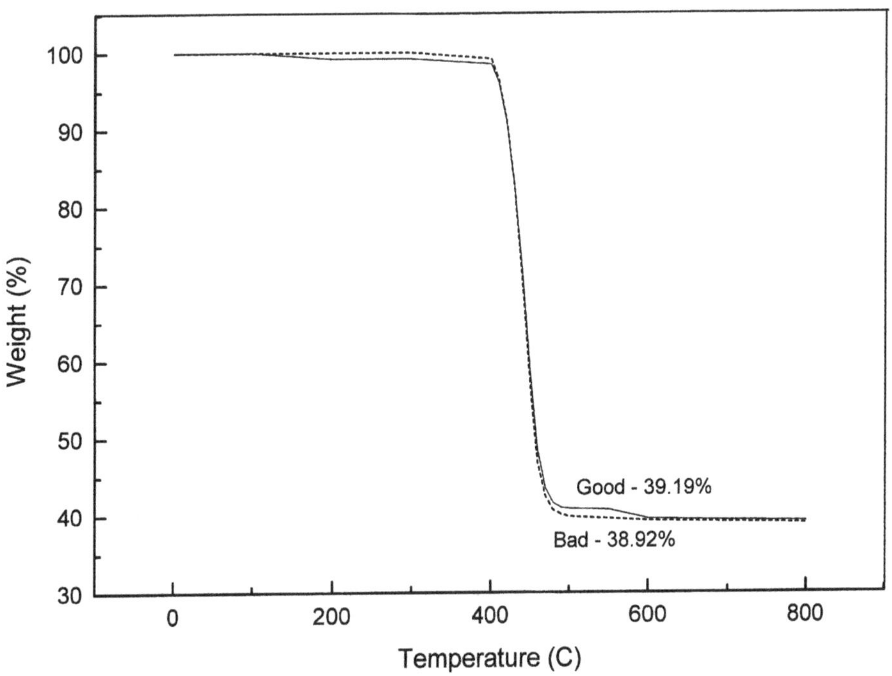

Figure 13.23 TGA scan comparing two nylon 6 compounds filled with a combination of glass fiber and mineral.

Talc and Carbon Carbonate

Some fillers are partially organic and will therefore undergo partial decomposition to form a new product. This is one instance where TGA tests are preferable to furnace results. The two most commonly used fillers in polypropylene, for example, are talc and calcium carbonate. Talc is a silicate and loses no mass at the temperatures normally used in determining filler content. As the name would suggest, however, calcium carbonate is a carbonate. Although silicon and oxygen stay bonded to each other at 850 to 900°C, carbon and oxygen do not. As a result, calcium carbonate breaks down near 740°C to produce calcium oxide, which remains as residue, and carbon dioxide, which is measured as a weight loss. Figure 13.24 shows a comparison of a 20% talc-filled and a 20% calcium carbonate-filled polypropylene. While similar in cost and appearance, talc produces a stiffer compound while calcium carbonate imparts better impact strength. Table 13.1 gives a short-term property comparison to illustrate this.

In determining the composition of an unknown material, therefore, a determination of filler type is not a trivial matter. At first glance the talc-filled and calcium carbonate–filled materials are different, but a classical furnace test would only see that one material contained 20% filler and the other only contained 12%. Because the

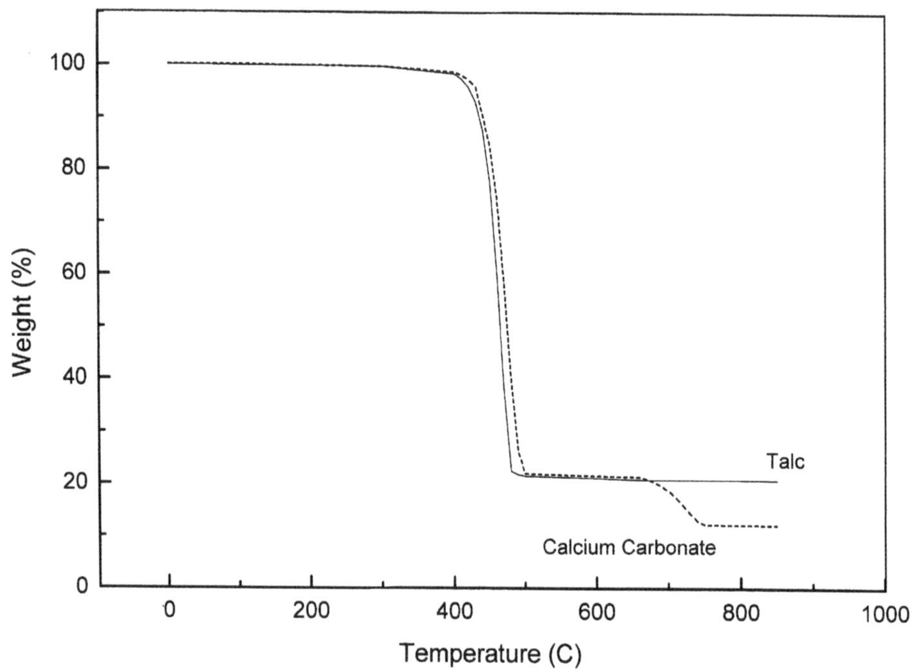

Figure 13.24 TGA comparing decomposition patterns for polypropylene filled with 20% of talc and calcium carbonate.

Table 13.1 Comparative Short-Term Properties of Talc and Calcium Carbonate–Filled PP

Property	Units	20% Talc	20% $CaCO_3$
Tensile strength @ yield	MPa (psi)	33.8 (4900)	27.6 (4000)
Elongation @ break	%	35.0	80.0
Flexural strength @ 5% deform.	MPa (psi)	44.8 (6500)	33.1 (4800)
Flexural modulus	MPa (kpsi)	2690 (390)	1790 (260)
Notched izod impact	J/m (ft-lb/in)	26 (0.50)	40 (0.75)

material with the 12% filler is, in fact, more flexible, the lack of strength and stiffness in the material filled with calcium carbonate would be blamed on a reduction in the filler level. The furnace test does not detect the fact that a little more than 9% of the weight loss occurs after the polypropylene has decomposed and is actually part of a different reaction. Once the calcium carbonate reaction is understood, the TGA results make perfect sense. Calcium oxide constitutes 56% of the calcium carbonate; the carbon dioxide accounts for the other 44%. If we look at the exact residue content we obtain 12.18%, whereas the late weight loss accounts for 9.36%, for a total of 21.54%. This breaks down to 56.5% and 43.5%, which is in excellent agreement with the theoretical.

Figure 13.25 shows the TGA scan for the calcium carbonate-filled polypropylene with the weight loss rate plot added. This derivative plot is often used in TGA analysis because it helps to pinpoint changes in weight loss rate that can provide clues to composition. In this case it is shown to emphasize the unique behavior of the calcium carbonate decomposition. Most decomposition rate curves are fairly symmetrical and bell-shaped, but the calcium carbonate breakdown is characteristically asymmetrical, starting slowly and ending more rapidly. This makes it easy to distinguish from other decomposition reactions that might occur in this temperature region.

With some prior knowledge of composition, TGA can be used for checking certain types of concentrates. Figure 13.26 shows a TGA scan on a polyethylene-based concentrate containing carbon black. The material is represented as containing 40% carbon black. If the type of polymer were not known in advance, it would not be possible to say with certainty that the entire weight loss above 600°C was due to carbon black; it could be partially due to char from polymer pyrolysis. Because we know that the material is based on polyethylene, however, we can state that the carbon burn off contains no contribution from the polymer and that the entire weight loss is attributable to the carbon black. This same type of determination can be very useful in some thermoset rubber compounds where carbon black is used as a primary filler. Thermoset compounds, such as EPDM, are chemically similar to polyethylene and polypropylene and produce virtually no char. In cases like this the high temperature weight loss can be assigned almost entirely to the carbon black. For other polymer types, such as nitriles, more sophisticated experimental techniques can be used to separate polymer char from carbon black [1].

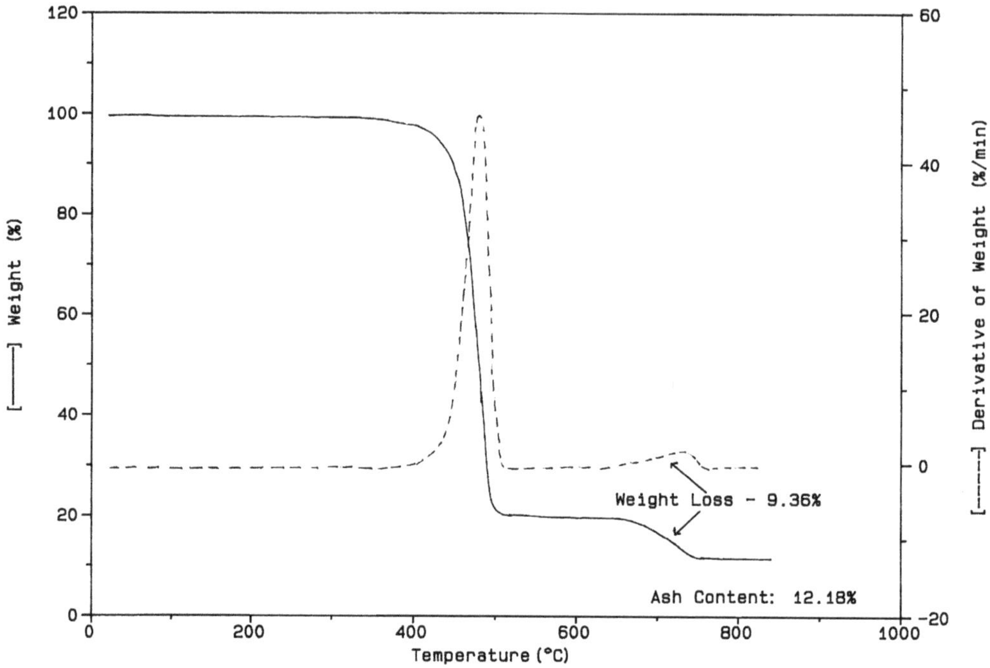

Figure 13.25 TGA scan of polypropylene filled with 20% calcium carbonate showing the derivative of the weight loss.

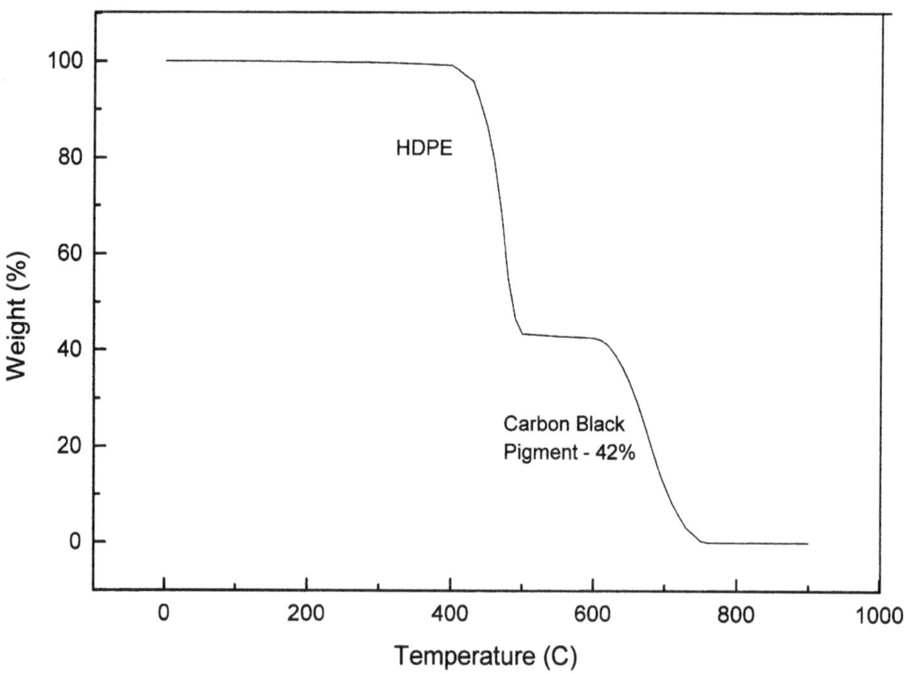

Figure 13.26 TGA showing determination of carbon black content in a high-density polyethylene concentrate.

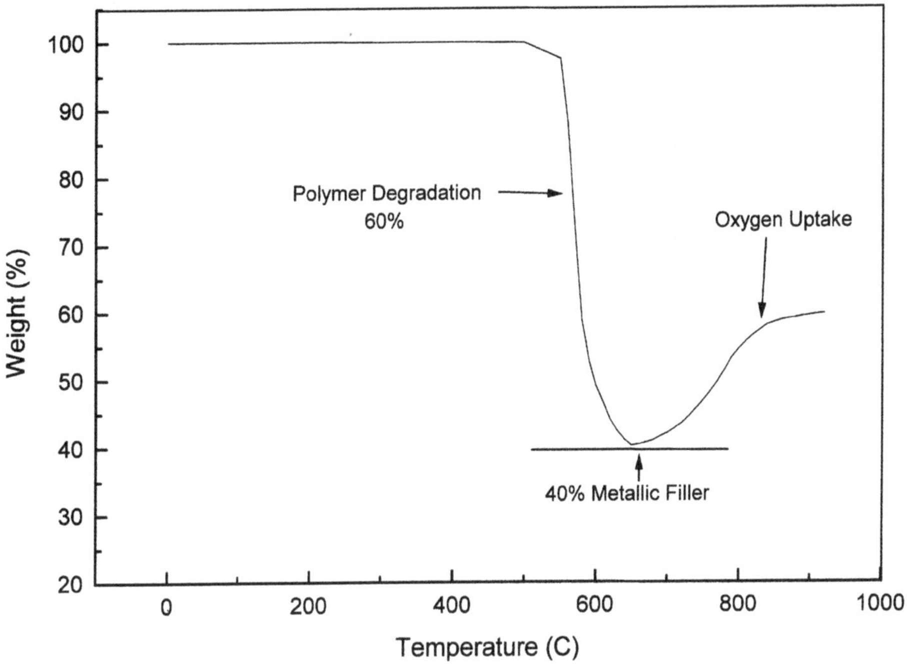

Figure 13.27 TGA scan for a Teflon filled with 40% bronze.

Finally, some novel fillers are metallic in nature. Bronze is added to some grades of PTFE to change wear characteristics, and stainless steel fiber is now employed in compounds designed to improve conductivity. In addition, some residues in samples are actually metallic inclusions that find their way into the molding process by mistake and plug up subgates and hot runner systems. Although metals will look like other fillers while in an inert atmosphere, they actually gain weight as they take on oxygen in air. Figure 13.27 shows a TGA result for an unknown material that had been shown by DSC to be based on PTFE, but had a very high specific gravity and a unique color. The sample lost 60% of its initial mass, then very quickly began to gain weight as air was introduced. This is a typical behavior for metallic additives and contaminants. When it is identified, the actual metals involved can be determined by running EDS tests on the residue.

13.1.3 Diagnostic Tools for Additives

Additives represent a very small but important part of any polymer compound. Even the most thermally stable polymers like polyethylene would have little chance of surviving a process like injection molding or extrusion without help from a class of stabilizers designed to prevent oxidation of the polymer. Other additives are employed to lubricate resin systems, make a material more flexible, prevent bacterial growth, protect against the deleterious effects of ultraviolet radiation, or inhibit the build-up of static electricity. Determining the presence of additives can be challenging because they are typically present in amounts that are beyond the detection limits of the more routine analytical techniques such as IR spectroscopy. In addition, techniques like IR do not provide quantitative information on the amount of an additive present in a material. In order to identify and quantify an additive, it is often necessary to extract it from the polymer matrix, isolate it, and then identify its exact structure.

Most additives are relatively volatile compared with the polymer matrix. Plasticizers are perhaps the easiest additives to detect because they are used in relatively high concentrations to increase the flexibility of a material. The most well-known material family in this category is PVC. The plasticizers that are traditionally used in PVC belong to a family of compounds known as *phthalates*. Figure 13.28 shows the chemical structure for one of the most commonly used of these compounds, di-isononyl phthalate (DINP). These compounds are readily detected by IR spectroscopy. They are such strong absorbers compared with the PVC polymer that they actually dominate the IR spectrum. Figure 13.29 shows a spectrum for a flexible PVC in which most of the significant bands belong to the plasticizer. To quantify the plasticizer, however, it is necessary to perform a solvent extraction or separate the materials by taking advantage of the relatively low molecular weight and boiling point of the plasticizer.

Running TGA tests under vacuum is an effective way of extracting and quantifying plasticizers. In order for this to be effective it is necessary that the temperature be kept low enough to prevent polymer degradation but high enough to allow for evaporation of the plasticizer. Employing a vacuum lowers the effective boiling point of the additive and widens the temperature window for conducting this experiment. Figure 13.30 shows the TGA result for an experiment where the plasticizer content in a flexible PVC was at issue. At 190°C under vacuum the PVC will not degrade, and the plasticizer can be extracted over a period of less than several hours. When the retained weight becomes constant, the separation is complete and the remainder of the material can be decomposed using a more standard heating rate. The determination of 46.08% agreed extremely well with the standard recipe that called for 46% plasticizer.

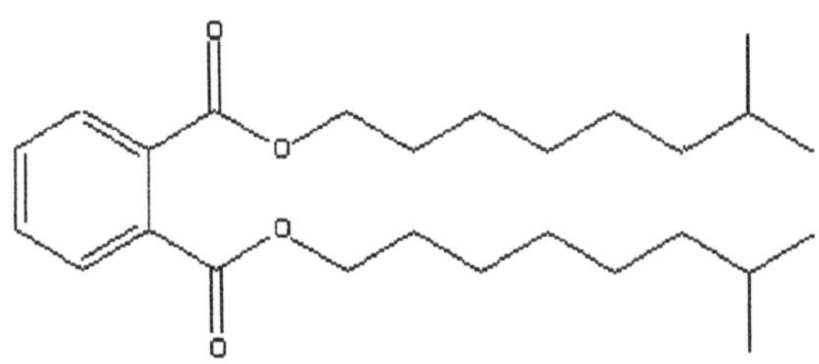

Figure 13.28 Chemical structure for di-isononyl phthalate, a common plasticizer in PVC.

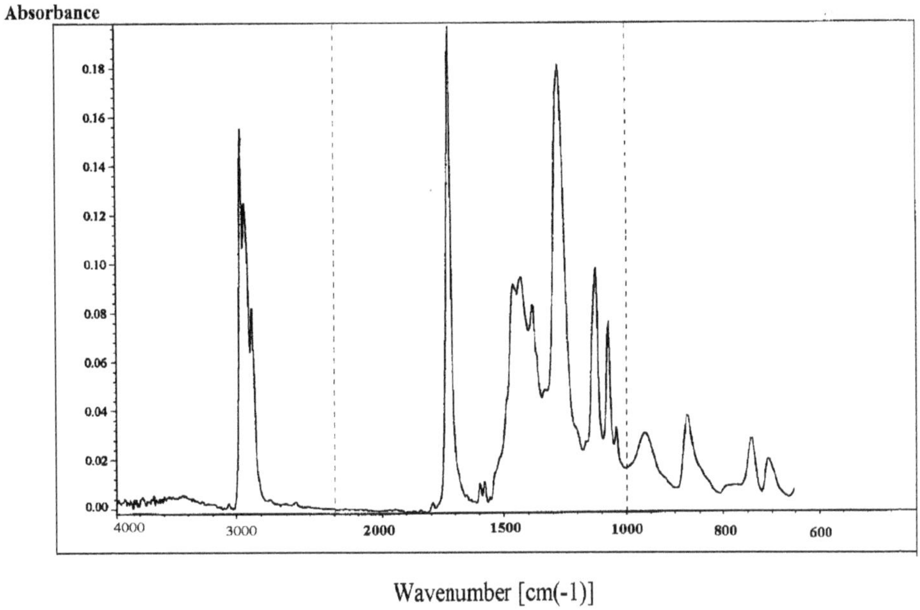

Figure 13.29 Infrared spectrum for flexible PVC showing characteristic absorption bands for the phthalate ester plasticizer.

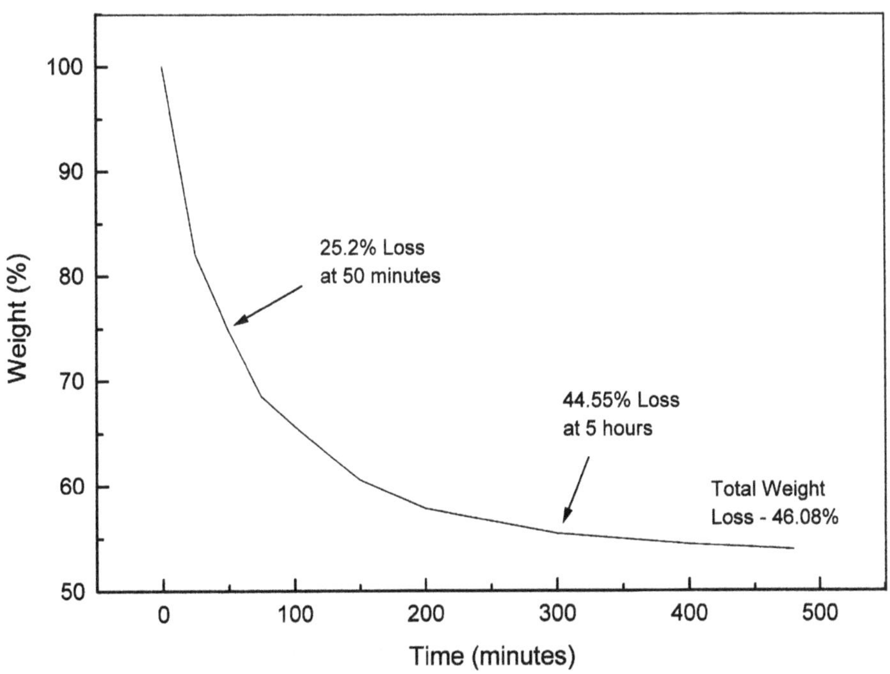

Figure 13.30 Plasticizer weight loss in flexible PVC by vacuum TGA.

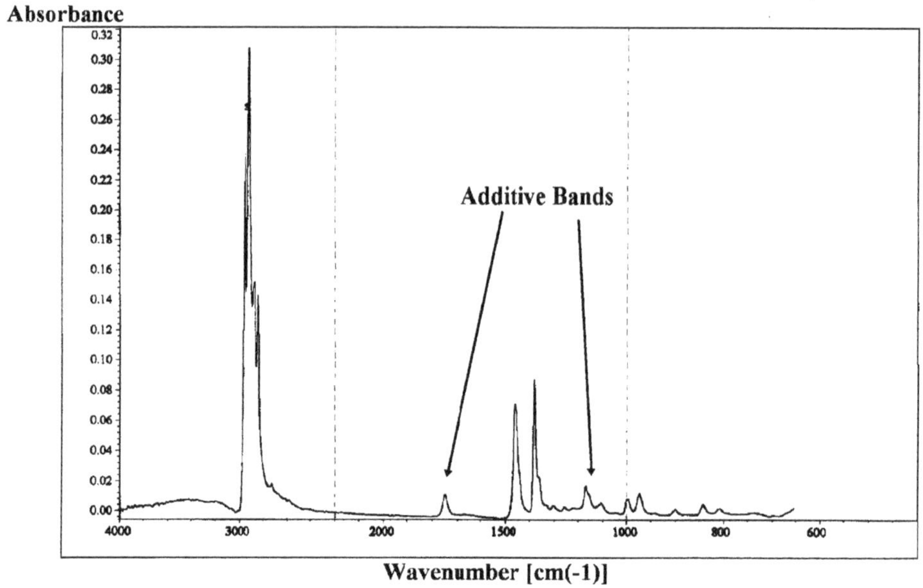

Figure 13.31 Infrared spectrum of polypropylene showing the presence of an additive.

Some additives such as antistatic agents are designed to work by migrating to the surface of the polymer. These types of additives may ultimately be lost from the formulation, but the fact that they concentrate on the surface often allows them to be detected easily by methods such as IR spectroscopy. The additive will be detectable if the concentration of such an additive on the surface approaches 0.5% and is incorporated into a material with a relatively simple IR spectrum such as polyethylene or polypropylene. Figure 13.31 shows such an analysis on a polypropylene medical part where the presence of an antistatic agent is necessary to promote the complete dispensing of polar liquids such as water. These additives have strong absorbers in areas where the polymer shows no activity. In this case the additive can be detected and the family of materials can be identified. To establish the additive as a specific compound and quantify the additive, however, would require extraction by either gas chromatography (GC) or high-performance liquid chromatography (HPLC) followed by a technique such as mass spectroscopy (MS).

Extraction

To perform an extraction, some knowledge of the chemistry of the polymer and the additive is needed. The objective is to selectively dissolve either the polymer or the additive, so that one material can be filtered off and the other can be isolated as a solid. Once this is accomplished the additive can be examined by IR spectroscopy. Once the specific family that the additive belongs to is identified, the additive can be placed in a gas chromatograph that will separate materials, usually according to their boiling points. The various fractions can then be sent to the mass spectrometer, which measures the molecular weight of the compound. The combination of the IR spectrum, a boiling point determination, and a molecular weight, will usually allow for a positive identification of the substance. Prior product knowledge regarding materials customarily used by the industry helps greatly in this endeavor. Because gas chromatography relies on converting the material of interest to a gas, as the name would suggest, it has its limits when we deal with chemicals that may be unstable when heated above the boiling point. Certain antioxidants fall into this category. This is where HPLC comes in. In this method the material does not need to be heated to temperatures where unwanted chemical changes may occur. This preserves the integrity of the additive and also allows for complex separations of multicomponent systems.

These are wonderful techniques for the research scientist, but they are sophisticated techniques that can take long periods of time and cost thousands of dollars depending upon the complexity of the additive package. Their practical use may be limited for most manufacturing applications. It may be possible, however, to gauge the practical influence of an additive by conducting tests that focus on the performance of the compound. This type of approach is useful for antioxidants. Antioxidants are incorporated to protect the polymer during processing, to provide long-term protection from sustained elevated temperatures in applications, and to prevent degradation during important secondary operations such as radiation sterilization. Although the mechanisms for various antioxidants may vary in their specific chemistries, the overall objective is the same—to prevent the polymer from oxidizing. Oxidation rapidly leads to deterioration of polymer molecular weight and properties.

Oxidation can be detected by DSC as an exothermic process. This can either be identified under isothermal conditions over time and is known as the oxidation induction time (OIT), or it can be established in dynamic heating conditions at a particular temperature. Different polymers respond better to one approach or the other, and standard test procedures make provisions for both types of tests. Figure 13.32 shows the results of a DSC test performed on a polypropylene material that was turning brittle after being sterilized by gamma irradiation. The test is an accelerated technique; it is not intended to establish an actual lifetime for the product, but it is meant to be a relative comparison of materials. In this result, the molded part before irradiation had a lifetime of approximately 50 minutes before the antioxidant package was consumed and the material began to break down rapidly. After irradiation, the stabilizer package was so deficient that the material began to degrade before the DSC could establish the temperature equilibrium.

Dynamic Heating Approach

Figure 13.33(a,b) illustrate the dynamic heating approach. In this case the rate of decomposition for the material did not allow for a meaningful measurement at

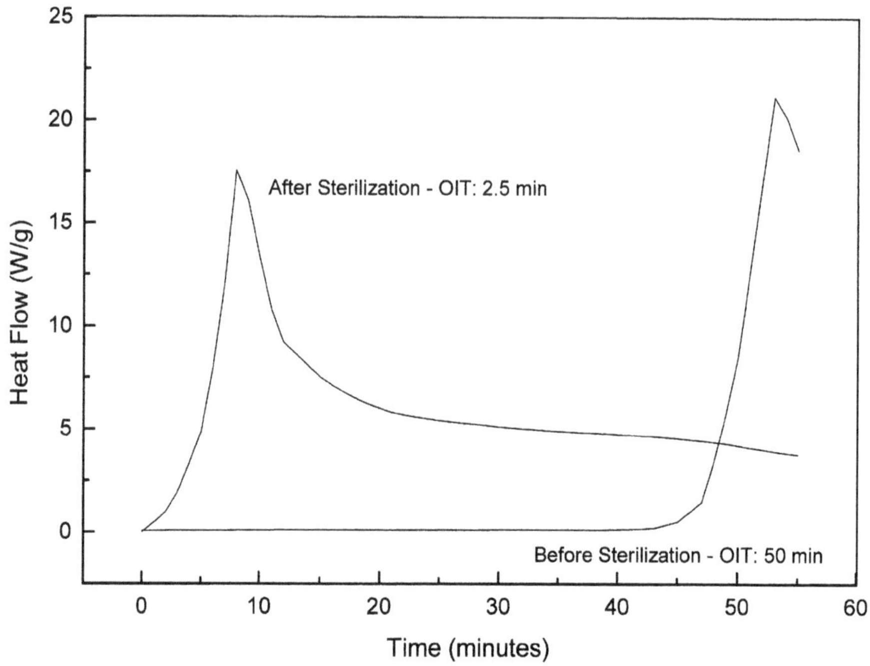

Figure 13.32 Oxidation induction time tests on a polypropylene with poor stabilization before and after radiation sterilization.

constant temperature. Below a threshold temperature the material would resist oxidation for extended periods of time and then degrade so gradually that the event did not register on the DSC. Once above this temperature, the degradation occurred so rapidly that it could not be captured. By heating at a constant rate, however, the onset temperature of oxidation can be established. In this case, the compound was degrading during processing. An examination of the DSC scan for the unmodified material shows why. Figure 13.33(a) shows the melting point of the material at 164°C followed almost immediately by an exotherm associated with oxidative degradation. There is no temperature window at which the material can be molten without rapidly decomposing. The user of the material added known amounts of antioxidant to various samples and then evaluated the relative stability of the compound using the DSC test. Figure 13.33(b) shows the behavior of the different formulations. As the antioxidant levels increased, the temperature gap between melting and oxidation increased until a manageable processing window was established. Although it is possible to extract, identify, and measure the level of the antioxidant after the fact, it is much simpler and of greater practical use to evaluate any future variations in the formulation by performing the oxidation induction test illustrated here.

This theme of choosing the practical over the best available is one that will come up again in future sections. Simple techniques can be very powerful in the right hands

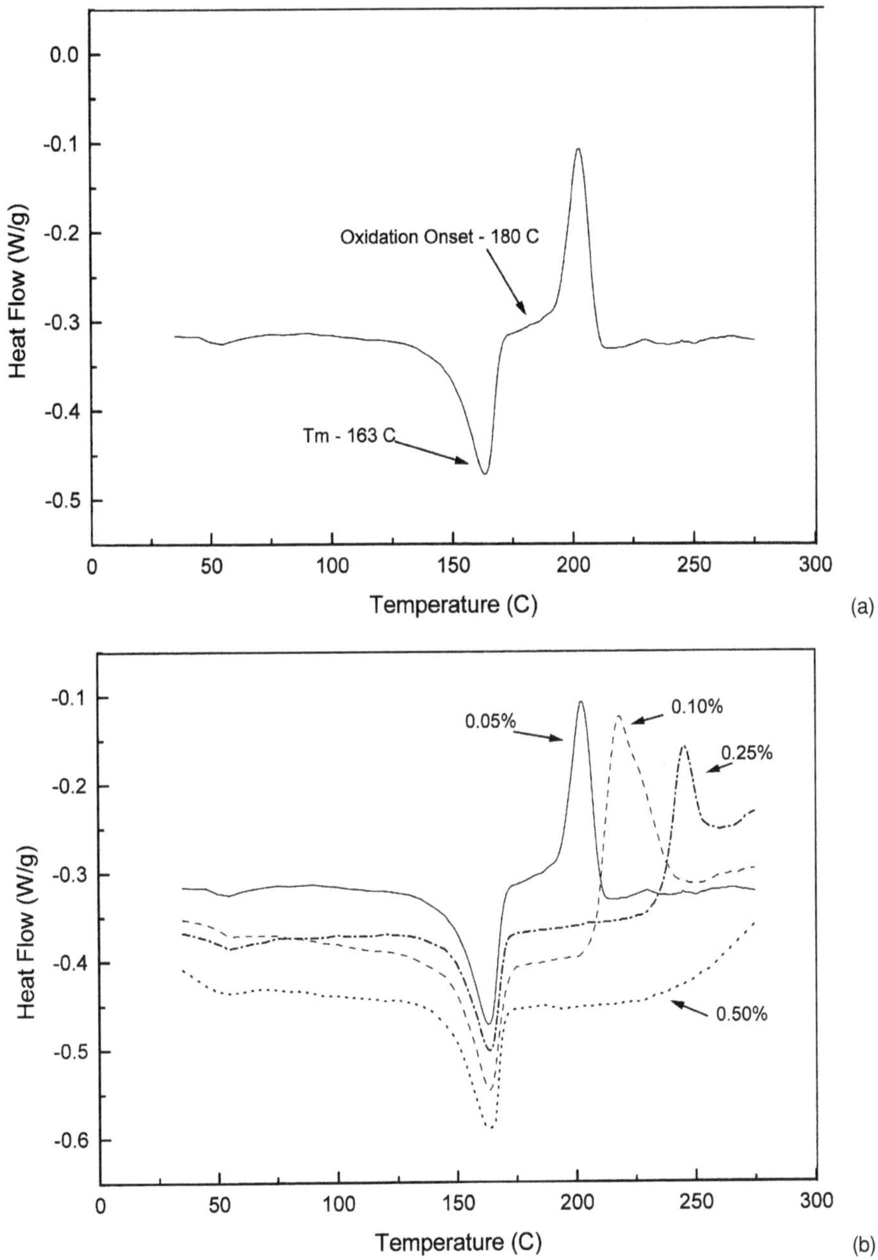

Figure 13.33 (a) DSC scan showing the melting and oxidation processes in a nylon-based elastomer. (b) DSC scans of the nylon-based elastomer showing improved resistance to oxidation as a function of increasing stabilizer content.

and they can save a great deal of money and time in solving a problem. Simplicity should always be emphasized as long as accuracy is not compromised. On the other hand, Einstein once commented that things should be made as simple as possible, but no simpler. An awareness of the limits of the easy methods and knowledge of where to go once those limits are reached is critical to good troubleshooting. Nowhere is this principle illustrated better than in the task of measuring molecular weight.

13.2 Molecular Weight Problems

Even though composition issues provide a fascinating array of possible problems and solutions, the reality is that most performance problems in polymers are related to an inappropriately low molecular weight. More than 80 years ago Hermann Staudinger proposed that the unique properties found in polymers were related to the large size of the individual molecules that made up the material. As much as this idea is accepted as fundamental today, it was scorned by most at the time of its first advancement, including Staudinger's own professors. In the early 1860s Thomas Graham had proposed that polymers were a completely different state of matter with unique types of chemical bonds that accounted for their unusual properties. Graham's work had led to the development of dialysis, so his place in history and in the eyes of the scientists who came after him was assured. When it came to explaining polymer behavior, however, Graham was wrong and Staudinger was right.

Fortunately, some people listened to the upstart. The most noteworthy among them may have been Wallace Carothers, the DuPont researcher who ultimately invented both polyester and nylon. Carothers used the principle of high-molecular weight to guide his development efforts. Eighty years later, despite our long-standing awareness of this important property, we still turn out products every day that are made of material whose molecular weight has been reduced below a safe limit. The process by which a polymer chain of adequate size is turned into one of inadequate size is commonly referred to as *degradation*.

Not all properties are affected equally as molecular weight declines. If we focus on the wrong property we may be convinced that reductions in molecular weight are inconsequential. In particular, properties such as heat deflection temperature and even tensile strength or flexural modulus are almost insensitive to changes in molecular weight. The most sensitive indicator of alterations in molecular weight is the ductility of the material. This is commonly measured with a standardized impact test such as the notched Izod test. This type of test does not always capture the true importance of the relationship between impact resistance and molecular weight unless it is viewed across another variable such as temperature. Figure 13.34 shows a plot of notched Izod impact results for five different grades of polycarbonate that are distinguished by their molecular weights. The impact resistance is shown as a function of temperature. At room temperature the differences in impact resistance between the highest and the lowest molecular weight material do not appear to be that

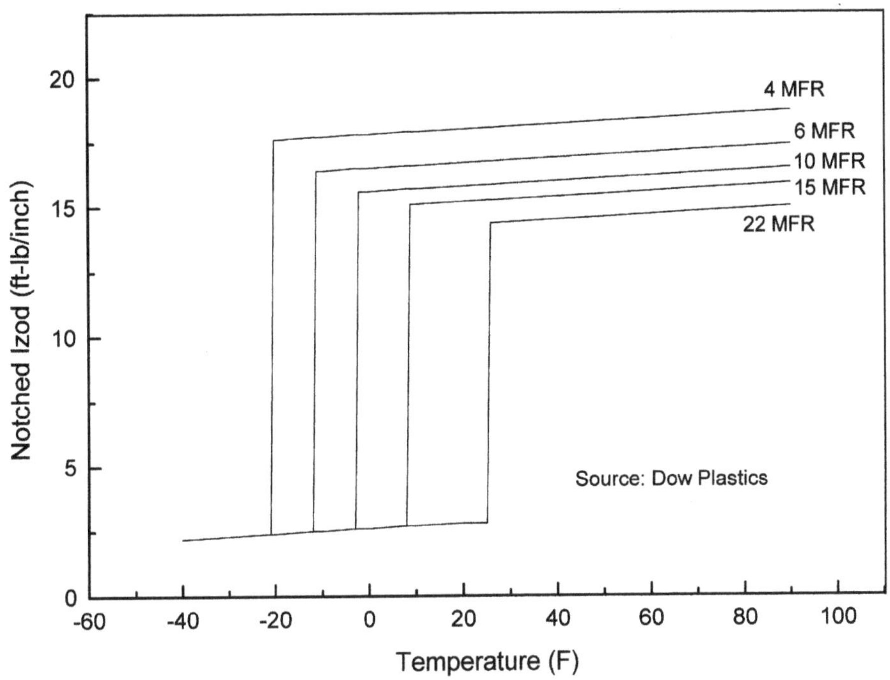

Figure 13.34 Notched Izod test results on various polycarbonates showing the effect of molecular weight on the ductile-to-brittle transition temperature.

significant. Even though everyone would agree that 18 ft-lbs/inch is better than 14 ft-lbs/inch, it can sometimes be difficult to distinguish these results in practical terms. By the same token, if we look at the low-temperature side of the figure we can see that the differences between grades are even smaller.

Things get interesting between room temperature and these low-temperature extremes. All five of the materials exhibit a rather sudden change in impact resistance at some specific temperature. This event is called the *ductile-to-brittle transition temperature* (DBTT), and it is a very important but seldom discussed attribute of polymer behavior. The material with the highest molecular weight in this grouping becomes brittle at a temperature 45°F (25°C) below the point where the lowest molecular weight material undergoes its transition. This cannot be captured in tabular data.

A table will sometimes manage to capture the importance of molecular weight. Table 13.2 shows the relationship of molecular weight to Izod impact for three different polycarbonates in transparent and opaque colors. The differences are once again small in transparent colors, but the lowest molecular weight grade is affected in an unexpected way in opaque colors. The range of impact values in this material ranges from a high end of 12 ft-lbs/inch, which is very ductile, to a low of 2 ft-lbs/inch, which is quite brittle. The wide range is due to the variation that the material exhibits

Table 13.2 Effect of Pigments on Impact Resistance of Polycarbonate

Material melt flow rate (g/10 minutes)	Notched Izod impact—clear and transparent colors	Notched Izod impact—opaque colors
10.5	15.0	15.0
17.5	13.0	13.0
25.0	12.0	2–12

as different pigment systems are employed. The higher molecular weight materials are not affected in this way.

Some materials are inherently somewhat brittle when tested according to the notched Izod method. For example, acetal copolymer will give average results that never even reach 2 ft-lbs/inch. In a case like this, alternate methods may be more useful to capture differences in ductility. In Table 13.3 we can see that an Izod test that makes use of unnotched specimens is more useful and shows a greater spread of values. Alternatively, the elongation at break values from tensile tests can be used, as Table 13.3 also shows. This is a measure of the distance that a specimen can be stretched before it fails outright. The highest molecular weight material shows the greatest tolerance for this type of stress.

Even though we have been discussing the importance of molecular weight, we have yet to refer directly to the property of molecular weight. We have instead been citing melt flow rate values. This brings us to the discussion of how we talk about the molecular weight of polymers. There are four primary techniques that provide insight into the molecular weight of a polymeric material. Two of these involve characterizing the material in the melt and two involve what are referred to as *solution techniques*, where the polymer is dissolved in a solvent. In all four methods the common thread is a measurement of viscosity. *Viscosity* is defined as resistance to flow, and there is a fundamental relationship between viscosity and molecular weight. The exact mathematical relationship varies depending upon the method of measurement, but the general correlation is that viscosity increases as molecular weight increases.

This is the tension that defines the injection molding process. Injection molding is arguably the most demanding of the melt processing techniques in that it requires both the elevated temperatures needed to melt the material as well as the application of considerable pressure and velocity to move the material through the flow

Table 13.3 Effect of Molecular Weight on Impact Properties of Acetal Copolymer

Material melt flow rate (g/10 min)	Notched Izod impact (ft-lbs/inch)	Unnotched Izod (ft-lbs/inch)	Tensile elongation @ break (%)
2.5	1.5	25	75
9.0	1.3	20	60
27.0	1.0	17	40

paths defined by the mold and part. Compared with extrusion, blow molding, and rotational molding, injection molding represents rough treatment of the raw material. As parts become more intricate and reductions in manufacturing costs become more important, these flow paths become increasingly demanding. Molders often cope with these demands by seeking out materials with lower viscosities. The viscosity of a material is often referred to in terms of the melt flow rate (MFR) or melt index (MI) of the material.

In order to appreciate when this measurement is useful and when it is not, it is important to gain an understanding of how it is developed. Figure 13.35 shows a schematic of a melt flow rate test apparatus. It consists of a heating device surrounding a block. In this block is a hole with a diameter of 9.55 mm (0.376 in). A removable insert is placed at the bottom of this hole and this insert contains an even smaller opening with a diameter of 2.095 mm (0.0825 in). Material is placed in this heated cylinder and brought to a temperature specified by a standard ASTM or ISO method. It is then forced through the orifice under a load that is also specified by the standard method. This test can be run in a constant volume mode where the same amount of material is extruded each time and the time required to perform this operation is measured. Alternatively, the test can be run for a fixed time and the mass of

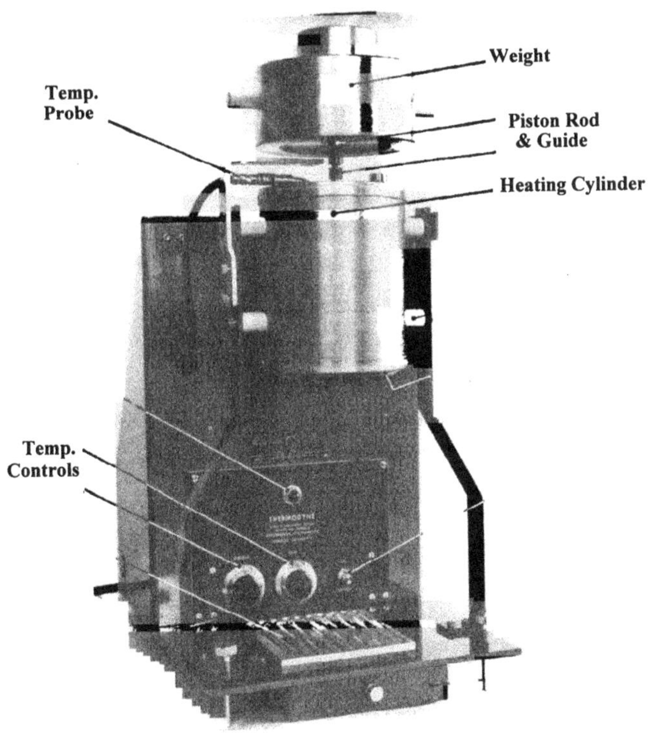

Figure 13.35 Schematic for a melt flow rate testing device.

the material extruded during that time period is measured. In either case the calculations produce a value that is expressed in the units of grams/10 minutes. This property has been refered to in our previous tables. Higher melt flow rate values are associated with materials that flow more easily, that is they have lower viscosities. Melt flow rate, therefore, is inversely related to viscosity.

Anyone who has spent any time processing knows this intuitively. It is easier to fill a mold cavity with a material that has a melt flow rate of 15 g/10 minutes than it is using a material with a melt flow rate of 5 g/10 minutes. The downside of using the easier flowing material is that the molecular weight of the polymer molecules in the 15-melt grade is lower, so the properties will be correspondingly lower. Thus, there is a constant balance being established between the degree of difficulty in processing and the performance of the molded product.

Unfortunately, the use of a higher molecular weight material does not guarantee a successful product with the anticipated properties. The rigors of the molding process act to reduce the average molecular weight of the polymer. The primary agents of this reduction are temperature, residence time, and shear. In addition, some materials are sensitive to the effects of moisture at the elevated temperatures associated with melt processing. Many materials are hygroscopic; they will absorb small amounts of moisture from the surrounding atmosphere and must be dried before they are processed. Failure to remove the excess moisture may result in cosmetic problems. But polyesters, polycarbonates, polyurethanes, and polyamides (nylon) will react with the excess water and undergo a chemical reaction known as hydrolysis. This process involves the breaking of the chemical bonds within the polymer chains, which results in shorter chains and a lower molecular weight. The preservation of the molecular weight of the polymer is of paramount importance in any process, and it is perhaps more challenging for injection molding than for any other type of polymer processing technique.

Because retention of molecular weight is so important, it is essential that tests be available that can detect problems that may occur during processing. It is not always possible to rely on visual cues to determine whether or not a material has been degraded. Materials like PET polyester may appear normal when they have actually been compromised significantly. The simplest and most economical method for evaluating the effects of processing on the polymer is the melt flow rate test. There are a lot of things wrong with the melt flow rate test, but most of the shortcomings arise from a misuse of the technique. Processors have viewed the melt flow rate test as a gauge of processability for many years; however, the differences in viscosity between materials are exaggerated by the melt flow rate test.

To illustrate this we must look at the behavior of viscosity in a polymer. Viscosity is influenced by two key processing parameters. One is the temperature of the melt and the other is the shear rate to which the material is subjected. Shear rate is a function of flow rate (processors read injection speed) and the size of the flow path (processors read nozzle, sprue bushing, runner, gate, and wall thickness dimensions). The tooling dimensions are essentially fixed unless the mold is modified, but the injection speed is a process variable that provides a means of modifying the viscosity of the material. The effect of flow rate on viscosity is not intuitive because it is not reflected in our experience with fluid flow. Low molecular weight fluids like water have an essentially constant viscosity, regardless of the flow rate. This independence of viscosity from flow rate is described as Newtonian behavior because Isaac Newton was one of the first people to wrestle with this concept of viscosity and he used the materials that were available for study. Polymers were not among those materials; if they had been Newton would have been surprised to see that some fluids actually exhibited a variable viscosity that depended greatly upon the flow rate.

Those of us who learned how to mold by trial and error grew up firmly convinced that the higher the melt flow rate was for a given material, the more easily it would fill an injection mold or the farther it would flow before the flow front would freeze. Although this is technically true, the melt flow numbers suggest that the differences in processability are much greater than they end up being in actual experience. For example, the melt flow rate numbers for a 4-melt and a 22-melt material suggest that the 22-melt material would flow more than five times farther or with less than one-fifth of the effort required for the 4-melt material. In actual practice, the differences are smaller than the numbers would suggest. The reason for the poor correlation can be seen if we examine the graph in Fig. 13.36. This shows the relationship of viscosity to shear rate for two polypropylene materials at a constant temperature of 227°C (440°F). The viscosity is measured in units called *Pascal-seconds* (Pa-s). If you are used to poise, there are 10 poise in a Pascal-second so simply add a zero to all the numbers on the vertical axis and you will be working in poise. The shear rate is given in an odd unit called *reciprocal seconds.* If we examine the equations for calculating

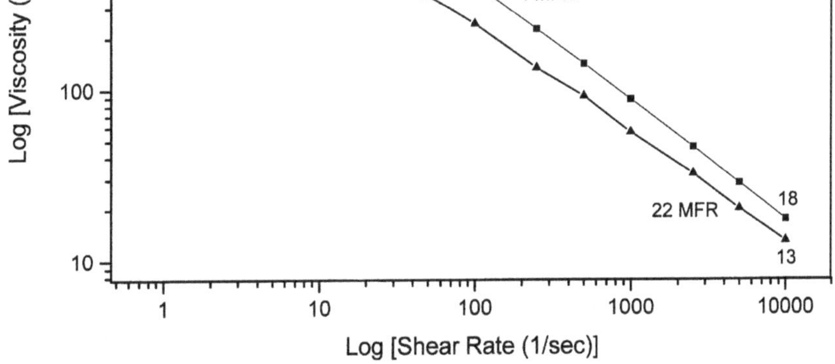

Figure 13.36 Viscosity versus shear rate comparison for two polypropylenes with melt flow rates of 4 g/10 minutes and 22 g/10 minutes.

shear rate values, which we will do later, these units make more sense. These graphs are developed using a device called a capillary rheometer. The instrument looks a lot like a melt flow tester but it has a drive motor on it that allows the speed of the piston to be controlled. Thus, it is a controlled shear rate device. In the melt flow test we rely on a fixed load and the speed is a result, not an input. It is therefore a controlled shear stress device. The difference may seem trivial, but the implications are important.

Data Analysis

First, let us look at the low shear rate region of the graph. The axes on this graph are logarithmic. This allows us to display large changes on a single piece of paper, but it also gives a distorted picture of what is actually happening to the material. Notice that at very low shear rates such as 1 reciprocal second (s^{-1}), the viscosity of the 4-melt flow polypropylene is 3460 Pa-s, whereas the 22-melt flow material has a viscosity of 1085 Pa-s. This is not quite the 5.5 : 1 ratio we might expect. As we increase the shear rate by moving the material at greater speeds or pushing it through smaller orifices, the viscosity of both materials drops dramatically. More importantly, the viscosities begin to converge. By the time we reach the shear rate of 10,000 s^{-1} the viscosity of the 4-melt material is only 18 Pa-s and the viscosity of the 22-melt material is down to 13 Pa-s. What was a difference of 70% at one end of the scale has been reduced to a 28% difference at the other end.

What does all of this have to do with the melt flow test and processability? During first-stage injection, shear rates of 10,000 reciprocal seconds and much more are common. The melt flow test for these two materials, however, operates in the range between 5 and 50 reciprocal seconds. The melt flow test therefore tends to magnify and exaggerate differences in viscosity, but the problem is even worse than it first appears. Remember that the melt flow test is a constant stress test. The shear rate, or rate of material flow, is a result, not a controlled input. By definition, a 22-melt flow rate polypropylene flows faster than 4-melt flow rate grade under this test. This means that the shear rate used in the test on the 4-melt material is lower than the shear rate on the 22-melt resin. How much lower? Five and a half times lower. The melt flow rate result itself means that in a constant volume test the same amount of material exits the melt flow test machine 5.5 times faster. In a constant time test, 5.5 times more material exits the orifice in the same amount of time. Either way, the shear rate for the high flow material is higher than it is for the stiffer flowing resin.

Shear Rates

We are not comparing the two materials at the same shear rate, and we are therefore exaggerating the differences in viscosity on two counts. First, we are measuring the viscosity in a shear rate region that bears little resemblance to processing conditions. Second, we are using a higher shear rate on the lower viscosity material. The shear rates that result in each test can be calculated by referring to a simplified equation that relates apparent shear rate to flow rate in a round cross-section.

$$S = 4Q/\pi r^3 \tag{13.1}$$

Rounded to the nearest whole number, the shear rate during the test of the 4-melt material is $8\,s^{-1}$, whereas the shear rate for the 22-melt grade is $44\,s^{-1}$. If we locate the viscosity for the 4-melt material at $8\,s^{-1}$ we will find it to be just under 2000 Pa-s (1850 to be exact). Now look at the plot for the 22-melt flow material, but do not pick the point corresponding to $8\,s^{-1}$. Instead move over to the point about midway between the lines representing 40 and $50\,s^{-1}$ and find the viscosity in that area. If you look here you will find a value close to 340 Pa-s. Take the ratio of 1850 to 340 and you get 5.44:1, which agrees very well with the 5.5:1 ratio suggested by the melt flow rate test.

This illustration shows why the melt flow rate test is so bad at estimating comparative processing characteristics for different resins; however, it also illustrates something that the melt flow rate test does very well. It distinguishes between different average molecular weights. The low shear rate region of the curve in Fig. 13.36 is where the differences in molecular weight are most apparent, and this is the primary value of the melt flow rate test. It is useful for verifying the consistency of the incoming material, and it is extremely valuable for assessing the effects of processing on average molecular weight.

In order to use this measurement tool, it is necessary to develop some guidelines for acceptable processing. The key here is the relative change in melt flow rate between the pellets at the beginning of the process and the molded parts at the end. Although there is some disagreement about the exact numbers, a large body of empirical data indicates that if the melt flow rate of the molded part produced in an unfilled material has increased by more than 40% over that of the pellets, then the process should be called into question. As part design pushes the envelope of a material's capabilities this threshold may need to be reduced. Some users of thin-walled parts are starting to look for increases of less than 20% as the dividing line between good and bad product. Other advocates of this evaluation method have quoted 30 to 35% as an acceptable increase. Studies show that depending upon the polymer, a well-controlled process can limit the increase in melt flow rate to 5–15%. The values of 20%, 30%, or 40% are not intended to represent thresholds below which all parts function as expected and above which all parts fail; however, there will be a measurable increase in product failures as these values rise because they relate inversely to the molecular weight of the polymer.

To perform an evaluation of this type properly, it is important to have virgin material and molded parts that were produced from that same lot of virgin material. There are a couple of reasons for this. First, if the parts are found to have an unacceptably high melt flow rate, it is important to determine that the raw material was within specification in the first place. Even within the range of acceptable melt flow rate measurements, the raw material may vary enough to affect the assessment of the molded parts. For example, a polycarbonate part that was failing during a snap-fit assembly was found to have a melt flow rate of 22 g/10 minutes. The specification for the base resin was 14–18 g/10 minutes. The evaluation of the integrity of the process depends greatly on the melt flow rate of the specific lot used to make the parts. If the value was nominal at 16 g/10 minutes, then the increase of 6 g/10 minutes constitutes an increase of 37.5% and is considered marginal. If the material started at the low end of 14, then the change is 57% and the process is considered to be beyond the normal conditions for proper handling of the polycarbonate. If the raw material starts at 18, then the change is only 22% and the process can be considered well controlled. In this case, when the raw material was obtained it was found to be 17.5 g/10 minutes and the shift of 26% is considered acceptable. When parts fail to perform in such a situation, it should immediately trigger a reassessment of the resin specification or the part design. These types of occurrences indicate that material made within specification and processed according to normal guidelines has a likelihood of failure. Either the design must be made more robust or the molecular weight of the material being used to produce the parts must be increased.

With the constant pressure to increase productivity and make more complex parts, there is an increasing chance that the molecular weight of a molded product will be a problem by design rather than due to poor processing. A good example is a part molded in high-density polyethylene. The original application used a 25-melt flow rate material in a 16-cavity mold and produced parts that were free of performance problems. When a new 32-cavity mold was constructed, the flow paths were much longer to the exterior cavities and the 25-melt material no longer worked. A new material with a melt flow rate of 40 g/minutes was specified and solved the processing problems. Within 6 months broken components began to come back to the original equipment manufacturer, who in turn brought them back to the molder for review. A melt flow rate test showed that the broken parts had a melt flow rate of 44 g/10 minutes, a 10% increase over the nominal value for the raw material and a 22% increase over low end of the spectrum. This was not a processing problem. It was a material specification problem, one that occurs with increasing frequency as material manufacturers reduce the molecular weight of their raw materials in an effort to fill the demand for thin-walled products in rapidly growing markets such as electronics and medical components.

The effects of processing can be easily captured by measuring the changes in average molecular weight using this test. Figure 13.37 shows the increase in melt flow rate for medical parts molded from a 4-melt flow rate polypropylene at two different melt temperatures. This material was purposely recycled back to the molding process repeatedly to assess the cumulative effects of using regrind as a cost saving measure. The melt temperatures were 205°C (400°F) and 250°C (480°F). Even though polypropylene is generally considered to be a material with good thermal stability, the results of this study were striking. They are summarized in Table 13.4 in terms of the percent increase in melt flow rate using the virgin material as the basis for the calculations.

The values in Table 13.4 represent the cumulative effect on the melt flow rate. They show that the change in molecular weight after one pass through the molding process at 250°C results in a greater reduction in molecular weight than do five passes through the process at 205°C. A visual and a physical evaluation of the product shows the onset of brittle behavior and a yellowing of the resin at the second regrind for the material processed at the higher melt temperature.

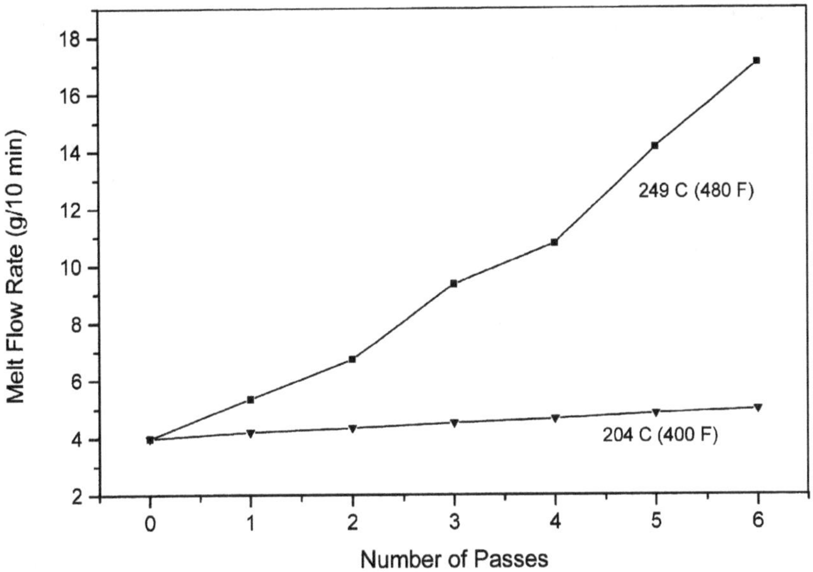

Figure 13.37 Change in melt flow rate of a polypropylene with repeated processing at two different melt temperatures.

The same effect will be observed if a material that is prone to hydrolysis is molded with excess moisture. Polycarbonate is a material used frequently in the medical industry for its clarity, toughness, and ability to resist certain types of sterilization, but many of the application environments to which polycarbonate is exposed can produce stress cracking. Resistance to stress cracking is significantly affected by molecular weight. One notable product failure involved parts that had been produced from a material with a nominal melt flow rate of 25 g/10 minutes. Reports of cracking from the field led to attempts to duplicate the problem in the manufacturer's laboratory using the chemical agents that were a known part of the application envi-

Table 13.4 Effect of Melt Temperature on the Melt Flow Rate of Polypropylene

Material used	MFR shift @ 205°C (%)	MFR shift @ 250°C (%)
Virgin	5.3	34.4
First regrind	8.8	69.1
Second regrind	13.3	135.4
Third regrind	16.8	170.9
Fourth regrind	21.6	255.8
Fifth regrind	25.4	329.9

ronment. No failures resulted. The failed parts were finally returned containing obvious failures and were found to have a melt flow rate of 50.6 g/10 minutes, an increase of more than 100%. Parts tested in the lab had a melt flow rate of 30.5, or a 22% shift. This explained why the lab tests could not duplicate the field failures. A review of manufacturing logs indicated that dryer problems had coincided with the times when the discrepant parts were manufactured, even though the parts did not show obvious signs of excess moisture.

Melt processing is not the only step in the process where molecular weight reduction can occur. Some materials are sensitive to prolonged drying at elevated temperatures. The most notable example of this is the polyamide (nylon) family. Nylon will degrade in the presence of excess moisture and must be dried to a maximum moisture level of 0.2% by weight of polymer to avoid going through the same hydrolysis process that plagues polycarbonate and PET polyester. The amide linkage, however, is also susceptible to oxidation, a process that occurs when the material is exposed to elevated temperatures in the presence of air. Most nylon suppliers suggest a maximum drying temperature of 80°C (176°F) or lower if the material is to be left in the dryer for an extended period of time. Because nylon that has picked up moisture can be difficult to dry in a timely fashion, however, there is a great temptation to elevate the drying temperature in order to hasten the process.

When nylon materials are dried at elevated temperatures, they tend to produce parts that are later found to be brittle. This is often blamed on overdrying, the removal of too much moisture from the material. It also may be attributed to the fact that the parts have not been conditioned to an equilibrium point with atmospheric moisture. A review of field failures, however, often shows that the brittle product contains 1.5 to 2% of moisture by weight. Studies have shown that simply remoisturizing the raw material to a higher moisture level before molding does not repair the damage done by the extended aggressive drying cycle [2].

Figure 13.38 shows the results of a melt flow rate evaluation performed on general-purpose nylon 6 dried at 80°C and 120°C. The material dried at the appropriate temperature maintains the same melt flow rate indefinitely and does produce usable product. The material dried at the elevated temperature shows an immediate and steady upward trend in the melt flow rate as it sits in the dryer. Within 4 days the melt flow rate has already increased by more than 40% without ever having been melted. Continued exposure raises the melt flow rate by as much as 200%. This change will result in the same symptoms of lost performance as in the previous examples. Figure 13.39 shows an infrared spectrum comparing the raw material that was properly dried with the material that was dried at elevated temperatures. The aggressively dried material shows an additional absorption band near 1740 cm^{-1} that verifies that the material has been oxidized even before molding.

Not all materials decline in molecular weight as they degrade. There are a few cases where degradation is accompanied by an increase in viscosity, which translates to a decrease in melt flow rate. This is usually caused by the cross-linking of the polymer or another ingredient in the compound. Cross-linking usually results in an embrittlement of the matrix, but the mechanism is different from what has been discussed to this point. Sulfone-based polymers are typical of materials that increase in

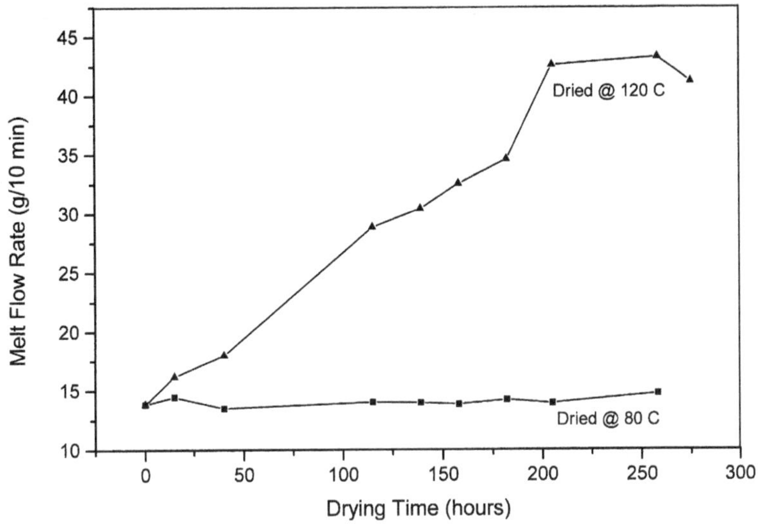

Figure 13.38 Effect of drying time and temperature on the melt flow rate of a nylon 6.

viscosity each time they are melt processed. Certain types of impact modifiers with unsaturated sites such as EPDM rubber and butadiene rubber can also cross-link and build viscosity. Incorporating a chain extender into a recycled material to offset the effects of degradation during reprocessing can also lead to an increase in viscosity that may not be controllable. The melt flow rate test will also detect these types of changes, although it may not offer insights into the mechanism.

The melt flow rate test can also act as a rapid method for determining the thermal stability of a compound in the melt. This test was actually devised by manufacturers

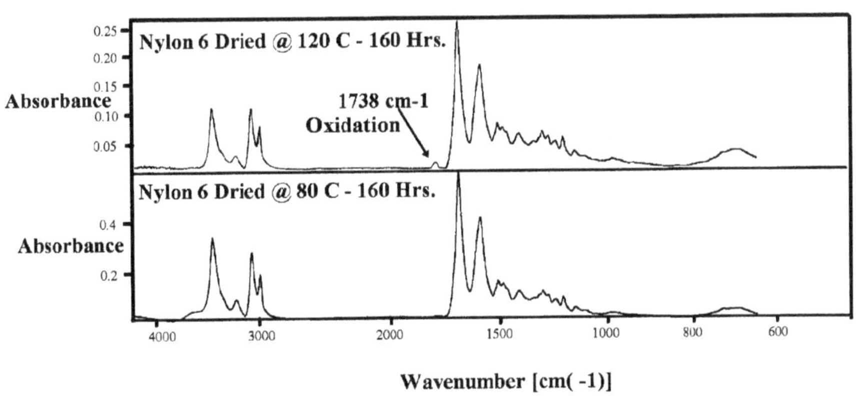

Figure 13.39 Comparison of effects of drying temperature on the structure of nylon 6. The elevated drying temperature results in oxidation of the polymer.

Table 13.5 Effect of Pigments on the Thermal Stability of Polycarbonate Raw Materials

Color	MFR—5 minutes	MFR—20 minutes	Percent shift
Clear	11.71	11.95	2.0
Black	12.37	13.86	12.0
Green	9.83	11.21	14.0
Blue	11.11	14.22	28.0
Red	10.18	13.15	29.2

of thermally sensitive materials that cross-link during processing when not properly stabilized. The test consists of running the standard melt flow test with the normal preheat time of 5 to 6 minutes, and then repeating the test using a longer preheat time to simulate the effects of an extended residence time in the processing machinery. This is not a perfect simulation of the molding conditions because the melt flow rate test does not include any shear from screw rotation or rapid injection, and it operates at the standard test temperatures for the melt flow rate, which are fairly mild compared with some processing conditions. Potential problems with processing can be screened by performing this expanded test despite these limitations.

The same test can be used to identify products that have limited thermal stability and may degrade during normal good processing. Table 13.5 shows this test applied to different colors of a polycarbonate used in making thin-walled parts for an electronics assembly. In this case the molder was having difficulty maintaining a very tight tolerance on the allowable increase in melt flow rate. The end user was attempting to hold the processor to a maximum increase of 20% in the melt flow rate. The clear and the black product were being molded successfully, but the clear tint colors were exhibiting an unusually high level of degradation given the customary thermal stability of polycarbonate. When a variety of process changes failed to produce the desired result, the melt stability test was conducted on the five different colors using a 5-minute preheat and a 20-minute preheat. The results show that two of the three transparent colors were inherently less stable than the black and clear product. This is a phenomenon that was attributed to an interaction between the polycarbonate and the pigments used to achieve the desired colors.

With these materials shifting by almost 30% during exposure in the melt flow tester, the processor had little chance of controlling the change to less than 20% during the rigors of the molding process. In this case the melt temperatures and levels of shear were too aggressive for the transparent blue and red. Changes in runners and gates reduced the shear on the material and permitted the selection of a lower melt temperature. Selecting a machine with more available pressure also made the reduced melt temperature possible and brought the product into line despite some inherent disadvantages in the material composition.

Our discussion of monitoring molecular weight with the melt flow rate test has focused on unfilled materials to this point. The technique can be applied to filled and reinforced materials, but the guidelines for judging when degradation has occurred are less easily defined because the size of reinforcing fibers and filler particles is also reduced during the molding process. This adds to the viscosity reduction that occurs when the polymer chains are shortened. As a result, the shift in melt flow rate becomes dependent upon the amount of filler as well as the type. Fibers undergo the most significant reduction in size because their long, thin geometry is the most vulnerable to breakage. In addition, fibers in a raw material may have a length that approximates the diameter of the orifice in the melt flow tester. At the low shear rates used in the melt flow rate test there is no significant degree of orientation, so the fibers do not align the way they do in the injection molding process.

Some fibers may line up in the direction of flow, but many others will lie in a variety of directions, including directly across the orifice in the transverse direction. This can significantly increase the viscosity of the material, especially when measured at low shear rates. After processing, the glass fibers are reduced in length by an average of 35 to 50%. This has a dramatic effect on the low shear viscosity. As a result, an increase in melt flow rate of 75 to 100% may occur in a material with 10 to 15% glass fiber. A material with 30 to 40% glass fiber may exhibit an increase in melt flow rate of up to 300% and still have the required integrity, but these determinations become much more approximate and it is more critical to run actual physical tests to correlate viscosity changes to performance.

One way to combat this uncertainty and still evaluate the material in the melt state is to employ the capillary rheometer to measure viscosity. This instrument is typically used to characterize the relationship between viscosity and shear rate, which is a necessary input for flow simulation. It can also be used, however, to run many of the same evaluations that the melt flow rate tester performs. Tests for degradation essentially look for the same responses detected by the melt flow rate test, but because higher shear rates are employed, the degree of difference between good and bad product is smaller. In addition, the shear rate that is used in the evaluations will influence the degree of reduction in viscosity that is measured.

To understand this mechanism, refer back to the comparison of viscosity versus shear rate behavior for the 4- and 22-melt polypropylenes shown in Fig. 13.36. For purposes of this illustration the 22-melt material can be considered as a degraded version of the 4-melt material. Although this would be a large shift in melt flow rate for an unfilled material, it would not be unheard of for a highly filled material and would characterize results where the higher value would represent a part that was just outside the realm of good product. As mentioned previously, the relative difference between the 4- and the 22-melt material varies with shear rate. At intermediate shear rate values the differences between the two materials will fall somewhere between the two extremes. An evaluation conducted at $400\,s^{-1}$ will therefore produce a result that would not be applicable to another set of data generated at $1000\,s^{-1}$.

Another important point is the change in units employed by the capillary test. The melt flow rate is expressed in grams/10 minutes, the higher the value the more degraded the material is considered to be. In addition, because the basis for the shift calculation is the lower number, it is possible to have changes of more than 100%. The capillary test measures viscosity directly. Melt viscosity decreases as melt flow rate increases. In this case the larger number is the basis for the calculation, and the change is expressed as a decrease. As a result, a 50% increase in melt flow rate and a 50% reduction in melt viscosity are not quantitatively the same. For this reason, the rules for judging a good and a bad product must change, even if a low shear rate such as $50\,s^{-1}$ is used. In general, the amount of empirical data available using this method is not as plentiful as it is for the melt flow rate technique. Qualifications relating performance and viscosity changes are often required in order to make good use of the data from the capillary rheometer technique. Nevertheless, the higher and controlled shear rates available on this instrument may produce more repeatable results for filled materials once a technique has been established.

The use of this technique to document degradation has been used effectively in studying the effects of excess moisture in PBT and PET polyesters at the time of processing [3]. In this study, 30% glass-filled grades of PBT and PET polyester were molded with moisture content values ranging from 50 ppm (0.005%) to 2000 ppm (0.2%). The molded parts were then assessed for a variety of physical properties as well as for melt viscosity by using a single shear rate of $400\,s^{-1}$ in a capillary rheometer.

Figure 13.40 shows the relationship between the viscosity of the final molded product made from 30% glass-filled PET and the moisture content of the base resin during the molding process. The test captures the decline in viscosity, and an evaluation based on this method can be constructed to document degradation using rules that are appropriate for the particular shear rate chosen for the test.

Figure 13.41 plots the results of unnotched Izod impact tests performed on parts molded from the various moisture contents and shows that the decline in this property reflects the reduction in melt viscosity. Figure 13.42 plots tensile strength as a

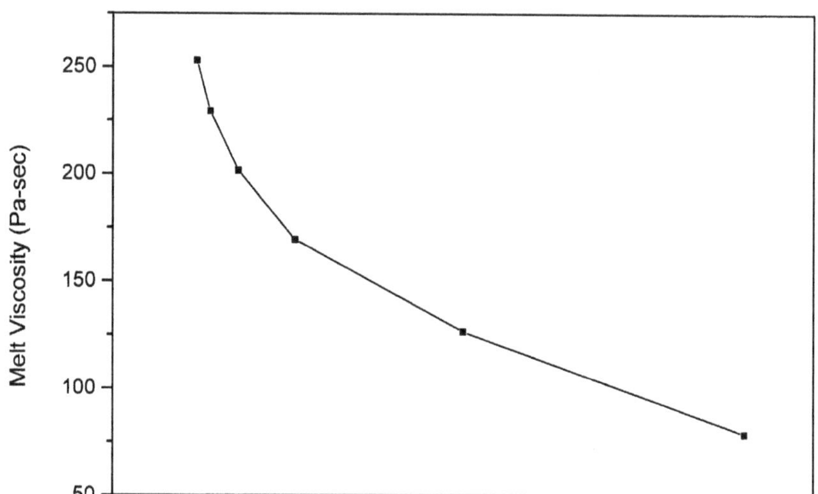

Figure 13.40 Effect of pellet moisture content on the melt viscosity of a 30% glass-filled PET polyester.

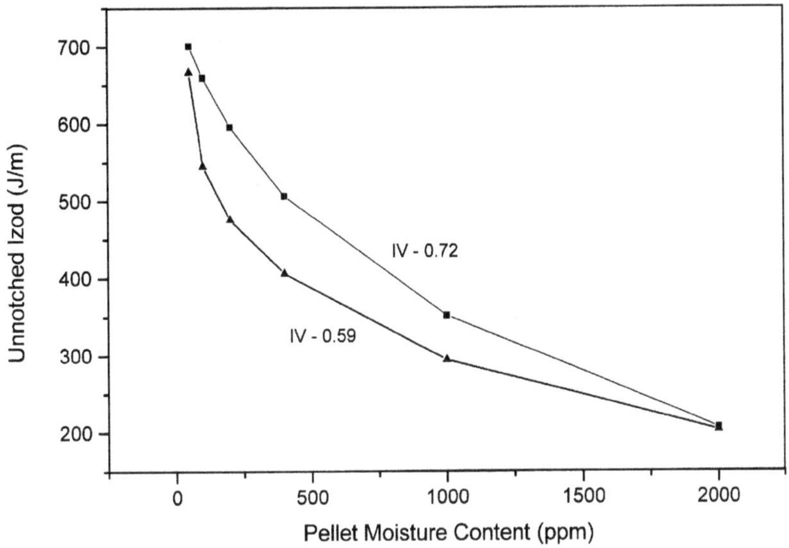

Figure 13.41 Effect of pellet moisture content on the unnotched Izod impact resistance of parts molded from two grades of 30% glass-filled PET polyester.

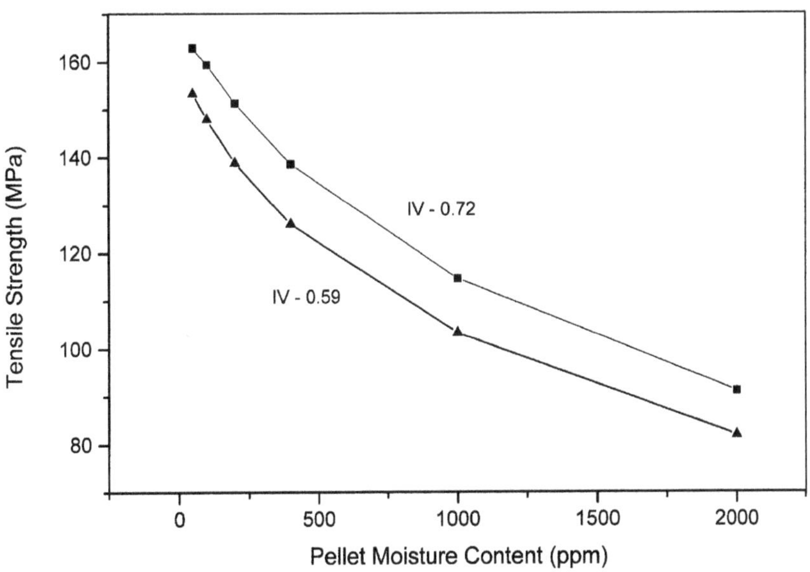

Figure 13.42 Effect of pellet moisture content on the tensile strength of parts molded from two grades of 30% glass-filled PET polyester.

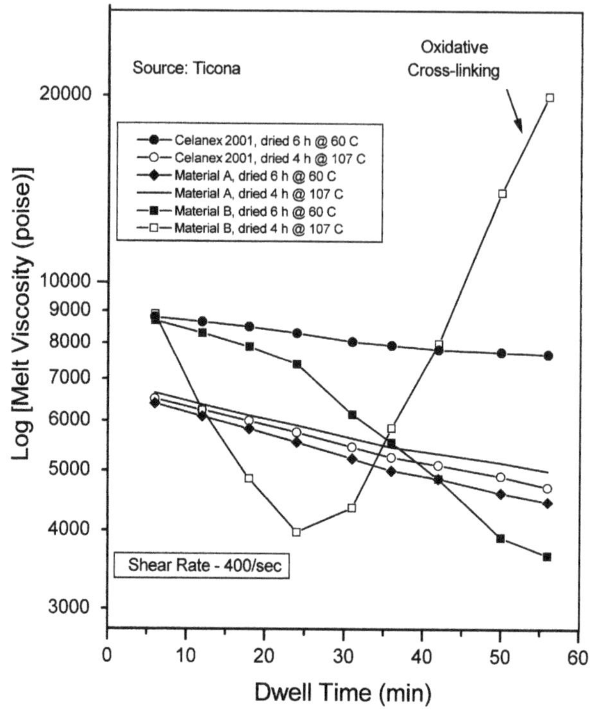

Figure 13.43 Thermal stability test results on various PBT polyesters by capillary rheometry.

function of the moisture content. Although the tensile strength is affected less than the impact resistance, the negative effects of running wet material are apparent. A major point of this study is the finding that moisture removal to levels below the recommended maximum value of 200 ppm (0.02%) has considerable benefits in improving physical properties. This tends to counter concerns over the notion of overdrying.

The capillary rheometer can also be used effectively to replicate the melt stability test discussed above. Figure 13.43 shows the results of a melt stability study conducted on different grades of PBT polyester. It highlights the use of different drying routines and the effects of prolonged exposure to process temperatures as a function of time using a constant shear rate of $400\,s^{-1}$. This study shows a variety of interesting responses. The material identified as Celanex 2001 shows a very small change in viscosity over time, which indicates good thermal stability. A second resin system exhibits slightly less stable behavior over time. The third resin shows poor stability, but more importantly it documents the cross-linking behavior that was discussed earlier. The material loses viscosity, but the viscosity of this system then starts to build rapidly as the shortened chains begin to connect using sites created by the initial degradation process. This represents an excellent use of the capillary rheometer as an analytical tool.

Solution Techniques for Measuring Viscosity

Solution techniques for measuring the viscosity of a polymer are an alternative to the capillary technique. Rather than overcoming the effect of a filler, this technique dissolves the compound, isolating the filler as a solid residue and allowing for its removal from the system before a viscosity measurement is made. The viscosity of the solution is then compared with the viscosity of the solvent to arrive at a value that is referred to as a relative viscosity. In its most basic form the measurement is a comparison of the time required for a fixed volume of pure solvent and a dilute solution of the polymer in that solvent to flow through a specialized glass tube known as a *viscometer*. The measurement is dependent upon the concentration of the polymer in the solvent. By a series of calculations, this relative viscosity is converted into a value known as the *intrinsic viscosity*. This is a value that is independent of concentration and is therefore more useful, but in order to obtain this improved result it is necessary to make several measurements at different concentrations of polymer.

The purpose of these methods is to determine in a relative way the molecular weight of the polymer using a solution technique rather than a melt technique. As a practical matter, the equipment involved in this method is more costly and sophisticated and involves the use of solvents that can be very aggressive and dangerous to handle. The method is also more demanding on the analyst because the results are extremely sensitive to temperature and therefore must be conducted in a water bath that can be controlled to within 0.1°C. In addition, the results will be influenced by the solvent that is used. This is an important consideration in comparing results between laboratories; however, the method is excellent for separating the effects of processing on the polymer from that of the filler. Solution techniques also have a distinct advantage over melt techniques in that they require a much smaller sample size. When working with a limited number of failures in small parts, melt techniques are simply not practical because they require too much material. The melt flow rate test uses 15 to 20 g of material and the capillary rheometry determinations can require ten times that amount. Solution techniques can be run with a fraction of a gram of material.

Molecular weight and intrinsic viscosity are related by a different mathematical relationship than the one relating molecular weight to melt viscosity. Where a melt viscosity reduction of 30% for an unfilled polymer is considered as a borderline between good and poor processing, the allowable shift for intrinsic viscosity is only 10%. Just as in the case of the melt flow rate and melt viscosity techniques, this value is not a firm dividing line between good and bad product. The greater the viscosity

Table 13.6 Intrinsic Viscosity Test Results on PBT Polyester

Sample	Intrinsic viscosity (dl/g)	Percent shift from virgin by I.V.	Percent shift from virgin by M.V.
Virgin raw material	0.887	—	—
Parts molded at 250°C	0.771	13.1	45.7
Parts molded at 265°C	0.648	26.9	59.4
Field returns	0.538	39.3	75.6

reduction becomes, however, the more likely the product is to fail. Table 13.6 summarizes the results of an assessment performed on parts molded from an unfilled flame-retardant PBT polyester. The parts that were functioning properly were molded at a melt temperature of 250°C. Even though they exhibited a shift in intrinsic viscosity that exceeded the 10% limit, the product still met the performance requirements. Parts molded at 265°C exhibited a much larger shift and displayed a moderate level of failure over a short period of time in the field characterized by brittle behavior. A field return processed at conditions that can be extrapolated to be very aggressive, originated from a batch of parts that failed nearly 100% of the time. These types of studies are also extremely useful in evaluating the relative stability of a particular resin system as a function of orchestrated process changes. PBT polyester is one of the most thermally sensitive commercial polymers commonly used, and these results quantify that sensitivity. The table also shows the percentage reductions in melt viscosity that were run in parallel to the solution technique. The same pattern is reflected even though the absolute numbers are different by this method.

In spite of the inherent degree of difficulty involved with solution viscosity techniques, modern equipment has simplified the process by making the measurement as a function of a pressure drop across a capillary tube. The pressure drop in the pure solvent is compared with that of the polymer solution and the viscosity is calculated by the software. It may be of passing interest to note the units used to express the intrinsic viscosity. Whereas melt viscosity is expressed in poise or Pa-s, which directly express the resistance to flow of the system, intrinsic viscosity is expressed in deciliters/gram (dl/g), which is not a viscosity term at all. Instead, it is a measurement of specific volume or how much room a given amount of material occupies. What is actually being measured is molecular size, the bigger the molecules are, the more space each one will occupy. The assumption is made that molecular size is related to molecular weight.

The gold standard in molecular weight measurements is known as *gel permeation chromatography* (GPC). GPC starts with the same polymer solution that is used in the intrinsic or relative viscosity measurement, but intrinsic viscosity is essentially a statistical average of the effects of all the various polymer chains in the sample. If all of these chains were of the same size, this measurement would be all that is needed to characterize the material. One of the fascinating consequences of synthesizing polymers, however, is that we never obtain a material with molecules of uniform size. Instead, we obtain a wide distribution of sizes that tend to follow a bell curve relationship. Whereas the intrinsic viscosity gives us an average value, GPC has the ability to show us the histogram behind that average.

Figure 13.44 gives a schematic of the principle on which GPC operates. The polymer solution is injected into a column that contains a packing material that is capable of adsorbing the polymer material onto its surface. The rate at which the polymer chains make their way through the column is related to their molecular size. The largest chains have the lowest affinity with the column material and therefore make it through in the shortest time, called the *elution time*. The smaller chains are retained for longer periods of time. The instrument therefore separates the material into its various molecular sizes as a function of time. Figure 13.45 shows a plot of an ideal result from such a test. The large molecular sizes of the polymer come out first,

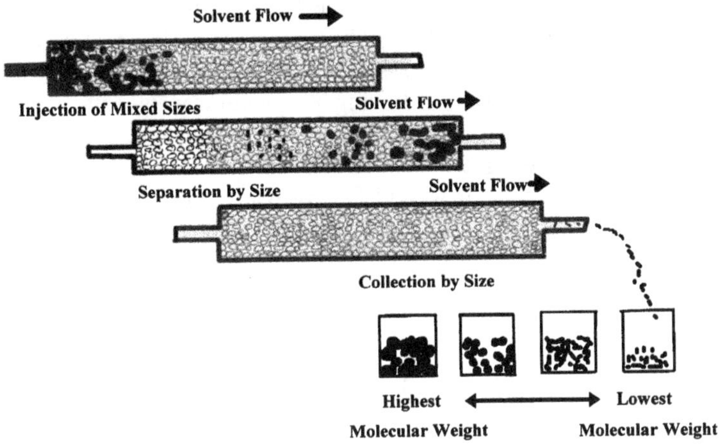

Figure 13.44 Schematic of the principle of operation in a gel permeation chromatograph.

followed by much shorter chains called *oligomers*. These are then followed by additives and residual monomer. This last phase can be broken down further by a variety of techniques including HPLC, GC-MS, and GC-IR, to identify and quantify the additive package.

Molecular Size

Let us return to the main event. As with intrinsic viscosity, the measurement being made here is molecular size. With the appropriate mathematics and attention to detail, however, these raw data are translatable into molecular weight. GPC is the only technique that is able to determine absolute molecular weight and provide a complete picture of all of the components that make up that molecular weight. Although the shape of the entire curve is important, there are key values that are an

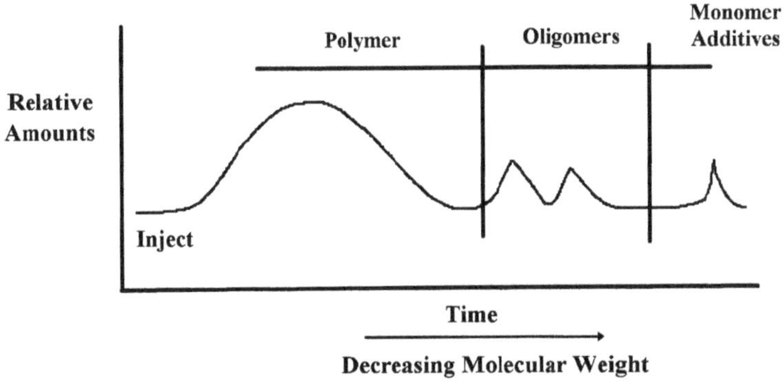

Figure 13.45 Idealized test result from a gel permeation chromatography experiment.

attribute of each curve that help to summarize the overall distribution. The first of these is the number average molecular weight, M_n, which is the most commonly used value when referring to the average molecular weight of a polymer. If a manufacturer refers to one of their materials as having a molecular weight of 23,500, and they do not specify a particular subscript, it is almost certain that they are referring to the number average molecular weight.

The number average molecular weight is expressed mathematically as

$$M_n = \frac{\sum_{i=1}^{N} N_i M_i}{\sum_{i=1}^{N} N_i} \tag{13.2}$$

where N_i = the number of moles of molecules with a molecular weight of M_i.
If the distribution curve is perfectly symmetrical, then the number average value will coincide with the peak of the curve because it represents a simple average of all of the various molecular weights divided by the number of molecules in the population.

The weight average molecular weight, M_w, is literally a weighted average. It replaces M_i in the numerator of the preceding equation with M_i^2 and incorporates M_i into the denominator. This has the effect of accentuating the contribution of the higher molecular weight fraction in the distribution. If all of the molecules in the polymer system were the same, as they are with classical low molecular weight materials, then M_w would equal M_n and we would refer to the system as monodisperse. Even though it is possible to manufacture some materials such as polystyrene with a nearly monodisperse distribution under controlled laboratory conditions, most commercial materials will be polydisperse. M_w, therefore, will always be greater than M_n and the ratio M_w/M_n is called the *polydispersity index* (PDI). It is a relative measure of the breadth of the distribution; the higher the PDI the broader the distribution. We refer to this characteristic of the material as the *molecular weight distribution* (MWD).

Although this may all seem very arcane, there are a lot of practical reasons for paying attention to this characteristic of a polymer. Whereas most of the polymer molecules obviously reside in the center portion of the curve, the molecules out at the extremities of the distribution have a substantial influence on certain properties. The low molecular weight fraction out at the far right of the main distribution curve greatly enhances polymer flow as the smaller chains act as a sort of plasticizer for the higher molecular weight fraction. At the other end of the distribution, the high molecular weight fraction provides a contribution to the impact resistance of the material that is disproportionately large compared with the number of chains of this size. In addition, these larger chains are extremely important in reducing the permeability of a material. If a manufacturer was selecting a polyethylene for a single-layer gasoline container and needed to minimize the loss of vapors through the container wall, a large high molecular weight fraction would be an important part of the specification and a large polydispersity would be desirable.

A third measurement, known as M_z further emphasizes the role of the high molecular weight chains. This value is calculated using the same equation form given earlier. In the M_z calculation, however, M_i is replaced by M_i^3 in the numerator; the denominator contains the term M_i^2. In cases where the molecular weight distribution curve is not perfectly symmetrical, M_n will not coincide with the peak of the distribution curve. In such cases, the peak point is picked out as a fourth measurement and is designated M_p. The difference between M_n and M_p is an indication of the degree of symmetry in the distribution curve.

Complexity and Cost

If all of this looks complicated and expensive, it is. First, as with the intrinsic viscosity technique, the polymer must be dissolved in a solvent. Some of these solvents are very aggressive to their users if mishandled, and they are very expensive. In addition, some polymers such as polyethylene will not dissolve at room temperature; therefore, they require a high-temperature system. This adds to the cost and complexity of the instrument. In addition, calculating absolute molecular weight values requires the use of a calibrated set of standards that are typically used with a specific solvent. Because polymer–solvent interactions influence the time it takes for the solution to pass through the column, it is common that errors creep into the calculations because of failure to take the solvent interaction factor into account. In addition, different detectors can be used to look at the various molecular weight fractions as they elute. A detector such as an in-line viscometer will be sensitive to high molecular-weight fractions, whereas a detector that measures the refractive index of the solution is more likely to miss the presence of such a fraction. For most manufacturing operations, GPC is beyond the scope of practical problem solving. It is no surprise, then, that the other techniques mentioned previously are often used in an attempt to circumvent the need for chromatography.

Table 13.7 shows results from melt flow rate and GPC evaluations of several polyethylene materials. The melt index is presented in the standard form of grams/10 minutes; however, in addition the melt flow rate test is used in a different way to assess the effects of increased shear rate, even though the melt flow rate test does not control the shear rate. This method was developed by Dow Chemical when it became apparent that the new polyethylene materials produced with metallocene or single-site catalysts did not flow in the same way that more traditional materials did.

Table 13.7 Molecular Weight & Viscosity Characterizations for HDPE

Property	Material A	Material B	Material C	Material D	Material E
Melt Index	6.2	3.3	2.0	5.8	6.8
MFR Ratio	15.4	24.0	24.0	15.5	24.8
M_n	32,100	23,680	28,960	34,440	18,800
M_w	79,890	86,320	103,190	71,680	69,600
M_z	358,040	255,460	293,530	124,670	204,700
M_w/M_n	2.49	3.65	3.56	2.08	3.70
M_z/M_w	4.48	2.96	2.84	1.74	2.94

Early users of the new materials quickly noticed that the melt index did a poor job of characterizing the anticipated flow characteristics of the materials during actual processing. A metallocene-based material with a melt index of 6g/10 minutes was more difficult to process than was a traditional material with the same melt index. More pressure and/or more heat was required to produce the same result.

Shear-Thinning Effects

This phenomenon points out one of the more significant effects of molecular weight distribution: shear-thinning effects are less significant as the molecular weight distribution of a polymer becomes narrower. Traditional polyethylene materials tend to have relatively broad molecular weight distributions. The new materials catalyzed with metallocenes exhibit a narrower distribution. Even though this property is more controllable with the new catalysts and allows for the tailoring of certain properties within a tighter range, it reduces the responsiveness of the material viscosity to the effects of increased shear. To better characterize the material, Dow expanded the melt index test to run at two different loads. The standard test uses the load of 2.16kg while the second test uses 10kg. A higher load of 21.6kg has become more popular, in part because it represents the standard load for testing high molecular weight polyethylenes—the so-called high-load melt index (HLMI). The 21.6kg load is used in the results shown in Table 13.7. The GPC data include actual molecular weight values for M_n, M_w, and M_z along with calculations of the polydispersity index and another ratio expressed as M_z/M_w.

There are several interesting correlations that show the utility of the simpler melt flow rate/melt index test. First, there is a reasonably good correlation between melt index and M_w, not M_n as is often assumed. Second, there is an excellent correlation between the MFR ratio and the polydispersity index (M_w/M_n). This test has become known as the poor man's GPC because it provides one of the essential pieces of data from the chromatography method with the use of the simple melt flow rate apparatus. Polyethylene materials with very broad molecular weight distributions have been known to produce MFR ratios of as high as 100, illustrating the broad range of results that can be obtained from such a test.

There are some details, however, that only the GPC can reveal. For example, the ratio M_z/M_w does not necessarily reflect the ratio M_w/M_n. This shows the importance of a comprehensive characterization, particularly for specialized applications where the high molecular weight fraction is critical. In comparing Materials A and D by melt flow techniques, they appear to have the same properties. Their nominal melt index values vary by less than 7%, and their MFR ratios are virtually the same. Even much of the GPC work does little to distinguish these materials from one another. The polydispersity indexes follow the MFR ratio results and are therefore very comparable. Even the absolute M_w and M_n values that make up the polydispersity values are close; however, the M_z values are very different, resulting in substantial differences in properties like fatigue resistance and hydrocarbon permeation rates. The shape of the molecular weight distribution curves associated with these tabular summaries will show these differences at a glance, even without the benefit of an exact calculation of the molecular weight.

There are other applications where there is no substitute for a full assessment of the molecular weight distribution. Many of these involve competing reactions that result in a small net change in viscosity. Refer back to Fig. 13.43 and note once again the curve for the grade of PBT that declines rapidly in viscosity and then recovers to an ultimate value that is far above its initial viscosity. There are a range of times during this process when an overall viscosity measurement, by either melt or solution techniques, would reveal little or no change; however, Fig. 13.46 illustrates the process by GPC. This test compares reference pellets to multiple determinations of a very brittle material that exhibits only a very small change in viscosity. At the high molecular weight end of the spectrum is a significant secondary population of very large chains. This is accompanied by a shift in the overall distribution to a lower average molecular weight. In addition, there is a significant change in the oligomer population located at the far right on the graph. This result shows a process that is transparent to other techniques. The bulk of the population of polymer chains in this system is showing a reduction in molecular weight. At the same time, however, some

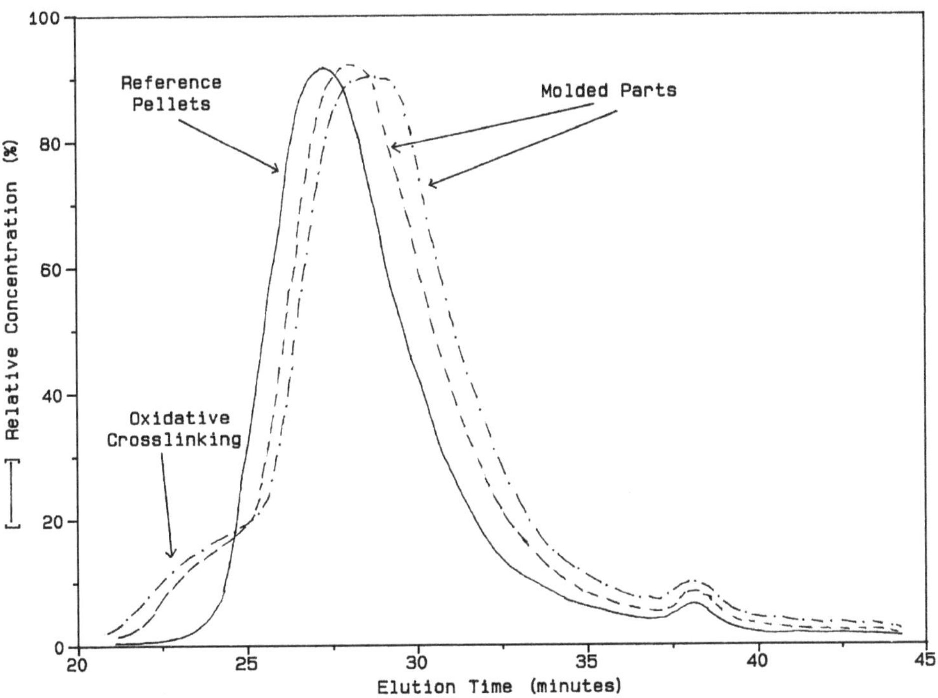

Figure 13.46 GPC test results showing simultaneous degradation and crosslinking in a PBT polyester.

of the smaller chains are forming cross-links that build molecular weight rapidly. The asymmetry in the curve is an indication of an extraordinary mechanism such as cross-linking as opposed to a general increase in the size of all the polymer chains. Finally, many of the by-products of the initial chain scission process are detected as chain fragments in an increasing population of oligomer.

In contrast, Fig. 13.47 shows an instance of increasing molecular weight through a mechanism known as solid-state polymerization. In this case a heat-stabilized nylon 6/6 had been exposed to elevated temperatures for a prolonged period of time in the solid state. Melt flow rate measurements had detected a decrease in the melt flow rate from 22 g/10 minutes to 6 g/10 minutes. Intrinsic viscosity measurements showed a 50% increase. Impact tests documented a very large increase in ductility and total energy to failure. Without a picture of the actual molecular weight distribution, however, it was difficult to determine the root cause for these changes. The GPC results show little change in the overall distribution of molecular sizes, the entire population is shifted almost uniformly to a higher molecular weight and no asymmetry appears in the curve to suggest the presence of gels or cross-links. There is also no change in the oligomer fraction.

Figure 13.48 shows the effects of the same type of extended exposure to elevated temperatures on a nylon 6/6 without the addition of the heat stabilizer. In this case the melt flow rate of the material had increased by 24% and intrinsic viscosity tests

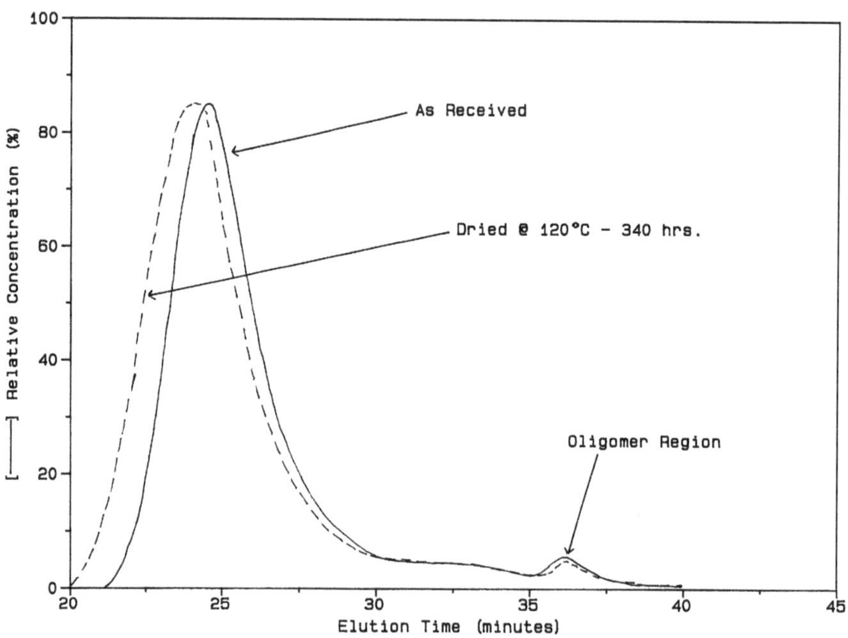

Figure 13.47 GPC results showing the effects of drying on the molecular weight of a nylon 6/6 containing a heat stabilizer.

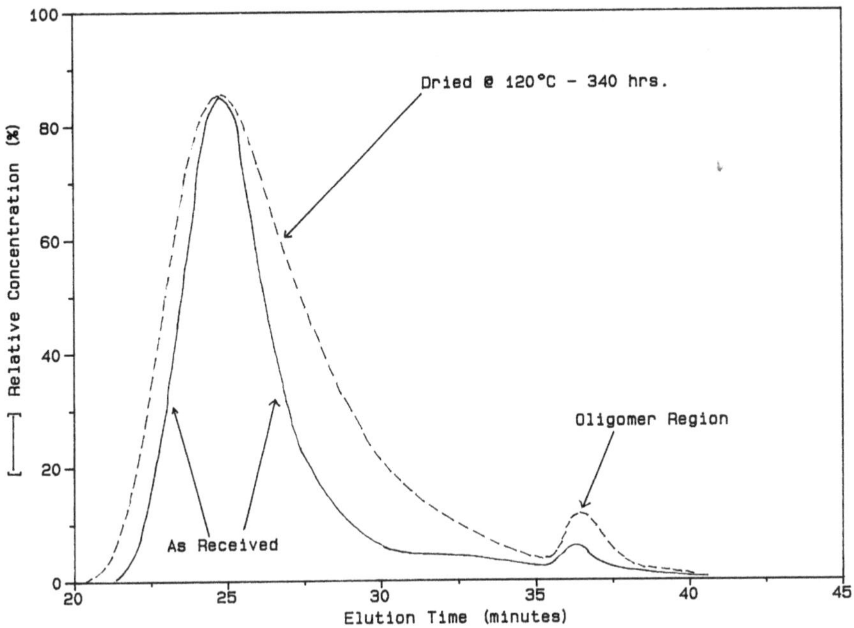

Figure 13.48 GPC results showing the effects of drying on the molecular weight distribution of a nylon 6/6 without heat stabilizer.

had detected a decrease of 10.4%. The product was extremely brittle after aging, however, and it failed to regain its ductility when conditioned with moisture. In this case the GPC results were the only method of evaluation that accurately illustrated the change in the polymer. The results show that the heat aging produced an increase in the size of some chains and a decrease in the size of others. The molecular weight distribution broadens substantially and produces an asymmetric curve that favors the low molecular weight side. The impact strength of the material is disproportionately affected by the lower molecular weight fraction. Neither the melt viscosity nor the intrinsic viscosity measurements can adequately account for the dramatic loss in properties in the polymer. This type of mechanism is frequently observed in nylon that has become brittle under application conditions that involve sustained exposure to elevated temperatures. Because there are many possible causes for the embrittlement of nylon polymers, this type of evaluation is indispensable to gaining an understanding of the root cause.

At this point we have reviewed the importance of molecular weight to the integrity of the polymer and have outlined the various tools available to diagnose the problems that can arise. They cover a range from simple to highly sophisticated, and they all have a place in determining how a material has changed and in connecting those changes to observed performance. Because the majority of problems arise in a manufacturing environment, the bias is for the rapid and low-cost approach, and this practical approach is a good one. It is important to understand the value of all the available tools, however, and to know when the simple methods have reached the limits of their usefulness. At that point the simple techniques should be abandoned in favor of a less familiar but more discerning approach.

In the last section of this chapter we will review some aspects of polymer behavior that bring the elements of composition and molecular weight together with the application environment. We will illustrate some of the common application failures and examine some of the common causes for their occurrence.

13.3 Performance Problems

At the beginning of this chapter, we listed the four aspects of a successful molded product: part design, mold design and construction, processing, and material selection. When a product fails to perform in the field as anticipated, rules involving one or more of these categories have been violated. The problem often represents a combination of problems that cross the lines between these disciplines; however, the problems involving material selection are perhaps the most difficult to diagnose because the rules governing the selection process are not as well defined as they are for the other three categories. In this section we will review some of the more common mistakes that are made in selecting the correct material and provide a brief overview of some systematic steps that can be used to prevent costly problems from arising later.

13.3.1 Material Selection Process

It sounds like the height of oversimplification, but the hard reality is that a good material selection process begins with an accurate assessment of the application conditions. This is sometimes more difficult than it sounds because the testing required to evaluate the application environment is often seen as an unnecessary and costly step in the development process. A good example of this is the environment under the hood of an automobile. Plastic products have been replacing metal components in the engine compartment for several decades, but it was almost impossible in the past to obtain scientific measurements of temperature in the various locations around the compartment. This presents special difficulties in a climate where cost reduction is a major focus because cost reduction in raw materials almost always involves a reduction in heat resistance.

Perhaps the best way to approach the problem of material selection errors is to highlight the areas of most frequent failure. An extensive study has been performed by the Rubber and Plastics Research Association (RAPRA) involving more than 5000 field failures spanning twenty-five years. They found that nearly one third of the failures were caused by environmental stress cracking (ESC) and that an additional

20% involved fatigue [4]. The good news in this finding is that problems with ESC and fatigue primarily follow the distinction between semi-crystalline and amorphous material families. An objective evaluation of ESC problems shows that the vast majority of them involve amorphous resins such as ABS, polycarbonate, polystyrene, and polysulfone. Good data on fatigue can be difficult to obtain because tests performed on various materials are rarely performed under comparable conditions. A pattern can be established, however, that shows that semi-crystalline materials generally tend to reach a limiting stress that is independent of the number of cycles involved in the fatigue evaluation. By contrast, amorphous materials undergo a slow but steady decline in limiting stress as a function of the increasing number of cycles. Figure 13.49 illustrates this behavior for unfilled grades of PBT polyester and polycarbonate. Even though the polycarbonate has superior performance over a relatively small number of cycles, extended testing shows that the semi-crystalline PBT polyester will eventually "catch up" to the polycarbonate.

The behavior of amorphous materials in fatigue is often worse than these standard tests indicate because the standard test frequencies are relatively high and can result in an uncontrolled temperature rise in the test specimens. This often results in a ductile failure mode at an extended number of cycles. Further work performed by RAPRA in this area has focused on low-cycle fatigue tests that often do a better job of simulating mechanical stress application in a real-world environment. Figure 13.50

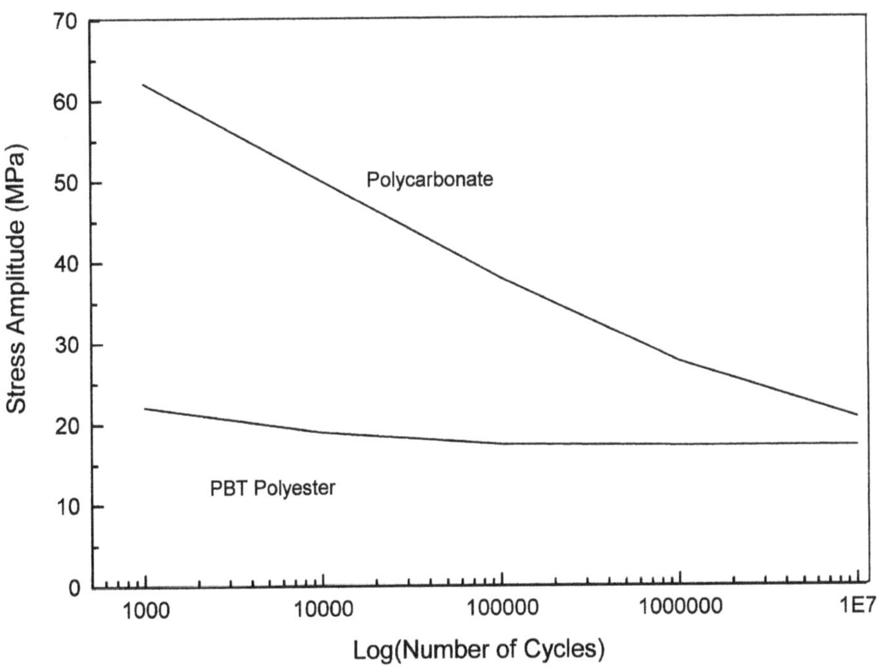

Figure 13.49 Fatigue resistance for PBT polyester and polycarbonate tested at a cycle rate of 30 Hz.

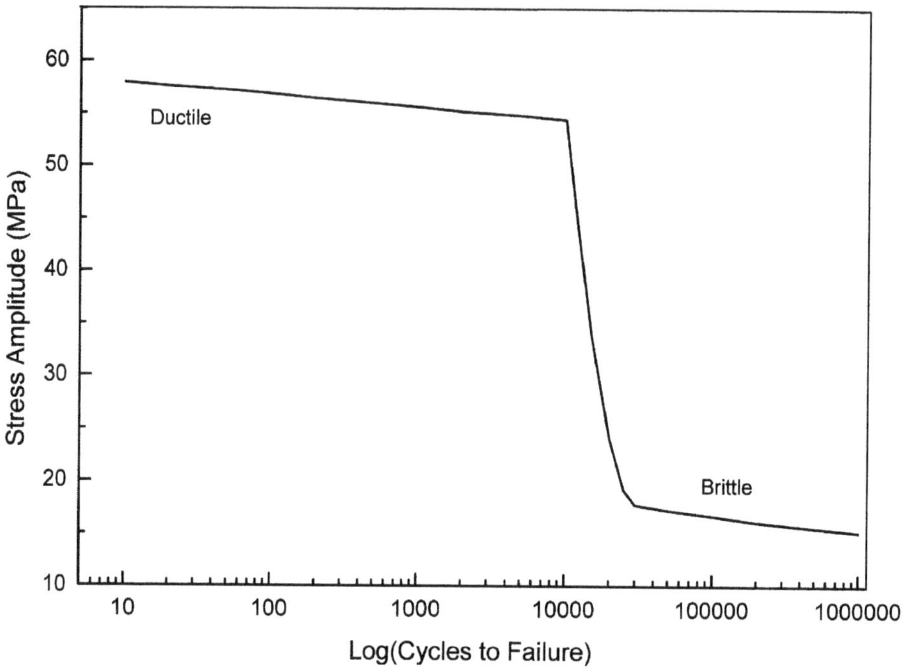

Figure 13.50 Fatigue behavior for polycarbonate tested at 0.5 Hz.

shows the results from a fatigue test performed on a polycarbonate at a frequency of 0.5 Hz. The shape of the S/N curve plotting stress amplitude versus number of cycles is distinctly different from the classical exponential decay behavior that is considered to be typical in plastics. The material instead exhibits a stress tolerance at a low number of cycles that approaches the yield tensile strength of the material. In this region the failure mode of the samples is ductile. At some point, however, the material transitions rapidly from a ductile to a brittle failure mode and the allowable stress levels decline by approximately 80%. This behavior is reminiscent of the ductile-to-brittle behavior observed when impact performance is evaluated as a function of temperature. The exact location of this transition can be influenced by part design, secondary application influences such as temperature and chemical exposure, and the molecular weight of the material. This type of evaluation goes a long way toward explaining the reasons for the catastrophic failure of materials that appear to be very ductile and strong when tested over short duration.

The inherent weaknesses of amorphous materials in the area of ESC and fatigue do not preclude their use in applications where these influences are factors. Thorough testing is needed, however, and everything must be done in the area of design and material selection to minimize the chances of failure. Design features that act as stress concentrators will reduce the time to failure. Molding techniques that build in high levels of stress in the part will also contribute to an early demise. As indicated earlier, a higher molecular weight material will extend the natural life of a product that might otherwise be prone to failure. The incorporation of a reinforcement such as glass fiber can also reduce the sensitivity of a material to the influences of fatigue and ESC; however, amorphous resins are often selected due to their superior ductility and/or their transparency. The use of reinforcements compromises both of these properties, so the trade-offs in esthetics and performance need to be well understood.

The use of semi-crystalline materials will improve the performance of a product in the applications where fatigue and ESC are factors. The decision to use a semi-crystalline material, however, is best made before the mold is designed and built. Semi-crystalline materials inherently shrink more because of the structural order that they achieve during the cooling process. Although unfilled amorphous materials all shrink by approximately 0.5% as they cool from the melt, unfilled semi-crystalline materials may shrink anywhere from 1 to 3%. In addition, semi-crystalline materials are prone to greater dimensional variation as a function of fluctuations in process conditions. A change in packing pressure may cause an ABS product to shrink 0.7% instead of 0.5%, but the shrinkage of an acetal part may increase from 2 to 3% under the same type of adjustment. If the materials under consideration are reinforced, there will be noticeable differences in the tendency for warpage and other manifestations of differential shrinkage. Because amorphous polymers naturally have lower shrinkage rates, oriented fiber in the part has only a small effect on the shrinkage rate. In an unfilled semi-crystalline material like nylon, however, where the natural shrinkage may be 2%, the presence of oriented fiber can substantially reduce the shrinkage of the part in the flow direction while having only a minor effect in the direction transverse to flow. This differential shrinkage sets up stresses in the molded part that can significantly increase the tendency for the part to warp and distort. In addition, part features that push the limits of good design principles can often be tolerated in amorphous materials because of their relatively isotropic behavior. These same features can cause substantial difficulties in semi-crystalline materials, where a sudden change in wall thickness may cause a significant increase in material shrinkage.

In cases where material selection is performed after the mold is already built, there is a tendency to modify the candidate materials to reduce shrinkage in an effort to avoid the need for tool changes. A classical case is a conversion from high-impact polystyrene to polypropylene. Over the years, the logic for such a change has included cost and availability issues as well as performance considerations. From a performance standpoint the semi-crystalline polypropylene has much better overall chemical resistance than does the polystyrene. This superiority also extends to the fatigue and ESC considerations mentioned earlier, but the shrinkage rate of high-impact polystyrene is 0.5 to 0.6%, which is typical for an amorphous material. Polypropylenes may shrink anywhere from 1 to 2.5% depending upon part design as well as the nuances of material composition. In addition, the modulus of polypropylene tends to be lower, particularly for those grades where the impact performance will approach that of the modified polystyrene. This lower modulus dictates that stresses in the part will result in substantial amounts of warpage.

In most part geometries, the conversion from polystyrene to polypropylene inevitably results in a part that is smaller and at the same time exhibits warpage in areas where the polystyrene was flat. The simplest way to limit the shrinkage of the polypropylene while combating the problems with warpage is to add a filler. Particulate fillers such as talc and calcium carbonate are preferred to glass fiber, where the geometry of the fiber accentuates the tendency for warpage. If weld-line integrity is a consideration, then calcium carbonate is preferred over talc because the shape of the calcium carbonate particles interferes only minimally with the knitting of the polypropylene flow fronts. In order to reduce the shrinkage of the polypropylene to the point where it is comparable to the polystyrene, however, it is frequently necessary to add the filler at a level of 40%. This changes a number of key properties, including strength, stiffness, and impact resistance. It also increases the weight of the part. In addition, the heavy loading of an opaque white filler can make color matches with the old material difficult. Changes in pigment type or pigment loading may be necessary to maintain esthetics. All of these factors make it likely that a change of this type, made after the mold is constructed and initial qualification has begun, will cause significant delays in product introduction, costly tool revisions, and lingering unresolvable conflicts between competing requirements. Understanding the nature of semi-crystalline and amorphous materials and all of the implications that they have for processing and performance is critical to the material selection process.

Once this fundamental has been mastered, the primary challenge is avoiding the use of short-term property data to predict long-term performance. Unlike metals, plastics are affected by a variety of environmental influences that are a continual surprise to the uninitiated. Part of the problem arises from the fact that there is a severe shortage of long-term performance information on plastic materials. More than 95% of the available information comes in the form of a short-term property table. This table is typically devoted to a few key thermal and mechanical properties. Some expanded treatments will include electrical properties, but even here there is extreme reliance on standardized tests that emphasize short-term behavior. With few exceptions, the short-term behavior is characterized at room temperature.

Temperature and time unfortunately change everything when it comes to polymer behavior. In addition, the effects of temperature and time are difficult to predict because the behavior of most properties with respect to these influences is not linear. Figure 13.51 shows a plot of elastic modulus versus temperature for two unfilled thermoplastics, polycarbonate and nylon 6, by a technique known as *dynamic mechanical analysis* (DMA). At room temperature both materials are reasonably stiff and strong. Property charts list the tensile strength at yield for the nylon 6 at 80 MPa (11,500 psi) and the modulus at 2.83 GPa (410,000 psi). The polycarbonate values are 62 MPa (9000 psi) for tensile strength at yield and 2.35 GPa (340,000 psi) for modulus. Even though DMA cannot track strength properties, it does provide a continuous measurement of modulus as a function of temperature. The effects of increasing temperature on the two materials are markedly different and are related to their structure. Polycarbonate is an amorphous material. Its modulus is relatively constant as a function of temperature, declining gradually by less than 20% as the temperature rises from 25°C to approximately 130°C. At this point the modulus begins to decline more rapidly, and in an interval of 30°C it has declined by over two orders of magnitude. This sudden change represents the glass transition temperature for the poly-

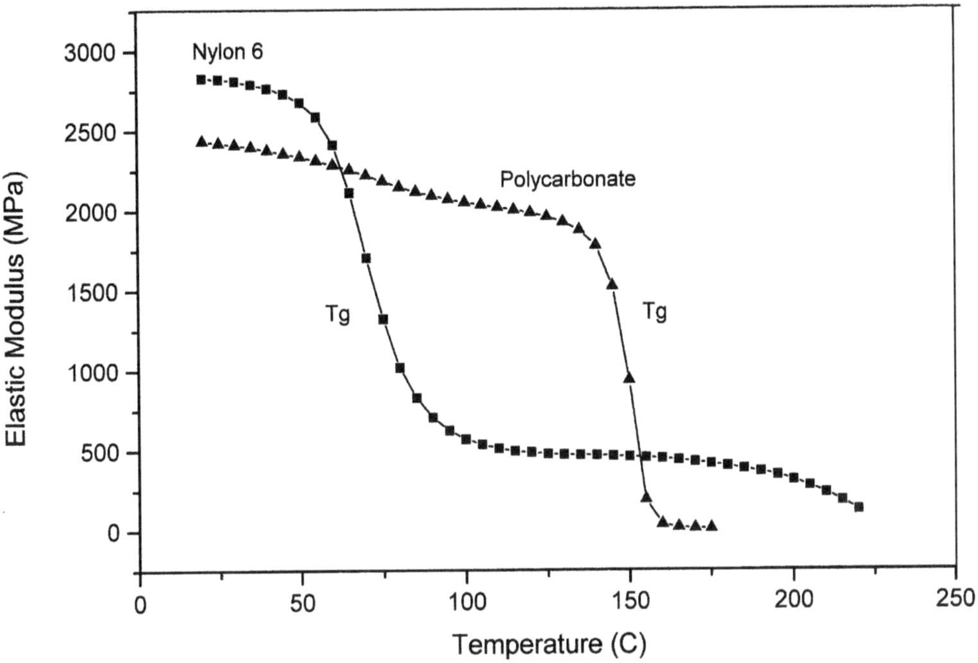

Figure 13.51 Dynamic mechanical analysis (DMA) showing the effect of temperature on the elastic modulus of an unfilled nylon 6 and polycarbonate.

carbonate. This temperature represents the onset of motion along extended segments of the polymer chain backbone and can be thought of as a sort of softening temperature for the amorphous regions of the polymer. Because polycarbonate is wholly amorphous, this results in a complete change from rigid and solid to very soft and compliant. As a useful load-bearing member, any product made from polycarbonate ceases to be useful over any significant duration once the temperature reaches 140°C.

The effect of temperature on the nylon 6 is significantly different. Nylon 6 is a semi-crystalline material; with a structure that consists of crystalline regions surrounded by amorphous material. There is no such thing in normal manufacturing processes as a 100% crystalline polymer. Because materials like nylon contain both crystalline and amorphous regions, they have two transition temperatures: a glass transition for the amorphous structure and a melting point for the crystals. In general, the melting point for a semi-crystalline polymer will be 150 to 175°C (270 to 315°F) above the glass transition temperature. The melting point for nylon 6 is 220 to 225°C (428 to 437°F) and the glass transition temperature is 65°C (149°F). As the nylon 6 is heated in the DMA, its modulus begins to decline almost immediately. Between 50 and 100°C the modulus declines to approximately 20% of its room-temperature value. This is the quantitative effect of the relaxation of the amorphous regions on the bulk properties of the nylon polymer. Unlike the polycarbonate, however, the

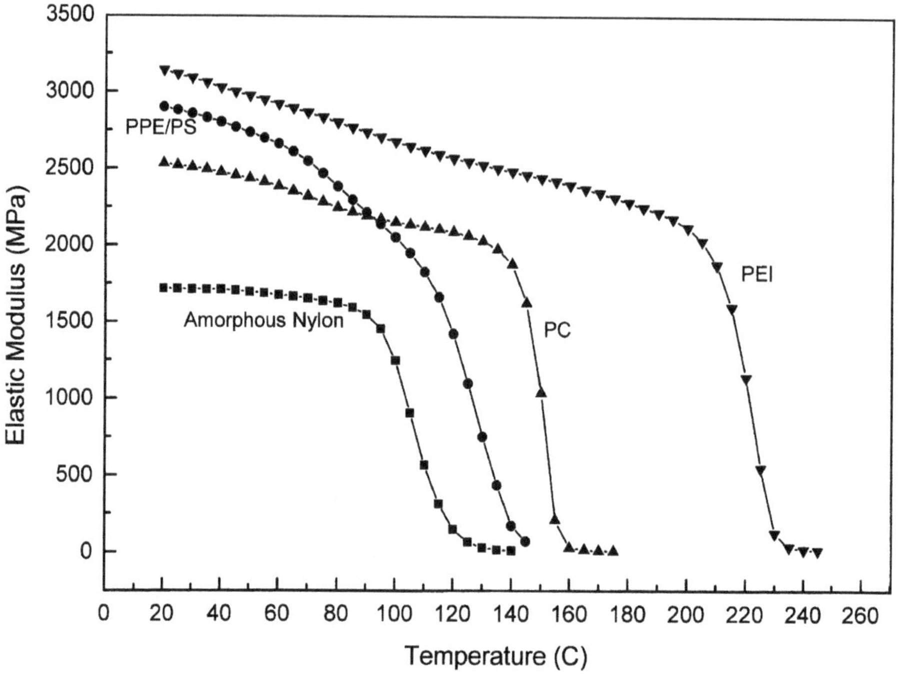

Figure 13.52 DMA scans for four unfilled amorphous polymers.

nylon 6 does not soften completely due to the presence of the crystalline structure. This crystalline material maintains the modulus at a reduced and almost constant level until the material approaches its crystalline melting point. At this point the modulus again declines as the material changes from solid to liquid.

It is quite apparent that in polymers the modulus is not an inherent property that can be quoted without reference to the temperature. It is also apparent that the changes in modulus are not linear with temperature, and that the degree of change associated with a particular transition is very much dependent upon the details of the polymer structure. An amorphous material with a more rigid polymer backbone, such as polyetherimide, will have a higher glass transition temperature than the polycarbonate. The effect of the glass transition on the modulus of the polyetherimide, however, will be comparable to the effect observed for the polycarbonate. Figure 13.52 compares the elastic modulus versus temperature behavior for several amorphous polymers. Semi-crystalline polymers also display a range of glass transition temperatures and melting points. For unfilled materials, many of these polymers have similar room temperature properties, but they exhibit a wide range of transition temperatures. The degree of modulus decline associated with the glass transition is a direct reflection on the degree of crystallinity, the amount of the polymer matrix that organizes into crystals. It is common, however, for an unfilled semi-crystalline polymer to lose 70 to 80% of its modulus as it passes through the glass transition region.

13.3.2 Use of Fillers

Because semi-crystalline materials do not completely soften above the glass transition temperature, they are expected to be functional in the temperature region between the glass transition and the melting point. Many applications, however, cannot tolerate a decline of 70 to 80% from room temperature performance. For this reason, it is common to find fillers and reinforcements added to these materials. If we examine property tables for various reinforced grades of a material such as the nylon 6 in our discussion, we will observe that the strength and modulus of the material increase with increasing filler content. Table 13.8 shows this summary for an unfilled nylon 6 and four analogs with various levels of filler and reinforcement.

The primary purpose of incorporating the reinforcement, however, has nothing to do with room-temperature properties. Instead its primary purpose is the improvement it provides in the temperature region between the glass transition and the melting point. Figure 13.53 illustrates the relative effect of the various reinforcements and reinforcement levels on the modulus of the material as a function of temperature. Table 13.9 summarizes some of the key features of these curves.

Note that the glass transition temperature (T_g) is affected very little by the presence of the filler; however, the effect of the T_g on the modulus is altered substantially. Even though the modulus of the unreinforced material declines by 80% as it passes through the glass transition, even the addition of a small amount of glass fiber changes this reduction to 56%. This reduction appears to reach a limiting value of 49% as the glass-fiber level increases. This illustration also points out the difference between a reinforcement and a filler. The 40% glass-fiber and mineral system consists of 15% glass fiber and 25% mineral. By weight, this material contains almost as much filler as the 44% glass fiber system; however, the room-temperature modulus of the 40% mixed filler system does not even reach the level attained by the 33% glass-fiber compound. It is perhaps more important that the glass and mineral system shows a decline

Table 13.8 Effect of Filler Content and Type on Short-Term Properties of Nylon 6

Material	Tensile strength @ yield (MPa)	Elongation @ break (%)	Flexural modulus (MPa)	Notched Izod impact (J/m)
Unfilled	79	70	2,829	48
14% Glass	124	3.5	5,520	59
33% Glass	200	3.0	9,384	117
44% Glass	228	2.0	11,868	133
40% Gl/Mineral	138	4.0	7,659	48

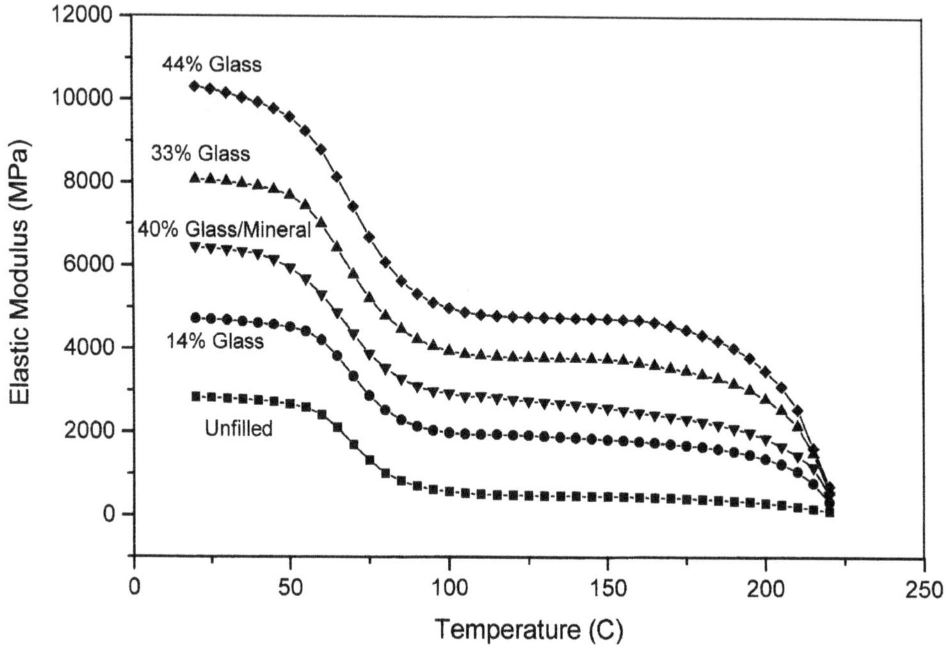

Figure 13.53 DMA scans showing the effect of filler content and type on the modulus of nylon 6 at various temperatures.

in modulus of 58% with the glass transition, which is the largest change of any of the filled materials. The mineral system is particulate as opposed to fibrous; therefore, it does not contribute proportionally to improving the properties of the polymer matrix.

Finally, the table and the graph both show that the fillers and reinforcements have a much greater effect on improving the properties of the material above the T_g than below it. At room temperature, where the standard property charts put all of the focus, the addition of 14% glass fiber increases the modulus by 59%. Above the T_g, however, the modulus improves by 250%. At the 44% loading the room-temperature modulus increases by a factor of 3.6, but above the T_g we measure a ninefold improve-

Table 13.9 Summary of Temperature-Dependent Behavior of Nylon 6

Filler	T_g (°C)	E' before T_g (GPa)	E' after T_g (GPa)	DTUL @ 1.82 Mpa (°C)
None	65	2.81	0.56	65
14% Glass	69	4.46	1.96	200
33% Glass	70	7.87	3.99	210
44% Glass	71	10.04	5.13	210
40% Glass/Mineral	69	6.44	2.69	206

ment. Many material selection decisions are unfortunately made without the benefit of this more detailed performance picture.

13.3.3 Deflection Temperature Under Load

There is an additional property listed in Table 13.9: the deflection temperature under load (DTUL). In most property charts, this attribute represents the only attempt to deal with the subject of performance at elevated temperatures. A review of the test method used to produce these values shows that this property is essentially a measurement of the temperature at which a specific stress produces a specific strain. Because modulus is defined as stress divided by strain, the DTUL is the temperature at which a material achieves a certain modulus. Both the ASTM and the ISO method dedicated to this measurement use various fixed stresses, 1.82 MPa (264 psi) and 455 kPa (66 psi). In addition, the ISO technique uses a third stress level of 8.00 MPa (1160 psi). Takemori [5] has shown that the modulus associated with the stress level of 1.82 MPa is 800 MPa (116,000 psi) and the modulus associated with the stress level of 455 kPa is 200 MPa (29,000 psi). Using this calculation method it can be shown that the newly employed ISO stress level of 8 MPa pinpoints an elastic modulus of 3520 MPa (510,000 psi). Because very few unfilled materials achieve a modulus in excess of 3520 MPa at room temperature, it is apparent that this last stress level is designed for re-evaluating filled and reinforced systems.

The value of this more stringent evaluation can be observed in Fig. 13.54. This illustration repeats the image from Fig. 13.53 but adds horizontal lines that correspond to the three modulus values cited above. If we examine Table 13.9 we can see that once the nylon 6 polymer is filled, there is no significant difference between the DTUL values for all of the compounds, despite the fact that the graph shows clear differences in the temperature-dependent behavior. The reason for this is clear from the graph. The modulus of 800 MPa that is required to establish the DTUL value is easily attained for any nylon 6 with an appreciable filler content. This modulus line is not crossed until just before the materials melt. However, at the higher stress level the DTUL is measured as the temperature at which the modulus drops below 3520 MPa or 3.52 GPa. At this level the four filled materials are easily distinguishable.

Although these attempts to make short-term single-point test values more relevant is laudable, it is no substitute for the graphical data that readily illustrates important aspects of polymer behavior. Temperature dependence is among the most important of these. It would be reasonable to assume that changes in temperature also influence the strength of the material. Figure 13.55 shows tensile stress–strain curves for a 15% glass-fiber–reinforced polypropylene measured at room temperature, 43°C (110°F), and 55°C (130°F). Although these represent relatively minor differences in temperature, the effect on the polypropylene is significant. Table 13.10 summarizes key values from each temperature at which the tests were conducted.

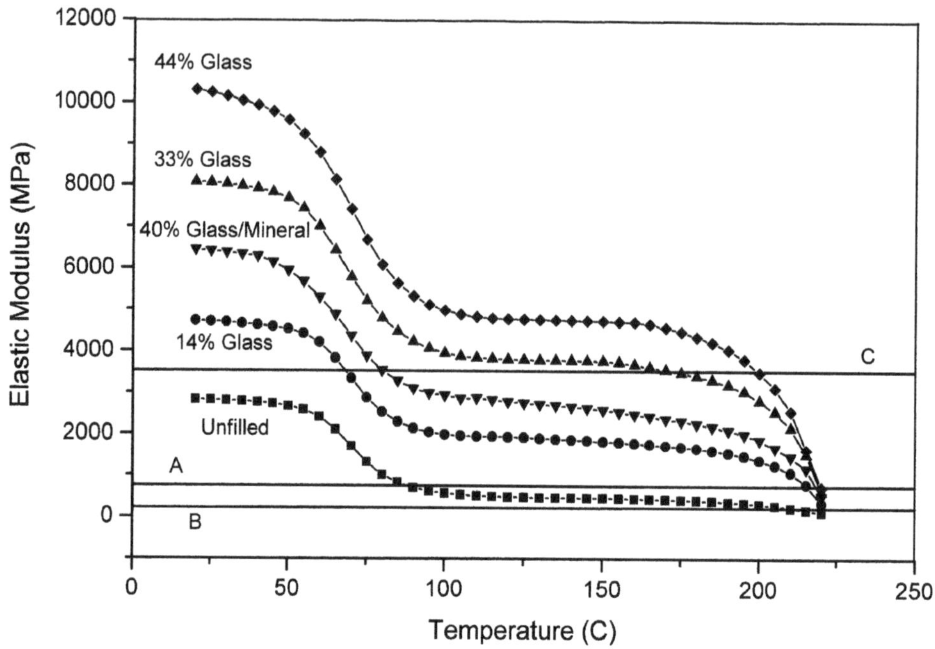

Figure 13.54 Materials from Figure 13.53 with lines added to indicate the modulus values associated with each of the three loads employed in the deflection temperature under load test by ISO 75.

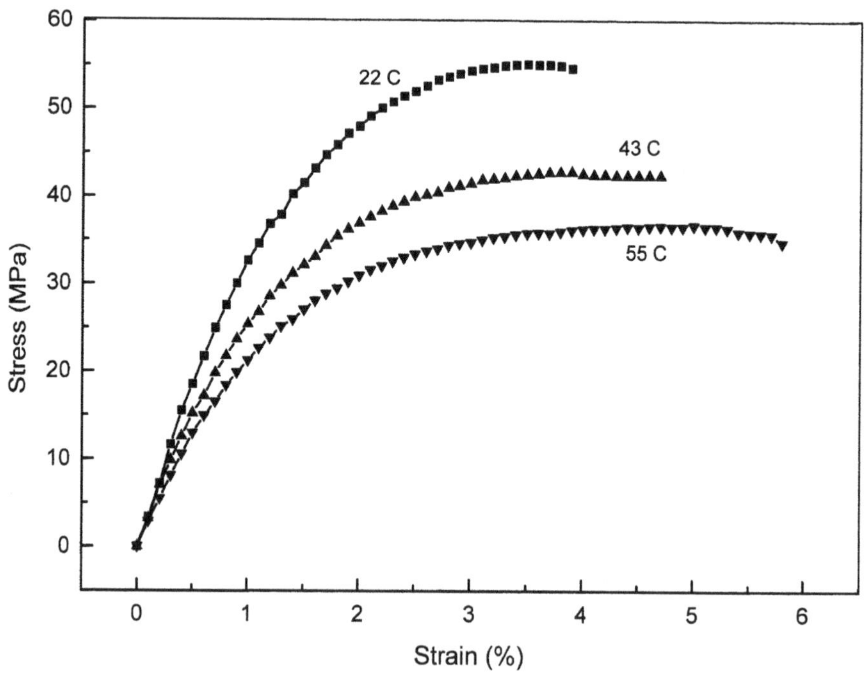

Figure 13.55 Tensile stress-strain curves at various temperatures for a polypropylene filled with 15% glass fiber.

Table 13.10 Summary of Tensile Properties of a Glass-Filled Polypropylene

Temperature (°C)	Tensile strength @ yield (Mpa)	Tensile modulus (GPa)	Elongation @ yield (%)	Elongation @ break (%)
22	55.43	4.07	3.57	3.86
43	42.09	3.12	4.10	4.98
55	36.66	2.61	4.79	5.82

The results show that the temperature rise substantially reduces the strength and stiffness of the material. In addition, the increase in the elongation at yield and failure show that the material is becoming more ductile. This is especially apparent from the difference between the two elongation values. Brittle materials yield and fail at essentially the same point. Tough materials elongate significantly beyond the yield point. The difference between yield strain and strain at break, therefore, is an indicator of ductility. Although the preceding table represents a convenient way of summarizing particular data points, the real value comes from the curves themselves. As with the DMA plots of modulus versus temperature, the stress–strain curves supply a full range of data points that help to capture the behavior of the material over a variety of application conditions. This is critical in assessing the fitness-for-use of any plastic material and it is all too often information that is not available. This is an element in the failure of plastic parts.

When temperature is combined with other possible application conditions, the effects can be even more surprising. This is a common problem in evaluating chemical resistance. Chemical resistance tests are too often performed at room temperature on unstressed specimens. When the temperature is elevated and an external stress is applied to the test samples, the assessment of the material can change dramatically. For example, ABS in contact with propanol produces no change in the material after 1 month, and the projection from this result is that the ABS should last indefinitely. When the temperature of the test is increased to 52°C (125°F), however, the ABS dissolves in a matter of hours.

A case study involving cracking in some molded ABS parts illustrates the effects of combining stress and a chemical agent, which brings us back to the phenomenon of environmental stress cracking. In this instance two ABS parts were assembled with a snap-fit feature. Between the two parts a rubber gasket was placed to prevent water from leaking around the assembly point. To ease insertion of the gasket, it was coated in silicone grease. The application had worked for years when it was suddenly discovered during one assembly run that areas adjacent to the snap-fit region were cracking approximately 24 hours after the components were put together. The investigation eventually focused on the lubricated gasket, and it was discovered by IR spectroscopy that the supplier of the lubricant had substituted poly(propylene glycol) for silicone. Chemical exposure tests incorporating a series of increasing loads showed that poly(propylene glycol) caused the ABS material to crack at a threshold stress of approximately 1000 psi, whereas the silicone did not produce problems until the stress exceeded 3000 psi. The snap-fit features in the part generated stresses that fell between these two values. The value of including time, temperature, and load in any chemical resistance evaluation is very important to long-term success of a product. This instance also emphasizes the tendency of amorphous resins to exhibit these subtle aspects of polymer behavior that can lead to poor performance.

13.3.4 Impact Properties

Even though a complete treatment of the inadequacies of the short-term property chart cannot possibly be addressed within the context of one chapter, it is important that impact performance be considered. We illustrated some of the temperature-dependent aspects of impact performance in the section on molecular weight. Figure 13.34 showed the interaction of molecular weight and temperature on the notched Izod impact strength of polycarbonate. The principle of ductile-to-brittle transition is an important one for predicting performance in an application. We have seen in the preceding discussion of the glass transition and crystal melting that strength and stiffness fall off with temperature in a nonlinear manner that is associated with the onset of certain types of molecular motion. Impact properties exhibit the same type of apparently unpredictable behavior as the temperature declines. In some cases, as with polycarbonate, the transition is very sudden. In other materials, such as ABS, the change from low- to high-impact resistance may cover a range of 15 to 20°C. Even more important than the energy required to produce failure, the failure mode itself will change from brittle to ductile. In applications this is a primary consideration in assessing overall resistance to damage, probable repair costs, and collateral injury due to contact with sharp edges and flying shards of material.

The standard test method for reporting impact resistance is the notched Izod test, which is a technique adapted from the metals industry. Other similar tests have been developed, but they all essentially rely on a notched specimen struck by a swinging pendulum, as shown in Fig. 13.56. The difference in the height of the pendulum at the beginning and the end of the test is a function of the resistance provided by the specimen. There are many difficulties associated with this test. First, all materials are notch sensitive. That is, sharp corners and similar stress concentrators will change an apparently ductile material into a brittle one; however, the threshold severity varies with each material. For example, polycarbonate does not begin to exhibit brittle behavior until the radius of a machined notch reaches 0.15 mm (0.006 in). A material like acetal may require a radius of 1.5 mm (0.059 in) before it begins to exhibit signs of ductility. The notch radius for the standard test is 0.25 mm (0.010 in), which tends to exaggerate differences between materials that fall on either side of this threshold.

Second, the test only captures the total energy to break; it does not delineate between the energy required to initiate a failure and the energy required to propagate that failure to completion. This can be a problem with any impact test evaluation, but it is inherent in the notched Izod test because the measurement system itself is not set up to provide such data. Any impact-resistance measurement that evalu-

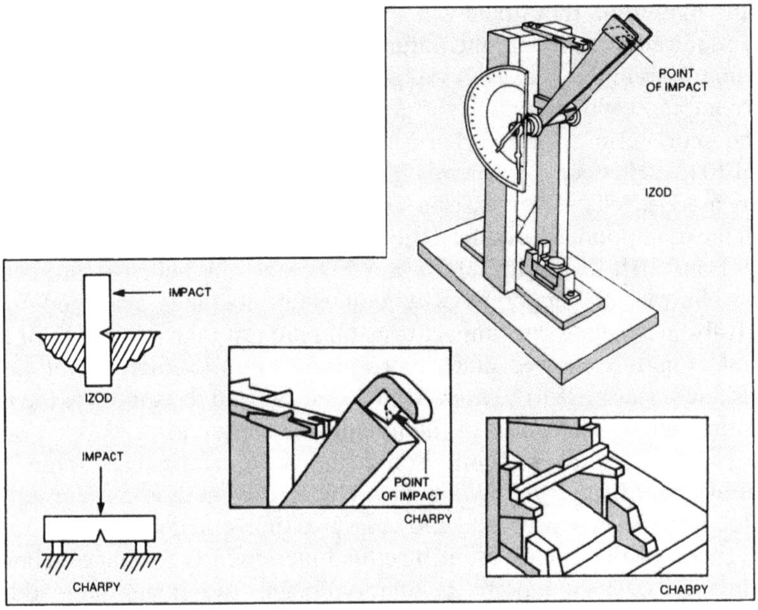

Figure 13.56 Apparatus for Izod and Charpy pendulum impact tests.

ates the total energy to failure is characterizing a composite of the energy required to produce the initial crack and the energy required to extend that initial crack to the point of failure. The initiation phase is primarily a function of the stiffness of the material. A very rigid material will frequently require more energy to produce crack initiation than will a softer, more compliant compound. Once failure is initiated, however, the ability of the material to manage the energy beyond the yield point is the other part of the impact-resistance calculation.

The failure to adequately capture the full effect of this second stage in the notched Izod test leads to some measured properties that do not accurately reflect performance in the application. One of the best examples of this is the effect of adding glass fiber to a neat resin with poor notched Izod performance values, such as nylon or polypropylene. Most nylon products that have not been conditioned to an equilibrium moisture content and do not contain additional impact modification will produce notched Izod values near 50 to 55 J/m (approximately 1 ft-lb/in). The addition of glass fiber to this material will more than double this value. This can be observed in Table 13.8; however, it can be demonstrated that the practical toughness of these glass-reinforced analogs is significantly lower than that of the unfilled material. The test emphasizes the stiffening effects of the glass, which increase faster than the ductility declines. In addition, an end-gated specimen will allow for an alignment of the glass fiber in such a way that the test must break through these aligned fibers.

This type of evaluation reaches almost absurd conclusions when tests are performed on long-fiber systems. In these materials, the increased length of the fiber increases the likelihood that fibers will span the area of crack propagation, making the energy required to complete the failure extremely high. The introduction of long glass fiber materials in the late 1980s was accompanied by colossal claims of increased toughness combined with increased stiffness. The supporting information came in the form of the short-term property chart, which showed an increase in strength and modulus of 10 to 20% over equivalently filled short glass analogs. It also documented as much as a 500 to 700% increase in notched Izod impact resistance. Long-glass polypropylene compounds produced notched Izod test results of 320, 425, and 530 J/m (6, 8, and 10 ft-lbs/in) for 30, 40, and 50% long-glass-fiber–reinforced grades, respectively. Short-glass materials with equivalent loadings struggled to exceed 107 J/m (2 ft-lbs/in). Falling dart impact tests that did not take advantage of fiber orientation unfortunately showed much more modest improvements over short glass compounds. Increases of 25 to 50% were more typical, and the source of the improvements was the increased stiffness of the long fiber compound.

Figure 13.57(a,b) show a comparison of impact performance by an alternate method known as *instrumented falling dart*. Under this method a radiused tup with an implanted transducer senses load and measures time and displacement to develop a graphic version of the impact event. The load increases as resistance from the test specimen increases. At some point the material begins to fail and the load begins to decline, eventually returning to zero as failure is completed. The energy collected during the event is a function of the load versus time relationship. The materials being compared are a long-glass fiber reinforced nylon 6 and a short-glass fiber reinforced nylon 6 with an impact modifier. In Fig. 13.57(a) the load curves develop in distinctly different ways. The load in the long-glass material increases more rapidly and reaches a higher maximum due to its superior stiffness. The load curve is slightly jagged on the increasing side, which indicates that the material is undergoing multiple microfractures during the overall impact event. Once maximum load is attained it drops off rapidly, which indicates that the material fails immediately beyond the point of maximum resistance. The impact-modified material does not reach the same peak load, but it does show a smoother and more attenuated load curve with significant energy collecting capabilities beyond maximum load.

Figure 13.57(b) shows the resulting energy values for the test. Even with a more sophisticated measurement system, such as the instrumented falling dart, this total value is still the focus of most evaluations. The result shows that the total energy for the long-glass material is just over 26 J (19 ft-lbs), whereas the short-glass impact modified material absorbs 32 J (23.5 ft-lbs). This is significant because the notched Izod value for the long-glass material is 480 J/m (9 ft-lbs/in), whereas the notched Izod result for the short-glass material is 240 J/m (4.5 ft-lbs/in).

These types of contradictions are plentiful in impact evaluations. Impact-modified polycarbonates were developed to improve the low-temperature impact resistance of the base polymer and increase damage resistance from repeated impact; however, the modifying agents reduce modulus. An impact-modified polycarbonate consequently will typically produce notched Izod values that are 30% lower than are those of general-purpose materials. A material selection process that focuses on short-term values will tend to neglect the materials with greater practical toughness.

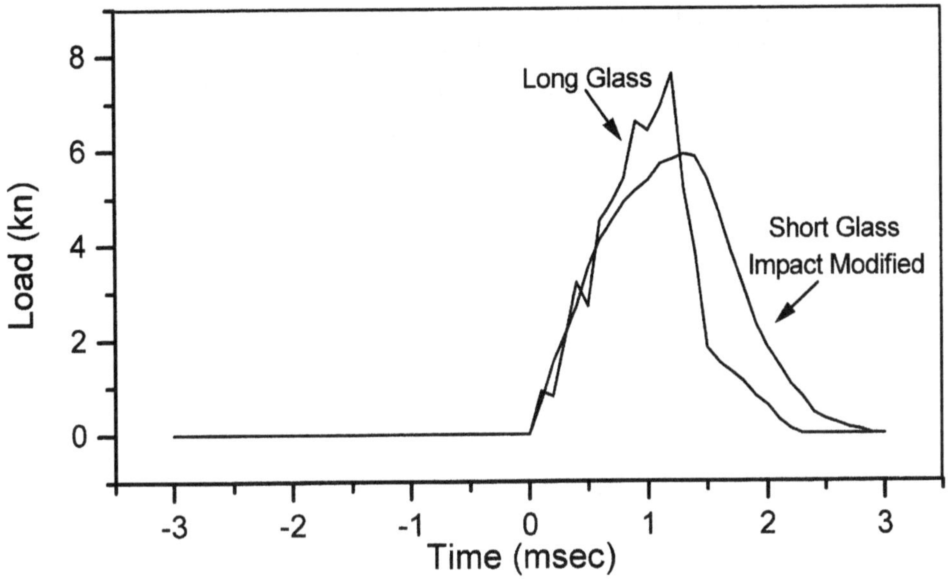

(a)

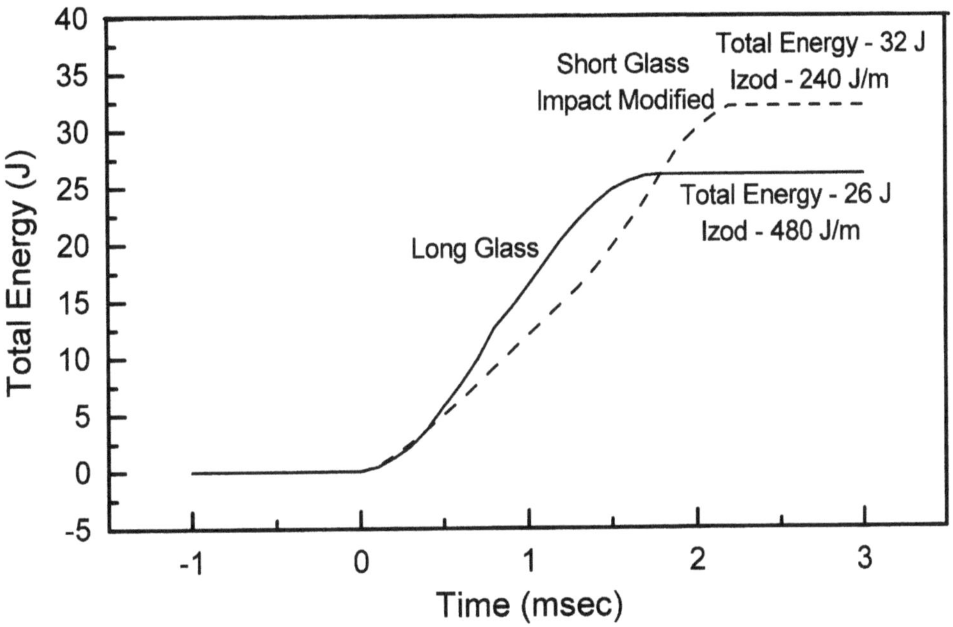

(b)

Figure 13.57 (a) Load versus time curves from instrumented falling dart impact tests on long and short glass reinforced nylon 6. (b) Total energy versus time curves from instrumented falling dart impact tests on long and short glass reinforced nylon 6.

Table 13.11 Impact Resistance of Polypropylene Materials by Falling Dart and Notched Izod

Material I. D.	Falling dart impact (J)	Notched Izod impact (J/m)
1	37.07	101
2	35.22	112
3	34.45	91
4	25.85	48
5	4.92	32
6	2.59	27

The instrumented falling dart test has several advantages over the notched Izod test. First, as we have seen above, it provides a graphical output. It also quantifies several different tabular values that allow for a more complete assessment of the impact event, such as total sample deflection and time to failure. Second, it does not rely on a prestressed specimen. The absence of the precut notch tends to result in a more accurate assessment of true differences in ductility. Table 13.11 shows instrumented falling dart results and notched Izod results for six different polypropylene materials. These results clearly show that even when we limit the reporting to tabular data, the falling dart test does a much better job of distinguishing between resins.

The instrumented impact test results in Table 13.11 are average values from five individual specimens. The top three materials consistently fail in a fully ductile manner. Materials #5 and #6 fail in a fully brittle manner. In these five materials the variation from specimen to specimen is very small. The fourth material represents an average made up of specimens that exhibit a wide range of behavior from fully ductile to fully brittle. There are actually no individual specimens that fail at 25 J/m. Some specimens instead achieve 35 J/m, whereas others in the same group fail near 5 J/m. This is an indication that a material is at a critical transition, and the apparent inconsistencies in the individual specimens are a reflection of that condition. Repeated testing of additional sets of five specimens produces the same type of data scatter. This table is a good illustration of the effects of the ductile-to-brittle transition. Three of these materials have a ductile-to-brittle transition temperature (DBTT) that is below room temperature, whereas two have a DBTT that is above room temperature. Material #4 has a DBTT that appears to be right at room temperature.

The effect of temperature on impact resistance is emphasized by a comparison of impact performance at room temperature and at 5°C (41°F) on the top three performers in Table 13.11. Table 13.12 shows the results of this evaluation and underscores the importance of evaluating materials at conditions that are relevant to the application. Materials #2 and #3 still failed in a ductile mode at 5°C, but material #1 was extremely brittle even though the temperature change was relatively small. This is a clear sign that material #1 has passed through its DBTT, but the transition temperature for the other two materials is still lower than the test temperature.

Table 13.12 Effect of Temperature on Instrumented Impact Resistance of Polypropylenes

Material I. D.	Impact strength @ 23°C (J/m)	Impact strength @ 5°C (J/m)
#1	37.07	2.63
#2	35.22	33.02
#3	34.45	37.42

Without a means of reporting the failure mode of an impact specimen, even the improved methods of evaluation are not as valuable. This makes an assessment of the effects of temperature much more difficult. Table 13.13 summarizes the results of instrumented falling dart impact tests performed on a polycarbonate/PET polyester blend at five different temperatures. The objective of this type of evaluation is to establish a minimum temperature for fully ductile failure, which is the criterion for acceptable performance. The actual values for energy absorption are secondary as long as they exceed a certain minimum value, in this case 50 J.

These results illustrate the role that increased modulus can have in inflating the total energy values for an impact test. The highest values are achieved at the lowest temperatures. A casual analysis of these data would lead to the conclusion that the impact resistance of this compound improves with declining temperature; however, attention to the failure mode avoids a crucial mistake in material evaluation that often occurs when too much credence is given to tabular data.

13.4 A Brief Discussion of Viscoelasticity

This discussion has attempted to illustrate the complexity behind the behavior of polymers and the need for more complete methods of evaluation to avoid material selection errors. In this final section, we will briefly cover one of the most important

Table 13.13 Effect of Temperature on the Impact Resistance of A PET/PC Alloy

Test temperature (°C)	Total energy to break (J)	Ductile failures
–29	67.38	1/5
–18	68.97	3/5
0	59.51	5/5
22	56.72	5/5
85	55.18	5/5

principles of polymer behavior: viscoelasticity. Viscoelastic behavior is at the root of virtually every phenomenon in polymer properties in both the solid state and the melt state. It is therefore exceedingly important that professionals working with plastics have some appreciation for the implications of viscoelastic behavior. Most treatments of the topic are unfortunately highly mathematical and seek to treat the subject completely. This leaves many in the field who seek a practical working knowledge of the topic without any resources.

The importance of understanding the effects of time-dependent behavior on the load-bearing properties of a material can be understood in practical terms if we return to the RAPRA study of failed applications mentioned earlier. Recall that more than half of these failures came from either environmental stress cracking or fatigue. The next largest pieces of that pie came from the combined effects of creep rupture and excessive deformation. These represent the brittle and ductile aspects of failure from mechanical overload, and they accounted for 18% and 10% of the failures, respectively. When added to the ESC and fatigue problems, these causes account for 80% of the problems evaluated. Most overload conditions come as a surprise because they do not look like candidates for overload until time-dependent behavior is factored in.

The complexity of polymer behavior ultimately relates to viscoelasticity. Most classical materials exhibit either elastic or viscous behavior in response to an applied stress. Elastic responses are typical in solid materials. When a stress is applied to an elastic system it deforms proportionally by a quantity identified as the strain. We can quantitatively express the relationship between the applied stress and the resulting strain as:

$$\tau = G \cdot \gamma \tag{13.3}$$

where τ = the stress in shear
γ = the strain
G = the shear modulus

The same equation can be written for other modes of stress such as tension. The response of an elastic system to applied stress is instantaneous and completely recoverable. We say that the system stores the energy and can return it to the system completely when the stress is removed. The preceding equation is familiar to us as Hooke's Law and a spring is used as the model for materials governed by this law. Figure 13.58 illustrates the behavior of an elastic system in time.

Viscous behavior is a characteristic of fluids, materials where the bond energies necessary for long-range translational order have been overcome. In these systems an applied stress results in a strain that increases proportionally with time until the stress is removed. The strain is not recoverable; when the stress is removed the deformation is completely retained. We say that the energy has been lost to the system. The model of a dashpot is frequently used as an analogy. Figure 13.59 shows this behavior graphically. Newton first defined the mathematical relationship between the

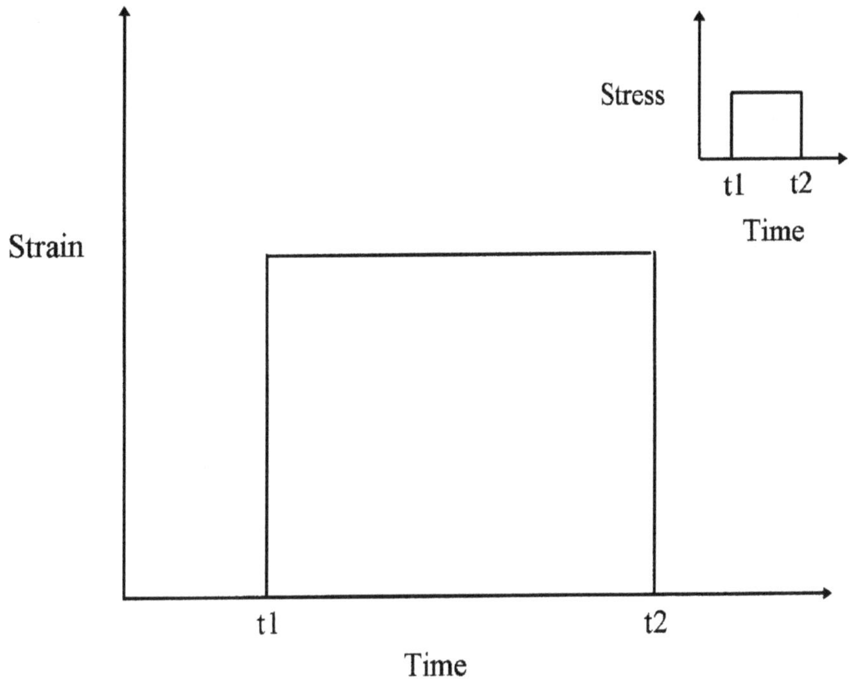

Figure 13.58 Schematic of stress-strain-time relationship for a perfect elastic material.

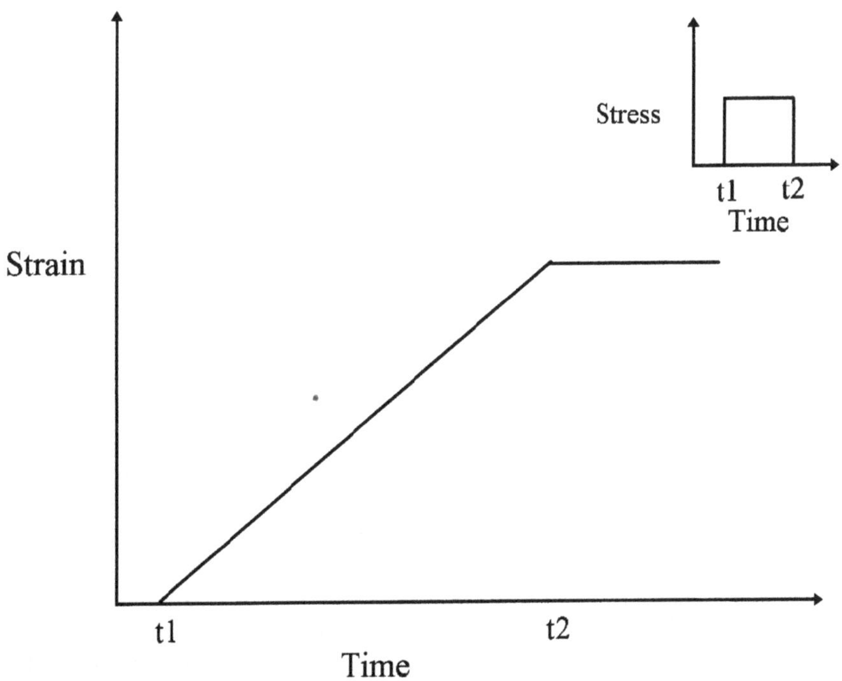

Figure 13.59 Schematic of stress-strain-time relationship for a perfect viscous material.

applied stress and the resulting strain rate in a fluid and termed the resulting ratio the *viscosity*,

$$\tau = \eta \cdot \dot{\gamma} \tag{13.4}$$

where τ = the stress
$\dot{\gamma}$ = the strain rate
η = the viscosity

In previous discussions involving polymer flow in the fluid state, we have already observed that the viscosity of a polymer is not an absolute property of the material. It varies with temperature, but perhaps more importantly it varies with strain rate, becoming lower as the strain rate increases. This demonstrates a general principle of viscoelastic behavior: As the time scale of the applied stress decreases the elastic behavior of the material becomes more dominant and the viscous aspects become less apparent. This behavior is a direct result of the large size and conformational variety of polymer molecules, which prevent these materials from forming the fully ordered systems that we normally associate with solid materials, but also hinder the complete independence of molecules that is normally associated with pure liquids. In both the solid and the fluid state, these materials exhibit a combination of elastic and viscous responses when placed under stress.

When we apply this model to the solid-state behavior of plastic materials, we find that it has very important implications for determining the long-term performance of our molded products. In a solid plastic beam we can perform classical measurements of stress versus strain that allow us to calculate a modulus. If we maintain a constant applied load, however, we find that the resulting strain is not constant: It continues to increase as a function of time. In engineering terms we refer to this as *creep* or *cold flow*, and it is actually a manifestation of viscous flow in the apparently solid polymer. A counterpart to this behavior is known as *stress relaxation*. Here the strain is held as the constant and the stress required to maintain that strain is measured as a function of time. In a viscoelastic system the stress decreases with time. Because *modulus* is defined as the ratio of stress to strain, it can be seen that the modulus calculation in viscoelastic systems must incorporate a time function and cannot be considered as an immutable property independent of the period over which the measurement is made. At a structural level, the polymer chains are slowly rearranging in response to the applied stress. Knowledge of the rate at which this occurs is critical to an accurate determination of a material's fitness-for-use in a particular application.

To further complicate the picture, the balance between the elastic and viscous response changes for a given material as a function of temperature, just as it does in the fluid state. In the solid state this balance is reflected in terms of load-bearing properties—time-dependent behavior such as creep and stress relaxation, as well as impact properties. In the fluid state, viscoelasticity provides information on molecular weight, molecular weight distribution, thermal stability, and cross-linking. The equation relating stress and strain in a viscoelastic system introduces the aspect of time dependency:

$$\tau = G(t) \cdot \gamma \tag{13.5}$$

where $G(t)$ = the stress relaxation modulus

The material initially responds in an elastic manner, then as a viscous fluid. When the stress is removed, the elastic portion recovers over an extended period of time. Figure 13.60 provides a generalized illustration of this compound behavior.

Determining the proportion of the elastic and viscous components in a polymer, and the factors that cause that balance to change, is crucial to understanding how a material will perform in a given application environment. It can also provide valuable information regarding structure and composition. Dynamic mechanical analysis, the same method we used to gain a broader understanding of temperature-dependent behavior, accomplishes this resolution. Although it is possible to perform dynamic mechanical measurements on solids and fluids, the focus of

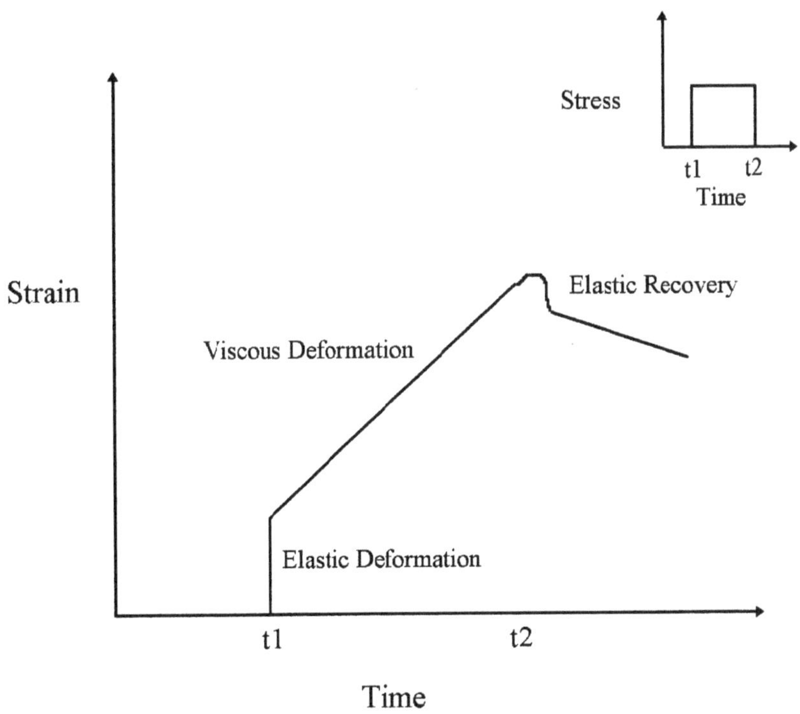

Figure 13.60 Schematic of the stress-strain-time relationship for a viscoelastic material.

this work is improved material selection for end-use applications; therefore, this work will concentrate on solid-state properties.

In the dynamic mode of operation, the DMA instrument applies an oscillatory stress with a controlled frequency. Dynamic modulus values obtained using this method are a function of the temperature of the material and the frequency of the oscillation. The stress function is sinusoidal. In a perfectly elastic system the applied stress and the resulting strain will be in phase as shown in Fig. 13.61. For an ideal fluid the stress will lead the strain by 90 degrees ($\pi/2$ radians), as illustrated in Fig. 13.62. A viscoelastic material will give some hybrid of these two responses. The stress and strain will be out of phase by some quantity known as the *phase angle* and commonly referred to as *delta* (δ). A small phase angle indicates high elasticity, whereas a large phase angle is associated with highly viscous properties. The complex response of the material is resolved into the elastic or storage modulus (G') and the viscous or loss modulus (G'') if the deformation is in the shear mode. If the deformation is in the tensile or flexural mode, then E' and E'' are used. Table 13.14 provides a summary of the key terms.

When tensile, flexural, or shear modulus are measured by traditional methods, the result of the test is actually a quantity known as the *complex modulus*. It is defined as the slope of the stress–strain curve in the linear region. The DMA resolves this complex modulus into the storage and loss components. The smaller the phase angle

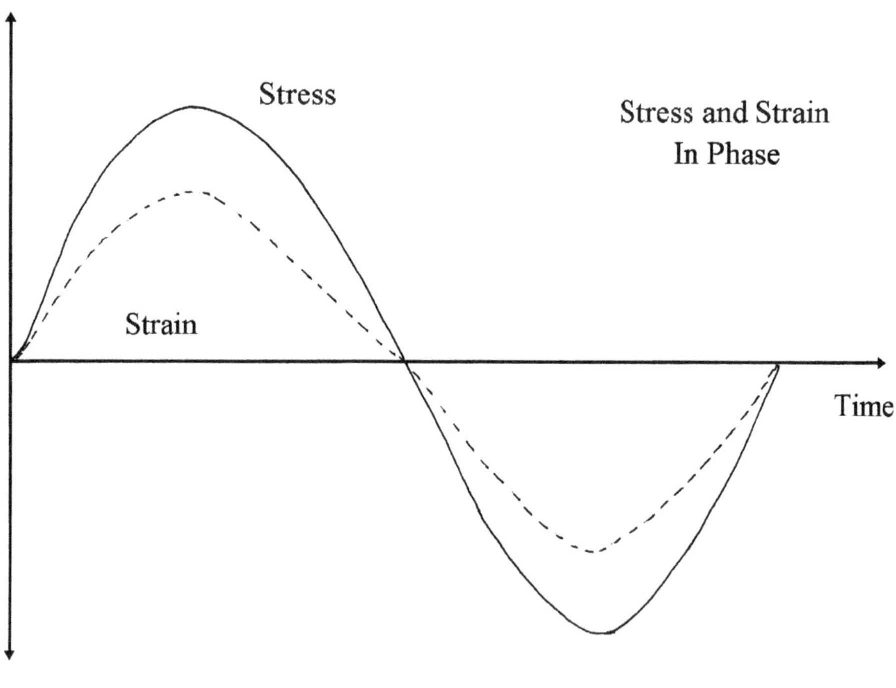

Figure 13.61 Response of elastic material to a sinusoidal applied stress.

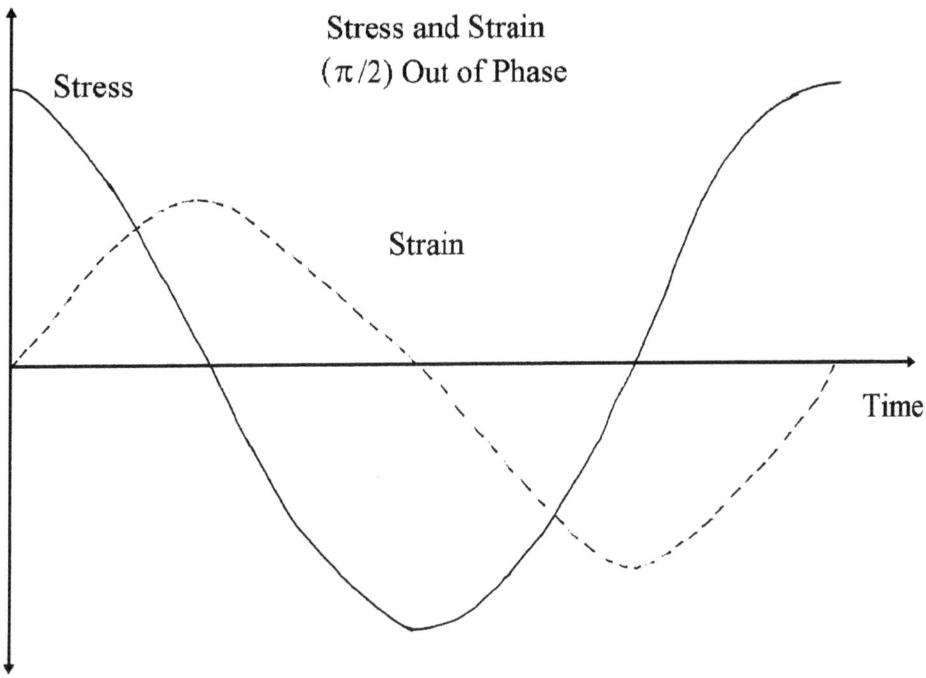

Figure 13.62 Response of viscous fluid to a sinusoidal applied stress.

is, the closer the elastic modulus is to the complex modulus. It is convenient to think of the elastic and viscous component in the vector terms illustrated in Fig. 13.63(a–c). Figure 13.63(a) shows the relationship between the stress and strain vectors. Figure 13.63(b) shows the stress vectors resolved into their storage and loss component. The storage component is in phase with the strain. Figure 13.63(c) expresses the vectors in terms of the modulus. The transposed loss modulus shows that the complex modulus can be thought of as the hypotenuse of a right triangle and the storage and loss components as the two shorter legs that are perpendicular to each other. The tangent of the phase angle, often referred to as *tan δ*, can be used to deduce the shape

Table 13.14 Key Viscoelastic Terms

Complex modulus	G^* or $E^* = \sigma^*/\gamma$
Elastic modulus	G' or $E' = \sigma'/\gamma = (\sigma^*/g)\cos$
Viscous modulus	G'' or $E'' = \sigma''/\gamma = (\sigma^*/\gamma)\sin$
Complex viscosity	$n^* = G^*/\dot{\gamma}$
Loss tangent	$\tan\delta = G''/G'$ or E''/E'

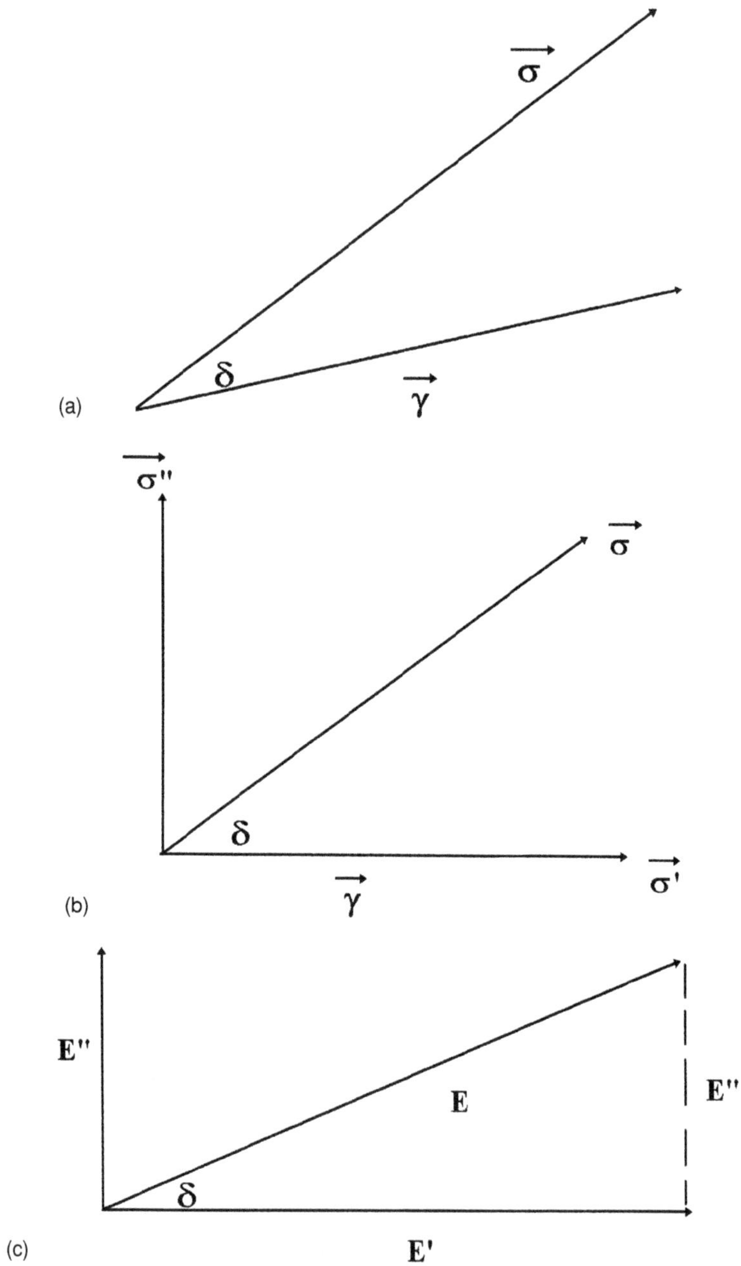

Figure 13.63 (a) Vector illustration of stress and strain in a viscoelastic material. The angle represents the phase lag between the application of the stress and the strain response. (b) Vector illustration from Figure 13.63(a) showing the stress resolved into storage and loss components. (c) Resolved vectors from Figure 13.63(b) expressed in terms of modulus.

Table 13.15 Effect of Tan δ on Variance between Complex and Storage Modulus

Tan δ	Variance (E^*/E')
0	1.00000
0.01	1.00005
0.03	1.00045
0.05	1.00125
0.10	1.00499
0.20	1.01980
0.30	1.04403
0.50	1.11803
0.75	1.25000
1.00	1.41421

of the right triangle. In the solid state, tan δ for a polymeric material rarely rises higher than 0.1 until the material approaches the softening temperature. A tan δ of 0.1 is analogous to a right triangle with a long side of 10 units and a short side of 1 unit. A triangle of these dimensions will have a hypotenuse 10.05 units long. This quantifies the relationship between the complex modulus measured by a classical stress–strain test and the elastic modulus measured by DMA. For the vast majority of the conditions at which DMA measurements are made on solid polymers, the complex modulus and the elastic modulus can be considered equivalent. Table 13.15 shows the relationship between tan δ and the degree of variation between the elastic and complex modulus.

Up to this point, when we have illustrated the results of dynamic mechanical experiments we have focused strictly on the elastic modulus because it is the most easily understood measurement and seems to have the greatest effect on our decisions about material properties. The full spectrum of dynamic mechanical properties will be shown here for representative materials from each important classification. The most common graphic presentation involves plotting the elastic or storage modulus (E' or G'), the viscous or loss modulus (E'' or G''), and tan δ as a function of temperature. From an engineering standpoint, E values are more useful than G values for evaluating performance in an application. The y-axis data is plotted conventionally on a logarithmic scale. This can be particularly useful for amorphous polymers where the glass transition may reduce the storage modulus of the material by two to three orders of magnitude and obscure changes related to molecular weight that may occur above the glass transition. In semi-crystalline systems, however, the changes in storage modulus are typically less than an order of magnitude until the material approaches the melting point. If the softening of the material is included in the plot, it can obscure the effects of the glass transition. In addition, logarithmic scales tend to obscure differences between materials in a comparative plot. For loss properties, logarithmic scales tend to diminish the visual impact of transitions. For data that focuses on solid-state performance, clarity is enhanced by utilizing a linear scale for all y-axis data, and this convention has been chosen for the graphs here.

In order to make the best use of DMA data, it is useful to relate representative plots to the structural characteristics of different polymer families. Two of these four examples, the polycarbonate and the nylon 6, have already been examined; however, this treatment will add the other important viscoelastic measurements. Figure 13.64(a) shows a typical DMA result for polycarbonate, an amorphous thermoplastic. The full-scale plot begins at –60°C and ends at 175°C. As we have already seen, there is little change in the storage modulus between the initial temperature and 130°C. Between 130 and 160°C, however, the storage modulus drops by more than two orders of magnitude, and the material has lost its usefulness as a structural material. This abrupt change in physical properties is associated with the glass transition. The amorphous structure in a polymer is often likened to that of glass because there is structural rigidity without the presence of a well-organized intermolecular structure. In an amorphous polymer the glass transition can be thought of as a softening temperature.

Figure 13.64(b) expands the graph to show the glass transition in more detail. We can see that the loss modulus rises to a maximum as the storage modulus is in its most rapid rate of descent. The peak of the loss modulus is conventionally identified as the glass transition temperature (T_g), even though the DMA plot clearly shows that the transition is a process that spans a temperature range. In most amorphous polymers the temperature range is relatively narrow, 25 to 40°C for materials that do not contain polymeric modifiers such as elastomeric toughening agents. The tan delta curve follows the loss modulus curve closely and provides a running tally for the ratio of the elastic and viscous phases in the polymer. Tan δ is well below 0.1 at low temperatures leading up to the glass transition. The rapid rise in the tan delta curve coincides with the rapid decline in the storage modulus. Above 150°C the tan δ curve rises rapidly and reaches a peak greater than 2.0. In this region the contribution of the loss modulus to the complex modulus is equal to or greater than that of the storage modulus. Once the glass transition is complete, the loss modulus drops back to a level close to the pretransition values. Because of the drastic reduction in elastic properties, however, the tan δ values do not decline significantly. The low storage modulus indicates that the material is easily deformed by an applied load. More significantly, the high tan δ values mean that once the deformation is induced, the material will not recover its original shape. It is considered to be soft and pliable. The pattern observed here for polycarbonate is typical of all amorphous materials. The key difference lies in the glass transition temperature (T_g) and the storage modulus below T_g.

Figure 13.65 shows a DMA plot for nylon 6, a semi-crystalline polymer. As we have already discussed, these materials are actually a mixture of amorphous and crystalline regions. They consequently exhibit both a melting point and a glass transition. The glass transition can be readily identified in the DMA plot. The storage modulus declines rapidly and the loss modulus and the tan δ curve rise to maximum values. Because of the presence of a crystalline matrix, however, the material does not soften above the glass transition. The new mobility of the amorphous regions causes a

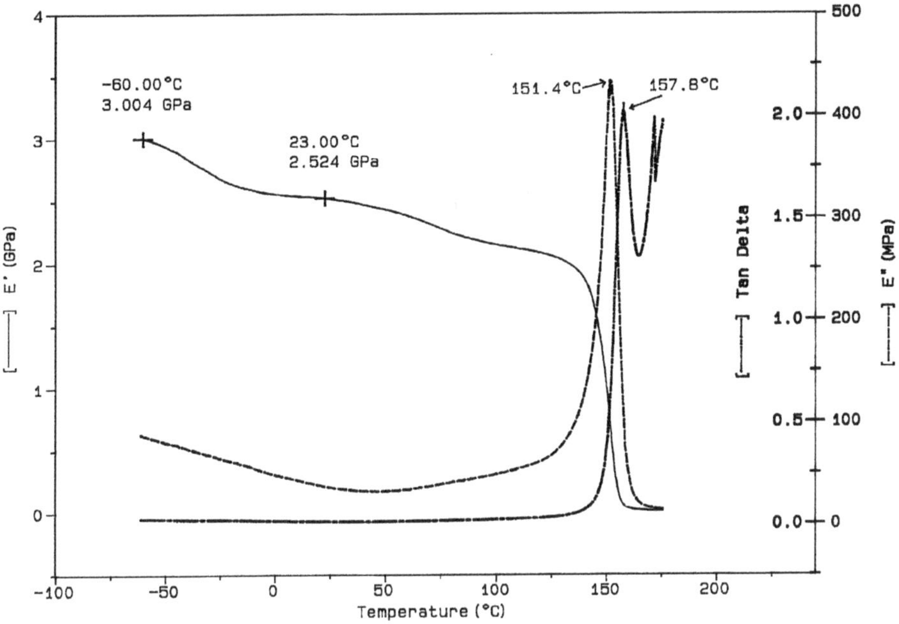

(a)

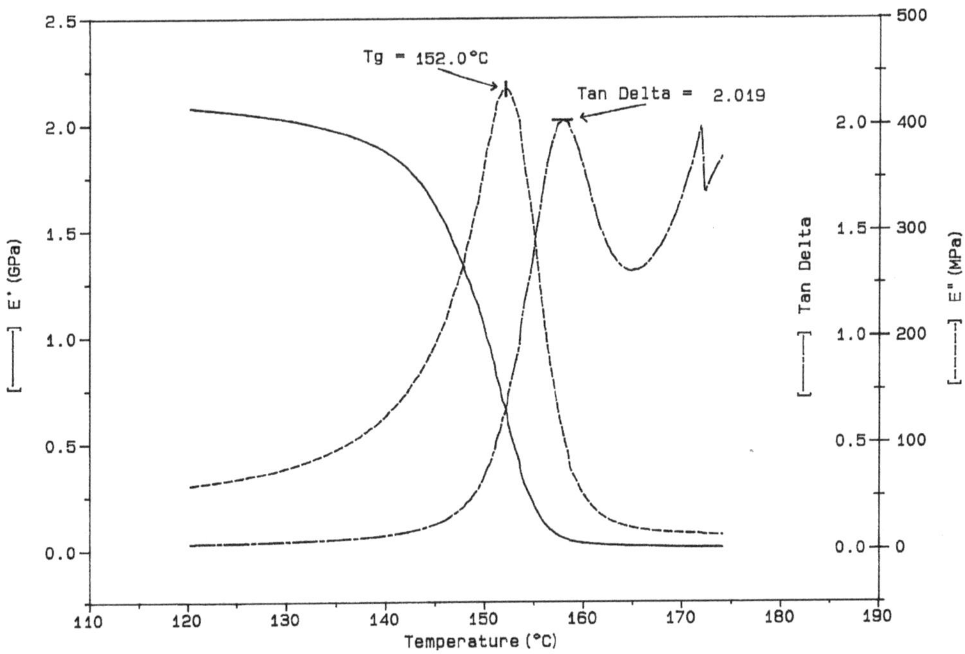

(b)

Figure 13.64 (a) DMA scan showing viscoelastic properties of unfilled polycarbonate. (b) DMA scan for an unfilled polycarbonate with glass transition region magnified for improved resolution.

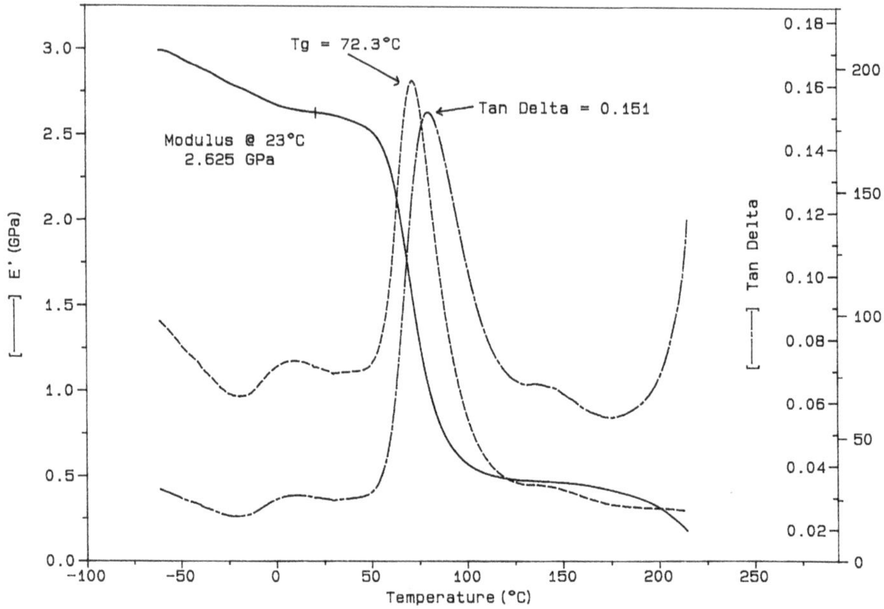

Figure 13.65 DMA scan showing the viscoelastic properties for an unfilled nylon 6.

reduction in the storage modulus, but the material exhibits useful solid-state properties until the material approaches the melting point, which is some 150°C greater than the glass transition. The diminished effect of the glass transition on the properties of the semi-crystalline material can also be seen in the tan δ peak value. Instead of rising above 1.0 as in most amorphous materials, the peak height for this material barely exceeds 0.15. Nylon 6 gives a result that is typical for a semi-crystalline polymer. The primary differences between semi-crystalline materials are in the actual glass transition temperatures, melting points, and degree of storage modulus decline associated with the glass transition.

Cross-linked systems, such as rigid thermosets, produce DMA results that are somewhat unique to the type of matrix polymer. Epoxies and phenolics, for example, have distinct temperature-dependent behaviors that make them easily distinguishable. In general, however, these materials all have a well-defined glass transition that produces the typical behavior of a declining storage modulus coincident with a rising loss modulus and tan δ. Figure 13.66 shows the storage and loss properties for an epoxy material used in printed circuit boards. Because the material is cross-linked, it has no melting point and in this respect it resembles an amorphous material. Due to the cross-linking, however, the plateau modulus beyond the glass transition does not decline to near zero. Instead, the material will still exhibit useful load-bearing characteristics even 50 to 75°C above T_g. Note also that in cross-linked systems the tan δ values above T_g return to pre-T_g levels.

Elastomers have glass transition temperatures below room temperature and their storage modulus properties are typically very low at ambient conditions. In this respect, they resemble a rigid amorphous material that has been heated above T_g. Unlike the amorphous materials, however, elastomers exhibit relatively low tan δ properties above T_g, which indicates that even though little force is required to deform the material, recovery will be good once the applied load is removed. This intuitively confirms our physical experience with elastomeric compounds. When the temperature is lowered, the material passes through the glass transition and presents itself as a rigid system. If the material is a cross-linked elastomer then it will have a low but measurable modulus to very high temperatures, whereas a thermoplastic elastomer will exhibit a second modulus decline associated with the melting point. This difference is most easily observed by plotting the storage modulus on a logarithmic scale. In addition, the tan δ values will be much higher for the melted thermoplastic system than for the crosslinked thermoset elastomer. Figure 13.67 shows a typical DMA result for a cross-linked elastomer.

Why does all of this matter so much? Simply because one of the very powerful principles of viscoelastic behavior is the equivalency of time and temperature on the structural characteristics of a material. Up to this point we have focused on measurements of modulus as a function of temperature in a constant time frame (i.e., at a constant frequency). Frequency is the inverse of time and the two can be related by the equation

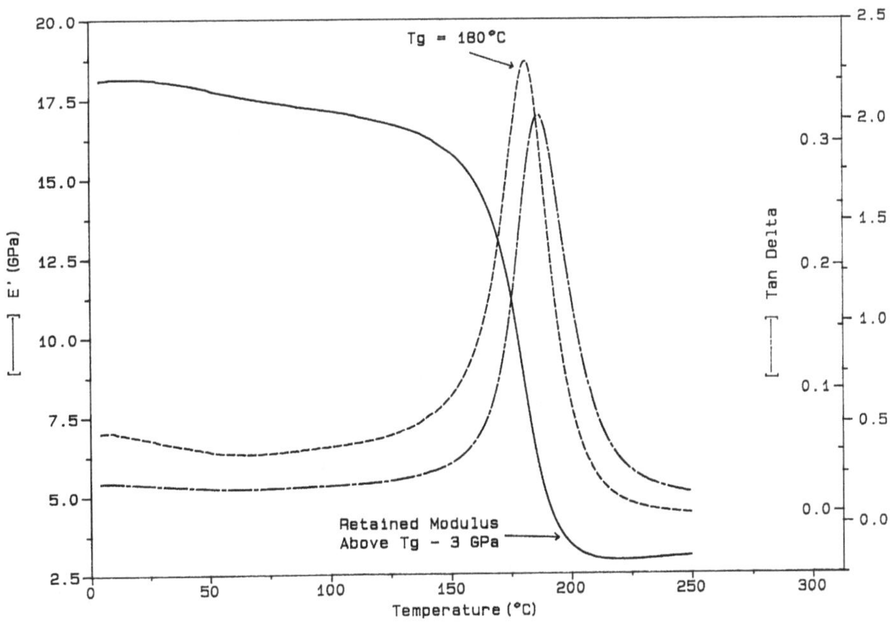

Figure 13.66 DMA scan showing viscoelastic properties for a glass fiber reinforced thermoset epoxy.

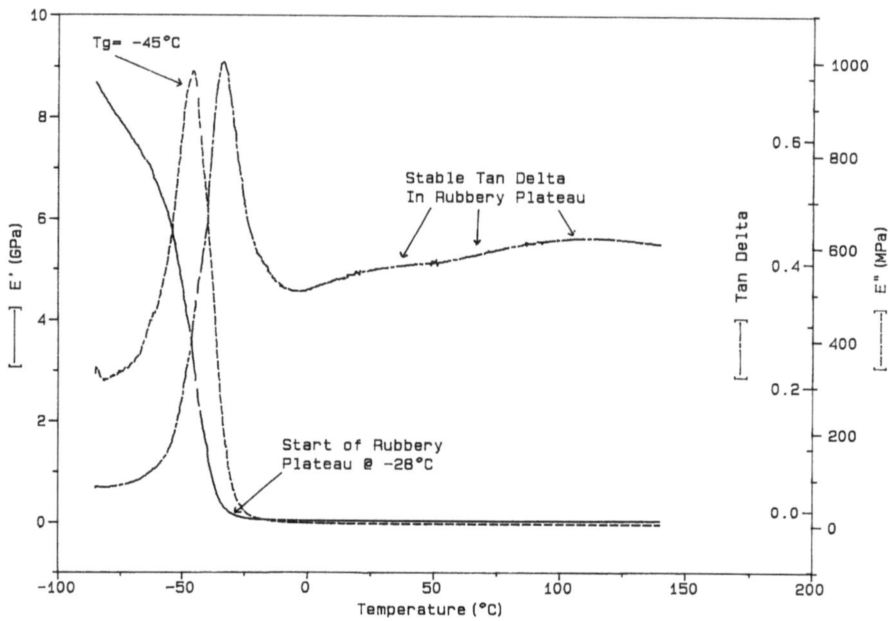

Figure 13.67 DMA scan showing viscoelastic properties for a crosslinked elastomer.

$$t = \frac{1}{2\pi f} \tag{13.6}$$

Modern DMA instruments, however, can also be programmed to operate in a constant stress or constant strain mode at isothermal conditions, allowing the measurement of modulus as a function of time at constant temperature. Figure 13.68 shows a plot of a 100-hour creep test (constant stress) with apparent modulus plotted as a function of time. Because the change in apparent modulus is very rapid in the early stages of the test and becomes more protracted at longer time frames, it is convention that this relationship is shown in a semi-logarithmic plot of apparent modulus versus time, as shown in Fig. 13.69. An alternate method of data presentation is to place both apparent modulus and time on a logarithmic scale, as shown in Fig. 13.70.

It is possible to observe graphically the equivalency of time and temperature if this is done for short-term measurements at multiple temperatures. Figure 13.71 shows a series of 30-minute creep tests conducted on a cross-linked vinyl ester between 101 and 136°C at 5°C intervals. As expected, the initial or zero-time modulus declines as the temperature is increased. In addition, the modulus at any given temperature decreases as the increasing strain is measured over the 30-minute period of each test. We can quantify the equivalency of the relationship between time and temperature for this particular material over this specific temperature range using this data. For example, we can see that the modulus declines in 30 minutes at 111°C by

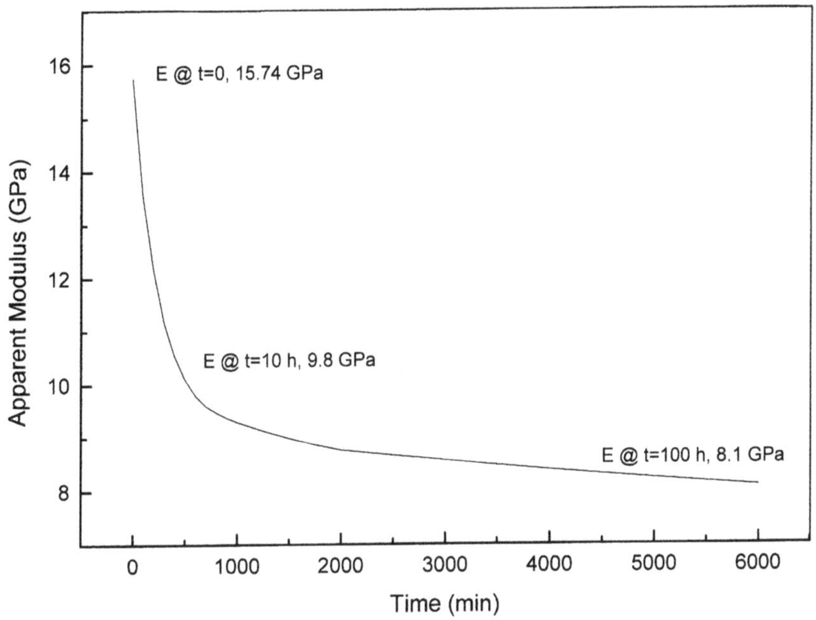

Figure 13.68 Linear plot of 100-hour creep test performed on a glass-filled nylon at 23°C.

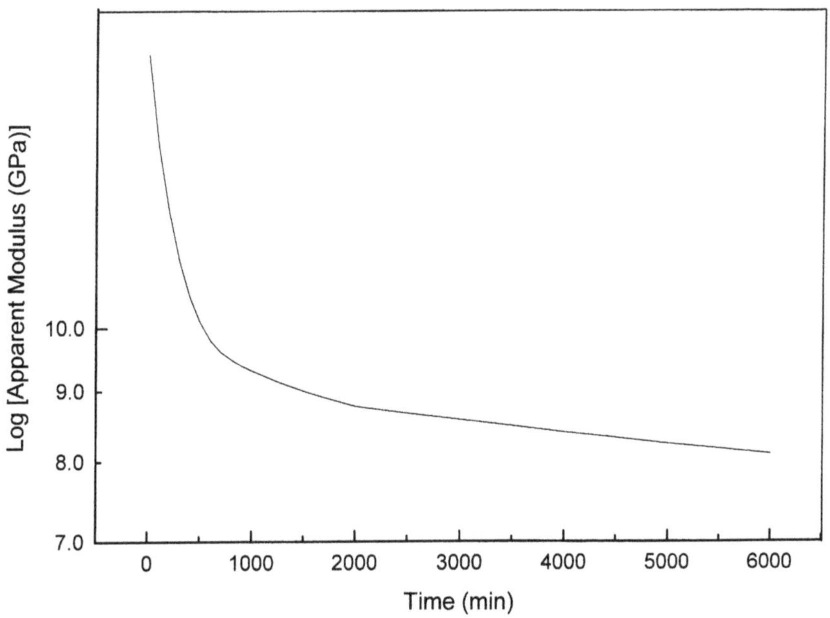

Figure 13.69 Data from Figure 13.68 plotted in semilogarithmic form.

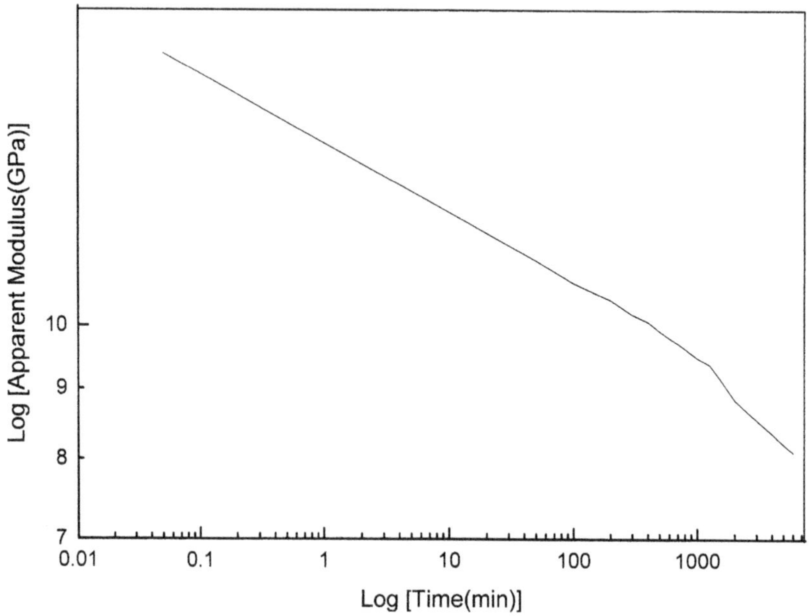

Figure 13.70 Data from Figure 13.68 plotted in logarithmic form.

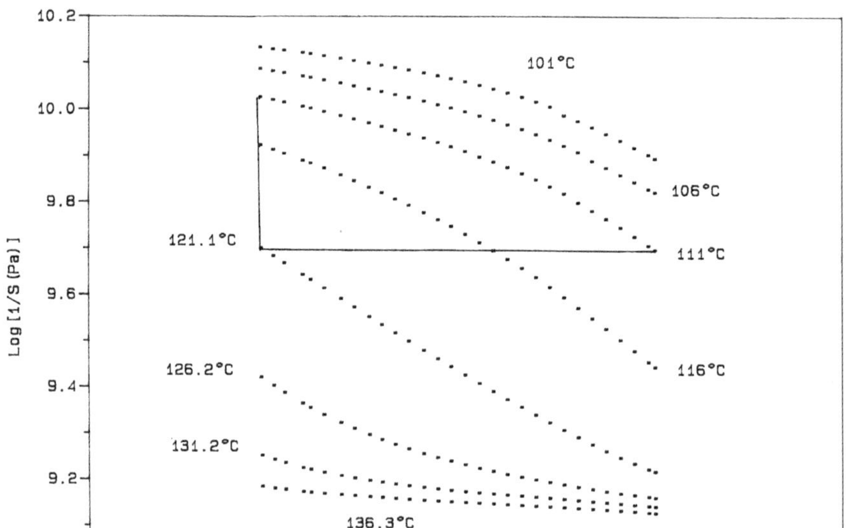

Figure 13.71 Results of a series of 30-minute creep tests on a vinyl ester composite showing a particular instance of the equivalence of time and temperature.

an amount that is equivalent to the decline in the zero-time modulus if the temperature is raised from 111 to 121°C. We can also see that this quantitative relationship changes due to the nonlinear behavior of modulus with temperature.

It is apparent that during this experiment the material has undergone a significant change in properties. The apparent modulus plots at the lower test temperatures are clustered together, which is an indication of relative structural stability. As the temperature is increased the zero-time modulus values begin to decline more rapidly. At the same time, the effect of time at any given temperature becomes more significant. This shows graphically that there is a correspondence between the effect of temperature and the effect of time. Near the end of the test, the zero-time values once again cluster together at a reduced level, and the time-dependent effects also become less significant. If we were to examine a modulus-temperature plot for this same temperature range we would see that the material has undergone a significant transition. The loss modulus and tan δ curves would show peaks typical of such a transition.

We can look at this time–temperature relationship in another way. Figure 13.72 shows the viscoelastic properties of a glass-fiber-reinforced poly(ether ether ketone) (PEEK). The glass transition is readily identified by the sharp decline in the storage modulus and the rapid rise of the loss modulus and tan δ to a maximum. We can conduct a series of 30-minute creep tests on a sample of the same material and plot

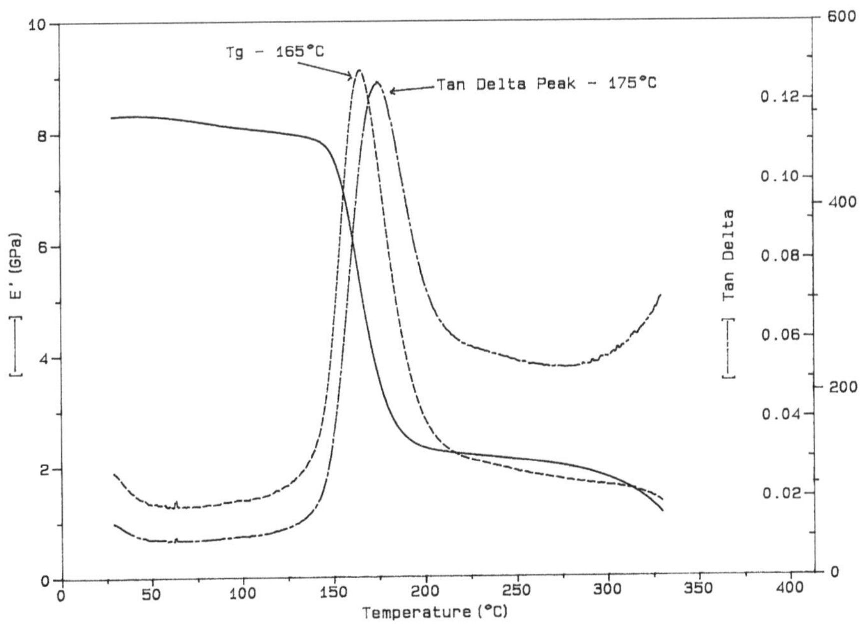

Figure 13.72 DMA scan showing viscoelastic properties for a 30% glass fiber reinforced poly(ether ether ketone).

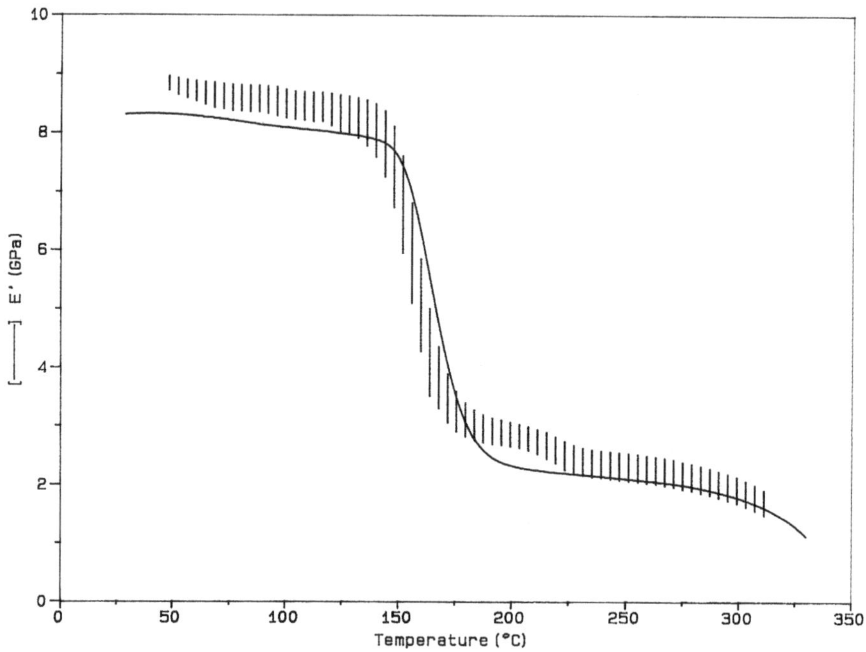

Figure 13.73 Elastic modulus versus temperature plot from Figure 13.72 superimposed on results of 30-minute creep tests performed on the same material.

the apparent modulus as a function of each temperature step on the same linear scale we used for the temperature scan. In Fig. 13.73 we superimpose the apparent modulus from the creep test on the storage modulus from the temperature scan. Because the x-axis is temperature in this graph, the apparent modulus plots measured at each temperature appear as vertical lines. Short vertical lines indicate low levels of time-dependent deformation, whereas longer lines denote regions where significant creep occurs. Note that in the temperature region below T_g the storage modulus is very stable with respect to temperature. In this same region, the changes in apparent modulus at any given temperature are small and the decline in zero-time modulus at each successive temperature is also small. As the material approaches the glass transition, however, the changes in apparent modulus become more substantial. Even before the zero-time modulus values begin to decline appreciably, the time-dependent behavior is already showing signs of a relaxation that is occurring over an ever-shorter time scale. There is again an obvious correspondence between time-dependent behavior at constant temperature and temperature-dependent behavior at constant time.

In qualitative terms, the storage modulus-temperature plot is a predictor of time-dependent behavior. If a projection of time-dependent behavior is sought, it can be estimated by selecting the appropriate temperature and then examining the behavior of the storage modulus as the temperature is increased above that reference point.

Thus, if a material is to be used at a temperature, and above that temperature the modulus is very stable, then time-dependent deformation will be small. If a material is evaluated just below the glass transition, however, a large reduction in apparent modulus can be expected in a short period of time even if the material is very rigid at the beginning of the evaluation. As a result, it is possible to make qualitative comparisons of creep resistance or stress decay between materials by examining the storage modulus-temperature curve. As a simple example, Fig. 13.74 shows a modulus plot for two amorphous materials, ABS and polycarbonate. For any temperature we wish to select, the lower T_g of the ABS and the tendency for the modulus of the ABS to fall off more rapidly with temperature allows us to conclude that polycarbonate will have superior creep resistance despite the higher modulus of the ABS at room temperature.

These qualitative determinations can be made quantitative by using a technique known as time-temperature superpositioning. This tool capitalizes on the principle that in viscoelastic materials a relaxation process that occurs rapidly at elevated temperatures will occur to the same degree over longer periods of time at lower temperatures. As a result, there are two experimental options for observing the time-dependent behavior of a polymer. The conventional method involves directly measuring the time-dependent response over longer time periods. This is obviously time-consuming, and in the current climate of rapid product development and com-

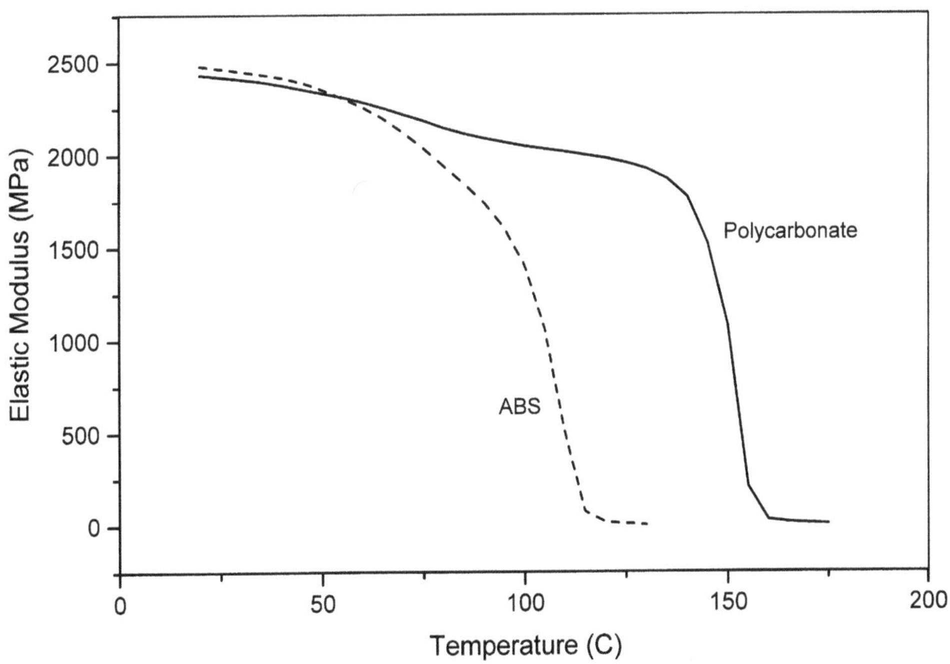

Figure 13.74 Comparison by DMA of the elastic modulus of a polycarbonate and an ABS.

pression in the time-to-market cycle, long-term testing is considered undesirable. The increased use of plastic materials in critical engineering applications, however, makes it unwise to forego the characterization of long-term behavior. The second option involves running the short-term experiment, whether it is constant stress (creep) or constant strain (stress decay), at progressively higher temperatures. The higher temperature data sets are then shifted to the right (to longer times) until they fall on the same line with the reference temperature. The resulting plot represents a prediction of time-dependent behavior called a *master curve*. This analysis is carried out on data plotted as apparent modulus versus time on a logarithmic scale.

Figure 13.75(a) shows raw data from a creep experiment for the cross-linked vinyl ester shown previously in Fig. 13.70. Temperatures between 100 and 135°C at 5°C increments were used, and each step in the experiment took 30 minutes. A 15-minute relaxation period was incorporated at the end of each stress period, and an additional 15 minutes was allocated between steps to allow the sample to equilibrate at each temperature step. Thus, the entire test took 10 hours to conduct. Figure 13.75(b) shows the master curve in its early stages of construction. At this point the first three temperature steps above the reference curve of 100°C have been moved into position so that they fall on the same line. As additional temperature steps are shifted, the curve is extended to increasingly longer times. Figure 13.75(c) shows the completed master curve extending to more than 100,000 hours.

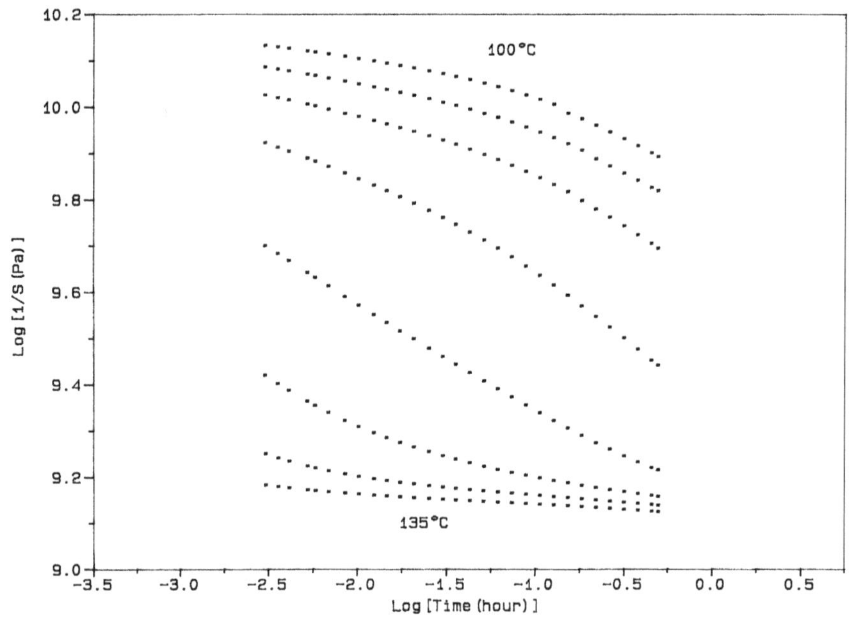

(a)

Figure 13.75 (a) Raw data from a series of 30-minute creep tests performed from 100 to 135°C on a vinyl ester composite. (b) Creep master curve for the vinyl ester composite in the process of construction at a reference temperature of 100°C. (c) Completed creep master curve for vinyl ester composite at a reference temperature of 100°C.

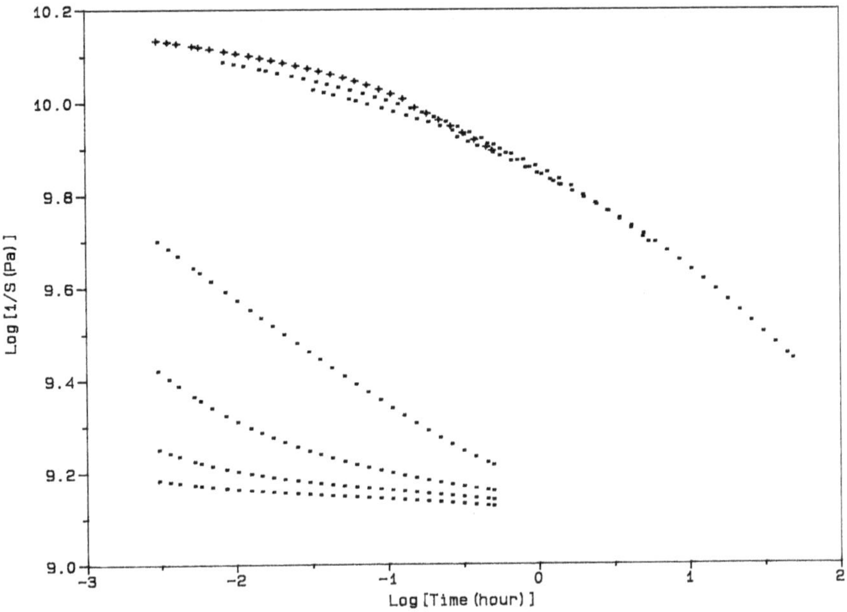

(b)

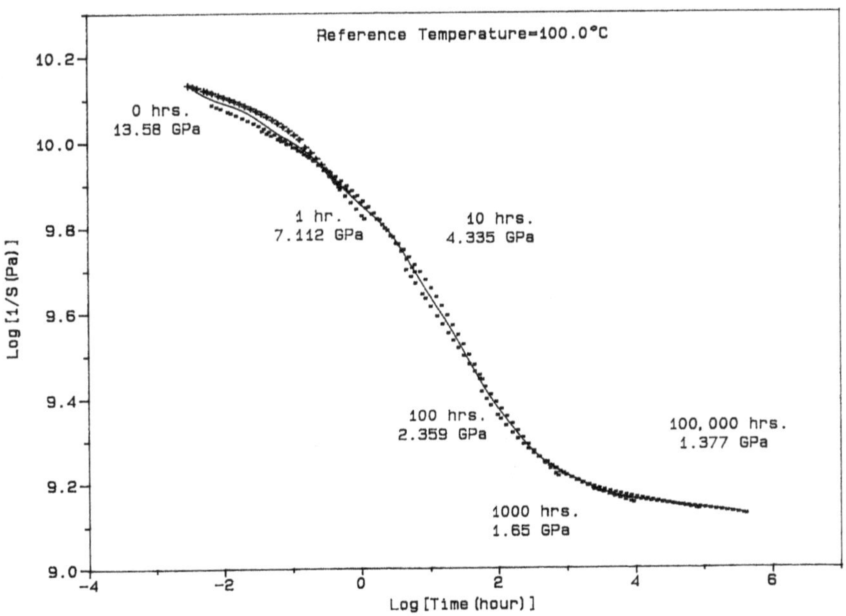

(c)

Figure 13.75 *(Continued)*

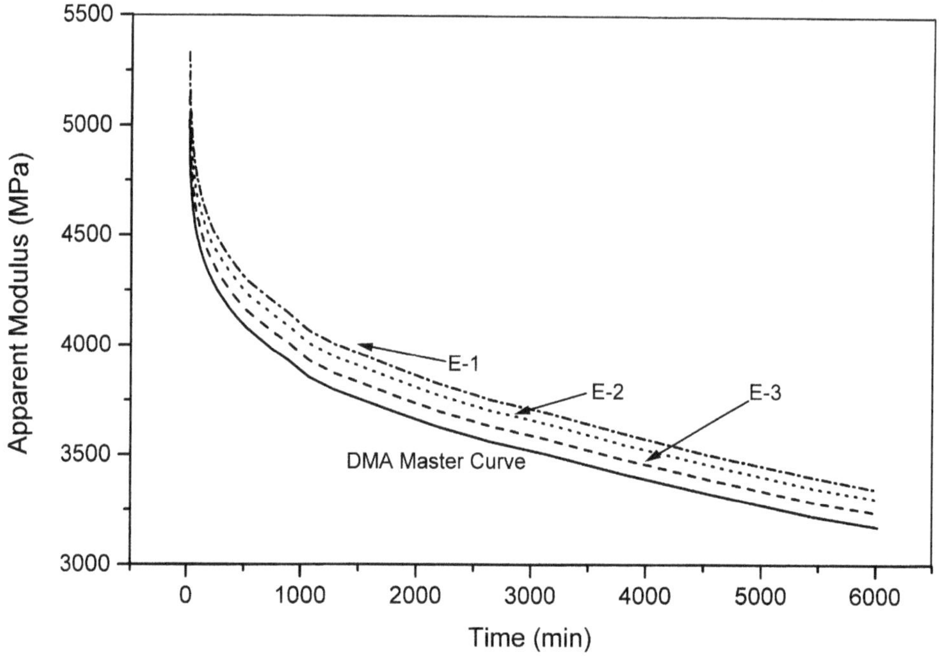

Figure 13.76 Comparison of creep master curve generated by DMA with three creep tests conducted in real time by classical methods.

Whenever accelerated testing of this type is conducted, it is natural to inquire about the agreement of such results with actual long-term testing. Figure 13.76 shows a comparison of the first 100 hours of the master curve for a cross-linked polyurethane developed at room temperature with three conventional creep tests conducted on the same material using a standard tensile testing machine. The plot is placed on a linear scale in order to maximize the visual appearance of discrepancies. Even with this treatment, the DMA master curve shows excellent agreement with classical creep test results. The difference between the master curve and creep test #3 (E-3) is smaller than the random variation that exists between the triplicate creep tests.

Figure 13.77(a) shows the raw data for a stress-relaxation test conducted on a polycarbonate. Figure 13.77(b) shows the completed master curve for a reference temperature of 135°C. Any test temperature can be used as a reference temperature for constructing a master curve; however, the extent of the projection will be limited by the number of tests run at temperatures higher than the reference temperature. In this case, 7 temperature steps comprising a 9-hour test provide a projection that extends to 30,000 hours.

Despite the power and success of the master curve in predicting long-term time-dependent behavior, some precautions are necessary. Some of these considerations,

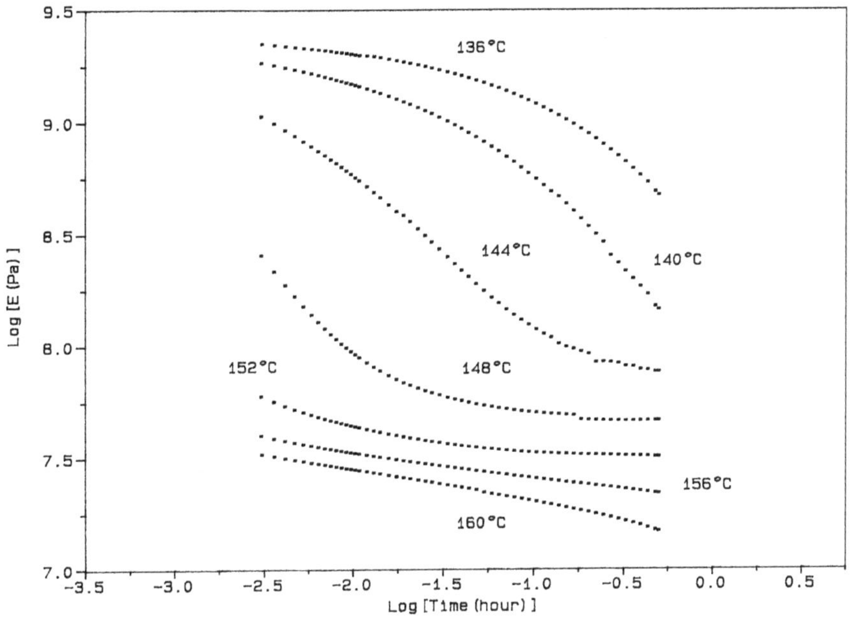

(a)

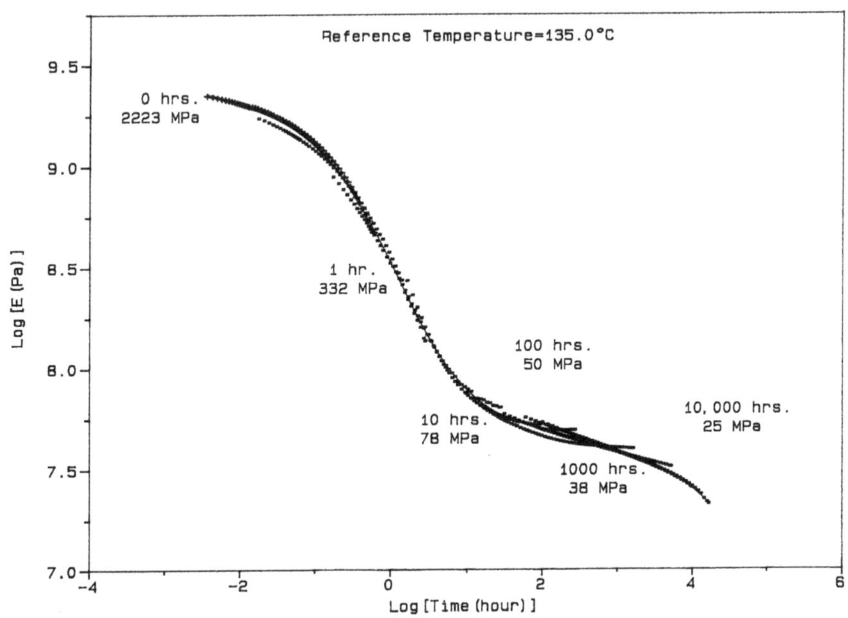

(b)

Figure 13.77 (a) Raw data from a stress relaxation experiment on an unfilled polycarbonate. (b) Stress relaxation master curve for polycarbonate at a reference temperature of 135°C.

such as corrections for changing temperature and density, result in minor changes in actual test results. These can be accounted for by incorporating material-specific data produced by other thermal analysis methods into the analysis software. Of much greater importance is the effect that irreversible structural changes can have on the accuracy of master curves. Events such as solid-state crystallization, postcuring, oxidative degradation, stress relief, or the melting of imperfect crystals can occur within the time frame of the short-term tests at elevated temperatures; however, these same events may never occur at the reference temperature. Incorporating the results of these structural changes into the long-term predictions can introduce serious error into the test results and accounts for poor results. These errors will be far more serious than subtle theoretical considerations based on correction factors and model selection.

As stated earlier, frequency and time are inversely related. Because viscoelastic responses are time-dependent they will also be frequency-dependent. Multiple-frequency experiments, which are often referred to as frequency *sweeps*, are capable of generating data similar to that obtained through short-term creep experiments. In addition, multiple frequency experiments provide information on loss properties, whereas creep experiments only supply data on the load-bearing component. Figures 13.78 and 13.79 show loss modulus versus temperature plots in the glass transition region for a 50% glass-fiber-reinforced nylon 6 and a 40% glass-fiber-reinforced

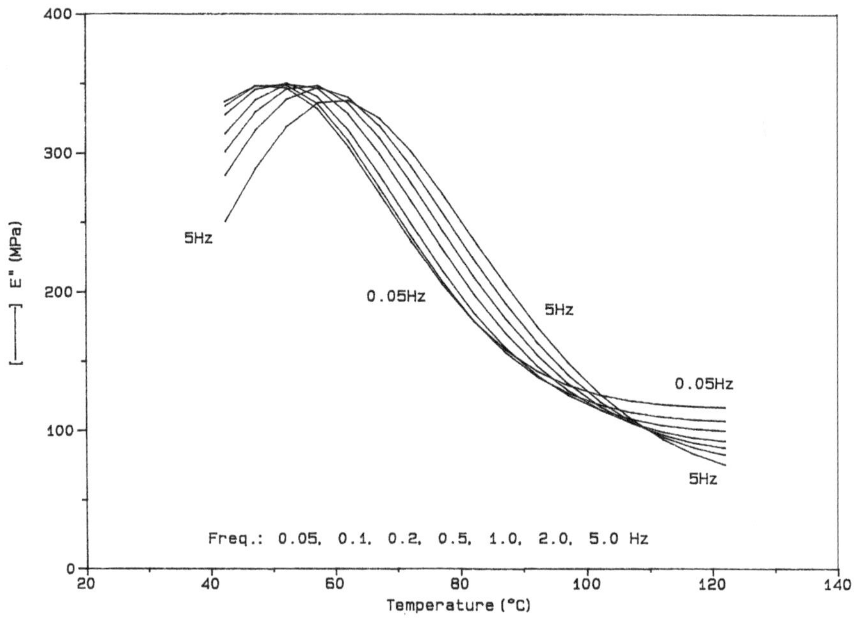

Figure 13.78 Multiple frequency scan by DMA showing loss modulus versus temperature through the glass transition of a nylon 6.

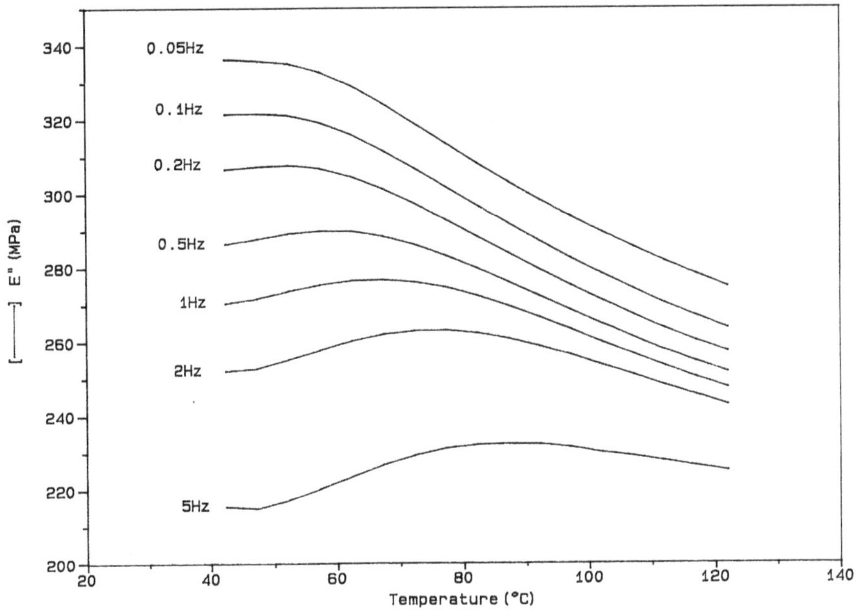

Figure 13.79 Multiple frequency scan of loss modulus versus temperature for a polypropylene.

polypropylene, respectively. In Fig. 13.78, the glass transition for the nylon 6 is well defined by the peaks in the loss modulus curves. The curves were generated at seven frequencies covering two orders of magnitude between 0.05 and 5.0 Hz. As the frequency is increased (time scale is decreased), the glass transition temperature increases slightly from 45 to 59°C. In Fig. 13.79, the polypropylene glass transition is barely perceptible as a maximum in the loss modulus. Nevertheless, it can be seen that the T_g is below the initial temperature of the test (40°C) at 0.05 Hz, whereas the peak is near 80°C at 5 Hz. Thus, the polypropylene is much more sensitive than is the nylon to the effects of the time scale of the measurement.

13.4.1 Glass Transition

It is important to remember that the glass transition is a region where loss properties increase and storage properties decrease. The T_g can be thought of as the temperature at which the elastic modulus is declining at the maximum rate. An increase in the T_g therefore represents a retardation of viscous flow. We would expect, as a result, that the storage modulus of a polymer would increase as the frequency of the measurement increases because a higher frequency equates to a shorter measurement time frame. Figure 13.80 shows a multiple frequency sweep from 0.02 to 2.0 Hz for the storage modulus of a polycarbonate. The temperature range is 97 to 175°C.

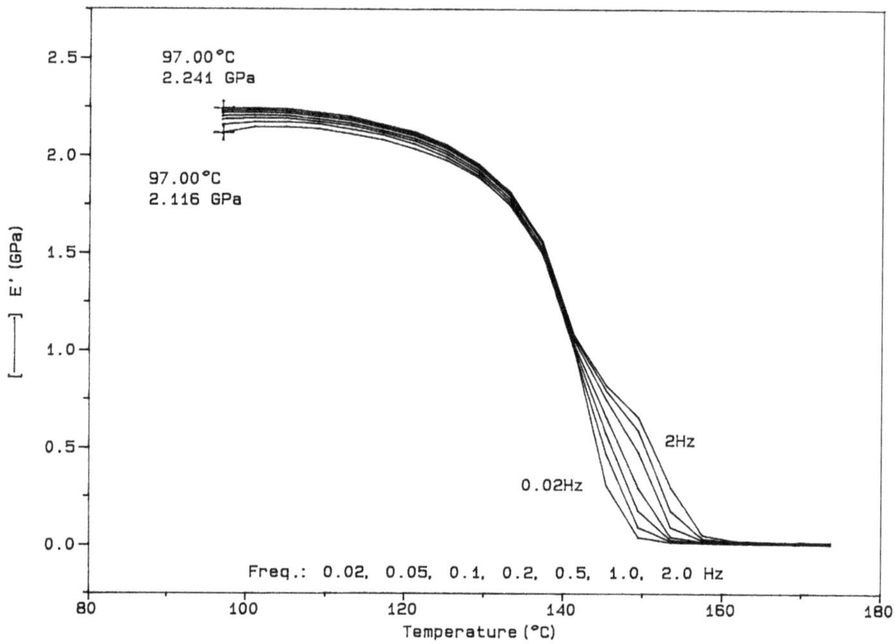

Figure 13.80 Multiple frequency scan of elastic modulus for a polycarbonate showing a dispersion of responses in the glass transition region.

As expected, the storage modulus increases with increasing frequency. The effects are small below the glass transition. At 97°C the modulus only increases by 6% across the frequency range used in the experiment. Above the glass transition, the storage modulus also appears to be affected very little by frequency. In the glass transition region, however, the effect of the measurement frequency is pronounced. The time scale of the measurement has a profound effect on how the transition is perceived. The shift to higher apparent stiffness with increasing frequency correlates to the increase in T_g that is measured by the loss modulus. Figure 13.81 shows the same phenomenon for another grade of polycarbonate evaluated between 125 and 165°C. The smaller temperature range provides more detail. At 125°C the modulus increases by 4% across the two decades of frequency. However, the curves begin to separate at 135°C. At 155°C, when the modulus measured at 0.02 Hz has reached the rubbery plateau, the modulus at 2 Hz is more than 15 times higher. At 2 Hz, the rubbery plateau is not attained until the test temperature reaches 162°C.

This may all seem somewhat esoteric when it comes to the practical world of material selection and accurately assessing real-world performance, but there is practical significance. These types of tests demonstrate that the viscoelastic balance in materials like polypropylene is shifted significantly as a function of the rate at which stress is applied. This behavior is measurable in physical terms that are easily understood at the engineering level. If a classical tensile stress–strain test is conducted on a rate-sensitive material like polypropylene, the peak stress and the modulus increase as a function of increasing strain rate. At the same time, the ultimate elongation, which is a relative measure of toughness, decreases with increasing strain rate. Figure 13.82(a) shows this behavior for the full scale of a series of tensile tests. Figure 13.82(b) expands the plot to show the detail of the yield section of the test. In Fig. 13.82(a) it can be seen that the yield stress increases from 26 MPa (3770 psi) at 5 mm/minute to 36 MPa at 500 mm/minute. Even though the elongation at yield is virtually unaffected, the ultimate elongation drops from 290% to less than 30%. The slope of the stress–strain plot in the linear region also increases with increasing strain rate. In engineering terms, the material behaves as a stronger, stiffer, and less impact-resistant system at higher strain rates; in other words the elastic properties are more dominant. At lower strain rates the material is weaker, more flexible, and tougher; the loss properties become more important. This shift in properties is related to the shift to a lower T_g at lower frequencies (lower strain rates) and a higher T_g at higher frequencies (higher strain rates). A similar family of stress–strain curves can be developed by holding the strain rate constant and varying the temperature. This again illustrates the equivalent effects of time (rate) and temperature.

The reduction in elongation with higher strain rates is analogous to the change in failure mode that some materials exhibit when subjected to an impact test at different velocities. Higher impact velocities will result in a more brittle failure mode, whereas lower velocities will produce a more ductile break. Here again, the rela-

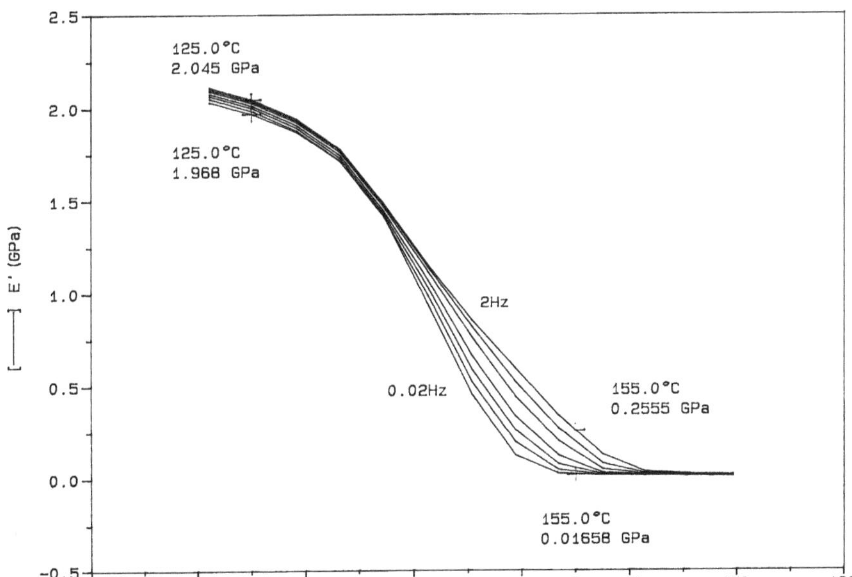

Figure 13.81 Multiple frequency scan of elastic modulus for a polycarbonate with the glass transition region magnified.

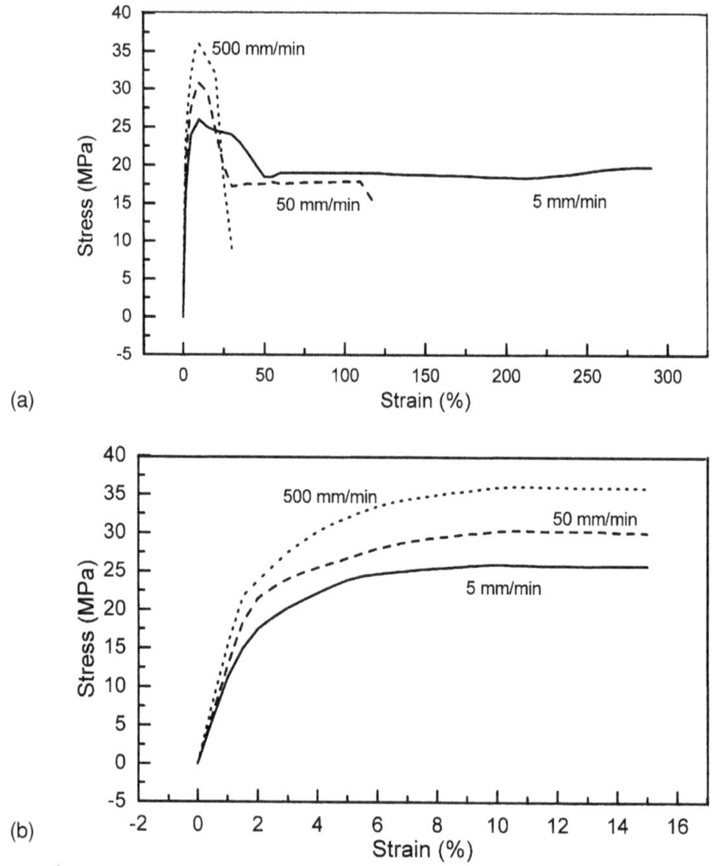

Figure 13.82 (a) Effect of strain rate on the tensile properties of a polypropylene. (b) Magnified view of the low-strain region of the stress-strain curves from Figure 13.82(a).

tionship between time and temperature is apparent. It is well known that impact testing at lower temperatures is more likely to produce a brittle failure, whereas tests conducted at higher temperatures will result in a more ductile failure. Thus, reducing the test temperature has the same effect as increasing the strain rate (decreasing the time scale) of the experiment, whereas increasing the test temperature produces the same result as reducing the strain rate (increasing the time scale) of the test. Figure 13.83 shows the result for an instrumented falling dart impact test on a grade of poly(vinyl chloride) (PVC) conducted at a velocity of 3.39 m/s (11.11 ft/s). The shape of the load curve shows a sharp drop as soon as the point of maximum load is achieved, indicating a brittle failure mode. The energy absorption of 20.7 J (15.1 ft-lbs) is considerable, but an inspection of the test specimens confirms the presence of brittle failure features. Performing the test at the same temperature but at a rate of 1.53 m/s (5 ft/s) changes the mode of failure from brittle to fully ductile, as shown in Fig. 13.84. This is another very practical demonstration of viscoelastic

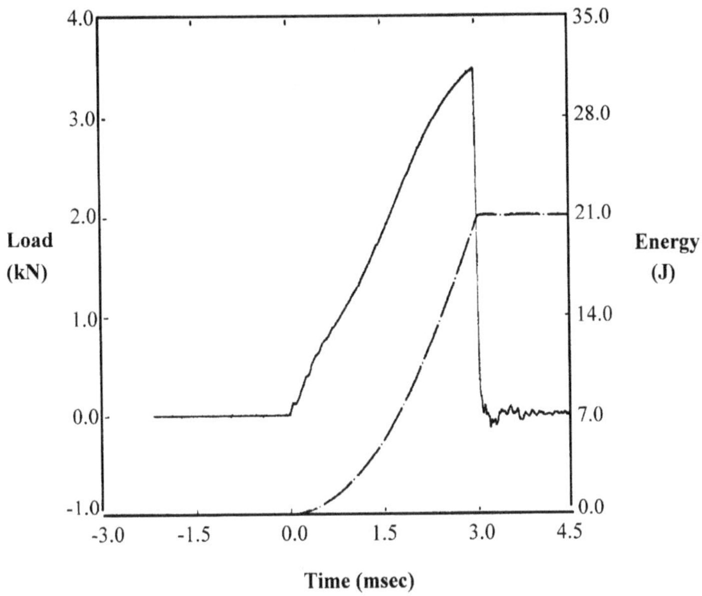

Figure 13.83 Load and total energy versus time results for instrumented impact tests on a PVC conducted at high velocity showing tendency for brittle failure.

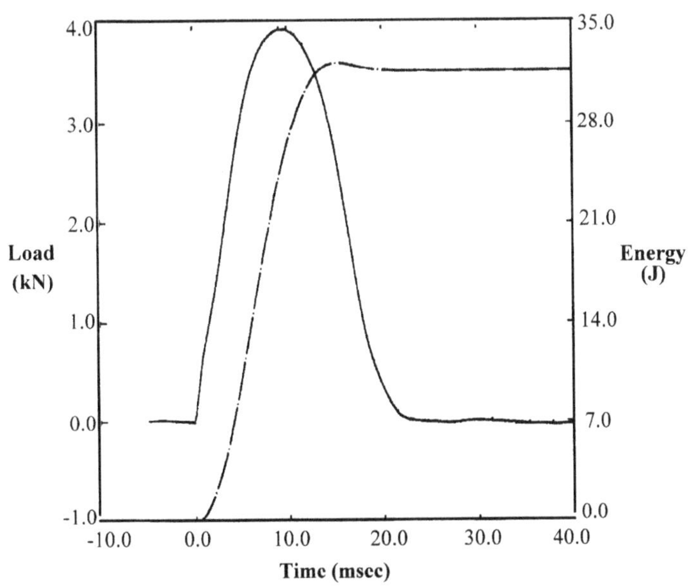

Figure 13.84 Load and total energy versus time results for instrumented impact tests on a PVC conducted at lower velocity showing fully ductile failure.

behavior that substantially affects the judgment about the fitness-for-use of a material for a given application. In this particular case, this behavior was discovered when an attempt to increase the insertion speed of metal hardware into the molded PVC parts resulted in a sudden increase in brittle failures.

13.5 Conclusion

Material properties play a critical role in the performance of molded products. These properties are the result of inherent chemistry, attention to detail during process, good part and mold design, and application environment. The many faces of plastic part performance reduce down to some fundamental aspects of composition, molecular weight, and structure. These in turn give rise to the unique behaviors captured by the property of viscoelasticity. In this chapter we have attempted to cover those critical aspects in the context of the tests required to arrive at the correct diagnosis for a given problem. As in every discipline, the increased understanding of the fundamentals at the beginning of an endeavor decreases the need for analysis of failures later in the life of a product.

References

1. Sircar, A. K., "Analysis of Elastomer Vulcanizate Composition by TG-DTG-Techniques", American Chemical Society, Toronto, Ontario, Canada, May 21–24, 1991.
2. Sepe, Michael, SPE Antec, Vol. 46, page 3198–3204, May 2000.
3. Nangrani, K. J., Wenger, R. M., Nell, P., SPE Antec, Vol. 32, page 79–83, May, 1986.
4. Wright, D., Unpublished presentation on Knowledge Based Systems for Material Selection given at SPE Antec, May, 1995.
5. Takemori, M., SPE Antec, Vol. 24, page 216, May 1978.

Index

www.ingramcontent.com/pod-product-compliance
Lightning Source LLC
LaVergne TN
LVHW081928200926
848484LV00009B/2390
9781569904206